PROOF THEORY

SECOND EDITION

Gaisi Takeuti
Professor Emeritus of Mathematics
The University of Illinois at Urbana-Champaign

Dover Publications, Inc.
Mineola, New York

Bibliographical Note

This Dover edition, first published in 2013, is an unabridged republication of the work originally published in 1987 by North-Holland, Amsterdam, as Volume 81 in the series *Studies in Logic and the Foundations of Mathematics.* The first edition was published in 1975.

Library of Congress Cataloging-in-Publication Data

Takeuti, Gaisi, 1926-
Proof theory / Gaisi Takeuti. — Dover edition.
pages cm. — (Dover books on mathematics)
Summary: "This comprehensive monograph is a cornerstone in the area of mathematical logic and related fields. Focusing on Gentzen-type proof theory, the book presents a detailed overview of creative works by the author and other 20th-century logicians that includes applications of proof theory to logic as well as other areas of mathematics. 1975 edition"— Provided by publisher.
Includes bibliographical references and index.
ISBN-13: 978-0-486-49073-1 (pbk.)
ISBN-10: 0-486-49073-4 (pbk.)
1. Proof theory. I. Title.
QA9.54.T34 2013
511.3'6—dc23

2012030734

Manufactured in the United States by *LSC* Communications
49073404 2023
www.doverpublications.com

PREFACE

This book is based on a series of lectures that I gave at the Symposium on Intuitionism and Proof Theory held at Buffalo in the summer of 1968. Lecture notes, distributed at the Buffalo symposium, were prepared with the help of Professor John Myhill and Akiko Kino. Mariko Yasugi assisted me in revising and extending the original notes. This revision was completed in the summer of 1971. At this point Jeffery Zucker read the first three chapters, made improvements, especially in Chapter 2, and my colleagues Wilson Zaring provided editorial assistance with the final draft of Chapters 4–6.

To all who contributed, including our departmental secretaries, who typed versions of the material for use in my classes. I express my deep appreciation.

Gaisi Takeuti

Urbana, March 1975

PREFACE TO THE SECOND EDITION

In addition to the corrections of misprints and minor gaps, this edition includes the following new material.

In Chapter 1, I have added Rasiowa–Sikorski's Completeness Theorem for intuitionistic predicate calculus using Heyting algebras. The relationship between Kripke semantics and Heyting valued models is discussed. In the author's opinion, the present form of the Completeness theorems for intuitionistic predicate calculus is much weaker than the counterpart for classical logic. The author believes that the investigation of stronger completeness theorems for intuitionistic logic is a very attractive new area of investigation.

In Chapter 2, Wainer's theory on the Hardy class, Kirby–Paris' work on Goodstein sequences, Ketonen–Solovay–Quinsey's result on Paris–Harrington's theorem and a weak form of Friedman's work on Kruskal's theorem are discussed as applications of Gentzen's consistency-proof.

In Chapter 4, I have made the material more readable and added a proof theoretic form of Borel determinacy. It is an interesting open problem to prove Borel determinacy by means of cut elimination.

In Chapter 5, I have simplified the accessibility proof of ordinal diagrams and included Arai's improvement of my consistency-proof and cut elimination theorem using ordinal diagrams. The theory of quasi-ordinal diagrams are developed as a generalization of ordinal diagrams.

As a postscript, I have added references which are not mentioned in the course of the book.

Included as an Appendix are Kreisel's article on his work, Pohlers' article on the work of Schütte school, Simpson's article on reverse mathematics, and Feferman's article on his approach to proof theory. The order of these articles is the order of the date when I received the manuscripts. These articles will give the reader a good idea of many different aspects of proof theory. I am very grateful to Professors Georg Kreisel, Wolfram Pohlers, Stephen G. Simpson and Solomon Feferman for their contributions.

Many people helped me with this edition. I would especially like to mention, with my deep appreciation, Nobuyoshi Motohashi for his help on the postscript, Toshiyasu Arai for his help on Chapter 4 and also for proof-reading, and Mitsuhiro Okada and Noriko Honda for proof-reading.

Gaisi Takeuti

Urbana, August 1986

CONTENTS

INTRODUCTION

Mathematics is a collection of proofs. This is true no matter what standpoint one assumes about mathematics—platonism, anti-platonism, intuitionism, formalism, nominalism, etc. Therefore, in investigating "mathematics", a fruitful method is to formalize the proofs of mathematics and investigate the structure of these proofs. This is what proof theory is concerned with.

Proof theory was initiated by D. Hilbert in his attempt to prove the consistency of mathematics. Hilbert's approach was later developed by the brilliant work of G. Gentzen. This textbook is devoted to the proof theory inspired by Gentzen's work: so-called Gentzen-type proof theory.

Part I treats the proof theory of first order formal systems. Chapter 1 deals with the first order predicate calculus; Gentzen's cut-elimination theorem plays a major role here. There are many consequences of this theorem as for example the various interpolation theorems.

We prove the completeness of the classical predicate calculus, by the use of reduction trees (following Schütte), and then we prove the completeness of the intuitionistic predicate calculus, by adapting this method to Kripke semantics.

Chapter 2 deals with the theory of natural numbers; the main topics are Gödel's incompleteness theorem and Gentzen's consistency proof. Since the author believes that the true significance of Gentzen's consistency proof has not been well understood so far, a philosophical discussion is also presented in this chapter. The author believes that the Hilbert-Gentzen finitary standpoint, with "Gedankenexperimenten" involving finite (and concrete) operations on (sequences of) concretely given figures, is most important in the foundations of mathematics.

Part II concerns the finite order predicate calculi and infinitary languages. In Chapter 3, the semantics for finite order systems, and the cut-elimination theorem for them, due to Tait, Takahashi and Prawitz, are considered. Since the finite type calculus is not complete, and further, much of traditional mathematics can be formalized in it, we are anxious to see progress in investigating the significance of the cut-elimination theorem here. The significance of the systems with infinitary languages which are presented in Chapter 4 is that they are complete systems in which the cut-elimination theorem holds, while at the same time they are essentially second order systems. The situation is, however, quite uncertain for systems with heterogeneous quantifiers. Here we propose a basic system of heterogeneous quantifiers which seems reasonable and for which so-called "weak completeness" holds. It seems, however, that our system is far from complete. A system which is obtained from ours by a slight modification is closely related to the axiom of determinateness (AD). Therefore the problem of how to extend our system to a (sound and) complete system is related to the justification of the axiom of determinateness.

Let M be a transitive model of ZF+DC (the axiom of dependent choices) which contains $P(\omega)$. It has been shown that the following two statements are equivalent: (1) AD holds in M; and (2) the cut-elimination theorem holds for any M-definable determinate logic. This suggests an interesting direction for the study of infinitary languages.

Part III is devoted to consistency proofs for stronger systems on which the author has worked.

We have tried to avoid overlapping of material with other textbooks. Thus, for example, we do not present the material in K. Schütte's Beweistheorie, although much of it is Gentzen-type proof theory. Those who wish to learn other approaches to proof theory are advised to consult G. Kreisel's Survey of Proof Theory I and II, Journal of Symbolic Logic (1968), and Proceedings of the Second Scandinavian Logic Symposium, ed. J. E. Fenstad (North-Holland, Amsterdam, 1971), respectively. We have made special efforts to clarify our position on foundational issues. Indeed, it is our view that in the study of the foundations of mathematics (which is not restricted to consistency problems), it is philosophically important to study and clarify the structures of mathematical proofs.

Concerning the impact of foundational studies on mathematics itself, we remark that while set theory, for example, has already contributed essentially to the development of modern mathematics, it remains to be seen what influence proof theory will have on mathematics.

No attempt has been made to make the references comprehensive although some names are attached to the theorems. In addition to those given above, a few references are recommended in the course of the book.

PART I

FIRST ORDER SYSTEMS

CHAPTER 1

FIRST ORDER PREDICATE CALCULUS

In this chapter we shall present Gentzen's formulation of the first order predicate calculus **LK** (logistischer klassischer Kalkül), which is convenient for our purposes. We shall also include a formulation of intuitionistic logic, which is known as **LJ** (logistischer intuitionistischer Kalkül). We then proceed to the proofs of the cut-elimination theorems for **LK** and **LJ**, and their applications.

§1. Formalization of statements

The first step in the formulation of a logic is to make the formal language and the formal expressions and statements precise.

DEFINITION 1.1. A first order (formal) language consists of the following symbols.

1) *Constants*:
   1.1) Individual constants: $k_0, k_1, \ldots, k_j, \ldots$ $(j = 0, 1, 2, \ldots)$.
   1.2) Function constants with i argument-places $(i = 1, 2, \ldots)$: $f_0^i, f_1^i, \ldots, f_j^i, \ldots$ $(j = 0, 1, 2, \ldots)$.
   1.3) Predicate constants with i argument-places $(i = 0, 1, 2, \ldots)$: $R_0^i, R_1^i, \ldots, R_j^i, \ldots$ $(j = 0, 1, 2, \ldots)$.
2) *Variables*:
   2.1) Free variables: $a_0, a_1, \ldots, a_j, \ldots$ $(j = 0, 1, 2, \ldots)$.
   2.2) Bound variables: $x_0, x_1, \ldots, x_j, \ldots$ $(j = 0, 1, 2, \ldots)$.
3) *Logical symbols*:
   $\neg$ (not), $\wedge$ (and), $\vee$ (or), $\supset$ (implies), $\forall$ (for all) and $\exists$ (there exists). The first four are called propositional connectives and the last two are called quantifiers.
4) *Auxiliary symbols*:
   (,) and , (comma).

We say that a first order language L is given when all constants are given. In every argument, we assume that a language L is fixed, and hence we omit the phrase "of L".

There is no reason why we should restrict the cardinalities of various kinds of symbols to exactly $\aleph_0$. It is, however, a standard approach in elementary logic to start with countably many symbols, which are ordered with order type ω. Therefore, for the time being, we shall assume that the language consists of the symbols as stated above, although we may consider various other types of language later on. In any case it is essential that each set of variables is infinite and there is at least one predicate symbol. The other sets of constants can have arbitrary cardinalities, even 0.

We shall use many notational conventions. For example, the superscripts in the symbols of 1.2) and 1.3) are mostly omitted and the symbols of 1) and 2) may be used as meta-symbols as well as formal symbols. Other letters such as $g, h, \ldots$ may be used as symbols for function constants, while $a, b, c, \ldots$ may be used for free variables and $x, y, z, \ldots$ for bound variables.

Any finite sequence of symbols (from a language L) is called an *expression* (of L).

DEFINITION 1.2. *Terms* are defined inductively (recursively) as follows:

1) Every individual constant is a term.

2) Every free variable is a term.

3) If f^i is a function constant with i argument-places and $t_1, \ldots, t_i$ are terms, then $f^i(t_1, \ldots, t_i)$ is a term.

4) Terms are only those expressions obtained by 1)–3). Terms are often denoted by $t, s, t_1, \ldots$.

Since in proof theory inductive (recursive) definitions such as Definition 1.2 often appear, we shall not mention it each time. We shall normally omit the last clause which states that the objects which are being defined are only those given by the preceding clauses.

DEFINITION 1.3. If R^i is a predicate constant with i argument-places and $t_1, \ldots, t_i$ are terms, then $R^i(t_1, \ldots, t_i)$ is called an *atomic formula*. *Formulas* and their outermost logical symbols are defined inductively as follows:

1) Every atomic formula is a formula. It has no outermost logical symbol.

2) If A and B are formulas, then $(\neg A)$, $(A \wedge B)$, $(A \vee B)$ and $(A \supset B)$ are formulas. Their outermost logical symbols are $\neg$, $\wedge$, $\vee$ and $\supset$, respectively.

3) If A is a formula, a is a free variable and x is a bound variable not occurring in A, then $\forall x\, A'$ and $\exists x\, A'$ are formulas, where A' is the expression obtained from A by writing x in place of a at each occurrence of a in A. Their outermost logical symbols are $\forall$ and $\exists$, respectively.

4) Formulas are only those expressions obtained by 1)–3).

Henceforth, $A, B, C, \ldots, F, G, \ldots$ will be metavariables ranging over formulas. A formula without free variables is called a *closed formula* or a *sentence*. A formula which is defined without the use of clause 3) is called *quantifier-free*. In 3) above, A' is called the *scope* of $\forall x$ and $\exists x$, respectively.

When the language L is to be emphasized, a term or formula in the language L may be called an L-*term* or L-*formula*, respectively.

REMARK. Although the distinction between free and bound variables is not essential, and is made only for technical convenience, it is extremely useful and simplifies arguments a great deal. This distinction will, therefore, be maintained unless otherwise stated.

It should also be noticed that in clause 3) of Definition 1.3, x must be a variable which does not occur in A. This eliminates expressions such as $\forall x\,(C(x) \wedge \exists x\, B(x))$. This restriction does not essentially narrow the class of formulas, since e.g. this expression $\forall x\,(C(x) \wedge \exists x\, B(x))$ can be replaced by $\forall y\,(C(y) \wedge \exists x\, B(x))$, preserving the meaning. This restriction is useful in formulating formal systems, as will be seen later.

In the following we shall omit parentheses whenever the meaning is evident from the context. In particular the outermost parentheses will always be omitted. For the logical symbols, we observe the following convention of priority: the connective $\neg$ takes precedence over each of $\wedge$ and $\vee$, and each of $\wedge$ and $\vee$ takes precedence over $\supset$. Thus $\neg A \wedge B$ is short for $(\neg A) \wedge B$, and $A \wedge B \supset C \vee D$ is short for $(A \wedge B) \supset (C \vee D)$. Parentheses are omitted also in the case of double negations: for example $\neg\neg A$ abbreviates $\neg(\neg A)$. $A \equiv B$ will stand for $(A \supset B) \wedge (B \supset A)$.

DEFINITION 1.4. Let A be an expression, let $\tau_1, \ldots, \tau_n$ be distinct primitive symbols, and let $\sigma_1, \ldots, \sigma_n$ be any symbols. By

$$\left(A \frac{\tau_1, \ldots, \tau_n}{\sigma_1, \ldots, \sigma_n}\right)$$

we mean the expression obtained from A by writing $\sigma_1, \ldots, \sigma_n$ in place of $\tau_1, \ldots, \tau_n$, respectively, at each occurrence of $\tau_1, \ldots, \tau_n$ (where these symbols are replaced simultaneously). Such an operation is called the (*simultaneous*) *replacement of* $(\tau_1, \ldots, \tau_n)$ *by* $(\sigma_1, \ldots, \sigma_n)$ *in* A. It is not required that $\tau_1, \ldots, \tau_n$ actually occur in A.

PROPOSITION 1.5. (1) *If A contains none of $\tau_1, \ldots, \tau_n$, then*

$$\left(A \frac{\tau_1, \ldots, \tau_n}{\sigma_1, \ldots, \sigma_n}\right)$$

is A itself.

(2) *If* $\sigma_1, \ldots, \sigma_n$ *are distinct primitive symbols, then*

$$\left(\left(A\frac{\tau_1, \ldots, \tau_n}{\sigma_1, \ldots, \sigma_n}\right)\frac{\sigma_1, \ldots, \sigma_n}{\theta_1, \ldots, \theta_n}\right)$$

is identical with

$$\left(A\frac{\tau_1, \ldots, \tau_n}{\theta_1, \ldots \theta_n}\right).$$

DEFINITION 1.6. (1) Let A be a formula and $t_1, \ldots, t_n$ be terms. If there is a formula B and n distinct free variables $b_1, \ldots, b_n$ such that A is

$$\left(B\frac{b_1, \ldots, b_n}{t_1, \ldots, t_n}\right),$$

then for each i ($1 \leq i \leq n$) the occurrences of t_i resulting from the above replacement are said to be indicated in A, and this fact is also expressed (less accurately) by writing B as $B(b_1, \ldots, b_n)$, and A as $B(t_1, \ldots, t_n)$. A may of course contain some other occurrences of t_i; this happens if B contains t_i.

(2) We say that a term t is fully indicated in A, or every occurrence of t in A is indicated, if every occurrence of t is obtained by such a replacement (from some formula B as above, with $n = 1$ and $t = t_1$).

It should be noted that the formula B and the free variables from which A can be obtained by replacement are not unique; the indicated occurrences of some terms of A are specified relative to such a formula B and such free variables.

PROPOSITION 1.7. *If $A(a)$ is a formula (in which a is not necessarily fully indicated) and x is a bound variable not occurring in $A(a)$, then $\forall x\, A(x)$ and $\exists x\, A(x)$ are formulas.*

PROOF. By induction on the number of logical symbols in $A(a)$.

In the following, let Greek capital letters $\Gamma, \Delta, \Pi, \Lambda, \Gamma_0, \Gamma_1, \ldots$ denote finite (possibly empty) sequences of formulas separated by commas. In order to formulate the sequential calculus, we must first introduce an auxiliary symbol $\rightarrow$.

DEFINITION 1.8. For arbitrary Γ and Δ in the above notation, $\Gamma \rightarrow \Delta$ is called a *sequent*. Γ and Δ are called the *antecedent* and *succedent*, respectively, of the sequent and each formula in Γ and Δ is called a *sequent-formula*.

Intuitively, a sequent $A_1, \ldots, A_m \rightarrow B_1, \ldots, B_n$ (where $m, n \geqslant 1$) means: if $A_1 \wedge \ldots \wedge A_m$, then $B_1 \vee \ldots \vee B_n$. For $m \geqslant 1$, $A_1, \ldots, A_m \rightarrow$ means that $A_1 \wedge \ldots \wedge A_m$ yields a contradiction. For $n \geqslant 1$, $\rightarrow B_1, \ldots, B_n$ means that $B_1 \vee \ldots \vee B_n$ holds. The empty sequent $\rightarrow$ means there is a contradiction. Sequents will be denoted by the letter S, with or without subscripts.

§2. Formal proofs and related concepts

DEFINITION 2.1. An *inference* is an expression of the form

$$\frac{S_1}{S} \quad \text{or} \quad \frac{S_1 \quad S_2}{S},$$

where S_1, S_2 and S are sequents. S_1 and S_2 are called the *upper sequents* and S is called the *lower sequent* of the inference.

Intuitively this means that when S_1 (S_1 and S_2) is (are) asserted, we can infer S from it (from them). We restrict ourselves to inferences obtained from the following rules of inference, in which A, B, C, D, $F(a)$ denote formulas.

1) Structural rules:

1.1) *Weakening*:

$$\text{left:} \quad \frac{\Gamma \rightarrow \Delta}{D, \Gamma \rightarrow \Delta}; \qquad \text{right:} \quad \frac{\Gamma \rightarrow \Delta}{\Gamma \rightarrow \Delta, D}.$$

D is called the weakening formula.

1.2) *Contraction*:

$$\text{left:} \quad \frac{D, D, \Gamma \rightarrow \Delta}{D, \Gamma \rightarrow \Delta}; \qquad \text{right:} \quad \frac{\Gamma \rightarrow \Delta, D, D}{\Gamma \rightarrow \Delta, D}.$$

1.3) *Exchange*:

$$\text{left:} \quad \frac{\Gamma, C, D, \Pi \rightarrow \Delta}{\Gamma, D, C, \Pi \rightarrow \Delta}; \qquad \text{right:} \quad \frac{\Gamma \rightarrow \Delta, C, D, \Lambda}{\Gamma \rightarrow \Delta, D, C, \Lambda}$$

We will refer to these three kinds of inferences as "weak inferences", while all others will be called "strong inferences".

1.4) *Cut*:

$$\frac{\Gamma \rightarrow \Delta, D \qquad D, \Pi \rightarrow \Lambda}{\Gamma, \Pi \rightarrow \Delta, \Lambda}.$$

D is called the *cut formula* of this inference.

2) Logical rules:

2.1) $\neg$: left : $\dfrac{\Gamma \rightarrow \Delta, D}{\neg D, \Gamma \rightarrow \Delta}$; $\neg$: right : $\dfrac{D, \Gamma \rightarrow \Delta}{\Gamma \rightarrow \Delta, \neg D}$.

D and $\neg D$ are called the *auxiliary formula* and the *principal formula*, respectively, of this inference.

2.2) $\wedge$: left : $\dfrac{C, \Gamma \rightarrow \Delta}{C \wedge D, \Gamma \rightarrow \Delta}$ and $\dfrac{D, \Gamma \rightarrow \Delta}{C \wedge D, \Gamma \rightarrow \Delta}$;

$\wedge$: right : $\dfrac{\Gamma \rightarrow \Delta, C \qquad \Gamma \rightarrow \Delta, D}{\Gamma \rightarrow \Delta, C \wedge D}$.

C and D are called the auxiliary formulas and $C \wedge D$ is called the principal formula of this inference.

2.3) $\vee$: left : $\dfrac{C, \Gamma \rightarrow \Delta \quad D, \Gamma \rightarrow \Delta}{C \vee D, \Gamma \rightarrow \Delta}$;

$\vee$: right : $\dfrac{\Gamma \rightarrow \Delta, C}{\Gamma \rightarrow \Delta, C \vee D}$ and $\dfrac{\Gamma \rightarrow \Delta, D}{\Gamma \rightarrow \Delta, C \vee D}$.

C and D are called the auxiliary formulas and $C \vee D$ the principal formula of this inference.

2.4) $\supset$: left : $\dfrac{\Gamma \rightarrow \Delta, C \quad D, \Pi \rightarrow \Lambda}{C \supset D, \Gamma, \Pi \rightarrow \Delta, \Lambda}$;

$\supset$: right : $\dfrac{C, \Gamma \rightarrow \Delta, D}{\Gamma \rightarrow \Delta, C \supset D}$.

C and D are called the auxiliary formulas and $C \supset D$ the principal formula.

2.1)–2.4) are called *propositional inferences*.

2.5) $\forall$: left : $\dfrac{F(t), \Gamma \rightarrow \Delta}{\forall x\, F(x), \Gamma \rightarrow \Delta}$, $\forall$: right : $\dfrac{\Gamma \rightarrow \Delta, F(a)}{\Gamma \rightarrow \Delta, \forall x\, F(x)}$,

where t is an arbitrary term, and a does not occur in the lower sequent. $F(t)$ and $F(a)$ are called the auxiliary formulas and $\forall x\, F(x)$ the principal formula. The a in $\forall$: right is called the *eigenvariable* of this inference.

Note that in $\forall$: right all occurrences of a in $F(a)$ are indicated. In $\forall$: left,

$F(t)$ and $F(x)$ are

$$\left(F(a)\frac{a}{t}\right) \quad \text{and} \quad \left(F(a)\frac{a}{x}\right),$$

respectively (for some free variable a), so not every t in $F(t)$ is necessarily indicated.

2.6) $\exists$: left: $\dfrac{F(a), \Gamma \to \Delta}{\exists x\, F(x), \Gamma \to \Delta}$, $\exists$: right: $\dfrac{\Gamma \to \Delta, F(t)}{\Gamma \to \Delta, \exists x\, F(x)}$,

where a does not occur in the lower sequent, and t is an arbitrary term.
$F(a)$ and $F(t)$ are called the auxiliary formulas and $\exists x\, F(x)$ the principal formula. The a in $\exists$: left is called the eigenvariable of this inference.

Note that in $\exists$: left a is fully indicated, while in $\exists$: right not necessarily every t is indicated. (Again, $F(t)$ is $(F(a)\frac{a}{t})$ for some a.)

2.5) and 2.6) are called *quantifier inferences*. The condition, that the eigenvariable must not occur in the lower sequent in $\forall$: right and $\exists$: left, is called the *eigenvariable condition* for these inferences.

A sequent of the form $A \to A$ is called an *initial sequent*, or *axiom*.

We now explain the notion of formal proof, i.e., proof in **LK**.

DEFINITION 2.2. A *proof* P (in **LK**), or **LK**-*proof*, is a tree of sequents satisfying the following conditions:

1) The topmost sequents of P are initial sequents.
2) Every sequent in P except the lowest one is an upper sequent of an inference whose lower sequent is also in P.

The following terminology and conventions will be used in discussing formal proofs in **LK**.

DEFINITION 2.3. From Definition 2.2, it follows that there is a unique lowest sequent in a proof P. This will be called the *end-sequent* of P. A proof with end-sequent S is called a *proof ending with* S or a *proof of* S. A sequent S is called *provable* in **LK**, or **LK**-*provable*, if there is an **LK**-proof of it. A formula A is called **LK**-*provable* (or a *theorem of* **LK**) if the sequent $\to A$ is **LK**-provable. The prefix "**LK**-" will often be omitted from "**LK**-proof" and "**LK**-provable".

A proof without the cut rule is called *cut-free*.

It will be standard notation to abbreviate part of a proof by $\vdots$. Thus, for example,

$$\begin{array}{c}\vdots\\ S\end{array} \qquad \text{and} \qquad \begin{array}{c}S_1 \quad S_2\\ \vdots\\ S\end{array}$$

denote a proof of S, and a proof of S from S_1 and S_2, respectively. Proofs are mostly denoted by letters $P, Q, \ldots$. An expression such as $P(a)$ means that all the occurrences of a in P are indicated. (Of course such notation is useful only when replacement of a by another term is being considered.) Then $P(t)$ is the result of replacing all occurrences of a in $P(a)$ by t.

Let us consider some slightly modified rules of inference, e.g.,

$$J \; \frac{\Gamma \to \Delta, A \qquad \Pi \to \Lambda, B}{\Gamma, \Pi \to \Delta, \Lambda, A \wedge B}.$$

This is not a rule of inference of **LK**. However, from the two upper sequents we can infer the lower sequent in **LK** using several structural inferences and an $\wedge$: right :

$$(*) \qquad \wedge : \text{right}: \; \frac{\dfrac{\dfrac{\Gamma \to \Delta, A}{\text{several weakenings and exchanges}}}{\Gamma, \Pi \to \Delta, \Lambda, A} \qquad \dfrac{\dfrac{\Pi \to \Lambda, B}{\text{several weakenings and exchanges}}}{\Gamma, \Pi \to \Delta, \Lambda, B}}{\Gamma, \Pi \to \Delta, \Lambda, A \wedge B}$$

Conversely, from the sequents $\Gamma \to \Delta, A$ and $\Gamma \to \Delta, B$ we can infer $\Gamma \to \Delta, A \wedge B$ using several structural inferences and an instance of the inference-schema J:

$$\frac{\dfrac{J \; \dfrac{\Gamma \to \Delta, A \qquad \Gamma \to \Delta, B}{\Gamma, \Gamma \to \Delta, \Delta, A \wedge B}}{\text{several contractions and exchanges}}}{\Gamma \to \Delta, A \wedge B}.$$

Thus we may regard J as an abbreviation of (*) above. In such a case we will use the notation

$$\begin{array}{c}\Gamma \to \Delta, A \qquad \Pi \to \Lambda, B\\ \hline\hline \Gamma, \Pi \to \Delta, \Lambda, A \wedge B\end{array}.$$

As in this example we often indicate abbreviation of several steps by double lines.

Another remark we wish to make here is that the restriction on bound variables (in the definition of formulas) prohibits an unwanted inference such as

$$\cfrac{\cfrac{\cfrac{A(a), B(b) \rightarrow A(a) \wedge B(b)}{A(a), B(b) \rightarrow \exists x\,(A(x) \wedge B(b))}}{A(a), B(b) \rightarrow \exists x\, \exists x\,(A(x) \wedge B(x))}}{\exists x\, A(x), \exists x\, B(x) \rightarrow \exists x\, \exists x\,(A(x) \wedge B(x)).}$$

In our system this can never happen, since $\exists x\, \exists x\,(A(x) \wedge B(x))$ is not a formula.

The quantifier-free part of **LK**, that is, the subsystem of **LK** which does not involve quantifiers, is called the *propositional calculus*.

EXAMPLE 2.4. The following are **LK**-proofs.

1)

$$\begin{array}{rc}
 & A \rightarrow A \\ \cline{2-2}
\neg : \text{right} & \rightarrow A, \neg A \\ \cline{2-2}
\vee : \text{right} & \rightarrow A, A \vee \neg A \\ \cline{2-2}
\text{exchange} : \text{right} & \rightarrow A \vee \neg A, A \\ \cline{2-2}
\vee : \text{right} & \rightarrow A \vee \neg A, A \vee \neg A \\ \cline{2-2}
\text{contraction} : \text{right} & \rightarrow A \vee \neg A.
\end{array}$$

2) Suppose that a is fully indicated in $F(a)$.

$$\begin{array}{rc}
 & F(a) \rightarrow F(a) \\ \cline{2-2}
\exists : \text{right} & F(a) \rightarrow \exists x\, F(x) \\ \cline{2-2}
\neg : \text{right} & \rightarrow \exists x\, F(x), \neg F(a) \\ \cline{2-2}
\forall : \text{right} & \rightarrow \exists x\, F(x), \forall y \neg F(y) \\ \cline{2-2}
\neg : \text{left} & \neg \forall y \neg F(y) \rightarrow \exists x\, F(x) \\ \cline{2-2}
\supset : \text{right} & \rightarrow \neg \forall y \neg F(y) \supset \exists x\, F(x)
\end{array}$$

It should be noted that the lower sequent of $\forall$: right does not contain the eigenvariable a.

EXERCISE 2.5. Prove the following in **LK**.
1) $A \vee B \equiv \neg(\neg A \wedge \neg B)$.
2) $A \supset B \equiv \neg A \vee B$.
3) $\exists x\, F(x) \equiv \neg \forall y \neg F(y)$.
4) $\neg \forall y\, F(y) \equiv \exists x \neg F(x)$.
5) $\neg(A \wedge B) \equiv \neg A \vee \neg B$.

EXERCISE 2.6. Prove the following in **LK**.
1) $\exists x\,(A \supset B(x)) \equiv A \supset \exists x\, B(x)$.

2) $\exists x\,(A(x) \supset B) \equiv \forall x\, A(x) \supset B$, where B does not contain x.
3) $\exists x\,(A(x) \supset B(x)) \equiv \forall x\, A(x) \supset \exists x\, B(x)$.
4) $\neg A \supset B \rightarrow \neg B \supset A$.
5) $\neg A \supset \neg B \rightarrow B \supset A$.

EXERCISE 2.7. Construct a cut-free proof of $\forall x\, A(x) \supset B \rightarrow \exists x\,(A(x) \supset B)$, where $A(a)$ and B are atomic and distinct.

DEFINITION 2.8. (1) When we consider a formula, term or logical symbol together with the place that it occupies in a proof, sequent or formula respectively, we refer to it as a formula, term or logical symbol in the proof, sequent or formula, respectively.

(2) A sequence of sequents in a proof P is called a *thread* (of P) if the following conditions are satisfied:

2.1) The sequence begins with an initial sequent and ends with the end-sequent.

2.2) Every sequent in the sequence except the last is an upper sequent of an inference, and is immediately followed by the lower sequent of this inference.

(3) Let S_1, S_2 and S_3 be sequents in a proof P. We say S_1 is *above* S_2 or S_2 is *below* S_1 (in P) if there is a thread containing both S_1 and S_2 in which S_1 appears before S_2. If S_1 is above S_2 and S_2 is above S_3, we say S_2 is *between* S_1 and S_3.

(4) An inference in P is said to be *below a sequent* S (in P) if its lower sequent is below S.

(5) Let P be a proof. A part of P which itself is a proof is called a *subproof* of P. This can also be described as follows. For any sequent S in P, that part of P which consists of all sequents which are either S itself or which occur above S, is called a subproof of P (with end-sequent S).

(6) Let P_0 be a proof of the form

$$\begin{array}{c} \vdots \\ \Gamma \rightarrow \Theta \\ (*)\{ \quad \vdots \end{array}$$

where (*) denotes the part of P_0 under $\Gamma \rightarrow \Theta$, and let Q be a proof ending with $\Gamma, D \rightarrow \Theta$. By a copy of P_0 from Q we mean a proof P of the form

$$\begin{array}{c} Q\left\{ \begin{array}{c} \vdots \\ \Gamma, D \rightarrow \Theta \end{array}\right. \\ (**)\{ \quad \vdots \end{array}$$

where (**) differs from (*) only in that for each sequent in (*), say $\Pi \rightarrow \Lambda$, the corresponding sequent in (**) has the form $\Pi, D \rightarrow \Lambda$. That is to say, P

is obtained from P_0 by replacing the subproof ending with $\Gamma \rightarrow \Theta$ by Q, and adding an extra formula D to the antecedent of each sequent in (*). Likewise, a copy can be defined for the case of an extra formula in the succedent. We can also extend the definition to the case where there are several of these formulas.

The precise definition can be carried out by induction on the number of inferences in (*). However this notion is intuitive, simple, and will appear often in this book.

(7) Let $S(a)$, or $\Gamma(a) \rightarrow \Delta(a)$, denote a sequent of the form $A_1(a), \ldots, A_m(a) \rightarrow B_1(a), \ldots, B_n(a)$. Then $S(t)$, or $\Gamma(t) \rightarrow \Delta(t)$, denotes the sequent $A_1(t), \ldots, A_m(t) \rightarrow B_1(t), \ldots, B_n(t)$.

We can define: t is fully indicated in $S(t)$, or $\Gamma(t) \rightarrow \Delta(t)$, by analogy with Definition 1.6.

In order to prove a basic property of provability, i.e., that provability is preserved under substitution of terms for free variables, we shall first list some lemmas, which themselves assert basic properties of proofs. We first define an important concept.

DEFINITION 2.9. A proof in **LK** is called *regular* if it satisfies the condition that firstly, all eigenvariables are distinct from one another, and secondly, if a free variable a occurs as an eigenvariable in a sequent S of the proof, then a occurs only in sequents above S.

LEMMA 2.10. (1) *Let $\Gamma(a) \rightarrow \Delta(a)$ be an* (**LK**-)*provable sequent in which a is fully indicated, and let $P(a)$ be a proof of $\Gamma(a) \rightarrow \Delta(a)$. Let b be a free variable not occurring in $P(a)$. Then the tree $P(b)$, obtained from $P(a)$ by replacing a by b at each occurrence of a in $P(a)$, is also a proof and its end-sequent is $\Gamma(b) \rightarrow \Delta(b)$.*

(2) *For an arbitrary* **LK**-*proof there exists a regular proof of the same end-sequent. Moreover, the required proof is obtained from the original proof simply by replacing free variables* (*in a suitable way*).

PROOF. (1) By induction on the number of inferences in $P(a)$. If $P(a)$ consists of simply an initial sequent $A(a) \rightarrow A(a)$, then $P(b)$ consists of the sequent $A(b) \rightarrow A(b)$, which is also an initial sequent. Let us suppose that our proposition holds for proofs containing at most n inferences and suppose that $P(a)$ contains $n+1$ inferences. We treat the possible cases according to the last inferences in $P(a)$. Since other cases can be treated similarly, we consider only the case where the last inference, say J, is a $\forall$: right. Suppose the eigenvariable of J is a, and $P(a)$ is of the form

$$\begin{array}{l} Q(a)\left\{ \begin{array}{c} \vdots \\ \Gamma \rightarrow \Lambda, A(a) \end{array}\right. \\ J \quad \dfrac{\Gamma \rightarrow \Lambda, A(a)}{\Gamma \rightarrow \Lambda, \forall x\, A(x)}, \end{array}$$

where $Q(a)$ is the subproof of $P(a)$ ending with $\Gamma \to \Lambda, A(a)$. It should be remembered that a does not occur in Γ, Λ or $A(x)$. By the induction hypotheses the result of replacing all a's in $Q(a)$ by b is a proof whose end-sequent is $\Gamma \to \Lambda, A(b)$. Γ and Λ contain no b's. Thus we can apply a $\forall$: right to this sequent using b as its eigenvariable:

$$Q(b)\left\{\begin{array}{c} \vdots \\ \Gamma \to \Lambda, A(b) \end{array}\right. \over \Gamma \to \Lambda, \forall x\, A(x)$$

and so $P(b)$ is a proof ending with $\Gamma \to \Lambda, \forall x\, A(x)$. If a is not the eigenvariable of J, $P(a)$ is of the form

$$Q(a)\left\{\begin{array}{c} \vdots \\ \Gamma(a) \to \Lambda(a), A(a, c) \end{array}\right. \over \Gamma(a) \to \Lambda(a), \forall x\, A(a, x).$$

By the induction hypothesis the result of replacing all a's in $Q(a)$ by b is a proof and its end-sequent is $\Gamma(b) \to \Lambda(b), A(b, c)$.

Since by assumption b does not occur in $P(a)$, b is not c, and so we can apply a $\forall$: right to this sequent, with c as its eigenvariable, and obtain a proof $P(b)$ whose end-sequent is $\Gamma(b) \to \Lambda(b), \forall x\, A(b, x)$.

(2) By mathematical induction on the number l of applications of $\forall$: right and $\exists$: left in a given proof P. If $l = 0$, then take P itself. Otherwise, P can be represented in the form:

$$\begin{array}{c} P_1 \quad P_2 \ldots P_k \\ (*)\left\{ \begin{array}{c} \ddots \;\vdots\; \dots \\ S \end{array}\right. \end{array}$$

where P_i is a subproof of P of the form

$$I_i \; \frac{\begin{array}{c}\vdots\\ \Gamma_i \to \Delta_i, F_i(b_i)\end{array}}{\Gamma_i \to \Delta_i, \forall y_i\, F_i(y_i)} \quad \text{or} \quad I_i \; \frac{\begin{array}{c}\vdots\\ F_i(b_i), \Gamma_i \to \Delta_i\end{array}}{\exists y_i\, F_i(y_i), \Gamma_i \to \Delta_i}$$

and I_i is a lowermost $\forall$: right or $\exists$: left in P $(i = 1, \ldots, k)$, i.e., there is no $\forall$: right or $\exists$: left in the part of P denoted by (*).

Let us deal with the case where I_i is $\forall$: right. P_i has fewer applications of $\forall$: right or $\exists$: left than P, so by the induction hypothesis there is a regular proof P_i' of $\Gamma_i \to \Delta_i, F_i(b_i)$. Note that no free variable in $\Gamma_i \to \Delta_i, F(b_i)$ (including b_i) is used as an eigenvariable in P_i'. Suppose $c_1, \ldots, c_m$ are all the eigenvariables in all the P_i's which occur in P above $\Gamma_i \to \Delta_i, \forall y_i\, F_i(y_i)$, $i = 1, \ldots, k$. Then change $c_1, \ldots, c_m$ to $d_1, \ldots, d_m$, respectively, where

$d_1, \ldots, d_m$ are the first m variables which occur neither in P nor in P'_i, $i = 1, \ldots, k$. If b_i occurs in P below $\Gamma_i \to \Delta_i, \forall y_i\, F_i(y_i)$, then change it to d_{m+i}.

Let P''_i be the proof which is obtained from P'_i by the above replacement of variables. Then $P''_1, \ldots, P''_k$ are each regular. P' is defined to be

$$(*)\left\{\begin{array}{c} P''_1 \cdots \dfrac{P''_i}{\Gamma_i \to \Delta_i, \forall y_i\, F_i(y_i)} \cdots P''_n \\ \vdots \\ S \end{array}\right.$$

where (*) is the same as in P, except for the replacement of b_i by d_{m+i}. This completes the proof.

From now on we will assume that we are dealing with regular proofs whenever convenient, and will not mention it on each occasion.

By a method similar to that in Lemma 2.10 we can prove the following.

LEMMA 2.11. *Let t be an arbitrary term. Let $\Gamma(a) \to \Delta(a)$ be a provable (in* **LK***) sequent in which a is fully indicated, and let $P(a)$ be a proof ending with $\Gamma(a) \to \Delta(a)$ in which every eigenvariable is different from a and not contained in t. Then $P(t)$ (the result of replacing all a's in $P(a)$ by t) is a proof whose end-sequent is $\Gamma(t) \to \Delta(t)$.*

LEMMA 2.12. *Let t be an arbitrary term, $\Gamma(a) \to \Delta(a)$ a provable (in* **LK***) sequent in which a is fully indicated, and $P(a)$ a proof of $\Gamma(a) \to \Delta(a)$. Let $P'(a)$ be a proof obtained from $P(a)$ by changing eigenvariables (not necessarily replacing distinct ones by distinct ones) in such a way that in $P'(a)$ every eigenvariable is different from a and not contained in t. Then $P'(t)$ is a proof of $\Gamma(t) \to \Delta(t)$.*

PROOF. By induction on the number of eigenvariables in $P(a)$ which are either a or contained in t, using Lemmas 2.10 and 2.11.

We rewrite part of Lemma 2.11 as follows.

PROPOSITION 2.13. *Let t be an arbitrary term and $S(a)$ a provable (in* **LK***) sequent in which a is fully indicated. The $S(t)$ is also provable.*

We will point out a simple, but useful fact about the formal proofs of our system, which will be used repeatedly.

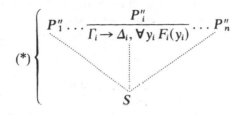

PROPOSITION 2.14. *If a sequent is provable, then it is provable with a proof in which all the initial sequents consist of atomic formulas. Furthermore, if a sequent is provable without cut, then it is provable without cut with a proof of the above sort.*

PROOF. It suffices to show that for an arbitrary formula A, $A \rightarrow A$ is provable without cut, starting with initial sequents consisting of atomic formulas. This, however, can be easily shown by induction on the complexity of A.

DEFINITION 2.15. We say that two formulas A and B are *alphabetical variants* (of one another) if for some $x_1, \ldots, x_n, y_1, \ldots, y_n$

$$\left(A \frac{x_1, \ldots, x_n}{z_1, \ldots, z_n}\right)$$

is

$$\left(B \frac{y_1, \ldots, y_n}{z_1, \ldots, z_n}\right),$$

where $z_1, \ldots, z_n$ are bound variables occurring neither in A nor in B: that is to say, if A and B are different, it is only because they have a different choice of bound variables. The fact that A and B are alphabetical variants will be expressed by $A \sim B$.

One can easily prove that the relation $A \sim B$ is an equivalence relation. Intuitively it is obvious that changing bound variables in a formula does not change its meaning. We can prove by induction on the number of logical symbols in A that if $A \sim B$, then $A \equiv B$ is provable without cut (indeed in **LJ**, which is to be defined in the next section). Thus two alphabetical variants will often be identified without mention.

§3. A formulation of intuitionistic predicate calculus

DEFINITION 3.1. We can formalize the intuitionistic predicate calculus as a sub-system of **LK**, which we call **LJ**, following Gentzen. (**J** stands for "intuitionistic".) **LJ** is obtained from **LK** by modifying it as follows (cf. Definitions 2.1 and 2.2 for **LK**):

1) A sequent in **LJ** is of the form $\Gamma \rightarrow \Delta$, where Δ consists of at most one formula.

2) Inferences in **LJ** are those obtained from those in **LK** by imposing the restriction that the succedent of each upper and lower sequent consists of at most one formula; thus there are no inferences in **LJ** corresponding to contraction : right or exchange : right.

The notions of proof, provable and other concepts for **LJ** are defined similarly to the corresponding notions for **LK**.

Every proof in **LJ** is obviously a proof in **LK**, but the converse is not true. Hence:

PROPOSITION 3.2. *If a sequent S of* **LJ** *is provable in* **LJ**, *then it is also provable in* **LK**.

Lemmas 2.10–2.12 and Propositions 2.13 and 2.14 hold, reading "**LJ**-provable" in place of "provable" or "provable (in **LK**)". We shall refer to these results (for **LJ**) as Lemma 3.3, Lemma 3.4, Lemma 3.5, Proposition 3.6 and Proposition 3.7, respectively. We omit the statements of these.)

EXAMPLE 3.8. The following are **LJ**-proofs.

1)

$$\begin{array}{rc}
 & A \to A \\ \cline{2-2}
\wedge:\text{left} & A \wedge \neg A \to A \\ \cline{2-2}
\neg:\text{left} & \neg A,\ A \wedge \neg A \to \\ \cline{2-2}
\wedge:\text{left} & A \wedge \neg A,\ A \wedge \neg A \to \\ \cline{2-2}
\text{contraction}:\text{left} & A \wedge \neg A \to \\ \cline{2-2}
\neg:\text{right} & \to \neg(A \wedge \neg A)
\end{array}$$

2) Suppose a is fully indicated in $F(a)$.

$$\begin{array}{rc}
 & F(a) \to F(a) \\ \cline{2-2}
\exists:\text{right} & F(a) \to \exists x\, F(x) \\ \cline{2-2}
\neg:\text{left} & \neg\exists x\, F(x),\ F(a) \to \\ \cline{2-2}
\text{exchange}:\text{left} & F(a),\ \neg\exists x\, F(x) \to \\ \cline{2-2}
\neg:\text{left} & \neg\exists x\, F(x) \to \neg F(a) \\ \cline{2-2}
\forall:\text{right} & \neg\exists x\, F(x) \to \forall y \neg F(y).
\end{array}$$

EXERCISE 3.9. Prove the following in **LJ**.

1) $\neg A \vee B \to A \supset B$.
2) $\exists x\, F(x) \to \neg\forall y \neg F(y)$.
3) $A \wedge B \to A$.
4) $A \to A \vee B$.
5) $\neg A \vee \neg B \to \neg(A \wedge B)$.
6) $\neg(A \vee B) \equiv \neg A \wedge \neg B$.
7) $(A \vee C) \wedge (B \vee C) \equiv (A \wedge B) \vee C$.
8) $\exists x \neg F(x) \to \neg\forall x\, F(x)$.

9) $\forall x\,(F(x) \wedge G(x)) \equiv \forall x\, F(x) \wedge \forall x\, G(x)$.
10) $A \supset \neg B \rightarrow B \supset \neg A$.
11) $\exists x\,(A \supset B(x)) \rightarrow A \supset \exists x\, B(x)$.
12) $\exists x\,(A(x) \supset B) \rightarrow \forall x\, A(x) \supset B$.
13) $\exists x\,(A(x) \supset B(x)) \rightarrow \forall x\, A(x) \supset \exists x\, B(x)$.

EXERCISE 3.10. Prove the following in **LJ**.
1) $\neg\neg(A \supset B), A \rightarrow \neg\neg B$.
2) $\neg\neg B \supset B, \neg\neg(A \supset B) \rightarrow A \supset B$.
3) $\neg\neg\neg A \equiv \neg A$.

EXERCISE 3.11. Define **LJ**$'$ to be the system which is obtained from **LJ** by adding to it, as initial sequents, all sequents $\neg\neg R \rightarrow R$, where R is atomic. Let A be a formula which does not contain $\vee$ or $\exists$. Then $\neg\neg A \rightarrow A$ is **LJ**$'$-provable. [*Hint*: By induction on the number of logical symbols in A; cf. Exercise 3.10.]

PROBLEM 3.12. For every formula A define A^* as follows.
1) If A is atomic, then A^* is $\neg\neg A$.
2) If A is one of the forms $\neg B$, $B \wedge C$, $B \vee C$ or $B \supset C$, then A^* is $\neg B^*$, $B^* \wedge C^*$, $\neg(\neg B^* \wedge \neg C^*)$ or $B^* \supset C^*$, respectively.
3) If A is of the form $\forall x\, F(x)$ or $\exists x\, F(x)$, then A^* is $\forall x\, F^*(x)$ or $\neg \forall x \neg F^*(x)$, respectively.
(Thus A^* does not contain $\vee$ or $\exists$.) Prove that for any A, A is **LK**-provable if and only if A^* is **LJ**-provable. [*Hint*: Follow the prescription given below.]
1) For any A, $A \equiv A^*$ is **LK**-provable.
2) Let S be a sequent of the form $A_1, \ldots, A_m \rightarrow B_1, \ldots, B_n$. Let S' be the sequent

$$A_1^*, \ldots, A_m^*, \neg B_1^*, \ldots, \neg B_n^* \rightarrow.$$

Prove that S is **LK**-provable if and only if S' is **LK**-provable.
3) $A^* \equiv \neg\neg A^*$ in **LJ**, from Exercise 3.11.
4) Show that if S is **LK**-provable, then S' is **LJ**-provable. (Use mathematical induction on the number of inferences in a proof of S.)
What must be proved is now a special case of 4).

§4. Axiom systems

DEFINITION 4.1. The basic system is **LK**.
1) A finite or infinite set $\mathscr{A}$ of sentences is called an *axiom system*, and each of these sentences is called an *axiom* of $\mathscr{A}$. Sometimes an axiom system is called a *theory*. (Of course this definition is only significant in certain contexts.)

2) A finite (possibly empty) sequence of formulas consisting only of axioms of $\mathscr{A}$ is called an *axiom sequence* of $\mathscr{A}$.

3) If there exists an axiom sequence Γ_0 of $\mathscr{A}$ such that $\Gamma_0, \Gamma \rightarrow \Delta$ is **LK**-provable, then $\Gamma \rightarrow \Delta$ is said to be *provable from* $\mathscr{A}$ (in **LK**). We express this by $\mathscr{A}, \Gamma \rightarrow \Delta$.

4) $\mathscr{A}$ is *inconsistent* (with **LK**) if the empty sequent $\rightarrow$ is provable from $\mathscr{A}$ (in **LK**).

5) If $\mathscr{A}$ is not inconsistent (with **LK**), then it is said to be *consistent* (with **LK**).

6) If all function constants and predicate constants in a formula A occur in $\mathscr{A}$, then A is said to be *dependent on* $\mathscr{A}$.

7) A sentence A is said to be *consistent* (*inconsistent*) if the axiom system $\{A\}$ is consistent (inconsistent).

8) $\mathbf{LK}_{\mathscr{A}}$ is the system obtained from **LK** by adding $\rightarrow A$ as initial sequents for all A in $\mathscr{A}$.

9) $\mathbf{LK}_{\mathscr{A}}$ is said to be *inconsistent* if $\rightarrow$ is $\mathbf{LK}_{\mathscr{A}}$-provable, otherwise it is *consistent*.

The following propositions, which are easily proved, will be used quite often.

PROPOSITION 4.2. *Let $\mathscr{A}$ be an axiom system. Then the following are equivalent*:

(a) *$\mathscr{A}$ is inconsistent* (*with* **LK**) (*as defined above*);

(b) *for every formula A* (*of the language*), *A is provable from $\mathscr{A}$*;

(c) *for some formula A, A and $\neg A$ are both provable from $\mathscr{A}$.*

PROPOSITION 4.3. *Let $\mathscr{A}$ be an axiom system. Then a sequent $\Gamma \rightarrow \Delta$ is $\mathbf{LK}_{\mathscr{A}}$-provable if and only if $\Gamma \rightarrow \Delta$ is provable from $\mathscr{A}$* (*in* **LK**).

COROLLARY 4.4. *An axiom system $\mathscr{A}$ is consistent* (*with* **LK**) *if and only if $\mathbf{LK}_{\mathscr{A}}$ is consistent.*

These definitions and the propositions hold also for **LJ**.

§5. The cut-elimination theorem

A very important fact about **LK** is the cut-elimination theorem, also known as Gentzen's Hauptsatz:

THEOREM 5.1 (the cut-elimination theorem: Gentzen). *If a sequent is* (**LK**)*-provable, then it is* (**LK**-)*provable without a cut.*

This means that any theorem in the predicate calculus can be proved without detours, so to speak. We shall come back to this point later. The purpose of the present section is to prove this theorem. We shall follow Gentzen's original proof.

First we introduce a new rule of inference, the mix rule, and show that the mix rule and the cut rule are equivalent. Let A be a formula. An inference of the following form is called a *mix* (with respect to A):

$$\frac{\Gamma \rightarrow \Delta \qquad \Pi \rightarrow \Lambda}{\Gamma, \Pi^* \rightarrow \Delta^*, \Lambda} \ (A)$$

where both Δ and Π contain the formula A, and Δ^* and Π^* are obtained from Δ and Π respectively by deleting all the occurrences of A in them. We call A the mix formula of this inference, and the mix formula of a mix is normally indicated in parentheses (as above).

Let us call the system which is obtained from **LK** by replacing the cut rule by the mix rule, **LK***. The following is easily proved.

LEMMA 5.2. **LK** *and* **LK*** *are equivalent, that is, a sequent S is* **LK**-*provable if and only if S is* **LK***-*provable.*

By virtue of the Lemma 5.2, it suffices to show that the mix rule is redundant in **LK***, since a proof in **LK*** without a mix is at the same time a proof in **LK** without a cut.

THEOREM 5.3 (cf. Theorem 5.1). *If a sequent is provable in* **LK***, *then it is provable in* **LK*** *without a mix.*

This theorem is an immediate consequence of the following lemma.

LEMMA 5.4. *If P is a proof of S* (*in* **LK***) *which contains* (*only*) *one mix, occurring as the last inference, then S is provable without a mix.*

The proof of Theorem 5.3 from Lemma 5.4 is simply by induction on the number of mixes occurring in a proof of S.

The rest of this section is devoted to proving Lemma 5.4. We first define two scales for measuring the complexity of a proof. The *grade* of a formula A (denoted by $g(A)$) is the number of logical symbols contained in A. The grade of a mix is the grade of the mix formula. When a proof P has a mix (only) as the last inference, we define the grade of P (denoted by $g(P)$) to be the grade of this mix.

Let P be a proof which contains a mix only as the last inference:

$$J\ \frac{\Gamma \rightarrow \Delta \qquad \Pi \rightarrow \Lambda}{\Gamma, \Pi^* \rightarrow \Delta^*, \Lambda} \ (A).$$

We refer to the left and right upper sequents as S_1 and S_2, respectively, and to the lower sequent as S. We call a thread in P a *left* (*right*) *thread* if it contains the left (right) upper sequent of the mix J. The ***rank*** of a thread $\mathscr{F}$ in P is defined as follows: if $\mathscr{F}$ is a left (right) thread, then the rank of $\mathscr{F}$ is the number of consecutive sequents, counting upward from the left (right) upper sequent of J, that contains the mix formula in its succedent (antecedent). Since the left (right) upper sequent always contains the mix formula, the rank of a thread in P is at least 1. The rank of a thread $\mathscr{F}$ in P is denoted by rank($\mathscr{F}$; P). We define

$$\text{rank}_l(P) = \max_{\mathscr{F}}(\text{rank}(\mathscr{F}; P)),$$

where $\mathscr{F}$ ranges over all the left threads in P, and

$$\text{rank}_r(P) = \max_{\mathscr{F}}(\text{rank}(\mathscr{F}; P)),$$

where $\mathscr{F}$ ranges over all the right threads in P. The rank of P, rank(P), is defined as

$$\text{rank}(P) = \text{rank}_l(P) + \text{rank}_r(P).$$

Notice that rank(P) is always $\geqslant 2$.

PROOF OF LEMMA 5.4. We prove the Lemma by double induction on the grade g and rank r of the proof P (i.e., transfinite induction on $\omega \cdot g + r$). We divide the proof into two main cases, namely $r = 2$ and $r > 2$ (regardless of g).

Case 1: $r = 2$, viz. $\text{rank}_l(P) = \text{rank}_r(P) = 1$.

We distinguish cases according to the forms of the proofs of the upper sequents of the mix.

1.1) The left upper sequent S_1 is an initial sequent. In this case we may assume P is of the form

$$J\,\frac{A \rightarrow A \qquad \Pi \rightarrow \Lambda}{A, \Pi^* \rightarrow \Lambda}.$$

We can then obtain the lower sequent without a mix:

$$\cfrac{\cfrac{\cfrac{\Pi \rightarrow \Lambda}{\text{some exchanges}}}{A, \ldots, A, \Pi^* \rightarrow \Lambda}}{\cfrac{\text{some contractions}}{A, \Pi^* \rightarrow \Lambda}}$$

1.2) The right upper sequent S_2 is an initial sequent. Similarly:

1.3) Neither S_1 nor S_2 is an initial sequent, and S_1 is the lower sequent of a structural inference J_1. Since $\text{rank}_l(P) = 1$, the formula A cannot appear in the succedent of the upper sequent of J_1, i.e., J_1 must be weakening: right, whose weakening formula is A:

$$J\frac{J_1\dfrac{\Gamma \rightarrow \Delta_1}{\Gamma \rightarrow \Delta_1, A} \qquad \Pi \rightarrow \Lambda}{\Gamma, \Pi^* \rightarrow \Delta_1, \Lambda}\ (A),$$

where Δ_1 does not contain A. We can eliminate the mix as follows:

$$\frac{\dfrac{\Gamma \rightarrow \Delta_1}{\Pi^*, \Gamma \rightarrow \Delta_1, \Lambda}\ \text{some weakenings}}{\Gamma, \Pi^* \rightarrow \Delta_1, \Lambda}\ \text{some exchanges}$$

1.4) None of 1.1)–1.3) holds but S_2 is the lower sequent of a structural inference. Similarly:

1.5) Both S_1 and S_2 are the lower sequents of logical inferences. In this case, since $\text{rank}_l(P) = \text{rank}_r(P) = 1$, the mix formula on each side must be the principal formula of the logical inference. We use induction on the grade, distinguishing several cases according to the outermost logical symbol of A. We treat here two cases and leave the others to the reader.

(i) The outermost logical symbol of A is $\wedge$. In this case S_1 and S_2 must be the lower sequents of $\wedge$: right and $\wedge$: left, respectively:

$$\frac{\dfrac{\Gamma \rightarrow \Delta_1, B \qquad \Gamma \rightarrow \Delta_1, C}{\Gamma \rightarrow \Delta_1, B \wedge C} \qquad \dfrac{B, \Pi_1 \rightarrow \Lambda}{B \wedge C, \Pi_1 \rightarrow \Lambda}}{\Gamma, \Pi_1 \rightarrow \Delta_1, \Lambda}\ (B \wedge C),$$

where by assumption none of the proofs ending with $\Gamma \rightarrow \Delta_1, B$; $\Gamma \rightarrow \Delta_1, C$ or $B, \Pi_1 \rightarrow \Lambda$ contain a mix. Consider the following:

$$\frac{\Gamma \rightarrow \Delta_1, B \qquad B, \Pi_1 \rightarrow \Lambda}{\Gamma, \Pi_1^{\#} \rightarrow \Delta_1^{\#}, \Lambda}\ (B),$$

where $\Pi_1^{\#}$ and $\Delta_1^{\#}$ are obtained from Π_1 and Δ_1 by omitting all occurrences of B. This proof contains only one mix, a mix that occurs as its last inference. Furthermore the grade of the mix formula B is less than $g(A)$ $(= g(B \wedge C))$. So by the induction hypothesis we can obtain a proof which contains no mixes and whose end-sequent is $\Gamma, \Pi_1^{\#} \rightarrow \Delta_1^{\#}, \Lambda$. From this we can obtain a proof without a mix with end-sequent $\Gamma, \Pi_1 \rightarrow \Delta_1, \Lambda$.

(ii) The outermost logical symbol of A is $\forall$. So A is of the form $\forall x\, F(x)$ and the last part of P has the form:

$$\dfrac{\dfrac{\Gamma \to \Delta_1, F(a)}{\Gamma \to \Delta_1, \forall x\, F(x)} \qquad \dfrac{F(t), \Pi_1 \to \Lambda}{\forall x\, F(x), \Pi_1 \to \Lambda}}{\Gamma, \Pi_1 \to \Delta_1, \Lambda} \quad (A)$$

(a being fully indicated in $F(a)$). By the eigenvariable condition, a does not occur in Γ, Δ_1 or $F(x)$. Since by assumption the proof ending with $\Gamma \to \Delta_1, F(a)$ contains no mix, we can obtain a proof without a mix, ending with $\Gamma \to \Delta_1, F(t)$ (cf. Lemma 2.12). Consider now

$$\dfrac{\Gamma \to \Delta_1, F(t) \qquad F(t), \Pi_1 \to \Lambda}{\Gamma, \Pi_1^{\#} \to \Delta_1^{\#}, \Lambda} \quad (F(t)),$$

where $\Pi_1^{\#}$ and $\Delta_1^{\#}$ are obtained from Π_1 and Δ_1 by omitting all occurrences of $F(t)$. This has only one mix. It occurs as the last inference and the grade of the mix formula is less than $g(A)$. Thus by the induction hypothesis we can eliminate this mix and obtain a proof ending with $\Gamma, \Pi_1^{\#} \to \Delta_1^{\#}, \Lambda$, from which we can obtain a proof, without a mix, ending with $\Gamma, \Pi_1 \to \Delta_1, \Lambda$.

Case 2. $r > 2$, i.e., $\text{rank}_l(P) > 1$ and/or $\text{rank}_r(P) > 1$.

The induction hypothesis is that from every proof Q which contains a mix only as the last inference, and which satisfies either $g(Q) < g(P)$, or $g(Q) = g(P)$ and $\text{rank}(Q) < \text{rank}(P)$, we can eliminate the mix.

2.1) $\text{rank}_r(P) > 1$.

2.1.1) Γ or Δ (in S_1) contains A. Construct a proof as follows.

$$\dfrac{\dfrac{\dfrac{\vdots}{\Pi \to \Lambda}}{\dfrac{\text{some exchanges and contractions}}{\dfrac{A, \Pi^* \to \Lambda}{\dfrac{\text{some weakenings and exchanges}}{\Gamma, \Pi^* \to \Delta^*, \Lambda}}}}}{} \qquad \dfrac{\dfrac{\dfrac{\vdots}{\Gamma \to \Delta}}{\dfrac{\text{some exchanges and contractions}}{\dfrac{\Gamma \to \Delta^*, A}{\dfrac{\text{some weakenings and exchanges}}{\Gamma, \Pi^* \to \Delta^*, \Lambda}}}}}{}$$

2.1.2) S_2 is the lower sequent of an inference J_2, where J_2 is not a logical inference whose principal formula is A. The last part of P looks like this:

$$\dfrac{\Gamma \to \Delta \qquad J_2\ \dfrac{\Phi \to \Psi}{\Pi \to \Lambda}}{\Gamma, \Pi^* \to \Delta^*, \Lambda} \quad (A),$$

where the proofs of $\Gamma \rightarrow \Delta$ and $\Phi \rightarrow \Psi$ contain no mixes and Φ contains at least one A. Consider the following proof P':

$$\text{mix}\ \frac{\overset{\vdots}{\Gamma \rightarrow \Delta} \qquad \overset{\vdots}{\Phi \rightarrow \Psi}}{\Gamma, \Phi^* \rightarrow \Delta^*, \Psi}\ (A).$$

In P', the grade of the mix is equal to $g(P)$, $\text{rank}_l(P') = \text{rank}_l(P)$ and $\text{rank}_r(P') = \text{rank}_r(P) - 1$. Thus by the induction hypothesis, $\Gamma, \Phi^* \rightarrow \Delta^*, \Psi$ is provable without a mix. Then we construct the proof

$$\begin{array}{c} \vdots \\ \Gamma, \Phi^* \rightarrow \Delta^*, \Psi \\ \hline \text{some exchanges} \\ \hline \Phi^*, \Gamma \rightarrow \Delta^*, \Psi \\ \hline \Pi^*, \Gamma \rightarrow \Delta^*, \Lambda \end{array}\ J_2$$

In case that the auxiliary formula in J_2 in P is a mix in Φ, we need an additional weakening before J_2 in the new proof.

2.1.3) Γ contains no A's, and S_2 is the lower sequent of a logical inference whose principal formula is A. Although there are several cases according to the outermost logical symbol of A, we treat only two examples here and leave the rest to the reader.

(i) A is $B \supset C$. The last part of P is of the form:

$$J\ \frac{\Gamma \rightarrow \Delta \qquad J_2\ \dfrac{\Pi_1 \rightarrow \Lambda_1, B \qquad C, \Pi_2 \rightarrow \Lambda_2}{B \supset C, \Pi_1, \Pi_2 \rightarrow \Lambda_1, \Lambda_2}}{\Gamma, \Pi_1^*, \Pi_2^* \rightarrow \Delta^*, \Lambda_1, \Lambda_2}\ (B \supset C)$$

Consider the following proofs P_1 and P_2:

$$P_1\ \frac{\overset{\vdots}{\Gamma \rightarrow \Delta} \qquad \overset{\vdots}{\Pi_1 \rightarrow \Lambda_1, B}}{\Gamma, \Pi_1^* \rightarrow \Delta^*, \Lambda_1, B}\ (B \supset C) \qquad P_2\ \frac{\overset{\vdots}{\Gamma \rightarrow \Delta} \qquad \overset{\vdots}{C, \Pi_2 \rightarrow \Lambda_2}}{\Gamma, C, \Pi_2^* \rightarrow \Delta^*, \Lambda_2}\ (B \supset C)$$

assuming that $B \supset C$ is in Π_1 and Π_2. If $B \supset C$ is not in Π_i ($i = 1$ or 2), then Π_i^* is Π_i and P_i is defined as follows.

$$P_1\ \begin{array}{c} \vdots \\ \Pi_1 \rightarrow \Lambda_1, B \\ \hline \text{weakenings and exchanges} \\ \hline \Gamma, \Pi_1^* \rightarrow \Delta^*, \Lambda_1, B \end{array} \qquad P_2\ \begin{array}{c} \vdots \\ C, \Pi_2 \rightarrow \Lambda_2 \\ \hline \text{weakenings and exchanges} \\ \hline \Gamma, C, \Pi_2^* \rightarrow \Delta^*, \Lambda_2 \end{array}$$

Note that $g(P_1) = g(P_2) = g(P)$, $\text{rank}_l(P_1) = \text{rank}_l(P_2) = \text{rank}_l(P)$ and $\text{rank}_r(P_1) = \text{rank}_r(P_2) = \text{rank}_r(P) - 1$. Hence by the induction hypothesis, the end-sequents of P_1 and P_2 are provable without a mix (say by P_1' and P_2'). Consider the following proof P':

$$
J\ \dfrac{\Gamma \rightarrow \Delta \qquad \dfrac{\begin{matrix} P_1' \\ \vdots \\ \Gamma, \Pi_1^* \rightarrow \Delta^*, \Lambda_1, B \end{matrix} \qquad \dfrac{\begin{matrix} P_2' \\ \vdots \\ \Gamma, C, \Pi_2^* \rightarrow \Delta^*, \Lambda_2 \end{matrix}}{C, \Gamma, \Pi_2^* \rightarrow \Delta^*, \Lambda_2}\ \text{some exchanges}}{B \supset C, \Gamma, \Pi_1^*, \Gamma, \Pi_2^* \rightarrow \Delta^*, \Lambda_1, \Delta^*, \Lambda_2}}{\Gamma, \Gamma, \Pi_1^*, \Gamma, \Pi_2^* \rightarrow \Delta^*, \Delta^*, \Lambda_1, \Delta^*, \Lambda_2}\ (B \supset C).
$$

Then $g(P') = g(P)$, $\text{rank}_l(P') = \text{rank}_l(P)$, $\text{rank}_r(P') = 1$, for Γ contains no occurrences of $B \supset C$ and $\text{rank}(P') < \text{rank}(P)$. Thus the end-sequent of P' is provable without a mix by the induction hypothesis, and hence so is the end-sequent of P.

(ii) A is $\exists x\, F(x)$. The last part of P looks like this:

$$
J\ \dfrac{\Gamma \rightarrow \Delta \qquad \dfrac{F(a), \Pi_1 \rightarrow \Lambda}{\exists x\, F(x), \Pi_1 \rightarrow \Lambda}}{\Gamma, \Pi_1^* \rightarrow \Delta^*, \Lambda}\ (\exists x\, F(x)).
$$

Let b be a free variable not occurring in P. Then the result of replacing a by b throughout the proof ending with $F(a), \Pi_1 \rightarrow \Lambda$ is a proof, without a mix, ending with $F(b), \Pi_1 \rightarrow \Lambda$, since by the eigenvariable condition, a does not occur in Π_1 or Λ (cf. Lemma 2.11).

Consider the following proof:

$$
J\ \dfrac{\begin{matrix} \vdots \\ \Gamma \rightarrow \Delta \end{matrix} \qquad \begin{matrix} \vdots \\ F(b), \Pi_1 \rightarrow \Lambda \end{matrix}}{\Gamma, F(b), \Pi_1^* \rightarrow \Delta^*, \Lambda}\ (\exists x\, F(x)).
$$

By the induction hypothesis, the end-sequent of this proof can be proved without a mix (say by P'). Now consider the proof

$$
J\ \dfrac{\begin{matrix} \vdots \\ \Gamma \rightarrow \Delta \end{matrix} \qquad \dfrac{\dfrac{\begin{matrix} P' \\ \vdots \\ \Gamma, F(b), \Pi_1^* \rightarrow \Delta^*, \Lambda \end{matrix}}{F(b), \Gamma, \Pi_1^* \rightarrow \Delta^*, \Lambda}\ \text{some exchanges}}{\exists x\, F(x), \Gamma, \Pi_1^* \rightarrow \Delta^*, \Lambda}}{\Gamma, \Gamma, \Pi_1^* \rightarrow \Delta^*, \Delta^*, \Lambda}\ (\exists x\, F(x)),
$$

where b occurs in none of $\exists x\, F(x)$, Γ, Π_1, Δ, Λ. This mix can then also be eliminated, by the induction hypothesis.

2.2) $\text{rank}_r(P) = 1$ (and $\text{rank}_l(P) > 1$).
This case is proved in the same way as 2.1) above.

This completes the proof of Lemma 5.4 and hence of the cut-elimination theorem.

It should be emphasized that the proof is constructive, i.e., a new proof is effectively constructed from the given proof in Lemma 5.2 and again in Lemma 5.4, and hence in Theorem 5.1.

The cut-elimination theorem also holds for **LJ**. Actually the above proof is designed so that it goes through for **LJ** without essential changes: we only have to keep in mind that there can be at most one formula in each succedent. The details are left to the reader; we simply state the theorem.

THEOREM 5.5. *The cut-elimination theorem holds for* **LJ**.

§6. Some consequences of the cut-elimination theorem

There are numerous applications of the cut-elimination theorem, some of which will be listed in this section, others as exercises. In order to facilitate discussion of this valuable, productive and important theorem, we shall first define the notion of subformula, which will be used often.

DEFINITION 6.1. By a *subformula* of a formula A we mean a formula used in building up A. The set of subformulas of a formula is inductively defined as follows, by induction on the number of logical symbols in the formula.

(1) An atomic formula has exactly one subformula, viz. the formula itself. The subformulas of $\neg A$ are the subformulas of A and $\neg A$ itself. The subformulas of $A \wedge B$ or $A \vee B$ or $A \supset B$ are the subformulas of A and of B, and the formula itself. The subformulas of $\forall x\, A(x)$ or $\exists x\, A(x)$ are the subformulas of any formula of the form $A(t)$, where t is an arbitrary term, and the formula itself.

(2) Two formulas A and B are said to be *equivalent* in **LK** if $A \equiv B$ is provable in **LK**.

(3) We shall say that in a formula A an occurrence of a logical symbol, say $\sharp$, is *in the scope* of an occurrence of a logical symbol, say $\natural$, if in the construction of A (from atomic formulas) the stage where $\sharp$ is the outermost logical symbol precedes the stage where $\natural$ is the outermost logical symbol (cf. Definition 1.3). Further, a symbol $\sharp$ is said to be in the left scope of a $\supset$ if $\supset$ occurs in the form $B \supset C$ and $\sharp$ occurs in B.

(4) A formula is called *prenex* (in prenex form) if no quantifier in it is in the scope of a propositional connective. It can easily be seen that any

formula is equivalent (in **LK**) to a prenex formula, i.e., for every formula A there is a prenex formula B such that $A \equiv B$ is **LK**-provable.

One can easily see that in any rule of inference except a cut, the lower sequent is no less complicated than the upper sequent(s); more precisely, every formula occurring in an upper sequent is a subformula of some formula occurring in the lower sequent (but not necessarily conversely). Hence a proof without a cut contains only subformulas of the formulas occurring in the end-sequent (the "subformula property"). So the cut-elimination theorem tells us that if a formula is provable in **LK** (or **LJ**) at all, it is provable by use of its subformulas only. (This is what we meant by saying that a theorem in the predicate calculus could be proved without detours.)

From this observation, we can convince ourselves that the empty sequent $\rightarrow$ is not **LK**- (or **LJ**-) provable. This leads us to the consistency proof of **LK** and **LJ**.

THEOREM 6.2 (consistency). **LK** *and* **LJ** *are consistent.*

PROOF. Suppose $\rightarrow$ were provable in **LK** (or **LJ**). Then, by the cut-elimination theorem, it would be provable in **LK** (or **LJ**) without a cut. But this is impossible, by the subformula property of cut-free proofs.

An examination of the proof of this theorem (including the cut-elimination theorem) shows that the consistency of **LK** (and **LJ**) was proved by quantifier-free induction on the ordinal ω^2. We shall not, however, go into the details of the consistency problem at this stage.

For convenience, we re-state the subformula property of cut-free proofs as a theorem.

THEOREM 6.3. *In a cut-free proof in* **LK** (*or* **LJ**) *all the formulas which occur in it are subformulas of the formulas in the end-sequent.*

PROOF. By mathematical induction on the number of inferences in the cut-free proof.

In the rest of this section, we shall list some typical consequences of the cut-elimination theorem. Although some of the results are stated for **LJ** as well as **LK**, we shall give proofs only for **LK**; those for **LJ** are left to the reader.

THEOREM 6.4 (1) (Gentzen's midsequent theorem for **LK**). *Let S be a sequent which consists of prenex formulas only and is provable in* **LK**. *Then* *there is a cut-free proof of S which contains a sequent (called a midsequent), say S′, which satisfies the following:*

1. *S′ is quantifier-free.*
2. *Every inference above S′ is either structural or propositional.*
3. *Every inference below S′ is either structural or a quantifier inference.*

Thus a midsequent splits the proof into an upper part, which contains the propositional inferences, and a lower part, which contains the quantifier inferences.

(2) (The midsequent theorem for **LJ** without ∨ : left.) *The above holds reading "***LJ*** without* ∨ : *left" in place of "***LK***".*

PROOF (outline). Combining Proposition 2.14 and the cut-elimination theorem, we may assume that there is a cut-free proof of S, say P, in which all the initial sequents consist of atomic formulas only. Let I be a quantifier inference in P. The number of propositional inferences under I is called the order of I. The sum of the orders for all the quantifier inferences in P is called the order of P. (The term "order" is used only temporarily here.) The proof is carried out by induction on the order of P.

Case 1: The order of a proof P is 0. If there is a propositional inference, take the lowermost such, and call its lower sequent S_0. Above this sequent there is no quantifier inference. Therefore, if there is a quantifier in or above S_0, then it is introduced by weakenings. Since the proof is cut-free, the weakening formula is a subformula of one of the formulas in the end-sequent. Hence no propositional inferences apply to it. We can thus eliminate these weakenings and obtain a sequent S_0' corresponding to S_0. By adding some weakenings under S_0' (if necessary), we derive S, and S_0' serves as the mid-sequent.

If there is no propositional inference in P, then take the uppermost quantifier inference. Its upper sequent serves as a midsequent.

Case 2: The order of P is not 0. Then there is at least one propositional inference which is below a quantifier inference. Moreover, there is a quantifier inference I with the following property: the uppermost logical inference under I is a propositional inference. Call it I'. We can lower the order by interchanging the positions of I and I'. Here we present just one example: say I is $\forall$: right.

P:

$$
\begin{array}{lc}
 & \vdots \\
I & \dfrac{\Gamma \rightarrow \Theta, F(a)}{\Gamma \rightarrow \Theta, \forall x\, F(x)} \\
(*)\ \{ & \vdots \\
I' & \overline{\qquad \Delta \rightarrow \Lambda \qquad}
\end{array},
$$

where the (*)-part of P contains only structural inferences and Λ contains

$\forall x\, F(x)$ as a sequent-formula. Transform P into the following proof P':

$$
\begin{array}{ll}
 & \Gamma \to \Theta, F(a) \\
\hline
 & \text{structural inferences} \\
\hline
 & \Gamma \to F(a), \Theta, \forall x\, F(x) \\
 & \quad \vdots \\
I' & \overline{\Delta \to F(a), \Lambda} \\
I & \overline{\overline{\Delta \to \Lambda, \forall x\, F(x)}} \\
 & \overline{\overline{\Delta \to \Lambda}} \\
 & \quad \vdots
\end{array}
$$

It is obvious that the order of P' is less than that of P.

Prior to the next theorem, Craig's interpolation theorem*, we shall first state and prove a lemma which itself can be regarded as an interpolation theorem for provable sequents and from which the original form of the interpolation theorem follows immediately. We shall present the argument for **LK** only, although everything goes through for **LJ** as well.

For technical reasons we introduce the predicate symbol T, with 0 argument places, and admit $\to \mathsf{T}$ as an additional initial sequent. (T stands for "true".) The system which is obtained from **LK** thus extended is denoted by **LK#**.

LEMMA 6.5. *Let $\Gamma \to \Delta$ be* **LK**-*provable, and let (Γ_1, Γ_2) and (Δ_1, Δ_2) be arbitrary partitions of Γ and Δ, respectively (including the cases that one or more of $\Gamma_1, \Gamma_2, \Delta_1, \Delta_2$ are empty). We denote such a partition by $[\{\Gamma_1; \Delta_1\}, \{\Gamma_2; \Delta_2\}]$ and call it a partition of the sequent $\Gamma \to \Delta$. Then there exists a formula C of* **LK#** *(called an interpolant of $[\{\Gamma_1; \Delta_1\}, \{\Gamma_2; \Delta_2\}]$ such that:*

(i) *$\Gamma_1 \to \Delta_1, C$ and $C, \Gamma_2 \to \Delta_2$ are both* **LK#**-*provable;*

(ii) *All free variables and individual and predicate constants in C (apart from T) occur both in $\Gamma_1 \cup \Delta_1$ and $\Gamma_2 \cup \Delta_2$.*

We will first prove the theorem (from this lemma) and then prove the lemma.

THEOREM 6.6 (Craig's interpolation theorem for **LK**). (1) *Let A and B be two formulas such that $A \supset B$ is* **LK**-*provable. If A and B have at least one predicate constant in common, then there exists a formula C, called an interpolant of $A \supset B$, such that C contains only those individual constants, predicate constants and free variables that occur in both A and B, and such*

---

* A strong general theory on interpolation theorems is established in N. Motohashi: Interpolation theorem and characterization theorem, Ann. Japan Assoc. Philos. Sci., **4** (1972) pp. 15–80.

that $A \supset C$ and $C \supset B$ are **LK**-*provable. If A and B contain no predicate constant in common, then either $A \rightarrow$ or $\rightarrow B$ is* **LK**-*provable.*

(2) *As above, with* **LJ** *in place of* **LK**.

PROOF. Assume that $A \supset B$, and hence $A \rightarrow B$, is provable, and A and B have at least one predicate constant in common. Then by Lemma 6.5, taking A as Γ_1 and B as Δ_1 (with Γ_2 and Δ_1 empty), there exists a formula C satisfying (i) and (ii). So $A \rightarrow C$ and $C \rightarrow B$ are **LK#**-provable. Let R be predicate constant which is common to A and B and has k argument places. Let R' be $\forall y_1 \ldots \forall y_k R(y_1, \ldots, y_k)$, where $y_1, \ldots, y_k$ are new bound variables. By replacing T by $R' \supset R'$, we can transform C into a formula C' of the original language, such that $A \rightarrow C'$ and $C' \rightarrow B$ are **LK**-provable. C' is then the desired interpolant.

If there is no predicate common to $\Gamma_1 \cup \Delta_1$ and $\Gamma_2 \cup \Delta_2$ in the partition described in Lemma 6.5, then by that lemma, there is a C such that $\Gamma_1 \rightarrow \Delta_1, C$ and $C, \Gamma_2 \rightarrow \Delta_2$ are provable, and C consists of T and logical symbols only. Then it can easily be shown, by induction on the complexity of C, that either $\rightarrow C$ or $C \rightarrow$ is provable. Hence either $\Gamma_1 \rightarrow \Delta_1$ or $\Gamma_2 \rightarrow \Delta_2$ is provable. In particular, this applies to $A \rightarrow B$ when A is taken as Γ_1 and B as Δ_2.

This method is due to Maehara and its significance lies in the fact that an interpolant of $A \supset B$ can be constructively formed from a proof of $A \supset B$. Note also that we could state the theorem in the following form: *If neither $\neg A$ nor B is provable, then there is an interpolant of $A \supset B$.*

A uniform form of interpolation theorem is given in
N. Motohashi: An Axiomatization Theorem, J. Math. Soc. Japan 34 (1982) 551–560.

PROOF OF LEMMA 6.5. The lemma is proved by induction on the number of inferences k, in a cut-free proof of $\Gamma \rightarrow \Delta$. At each stage there are several cases to consider; we deal with some examples only.

1) $k = 0$. $\Gamma \rightarrow \Delta$ has the form $D \rightarrow D$. There are four cases: (1) $[\{D; D\}, \{;\}]$, (2) $[\{;\}, \{D; D\}]$, (3) $[\{D;\}, \{; D\}]$, and (4) $[\{; D\}, \{D;\}]$. Take for C: $\neg\mathsf{T}$ in (1), T in (2), D in (3) and $\neg D$ in (4).

2) $k > 0$ and the last inference is $\wedge$: right:

$$\frac{\Gamma \rightarrow \Delta, A \qquad \Gamma \rightarrow \Delta, B}{\Gamma \rightarrow \Delta, A \wedge B}.$$

Suppose the partition is $[\{\Gamma_1; \Delta_1, A \wedge B\}, \{\Gamma_2; \Delta_2\}]$. Consider the induced partition of the upper sequents, viz. $[\{\Gamma_1; \Delta_1, A\}, \{\Gamma_2; \Delta_2\}]$ and $[\{\Gamma_1; \Delta_1, B\}, \{\Gamma_2; \Delta_2\}]$, respectively. By the induction hypotheses applied to the sub-proofs of the upper sequents, there exist interpolants C_1 and C_2 so that

$\Gamma_1 \to \Delta_1, A, C_1$; $C_1, \Gamma_2 \to \Delta_2$; $\Gamma_1 \to \Delta_1, B, C_2$; and $C_2, \Gamma_2 \to \Delta_2$ are all **LK#**-provable. From these sequents, $\Gamma_1 \to \Delta_1, A \wedge B, C_1 \vee C_2$ and $C_1 \vee C_2, \Gamma_2 \to \Delta_2$ can be derived. Thus $C_1 \vee C_2$ serves as the required interpolant.

3) $k > 0$ and the last inference is $\forall$: left:

$$\frac{F(s), \Gamma \to \Delta}{\forall x F(x), \Gamma \to \Delta}.$$

Suppose $b_1, \ldots, b_n$ are all the free variables and constants (possibly none) which occur in s. Suppose the partition is $[\{\forall x F(x), \Gamma_1; \Delta_1\}, \{\Gamma_2; \Delta_2\}]$. Consider the induced partition of the upper sequent and apply the induction hypothesis. So there exists an interpolant $C(b_1, \ldots, b_n)$ so that

$$F(s), \Gamma_1 \to \Delta_1, C(b_1, \ldots, b_n) \quad \text{and} \quad C(b_1, \ldots, b_n), \Gamma_2 \to \Delta_2$$

are **LK#**-provable. Let $b_{i_1}, \ldots, b_{i_m}$ be all the variables and constants among $b_1, \ldots, b_n$ which do not occur in $\{F(x), \Gamma_1; \Delta_1\}$. Then

$$\forall y_1 \ldots \forall y_m C(b_1, \ldots, y_1, \ldots, y_m, \ldots, b_n),$$

where $b_{i_1}, \ldots, b_{i_m}$ are replaced by the bound variables, serves as the required interpolant.

4) $k > 0$ and the last inference is $\forall$: right:

$$\frac{\Gamma \to \Delta, F(a)}{\Gamma \to \Delta, \forall x F(x)},$$

where a does not occur in the lower sequent.

Suppose the partition is $[\{\Gamma_1; \Delta_1, \forall x F(x)\}, \{\Gamma_2; \Delta_2\}]$. By the induction hypothesis there exists an interpolant C so that $\Gamma_1 \to \Delta_1, F(a), C$ and $C, \Gamma_2 \to \Delta_2$ are provable. Since C does not contain a, we can derive

$$\Gamma_1 \to \Delta_1, \forall x F(x), C,$$

and hence C serves as the interpolant.

All other cases are treated similarly.

EXERCISE 6.7. Let A and B be prenex formulas which have only $\forall$ and $\wedge$ as logical symbols. Assume furthermore that there is at least one predicate constant common to A and B. Suppose $A \supset B$ is provable.

Show that there exists a formula C such that

1) $A \supset C$ and $C \supset B$ are provable;
2) C is a prenex formula;

3) the only logical symbols in C are $\forall$ and $\wedge$;

4) the predicate constants in C are common to A and B.

[*Hint*: Apply the cut-elimination theorem and the midsequent theorem.]

DEFINITION 6.8. (1) A semi-term is an expression like a term, except that bound variables are (also) allowed in its construction. (The precise definition is left to the reader.) Let t be a term and s a semi-term. We call s a sub-semi-term of t if

(i) s contains a bound variable (that is, s is not a term),

(ii) s is not a bound variable itself,

(iii) some subterm of t is obtained from s by replacing all the bound variables in s by appropriate terms.

(2) A semi-formula is an expression like a formula, except that bound variables are (also) allowed to occur free in it (i.e., not in the scope of a quantifier).

THEOREM 6.9. *Let t be a term and S a provable sequent satisfying*:

(*) *There is no sub-semi-term of t in S.*

Then the sequent which is obtained from S by replacing all the occurrences of t in S by a free variable is also provable.

PROOF (outline). Consider a cut-free regular proof of S, say P. From the observation that if (*) holds for the lower sequent of an inference in P then it holds for the upper sequent(s), the theorem follows easily by mathematical induction on the number of inferences in P.

DEFINITION 6.10. Let $R_1, \ldots, R_m, R$ be predicate constants. Let $A(R, R_1, \ldots, R_m)$ be a sentence in which all occurrences of $R, R_1, \ldots, R_m$ are indicated. Let R' be a predicate constant with the same number of argument-places as R. Let B be $\forall x_1 \ldots \forall x_k\,(R(x_1, \ldots, x_k) \equiv R'(x_1, \ldots, x_k))$, where the string of quantifiers is empty if $k = 0$, and let C be $A(R, R_1, \ldots, R_m) \wedge A(R', R_1, \ldots, R_m)$. We say that $A(R, R_1, \ldots, R_m)$ defines (in **LK**) R implicitly in terms of $R_1, \ldots, R_m$ if $C \supset B$ is (**LK**-)provable and we say that $A(R, R_1, \ldots, R_m)$ defines (in **LK**) R explicitly in terms of $R_1, \ldots, R_m$ and the individual constants in $A(R, R_1, \ldots, R_m)$ if there exists a formula $F(a_1, \ldots, a_k)$ containing only the predicate constants $R_1, \ldots, R_m$ and the individual constants in $A(R, R_1, \ldots, R_m)$ such that

$$A(R, R_1, \ldots, R_m) \rightarrow \forall x_1 \ldots \forall x_k\,(R(x_1, \ldots, x_k) \equiv F(x_1, \ldots, x_k))$$

is **LK**-provable.

PROPOSITION 6.11 (Beth's definability theorem for **LK**). *If a predicate constant R is defined implicitly in terms of $R_1, \ldots, R_m$ by $A(R, R_1, \ldots, R_m)$, then R can be defined explicitly in terms of $R_1, \ldots, R_m$ and the individual constants in $A(R, R_1, \ldots, R_m)$.*

PROOF (outline). Let $c_1, \ldots, c_n$ be free variables not occurring in A. Then

$$A(R, R_1, \ldots, R_m), A(R', R_1, \ldots, R_m) \rightarrow R(c_1, \ldots, c_n) \equiv R'(c_1, \ldots, c_n)$$

and hence also

$$A(R, R_1, \ldots, R_m) \wedge R(c_1, \ldots, c_k) \rightarrow A(R', R_1, \ldots, R_m) \supset R'(c_1, \ldots, c_n)$$

are provable. Now apply Craig's theorem (i.e., part (1) of Theorem 6.6) to the latter sequent.

We now present a version of Robinson's theorem (for **LK**).

PROPOSITION 6.12 (Robinson). *Assume that the language contains no function constants. Let $\mathscr{A}_1$ and $\mathscr{A}_2$ be two consistent axiom systems. Suppose furthermore that, for any sentence A which is dependent on $\mathscr{A}_1$ and $\mathscr{A}_2$, it is not the case that $\mathscr{A}_1 \rightarrow A$ and $\mathscr{A}_2 \rightarrow \neg A$ (or $\mathscr{A}_1 \rightarrow \neg A$ and $\mathscr{A}_2 \rightarrow A$) are both provable. Then $\mathscr{A}_1 \cup \mathscr{A}_2$ is consistent. (See Definition 4.1 for the technical terms.)*

PROOF (outline). Suppose $\mathscr{A}_1 \cup \mathscr{A}_2$ is not consistent. Then there are axiom sequences Γ_1 and Γ_2 from $\mathscr{A}_1$ and $\mathscr{A}_2$ respectively such that $\Gamma_1, \Gamma_2 \rightarrow$ is provable. Since $\mathscr{A}_1$ and $\mathscr{A}_2$ are each consistent, neither Γ_1 nor Γ_2 is empty. Apply Lemma 6.5 to the partition $[\{\Gamma_1;\}, \{\Gamma_2;\}]$.

Let **LK**′ and **LJ**′ denote the quantifier-free parts of **LK** and **LJ**, respectively, viz. the formulations (in tree form) of the classical and intuitionistic propositional calculus, respectively.

THEOREM 6.13. *There exist decision procedures for* **LK**′ *and* **LJ**′.

PROOF (outline). The following decision procedure was given by Gentzen. A sequent of **LK**′ (or **LJ**′) is said to be reduced if in the antecedent the same formula does not occur at more than three places as sequent-formulas, and likewise in the succedent. A sequent S' is called a *reduct* of a sequent S is S' is reduced and is obtained from S by deleting some occurrences of formulas. Now, given a sequent S of **LK**′ (or **LJ**′), let S' be any reduct of S. We note the following.

1) S is provable or unprovable according as S' is provable or unprovable.

2) The number of all reduced sequents which contain only subformulas of the formulas in S is finite.

Consider the finite system of sequents as in 2), say $\mathscr{S}$. Collect all initial sequents in the systems. Call this set $\mathscr{S}_0$. Then examine $\mathscr{S} - \mathscr{S}_0$ to see if there is a sequent which can be the lower sequent of an inference whose upper sequent(s) is (are) one (two) sequent(s) from $\mathscr{S}_0$. Call the set of all sequents which satisfy this condition $\mathscr{S}_1$. Now see if there is a sequent in $(\mathscr{S} - \mathscr{S}_0) - \mathscr{S}_1$ which can be the lower sequent of an inference whose upper sequent(s) is (are) one (two) of the sequent(s) in $\mathscr{S}_0 \cup \mathscr{S}_1$. Continue this process until either the sequent S' itself is determined as provable, or the process does not give any new sequent as provable. One of the two must happen. If the former is the case, then S is provable. Otherwise S is unprovable. (Note that the whole argument is finitary.)

THEOREM 6.14 (1) (Harrop). *Let Γ be a finite sequence of formulas such that in each formula of Γ every occurrence of $\vee$ and $\exists$ is either in the scope of a $\neg$ or in the left scope of a $\supset$* (cf. *Definition* 6.1, *part* 3)). *This condition will be referred to as* (*) *in this theorem. Then*

1) *$\Gamma \to A \vee B$ is* **LJ**-*provable if and only if $\Gamma \to A$ or $\Gamma \to B$ is* **LJ**-*provable*,

2) *$\Gamma \to \exists x\, F(x)$ is* **LJ**-*provable if and only if for some term s, $\Gamma \to F(s)$ is* **LJ**-*provable*.

(2) *The following sequents* (*which are* **LK**-*provable*) *are not* (*in general*) **LJ**-*provable*.

$$\neg(\neg A \wedge \neg B) \to A \vee B; \qquad \neg\forall x \neg F(x) \to \exists x\, F(x);$$

$$A \supset B \to \neg A \vee B; \qquad \neg\forall x\, F(x) \to \exists x \neg F(x);$$

$$\neg(A \wedge B) \to \neg A \vee \neg B.$$

PROOF. (1) part 1): The "if" part is trivial. For the "only if" part, consider a cut-free proof of $\Gamma \to A \vee B$. The proof is carried out by induction on the number of inferences below all the inferences for $\vee$ and $\exists$ in the given proof. If the last inference is $\vee$: right, there is nothing to prove. Notice that the last inference cannot be $\vee$, $\neg$, or $\exists$: left.

Case 1: The last inference is $\wedge$: left:

$$\frac{C, \Gamma \to A \vee B}{C \wedge D, \Gamma \to A \vee B}.$$

It is obvious that C satisfies the condition (*). Thus the induction hypothesis applies to the upper sequent; hence either $C, \Gamma \to A$ or $C, \Gamma \to B$ is provable. In either case, the end-sequent can be derived in **LJ**.

Case 2: The last inference is $\supset$: left:

$$\frac{\Gamma \to C \qquad D, \Gamma \to A \vee B}{C \supset D, \Gamma \to A \vee B}.$$

It is obvious that D satisfies the condition (*); thus, by the induction hypothesis applied to the right upper sequent, $D, \Gamma \to A$ or $D, \Gamma \to B$ is provable. In either case the end-sequent can be derived.

Other cases are treated likewise. The proofs of (1) part 2), and (2), are left to the reader.

§7. The predicate calculus with equality

Definition 7.1. The predicate calculus with equality (denoted $\mathbf{LK_e}$) can be obtained from **LK** by specifying a predicate constant of two argument places (=: read equals) and adding the following sequents as additional initial sequents ($a = b$ denoting $=(a, b)$):

$$\to s = s;$$

$$s_1 = t_1, \ldots, s_n = t_n \to f(s_1, \ldots, s_n) = f(t_1, \ldots, t_n)$$

for every function constant f of n argument-places ($n = 1, 2, \ldots$);

$$s_1 = t_1, \ldots, s_n = t_n, R(s_1, \ldots, s_n) \to R(t_1, \ldots, t_n)$$

for every predicate constant R (including =) of n argument-places ($n = 1, 2, \ldots$); where $s, s_1, \ldots, s_n, t_1, \ldots, t_n$ are arbitrary terms.

Each such sequent may be called an equality axiom of $\mathbf{LK_e}$.

Proposition 7.2. *Let $A(a_1, \ldots, a_n)$ be an arbitrary formula. Then*

$$s_1 = t_1, \ldots, s_n = t_n, A(s_1, \ldots, s_n) \to A(t_1, \ldots, t_n)$$

is provable in $\mathbf{LK_e}$, *for any terms* s_i, t_i $(1 \leq i \leq n)$. *Furthermore,* $s = t \to t = s$ *and* $s_1 = s_2, s_2 = s_3 \to s_1 = s_3$ *are also provable.*

Definition 7.3. Let Γ_e be the set (axiom system) consisting of the following sentences:

$$\forall x (x = x),$$

$$\forall x_1 \ldots \forall x_n \forall y_1 \ldots \forall y_n [x_1 = y_1 \wedge \ldots \wedge x_n = y_n \supset f(x_1, \ldots, x_n) = f(y_1, \ldots, y_n)]$$

for every function constant f with n argument-places ($n = 1, 2, \ldots$),

$$\forall x_1 \ldots \forall x_n \forall y_1 \ldots \forall y_n [x_1 = y_1 \wedge \ldots \wedge x_n = y_n \wedge R(x_1, \ldots, x_n) \supset R(y_1, \ldots, y_n)]$$

for every predicate constant R of n argument-places ($n = 1, 2, \ldots$). Each such sentence is called an equality axiom.

PROPOSITION 7.4. *A sequent $\Gamma \rightarrow \Delta$ is provable in* **LK$_e$** *if and only if $\Gamma, \Gamma_e \rightarrow \Delta$ is provable in* **LK**.

PROOF. Only if: It is easy to see that all initial sequents of **LK$_e$** are provable from Γ_e. Therefore the proposition is proved by mathematical induction on the number of inferences in a proof of the sequent $\Gamma \rightarrow \Delta$.

If: All formulas of Γ_e are **LK$_e$**-provable.

DEFINITION 7.5. If the cut formula of a cut in **LK$_e$** is of the form $s = t$, then the cut is called inessential. It is called essential otherwise.

THEOREM 7.6 (the cut-elimination theorem for the predicate calculus with equality, **LK$_e$**). *If a sequent of* **LK$_e$** *is* **LK$_e$**-*provable, then it is* **LK$_e$**-*provable without an essential cut.*

PROOF. The theorem is proved by removing essential cuts (mixes as a matter of fact), following the method used for Theorem 5.1.

If the rank is 2, S_2 is an equality axiom and the mix formula is not of the form $s = t$, then the mix formula is of the form $P(t_1, \ldots, t_n)$. If S_1 is also an equality axiom, then it has the form

$$s_1 = t_1, \ldots, s_n = t_n, P(s_1, \ldots, s_n) \rightarrow P(t_1, \ldots, t_n).$$

From this and S_2, i.e.,

$$t_1 = r_1, \ldots, t_n = r_n, P(t_1, \ldots, t_n) \rightarrow P(r_1, \ldots, r_n),$$

we obtain by a mix

$$s_1 = t_1, \ldots, s_n = t_n, t_1 = r_1, \ldots, t_n = r_n, P(s_1, \ldots, s_n) \rightarrow P(r_1, \ldots, r_n).$$

This may be replaced by

$$s_i = t_i, t_i = r_i \rightarrow s_i = r_i \quad (i = 1, 2, \ldots, n);$$

$$s_1 = r_1, \ldots, s_n = r_n, P(s_1, \ldots, s_n) \rightarrow P(r_1, \ldots, r_n);$$

and then repeated cuts of $s_i = r_i$ to produce the same end-sequent. All cuts (or mixes) introduced here are inessential.

If $P(t_1, \ldots, t_n)$ in S_2 is a weakening formula, then the mix inference is:

$$\frac{s_1 = t_1, \ldots, s_n = t_n, P(s_1, \ldots, s_n) \rightarrow P(t_1, \ldots, t_n) \qquad P(t_1, \ldots, t_n), \Pi \rightarrow \Lambda}{s_1 = t_1, \ldots, s_n = t_n\ P(s_1, \ldots, s_n), \Pi \rightarrow \Lambda.}.$$

Transform this into:

$$\begin{array}{c} \Pi \rightarrow \Lambda \\ \hline\hline \text{end-sequent.} \end{array}$$

The rest of the argument in Theorem 5.1 goes through.

PROBLEM 7.7. A sequent of the form

$$s_1 = t_1, \ldots, s_n = t_n \rightarrow s = t \quad (n = 0, 1, 2, \ldots)$$

is said to be simple if it is obtained from sequents of the following four forms by applications of exchanges, contractions, cuts, and weakening left.

1) $\rightarrow s = s$.
2) $s = t \rightarrow t = s$.
3) $s_1 = s_2, s_2 = s_3 \rightarrow s_1 = s_3$.
4) $s_1 = t_1, \ldots, s_m = t_m \rightarrow f(s_1, \ldots, s_m) = f(t_1, \ldots, t_m)$.

Prove that if $s_1 = s_1, \ldots, s_m = s_m \rightarrow s = t$ is simple, then $s = t$ is of the form $s = s$. As a special case, if $\rightarrow s = t$ is simple, then $s = t$ is of the form $s = s$.

Let $\mathbf{LK}_e'$ be the system which is obtained from **LK** by adding the following sequents as initial sequents:

a) simple sequents,
b) sequents of the form

$$s_1 = t_1, \ldots, s_m = t_m, R(s_1', \ldots, s_n') \rightarrow R(t_1', \ldots, t_n'),$$

where $s_1 = t_1, \ldots, s_m = t_m \rightarrow s_i' = t_i'$ is simple for each i ($i = 1, \ldots, n$). First prove that the initial sequents of $\mathbf{LK}_e'$ are closed under cuts and that if

$$R(s_1, \ldots, s_n) \rightarrow R(t_1, \ldots, t_n)$$

is an initial sequent of $\mathbf{LK}_e'$ (where R is not $=$), then it is of the form $D \rightarrow D$. Finally, prove that the cut-elimination theorem (without the exception of inessential cuts) holds for $\mathbf{LK}_e'$.

PROBLEM 7.8. Show that if a sequent S without the $=$ symbol is $\mathbf{LK}_e$-provable, then it is provable in **LK** (without $=$).

PROBLEM 7.9. Prove that Theorems 6.2–6.4, 6.6, 6.9, and 6.14, Propositions 6.11 and 6.12 and Exercise 6.7 hold for $\mathbf{LK}_e$ when they are modified in the following way: References to **LK**- (or **LJ**-) provability are replaced throughout by references to $\mathbf{LK}_e$-provability, and further, when the statement demands that a formula can contain only certain constants, $=$ can be added as an exception.

The general technique of proof is to change a condition that a sequent $\Gamma \to \Delta$ be provable in **LK** to one that a sequent $\Pi, \Gamma \to \Delta$ be provable in **LK**, where Π is a set of equality axioms, and in this way to reduce the problem to **LK**.

§8. The completeness theorem

Although we do not intend to develop model theory in this book, we shall outline a proof of the completeness theorem for **LK**. The completeness theorem for the first order predicate calculus was first proved by Gödel. Here we follow Schütte's method, which has a close relationship to the cut-elimination theorem. In fact the cut-elimination theorem is a corollary of the completeness theorem as formulated below. (The importance of the proof of cut-elimination in §5 lies in its constructive nature.)

DEFINITION 8.1. (1) Let L be a language as described in §1. By a *structure* for L (an L-structure) we mean a pair $\langle D, \phi \rangle$, where D is a non-empty set and ϕ is a map from the constants of L such that

(i) if k is an individual constant, then ϕk is an element of D;

(ii) if f is a function constant of n arguments, then ϕf is a mapping from D^n into D;

(iii) if R is a predicate constant of n arguments, then ϕR is a subset of D^n.

(2) An *interpretation* of L is a structure $\langle D, \phi \rangle$ together with a mapping ϕ_0 from variables into D. We may denote an interpretation $(\langle D, \phi \rangle, \phi_0)$ simply by $\mathfrak{I}$. ϕ_0 is called an assignment from D.

(3) We say that an interpretation $\mathfrak{I} = (\langle C, \phi \rangle, \phi_0)$ *satisfies* a formula A if this follows from the following inductive definition. In fact we shall define the notion of "satisfying" for all semi-formulas (cf. Definition 6.8).

0) Firstly, we define $\phi(t)$, for every semi-term t, inductively as follows. We define $\phi(a) = \phi_0(a)$ and $\phi(x) = \phi_0(x)$ for all free variables a and bound variables x. Next, if f is a function constant and t is a semi-term for which ϕt is already defined, then $\phi(f(t))$ is defined to be $(\phi f)(\phi t)$.

1) If R is a predicate constant of n arguments and $t_1, \ldots, t_n$ are semi-terms, then $\mathfrak{I}$ satisfies $R(t_1, \ldots, t_n)$ if and only if $\langle \phi t_1, \ldots, \phi t_n \rangle \in \phi R$.

2) $\mathfrak{I}$ satisfies $\neg A$ if and only if it does not satisfy A; $\mathfrak{I}$ satisfies $A \wedge B$ if

and only if it satisfies both A and B; $\mathfrak{J}$ satisfies $A \vee B$ if and only if it satisfies either A or B; $\mathfrak{J}$ satisfies $A \supset B$ if and only if either it does not satisfy A or it satisfies B.

3) $\mathfrak{J}$ satisfies $\forall x\, B$ if and only if for every ϕ_0' such that ϕ_0 and ϕ_0' agree, except possibly on x, $(\langle D, \phi \rangle, \phi_0')$ satisfies B; $\mathfrak{J}$ satisfies $\exists x\, B$ if and only if for some ϕ_0' such that ϕ_0 and ϕ_0' agree, except possibly on x, $(\langle D, \phi \rangle, \phi_0')$ satisfies B.

If $\mathfrak{J} = (\langle D, \phi \rangle, \phi_0)$ satisfies a formula A, we say that A is satisfied in $\langle D, \phi \rangle$ by ϕ_0, or simply A is satisfied by $\mathfrak{J}$.

(4) A formula is called valid in $\langle D, \phi \rangle$ if and only if for every ϕ_0, $(\langle D, \phi \rangle, \phi_0)$ satisfies that formula. It is called valid if it is valid in every structure.

(5) A sequent $\Gamma \rightarrow \Delta$ is satisfied in $\langle D, \phi \rangle$ by ϕ_0 (or $\mathfrak{J} = (\langle D, \phi \rangle, \phi_0)$ satisfies $\Gamma \rightarrow \Delta$) if either some formula in Γ is not satisfied by $\mathfrak{J}$, or some formula in Δ is satisfied by $\mathfrak{J}$. A sequent is valid if it is satisfied in every interpretation.

(6) A structure may also be denoted as

$$\langle D; \phi k_0, \phi k_1, \ldots, \phi f_0, \phi f_1, \ldots, \phi R_0, \phi R_1, \ldots \rangle.$$

A structure is called a model of an axiom system Γ if every sentence of Γ is valid in it. It is called a counter-model of Γ if there is a sentence of Γ which is not valid in it.

THEOREM 8.2 (completeness and soundness). *A formula is provable in* **LK** *if and only if it is valid.*

NOTES. (1) The "if" part of the theorem is the statement of the completeness of **LK**. In general, a system is said to be complete if and only if every valid formula is provable in the system (for a suitable definition of validity).

Soundness means: all provable sequents are valid, i.e., the "only if" part of the theorem. Soundness ensures consistency.

(2) The theorem connects proof theory with semantics, where semantics means, very roughly, the study of the interpretation of formulas in a structure (of a language), and hence of their truth or falsity.

PROOF OF THEOREM 8.2. The "only if" part is easily proved by induction on the number of inferences in a proof of the formula. We prove the "if" part in the following generalized form:

LEMMA 8.3. *Let S be a sequent. Then either there is a cut-free proof of S, or there is an interpretation which does not satisfy S (and hence S is not valid).*

PROOF. We will define, for each sequent S, a (possibly infinite) tree, called the reduction tree for S, from which we can obtain either a cut-free proof of S or an interpretation not satisfying S. (This method is due to Schütte.) This reduction tree for S contains a sequent at each node. It is constructed in stages as follows.

Stage 0: Write S at the bottom of the tree.

Stage k ($k > 0$): This is defined by cases:

Case I. Every topmost sequent has a formula common to its antecedent and succedent. Then stop.

Case II. Not Case I. Then this stage is defined according as

$$k \equiv 0, 1, 2, \ldots, 11, 12 \pmod{13}.$$

$k \equiv 0$ and $k \equiv 1$ concern the symbol $\neg$; $k \equiv 2$ and $k \equiv 3$ concern $\wedge$; $k \equiv 4$ and $k \equiv 5$ concern $\vee$; $k \equiv 6$ and $k \equiv 7$ concern $\supset$; $k \equiv 8$ and $k \equiv 9$ concern $\forall$; and $k \equiv 10$ and $k \equiv 11$ concern $\exists$.

Since the formation of reduction trees is a common technique and will be used several times in this text, we shall describe these stages of the so-called reduction process in detail. In order to make the discussion simpler, let us assume that there are no individual or function constants.

All the free variables which occur in any sequent which has been obtained at or before stage k are said to be "available at stage k". In case there is none, pick any free variable and say that it is available.

0) $k \equiv 0$. Let $\Pi \rightarrow \Lambda$ be any topmost sequent of the tree which has been defined by stage $k-1$. Let $\neg A_1, \ldots, \neg A_n$ be all the formulas in Π whose outermost logical symbol is $\neg$, and to which no reduction has been applied in previous stages. Then write down

$$\Pi \rightarrow \Lambda, A_1, \ldots, A_n$$

above $\Pi \rightarrow \Lambda$. We say that a $\neg$: left reduction has been applied to $\neg A_1, \ldots, \neg A_n$.

1) $k \equiv 1$. Let $\neg A_1, \ldots, \neg A_n$ be all the formulas in Λ whose outermost logical symbol is $\neg$ and to which no reduction has been applied so far. Then write down

$$A_1, \ldots, A_n, \Pi \rightarrow \Lambda$$

above $\Pi \rightarrow \Lambda$. We say that a $\neg$: right reduction has been applied to $\neg A_1, \ldots, \neg A_n$.

2) $k \equiv 2$. Let $A_1 \wedge B_1, \ldots, A_n \wedge B_n$ be all the formulas in Π whose outermost logical symbol is $\wedge$ and to which no reduction has been applied yet. Then write down

$$A_1, B_1, A_2, B_2, \ldots, A_n, B_n, \Pi \rightarrow \Lambda$$

above $\Pi \rightarrow \Lambda$. We say that an $\wedge$: left reduction has been applied to

$$A_1 \wedge B_1, \ldots, A_n \wedge B_n.$$

3) $k \equiv 3$. Let $A_1 \wedge B_1,\ A_2 \wedge B_2, \ldots, A_n \wedge B_n$ be all the formulas in Λ whose outermost logical symbol is $\wedge$ and to which no reduction has been applied yet. Then write down all sequents of the form

$$\Pi \rightarrow \Lambda,\ C_1, \ldots, C_n,$$

where C_i is either A_i or B_i, above $\Pi \rightarrow \Lambda$. Take all possible combinations of such: so there are 2^n such sequents above $\Pi \rightarrow \Lambda$. We say that an $\wedge$: right reduction has been applied to $A_1 \wedge B_1, \ldots, A_n \wedge B_n$.

4) $k \equiv 4$. $\wedge$: left reduction. This is defined in a manner symmetric to 3).

5) $k \equiv 5$. $\vee$: right reduction. This is defined in a manner symmetric to 2).

6) $k \equiv 6$. Let $A_1 \supset B_1, \ldots, A_n \supset B_n$ be all the formulas in Π whose outermost logical symbol is $\supset$ and to which no reduction has been applied yet. Then write down the following sequents above $\Pi \rightarrow \Lambda$:

$$B_{i_1}, B_{i_2}, \ldots, B_{i_k}, \Pi \rightarrow \Lambda, A_{j_1}, A_{j_2}, \ldots, A_{j_{n-k}},$$

where $i_1 < i_2 < \ldots < i_k$, $j_1 < j_2 < \ldots < j_{n-k}$ and $(i_1, i_2, \ldots, i_k, j_1, j_2, \ldots, j_{n-k})$ is a permutation of $(1, 2, \ldots, n)$. Take all possible permutations: so there are 2^n such sequents above $\Pi \rightarrow \Lambda$. We say that an $\supset$: left reduction has been applied to $A_1 \supset B_1, \ldots, A_n \supset B_n$.

7) $k \equiv 7$. Let $A_1 \supset B_1, \ldots, A_n \supset B_n$ be all the formulas in Λ whose outermost logical symbol is $\supset$ and to which no reduction has been applied yet. Then write down

$$A_1, A_2, \ldots, A_n, \Pi \rightarrow \Lambda, B_1, B_2, \ldots, B_n$$

above $\Pi \rightarrow \Lambda$. We say that an $\supset$: right reduction has been applied to

$$A_1 \supset B_1, \ldots, A_n \supset B_n.$$

8) $k \equiv 8$. Let $\forall x_1\, A_1(x_1), \ldots, \forall x_n\, A_n(x_n)$ be all the formulas in Π whose outermost logical symbol is $\forall$. Let a_i be the first variable available at this stage which has not been used for a reduction of $\forall x_i\, A_i(x)$ for $1 \leq i \leq n$. Then write down

$$A_1(a_1), \ldots, A_n(a_n), \Pi \rightarrow \Lambda$$

above $\Pi \rightarrow \Lambda$. We say that a $\forall$: left reduction has been applied to

$$\forall x_1\, A_1(x), \ldots, \forall x_n\, A_n(x_n).$$

9) $k \equiv 9$. Let $\forall x_1\, A_1(x_1), \ldots, \forall x_n\, A_n(x_n)$ be all the formulas in Λ whose outermost logical symbol is $\forall$ and to which no reduction has been applied so far. Let $a_1, \ldots, a_n$ be the first n free variables (in the list of variables) which are *not* available at this stage. Then write down

$$\Pi \to \Lambda, A_1(a_1), \ldots, A_n(a_n)$$

above $\Pi \to \Lambda$. We say that a $\forall$: right reduction has been applied to $\forall x_1\, A_1(x_1), \ldots, \forall x_n\, A_n(x_n)$. Notice that $a_1, \ldots, a_n$ are new available free variables.

10) $k \equiv 10$. $\exists$: left reduction. This is defined in a manner symmetric to 9).

11) $k \equiv 11$. $\exists$: right reduction. This is defined in a manner symmetric to 8).

12) If Π and Λ have any formula in common, write nothing above $\Pi \to \Lambda$ (so this remains a topmost sequent). If Π and Λ have no formula in common and the reductions described in 0)–11) are not applicable, write the same sequent $\Pi \to \Lambda$ again above it.

So the collection of those sequents which are obtained by the above reduction process, together with the partial order obtained by this process, is the reduction tree (for S). It is denoted by $T(S)$. We will construct "reduction trees" like this again.

As an example of the case where the reduction process does not terminate, consider a sequent of the form $\forall x\, \exists y\, A(x, y) \to$, where A is a predicate constant.

Now a (finite or infinite) sequence $S_0, S_1, S_2, \ldots$ of sequents in $T(S)$ is called a branch if (i) $S_0 = S$; (ii) S_{i+1} stands immediately above S_i; (3) if the sequence is finite, say $S_1, \ldots, S_n$, then S_n has the form $\Pi \to \Lambda$, where Π and Λ have a formula in common.

Now, given a sequent S, let T be the reduction tree $T(S)$. If each branch of T ends with a sequent whose antecedent and succedent contain a formula in common, then it is a routine task to write a proof without a cut ending with S by suitably modifying T. Otherwise there is an infinite branch. Consider such a branch, consisting of sequents $S = S_0, S_1, \ldots, S_n, \ldots$.

Let S_i be $\Gamma_i \to \Delta_i$. Let $\bigcup \Gamma$ be the set of all formulas occurring in Γ_i for some i, and let $\bigcup \Delta$ be the set of all formulas occurring in Δ_j for some j. We shall define an interpretation in which every formula in $\bigcup \Gamma$ holds and no formula in $\bigcup \Delta$ holds. Thus S does not hold in it.

First notice that from the way the branch was chosen, $\bigcup \Gamma$ and $\bigcup \Delta$ have no atomic formula in common. Let D be the set of all the free variables. We consider the interpretation $\mathfrak{J} = (\langle D, \phi \rangle, \phi_0)$, where ϕ and ϕ_0 are defined as follows: $\phi_0(a) = a$ for all free variables a, $\phi_0(x)$ is defined arbitrarily for all bound variables x. For an n-ary predicate constant R, ϕR is any subset of D^n such that: if $R(a_1, \ldots, a_n) \in \bigcup \Gamma$, then $(a_1, \ldots, a_n) \in \phi R$, and $(a_1, \ldots, a_n) \notin \phi R$ otherwise.

We claim that this interpretation $\mathfrak{J}$ has the required property: it satisfies every formula in $\bigcup \Gamma$, but no formula in $\bigcup \Delta$. We prove this by induction on the number of logical symbols in the formula A. We consider here only the case where A is of the form $\forall x\, F(x)$ and assume the induction hypothesis:

Subcase 1. A is in $\bigcup \Gamma$. Let i be the least number such that A is in Γ_i. Then A is in Γ_j for all $j > i$. It is sufficient to show that all substitution instances $A(a)$, for $a \in D$, are satisfied by $\mathfrak{J}$, i.e., all these substitution instances are in $\bigcup \Gamma$. But this is evident from the way we construct the tree.

Subcase 2. A is in $\bigcup \Delta$. Consider the step at which A was used to define an upper sequent from $\Gamma_i \to \Delta_i$ (or $\Gamma_i \to \Delta_i^1, A, \Delta_i^2$). It looks like this:

$$\frac{\Gamma_{i+1} \to \Delta_{i+1}^1, F(a), \Delta_{i+1}^2}{\Gamma_i \to \Delta_i^1, A, \Delta_i^2}.$$

Then by the induction hypothesis, $F(a)$ is not satisfied by $\mathfrak{J}$, so A is not satisfied by $\mathfrak{J}$ either. This completes the proof.

PROBLEM 8.4 (Feferman). Let J be a non-empty set. Each element of J is called a *sort*. A many-sorted language for the set of sorts J, say $L(J)$, consists of the following.

1) Individual constants: $k_0, k_1, \ldots, k_i, \ldots$, where to each k_i is assigned one sort.

2) Predicate constants: $R_0, R_1, \ldots, R_i, \ldots$, where to each R_i is assigned a number n (≥ 0) (the number of arguments) and sorts $j_1, \ldots, j_n$. We say that $(n; j_1, \ldots, j_n)$ is assigned to R_i.

3) Function constants: $f_0, f_1, \ldots, f_i, \ldots$, where to each f_i is assigned a number n (≥ 1) (the number of arguments) and sorts $j_1, \ldots, j_n, j$. We say that $(n; j_1, \ldots, j_n, j)$ is assigned to f_i.

4) Free variables of sort j for each j in J: $a_0^j, a_1^j, \ldots, a_i^j, \ldots$.

5) Bound variables of sort j for each j in J: $x_0^j, x_1^j, \ldots, x_i^j, \ldots$.

6) Logical symbols: $\neg, \wedge, \vee, \supset, \forall, \exists$.

Terms of sort j for each j are defined as follows. Individual constants and free variables of sort j are terms of sort j; if f is a function constant with $(n; j_1, \ldots, j_n, j)$ assigned to it and $t_1, \ldots, t_n$ are terms of sort $j_1, \ldots, j_n$, respectively, then $f(t_1, \ldots, t_n)$ is a term of sort j.

If R is a predicate constant with $(n; j_1, \ldots, j_n)$ assigned to it and $t_1, \ldots, t_n$ are terms of sort $j_1, \ldots, j_n$, respectively, then $R(t_1, \ldots, t_n)$ is an atomic formula. If $F(a^j)$ is a formula and x^j does not occur in $F(a^j)$, then $\forall x^j\, F(x^j)$ and $\exists x^j\, F(x^j)$ are formulas; the other steps in building formulas of $L(J)$ are as usual. The sequents of $L(J)$ are defined as usual.

The rules of inference are those of **LK**, except that in the rules for $\forall$ and $\exists$, terms and free variables must be replaced by bound variables of the same sort.

Prove the following:

(1) The cut-elimination theorem holds for the system just defined.

Next, define Sort, Ex, Un, Fr, and Pr as follows. Sort(A) is the set of j in J such that a symbol of sort j occurs in A; Ex(A) and Un(A) are the sets of sorts of bound variables which occur in some essentially existential, respectively universal quantifier in A. (An occurrence of $\exists$, say $\#$, is said to be essentially existential or universal according to the following definition. Count the number of $\neg$ and $\supset$ in A such that $\#$ is either in the scope of $\neg$, or in the left scope of $\supset$. If this number is even, then $\#$ is essentially existential in A, while if it is odd then $\#$ is essentially universal. Likewise, we define, dually, an occurrence of $\forall$ to be essentially existential or universal.) Fr(A) is the set of free variables in A; Pr(A) is the set of predicate constants in A.

(2) Suppose $A \supset B$ is provable in the above system and at least one of $\mathrm{Sort}(A) \cap \mathrm{Ex}(B)$ and $\mathrm{Sort}(B) \cap \mathrm{Un}(A)$ is not empty. Then there is a formula C such that $\sigma(C) \subseteq \sigma(A) \cap \sigma(B)$, where σ stands for Fr, Pr or Sort, and such that $\mathrm{Un}(C) \subseteq \mathrm{Un}(A)$ and $\mathrm{Ex}(C) \subseteq \mathrm{Ex}(B)$. [*Hint*: Re-state the above theorem for sequents and apply (1), viz. the cut-elimination theorem.]

DEFINITION 8.5. We can define a structure for a many-sorted language as follows. Let L(J) be a many-sorted language. A structure for L(J) is a pair $\langle D, \phi \rangle$, where D is a set of non-empty sets $\{D_j; j \in J\}$ and ϕ is a map from the constants of L(J) into appropriate objects. We call D_j the domain of the structure of sort j. We leave the listing of the conditions on ϕ to the reader; we only have to keep in mind that an individual constant of sort j is a member of D_j. Let $\mathcal{M} = \langle D, \phi \rangle$ and $\mathcal{M}' = \langle D', \phi' \rangle$ be two structures for L(J). We say $\mathcal{M}'$ is an extension of $\mathcal{M}$ and write $\mathcal{M} \subseteq \mathcal{M}'$ if

(i) for each j in J, $D_j \subseteq D'_j$,

(ii) for each individual constant k, $\phi' k = \phi k$,

(iii) for each predicate constant R with $(n; j_1, \ldots, j_n)$ assigned to it,

$$\phi R = \phi' R \cap (D_{j_1} \times \cdots \times D_{j_n}),$$

(iv) for each function constant f with $(n; j_1, \ldots, j_n, j)$ assigned to it and $(d_1, \ldots, d_n) \in D_{j_1} \times \cdots \times D_{j_n}$,

$$(\phi' f)(d_1, \ldots, d_n) = (\phi f)(d_1, \ldots, d_n).$$

A formula is said to be existential if $U_n(A)$ is empty.

COROLLARY 8.6 (Łos-Tarski). *The following are equivalent: let A be a formula of an ordinary (i.e., single-sorted) language* L.

(i) *For any structure $\mathcal{M}$ (for* L*) and extension $\mathcal{M}'$, and any assignments ϕ, ϕ' from the domains of $\mathcal{M}$, $\mathcal{M}'$, respectively, which agree on the free variables of A, if $(\mathcal{M}, \phi)$ satisfies A, then so does $(\mathcal{M}', \phi')$.*

(ii) *There exists an (essentially) existential formula B such that $A \equiv B$ is provable and the free variables of B are among those of A.*

PROOF. (Feferman) We assume (for simplicity) that the language has no individual and function constants. The major task is to write down the conditions in (1) syntactically, by considering an extended language in which we can express the relation between two structures.

Let $\mathcal{M}$ and $\mathcal{M}'$ be two structures of the form

$$\mathcal{M} = \langle D_1, \{R_i\} i \in I \rangle, \qquad \mathcal{M}' = \langle D_2, \{R_i'\} i \in I \rangle.$$

Let J be $\{1, 2\}$. $(J, I, \langle k_i \rangle i \in I)$ will determine a 'type' of structures. Let L^+ be a corresponding language. It contains the original language L as the sublanguage of sort 1. For each bound variable u, the nth bound variable of sort 1, let u' be the nth bound variable of sort 2. If C is an L-formula, then C' denotes the result of replacing each bound variable u in C by u'; hence $\mathrm{Fr}(C) = \mathrm{Fr}(C')$. With this notation, define Ext to be the sentence of the form $\forall u' \exists u (u' = u)$. Then, Ext and $\exists u_i'(u_i' = b_i)$ for $i = 1, \ldots, n$ yield $A' \to A$. I.e., we have

$$\mathrm{Ext}, \{\exists u_i'(u_i' = b_i)\}_{i=1}^n, A' \to A.$$

Now apply the result of Problem 8.4. An interpolant B can be chosen so as to satisfy:

(i) $\mathrm{Fr}(B) \subseteq \mathrm{Fr}(A) = \{b_1, \ldots, b_n\}$,
(ii) $\mathrm{Pr}(B) \subseteq \mathrm{Pr}(A)$,
(iii) every bound variable in B is of sort 1 i.e., in L,
(iv) $U_n(B)$ is empty.

Hence B is an existential formula of L. Since

$$\mathrm{Ext}, \{\exists u_i'(u_i' = b_i)\}_{i=1}^n, A' \to B \text{ and } B \to A$$

are provable, we obtain that $A \equiv B$ is provable.

A general syntactic theory including Problem 8.6 is obtained by N. Motohashi: Interpolation Theorem and Characterization Theorem, Ann. Japan Assoc. Philos. Sci. 4 (1972) 15–80.

PROBLEM 8.7. Let $\mathscr{A}$ be an axiom system in a language L, $\forall x \exists y A(x, y)$ a sentence of L provable from $\mathscr{A}$, and f a function symbol not in L. Then any L-formula which is provable from $\mathscr{A} \cup \{\forall x\, A(x, f(x))\}$ is also provable from $\mathscr{A}$ in L. (That is to say, the introduction of f in this way does not essentially extend the system.) [*Hint* (Maehara's method): This is a corollary of the following lemma.]

LEMMA 8.8. *Let $\forall x \exists y\, A(x, y)$ be a sentence of* L, *f a function symbol not in* L, *and Γ and Θ finite sequences of* L-*formulas. If $\forall x\, A(x, f(x))$, $\Gamma \to \Theta$ is* (**LK**-) *provable, then $\forall x \exists y\, A(x, y)$, $\Gamma \to \Theta$ is provable in* L.

PROOF. Let P be a cut-free regular proof of $\forall x\, A(x, f(x)), \Gamma \to \Theta$. Let $t_1, \ldots, t_n$ be all the terms in P (i.e. proper terms, not semi-terms) whose outermost function symbol is f. These are arranged in an order such that t_i is not a subterm of t_j for $i < j$. Suppose t_i is $f(s_i)$ for $i = 1, \ldots, n$. P is transformed in three steps.

Step (1): Let $a_1, \ldots, a_n$ be distinct free variables not occurring in P. Transform P by replacing t_1 by a_1, then t_2 by a_2, and so on. The resulting figure P' has the same end-sequent as P, but is not, in general, a proof (as we will see below) and must be further transformed.

Step (2): Since P is cut-free and f does not occur in Γ or Θ, it can be seen that the only occurrences of f in P are in the context. $\forall x\, A(x, f(x))$, and further, all these $\forall x\, A(x, f(x))$ occur in antecedents of sequents in P', and the corresponding occurrences of $\forall x\, A(x, f(x))$ in P are introduced (in P) only by weakening : left or by some inferences of the form

$$I \quad \frac{A(s_i, f(s_i)), \Pi \to \Lambda}{\forall x\, A(x, f(x)), \Pi \to \Lambda}$$

(for some of the i, $1 \leq i \leq n$). Suppose the upper sequent of I is transformed into

$$A(s'_i, a_i), \Pi' \to \Lambda'$$

in P'. (So I is not transformed by step (1) into a correct inference in P'.) Now replace all occurrences of $\forall x\, A(x, f(x))$ in P' by

$$A(s'_1, a_1), \ldots, A(s'_n, a_n)$$

(where s'_i is formed by replacing all t_j in s_i by a_j). Then the lower sequent of (the transform of) I can be derived from the upper sequent by several weakenings.

The result (after applying some contractions etc.) is a figure P'' with end-sequent

$$A(s'_1, a_1), \ldots, A(s'_n, a_n), \Gamma \to \Theta.$$

However it may still not be a proof, as we now show, and must be transformed further.

Step (3): Consider a $\exists$: left in P:

$$J \quad \frac{B(b), \Delta \to \Psi}{\exists z\, B(z), \Delta \to \Psi}$$

and suppose this is transformed in P'' (by steps (1) and (2)) to

$$J' \quad \frac{B'(b), \Delta' \to \Psi'}{\exists z\, B'(z), \Delta' \to \Psi'}.$$

Now it may happen that for some i, the eigenvariable b occurs in s_i (and also s'_i), and further, the formula $A(s'_i, a_i)$ occurs in Δ' or Ψ'; so that the eigenvariable condition is no longer satisfied in J'.

So we transform all J' in P'' (arising from $\exists$: left inferences J in P) as follows:

$$\supset:\text{left}\ \frac{\exists z\, B'(z) \to \exists z\, B'(z) \qquad B'(b), \Delta' \to \Psi'}{\exists z\, B'(z) \supset B'(b), \exists z\, B'(z), \Delta' \to \Psi'}$$

and carry the extra formula $\exists z\, B'(z) \supset B'(b)$ down to the end-sequent.

For the same reason, for every $\forall$:right in P

$$J\ \frac{\Delta \to \Psi, B(b)}{\Delta \to \Psi, \forall z\, B(z)}$$

we replace its transform in P''

$$J'\ \frac{\Delta' \to \Psi', B'(b)}{\Delta' \to \Psi', \forall z\, B'(z)}$$

by

$$\supset:\text{left}\ \frac{\overset{\vdots}{\to \forall z\, B'(z), \exists z \neg B'(z)} \qquad \dfrac{\Delta' \to \Psi', B'(b)}{\neg B'(b), \Delta' \to \Psi'}}{\exists z \neg B'(z) \supset \neg B'(b), \Delta' \to \Psi', \forall z\, B'(z)}$$

(and carry the extra formula down to the end).

The result (after some obvious adjustments with structural inferences) is a proof, without $\exists$: left or $\forall$: right, whose end-sequent has the form

(S_1) $$\exists z\, B'(z) \supset B'(b), \ldots, A(s_i', a_i), \ldots, \Gamma \to \Theta.$$

Now apply $\exists$: left and $\forall$: left inferences in a suitable order (see below) (and contractions, etc.) to derive

(S_2) $$F, \ldots, \forall x\, \exists y\, A(x, y), \Gamma \to \Theta,$$

where F is the formula obtained from $\exists u\, (\exists z\, B'(z) \supset B'(u))$ by universal quantification over all its free variables.

Finally, applying cuts with sequents $\to F$, we obtain a proof, as desired, of

$$\forall x\, \exists y\, A(x, y), \Gamma \to \Theta.$$

We must still check that it is indeed possible to find a suitable order for applying the quantifier inferences in proceeding from (S_1) to (S_2) above, so that they all satisfy the eigenvariable condition. To this end, we use the following (temporary) notation. For terms s and t and a formula B, $s \subset t$ means that s is a (proper) subterm of t, $s \subseteq t$ means that s is a subterm of t or t itself, and $s \subset B$ means that s is contained in B.

Now note that the following condition (C) is satisfied for any of the auxiliary formulas $B(b)$ of P with eigenvariable b, considered above, and $1 \leq i \leq n$:

(C) If $b \subset t_i$, then $t_i \not\subset B(b)$.

(For suppose $b \subset t_i$ and also t_i, which we write as $f(s_i(b))$, occurs in $B(b)$. Then in the lower sequent of the inference J with auxiliary formula $B(b)$, f would occur in the principal formula $\exists z\, B(z)$ (or $\forall z\, B(z)$) in the context of the semiterm $f(s_i(z))$, and so, since P is cut-free, f would also occur (in a similar context) in all sequents of P below this, and hence in Γ or Θ.)

Now let $J_1, \ldots, J_m$ be all the $\exists$: left and $\forall$: right inferences in P, with eigenvariables $b_1, \ldots, b_m$ and auxiliary formulas $B_1(b_1), \ldots, B_m(b_m)$, respectively. Consider the partial order on $a_1, \ldots, a_n$, $b_1, \ldots, b_m$, generated by the relation $<$, which is defined by the following conditions:

(1a) If $t_j \subset t_i$, then $a_i < a_j$.

(1b) If $b_j \subset t_i$, then $a_i < b_j$.

(2a) If $t_j \subset B_i(b_i)$, then $b_i < a_j$.

(2b) If $b_j \subset B_i(b_i)$ ($j \neq i$), then $b_i < b_j$.

We will prove below that this does indeed generate a partial order, i.e., no circularities are formed. Assume this for the moment. Then, starting with sequent (S_1), we apply, in any $<$-increasing order, the quantifier inferences

$$\frac{\dfrac{A(s'_i, a_i), \ldots}{\exists : \text{left and } \forall : \text{left}}}{\forall x\, \exists y\, A(x, y), \ldots}$$

and

$$\frac{\dfrac{\exists z\, B_j(z, a_i, \ldots, b_k, \ldots) \supset B_j(b_j, a_i, \ldots, b_k, \ldots), \ldots}{\exists : \text{left and } \forall : \text{left}}}{\forall x \ldots \forall y \ldots \exists u\, (\exists z\, B_j(z, x, \ldots, y, \ldots) \supset B_j(u, x, \ldots, y, \ldots)), \ldots}$$

so as to obtain (S_2). We can see that the eigenvariable condition is satisfied throughout, from the way in which $<$ was defined (and since $a_j \subset s'_i \Rightarrow t_j \subset t_i$, $b_j \subset s'_i \Rightarrow b_j \subset t_i$, $a_j \subset B'_i(b_i) \Rightarrow t_j \subset B_i(b_i)$, and $b_j \subset B'_i(b_i) \Rightarrow b_j \subset B_i(b_i)$).

Finally we must show that the relation $<$ does generate a partial order. This follows from the following two sublemmas.

SUBLEMMA 8.9 (in the notation of Lemma 8.8). (a) *For any $<$-increasing sequence $b_i < \ldots < b_j$, J_i lies above J_j in P. (So $i \neq j$.)*

(b) *For any $<$-increasing sequence $a_i < \ldots < a_j$, we have $t_i \not\subseteq t_j$. (So, in particular, $i \neq j$.)*

PROOF OF (a). The proof is by induction on the length of this sequence.

(i) If the length is 2, i.e., $b_1 < b_j$, this follows from the definition of $<$ (part 2b) and the eigenvariable condition in P.

(ii) For the case $b_i < a_k < b_j$: we have $t_k \subset B_i(b_i)$ (by 2a) and $b_j \subset t_k$ (by 1b). Hence $b_j \subset B_i(b_i)$. Also $i \neq j$, by condition (C). So again $b_i < b_j$ (by 2b) and J_i is above J_j.

(iii) For the case $b_i < a_k < \ldots < a_l < b_j$ (with only a's between b_i and b_j): notice that $t_l \subset t_k$ (from 1a). The argument is now similar to that in (ii).

(iv) For the remaining case, $b_i < \ldots < b_k < \ldots < b_j$, use the induction hypothesis.

PROOF OF (b). The proof is by induction on the length of this sequence.

(i) If the length is 2, i.e. $a_i < a_j$, this follows from the definition (part 1a).

(ii) For the case $a_i < b_k < a_j$: we have $b_k \subset t_i$ and $t_j \subset B_k(b_k)$. So $t_i \subseteq t_j$ would imply $t_i \subset B_k(b_k)$, contradicting (C).

(iii) For the case $a_i < b_k < \ldots < b_l < a_j$ (with anything between b_k and b_l) we have $b_k \subset t_i$, $t_j \subset B_l(b_l)$ and J_k is above J_l (by Sublemma 8.12(a)). So $t_i \subseteq t_j$ would imply $b_k \subset B_l(b_l)$, contradicting the eigenvariable condition in P.

For the remaining two cases:

(iv) $a_i < a_k < \ldots < a_j$,

(v) $a_i < \ldots < a_k < a_j$,

use (1a) and the induction hypothesis.

This completes the proof of the sublemmas, and hence of Lemma 8.8.

A general syntactic theory including Lemma 8.8 and its analogue in **LJ**, is obtained by

N. Motohashi: Approximation Theory of Uniqueness Conditions by Existence Conditions, Fund. Math. 120 (1983) 29–44.

The following proposition is not strictly proof-theoretical in nature; however, it is useful for the next topic (in the proof of Proposition 8.13). We first give some definitions.

DEFINITION 8.10. Let R be a set and suppose a set W_p is assigned to every $p \in R$. If $R_1 \subseteq R$ and $f \in \prod_{p \in R_1} W_p$, then f is called a *partial function* (*over* R) with domain $\mathrm{Dom}(f) = R_1$. If $\mathrm{Dom}(f) = R$, then f is called a *total function* (*over* R). If f and g are partial functions and $\mathrm{Dom}(f) = D_0 \subseteq \mathrm{Dom}(g)$ and $f(x) = g(x)$ for every $x \in D_0$, then we call g an *extension* of f and write $f < g$ and $f = g \upharpoonright D_0$.

PROPOSITION 8.11 (a generalized König's lemma). *Let R be any set. Suppose a finite set W_p is assigned to every $p \in R$. Let* P *be a property of partial functions f over R (defined as above) satisfying the following conditions*:

1) $\mathrm{P}(f)$ *holds if and only if there exists a finite subset N of R satisfying* $\mathrm{P}(f \upharpoonright N)$,

2) $\mathrm{P}(f)$ *holds for every total function f.*

Then there exists a finite subset N_0 of R such that $\mathrm{P}(f)$ *holds for every f with* $N_0 \subseteq \mathrm{Dom}(f)$.

Note that R can have arbitrarily large cardinality. The case that R is the set of natural numbers is the original König's lemma.

PROOF. Let $X = \prod_{p \in R} W_p$, and give each W_p the discrete topology, and X the product topology. Since each W_p is compact, so is X (Tychonoff's theorem). For each g such that Dom(g) is finite, let

$$N_g = \{f \mid f \text{ is total and } g < f\}.$$

Let

$$C = \{N_g \mid \text{Dom}(g) \text{ is finite, and } \mathrm{P}(g)\}.$$

C is an open cover of X. Therefore C has a finite subcover, say

$$N_{g_1}, \ldots, N_{g_k}.$$

Let $N_0 = \text{Dom}(g_1) \cup \ldots \cup \text{Dom}(g_k)$. We will show that N_0 satisfies the condition of the theorem. If $N_0 \subseteq \text{Dom}(g)$, then let $g < f$, f total. Then $\mathrm{P}(f)$ and $f \in N_{g_1} \cup \ldots \cup N_{g_k}$. Say $f \in N_{g_i}$. So $g_i < f$, $\mathrm{P}(g_i)$ and $g_i < g$. Therefore $\mathrm{P}(g)$. This completes the proof.

What happens if we wish to apply to **LJ** the technique which has been used in proving completeness for **LK**? This leads us naturally to the study of Kripke models of **LJ**, relative to which one can prove the completeness of **LJ**. In order to simplify the discussion, we assume again that our language does not contain individual or function constants. Again, there should be no essential difficulty in extending the argument to the case where individual and function constants are included.

For technical reasons, we will deal with a system which is an equivalent modification of **LJ**. This system, invented by Maehara, will be called **LJ**′. **LJ**′ is defined by restricting **LK** (rather than **LJ**) as follows: The inferences $\neg$: right, $\supset$: right and $\forall$: right are allowed only when the principal formulas are the only formulas in the succedents of the lower sequents. (These are called the "critical inferences" of **LJ**′.) Thus, for instance, $\neg$: right will take a form:

$$\frac{D, \Gamma \rightarrow}{\Gamma \rightarrow \neg D}.$$

As is obvious from the definition, the sequents of **LJ**′ are those of **LK** (so the restriction on the sequents of **LJ**, that there can be at most one formula in the succedent of a sequent, is lifted here). It should be noted that all the other inferences are exactly those of **LK**. In particular, in $\vee$: right, the inference

$$\frac{\Gamma \rightarrow \Delta, A}{\Gamma \rightarrow \Delta, A \vee B}$$

is allowed even if Δ is not empty.

By interpreting a sequent of **LJ'**, say $\Gamma \rightarrow B_1, \ldots, B_n$, as $\Gamma \rightarrow B_1 \vee \ldots \vee B_n$, it is a routine matter to prove that **LJ'** and **LJ** are equivalent. Also, the cut-elimination theorem holds for **LJ'**. (Combine the proofs of cut-elimination for **LK** and **LJ**.)

The question now arises: Given a sequent of **LJ'**, say $\Gamma \rightarrow \Delta$, is there a cut-free proof of $\Gamma \rightarrow \Delta$ in **LJ'**?

Starting with a given $\Gamma \rightarrow \Delta$, we can carry out the reduction process which was defined for the classical case (cf. Lemma 8.3), except that we omit the stages 1) ($\neg$: right reduction), 7) ($\supset$: right reduction) and 9) ($\forall$: right reduction); in other words, all the reductions are as for the classical case, except those which concern the critical inferences of **LJ'**, which are simply omitted. We return to consider this point later.

The tree obtained by the above reduction process is (again) called the reduction tree for $\Gamma \rightarrow \Delta$.

In preparation for Kripke's semantics for intuitionistic systems and the completeness theorem for **LJ**, we will generalize the above reduction process to the case where Γ and/or Δ are infinite; i.e., we define reduction trees for infinite sequents $\Gamma \rightarrow \Delta$.

DEFINITION 8.12. Let Γ and Δ be well-ordered sequences of formulas, which may be infinite. We say that $\Gamma \rightarrow \Delta$ is provable (cut-free provable) (in **LJ'**) if there are finite subsequences of Γ and Δ, say $\tilde{\Gamma}$ and $\tilde{\Delta}$, respectively, such that $\tilde{\Gamma} \rightarrow \tilde{\Delta}$ is provable (cut-free provable).

(It is clear that $\Gamma \rightarrow \Delta$ is provable (in **LJ'**) if and only if it is provable without cut, even when Γ and/or Δ are infinite, by the cut-elimination theorem of §5, adapted to **LJ'**.)

The reduction process which has just been described can be generalized immediately to the case of infinite sequents. We shall only point out a few modifications in the stages. Note: for the reduction process, we assume that the language is augmented by uncountably many new free and bound variables (in a well-ordered sequence).

8) $k \equiv 8$. Let $\forall x_1 A_1(x_1), \ldots, \forall x_\alpha A_\alpha(x_\alpha), \ldots$ be all the formulas in Π whose outermost logical symbol is $\forall$. Let $a_1, \ldots, a_\beta, \ldots$ be all the free variables available at this stage. Then write down

$$A_1(a_1), \ldots, A_1(a_\beta), \ldots, A_\alpha(a_1), \ldots, A_\alpha(a_\beta), \ldots, \Pi \rightarrow \Lambda$$

above $\Pi \rightarrow \Lambda$.

10) $k \equiv 10$. Let $\exists x_1 A_1(x_1), \ldots, \exists x_\alpha A_\alpha(x_\alpha), \ldots$ be all the formulas in Π whose outermost logical symbol is $\exists$. Introduce new free variables $b_1, b_2, \ldots, b_\alpha, \ldots$. Then write down

$$A_1(b_1), \ldots, A_\alpha(b_\alpha), \ldots, \Pi \rightarrow \Lambda$$

above $\Pi \rightarrow \Lambda$.

PROPOSITION 8.13. (a) *If a sequent $\Gamma \to \Delta$ is provable (in* **LJ'**), *then every sequent of the reduction tree for $\Gamma \to \Delta$ is provable.*

(b) *If a sequent $\Gamma \to \Delta$ is unprovable, then there is a branch (in the tree for $\Gamma \to \Delta$) in which every sequent is unprovable.*

PROOF. (a) is obvious. In order to prove (b), we shall first prove the following: Let $\Pi \to \Lambda$ be a sequent in the tree and let $\Pi_\lambda \to \Lambda_\lambda$, $\lambda = 1, 2, \ldots, \alpha, \ldots$ be all its upper sequents, given by a reduction. If each is provable, then $\Pi \to \Lambda$ is provable. In other words, if for each λ, $\lambda = 1, 2, \ldots, \alpha, \ldots$, there are finite subsets of Π_λ and Λ_λ, say Π'_λ respectively, such that $\Pi'_\lambda \to \Lambda'_\lambda$ is provable, then there are finite subsets of Π and Λ, say Π' and Λ' respectively, such that $\Pi' \to \Lambda'$ is provable. We shall only deal with a few cases.

1) A $\exists$: left reduction has been applied to $\Pi \to \Lambda$. Then its upper sequent is of the form

$$A_1(b_1), \ldots, A_\alpha(b_\alpha), \ldots, \Pi \to \Lambda,$$

where $\exists x_\alpha A_\alpha(x_\alpha)$ is in Π for each α, and $b_1, \ldots, b_\alpha, \ldots$ are newly introduced free variables. By the hypothesis, there are finite subsets of $A_1(b_1), \ldots, A_\alpha(b_\alpha), \ldots$ (say $B_1(c_1), \ldots, B_n(c_n)$), of Π (say Π'), and of Λ (say Λ'), such that

$$B_1(c_1), \ldots, B_n(c_n), \Pi' \to \Lambda'$$

is provable. By repeated $\exists$: left and some weak inferences, we obtain $\Pi \to \Lambda'$, which is a subsequent of $\Pi \to \Lambda$. Notice that since $B_1(c_1), \ldots, B_n(c_n)$, $\Pi' \to \Lambda'$ is provable (with a finite proof), we may regard $c_1, \ldots, c_n$ as free variables of our original language.

2) An $\wedge$:right reduction has been applied to $\Pi \to \Lambda$. Then all upper sequents are of the form

$$\Pi \to \Lambda, C_1, \ldots, C_\alpha, \ldots,$$

where $A_1 \wedge B_1, \ldots, A_\alpha \wedge B_\alpha, \ldots$ are all the formulas of Λ whose outermost logical symbol is $\wedge$ and each C_α is A_α or B_α. We shall distinguish these cases by denoting C_α by $C_{0,\alpha}$ if C_α is A_α and by $C_{1,\alpha}$ if C_α is B_α. Then the upper sequents are the sequents

$$\Gamma \to \Lambda, C_{i_1,1}, \ldots, C_{i_\alpha,\alpha}, \ldots,$$

where $i_\alpha = 0$ or 1, for all possible combinations of values of $i_1, \ldots, i_\alpha, \ldots$. Let f denote any sequence $(i_1, \ldots, i_\alpha, \ldots)$. By assumption, there is a finite subsequent of each sequent, say $\Pi^f \to \Lambda^f, C_1^f, \ldots, C_{n_i}^f$, which is provable, where $C_1^f, \ldots, C_{n_i}^f$ is a finite subset of $C_{i_1,1}, \ldots, C_{i_\alpha,\alpha}, \ldots$.

In order now to exploit the generalized König's lemma (Proposition 8.14), we let R be a set with the order type of the sequence $C_1, C_2, \ldots, C_\alpha, \ldots$ (say $R = \{1, 2, \ldots, \alpha, \ldots\}$). Define $W_\alpha = 2\,(=\{0, 1\})$. For any subset $R_1 \subseteq R$ and any $f \in \prod_{\alpha \in R_1} W_\alpha$, we say that a finite sequence of formulas

$$(C_{f(\alpha_1), \alpha_1}, \ldots, C_{f(\alpha_n), \alpha_n})$$

(with $\alpha_1, \ldots, \alpha_n \in R_1$) is selected for f if there are finite subsets of Π and Λ, say Π' and Λ', respectively, such that

$$\Pi' \to \Lambda', C_{f(\alpha_1), \alpha_1}, \ldots, C_{f(\alpha_n), \alpha_n}$$

is provable. From the observation above, there is such a selected subset for any total function f. Now, for any $R_1 \subseteq R$ and any $f \in \prod_{\alpha \in R_1} W_\alpha$, we define

$$P(f) \Leftrightarrow \exists k\, \exists \alpha_1 \ldots \exists \alpha_k\, (\alpha_1, \ldots, \alpha_k \text{ are in the domain of } f \text{ and}$$

$$(C_{f(\alpha_1), \alpha_1}, \ldots, C_{f(\alpha_k), \alpha_k}) \text{ is selected}), \text{ where } k \text{ ranges}$$

$$\text{over the natural numbers.}$$

Then conditions 1 and 2 in the hypothesis of the generalized König's lemma are satisfied; hence by this lemma, there exists a finite subset of R, say $N_0 = \{\gamma_1, \ldots, \gamma_l\}$, such that if $\mathrm{Dom}(f)$ contains N_0, then $\mathrm{P}(f)$ holds.

Let

$$F = \{f \mid \mathrm{Dom}(f) = N_0\} = \prod_{j=1}^{l} W_{\gamma_j}.$$

F is a finite set and, for every f in F, $\mathrm{P}(f)$ holds, i.e., there is a subset of $\gamma_1, \ldots, \gamma_l$, say $\alpha_1, \ldots, \alpha_k$, such that $(C_{f(\alpha_1), \alpha_1}, \ldots, C_{f(\alpha_k), \alpha_k})$ is selected for f; i.e., there exist finite subsets of Π and Λ, say Π' and Λ' respectively, such that

$$\Pi' \to \Lambda', C_{f(\alpha_1), \alpha_1}, \ldots, C_{f(\alpha_k), \alpha_k}$$

is provable. Therefore, for every possible combination of values of $(i_1, \ldots, i_k)$ $(= i)$, there are finite subsets of Π and Λ, say Π^i and Λ^i respectively, such that

$$\Pi^i \to \Lambda^i, C_{i_1, \alpha_1}, \ldots, C_{i_k, \alpha_k}$$

is provable. Hence by weakenings and repeated $\wedge$: right, we obtain

$$\tilde{\Pi}' \to \tilde{\Lambda}', A_{\alpha_1} \wedge B_{\alpha_1}, \ldots, A_{\alpha_k} \wedge B_{\alpha_k},$$

where $\tilde{\Pi}'$ consists of all the Π''s for f in F, and likewise with $\tilde{\Lambda}'$.

Now, from the argument just completed, if the given sequent $\Gamma \to \Delta$ is not provable, then there is one branch in which every sequent is unprovable.

Having finished these preparations, we now define Kripke (intuitionistic) structures (for a language L).

DEFINITION 8.14. (1) A *partially ordered structure* $P = \langle O, \leq \rangle$ consists of a set O together with a binary relation $\leq$ satisfying the following:

a) $p \leq p$,

b) $p \leq q$ and $q \leq p$ imply $p = q$,

c) $p \leq q$ and $q \leq r$ imply $p \leq r$,

where p, q and r range over elements of O.

(2) A Kripke structure for a language L is an ordered triple $\langle P, U, \phi \rangle$ such that:

1) $P = \langle O, \leq \rangle$ is a partially ordered structure.

2) U is a map which assigns to every member of O, say p, a non-empty set, say U_p, such that, if $p \leq q$, then $U_p \subseteq U_q$ (where $\subseteq$ means set inclusion).

3) ϕ is a binary function $\phi(R, p)$, where R ranges over predicate constants in the language L and p ranges over members of O. Further:

3.1) Suppose the number of argument places of R is 0. Then $\phi(R, p) =$ T or F, and if $\phi(R, p) =$ T and $p \leq q$, then $\phi(R, q) =$ T.

3.2) Suppose R is an n-ary predicate ($n \geq 1$). Then $\phi(R, p)$ is a subset of

$$U_p^n = \underbrace{U_p \times \ldots \times U_p}_{n \text{ times}},$$

and $p \leq q$ implies $\phi(R, p) \subseteq \phi(R, q)$.

We define $U = \bigcup_{p \in O} U_p$. Then U can be thought of as the universe of the model or structure, and the elements of O as stages (see below).

Suppose that there is an assignment of objects of U to all the free variables; i.e., to each free variable a_i an object of U, say c_i, is assigned. Let $F(a_1, \ldots, a_n)$ be a formula with free variables $a_1, \ldots, a_n$ (at most). The *interpretation* of $F(a_1, \ldots, a_n)$ at (the stage) p (under this assignment) is defined as follows by induction on the number of logical symbols in $F(a_1, \ldots, a_n)$, and this interpretation is expressed as $\phi(F(c_1, \ldots, c_n), p)$. The value of such an interpretation is T or F.

a) $\phi(R(c_1, \ldots, c_n), p) =$ T if and only if $\langle c_1, \ldots, c_n \rangle \in \phi(R, p)$ (for $n > 0$).

b) $\phi(A \wedge B, p) =$ T if and only if $\phi(A, p) =$ T and $\phi(B, p) =$ T.

c) $\phi(A \vee B, p) =$ T if and only if $\phi(A, p) =$ T or $\phi(B, p) =$ T.

d) $\phi(A \supset B, p) =$ T if and only if for all q such that $p \leq q$, either $\phi(A, q) =$ F or $\phi(B, q) =$ T.

e) $\phi(\neg A, p) = \mathsf{T}$ if and only if for all q such that $p \leq q$, $\phi(A, q) = \mathsf{F}$.
f) $\phi(\exists x\, A(c_1, \ldots, c_n, x), p) = \mathsf{T}$ if and only if there is a c in U_p such that $\phi(A(c_1, \ldots, c_n, c), p) = \mathsf{T}$.
g) $\phi(\forall x\, F(c_1, \ldots, c_n, x), p) = \mathsf{T}$ if and only if for all q such that $p \leq q$, and for all c in U_q, $\phi(F(c_1, \ldots, c_n, c), q) = \mathsf{T}$.

We can generalize the definition of interpretation which has just been given to the case of sequents (finite or infinite). Let $\Gamma \to \Delta$ be a sequent. Then $\phi(\Gamma \to \Delta, p)$ is defined to be T if and only if, for all q such that $p \leq q$, either $\phi(A, q) = \mathsf{F}$ for some A in Γ or $\phi(B, q) = \mathsf{T}$ for some B in Δ.

A sequent $\Gamma \to \Delta$ is said to be valid in a Kripke structure $\langle P, U, \phi \rangle$ (with $P = \langle O, \leq \rangle$) if $\phi(\Gamma \to \Delta, p) = \mathsf{T}$ for all p in O.

PROPOSITION 8.15. *Suppose $\Gamma \to \Delta$ is provable in* **LJ**$'$, *and $\langle P, U, \phi \rangle$ is a Kripke structure. Then $\Gamma \to \Delta$ is valid in $\langle P, U, \phi \rangle$.*

PROOF. This is only a routine matter: by mathematical induction on the number of inferences in a proof of $\Gamma \to \Delta$ (or a subsequent of it).

Now, in order to finish the completeness proof for **LJ**$'$, we shall start with an unprovable sequent $\Gamma \to \Delta$ and construct a counter-model in the sense of Kripke. This will be constructed from the reduction tree for $\Gamma \to \Delta$. Let us call this tree T. (Remember, in the construction of T, the $\neg$: right, $\supset$: right and $\forall$: right reductions were omitted.) This situation, i.e., with just this tree present, is called stage 0. By Proposition 8.16, there is a branch of T, say B_0, containing (only) unprovable sequents. If B_0 is finite, let $\Gamma_0 \to \Delta_0$ be its uppermost sequent. If B_0 is infinite, let Γ_0 and Δ_0 be the union of all formulas in the antecedents and succedents respectively of the sequents in B_0 (each arranged in a well-ordered sequence), and consider the (possibly infinite) sequent $\Gamma_0 \to \Delta_0$. Single out all the formulas in Δ_0 whose outermost symbols are $\neg$, $\supset$ or $\forall$. (If there is no such formula, then stop.) Let the symbol p range over all such formulas. We call each such p an immediate successor of 0 (and 0 an immediate predecessor of p).

Case 1. p is a formula of the form $\neg A$. Then consider the sequent $A, \Gamma_0 \to$.

Case 2. p is $B \supset C$. Then consider the sequent $B, \Gamma_0 \to C$.

Case 3. p is $\forall x\, F(x)$. Let a be a free variable which does not belong to U_0.
(This can always be done by introducing a new symbol if necessary.) Then consider the sequent $\Gamma_0 \to F(a)$.

It is easily shown that (in each case) this new sequent is not provable, since otherwise $\Gamma_0 \to \Delta_0$ would be provable. Let us call this new sequent $\tilde{\Gamma}_p \to \tilde{\Delta}_p$, and let T_p be the reduction tree for $\tilde{\Gamma}_p \to \tilde{\Delta}_p$.

As before, let B_p be a branch of T_p containing unprovable sequents, and let $\Gamma_p \to \Delta_p$ be either the topmost sequent of B_p, or (if B_p is infinite) the "union" of all sequents in B_p, as before. Then follow exactly the same process as the preceding paragraph. Namely, let q range over all formulas in Δ_p whose outermost logical symbol is $\neg$, $\supset$ or $\forall$. (If there is no such formula, then stop.) Again, for all such q and p (with q in Δ_p as above), we call q an immediate successor of p and p an immediate predecessor of q. Then define as before the tree T_q and branch B_q.

Continue this procedure ω times. Let O be the set of all these p's, and let $\leqslant$ be the transitive reflexive relation on O generated by the immediate predecessor relation defined above. O is partially ordered by $\leqslant$. Now define U_p to be the set of all free variables occurring in B_p, for all $p \in O$, and define $U = \bigcup_{p\in O} U_p$. Notice the following.

1) If $p \leqslant q$, then $U_p \subseteq U_q$.

2) If q is an immediate successor of p, then all formulas in Γ_p occur in the antecedents of all sequents in T_q (and hence in B_q).

We now define the function ϕ as follows. For any n-ary predicate symbol R ($n > 0$), and any $p \in O$,

$$\phi(R, p) = \{\langle a_1, \ldots, a_n\rangle \mid a_1, \ldots, a_n \in U_p \text{ and } R(a_1, \ldots, a_n) \text{ occurs in } \Gamma_p\}$$

(and for $n = 0$, $\phi(R, p) = \mathsf{T}$ if and only if R occurs in Γ_p).

So we have defined a Kripke structure $\langle P, U, \phi\rangle$. We shall consider the interpretation of formulas in this structure relative to the (natural) assignment of each free variable to itself.

PROPOSITION 8.16 (with the above notation). *Let A be a formula in B_p. If A occurs in the antecedent of a sequent in B_p, then $\phi(A, p) = \mathsf{T}$; if it occurs in the succedent, then $\phi(A, p) = \mathsf{F}$.*

PROOF. By induction on the number of logical symbols in A. First it should be noticed that if a formula occurs in the antecedent of a sequent in B_p, then it does not occur in the succedent of any sequent in B_p. The same holds with "antecedent" and "succedent" interchanged. Also, once a formula appears on one side of a sequent, it will appear on the same side of all higher sequents of B_p, and hence of the sequent $\Gamma_p \rightarrow \Delta_p$.

1) A is an atomic formula $R(a_1, \ldots, a_n)$. If A occurs in an antecedent, hence in Γ_p, then by definition $\langle a_1, \ldots, a_n\rangle \in \phi(R, p)$, which implies, again by definition, that $\phi(A, p) = \mathsf{T}$. If A occurs in a succedent, then $\langle a_1, \ldots, a_n\rangle \notin \phi(R, p)$, so $\phi(A, p) = \mathsf{F}$.

2) A is $\neg B$. Suppose A occurs in the antecedent. Then A occurs in Γ_p. This implies that, given any q such that $p \leqslant q$, A occurs in the antecedent of all the sequents in B_q; hence B occurs in the succedent of a sequent in B_q; therefore, by the induction hypothesis, $\phi(B, q) = \mathsf{F}$. So $\phi(B, q) = \mathsf{F}$ for any q such that $p \leqslant q$. This means that $\phi(A, p) = \mathsf{T}$.

Suppose next that A occurs in the succedent of a sequent in B_p. Then there exists a next stage, say q. It starts with $B, \Gamma_p \rightarrow$. By the induction hypothesis, $\phi(B, q) = \mathsf{T}$. That is to say, there is a q such that $p \leqslant q$ and $\phi(B, q) = \mathsf{T}$. Therefore by definition $\phi(A, p) = \mathsf{F}$.

3) A is $B \wedge C$ or $B \vee C$. Those cases are easy; so they are left to the reader.

4) A is $\forall x\, F(x)$. Suppose A occurs in the antecedent of a sequent in B_p and suppose $p \leq q$. Then A occurs in the antecedent of a sequent in B_q. Let a be an element of U_q. Then $F(a)$ occurs in the antecedent of a sequent in B_q. Hence, by the induction hypothesis, $\phi(F(a), q) = \mathsf{T}$. So for any q such that $p \leq q$ and any a in U_q, $\phi(F(a), q) = \mathsf{T}$, which means that $\phi(A, p) = \mathsf{T}$.

Suppose next that A occurs in the succedent of a sequent in B_p. So the next stage, say q, starts with $\Gamma_p \rightarrow F(a)$, where a is a (new) variable in U_q. By the induction hypothesis, $\phi(F(a), q) = \mathsf{F}$. So there exists a q such that $p \leq q$, and a member a of U_q, such that $\phi(F(a), q) = \mathsf{F}$. This means that

$$\phi(\forall x\, F(x), p) = \mathsf{F}.$$

5) A is of the form $\exists x\, F(x)$. This case is left as an exercise.

6) A is of the form $B \supset C$. Suppose that A occurs in the antecedent of a sequent in B_p. Then either C occurs in Γ_p or B occurs in Δ_p. Let $p \leq q$. Then either C occurs in the antecedent or B occurs in the succedent of a sequent in B_q. So for any q, with $p \leq q$, either $\phi(C, q) = \mathsf{T}$ or $\phi(B, q) = \mathsf{F}$. So $\phi(B \supset C, p) = \mathsf{T}$.

Suppose next that A occurs in the succedent of a sequent in B_p. Then the next stage, say q, starts with $B, \Gamma_p \rightarrow C$. Hence there is a q such that $p \leq q$, $\phi(B, q) = \mathsf{T}$ and $\phi(C, q) = \mathsf{F}$; so $\phi(B \supset C, p) = \mathsf{F}$.

So now we can conclude that if $\Gamma \rightarrow \Delta$ is unprovable, then we can construct a Kripke structure $\langle P, U, \phi \rangle$ such that (under a suitable assignment to free variables) every formula in Γ assumes the value T and every formula in Δ assumes the value F; in other words, there is a Kripke counter-model for $\Gamma \rightarrow \Delta$. This ends the completeness proof. Thus we have obtained:

THEOREM 8.17 (completeness of the intuitionistic predicate calculus: a generalized version; cf. Theorem 8.2). *Let $\Gamma \rightarrow \Delta$ be a sequent (finite or infinite). If $\Gamma \rightarrow \Delta$ is valid in all Kripke structures, then $\Gamma \rightarrow \Delta$ is provable. In particular,* **LJ** *is complete.*

(Recall that the soundness of **LJ** was established by Proposition 8.18). Notice that the method which has been prescribed here for completeness of **LJ** works even when the language is not countable, while the method for **LK** works only for a countable language. Although we could in fact use a method for **LK** similar to this one for **LJ**′, we do not attempt to do so, since Henkin's simple method is sufficient for that purpose.

EXERCISE 8.18. Construct a Kripke counter-model for each of the following sequents.

1) $\rightarrow P \vee \neg P$, where P is a predicate symbol.

2) $\forall x\,(P(x) \vee Q) \rightarrow \forall x\, P(x) \vee Q$, where P and Q are predicate symbols of the indicated numbers of argument.

3) $\rightarrow \exists x\,(\exists y\, P(y) \supset P(x))$, where P is a unary predicate.

[*Hint* for 1): At stage 0:

$$\frac{\rightarrow P \vee \neg P,\, P,\, \neg P}{\rightarrow P \vee \neg P}.$$

Let p be $\neg P$. Then at stage p:

$$\frac{P \rightarrow}{\rightarrow \neg P}.$$

So define $O = \{0, p\}, 0 \leqslant p$, $U_0 = U_p = \{a\}$, $\phi(P, 0) = \mathsf{F}$. Then $\phi(P \vee \neg P, 0) = \mathsf{F}$ can be easily proved.]

In order to discuss the completeness theorem for the intuitionistic logic further, we first discuss the complete Heyting algebras. A complete Heyting algebra (abbreviated by **cHa**) is a special kind of complete lattice. Let Ω be a complete lattice and $A \subseteq \Omega$. The least upperbound of A and the greatest lower bound of A are denoted by $\bigvee A$ and $\bigwedge A$ respectively. If $A = \{a_i \mid i \in I\}$, then $\bigvee A$ is denoted by $\bigvee_{i \in I} a_i$ or $\bigvee_i a_i$ and $\bigwedge A$ is denoted by $\bigwedge_{i \in I} a_i$ or $\bigwedge_i a_i$. For a and b in Ω, $a \vee b$ denotes the join of a and b and $a \wedge b$ denotes the meet of a and b.

DEFINITION 8.19. A **cHa** is a complete lattice Ω satisfying the following $\wedge, \bigvee$-distributive law:

$$p \wedge \bigvee_{i \in I} q_i = \bigvee_{i \in I} (p \wedge q_i)$$

for all $p \in \Omega$ and all subsets $\{q_i \mid i \in I\} \subseteq \Omega$. We denote the greatest element and the least element of Ω by 1 and 0 respectively. Let Ω be a **cHa** and $p, q \in \Omega$. We define $p \supset q$ as $\bigvee\{r \in \Omega \mid p \wedge r \leqslant q\}$ and $\neg p$ as $p \supset 0$.

A complete Boolean algebra is a **cHa**. Another typical example of **cHa** is a topological space. Let X be a topological space and $\mathcal{O}(X)$ be the set of all open sets of X. Then $\mathcal{O}(X)$ is a **cHa** by the following definition: $p \wedge q = p \cap q$ and $\bigvee_i p_i = \bigcup_i p_i$, where p, q and p_i are open sets. It is easily seen that $\neg p = (X - p)^\circ$ and $\bigwedge_i p_i = (\bigcap_i p_i)^\circ$, where q° denotes the open kernel of q.

EXERCISE 8.20. Let Ω be a **cHa** and $p, q, r \in \Omega$. Then the following equivalences hold.

1) $p \wedge q \leqslant r$ iff $q \leqslant (p \supset r)$.

2) $p \wedge q = 0$ iff $q \leqslant \neg p$.

Let L be a language without function constants. For any non-empty set D, $L\{D\}$ is the extended language obtained from L by introducing a new individual constant $\bar{d}$ for every member d in D.

DEFINITION 8.21. (1) Let Ω be a **cHa**. By an Ω-valued structure for L we mean a pair $\langle D, \phi \rangle$, where D is a non-empty set and ϕ is a map from the constants of $L\{D\}$ such that:

(i) if k is an individual constant, then ϕk is an element of D;

(ii) if d is an element of D, then $\phi \bar{d}$ is d,

(iii) if R is a predicate constant of n arguments, then ϕR is a function from D^n into Ω.

(2) Let A be a sentence in $L\{D\}$. Then the truth value $[\![A]\!]$ of A is defined by an Ω-valued structure $\langle D, \phi \rangle$ as follows.

1) If R is a predicate constant of n arguments and $t_1, \ldots, t_n$ are individual constants in $L\{D\}$, then $[\![R(t_1, \ldots, t_n)]\!]$ is $\phi R(\phi t_1, \ldots, \phi t_n)$. $[\![R(t_1, \ldots, t_n)]\!]$ is a member of Ω.

2) If A is a sentence with a logical symbol, then $[\![A]\!]$ is defined according to its outermost logical symbols as follows.

(i) $[\![\neg A]\!] = \neg [\![A]\!]$,

$[\![A \wedge B]\!] = [\![A]\!] \wedge [\![B]\!]$,

$[\![A \vee B]\!] = [\![A]\!] \vee [\![B]\!]$,

$[\![A \supset B]\!] = [\![A]\!] \supset [\![B]\!]$,

where $\neg$, $\wedge$, $\vee$, and $\supset$ in the left-hand side of $=$ are logical symbols of the language and $\neg$, $\wedge$, $\vee$, and $\supset$ in the right-hand side of $=$ are operations on Ω.

(ii) $[\![\forall x\, A(x)]\!] = \bigwedge_{d \in D} [\![A(\bar{d})]\!]$,

$[\![\exists x\, A(x)]\!] = \bigvee_{d \in D} [\![A(\bar{d})]\!]$.

For every sentence A of $L\{D\}$, $[\![A]\!]$ is a member of Ω. Let $\mathscr{B}$ be a complete Boolean algebra. Since a complete Boolean algebra is a **cHa**, the definitions of a $\mathscr{B}$-valued structure $\langle D, \phi \rangle$ and its truth value $[\![\;]\!]$ are obtained as special cases of the above definition. If $\mathscr{B}$ is the complete Boolean algebra $\{0, 1\}$, then we can identify a $\mathscr{B}$-valued structure and a structure in Definition 8.1 as follows. If $\langle D, \phi \rangle$ is a structure in Definition 8.1, then we assign to $\langle D, \phi \rangle$ a $\{0, 1\}$-valued structure $\langle D, \tilde{\phi} \rangle$ defined as follows.

(i) For an individual constant k in L, $\tilde{\phi} k$ is ϕk.

(ii) For any element d of D, $\tilde{\phi} \bar{d}$ is d.

(iii) For a predicate constant R of n arguments, $\tilde{\phi} R$ is a function from D^n into $\{0, 1\}$ satisfying

$$\tilde{\phi} R(d_1, \ldots, d_n) = 1 \text{ iff } \langle d_1, \ldots, d_n \rangle \in \phi R.$$

It is easily proved that for every sentence A in L and for every interpretation $\mathscr{T} = (\langle D, \phi \rangle, \phi_0)$, $\mathscr{T}$ satisfies A iff $[\![A]\!] = 1$, where $[\![A]\!]$ is the truth value of A by $\langle D, \tilde{\phi} \rangle$.

Now let Ω be again a **cHa** and $\langle D, \phi\rangle$ be an Ω-valued structure. The truth value of a closed sequent in **LJ** is defined by

$$[\![A_1, \ldots, A_n \to B]\!] = [\![A_1 \wedge \ldots \wedge A_n \supset B]\!], \text{ and}$$

$$[\![A_1, \ldots, A_n \to \quad]\!] = [\![\neg(A_1 \wedge \ldots \wedge A_n)]\!].$$

A closed sequent $\Sigma \to \Delta$ of $L\{D\}$ is said to be valid in $\langle D, \phi\rangle$ if $[\![\Sigma \to \Delta]\!] = 1$, where $[\![\Sigma \to \Delta]\!]$ is the truth value of $\Sigma \to \Delta$ by $\langle D, \phi\rangle$. The sequent is said to be Ω-valid if it is valid in every Ω-valued structure $\langle D, \phi\rangle$.

Then Theorem 8.2 implies the following theorem.

THEOREM 8.22. *Let $\mathcal{B}$ be a complete Boolean algebra. A closed sequent $\Gamma \to \Delta$ is provable in* **LK** *if and only if it is $\mathcal{B}$-valid.*

The proof that provability implies $\mathcal{B}$-validity is routine. The converse direction is immediate from Theorem 8.2 since $\{0, 1\}$ is a subalgebra of $\mathcal{B}$ for every Boolean algebra $\mathcal{B}$.

Before Kripke, Rasiowa and Sikorski proved a completeness theorem for the intuitionistic predicate calculus by using the **cHa**. We shall discuss their completeness theorem and its relation with Kripke's completeness theorem. First we shall give the definition of Heyting algebra itself and another definition of **cHa**.

DEFINITION 8.23. A lattice H is said to be a Heyting algebra if $\supset$ satisfying the following condition is defined on H

$$p \wedge q \leqslant r \text{ iff } q \leqslant (p \supset r).$$

THEOREM 8.24. *A* **cHa** *is a Heyting algebra and a Heyting algebra is a* **cHa** *if it is complete.*

PROOF. The first part is 1) of Exercise 8.20. Now we assume that Ω is a Heyting algebra and complete. It suffices to show

$$a \wedge \bigvee_i b_i \leqslant \bigvee_i (a \wedge b_i).$$

This is proved as follows

$$a \wedge b_i \leqslant \bigvee_i (a \wedge b_i) \to b_i \leqslant \Big(a \supset \bigvee_i (a \wedge b_i)\Big)$$

$$\to \bigvee_i b_i \leqslant \Big(a \supset \bigvee_i (a \wedge b_i)\Big)$$

$$\to a \wedge \bigvee_i b_i \leqslant \bigvee_i (a \wedge b_i).$$

THEOREM 8.25. (Rasiowa and Sikorski). *Let $\Gamma \to \Delta$ be a (finite or infinite) sequent in* L. *If $\Gamma \to \Delta$ is not provable in* **LJ**, *then there exists a* **cHa** Ω, *and a Ω-valued structure $\langle D, \phi \rangle$, such that $[\![A]\!] = 1$ for each A in Γ and $[\![A]\!] \neq 1$ for each A in Δ.*

PROOF. We assume that there are infinitely many free variables which do not occur in Γ or Δ. Let D be the set of all terms in L and L(D) be the set of all formulas in L. We define an order $\leq$ on L(D) as follows.

$A \leq B$ iff $\Gamma, A \to B$ is provable in **LJ**. We also define $A \equiv B$ to be '$A \leq B$ and $B \leq A$'. Then $\equiv$ is an equivalence relation on L(D) and the equivalence class of A in L(D) is denoted by $[\![A]\!]$. Let H be the set of $[\![A]\!]$ namely $H = \{[\![A]\!] \mid A \in \mathrm{L}(D)\}$. The order $\leq$ on L(D) induces an order $\leq$ on H. It is easily seen that H is a Heyting algebra. The following properties hold.

1) $[\![\forall x\, A(x)]\!] = \bigwedge_{d \in D} [\![A(d)]\!]$.

2) $[\![\exists x\, A(x)]\!] = \bigvee_{d \in D} [\![A(d)]\!]$.

Proof of 1). For every $d \in D [\![\forall x\, A(x)]\!] \leq [\![A(d)]\!]$ since $\forall x\, A(x) \to A(d)$ is provable in **LJ**. Now let $[\![C]\!] \leq [\![A(d)]\!]$ for every $d \in D$. Then $\Gamma, C \to A(d)$ is provable for every $d \in D$. Take d to be a free variable which does not occur in $\Gamma, C, \forall x\, A(x)$. Then $\Gamma, C \to \forall x\, A(x)$ is provable in **LJ**.

The proof of 2) goes in the same way as 1).

For each n-ary predicate constant R, we define ϕR by $(\phi R)(d_1, \ldots, d_n) = [\![R(d_1, \ldots, d_n)]\!]$ for $d_1, \ldots, d_n \in D$. Then $\langle D, \phi \rangle$ is a H-valued structure, $[\![A]\!] = 1$ for every A in Γ, and $[\![A]\!] \neq 1$ for every A in Δ. Therefore the theorem immediately follows from the following Rasiowa–Sikorski's embedding lemma.

EXERCISE 8.26. For every Heyting algebra A there exists a **cHa** A^* and an isomorphism of A into A^*, preserving all infinite joins and meets in A.

[*Hint*: Follow the prescription given below.]

1). Let J be a subset of A. J is said to be an ideal of A if it satisfies the following conditions:

i) $a_1 \in J,\ a_2 \leq a_1 \to a_2 \in J$;

ii) $a_1 \in J,\ a_2 \in J \to (a_1 \vee a_2) \in J$.

An ideal J of A is said to be a prime ideal if $1 \notin J$ and the following condition is satisfied: $(a_1 \wedge a_2) \in J \to a_1 \in J \vee a_2 \in J$. The set of all prime ideals of A is called the Stone space of A.

Let X be the Stone space of A. Then there exists an embedding $h: \langle A, \wedge, \vee \rangle \to \langle \mathcal{P}(X), \cap, \cup \rangle$, defined by $h(a) = \{x \in X \mid a \notin x\}$.

2). Let B be the Boolean subalgebra of $\mathcal{P}(X)$ generated by $h(A)$. Every element of B is of the form $((X - a_1) \cup b_1) \cap \ldots \cap ((X - a_n) \cup b_n)$, where $a_1, b_1, \ldots, a_n, b_n \in h(A)$. We define an interior operation $I : B \to h(A)$ by the following equation

$$I(((X - a_1) \cup b_1) \cap \ldots \cap ((X - a_n) \cup b_n)) = (a_1 \supset b_1) \cap \ldots \cap (a_n \supset b_1),$$

where $h(a) \supset h(b)$ denotes $h(a \supset b)$. Then $I : B \to B$ satisfies the following conditions.

i) $I(b_1 \wedge b_2) = I(b_1) \wedge I(b_2)$.

ii) $I(b) \leqslant b$.

iii) $II(b) = I(b)$.

iv) $I(1) = 1$.

A Boolean algebra B together with I satisfying i)–iv) is called a topological Boolean algebra. Our topological Boolean algebra $\langle B, I\rangle$ satisfies $I(B) = h(A)$.

3). Let B^* be the complete Boolean algebra of all regular open sets of the Stone space of B. Then there exists a canonical embedding $h' : B \to B^*$ which preserves infinite joins and infinite meets. We extend the interior operation $I : B \to h(A)$ to $I : B^* \to h(A)$ by the following equation

$$I\left(\bigvee_i{}^{B^*} h'(b_i)\right) = \bigvee_i{}^{B^*} h'(Ib_i).$$

Then $\langle B^*, I\rangle$ is a topological Boolean algebra and $I(B^*)$ is a **cHa** satisfying Exercise 8.26.

Now we shall consider the relation between Kripke models and **cHa**-valued structures.

Let $\langle P, U, \phi\rangle$ be a Kripke structure for L, where $P = \langle P, \leqslant\rangle$. For $p \in P$, we denote $\phi(A, p) = T$ by $p \Vdash A$ and $\phi(A, p) = F$ by $p \nVdash A$. Then b)–g) in Definition 8.14 are rewritten as follows.

b) $p \Vdash A \wedge B$ iff $p \Vdash A$ and $p \Vdash B$.

c) $p \Vdash A \vee B$ iff $p \Vdash A$ or $p \Vdash B$.

d) $p \Vdash A \supset B$ iff $\forall q \geqslant p(q \nVdash A$ or $q \Vdash B)$.

e) $p \Vdash \neg A$ iff $\forall q \geqslant p(q \nVdash A)$.

f) $p \Vdash \exists x\, A(x)$ iff $\exists c \in U_p(p \Vdash A(c))$.

g) $p \Vdash \forall x\, A(x)$ iff $\forall q \geqslant p\, \forall c \in U_q(q \Vdash A(c))$.

First we extend the language L to L^E by adding a new unary predicate constant E and we define

$$p \Vdash E(c) \text{ iff } c \in U_p.$$

Then the following holds

$$p \Vdash E(c),\ p \leqslant q \to q \Vdash E(c)$$

and $\langle P, U, \phi\rangle$ becomes a Kripke structure for L^E.

Let A be a sentence of $\mathrm{L}(U)$. Then A^E is defined to be a relativization of A by E, namely, A^E is obtained from A by replacing all quantifiers $\forall x, \exists y, \ldots$ in A by $\forall x\,(Ex \supset), \exists y\,(Ey \wedge), \ldots$

THEOREM 8.27. *There exists a* **cHa**-*valued structure* $\langle U, [\![\ ]\!]\rangle$ *of* L^E, *where*

$U = \bigcup_{p \in P} U_p$, *such that for every sentence* A *of* L,

$$A \text{ is valid in } \langle P, U, \phi\rangle \text{ iff } [\![A^E]\!] = 1 \text{ in } \langle U, [\![\ ]\!]\rangle.$$

PROOF. Let $L^E(U)$ be obtained from L^E by adding all elements of U as constants. For a sentence A in $L^E(U)$, $[A]$ is defined by the following equation

$$[A] = \{p \in P \mid p \Vdash A\}.$$

Then we define H by the equation

$$H = \{[A] \mid A \text{ is a sentence of } L^E(U)\}.$$

The following properties hold.

1) $[A] \subseteq [B]$ iff $\forall p\,(p \Vdash A \supset B)$.

If $[A] \subseteq [B]$, then $\forall p\,(p \Vdash A$ implies $p \Vdash B)$. Therefore $q \geqslant p$, $q \Vdash A \to q \Vdash B$ namely

$$\forall p\,(p \Vdash A \supset B).$$

Now we assume $\forall p\,(p \Vdash A \supset B)$. Then $\forall p\,(p \Vdash A$ implies $p \Vdash B)$. Therefore $q \in [A]$ implies $q \in [B]$.

2) $[A] \cup [B] = [A \vee B]$.

The proofs of this and the following are obvious.

3) $[A] \cap [B] = [A \wedge B]$.

4) $[A] \cap [C] \subseteq [B]$ iff $[C] \subseteq [A \supset B]$.

$$\begin{aligned}[A] \cap [C] \subseteq [B] &\text{ iff } \forall p\,(p \Vdash A \wedge C \supset B) \text{ by 1), 3)}\\ &\text{ iff } \forall p\,(p \Vdash C \supset (A \supset B))\\ &\text{ iff } [C] \subseteq [A \supset B].\end{aligned}$$

5) $\bigcup_{u \in U} [Eu \wedge A(u)] = [\exists x\, A(x)]$.

$$\begin{aligned}p \in \bigcup_{u \in U} [Eu \wedge A(u)] &\text{ iff } \exists u \in U(p \Vdash Eu \wedge A(u))\\ &\text{ iff } \exists u \in U_p(p \Vdash A(u))\\ &\text{ iff } p \Vdash \exists x\, A(x).\end{aligned}$$

6) $\bigcap_{u \in U} [Eu \supset A(u)] = [\forall x\, A(x)]$.

$$p \in \bigcap_{u \in U} [Eu \supset A(u)] \text{ iff } \forall u \in U(p \Vdash Eu \supset A(u))$$

$$\text{iff } \forall u \in U\, \forall q \geqslant p(q \Vdash Eu \text{ implies } q \Vdash A(u))$$

$$\text{iff } \forall u \in U\, \forall q \geqslant p(u \in U_q \text{ implies } q \Vdash A(u))$$

$$\text{iff } \forall q \geqslant p\, \forall u \in U_q(q \Vdash A(u))$$

$$\text{iff } p \Vdash \forall x\, A(x).$$

From 1)–4), $\langle H, \cap, \cup \rangle$ is a Heyting algebra, where $1 = P$ and $0 = \phi$. For every n-ary predicate constant R we define

$$[\![R(u_1, \ldots, u_n)]\!] = [R(u_1, \ldots, u_n)].$$

Then for every sentence A in $\mathrm{L}(U)$, $[\![A^E]\!]$ is defined as a member of H.

7) $[\![A^E]\!] = [A]$.

If A is atomic, then this is obvious from the definition. The cases for $A \vee B$, $A \wedge B$, and $A \supset B$ are also immediate. We prove only the cases $\forall x\, A(x)$ and $\exists x\, A(x)$.

$$\begin{aligned} [\![(\forall x\, A(x))^E]\!] &= [\![\forall x\, (Ex \supset A^E(x))]\!] \\ &= \bigcap_{u \in U} ([\![Eu]\!] \supset [\![A^E(u)]\!]) \\ &= [\forall x\, A(x)] \qquad \text{by 6).} \end{aligned}$$

$$\begin{aligned} [\![(\exists x\, A(x))^E]\!] &= [\![\exists x\, (Ex \wedge A^E(x))]\!] \\ &= \bigcup_{u \in U} [\![Eu]\!] \wedge [\![A^E(u)]\!] \\ &= \bigcup_{u \in U} [Eu \wedge A(u)] \\ &= [\exists x\, A(x)] \qquad \text{by 5).} \end{aligned}$$

By Rasiowa–Sikorski's embedding lemma (Exercise 8.26) there exist a **cHa** Ω and an isomorphism $H \to \Omega$, in which all infinite joins and meets are preserved. Then $\langle U, [\![\ ]\!] \rangle$ can be considered an Ω-valued structure and we have

$$\begin{aligned} [\![A^E]\!] = 1 \text{ in } \langle U, [\![\ ]\!] \rangle &\text{ iff } [A] = P \\ &\text{ iff } \forall p \in P(p \Vdash A). \end{aligned}$$

Compared with the completeness theorem for **LK**, Theorem 8.26 for **LJ** is much weaker. It says only that $\Gamma \to \Delta$ is provable if and only if it is Ω-valid for every **cHa** Ω. Therefore the following open problems are very interesting.

1. Let Ω be a **cHa**. Formulate an extension $\mathbf{LJ}_\Omega$ of **LJ** such that the following equivalence holds: $\Gamma \to \Delta$ is provable in $\mathbf{LJ}_\Omega$ iff it is Ω-valid. An interesting special case is that Ω is the **cHa** of the n-dimensional Euclidean space.

2. Let $\mathscr{F}$ be a class of **cHa**. Formulate an extension $\mathbf{LJ}_\mathscr{F}$ of **LJ** such that the following equivalence holds: $\Gamma \to \Delta$ is provable in $\mathbf{LJ}_\mathscr{F}$ iff it is Ω-valid for every **cHa** Ω in $\mathscr{F}$. An interesting special case is that $\mathscr{F}$ is the class of $\mathcal{O}(X)$ where X is a certain type of topological spaces e.g. $\mathscr{F} = \{\mathcal{O}(X) \mid X$ is a finite topological space$\}$.

We have the impression that an axiomatization of $\mathbf{LJ}_\Omega$ might be impossible for almost all infinite **cHa** Ω. In other words, an axiomatization of $\mathbf{LJ}_\mathscr{F}$ is very likely only when $\mathscr{F}$ is a large class of **cHa**. In this sense, the following example $\mathbf{LJ}_I$ is very interesting since I is a single infinite **cHa** and an axiomatization of $\mathbf{LJ}_I$ is possible. It seems to the author very attractive to find many such examples and to classify in what kind of Ω or $\mathscr{G}$ $\mathbf{LJ}_\Omega$ or $\mathbf{LJ}_\mathscr{F}$ is axiomatizable.

In the rest of this section, we solve the first problem for a **cHa** of $[0, 1]$ (a closed interval of real numbers). Let I be the **cHa** of $[0, 1]$. The operations $\wedge$ and $\bigvee$ are defined as follows;

$a \wedge b$ is the minimum of a and b, and $\bigvee_i a_i$ is the least upper bound of a_i.

It is easily seen that I is a **cHa**. 0 and 1 are the smallest element and the greatest element of I respectively.

In order to formulate $\mathbf{LJ}_I$ we introduce propositional variables $\alpha_0, \alpha_1, \alpha_2, \ldots$. $\alpha, \beta, \gamma, \ldots$ may be used for propositional variables. The formation of formulas is supplemented by the following rule.

A propositional variable only is an atomic formula.

Then $\mathbf{LJ}_I$ is obtained from **LJ** by introducing the following extra axiomschemata and extra inference rule.

Extra axiomschemata for $\mathbf{LJ}_I$.

1. $\to (A \supset B) \vee ((A \supset B) \supset B)$,
2. $(A \supset B) \supset B \to (B \supset A) \vee B$,
3. $\forall x\,(C \vee A(x)) \to C \vee \forall x\,A(x)$,
4. $\forall x\,A(x) \supset C \to \exists x\,(A(x) \supset D) \vee (D \supset C)$.

Extra inference rule

$$\frac{\Gamma \to A \vee (C \supset \alpha) \vee (\alpha \supset B)}{\Gamma \to A \vee (C \supset B)},$$

where α does not occur in the lower sequent.

We express two sequents $A \to B$ and $B \to A$ by $A \leftrightarrow B$.

PROPOSITION 8.28. *The following sequents are provable in* $\mathbf{LJ}_I$.

(1) $\to (A \supset B) \vee (B \supset A)$,

(2) $A \supset B \vee C \leftrightarrow (A \supset B) \vee (A \supset C)$,

(3) $A \wedge B \supset C \leftrightarrow (A \supset C) \vee (B \supset C)$,

(4) $C \supset \exists x\, A(x) \to (C \supset D) \vee \exists x\, (D \supset A(x))$,

where x does not occur in D.

PROOF. (1) follows from Extra axioms 1 and 2.
($\leftarrow$) of (2) and (3) are provable in **LJ**.
($\to$) of (2)

$$A \supset B \vee C \to (A \supset B \vee C) \wedge ((B \supset C) \vee (C \supset B))$$
$$\to (A \supset C) \vee (A \supset B)$$

($\to$) of (3)

$$A \wedge B \supset C \to (A \wedge B \supset C) \wedge ((A \supset B) \vee (B \supset A))$$
$$\to (A \supset C) \vee (B \supset C)$$

(4): By the use of Extra axiom 1, it suffices to prove the following sequent.

(i) $C \supset \exists x\, A(x), (C \supset D) \supset D \to \exists x\, (D \supset A(x)) \vee (C \supset D)$.

Let Γ be $C \supset \exists x\, A(x)$, $(C \supset D) \supset D$. Then we have

(ii) $\Gamma \to (\exists x\, A(x) \supset D) \supset D$ and

(iii) $(\exists x\, A(x) \supset D) \supset D \to \forall x\, (A(x) \supset D) \supset D$.

By using (ii), (iii) and Extra axiom 4, we have

(iv) $\Gamma \to \exists x\, ((A(x) \supset D) \supset C) \vee (C \supset D)$.

On the other hand, from $\to (A(a) \supset D) \vee (D \supset A(a))$ we have $(A(a) \supset D) \supset C \to (D \supset A(a)) \vee C$. Hence from (iv) we have

$$\Gamma \to \exists x\, (D \supset A(x)) \vee C \vee (C \supset D).$$

Since $\Gamma, C \to \exists x\, (D \supset A(x))$, we have (i).

Since we extended the language by introducing propositional variables, we extend the definition of the truth value accordingly. We call a formula in this extended language a sentence even if it has propositional variables. So a formula is a sentence if no free variables occur in it.

DEFINITION 8.29. Let $\langle D, \phi\rangle$ be an Ω-valued structure. An interpretation of $L\{D\}$ is a structure $\langle D, \phi\rangle$ together with a mapping ρ from propositional variables into I. We may denote an interpretation $(\langle D, \phi\rangle, \rho)$ simply by $\mathcal{I}$. Here ρ is called an assignment from the set of all propositional variables. Now we extend the truth value of a sentence. The truth value of A depends on ρ and is denoted by $[\![A]\!]_\rho$ or simply by $[\![A]\!]$. The definition of $[\![A]\!]_\rho$ is obtained by rewriting $[\![A]\!]$ in the previous definition by $[\![A]\!]_\rho$ and adding the following rule.

$[\![\alpha_i]\!]_\rho = \rho\alpha_i$, where α_i is a propositional variable.

Obviously $[\![A]\!]_\rho$ does not depend on ρ if A does not have any propositional variable. We say that $\Gamma \to \Delta$ is valid (more precisely I-valid) if $[\![\Gamma \to \Delta]\!]_\rho = 1$ for every interpretation $(\langle D, \phi\rangle, \rho)$.

PROPOSITION 8.30. *If a closed sequent $\Gamma \to \Delta$ is provable in* $\mathbf{LJ}_I$, *then it is valid.*

PROOF. We prove only the validity of the extra inference rule, since it is clear for other cases.

Let $\mathcal{I} = (\langle D, \phi\rangle, \rho)$ be an interpretation and the lower sequent of

$$\frac{\Gamma \to A \vee (C \supset \alpha) \vee (\alpha \supset B)}{\Gamma \to A \vee (C \supset B)}$$

be not valid in $\mathcal{I}$. Then

$$[\![\Gamma \to A \vee (C \supset B]\!]_\rho \neq 1.$$

It follows that

$$[\![D]\!]_\rho \wedge [\![C]\!]_\rho > [\![B]\!]_\rho$$

for each $D \in \Gamma$, and so there exists $p \in [0, 1]$ such that

$$\bigvee_{D \in \Gamma} [\![D \wedge C]\!]_\rho > p > [\![B]\!]_\rho.$$

Let ρ' be an assignment which has the same values as ρ except $\rho'(\alpha) = p$. Then we have

$$[\![\Gamma \to A \vee (C \supset \alpha) \vee (\alpha \supset B)]\!]_{\rho'} \neq 1.$$

Therefore the upper sequent is not valid.

THEOREM 8.31 (Takeuti–Titani) (Completeness theorem for $\mathbf{LJ}_I$). *Let L* *be a countable language. A closed sequent $\Gamma \rightarrow \Delta$ is valid if and only if it is* $\mathbf{LJ}_I$*-provable.*

PROOF. We have already proved the 'if-part' in Proposition 8.33. Now suppose $\Gamma \rightarrow \Delta$ is not provable in $\mathbf{LJ}_I$. Without loss of generality we may assume that Γ is empty and Δ consists of one formula A. We enumerate all free variables $b_0, b_1, b_2, \ldots$ and all propositional variables $\beta_0, \beta_1, \beta_2, \ldots$ which do not occur in A. We write $\vdash C$ if $\rightarrow C$ is provable in $\mathbf{LJ}_I$ and $\nvdash C$ if $\rightarrow C$ is not provable in $\mathbf{LJ}_I$. If S is of the form $\rightarrow C$, we write $\vdash S$ for $\vdash C$ and $\nvdash S$ for $\nvdash C$. Enumerate the set $\mathscr{F}$ of all formulas of L, say $\mathscr{F} = \{B_1, B_2, B_3, \ldots\}$. We will construct the set M_k, $k = 0, 1, 2, \ldots$ of finite formulas by induction so that

$$\nvdash \bigvee M_k,$$

where $\bigvee M_k$ with $M_k = \{C_1, \ldots, C_n\}$ denotes $C_1 \vee \ldots \vee C_n$. Let $M_0 = \{A\}$. Then $\nvdash \bigvee M_0$. For the induction step, we assume that M_k has been defined and $\nvdash \bigvee M_k$.

Case 1. If $\vdash \bigvee M_k \vee B_{k+1}$, then put

$$M_{k+1} = M_k.$$

Case 2. If $\nvdash \bigvee M_k \vee B_{k+1}$, then put

$$M'_{k+1} = M_k \cup \{B_{K+1}, \beta_j, \beta_j \supset B_{k+1}\},$$

where β_j is the first β which does not occur in $\bigvee M_k \vee B_{k+1}$.

Case 2.1. If $\nvdash \bigvee M_k \vee B_{k+1}$ and B_{k+1} is neither of the form $C_1 \supset C_2$ nor of the form $\forall x\, \varphi(x)$, then let $M_{k+1} = M'_{k+1}$.

Case 2.2. If $\nvdash \bigvee M_k \vee B_{k+1}$ and B_{k+1} is of the form $C_1 \supset C_2$, then let

$$M_{k+1} = M'_{k+1} \cup \{C_1 \supset \beta_j, \beta_j \supset C_2\},$$

where β_j is the first β which does not occur in M'_{k+1}. It is clear that $\nvdash \bigvee M_{k+1}$.

Case 2.3. If $\nvdash \bigvee M \vee B_{k+1}$ and B_{k+1} is of the form $\forall x\, \varphi(x)$, then let

$$M_{k+1} = M'_{k+1} \cup \{\varphi(b_i)\},$$

where b_i is the first b which does not occur in M'_{k+1}. Then we have

$$\nvdash \bigvee M_{k+1}.$$

Now that $\{M_k\}_{k \in \omega}$ have been defined, put $M = \bigcup_{k \in \omega} M_k$.

Let $P = \{\gamma_0, \gamma_1, \gamma_2, \dots\}$ be the set of propositional variables such that $\gamma \vee \neg\gamma \in M$.

For the set M we have:

LEMMA 8.32. (1) *If $C \notin M$, then there exist $C_1, \dots, C_n \in M$ such that*

$$\vdash C_1 \vee \dots \vee C_n \vee C.$$

(2) *If $C_1, \dots, C_n \in M$, then $C_1 \vee \dots \vee C_n \in M$.*
(3) *If $C \in M$ and $\vdash E \to C$, then $E \in M$.*
(4) *If $C \in M$, then there exists $\gamma \in P$ such that $(\gamma \supset C) \in M$.*
(5) *If $C_1 \wedge C_2 \in M$, then $C_1 \in M$ or $C_2 \in M$.*
(6) *If $(C_1 \supset C_2) \in M$, then there exists $\gamma \in P$ such that $(C_1 \supset \gamma) \vee (\gamma \supset C_2) \in M$.*
(7) *If $\forall x\, \varphi(x) \in M$, then there exists a free variable a such that $\varphi(a) \in M$.*
(8) *If $\exists x\, \varphi(x) \notin M$ and $\gamma \in P$, then there exists a free variable a such that $(\gamma \supset \varphi(a)) \notin M$.*
(9) *For each formula C and $\gamma \in P$, $(C \supset \gamma) \in M$ or $(C \supset \gamma) \supset \gamma \in M$.*

PROOF. (1)–(7) are obvious from the construction of M.

(8) Assume $\exists x \varphi(x) \notin M$ and $\gamma \in P$. We have

$$\exists x\, (\gamma \supset \varphi(x)) \notin M, \exists x\, (\gamma \supset \varphi(x)) \supset \gamma \in M$$

and $\forall x\, ((\gamma \supset \varphi(x)) \supset \gamma) \in M$. Hence there exists a free variable a such that $(\gamma \supset \varphi(a)) \supset \gamma \in M$. Since $\vdash((\gamma \supset \varphi(a)) \supset \gamma) \vee (\gamma \supset (\gamma \supset \varphi(a))$ and $\vdash \gamma \supset (\gamma \supset \varphi(a)) \leftrightarrow \gamma \supset \varphi(a)$, we have $(\gamma \supset \varphi(a)) \notin M$.

(9) follows from $\vdash (C \supset \gamma) \wedge ((C \supset \gamma) \supset \gamma) \to \gamma$.

Now we shall construct an interpretation $(\langle D, \phi\rangle, \rho)$. Let D be the set of all terms in L. Define a relation $\leqslant$ on $P = \{\gamma_0, \gamma_1, \gamma_2, \dots\}$ by

$$\gamma_i \leqslant \gamma_j \text{ iff } \gamma_i \supset \gamma_j \notin M.$$

Then $\gamma_i < \gamma_j$ (i.e., $\gamma_i \leqslant \gamma_j \wedge \gamma_j \not\leqslant \gamma_i$) iff $\gamma_j \supset \gamma_i \in M$.

LEMMA 8.33. *P is a countable ordered set satisfying*
a) *the order $<$ is dense;*
b) *P has no least nor greatest element.*

PROOF. 1) P is clearly ordered.
1.1) $\gamma \not< \gamma$ since $\gamma \supset \gamma \notin M$.
1.2) $\gamma_i < \gamma_j$ and $\gamma_j < \gamma_k$ implies $\gamma_i < \gamma_k$.

$$\begin{aligned}\gamma_i < \gamma_j \wedge \gamma_i \not< \gamma_k &\rightarrow \gamma_i \supset \gamma_j \notin M \wedge \gamma_k \supset \gamma_i \notin M\\ &\rightarrow \gamma_k \supset \gamma_j \notin M\\ &\rightarrow \gamma_j \not< \gamma_k.\end{aligned}$$

$\gamma_i \not< \gamma_j$ implies $\gamma_j < \gamma_i$ or $\gamma_j = \gamma_i$.

(a) If $\gamma_i < \gamma_j$, i.e., $\gamma_j \supset \gamma_i \in M$, then there exists α such that $(\gamma_j \supset \alpha) \vee (\alpha \supset \gamma_i) \in M$.
Since $\vdash \alpha \vee \neg\alpha \supset (\gamma_j \supset \alpha) \vee (\alpha \supset \gamma_i)$, $\alpha \vee \neg\alpha \in M$.
Therefore $\alpha \in P$ and $\gamma_j > \alpha > \gamma_i$.

(b) If $\gamma_i \in P$, then $\gamma_i \vee \neg\gamma_i \in M$, that is, $\gamma_i \rightarrow O \in M$, where O is a formula of the form $C \wedge \neg C$. By Lemma 8.35, (6), there exists $\gamma \in P$ such that $(\gamma_i \supset \gamma) \vee (\gamma \supset O) \in M$. Hence $\gamma_i > \gamma$.

If $\gamma_i \in P$, then $1 \supset \gamma_i \in M$, where 1 is a provable formula. By Lemma (6), there exists a $\gamma \in P$ such that $(1 \supset \gamma) \vee (\gamma \supset \gamma_i) \in M$. So $\gamma > \gamma_i$.

By the lemma, there exists an isomorphism τ of P onto the set of rational numbers r such that $0 < r < 1$, which is denoted by $(0, 1)_Q$. Extend the isomorphism $\tau : P \rightarrow (0, 1)_Q$ to $\rho : \{\alpha_0, \alpha_1 \ldots\} \rightarrow [0, 1]$ by

$$\rho(\alpha_i) = \bigwedge\{\tau(\gamma) \mid \gamma \in P \quad \text{and} \quad (\gamma \supset \alpha) \in M\}.$$

We also define ϕ by $\phi(d) = d$ for any element d of D. Finally we define ϕR for a predicate constant R with n arguments by the following equation.

$$\phi R(d_1, \ldots, d_n) = \bigwedge\{\tau(\gamma) \mid \gamma \in P \quad \text{and} \quad (\gamma \supset R(d_1, \ldots, d_n)) \in M\}.$$

We will prove that A is not valid in the interpretation we defined. First we prove the following lemma.

LEMMA 8.34. *For a formula φ of* L *and $\gamma \in P$,*

$$[\![\varphi]\!] < [\![\gamma]\!] \text{ iff } (\gamma \supset \varphi) \in M.$$

PROOF. We prove this by induction on the complexity of φ.

(1) If φ is atomic, i.e., propositional variable α or of the form $R(t_1, \ldots, t_n)$, then it is obvious from the definition.

(2) If φ is $B \vee C$, then

$$\begin{aligned}(\gamma \supset \varphi) \in M &\text{ iff } (\gamma \supset B) \vee (\gamma \supset C) \in M\\ &\text{ iff } [\![\varphi]\!] < [\![\gamma]\!],\end{aligned}$$

by using induction hypothesis.

(3) If φ is $B \wedge C$, then $(\gamma \supset \varphi) \in M$ iff $[\![\varphi]\!] < [\![\gamma]\!]$. The proof is similar as in (2).

(4) Let φ be $B \supset C$. Then $(\gamma \supset \varphi) \in M$ iff $(\gamma \supset C) \vee (B \supset C) \in M$, since $\gamma \supset (B \supset C) \leftrightarrow (\gamma \wedge B) \supset C \leftrightarrow (\gamma \supset C) \vee (B \supset C)$. Hence $(\gamma \supset \varphi) \in M$ iff there exists $\gamma_j \in P$ such that $(\gamma \supset C) \vee (B \supset \gamma_j) \vee (\gamma_j \supset C) \in M$. It follows, by the induction hypothesis, that

$$(\gamma \supset \varphi) \in M \text{ iff } [\![\gamma]\!] > [\![C]\!] = [\![\varphi]\!].$$

(5) Let φ be $\forall x\, B(x)$. Then

$$[\![\varphi]\!] = \bigwedge_{d \in D} [\![B(d)]\!].$$

If $[\![\varphi]\!] < [\![\gamma]\!]$, then there exists $d \in D$ such that $[\![B(d)]\!] < [\![\gamma]\!]$ and hence, by the induction hypothesis, $(\gamma \supset \forall x\, B(x)) \in M$. Conversely, if $(\gamma \supset \forall x\, B(x)) \in M$, then $\forall x\, (\gamma \supset B(x)) \in M$ and so there exists a free variable a such that $\gamma \supset B(a) \in M$. Therefore

$$[\![\forall x\, B(x)]\!] \leqslant [\![B(a)]\!] < [\![\gamma]\!].$$

(6) Let φ be $\exists x\, B(x)$. Then $[\![\varphi]\!] = \bigvee_{d \in D} [\![B(d)]\!]$. If $[\![\varphi]\!] < [\![\gamma]\!]$, then there exists $\gamma_j \in P$ such that $[\![\varphi]\!] < [\![\gamma_j]\!] < [\![\gamma]\!]$ and hence $\gamma \supset (\gamma_j \supset B(d)) \in M$ for each $d \in D$, since

$$\gamma \supset (\gamma_j \supset B(d)) \leftrightarrow \gamma \wedge \gamma_j \supset B(d) \leftrightarrow (\gamma \supset B(d)) \vee (\gamma_j \supset B(d)).$$

Therefore by (8) of Lemma 8.35

$$\exists x\, (\gamma_j \supset B(x)) \in M \quad \text{and} \quad \gamma \supset \gamma_j \in M.$$

Consequently $(\gamma \supset \gamma_j) \vee \exists x\, (\gamma_j \supset B(x)) \in M$. So, by Proposition 8.31 (4) $\gamma \supset \exists x\, B(x) \in M$. Conversely, let $\gamma \supset \exists x\, B(x) \in M$. Then there exists γ_j such that

$$(\gamma \supset \gamma_j) \vee (\gamma_j \supset \exists x\, B(x)) \in M.$$

If $[\![\exists x\, B(x)]\!] = \bigvee_{d \in D} [\![B(d)]\!] > [\![\gamma_j]\!]$, then there exists $d \in D$ such that $[\![B(d)]\!] > [\![\gamma_j]\!]$, hence, by the induction hypothesis, $\gamma_j \supset B(d) \notin M$. This is a contradiction, since $\gamma_j \supset \exists x\, B(x) \in M$. Therefore, $[\![\gamma_j]\!] \geqslant [\![\exists x\, B(x)]\!]$, and so $[\![\gamma]\!] > [\![\exists x\, B(x)]\!]$.

PROOF OF $[\![A]\!] \neq 1$. By the construction of M, A is in M. Hence there exists $\gamma \in P$ such that

$$\gamma \supset A \in M,$$

thus $[\![A]\!] \neq 1$.

CHAPTER 2

PEANO ARITHMETIC

In this chapter we shall formulate first-order Peano arithmetic, prove Gödel's incompleteness theorem, develop a constructive theory of ordinals up to the first ε-number ε_0, and then present a consistency proof of the system, due to Gentzen.

§9. A formulation of Peano arithmetic

DEFINITION 9.1. The language of the system, which will be called Ln, contains finitely many constants, as follows. (See also Definition 1.1.)

Individual constant: 0;

Function constants: $'$, $+$, $\cdot$;

Predicate constant: $=$;

where $'$ is unary while the other constants are binary.

The intended interpretation of the above constants should be obvious. We shall use expressions like $s = t$, $s + t$, $s \cdot t$ and s' rather than formal expressions like $+(s, t)$.

A numeral is an expression of the form $0'^{\cdots\prime}$, i.e., zero followed by n primes for some n, which is used as a formal expression for the natural number n, and is denoted by $\bar{n}$. Further, if s is a closed term of Ln denoting a number m (in the intended interpretation), then $\bar{s}$ denotes the numeral $\bar{m}$ (e.g., if s is $\bar{2} + \bar{3}$, then $\bar{s}$ denotes $\bar{5}$).

DEFINITION 9.2. The first axiom system of Peano arithmetic which we consider, CA, consists of Γ_e for Ln in Definition 7.3 and the following sentences.

A1 $\forall x \forall y\, (x' = y' \supset x = y)$;

A2 $\forall x\, (\neg x' = 0)$;

A3 $\forall x\, (x + 0 = x)$;

A4 $\forall x \forall y\, (x + y' = (x + y)')$;

A5 $\forall x\, (x \cdot 0 = 0)$;

A6 $\forall x \forall y\, (x \cdot y' = x \cdot y + x)$.

The second axiom system of Peano arithmetic which we consider, VJ, consists of all sentences of the form

$$\forall z_1 \dots \forall z_n \forall x\, (F(0, z) \wedge \forall y\, (F(y, z) \supset F(y', z)) \supset F(x, z)),$$

where z is an abbreviation for the sequence of variables $z_1, \dots, z_n$; and all the variables which are free in $F(x, z)$ are among x, z.

The basic logical system of Peano arithmetic is **LK**. Then CA ∪ VJ is an axiom system with equality, regarding $=$ as the distinguished predicate constant in §7. Furthermore, $\forall x\, \forall y\, (x = y \supset (F(x) \equiv F(y)))$ is provable for every formula of Ln (cf. Proposition 7.2).

As an example of the strength of CA ∪ VJ, we mention that the theory of primitive recursive functions can be developed in this system. Although this point will not be discussed further here, such knowledge is assumed.

DEFINITION 9.3. The system **PA** (Peano arithmetic) is obtained from **LK** (in the language Ln) by adding extra initial sequents (called the *mathematical initial sequents*) and a new rule of inference called "*ind*", stated below.

1) Mathematical initial sequents: additional initial sequents of $\mathbf{LK}_e$ for Ln in Definition 7.1 and the following sequents

$$s' = t' \rightarrow s = t;$$

$$s' = 0 \rightarrow;$$

$$\rightarrow s + 0 = s;$$

$$\rightarrow s + t' = (s + t)';$$

$$\rightarrow s \cdot 0 = 0;$$

$$\rightarrow s \cdot t' = s \cdot t + s,$$

where s, t, r are arbitrary terms of Ln.

2) Ind:

$$\frac{F(a), \Gamma \rightarrow \Delta, F(a')}{F(0), \Gamma \rightarrow \Delta, F(s)}$$

where a is not in $F(0)$, Γ or Δ; s is an arbitrary term (which may contain a); and $F(a)$ is an arbitrary formula of Ln.

$F(a)$ is called the *induction formula*, and a is called the *eigenvariable* of this inference. Further, we call $F(a)$ and $F(a')$ the *left* and *right auxiliary formula*, respectively, and $F(0)$ and $F(s)$ the *left* and *right principal formula*, respectively, of this inference.

The initial sequents of the form $D \rightarrow D$ are called *logical* initial sequents (in contrast to the mathematical initial sequents defined above).

To summarize, then: there are two kinds of initial sequents of **PA**: logical and mathematical; and three kinds of inference rules: structural, logical and ind (cf. Definition 2.1).

Finally, a *weak inference* is a structural inference other than cut.

We shall adapt the concepts concerning proofs which were defined in Chapter 1 with some modifications; the new inference "ind" must be taken into account in every definition. In particular, the successor of $F(a)$ (respectively, $F(a')$) in ind is $F(0)$ (respectively, $F(s)$). Otherwise all definitions in Chapter 1 are relevant here.

As an easy corollary of the definitions we have

PROPOSITION 9.4. *A sequent is provable from* $\mathrm{CA} \cup \mathrm{VJ}$ (*in* **LK**) *if and only if it is provable in* **PA**. *Hence the axiom system* $\mathrm{CA} \cap \mathrm{VJ}$ *is consistent if and only if* $\rightarrow$ *is not provable in* **PA**.

Thus we can restrict our attention to the system **PA**. In the rest of this chapter, "provability" means provability in **PA**. It was Gentzen's great development to formulate first-order arithmetic in the form of **PA**.

Similarly to Lemma 2.11, we can prove the following proposition, which we shall use without mention.

PROPOSITION 9.5. *Let P be a proof in* **PA** *of a sequent $S(a)$, where all the occurrences of a in $S(a)$ are indicated. Let s be an arbitrary term. Then we may construct a* **Pa***-proof P' of $S(s)$ such that P' is regular* (*cf. Lemma* 2.9, *part* (2)) *and P' differs from P only in that some free variables are replaced by some other free variables and some occurrences of a are replaced by s.*

The following lemma will be used later.

LEMMA 9.6. (1) *For an arbitrary closed term s, there exists a unique numeral $\bar{n}$ such that $s = \bar{n}$ is provable without an essential cut and without ind.* (*See Definition* 7.5 *for "essential cut".*)

(2) *Let s and t be closed terms. Then either $\rightarrow s = t$ or $s = t \rightarrow$ is provable without an essential cut or ind.*

(3) *Let s and t be closed terms such that $s = t$ is provable without an essential cut or ind and let $q(a)$ and $r(a)$ be two terms with some occurrences of a (possibly none). Then $q(s) = r(s) \rightarrow q(t) = r(t)$ is provable without an essential cut or ind.*

(4) *Let s and t be as in* (3). *For an arbitrary formula $F(a)$: $s = t$, $F(s) \rightarrow F(t)$ is provable without an essential cut or ind.*

PROOF. (1) By induction on the complexity of s.

We defined some notions concerning formal proofs in §2. In order to carry out the consistency proof for **PA**, however, we need some more of these. We shall list them all here.

DEFINITION 9.7. When we consider a formula or a logical symbol together with the place that it occupies in a proof, in a sequent or in a formula, we refer to it (respectively) as a formula or a logical symbol in the proof, in the sequent or in the formula. A formula in a sequent is also called a *sequent-formula*.

(1) Successor. If a formula E is contained in the upper sequent of an inference using one of the rules of inference in §1 or "ind", then the *successor* of E is defined as follows:
   (1.1) If E is a cut formula, then E has no successor.
   (1.2) If E is an auxiliary formula of any inference other than a cut or exchange, then the principal formula is the successor of E. (For the case of ind, see above.)
   (1.3) If E is the formula denoted by C (respectively, D) in the upper sequent of an exchange (in Definition 2.1), then the formula C (respectively, D) in the lower sequent is the successor of E.
   (1.4) If E is the kth formula of Γ, Π, Δ, or Λ in the upper sequent (in Definition 2.1), then the kth formula of Γ, Π, Δ or Λ, respectively, in the lower sequent is the successor of E.
(2) Thread. The notion of a sequence of sequents in a proof, called a *thread*, has been defined in Definition 2.8.
(3) The notions of a sequent being above or below another, and of a sequent being between two others, were defined in Definition 2.8; so was the notion of an inference being below a sequent.
(4) A sequent formula is called an *initial formula* or an *end-formula* if it occurs, respectively, in an initial sequent or an end-sequent.
(5) Bundle. A sequence of formulas in a proof with the following properties is called a *bundle*:
   (5.1) The sequence begins with an initial formula or a weakening formula.
   (5.2) The sequence ends with an end-formula or a cut-formula.
   (5.3) Every formula in the sequence except the last is immediately followed by its successor.
(6) Ancestor and decendant. Let A and B be formulas. A is called an *ancestor* of B and B is called a *descendent* of A if there is a bundle containing both A and B in which A appears above B.
(7) Predecessor. Let A and B be formulas. If A is the successor of B, then B is called a *predecessor* of A.

Some principal formulas, e.g. of $\wedge$: right, has two predecessors. In such cases we call a predecessor the *first* or the *second* predecessor of A, according as it is in the left or right upper sequent.

(8) The concepts of explicit and implicit.
   (8.1) A bundle is called *explicit* if it ends with an end formula.

(8.2) It is called *implicit* if it ends with a cut-formula.

A formula in a proof is called explicit or implicit according as the bundles containing the formula are explicit or implicit.

A sequent in a proof is called implicit or explicit according as this sequent contains an implicit formuula or not.

A logical inference in a proof is called explicit or implicit according as the principal formula of this inference is explicit or implicit.

(9) End-piece. The *end-piece* of a proof is defined as follows:

(9.1) The end-sequent of the proof is contained in the end-piece.

(9.2) The upper sequent of an inference other than an implicit logical inference is contained in the end-piece if and only if the lower sequent is contained in it.

(9.3) The upper sequent of an implicit logical inference is not contained in the end-piece.

We can rephrase this definition as follows: A sequent in a proof is in the end-piece of the proof if and only if there is no implicit logical inference below this sequent.

(10) An inference of a proof is said to be *in the end-piece* of the proof if the lower sequent of the inference is in the end-piece.

(11) Boundary. Let J be an inference in a proof. We say *J belongs to the boundary* (or J is a *boundary inference*) if the lower sequent of J is in the end-piece and the upper sequent is not. It should be noted that if J belongs to the boundary, then it is an implicit logical inference.

(12) Suitable cut. A cut in the end-piece is called *suitable* if each cut formula of this cut has an ancestor which is the principal formula of a boundary inference.

(13) Essential and inessential cuts. A cut is called *inessential* if the cut formula contains no logical symbol; otherwise it is called essential.

In **PA**, the cut formulas of inessential cuts are of the form $s = t$.

(14) A proof P is *regular* if: (i) the eigenvariables of any two distinct inferences ($\forall$: right, $\exists$: left or induction) in P are distinct from each other; and (ii) if a free variable a occurs as an eigenvariable of a sequent S of P, then a only occurs in sequents above S.

PROPOSITION 9.8. *For an arbitrary proof of* **PA**, *there exists a regular proof of the same end-sequent, which can be obtained from the original proof by simply replacing free variables.*

PROOF. The proof is as for Lemma 2.10, part (2).

§10. The incompleteness theorem

In this section we shall prove the incompleteness of **PA**. This is a celebrated result of Gödel. We shall actually consider any axiomatizable system which contain **PA** as a subsystem.

DEFINITION 10.1. An axiom system $\mathscr{A}$ (cf. §4) is said to be *axiomatizable* if there is a finite set of schemata such that $\mathscr{A}$ consists of all the instances of these schemata. A formal system **S** is called axiomatizable if there is an axiomatizable axiom system $\mathscr{A}$ such that **S** is equivalent to $\mathbf{LK}_{\mathscr{A}}$ (cf. §4). (Two systems are called equivalent if they have exactly the same theorems.)

A system **S** is called an extension of **PA** if every theorem of **PA** is provable in **S**. Throughout this section we deal with axiomatizable systems which are extensions of **PA**. They are denoted by **S**. Such an **S** is arbitrary but fixed; so is the language of **S**, say L (which will always extend Ln).

DEFINITION 10.2. The class of primitive recursive functions is the smallest class of functions generated by the following schemata. (These can be thought of as the clauses of an inductive definition, or as the defining equations of the function being defined.)

(i) $f(x) = x'$, where $'$ is the successor function.

(ii) $f(x_1, \ldots, x_n) = k$, where $n \geqslant 1$ and k is a natural number.

(iii) $f(x_1, \ldots, x_n) = x_i$, where $1 \leqslant i \leqslant n$.

(iv) $f(x_1, \ldots, x_n) = g(h_1(x_1, \ldots, x_n), \ldots, h_m(x_1, \ldots, x_n))$, where $g, h_1, \ldots, h_m$ are primitive recursive functions.

(v) $f(0) = k$, $f(x') = g(x, f(x))$, where k is a natural number and g is a primitive recursive function.

(vi) $f(0, x_2, \ldots, x_n) = g(x_2, \ldots, x_n)$, $f(x', x_2, \ldots, x_n) = h(x, f(x, x_2, \ldots, x_n), x_2, \ldots, x_n)$, where g and h are primitive recursive functions.

This formulation is due to Kleene.

An n-ary relation R (of natural numbers) is said to be primitive recursive if there is a primitive recursive function f which assumes values 0 and 1 only such that $R(a_1, \ldots, a_n)$ is true if and only if $f(a_1, \ldots, a_n) = 0$.

EXERCISE 10.3. We define $+$ and $\cdot$ as follows:

$$a + 0 = a, \qquad a \cdot 0 = 0,$$

$$a + b' = (a + b)', \qquad a \cdot b' = a \cdot b + a.$$

Prove the following from the above equations in **PA**.

(1) $a + b = b + a$.

(2) $a \cdot b = b \cdot a$.

(3) $a \cdot (b + c) = a \cdot b + a \cdot c$.

EXERCISE 10.4. Prove that $=$ and $<$ are primitive recursive relations of natural numbers.

Here we shall state a basic metamathematical lemma without proof, which we shall use later.

LEMMA 10.5. *The consistency of* **S** (*i.e.*, **S**-*unprovability of* $\rightarrow$) *is equivalent to the* **S**-*unprovability of* $0=1$. *In other words*, $0=1$ *is* **S**-*provable if and only if every formula of* L *is* **S**-*provable*. (Cf. Proposition 4.2.)

PROPOSITION 10.6 (Gödel). (1) *The graphs of all the primitive recursive functions can be expressed in* Ln, *so that* (*the translations of*) *their defining equations are provable in* **PA**.

Thus the theory of primitive recursive functions can be translated into our formal system of arithmetic. We may therefore assume that **PA** (*or any of its extensions*) *actually contains the function symbols for primitive recursive functions and their defining equations, as well as predicate symbols for the primitive recursive relations.*

We must distinguish between informal objects and their formal expressions (although this will lead to notational complications). For example, the formal expression (function symbol) for a primitive recursive function f will be denoted by $\bar{f}$; if R is a predicate (of natural numbers) which can be expressed in the formal language, then its formal expression will be denoted by $\bar{R}$. Also, as stated earlier, for any closed term t, $\bar{t}$ is the numeral of the number denoted by t. Although in later sections we may omit such a rigorous distinction between formal and informal expressions, it is essential in this section.

(2) *Let R be a primitive recursive relation of n arguments. It can be represented in* **PA** *by a formula* $\bar{R}(a_1, \ldots, a_n)$, *namely* $\bar{f}(a_1, \ldots, a_n)=\bar{0}$, *where f is the characteristic function of R. Then, for any n-tuple of numbers* $(m_1, \ldots, m_n)$, *if* $R(m_1, \ldots, m_n)$ *is true, then* $\bar{R}(\bar{m}_1, \ldots, \bar{m}_n)$ *is* **PA**-*provable*.

PROOF. The proof of (1) is by induction on the inductive definition of the primitive recursive functions (i.e., by induction on their construction).

The proof of (2) is carried out as follows. We prove that for any primitive recursive function f (of n arguments) and any numbers $m_1, \ldots, m_n, p$, if $f(n_1, \ldots, m_n)=p$, then $\bar{f}(\bar{m}_1, \ldots, \bar{m}_n)=\bar{p}$ is **PA**-provable. The proof is by induction on the construction of f (according to its defining equations). Then, finally, if f is a primitive recursive function which is the characteristic function of R, we have, for all $m_1, \ldots, m_n$, if $R(m_1, \ldots, m_n)$ is true, then $\bar{f}(\bar{m}_1, \ldots, \bar{m}_n)=\bar{0}$ is **PA**-provable.

Since the rest of the argument depends heavily on this proposition, we shall use it without quoting it each time.

Note that the converse proposition (i.e., for primitive recursive R, if $\bar{R}(\bar{m}_1, \ldots, \bar{m}_n)$ is **PA**-provable, then $R(m_1, \ldots, m_n)$ is true) follows from the consistency of **PA**.

DEFINITION 10.7 (Gödel numbering). We shall define a one-to-one map from the formal expressions of the language L, such as symbols, terms, formulas, sequents and proofs, to the natural numbers. (The following is only one example of a suitable map.) For an expression X, we use $\ulcorner X \urcorner$ to denote the corresponding number, which we call the Gödel number of X.

(1) First assign different odd numbers to the symbols of Ln. (We include $\rightarrow$ and $-$ among the symbols of the language here.)

(2) Let X be a formal expression $X_0 X_1 \ldots X_n$, where each X_i, $0 \leq i \leq n$, is a symbol of L. Then $\ulcorner X \urcorner$ is defined to be $2^{\ulcorner X_0 \urcorner} 3^{\ulcorner X_1 \urcorner} \ldots p_n^{\ulcorner X_n \urcorner}$, where p_n is the nth prime number.

(3) If P is a proof of the form

$$\frac{Q}{S} \quad \text{or} \quad \frac{Q_1 \quad Q_2}{S}$$

then $\ulcorner P \urcorner$ is $2^{\ulcorner Q \urcorner} 3^{\ulcorner - \urcorner} 5^{\ulcorner S \urcorner}$ or $2^{\ulcorner Q_1 \urcorner} 3^{\ulcorner Q_2 \urcorner} 5^{\ulcorner - \urcorner} 7^{\ulcorner S \urcorner}$, respectively.

If an operation or relation defined on a class of formal objects (e.g., formulas, proofs, etc.) is thought of in terms of the corresponding number-theoretic operation or relation on their Gödel numbers, we say that the operation or relation has been *arithmetized*. More precisely, suppose ψ is an operation defined on n-tuples of formal objects of a certain class, and f is a number-theoretic function such that for all formal objects $X_1, \ldots, X_n, X$ (of the class considered), if ψ applied to $X_1, \ldots, X_n$ produces X, then $f(\ulcorner X_1 \urcorner, \ldots, \ulcorner X_n \urcorner) = \ulcorner X \urcorner$. Then f is called the *arithmetization* of ψ. Similarly with relations.

LEMMA 10.8. (1) *The operation of substitution can be arithmetized primitive recursively, i.e., there is a primitive recursive function* sb *of two arguments such that if* $X(a_0)$ *is an expression of* L (*where all occurrences of* a_0 *in* X *are indicated*), *and* Y *is another expression, then* $\mathrm{sb}(\ulcorner X(a_0) \urcorner, \ulcorner Y \urcorner) = \ulcorner X(Y) \urcorner$, *where* $X(Y)$ *is the result of substituting* Y *for* a_0 *in* X.

(2) *There is a primitive recursive function* ν *such that* $\nu(m) = \ulcorner$the mth numeral$\urcorner$. *In terms of our notation,* $\nu(m) = \ulcorner \bar{m} \urcorner$.

(3) *The notion that P is a proof* (*of the system* **S**) *of a formula A* (*or a sequent S*) *is arithmetized primitive recursively*; *i.e., there is a primitive recursive relation* $\mathrm{Prov}(p, a)$ *such that* $\mathrm{Prov}(p, a)$ *is true if and only if there is a proof P and a formula A* (*or a sequent S*) *such that* $p = \ulcorner P \urcorner$, $a = \ulcorner A \urcorner$ (*or* $a = \ulcorner S \urcorner$) *and P is a proof of A* (*or S*).

(4) Prov *may be written as* $\mathrm{Prov}_{\mathbf{S}}$ *to emphasize the system* **S**.

(5) *As was mentioned before, the formal expression for* Prov *will be denoted by* $\overline{\mathrm{Prov}}$.

We shall not prove this lemma. It is important to note that the axiomatizability of **S** is used in (3); (3) is crucial in the subsequent argument. We also use the following fact about Gödel numbering: we can go effectively from formal objects to their Gödel numbers, and back again (i.e., decide effectively whether a given number is a Gödel number, and if so, of what formal object).

$\exists x\, \overline{\text{Prov}}(x, \overline{\ulcorner A \urcorner})$ is often abbreviated to $\overline{\text{Pr}}(\overline{\ulcorner A \urcorner})$ or $\vdash \overline{\ulcorner A \urcorner}$.

PROPOSITION 10.9. (1) *If A is* **S**-*provable, then* $\vdash \overline{\ulcorner A \urcorner}$ *is* **S**-*provable.*

(2) *If $A \leftrightarrow B$ is* **S**-*provable, then* $\overline{\text{Pr}}(\overline{\ulcorner A \urcorner}) \leftrightarrow \overline{\text{Pr}}(\overline{\ulcorner B \urcorner})$, *i.e.,* $\vdash \overline{\ulcorner A \urcorner} \leftrightarrow \vdash \overline{\ulcorner B \urcorner}$, *is* **S**-*provable.*

(3) $\vdash \overline{\ulcorner A \urcorner} \rightarrow (\vdash \overline{\ulcorner \vdash \overline{\ulcorner A \urcorner} \urcorner})$ *is* **S**-*provable.*

PROOF. (1) Suppose A is provable with a proof P. Then by (3) of Lemma 10.5, $\text{Prov}(\ulcorner P \urcorner, \ulcorner A \urcorner)$ is true, which implies, by (2) of Proposition 10.6, that $\exists x\, \overline{\text{Prov}}(x, \overline{\ulcorner A \urcorner})$, i.e., $\vdash \overline{\ulcorner A \urcorner}$, is **S**-provable.

(2) Suppose $A \equiv B$ is provable with a proof P and A is provable with a proof Q. There is a prescription for constructing a proof of B from P and Q, uniform in P and Q, which can be arithmetized by a primitive recursive function f. Thus $\text{Prov}(q, \ulcorner A \urcorner) \rightarrow \text{Prov}(f(p, q), \ulcorner B \urcorner)$ is true, from which it follows by (2) of Proposition 10.6 that $\vdash \overline{\ulcorner A \urcorner} \rightarrow \vdash \overline{\ulcorner B \urcorner}$ is provable. The same argument works for $\vdash \overline{\ulcorner B \urcorner} \rightarrow \vdash \overline{\ulcorner A \urcorner}$.

(3) If P is a proof of A, then we can construct a proof Q of $\vdash \overline{\ulcorner A \urcorner}$ by (1). This process is uniform in P; in other words, there is a uniform prescription for obtaining Q from P. Thus

$$\text{Prov}(p, \ulcorner A \urcorner) \rightarrow \text{Prov}(f(p), \ulcorner \overline{\text{Pr}}(\overline{\ulcorner A \urcorner}) \urcorner)$$

is true for some primitive recursive function f, from which it follows that $\vdash \overline{\ulcorner A \urcorner} \rightarrow \vdash \overline{\ulcorner \vdash \overline{\ulcorner A \urcorner} \urcorner}$ is provable.

We shall now consider the notion of truth definition and Tarski's theorem concerning it.

DEFINITION 10.10. A formula of L (the language of **S**) with one free variable, say $T(a_0)$, is called a *truth definition* for **S** if every sentence A of L,

$$T(\overline{\ulcorner A \urcorner}) \equiv A$$

is **S**-provable.

THEOREM 10.11 (Tarski). *If* **S** *is consistent, then it has no truth definition.*

PROOF. Suppose otherwise. Then there is a formula $T(a_0)$ of L such that for every sentence A of L, $T(\overline{\ulcorner A \urcorner}) \equiv A$ is provable in **S**. Consider the formula $F(a_0)$, with sole free variable a_0, defined as: $\neg T(\overline{\text{sb}}(a_0, \bar{\nu}(a_0)))$. Put $p = \ulcorner F(a_0) \urcorner$, and let A_T be the sentence $F(\bar{p})$. Then by definition:

$$A_T \equiv \neg T(\overline{\text{sb}}(\bar{p}, \bar{\nu}(\bar{p}))). \tag{1}$$

Also, since $\ulcorner A_T \urcorner = \mathrm{sb}(p, \nu(p))$, we can prove in **S** the equivalences:

$$\begin{aligned} A_T &\equiv T(\overline{\ulcorner A_T \urcorner}) \qquad \text{(by assumed property of } T\text{)} \\ &\equiv T(\overline{\mathrm{sb}}(\bar{p}, \bar{\nu}(\bar{p}))). \end{aligned} \tag{2}$$

(1) and (2) together contradict the consistency of **S**.

An interesting consequence of Theorem 10.11 is the following. First note that in the proof of Theorem 10.11, we need *not* assume that **S** is axiomatizable (cf. Def. 10.1). So we may take as the axioms of **S** the set of all sentences of Ln which are *true* in the intended interpretation (or standard model) $\mathfrak{M}$ of **PA** (using the ordinary semantic or model-theoretic definition of truth in $\mathfrak{M}$). We then obtain that there is no formula $T(a_0)$ of Ln such that for any sentence A of Ln:

$$A \text{ is true} \Leftrightarrow T(\overline{\ulcorner A \urcorner}) \text{ is true}$$

(i.e., true in $\mathfrak{M}$). This corollary of Theorem 10.11 can be stated in the form: "The notion of arithmetical truth is not arithmetical" (i.e., cannot be expressed by a formula of Ln). This is often taken as the statement of Tarski's theorem.

DEFINITION 10.12. **S** is called *incomplete* if for some sentence A, neither A nor $\neg A$ is provable in **S**.

Next we introduce "Gödel's trick" for use in Theorem 10.16.

DEFINITION 10.13. Consider a formula $F(\alpha)$ with a metavariable α (i.e., a new predicate variable, not in L, which we only use temporarily for notational convenience), where α is regarded as an atomic formula in $F(\alpha)$ and $F(\alpha)$ is closed. $F(\vdash \overline{\mathrm{sb}}(a_0, \bar{\nu}(a_0)))$ is a formula with a_0 as its sole free variable. Define $p = \ulcorner F(\vdash \overline{\mathrm{sb}}(a_0, \bar{\nu}(a_0)))\urcorner$ and A_F as $F(\vdash \overline{\mathrm{sb}}(\bar{p}, \bar{\nu}(\bar{p})))$. Note that A_F is a sentence of L.

LEMMA 10.14. $A_F \equiv F(\vdash \overline{\ulcorner A_F \urcorner})$ *is provable in* **S**.

PROOF. Since $\ulcorner A_F \urcorner = \mathrm{sb}(p, \nu(p))$ by definition,

$$\overline{\ulcorner A_F \urcorner} = \overline{\mathrm{sb}}(\bar{p}, \bar{\nu}(\bar{p})) \text{ is provable in } \mathbf{S}.$$

Hence $A_F \equiv F(\vdash \overline{\ulcorner A_F \urcorner})$ is provable in **S**.

From now on, we shall use the abbreviation $\vdash A$ for $\vdash \overline{\ulcorner A \urcorner}$.

DEFINITION 10.15. **S** is called *ω-consistent* if the following condition is satisfied. For every formula $A(a_0)$, if $\neg A(\bar{n})$ is provable in **S** for every $n = 0, 1, 2, \ldots$, then $\exists x\, A(x)$ is not provable in **S**. Note that ω-consistency of **S** implies consistency of **S**.

THEOREM 10.16 (Gödel's first incompleteness theorem). *If* **S** *is ω-consistent, then* **S** *is incomplete.*

PROOF. There exists a sentence A_G of L such that $A_G \equiv \neg\vdash A_G$ is provable in **S**. (Any such sentence will be called a Gödel sentence for **S**.) This follows from Lemma 10.14, by taking $F(\alpha)$ in Definition 10.13 to be $\neg\alpha$. Then $A_G \equiv \neg\vdash A_G$ is provable in **S**. First we shall show that A_G is not provable in **S**, assuming only the consistency of **S** (but without assuming the ω-consistency of **S**). Suppose that A_G were provable in **S**. Then by (1) of Proposition 10.9, $\vdash A_G$ is provable in **S**; thus by the definition of Gödel sentence, $\neg A_G$ is provable in **S**, contradicting the consistency of **S**.

Next we shall show that $\neg A_G$ is not provable in **S**, assuming the ω-consistency of **S**. Since we have proved that A_G is not provable in **S**, for each $n = 0, 1, 2, \ldots \neg\overline{\mathrm{Prov}}(\bar{n}, \overline{\ulcorner A_G \urcorner})$ is provable in **S**. By the ω-consistency of **S**, $\exists x\, \overline{\mathrm{Prov}}(x, \overline{\ulcorner A_G \urcorner})$ is not provable in **S**. Since $\neg A_G \equiv \vdash A_G$ is provable in **S**, $\neg A_G$ is not provable in **S**.

REMARK. In fact, A_G, although unprovable, is (intuitively) true, since it asserts its own unprovability.

DEFINITION 10.17. $\overline{\mathrm{Consis}}_{\mathbf{S}}$ is the sentence $\neg\vdash 0 = 1$. (So $\overline{\mathrm{Consis}}_{\mathbf{S}}$ asserts the consistency of **S**.)

THEOREM 10.18. (Gödel's second incompleteness theorem). *If* **S** *is consistent, then* $\overline{\mathrm{Consis}}_{\mathbf{S}}$ *is not provable in* **S**.

PROOF. Let A_G be a Gödel sentence. In the proof of Theorem 10.16, we proved that A_G is not provable, assuming only consistency of **S**. Now we shall prove a stronger theorem: that $A_G \equiv \overline{\mathrm{Consis}}_{\mathbf{S}}$ is provable in **S**.

(1) To show $A_G \to \overline{\mathrm{Consis}}_{\mathbf{S}}$ is provable in **S**. By Lemma 10.5, $\neg\overline{\mathrm{Consis}}_{\mathbf{S}} \equiv \forall^{\ulcorner} A^{\urcorner}\ (\vdash A)$ is provable (where $\forall^{\ulcorner} A^{\urcorner}$ means: for all Gödel numbers of formulas A). Therefore, $A_G \to \neg\vdash A_G \to \neg\forall^{\ulcorner} A^{\urcorner}\ (\vdash A) \to \overline{\mathrm{Consis}}_{\mathbf{S}}$.

(2) To show $\overline{\mathrm{Consis}}_{\mathbf{S}} \to A_G$ is provable in **S**. Again by Lemma 10.5, $\overline{\mathrm{Consis}}_{\mathbf{S}}, \vdash A_G \to \neg\vdash\neg A_G \to \neg\vdash\vdash A_G$, since $\neg A_G \equiv \vdash A_G$ (of (3) of Lemma 10.8). But $\vdash A_g \to \vdash\vdash A_g$, by Proposition 10.9. So $\overline{\mathrm{Consis}}_{\mathbf{S}}, \vdash A_G \to \neg\vdash\vdash A_G \wedge \vdash\vdash A_G$, and so $\overline{\mathrm{Consis}}_{\mathbf{S}} \to \neg\vdash A_G \to A_G$

EXERCISE 10.19. Define the system **QA** as the quantifier-free part of **PA**.

Show that the following are provable in **QA** for free variables a, b, c.

(1) $a = a$,

(2) $a = b \rightarrow b = a$,

(3) $a + b = b + a$,

(4) $a \cdot b = b \cdot a$,

(5) $a \cdot (b + c) = a \cdot b + a \cdot c$.

EXERCISE 10.20. In Gödel's trick (cf. Definition 10.13) we may replace $\mathrm{sb}(a_0, \nu(a_0))$ by $e(\mathrm{sb}(a_0, \nu(a_0)))$ for some primitive recursive function e which satisfies that if A is a formula then $e(\ulcorner A \urcorner)$ is Gödel number of a formula obtained from A by adding some more stages of the definition of formula; for example, $e(\ulcorner A \urcorner) = \ulcorner \neg A \urcorner$. Show that if $e(\ulcorner A \urcorner) = \ulcorner \neg A \urcorner$, $p = \ulcorner F(\vdash \bar{e}(\overline{\mathrm{sb}}(a_0, \bar{\nu}(a_0)))) \urcorner$ and B_F is $F(\vdash \bar{e}(\overline{\mathrm{sb}}(\bar{p}, \bar{\nu}(\bar{p}))))$, then $\ulcorner B_F \urcorner = \mathrm{sb}(p, \nu(p))$, i.e., B_F is $F(\vdash \neg B_F)$.

PROBLEM 10.21 (Löb). Show that for any sentence A, if $(\vdash A) \rightarrow A$ is **PA**-provable, then A is itself provable. [*Hint*: By Gödel's trick there is a sentence B such that $B \equiv (\vdash B \supset A)$. For such B, if B is provable then $\vdash B$ is provable (cf. Proposition 10.9) and $(\vdash B) \rightarrow A$ is provable; thus A is provable. This procedure is uniform in the proofs of B; hence by formalizing the entire process we obtain $(\vdash B) \rightarrow (\vdash A)$. This and the assumption $(\vdash A) \rightarrow A$ imply $(\vdash B) \rightarrow A$. But, by the definition of B, the last sequent implies B itself, and hence $\vdash B$ (Proposition 10.9). So, since $\vdash B$ and $(\vdash B) \rightarrow A$ are both provable, so is A.]

PROBLEM 10.22 (Rosser). Let e be a primitive recursive function satisfying $e(\ulcorner A \urcorner) = \ulcorner \neg A \urcorner$ as in Exercise 10.20. Let $F(a_{10})$ be

$$\forall x_1 (\overline{\mathrm{Prov}}(x_1, \overline{\mathrm{sb}}(a_0, \bar{\nu}(a_0))) \supset \exists x_2 (x_2 \leqslant x_1 \vee \overline{\mathrm{Prov}}(x_2, \bar{e}(\overline{\mathrm{sb}}(x_1, \bar{\nu}(x_0)))))).$$

Define $p = \ulcorner F(a_0) \urcorner$ and A_R as $F(\bar{p})$. Prove that if **S** is consistent, then neither A_R nor $\neg A_R$ is provable in **S**.

REMARK. This strengthens Gödel's first incompleteness theorem. Namely, the hypothesis of the ω-consistency in Theorem 10.16 is weakened to the consistency.

§11. A discussion of ordinals from a finitist standpoint

When one is concerned with consistency proofs, their philosophical interpretation is always a paramount problem. There is no doubt that Hilbert's "finitist standpoint" which considers only a finite number of symbols concretely given and arguments concretely given about finite sequences of these symbols (called expressions) is an ideal standpoint in proving consistency. From this standpoint, one defines expressions in the following way (as we have, in fact, done already).

(0) Firstly, we give a finite set of symbols, called an alphabet.

(1) Next, we give a finite set of finite sequences of these symbols, called initial expressions.

(2) Next, we give a finite set of concrete operations, for constructing or generating expressions from expressions already obtained.

(3) Finally, we restrict ourselves to considering only expressions obtained by starting with step (1) and iterating step (2).

As a special case of the above, let us suppose that we are given symbols $a_1, \ldots, a_n$ by (1), and concrete operations $f_1, \ldots, f_j$, to obtain new expressions from expressions we already have, and let $\mathscr{D}$ be the collection of all expressions thus obtained. Then the definition of $\mathscr{D}$ is as follows:

(0) The alphabet consists of $\{a_1, \ldots, a_n\}$.

(1) $a_1, \ldots, a_n$ (considered as sequences of length 1) are in $\mathscr{D}$.

(2) If $x_1, \ldots, x_{k_i}$ are in $\mathscr{D}$, then $f_i(x_1, \ldots, x_{k_i})$ is in $\mathscr{D}$ $(i = 1, \ldots, j)$.

(3) $\mathscr{D}$ consists of only these objects (expressions) obtained by (1) and (2).

This is called a *recursive* or *inductive definition* of the class $\mathscr{D}$. Corresponding to this inductive definition, we have a principle of "*proof by induction*" on (the elements of) $\mathscr{D}$, namely, let A be any property (of expressions), and suppose we can do the following.

(1) Prove that $A(a_1), \ldots, A(a_n)$ hold;

(2) Assuming $A(x_1), \ldots, A(x_k)$ hold for $x_1, \ldots, x_k$ in $\mathscr{D}$, infer that

$$A(f_1(x_1, \ldots, x_{k_1})), \ldots, A(f_j(x_1, \ldots, x_{k_j}))$$

hold.

Then we conclude that $A(x)$ holds for all x in $\mathscr{D}$. This follows since for any x in $\mathscr{D}$ that is concretely given, one can show that $A(x)$ holds by following the steps in constructing this x, by applying (1) and (2) above step by step. According to this viewpoint, we can regard "induction" simply as a general statement of a concrete method of proof applicable for any given expression x, and not as an axiom that is accepted a priori.

Though nobody denies that the above way of thinking is contained in Hilbert's standpoint, there are many opinions about where to set the boundary of this standpoint: for example, assuming that transfinite induction up to each of $\omega, \omega \cdot 2, \omega \cdot 3, \ldots$ is accepted, whether transfinite induction up to ω^2 should also be accepted; or, assuming that transfinite induction up to each of $\omega, \omega^\omega, \omega^{\omega^\omega}, \ldots$ is accepted, whether transfinite induction up to the first ε-number (denoted by ε_0) should be. If we consider each concretely given expression (in this case an ordinal less than ε_0), then it must be less than some ω_n, and so should be accepted—or should it? Here ω_n denotes the ordinal

$$\left.\omega^{\omega^{\cdot^{\cdot^{\omega}}}}\right\} n.$$

When one thinks about this in a very skeptical way, how far can one accept induction? One might even perhaps doubt whether induction up to ω itself is already beyond Hilbert's standpoint.

However, if we interpret Hilbert's finitist standpoint in an extremely pure and restricted way so as to forbid both transfinite induction and all abstract notions such as Gödel's primitive recursive functionals of finite types, then by Gödel's incompleteness theorem, it is clear that *the consistency of* **PA** *cannot be proved if one adheres to this standpoint*, since (presumably) such strictly finitist methods can be formalized in **PA** (in fact, in "primitive recursive arithmetic": see below).

Therefore in a consistency proof it is always very interesting to see what is used that goes beyond Hilbert's finitist standpoint, and on what basis it can be justified.

At present, the methods used mainly for consistency proofs are firstly those using transfinite induction (initiated by Gentzen), and, secondly, those using higher type functionals (initiated by Gödel).

We explain the first method, that of Gentzen. First, in order to make sure of our standpoint, let us consider an inductive definition of natural numbers that adheres most closely to the above scheme:

N 1 1 is a natural number.

N 2 If a is a natural number, then $a1$ is a natural number.

N 3 Only those objects obtained by N 1 and N 2 are natural numbers.

Although we normally consider a definition like this to be obvious, it seems that this is because much knowledge is often unconsciously presupposed. In order to clarify our unconsciously-arrived-at standpoint, let us ask ourselves questions that a person E who has no understanding of N 1–N 3 might ask.

First, E might say he did not understand N 2 and N 3. For E it is impossible to understand N 2 using the notion of natural number when one does not understand "natural numbers" (a "vicious circle"). Moreover, E cannot understand in N 3 what "those objects obtained by N 1 and N 2" means. There are many possible answers to these doubts. The most practical one from the didactic point of view will be as follows: 1 is a natural number by N 1. Now that we know 1 is a natural number, 11 is a natural number by N 2; now that we know 11 is a natural number, 111 is a natural number by N 2. Everything obtained in this way by starting with N 1 and iterating the operation N 2 is a natural number. N 3 says on the other hand, that only those things obtained in this way are natural numbers. Of course E might ask more questions about the above explanation: "What do you mean by 'iterating the operation N 2'?", "What do you mean by 'everything obtained in this way'?" etc., and this kind of discussion can be continued endlessly. I hope that E will finally get the idea. The important fact is that the general concept of a (potentially) infinite process of creating new things by iterating a concrete operation a finite number of times is presupposed in order to understand the definition N 1–N 3 of natural numbers, and that the purpose of the definition N 1–N 3 is to specify the process of defining natural numbers by such a procedure.

When we analyze precisely the discussion repeated endlessly with E, we will realize that we must accept or presuppose to some extent the notion of finite sequence (or finite iteration of an operation) as our basic notion. Here an important remark should be made: this does not mean that we must accept large amounts of knowledge about sequences and finiteness separately; only that which seems absolutely necessary to understand the single notion of finite sequence.

In order to clarify our standpoint further, let us consider the inductive definition of the finite (non-empty) sequences of natural numbers:

S 1 If n is a natural number, then n itself is a finite sequence of natural numbers.

S 2 If m is a natural number and s is a finite sequence of natural numbers, then $s * m$ is a finite sequence of natural numbers.

S 3 Only those objects obtained by S 1 and S 2 are finite sequences of natural numbers.

It should be realized that this kind of definition is regarded as basic and clear, no matter what standpoint one assumes.

We shall present some more examples of such inductively defined classes of concrete objects, and properties of them.

For instance, the notion of length of a finite sequence of natural numbers is defined inductively as follows:

L 1 If s is a sequence of natural numbers consisting of a natural number n only, then the length of s is 1.

L 2 If s is a sequence of natural numbers of the form $s_0 * n$, and the length of s_0 is l, then the length of s is $l+1$.

We can certainly take an alternative definition: given a sequence of natural numbers, say s, examine s and count the number of $*$'s in it. If the number of $*$'s is l, then the length of s is $l+1$. (Each of these definitions presents an operation which applies to the concretely given figures in a general form.)

These finitist inferences often present striking similarities to the arguments in the following formalism, which we call primitive recursive arithmetic.

(1) The basic logical system is the propositional calculus.

(2) The defining equations of primitive recursive functions are assumed as axioms.

(3) No quantifiers are introduced.

(4) Mathematical induction (for quantifier-free formulas) is admitted:

$$\frac{A(a), \Gamma \rightarrow \Delta, A(a')}{A(0), \Gamma \rightarrow \Delta, A(t)},$$

where a does not occur in $A(0)$, Γ or Δ, and t is an arbitrary term.

From the above discussion, it seems quite reasonable to characterize Hilbert's finitist standpoint as that which can be formalized in *primitive recursive arithmetic*. This standpoint shall be called the "purely finitist standpoint". It is therefore of paramount importance to clarify where a consistency proof exceeds this formalism, i.e., the purely finitist standpoint. (Thus, in the following, we shall not bother with arguments which can be carried out within the above formalism.) In order to pursue this point, we shall first present the recursive definition of ordinal numbers up to ε_0 (the first ε-number); temporarily, by "ordinal" we mean: ordinal less than ε_0.

O 1 0 is an ordinal.

O 2 Let μ and $\mu_1, \mu_2, \ldots, \mu_n$ be ordinals. Then $\mu_1 + \mu_2 + \ldots + \mu_n$ and ω^{μ} are ordinals.

O 3 Only those objects obtained by O 1 and O 2 are ordinals.

ω^0 will be denoted by 1. Regarding 1 as the natural number 1, $1+1$ as 2, etc., we may assume that the natural numbers are included in the ordinals. (We may also include 0 among the natural numbers if we wish.)

We can now define the relations $=$ and $<$ on ordinals so that they match the notions of equality and the natural ordering of ordinals which we know from set theory, and develop the theory of ordinals for these relations within the purely finitist standpoint. We can actually inductively define $=$, $<$, $+$, and $\cdot$ simultaneously so that they satisfy the following.

(1) $<$ is a linear ordering and 0 is its least element.

(2) $\omega^{\mu} < \omega^{\nu}$ if and only if $\mu < \nu$.

(3) Let μ be an ordinal containing an occurrence of the symbol 0 but not 0 itself, and let μ' be the ordinal obtained from μ by eliminating this occurrence of 0 as well as excessive occurrences of $+$. Then $\mu = \mu'$.

As a consequence of (3) it can be easily shown that

(4) Every ordinal which is not 0 can be expressed in the form

$$\omega^{\mu_1} + \omega^{\mu_2} + \ldots + \omega^{\mu_n},$$

where each of $\mu_1, \mu_2, \ldots, \mu_n$ which is not 0 has the same property. (Each term ω^{μ_i} is called a monomial of this ordinal.)

(5) Let μ and ν be of the forms

$$\omega^{\mu_1} + \omega^{\mu_2} + \ldots + \omega^{\mu_k} \quad \text{and} \quad \omega^{\nu_1} + \omega^{\nu_2} + \ldots + \omega^{\nu_l},$$

respectively. Then $\mu + \nu$ is defined as

$$\omega^{\mu_1} + \omega^{\mu_2} + \ldots + \omega^{\mu_k} + \omega^{\nu_1} + \omega^{\nu_2} + \ldots + \omega^{\nu_l}.$$

(6) Let μ be an ordinal which is written in the form of (4) and contains two consecutive terms ω^{μ_j} and $\omega^{\mu_{j+1}}$ with $\mu_j < \mu_{j+1}$, i.e., μ is of the form

$$\ldots + \omega^{\mu_j} + \omega^{\mu_{j+1}} + \ldots,$$

and let μ' be an ordinal obtained from μ by deleting "$\omega^{\mu_j} +$", so that μ' is of the form

$$\ldots \omega^{\mu_{j+1}} + \ldots.$$

Then $\mu = \mu'$.

As a consequence of (6) we can show that

(7) For every ordinal μ (which is not 0) there is an ordinal of the form

$$\omega^{\mu_1} + \omega^{\mu_2} + \ldots + \omega^{\mu_n},$$

where $\mu_1 \geqslant \ldots \geqslant \mu_n$ such that $\mu = \omega^{\mu_1} + \ldots + \omega^{\mu_n}$, where $\mu \geqslant \nu$ means: $\nu < \mu$ or $\nu = \mu$. The latter is called the normal form of μ. (This normal form of μ is unique, since the same holds for every ordinal which is used in constructing μ: see O 2.)

Suppose $\mu = \omega^{\mu_1} + \ldots + \omega^{\mu_n}$ and $\nu = \omega^{\nu_1} + \ldots + \omega^{\nu_m}$ are in the normal form. Then, $\mu < \nu$ if and only if $\omega^{\mu_i} < \omega^{\nu_i}$ for some i and $\omega^{\mu_j} = \omega^{\nu_j}$ for all $j < i$, or $n < m$ and $\omega^{\mu_i} = \omega^{\nu_i}$ for all $i \leqslant n$.

(8) Let μ have the normal form

$$\omega^{\mu_1} + \omega^{\mu_2} + \ldots + \omega^{\mu_n}$$

and ν be > 0. Then $\mu \cdot \omega^{\nu} = \omega^{\mu_1 + \nu}$.

(9) Let μ and ν be as in (5). Then

$$\mu \cdot \nu = \mu \cdot \omega^{\nu_1} + \mu \cdot \omega^{\nu_2} + \ldots + \mu \cdot \omega^{\nu_l}.$$

(10) $(\omega^{\mu})^n$ is defined as $\omega^{\mu} \ldots \omega^{\mu}$ (n times) for any natural number n. Then $(\omega^{\mu})^n = \omega^{\mu \cdot n}$.

As a consequence of our definitions, it can easily be shown that for an arbitrary ordinal μ an ordinal of the form ω_n which satisfies $\mu < \omega_n$ can be constructed.

It is obvious that for any given natural number n the length of a strictly decreasing sequence of ordinals which starts with n is at most $n+1$; in other words, there can be no strictly decreasing sequence of ordinals which starts with n and has length $n+2$. This fact tells us that the notion of arbitrary, strictly decreasing sequences of ordinals which start with n is a clear notion.

At this point it is not very meaningful to object to this on the grounds that if we write the statement that a strictly decreasing sequence terminates, in terms of expressions in the Kleene hierarchy, it turns out to belong to the Π_1^1-class. The important fact is not to which class of the hierarchy it belongs but how evident it is. We shall come to this point later.

In the following section, a consistency proof (for **PA**) will be given in the following way. In order to emphasize the concrete or "figurative" aspect of the arguments, we say "proof-figure" for formal proof.

1) We present a uniform method such that, if a proof-figure P is concretely given, then the method enables us to concretely construct another proof-figure P'; furthermore, the end-sequent of P' is the same as that of P if the end-sequent of P does not contain quantifiers. The process of constructing P' from P is called the "reduction" (of P) and may be denoted by r. Thus $P' = r(P)$.

2) There is a uniform method by which every proof-figure is assigned an ordinal $< \varepsilon_0$. The ordinal assigned to P (the ordinal of P) may be denoted by $o(P)$.

3) o and r satisfy: whenever a proof-figure P contains an application of ind or cut, then $o(P) > \omega$ and $o(r(P)) < o(P)$, and if P does not contain any such application, then $o(P) < \omega$.

Suppose we have concretely shown that any strictly decreasing sequence of natural numbers is finite, and that whenever a concrete method of constructing decreasing sequences of ordinals $< \varepsilon_0$ is given it can be recognized that any decreasing sequence constructed this way is finite (or such a sequence terminates). (By "decreasing sequence" we will always mean strictly decreasing sequence.) We can then conclude, in the light of 1)–3) above, that, for any given proof-figure P whose end-sequent does not contain quantifiers, there is a concrete method of transforming it into a proof-figure with the same end-sequent and containing no applications of the rules cut and ind. It can be easily seen, on the other hand, that no proof-figure without applications of a cut or ind can be a proof of the empty sequent. Thus we can claim that the consistency of the system has been proved.

The crucial point in the process described above is to demonstrate:

(*) Whenever a concrete method of constructing decreasing sequences of ordinals is given, any such decreasing sequence must be finite.

We are going to represent a version of such a demonstration, which the author believes represents the most illuminating approach to the consistency proof.

Suppose $a_0 > a_1 > \dots$ is a decreasing sequence concretely given.

(I) Assume $a_0 < \omega$, or a_0 is a natural number.

Consider a decreasing sequence which starts with a concretely given natural number. As soon as one writes down its first term n, one can recognize that its length must be at most $n+1$. Hence we can assume that a_0 is not a natural number.

In order to deal with all ordinals $< \varepsilon_0$, we shall define the concept of α-sequence and α-eliminator for all $\alpha < \varepsilon_0$. We start, however, with a simple example rather than the general definition.

(II) Suppose each a_i in $a_0 > a_1 > \ldots$ is written in the canonical form; a_i has the form

$$\omega^{\mu_1^i} + \omega^{\mu_2^i} + \ldots + \omega^{\mu_{n_i}^i} + k_i,$$

where $\mu_j^i > 0$ and k_i is a natural number. (This includes the case where $+ k_i$ does not actually appear.) A sequence in which k_i does not appear for any a_i will be called a 1-sequence. We call $\omega^{\mu_1^i} + \omega^{\mu_2^i} + \ldots + \omega^{\mu_{n_i}^i}$ in a_i the 1-major part of a_i. We shall give a concrete method (M_1) which enables us to do the following: given a descending sequence $a_0 > a_1 > \ldots$, where each a_i is written in its canonical form, the method M_1 concretely produces a (decreasing) 1-sequence $b_0 > b_1 > \ldots$ so as to satisfy the condition

(C_1) b_0 is the 1-major part of a_0, and we can concretely show that if $b_0 > b_1 > \ldots$ is a finite sequence, then so is $a_0 > a_1 > \ldots$.

This method M_1 (a 1-eliminator) is defined as follows. Put $a_i = a_i + k_i$, where a_i' is the 1-major part of a_i. Then $a_0 > a_1 > a_2 > \ldots$ can be expressed as $a_0' + k_0 > a_1' + k_1 > a_2' + k_2 > \ldots$.

Put $b_0 = a_0'$. Suppose $b_0 > b_1 > \ldots > b_m$ has been constructed in such a manner that b_m is a_j' for some j. Then either $a_j' = a_{j+1}' = \ldots = a_{j+p}'$ for some p and a_{j+p} is the last term in the sequence, or $a_j' = a_{j+1}' = \ldots = a_{j+p}' > a_{j+p+1}'$. This is so, since $a_j' = a_{j+1}' = \ldots = a_{j+p}' = \ldots$ implies $k_j > k_{j+1} > \ldots > k_{j+p} > \ldots$, but such a sequence (of natural numbers) must stop (cf. (I)). Therefore, as stated above, either the whole sequence stops, or $a_{j+p}' > a_{j+p+1}'$ for some p. If the former is the case, then stop. If the latter holds, then put $b_{m+1} = a_{j+p+1}'$.

From the definition, it is obvious that $b_0 > b_1 > \ldots > b_m > \ldots$. Suppose this sequence is finite, say $b_0 > b_1 > \ldots > b_m$. Then according to the prescribed construction of b_{m+1} the original sequence is finite. Thus the sequence $b_0 > b_1 > \ldots$ satisfies (C_1), and we have completed the definition of M_1.

(III) Suppose we are given a decreasing sequence $a_0 > a_1 > \ldots$, in which $a_0 < \omega^2$. Then by a 1-eliminator M_1 applied to this sequence, we can construct a 1-sequence $b_0 > b_1 > \ldots$, where $b_0 \leq a_0$. Then $b_0 > b_1 > \ldots$ can be written in the form $\omega \cdot k_0 > \omega \cdot k_1 > \ldots$, which implies $k_0 > k_1 > \ldots$. Then by (I), $k_0 > k_1 > \ldots$ must be finite, which successively implies that $b_0 > b_1 > \ldots$ and $a_0 > a_1 > \ldots$ are finite.

(IV) We now define "n-sequences" as follows. Let $a_0 > a_1 > \ldots$ be a descending sequence which is written in the form $a_0' + c_0 > a_1' + c_1 > \ldots$, where if $a_i = a'_i + c_i$ then each monomial in a'_i is $\geqslant \omega^n$ and each monomial in c_i is $< \omega^n$. (a'_i is called the n-major part of a_i.) Such a sequence is called an n-sequence if every c_i is empty.

Now assume (as an induction hypothesis) that any descending sequence $d_0 > d_1 > \dots$, with $d_0 < \omega^n$, is finite. We shall define a concrete method M_n (an n-eliminator) such that, given a decreasing sequence $a_0 > a_1 > \dots$, M_n concretely produces an n-sequence, say $b_0 > b_1 > \dots$, which satisfies:

(C_n) b_0 is the n-major part of a_0, and if $b_0 > b_1 > \dots$ is finite then we can concretely show that $a_0 > a_1 > \dots$ is also finite.

The prescription for M_n is as follows. Write each a_i as $a'_i + c_i$, where a'_i is the n-major part of a_i. The definition now proceeds very much like that for 1-sequences in (II). Namely, put $b_0 = a'_0$. Suppose $b_0 > b_1 > \dots > b_m$ has been constructed and b_m is a'_j. If $a'_j = a'_{j+1} = \dots = a'_{j+p}$ and a'_{j+p} is the last term in the given sequence, then stop. Otherwise $a'_j = a'_{j+1} = \dots = a'_{j+p} > a'_{j+p+1}$ for some p, since $a'_j = a'_{j+1} = \dots = a'_{j+p}$ implies that $c_j > c_{j+1} > \dots > c_{j+p}$, which, by the induction hypothesis, is finite; hence for some p, $c_{j+p+1} \geqslant c_{j+p}$, which implies $a'_{j+p} > a'_{j+p+1}$. Then define $b_m = a'_{j+p+1}$. Then the sequence $b_0 > b_1 > \dots$ satisfies (C_n), and so we have successfully defined M_n.

(V) By means of the n-eliminator M_n, we shall prove that a decreasing sequence $a_0 > a_1 > \dots$, where $a_0 < \omega^{n+1}$, must be finite. By applying M_n to $a_0 > a_1 > \dots$, we can construct concretely an n-sequence, say $b_0 > b_1 > \dots$, where $b_0 \leqslant a_0$. Moreover, b_i can be written as $\omega^n \cdot k_i$, where k_i is a natural number. So, $\omega^n \cdot k_0 > \omega^n \cdot k_1 > \dots$, and this implies $k_0 > k_1 > \dots$, which is a finite sequence by (I), hence $b_0 > b_1 > \dots$ is finite, which in turn implies that $a_0 > a_1 > \dots$ is finite.

(VI) From (III) and (V) we conclude: given (concretely) any natural number n, we can concretely demonstrate that any decreasing sequence $a_0 > a_1 > \dots$ with $a_0 < \omega^n$ is finite.

(VII) Any decreasing sequence $a_0 > a_1 > \dots$ is finite if $a_0 < \omega^\omega$, for this means that $a_0 < \omega^n$ for some n, and hence (VI) applies.

(VIII) Now the general theory of α-sequences and (α, n)-eliminators will be developed, where α ranges over all ordinals $< \varepsilon_0$ and n ranges over natural numbers > 0. A descending sequence $d_0 > d_1 > \dots$ is called an α-sequence if in each d_i all the monomials are $\geqslant \omega^\alpha$. If $a = a' + c$ where each monomial in a' is $\geqslant \omega^\alpha$ and each monomial in c is $< \omega^\alpha$, then we say that a' is the α-major part of a. An α-eliminator has the property that given any concrete descending sequence, say $a_0 > a_1 > \dots$, it concretely produces an α-sequence $b_0 > b_1 > \dots$ such that

(i) b_0 is the α-major part of a_0,

(ii) if $b_0 > b_1 > \dots$ is a finite sequence then we can concretely demonstrate that $a_0 > a_1 > \dots$ is finite.

(Clearly $a_0 \geq b_0$.)

We delay the definition of α-eliminators. Assuming that an α-eliminator has been defined for every α, we can show that any decreasing sequence is finite. For consider $a_0 > a_1 > \ldots$. There exists an α such that $a_0 < \omega^{\alpha+1}$. An α-eliminator concretely gives an α-sequence $b_0 > b_1 > \ldots$ satisfying (i) and (ii) above. Since $b_0 \leq a_0$, each b_i can be written in the form $\omega^\alpha \cdot k_i$; thus $\omega^\alpha \cdot k_0 > \omega^\alpha \cdot k_1 > \ldots$, which implies $k_0 > k_1 > \ldots$. By (I) this means that $k_0 > k_1 > \ldots$ is finite, hence so is $b_0 > b_1 > \ldots$; so $a_0 > a_1 > \ldots$ is finite. This proves our objective (*). Therefore, what must be done is to define (construct) α-eliminators for all $\alpha < \varepsilon_0$.

(IX) We rename an α-eliminator to be an $(\alpha, 1)$-eliminator. Suppose that (α, n)-eliminators have been defined. A $(\beta, n+1)$-eliminator is a *concrete* method for constructing an $(\alpha \cdot \omega^\beta, n)$-eliminator from any given (α, n)-eliminator. We must go through the following procedure.

(X) Suppose $\{\mu_m\}_{m<\omega}$ is an increasing sequence of ordinals whose limit is μ (where there is a concrete method for obtaining μ_m for each m), and suppose g_m is a μ_m-eliminator. Then the g defined as follows is a μ-eliminator. Suppose $a_0 > a_1 > \ldots$ is a concretely given sequence. If a_0 is written as $a_0' + c_0$, where a_0' is the μ-major part of a_0, then there exists an m for which $c_0 < \omega^{\mu_m}$, so we may assume that each a_i is written as $a_i' + c_i$, where a_i' is the μ_m-major part of a_i. Then g_m can be applied to the sequence $a_0 > a_1 > \ldots$ and hence it concretely produces a μ_m-sequence

$$b_{10} > b_{11} > b_{12} > \ldots \tag{1}$$

satisfying (i) and (ii) above (with μ_m in place of α), with $b_{10} = a_0'$, so that in fact b_{10} is the μ-major part of a_0. Write $b_0 = b_{10}$.

Now consider the sequence $b_{11} > b_{12} > \ldots$. Suppose $b_{11} \geq \omega^\mu$. Then repeat the above procedure: i.e., for the sequence (1), write $b_{10} = b_{10}' + c_{10}$, where b_{10}' is the μ-major part of b_{10}. Then there exists an m_1 such that $c_{10} < \omega^{\mu_{m_1}}$. So apply g_{m_1} to the sequence $b_{11} > b_{12} > b_{13} > \ldots$, to obtain a μ_{m_1}-sequence

$$b_{21} > b_{22} > b_{23} > \ldots$$

satisfying (i) and (ii) (with μ_{m_1} in place of α), with b_{21} the μ-major part of b_{10}. Put $b_1 = b_{21}$. Suppose $b_{22} \geq \omega^\mu$. Then repeat this procedure with the sequence $b_{22} > b_{23} > \ldots$ to obtain a sequence

$$b_{32} > b_{33} > b_{34} > \ldots,$$

and put $b_2 = b_{32}$. Continuing in this way, we obtain a μ-sequence

$$b_0 > b_1 > b_2 > \ldots.$$

If this sequence is finite with last term (say) $b_l = b_{l+1,l}$, then it follows that in the sequence

$$b_{l+1,l} > b_{l+1,l+1} > b_{l+1,l+2} > \dots \qquad (2)$$

we must have $b_{l+1,l+1} < \omega^{\mu}$. So $b_{l+1,l+1} < \omega^{\mu_{m'}}$ for some m'. Apply $g_{m'}$ to the sequence (2); we then obtain a finite $\mu_{m'}$-sequence with only the term 0; hence the sequence (2) is finite (by definition of $\mu_{m'}$-eliminator); hence the sequence $b_{l,l-1} > b_{l,l} > \dots$ is finite; and so on (backwards), until we deduce that the original sequence $a_0 > a_2 > \dots$ is finite.

(XI) Suppose $\{\mu_m\}_{m<\omega}$ is a sequence of ordinals whose limit is μ and suppose for each m, a $(\mu_m, n+1)$-eliminator is concretely given. Then we can define a $(\mu, n+1)$-eliminator g as follows. The definition is by induction on n. For $n = 0$ (so $n+1 = 1$), (X) applies. Assume (XI) for n; so there is an operation k_n such that for any sequence $\{\gamma_m\}_{m<\omega}$ with limit γ and (γ_m, n)-eliminator g'_m, k_n applied to g'_m concretely produces a (γ, n)-eliminator. Now for $n+1$, suppose a sequence $\{\beta_m\}_{m<\omega}$ with limit β and an (α, n)-eliminator p are given. Since g_m is a $(\beta_m, n+1)$-eliminator, it produces concretely an $(\alpha \cdot \omega^{\beta_m}, n)$-eliminator from p, which we denote by $g_m(p)$. So, by taking $\alpha \cdot \omega^{\beta_m}$ for γ_m, $g_m(p)$ for g'_m and $\alpha \cdot \omega^{\beta}$ for γ, we can apply the induction hypothesis; thus k_n applied to $\{g'_m\}$ defines an $(\alpha \cdot \omega^{\beta}, n)$-eliminator q. This procedure for defining q from p is concrete, and so serves as a $(\beta, n+1)$-eliminator.

(XII) Suppose g is a $(\mu, n+1)$-eliminator. Then we will construct a $(\mu \cdot \omega, n+1)$-eliminator. In virtue of (XI) it suffices to show that we can concretely construct (from g) a $(\mu \cdot m, n+1)$-eliminator for every $m < \omega$. Suppose an (α, n)-eliminator, say f, is given. Note that

$$\alpha \cdot \omega^{\mu \cdot m} = \alpha \cdot \underbrace{\omega^{\mu} \cdot \omega^{\mu} \dots \omega^{\mu}}_{m}.$$

Since g is a $(\mu, n+1)$-eliminator, g concretely constructs an $(\alpha \cdot \omega^{\mu}, n)$-eliminator from f, which we denote by $g(f)$. Now apply g to this, to obtain an $(\alpha \cdot \omega^{\mu} \cdot \omega^{\mu}, n)$-eliminator $g(g(f))$. Repeating this procedure m times, we obtain the $(\alpha \cdot \omega^{\mu \cdot m}, n)$-eliminator $g(g(\dots g(f)\dots))$.

(XIII) We can now construct a $(1, m+1)$-eliminator for every $m \geq 0$. The construction is by induction on m. We may take M_1 as a $(1, 1)$-eliminator. For $m = 1$, the construction of a $(1, 2)$-eliminator is reduced to the construction of an $(\alpha + \alpha)$-eliminator from an α-eliminator. Given $a_0 > a_1 > \dots$, apply an α-eliminator to obtain $b_0 > b_1 > \dots$, where $\{b_i\}$ is an α-sequence, b_0 is the α-major part of a_0, and if $\{b_i\}$ is finite, then so is $\{a_i\}$. Each b_i can be written in the form $\omega^{\alpha} \cdot c_i$, where $\{c_i\}$ is decreasing and, if $\{c_i\}$ is finite, then so is $\{b_i\}$. $a_0 = b_0 + e_0$ where $e_0 < \omega^{\alpha}$. Apply and α-eliminator to $\{c_i\}$ to obtain $d_0 > d_1 > \dots$, where $\{d_i\}$ is an α-sequence, d_0 is the α-major part of c_0 and, if $\{d_i\}$ is finite, then so is $\{c_i\}$. $\{\omega^{\alpha} \cdot d_i\}$ is an

$(\alpha+\alpha)$-sequence and decreasing. If $\{\omega^\alpha d_i\}$ is finite, then so are $\langle d_i\}$, $\{c_i\}$, $\{b_i\}$, $\{a_i\}$ successively, and

$$\begin{aligned}\omega^\alpha \cdot d_0 &= \omega^\alpha \cdot (\text{the } \alpha\text{-major part of } c_0)\\ &= (\alpha+\alpha)\text{-major part of } b_0\\ &= (\alpha+\alpha)\text{-major part of } a_0.\end{aligned}$$

So $\{\omega^\alpha d_i\}$ is the $(\alpha+\alpha)$-sequence which was desired for $\{a_i\}$.

For $m>1$, suppose f is an (α, m)-eliminator. Then, by (XII) (with $n+1=m$), we can construct an $(\alpha \cdot \omega, m)$-eliminator concretely from f. Hence we have a $(1, m+1)$-eliminator.

(XIV) Conclusion: An (α, n)-eliminator can be constructed for every α of the form ω_m, i.e.,

$$\left.\omega^{\cdot^{\cdot^{\omega}}}\right\}m.$$

The construction is by induction on m. If $m=0$, then we define α to be $1=\omega^0$. Then an (α, n)-eliminator has been defined in (XIII) for every n. Suppose f is a $(1, n)$-eliminator, and g is an $(\alpha, n+1)$-eliminator, which we assume to have been defined. Then g operates on f and produces the required $(1 \cdot \omega^\alpha, n) = (\omega^\alpha, n)$-eliminator. This completes the proof.

NOTE. We can also develop the theory of eliminators if we define a $(\beta, n+1)$-eliminator to be a concrete method for constructing an $(\alpha + \omega^\beta, n)$-eliminator from any given (α, n)-eliminator. (Compare this with (IX).)

Our standpoint, which has been discussed above, is like Hilbert's in the sense that both standpoints involve "Gedankenexperimente" only on clearly defined operations applied to some concretely given figures and on some clearly defined inferences concerning thess operations. An α-eliminator is a concrete operation which operates on concretely given figures. A $(\beta, 2)$-eliminator is a concrete method which enables one to exercise a Gedanken-experiment in constructing an $\alpha \cdot \omega^\beta$-eliminator from any concretely given α-eliminator. So if an ordinal, say ω_k is given, then we have a method for concretely constructing an ω_k-eliminator.

We believe that the most illuminating way to view the consistency proof of **PA**, to be described in §12, is in terms of the notion of eliminators, as described above. (In fact, it is not difficult to generalize this notion, so as to include, say, the concept of (α, ω)-eliminator, and so on; however, this is unnecessary for the consistency proof for **PA**.)

The ideas we have presented are normally formulated in terms of the notion of accessibility. It may be helpful to reformulate our ideas in terms of this notion, which (we believe) is a rough but convenient way of expressing the idea of eliminators.

We say that an ordinal μ is accessible if it has been demonstrated that every strictly decreasing sequence starting with μ is finite. More precisely, we consider the notion of accessibility only when we have actually seen, or demonstrated constructively, that a given ordinal is accessible. Therefore we never consider a general notion of accessibility, and hence we do not define the negation of accessibility as such. If we mention "the negation of accessibility", it means that we are concretely given an infinite, strictly decreasing sequence.

First, we assume we have arithmetized the construction of the ordinals (less than ε_0) given by clauses O 1–O 3. In other words, we assume a Gödel numbering of these (expressions for) ordinals, with certain nice properties: namely, the induced number-theoretic relations and functions corresponding to the ordinal relations and functions $=$, $<$, $+$, $\cdot$, and exponentiation by ω (which we will often continue to denote by the same symbols) are primitive recursive; also we can primitive recursively represent any (Gödel number of an) ordinal in its normal form, and hence decide primitive recursively whether it represents a limit or successor ordinal, etc. The ordering of the natural numbers corresponding to $<$ (on the ordinals) will be called a "standard well-ordering of type ε_0", or just "standard ordering of ε_0".

Our method for proving the accessibility of ordinals will be as follows. (We work with our standard well-ordering of type ε_0.)

(1) When it is known that $\mu_1 < \mu_2 < \mu_3 \ldots \to \nu$ (i.e., ν is the limit of the increasing sequence $\{\mu_i\}$) and that every μ_i is accessible, then ν is also accessible.

(2) A method is given by which, from the accessibility of a subsystem, one can deduce the accessibility of a larger system.

(3) By repeating (1) and (2), we show that every initial segment of our ordering is accessible, and hence so is the whole ordering.

The fact that every decreasing sequence which starts with a natural number is finite can be proved as in (I) above.

Let us proceed to the next stage: decreasing sequences of ordinals less than $\omega+\omega$. Here we can again see that every decreasing sequence terminates. This is done as follows. Consider the first term μ_0 of such a sequence. We can effectively decide whether it is of the form n or of the form $\omega+n$, where n is a natural number. If it is of the form n, then it suffices to repeat the above argument for natural numbers. If it is of the form $\omega+n$, consider the first $n+2$ terms of the sequence

$$\mu_{n+1} < \ldots < \mu_2 < \mu_1 < \mu_0.$$

It is easily seen that μ_{n+1} cannot be of the form $\omega+m$ for any natural number m and hence must be a natural number, so we now repeat the proof for natural numbers. This method can be extended to the cases of decreasing sequences of ordinals less than $\omega \cdot n$, less than ω^2, less than ω^ω, etc.

A more mathematical presentation of this idea now follows.

LEMMA 11.1. *If μ and ν are accessible, then so is $\mu + \nu$.*

PROOF. We just generalize the proof that $\omega + \omega$ is accessible and make use of the following fact which is easily seen: given ordinals μ, ξ, ν such that $\mu \leq \xi < \nu$, we can effectively find a ν_0 such that $\nu_0 < \nu$ and $\xi = \mu + \nu_0$.

LEMMA 11.2. *If μ is accessible, then so is $\mu \cdot \omega$.*

PROOF. We use the following fact, which is easy to show: if $\nu < \mu \cdot \omega$, then we can find an n such that $\nu < \mu \cdot n$.

With these lemmas, let us prove that all ordinals less than ε_0 are accessible. First we introduce the technical term: "n-accessible", for every n, by induction on n.

DEFINITION 11.3. μ is said to be 1-*accessible* if μ is accessible. μ is said to be $(n+1)$-*accessible* if for every ν which is n-accessible, $\nu \cdot \omega^\mu$ is n-accessible.

It should be emphasized that "ν being n-accessible" is a clear notion only when it has been concretely demonstrated that ν is n-accessible.

LEMMA 11.4. *If μ is n-accessible and $\nu < \mu$, then ν is n-accessible.*

LEMMA 11.5. *Suppose $\{\mu_m\}$ is an increasing sequence of ordinals with limit μ. If each μ_m is n-accessible, then so is μ.*

LEMMA 11.6. *If ν is $(n+1)$-accessible, then so is $\nu \cdot \omega$.*

PROOF. We must show that for any n-accessible μ, $\mu \cdot \omega^{\nu \cdot \omega}$ is n-accessible. For this purpose it suffices so show that $\mu \cdot \omega^{\nu \cdot m}$ is n-accessible for each m (cf. Lemma 11.5). This is, however, obvious, since

$$\mu \cdot \omega^{\nu \cdot m} = \mu \cdot (\omega^\nu)^m = \mu \cdot \omega^\nu \ldots \omega^\nu$$

and ν is $(n+1)$-accessible.

PROPOSITION 11.7. 1 *is $(n+1)$-accessible.*

PROOF. Suppose μ is n-accessible. Then by Lemma 11.6, $\mu \cdot \omega = \mu \cdot \omega^1$ is n-accessible, which means by definition that 1 is $(n+1)$-accessible.

DEFINITION 11.8. $\omega_0 = 1$; $\omega_{n+1} = \omega^{\omega_n}$.

PROPOSITION 11.9. *ω_k is $(n-k)$-accessible for an arbitrary $n > k$.*

PROOF. By induction on k. If $k = 0$, then $\omega_k = 1$ and hence is n-accessible for all n (cf. Proposition 11.7). Suppose ω_k is $(n-k)$-accessible. Since 1 is $[n-(k+1)]$-accessible, $1 \cdot \omega^{\omega^k}$ is $[n-(k+1)]$-accessible by Definition 11.3, i.e., ω_{k+1} is $[n-(k+1)]$-accessible.

As a special case of Proposition 11.9 we have:

PROPOSITION 11.10. *ω_k is accessible for every k.*

Given any decreasing sequence of ordinals (less than ε_0), there is an ω_k such that all ordinals in the sequence are less than ω_k. Therefore the sequence must be finite by Proposition 11.10. Thus we can conclude:

PROPOSITION 11.11. *ε_0 is accessible.*

An important point to note is this. Our proof of the accessibility of ε_0 (by the method of eliminators, (I)–(XIV), or by the method of Proposition 11.11) depends essentially on the fact that we are using a standard well-ordering of type ε_0, for which the successive steps in the argument are evident. Of course this is not so for an arbitrary well-ordering of type ε_0, nor for the general notion of well-ordering or ordinal.

Comparison of our standpoint with some other standpoints may help one to understand our standpoint better. First, consider set theory. Our standpoint does not assume the absolute world as set theory does, which we can think of as being based on the notion of an "infinite mind". It is obvious that, on the contrary, it tries to avoid the absolute world of an "infinite mind" as much as possible. It is true that in the study of number theory, which does not involve the notion of sets, the absolute world of numbers 0, 1, 2, ... is not such a complicated notion; to an infinite mind it would be quite clear and transparent. Nevertheless, our minds being finite, it is, after all, an imaginary world to us, no matter how clear and transparent it may appear. Therefore we need reassurance of such a world in one way or another.

Next, consider intuitionism. Although our standpoint and that of intuitionism have much in common, the difference may be expressed as follows.

Our standpoint avoids abstract notions as much as possible, except those which are eventually reduced to concrete operations or Gedankenexperimente on concretely given sequences. Of course we also have to deal with operations on operations, etc. However, such operations, too, can be thought of as Gedankenexperimente on (concrete) operations.

By contrast, intuitionism emphatically deals with abstract notions. This is seen by the fact that its basic notion of "construction" (or "proof") is absolutely abstract, and this abstract nature also seems necessary for its impredicative concept of "implication". It is not the aim of intuitionism to reduce these abstract notions to concrete notions as we do.

We believe that our standpoint is a natural extension of Hilbert's finitist standpoint, similar to that introduced by Gentzen, and so we call it the Hilbert-Gentzen finitist standpoint.

Now a Gentzen-style consistency proof is carried out as follows:

(1) Construct a suitable standard ordering, in the strictly finitist standpoint.

(2) Convince oneself, in the Hilbert-Gentzen standpoint, that it is indeed a well-ordering.

(3) Otherwise use only strictly finitist means in the consistency proof.

We now present a consistency proof of this kind for **PA**.

§12. A consistency proof of PA

We assume from now on that **PA** is formalized in a language which includes a constant f for every primitive recursive function f. We call this language L.

As initial sequents of **PA** we will also take from now on the defining equations for all primitive recursive functions, as well as all sequents $\rightarrow s = t$, where s, t are closed terms of L denoting the same number, and all sequents $s = t \rightarrow$, where s, t are closed terms of L denoting different numbers.

We shall follow Gentzen's second version of his consistency proof for first order arithmetic. This involves a "reduction method". Since this method will recur often, we shall abstract the concept here. (We assume that the ordinals less than ε_0 are represented as notations in a fixed standard well-ordering, as described in §11.)

First, suppose that ordinals less than ε_0 are effectively assigned to proofs. Now let R be a property of proofs such that:

(*) For any proof P satisfying R, we can find (effectively from P) a proof P' satisfying R such that P' has a smaller ordinal than P.

We can then infer from (*), and the accessibility of ε_0:

(**) No proof satisfies R.

The procedure of finding (or constructing) P' from P in (*) is called: a *reduction of P to P'* (for the property R).

The property R of proofs that we will be interested in, is the property of having $\rightarrow$ as an end-sequent.

By giving a uniform reduction procedure for this property (Lemma 12.8), we will have shown (by (**)) that no proof of **PA** ends with $\rightarrow$; in other words:

THEOREM 12.1. *The system* **PA** *is consistent.*

Of course the importance of this theorem exists in its proof, which, apart from the assumption of the accessibility of ε_0, is strictly finitist. (Nobody suspects the consistency of Peano arithmetic!)

Theorem 12.1 follows from Lemma 12.8 (as just stated). First, we need:

DEFINITION 12.2. A proof in **PA** is *simple* if no free variables occur in it, and it contains only mathematical initial sequents, weak inferences and inessential cuts.

(Recall that a weak inference is a structural inference other than a cut. Cf. §9 for other definitions.)

LEMMA 12.3. *There is no simple proof of* $\rightarrow$.

PROOF. Let P be any simple proof. All the formulas in P are of the form $s = t$ with s and t closed. Note that with the natural interpretation of the constants, it can be determined (finitistically) whether $s = t$ is true or false (since this only involves the evaluation of certain primitive recursive functions). A sequent in P is then given the value T if at least one formula in the antecedent is false, or at least one formula in the succedent is true, and it is given the value F otherwise. It is easy to see that all mathematical initial sequents take the value T, and weak inferences and inessential cuts preserve the value T downward for sequents. So all sequents of P have the value T. But $\rightarrow$ has the value F.

DEFINITION 12.4. (1) The *grade of a formula*, is (as defined in §5) the number of logical symbols it contains. The *grade of a cut* is the grade of the cut formula; the *grade of an ind inference* is the grade of the induction formula.

(2) The *height of a sequent S in a proof P* (denoted by $h(S; P)$ or, for short, $h(S)$) is the maximum of the grades of the cuts and ind's which occur in P below S.

PROPOSITION 12.5. (1) *The height of the end-sequent of a proof is* 0.

(2) *If S_1 is above S_2 in a proof, then $h(S_1) \geq h(S_2)$; if S_1 and S_2 are the upper sequents of an inference, then $h(S_1) = h(S_2)$.*

Before defining the assignment of ordinals to proofs, we introduce the following notation. For any ordinal α and natural number n, $\omega_n(\alpha)$ is defined by induction on n; $\omega_0(\alpha) = \alpha$, $\omega_{n+1}(\alpha) = \omega^{\omega_n(\alpha)}$. So

$$\omega_n(\alpha) = \underbrace{\omega^{\cdot^{\cdot^{\omega^{\alpha}}}}}_{n}$$

DEFINITION 12.6. Assignment of ordinals (less than ε_0) to the proofs of **PA**. First we assign ordinals to the sequents in a proof. The ordinal assigned to a sequent S in a proof P is denoted by $o(S; P)$ or $o(S)$. Now suppose a proof P is given. We shall define $o(S) = o(S; P)$, for all sequents S in P.

We shall henceforth assume that the ordinals are expressed in normal form (cf. §11). If μ and ν are ordinals of the form $\omega^{\mu_1} + \omega^{\mu_2} + \ldots + \omega^{\mu_n}$ and $\omega^{\nu_1} + \omega^{\nu_2} + \ldots + \omega^{\nu_n}$ respectively (so that $\mu_1 \geq \mu_2 \geq \ldots \geq \mu_m$ and $\nu_1 \geq \nu_2 \geq \ldots \geq \nu_n$), then $\mu \# \nu$ denotes the ordinal $\omega^{\lambda_1} + \omega^{\lambda_2} + \ldots + \omega^{\lambda_{m+n}}$, where $\{\lambda_1, \lambda_2, \ldots, \lambda_{m+n}\} = \{\mu_1, \mu_2, \ldots, \nu_1, \nu_2, \ldots\}$ and $\lambda_1 \geq \ldots \geq \lambda_{m+n}$. $\mu \# \nu$ is called the natural sum of μ and ν.

(1) An initial sequent (in P) is assigned the ordinal 1.

(2) If S is the lower sequent of a weak inference, then $o(S)$ is the same and the ordinal of its upper sequent.

(3) If S is the lower sequent of $\wedge$: left, $\vee$: right, $\supset$: right, $\neg$: right, $\neg$: left or an inference involving a quantifier, and the upper sequent has the ordinal μ, then $o(S) = \mu + 1$.

(4) If S is the lower sequent of $\wedge$: right, $\vee$: left, or $\supset$: left and the upper sequents have ordinals μ and ν, then $o(S) = \mu \# \nu$.

(5) If S is the lower sequent of a cut and its upper sequents have the ordinals μ and ν, then $o(S)$ is $\omega_{k-l}(\mu \# \nu)$, i.e.,

$$\left.\omega^{\cdot^{\cdot^{\omega^{\mu \# \nu}}}}\right\} k-l,$$

where k and l are the heights of the upper sequents and of S, respectively.

(6) If S is the lower sequent of an ind and its upper sequent has the

ordinal μ, then $o(S)$ is $\omega_{k-l+1}(\mu_1+1)$, i.e.,

$$\left.\omega^{\omega^{\cdot^{\cdot^{\omega^{\mu_1+1}}}}}\right\}(k-l)+1,$$

where μ has the normal form $\omega^{\mu_1}+\omega^{\mu_2}+\ldots+\omega^{\mu_n}$ (so that $\mu_1\geq\mu_2\geq\ldots\geq\mu_n$), and k and l are the heights of the upper sequent and of S, respectively.

(7) The ordinal of a proof P, $o(P)$, is the ordinal of its end-sequent.

We use the notation

$$P:\quad \begin{array}{c}\vdots\\ \Gamma\overset{\mu}{\rightarrow}\Delta\end{array}$$

to denote a proof P of $\Gamma\rightarrow\Delta$ such that $o(\Gamma\rightarrow\Delta;P)=o(P)=\mu$.

LEMMA 12.7. *Suppose P is a proof containing a sequent S_1, there is no ind below S_1, P_1 is the subproof of P ending with S_1, P_1' is any other proof of S_1, and P' is the proof formed from P by replacing P_1 by P_1':*

$$P:\quad P_1\left\{\begin{array}{c}\vdots\\ S_1\end{array}\right.\\ \vdots \qquad\qquad P':\quad P_1'\left\{\begin{array}{c}\vdots\\ S_1\end{array}\right.\\ \vdots$$

Suppose also that $o(S_1;P')<o(S_1;P)$. Then $o(P')<o(P)$.

PROOF. Consider a thread of P passing through S_1. We show that for any sequent S of this thread at or below S_1: if S' is the sequent "corresponding to" S in P', then

$$(*)\qquad o(S';P')<o(S;P).$$

This is true for $S=S_1$ by assumption, and this property (*) is preserved downwards by all the inference rules, as can be checked. (We use the fact that the natural sum is strictly monotonic in each argument, i.e., $\alpha<\beta\Rightarrow\alpha\,\#\,\gamma<\beta\,\#\,\gamma$, etc.) Finally, letting S be the end-sequent of P, we obtain the desired conclusion.

This lemma is used repeatedly in the consistency proof.

Now let R be the property of proofs of ending with the sequent $\rightarrow$; i.e., for any proof P, R(P) holds if and only if P is a proof of $\rightarrow$.

Notice first that if P is a proof of $\rightarrow$, then every logical inference of P is implicit! (cf. Definition 9.7) (since otherwise a bundle containing the principal formula of this inference would end with an end formula).

Hence the definition of end-piece for such proofs can be simply stated as follows.

The end-piece of a proof of $\rightarrow$ consists of all those sequents that are encountered as we ascend each thread from the end-sequent and stop as soon as we arrive at a logical inference. (Then the upper sequent of this inference no longer belongs to the end-piece, but the lower sequent, and all sequents below it, do.) This inference belongs to the boundary.

LEMMA 12.8. *If P is a proof of $\rightarrow$, then there is another proof P' of $\rightarrow$ such that $o(P') < o(P)$.*

PROOF. Let P be a proof of $\rightarrow$. We can assume, by Proposition 9.8, that P is regular. We describe a "reduction" of P to obtain the desired P'. The reduction consists of a number of steps, described below. Each step is performed, perhaps finitely often (as will be clear), and at each step, we assume that the previous steps have been performed (as often as possible).

At each step, the ordinal of the resulting proof does not increase, and at least at one step, the ordinal decreases.

Step 1. Suppose the end-piece of P contains a free variable, say a, which is not used as an eigenvariable. Then replace a by the constant 0. This results in a proof of $\rightarrow$ (using the analogue of Lemma 2.10 for **PA**), with the same ordinal.

Step 1 is performed repeatedly until there is no free variable in the end-piece which is not used as an eigenvariable.

Step 2. Suppose the end-piece of P contains an ind. Then take a lowermost one, say I. Suppose I is of the following form:

$$\begin{array}{ll} P_0(a) & \left\{ \begin{array}{c} \vdots \\ F(a), \Gamma \overset{\mu}{\rightarrow} \Delta, F(a') \end{array} \right. \\ I & \quad \overline{\quad F(0), \Gamma \rightarrow \Delta, F(s) \quad}, \\ & \qquad\qquad \vdots \\ & \qquad\qquad \rightarrow \end{array}$$

where P_0 is the subproof ending with $F(a), \Gamma \rightarrow \Delta, F(a')$, and let l and k be the heights of the upper sequent (call it S) and the lower sequent (call it S_0) of I, respectively. Then

$$o(S_0) = \omega_{l-k+1}(\mu_1 + 1),$$

where $\mu = o(S) = \omega^{\mu_1} + \omega^{\mu_2} + \ldots + \omega^{\mu_n}$ and $\mu_n \leqslant \ldots \leqslant \mu_2 \leqslant \mu_1$. Since no free variable occurs below I, s is a closed term and hence there is a number m such that $\rightarrow s = \bar{m}$ is **PA**-provable without an essential cut or ind (cf. Lemma 9.6); hence there is a proof Q of $F(\bar{m}) \rightarrow F(s)$ without an essential cut or ind (cf. Lemma 9.6). Let $P_0(\bar{n})$ be the proof which is obtained from P_0 by replacing a by $\bar{n}$ throughout. Consider the following proof P'.

$$
\begin{array}{ll}
 & \quad P_0(\bar{0}) \qquad\qquad\qquad\qquad P_0(\bar{1}) \qquad\qquad\qquad\qquad P_0(\bar{2}) \\
S_1 & \dfrac{F(0), \Gamma \rightarrow \Delta, F(0') \qquad F(0'), \Gamma \rightarrow \Delta, F(0'')}{} \\
S_2 & \dfrac{F(0), \Gamma \rightarrow \Delta, F(0'') \qquad\qquad F(0''), \Gamma \rightarrow \Delta, F(0''')}{} \\
S_3 & \qquad F(0), \Delta \rightarrow \Delta, F(0''') \\
 & \qquad\qquad\qquad\qquad\qquad\qquad\qquad\qquad Q \\
S_m & F(0), \Gamma \rightarrow \Delta, F(\bar{m}) \quad F(\bar{m}) \rightarrow F(s) \\
S_0 & \dfrac{\qquad}{F(0), \Gamma \rightarrow \Delta, F(s)} \\
 & \qquad\qquad \rightarrow
\end{array}
$$

where $S_1, S_2, \ldots, S_0$ denote the sequents shown on their right, $S_1, \ldots, S_m$ all have height l, since the formulas $F(\bar{n})$, $n = 0, \ldots, m$, all have the same grade. Therefore,

$$o(F(\bar{n}), \Gamma \rightarrow \Delta, F(\bar{n}'); P') = \mu \quad \text{for} \quad n = 0, 1, \ldots, m.$$

Since Q has no essential cut or ind, $o(F(\bar{m}) \rightarrow F(s); P') = q$ (say) $< \omega$, $o(S_2) = \mu \# \mu$; $o(S_3) = \mu \# \mu \# \mu$; $, \ldots,$ and in general, writing $\mu * n = \mu \# \mu \# \ldots \# \mu$ (n times), $o(S_n) = \mu * n$ for $n = 1, 2, \ldots, m$. Thus

$$o(S_0) = \omega_{l-k}(\mu * m + q)$$

and $\mu * m + q < \omega^{\mu_1 + 1}$, since $q < \omega$. Therefore

$$o(S_0; P') = \omega_{l-k}(\mu * m + q) < \omega_{l-k+1}(\mu_1 + 1) = o(S_0; P).$$

Thus $o(S_0; P') < o(S_0; P)$, and hence by Lemma 12.7, $o(P') < o(P)$.

Thus, if P has an ind in the end-piece, we are done: we have reduced P to a proof P' of $\rightarrow$ with $o(P') < o(P)$. Otherwise, we assume from now on that P has no ind in its end-piece, and go to Step 3.

Step 3. Suppose the end-piece of P contains a logical initial sequent $D \rightarrow D$. Since the end-sequent is empty, both D's (or more strictly, descendants of both D's) must disappear by cuts. Suppose that (a descendant of) the D in the antecedent is a cut formula first (viz. in the following figure a descendant of the D in the succedent of $D \to D$ occurs in Ξ).

$$\begin{array}{ccc}
 & & D \to D \\
 & \vdots & \vdots \\
 & \Gamma \to \Delta, D & D, \Pi \to \Xi \\
\cline{2-3}
S & \multicolumn{2}{c}{\Gamma, \Pi \to \Delta, \Xi} \\
 & \multicolumn{2}{c}{\vdots} \\
 & \multicolumn{2}{c}{\to}
\end{array}$$

P is reduced to the following P':

$$\begin{array}{cc}
 & \vdots \\
 & \Gamma \to \Delta, D \\
\hline
 & \text{weakenings and exchanges} \\
\hline
S' & \Gamma, \Pi \to \Delta, \Xi \\
 & \vdots \\
 & \to
\end{array}.$$

Note that there is a cut whose cut formula is D below S since both D's in $D \to D$ must disappear by cuts. Hence, the height of $\Gamma \to \Delta, D$ does not change when we transform P into P': $o(S'; P') < o(S; P)$.

Hence, by Lemma 12.7, $o(P') < o(P)$.

The other case is proved likewise.

So, if the end-piece of P contained a logical initial sequent, we have found a P' as desired. Otherwise, we assume from now on that the end-piece of P contains no logical initial sequents, and go on to Step 4.

Step 4. Suppose there is a weakening in the end-piece. Let I be the lower most weakening inference in the end-piece. Since the end-sequent is empty, there must exist a cut, J, below I and the cut formula is the descendent of the principal formula of I.

$$\begin{array}{ccc}
 & & \vdots \\
 & & \Pi' \to \Xi' \\
\cline{3-3}
 & & D, \Pi' \to \Xi' \quad I \\
 & \vdots & \vdots \\
 & \Gamma \to \Delta, D & D, \Pi \to \Xi(k) \\
\cline{2-3}
J & \multicolumn{2}{c}{\Gamma, \Pi \to \Delta, \Xi(l)} \\
 & \multicolumn{2}{c}{\vdots} \\
 & \multicolumn{2}{c}{\to}
\end{array}$$

Case (1)

If no contraction is applied to D from the inference I through J, by deleting some exchanges from P if necessary, reduce P into the following proof P':

$$\begin{array}{c} \vdots \\ \Pi' \to \Xi' \\ \vdots \\ \Pi \to \Xi \quad (l) \\ \hline\hline \text{weakenings and exchanges} \\ \hline \Gamma, \Pi \to \Delta, \Xi \quad (l) \\ \vdots \\ \to \end{array}$$

Let $h(\Gamma, \Pi \to \Delta, \Xi; P) = l$ and $h(D, \Pi \to \Xi; P) = k$. Then, $l \leqslant k$ and $h(\Pi \to \Xi; P') = h(\Gamma, \Pi \to \Delta, \Xi; P') = l$. Let S be a sequent in P above $D, \Pi \to \Xi$, and let S' be the corresponding sequent in P'. Then, by the induction on number of inferences up to $D, \Pi \to \Xi$, we can show

$$\omega_{k_1 - k_2}(o(S; P)) \geqslant o(S'; P'),$$

where $k_1 = h(S; P)$ and $k_2 = h(S'; P')$. Hence, if $o(\Gamma \to \Delta, D; P) = \mu_1$, $o(D, \Pi \to \Xi; P) = \mu_2$, $o(\Gamma, \Pi \to \Delta, \Xi) = \nu$, $o(\Pi \to \Xi; P) = \mu_2'$ and $o(\Gamma, \Pi \to \Delta, \Xi) = \nu'$, then

$$\omega_{k-l}(\mu_2) \geqslant \mu_2'$$

and further,

$$\nu = \omega_{k-l}(\mu_2 \# \mu_1) > \omega_{k-l}(\mu_2) \geqslant \mu_2' = \nu'.$$

Thus, $o(P) > o(P')$.

Case (2)

If not the Case (1), let the uppermost contraction applied to D be I'. Reduce P into the following proof Q:

P: Q:

$$\begin{array}{c} \vdots \\ \Pi' \to \Xi' \\ \hline D, \Pi' \to \Xi' \\ \vdots \\ D, D, \Pi'' \to \Xi'' \\ \hline D, \Pi'' \to \Xi'' \\ \vdots \\ D, \Pi \to \Xi \end{array} \qquad\qquad \begin{array}{c} \vdots \\ \Pi' \to \Xi' \\ \vdots \\ D, \Pi'' \to \Xi'' \\ \vdots \\ D, \Pi \to \Xi. \end{array}$$

Apparently, $o(P) = o(P')$. Hence, we can assume that the end-piece of P contains no weakening.

Step 5. We can now assume that P is not its own end-piece, since otherwise it would be simple (Definition 12.2), as is easily seen, and hence by Lemma 12.3, could not end with $\rightarrow$.

Under these assumptions, we shall prove that the end-piece of P contains a suitable cut (cf. Definition 9.7). We actually prove a stronger result, which is used again later (for Problem 12.11):

SUBLEMMA 12.9. *Suppose that a proof in* **PA**, *say P, satisfies the following.*

(1) *P is not its own end-piece.*

(2) *The end-piece of P does not contain any logical inference, ind or weakening.*

(3) *If an initial sequent belongs to the end-piece of P, then it does not contain any logical symbol.*

Then there exists a suitable cut in the end-piece of P.

(Notice that we do not assume here that the end-sequent is $\rightarrow$.)

PROOF. This is proved by induction on the number of essential cuts in the end-piece of P. The end-piece of P contains an essential cut, since P is not its own end-piece. Take a lowermost such cut, say I. If I is a suitable cut, then the sublemma is proved. Otherwise, let P be of the form

$$I \quad \frac{P_1\left\{\begin{array}{c}\vdots\\ \Gamma \rightarrow \Delta, D\end{array}\right. \qquad P_2\left\{\begin{array}{c}\vdots\\ D, \Pi \rightarrow \Lambda\end{array}\right.}{\Gamma, \Pi \rightarrow \Delta, \Lambda}\ .$$

Since I is not a suitable cut, one of two cut formulas of I is not a descendant of the principal formula of a boundary inference. Suppose that D in $\Gamma \rightarrow \Delta, D$ is not a descendant of the principal formula of a boundary inference. Now we prove:

(i) P_1 contains a boundary inference of P.

Suppose otherwise. Then D in $\Gamma \rightarrow \Delta, D$ is a descendant of D in an initial sequent in the end-piece of P, by (2). This contradicts the assumption that I is an essential cut, by (3).

(ii) If an inference J in P_1 is a boundary inference of P, then J is a boundary inference of P_1.

This is easily seen by the fact that I is a lowermost essential cut of P and D is not a descendant of the principal formula of a boundary inference.

(iii) P_1 is not its own end-piece and the end-piece of P_1 is the intersection of P_1 and the end-piece of P.

This follows immediately from (i), (ii) and (1).

Now from the induction hypothesis, the end-piece of P_1 has a suitable cut. This cut is a suitable cut in the end-piece of P.

Returning to our proof P of $\rightarrow$ which satisfies the conclusion of steps 1–4, we have, as an immediate consequence of Sublemma 12.9, that the end-piece of P contains a suitable cut. We now define an *essential reduction* of P.

Take a lowermost suitable cut in the end-piece of P, say I.

Case 1. The cut formula of I is of the form $A \wedge B$. Suppose P is of the form

$$
I \; \dfrac{\dfrac{\begin{array}{c}\vdots\\ \Gamma' \rightarrow \Theta', A \qquad \Gamma' \rightarrow \Theta', B\end{array}}{\Gamma' \rightarrow \Theta', A \wedge B}\; I_1 \quad \vdots \quad \Gamma \overset{\mu}{\rightarrow} \Theta, A \wedge B \qquad \dfrac{\begin{array}{c}\vdots\\ A, \Pi' \rightarrow \Lambda'\end{array}}{A \wedge B, \Pi' \rightarrow \Lambda'}\; I_2 \quad \vdots \quad A \wedge B, \Pi \overset{\nu}{\rightarrow} \Lambda \quad (l)}{\Gamma, \Pi \rightarrow \Theta, \Lambda}
$$

$$
\begin{array}{c}
\Gamma, \Pi \rightarrow \Theta, \Lambda\\
\vdots\\
\Delta \overset{\lambda}{\rightarrow} \Xi \quad (k)\\
\vdots\\
\rightarrow
\end{array}
$$

where $\Delta \rightarrow \Xi$ denotes the uppermost sequent below I whose height is less than that of the upper sequents of I. Let l be the height of each upper sequent of I, and k that of $\Delta \rightarrow \Xi$. Then $k < l$. Notice that $\Delta \rightarrow \Xi$ may be the lower sequent of I, or the end-sequent. The existence of such a sequent follows from Proposition 12.5.

$\Delta \rightarrow \Xi$ must be the lower sequent of a cut J (since there is no ind below I). Let $\mu = o(\Gamma \rightarrow \Theta, A \wedge B)$, $\nu = o(A \wedge B, \Pi \rightarrow \Lambda)$, $\lambda = o(\Delta \rightarrow \Xi)$ as shown. Consider the following proofs:

P_1:

$$
\begin{array}{c}
\vdots\\
\Gamma' \rightarrow \Theta', A\\
\hline\hline
\Gamma' \rightarrow A, \Theta'\\
\hline
\Gamma' \rightarrow A, \Theta', A \wedge B
\end{array}
\quad \text{(weakening : right)}
$$

$$
\begin{array}{c}
\vdots\\
J_1 \quad \dfrac{\Gamma \overset{\mu_1}{\rightarrow} A, \Theta, A \wedge B \qquad A \wedge B, \Pi \overset{\nu_1}{\rightarrow} \Lambda \quad (l)}{\Gamma, \Pi \rightarrow A, \Theta, \Lambda}\\
\vdots\\
\Delta \overset{\lambda_1}{\rightarrow} A, \Xi\\
\hline\hline
\Delta \rightarrow \Xi, A \quad (m)
\end{array}
$$

P_2:

$$\begin{array}{c}
\vdots \\
A, \Pi' \rightarrow \Lambda' \\
\hline\hline
\Pi', A \rightarrow \Lambda' \\
\hline
A \wedge B, \Pi', A \rightarrow \Lambda'
\end{array}\quad \text{(weakening : left)}$$

$$J_2\quad \begin{array}{c}
\vdots \qquad\qquad\qquad\qquad \vdots \\
\Gamma \xrightarrow{\mu_2} \Theta, A \wedge B \qquad A \wedge B, \Pi, A \xrightarrow{\nu_2} \Lambda \quad (l) \\
\hline
\Gamma, \Pi, A \rightarrow \Theta, \Lambda \\
\vdots \\
\Delta, A \xrightarrow{\lambda_2} \Xi \quad (m) \\
\hline\hline
A, \Delta \rightarrow \Xi
\end{array}$$

(where l and m are the heights of the sequents shown, not in P_1 and P_2, but in P', defined below, which contains these as subproofs).

Define P' to be the proof:

$$\begin{array}{c}
P_1 \qquad\qquad\qquad\qquad P_2 \\
\vdots \qquad\qquad\qquad\qquad \vdots \\
\Delta \xrightarrow{\lambda_1} \Xi, A \quad (m) \qquad A, \Delta \xrightarrow{\lambda_2} \Xi \quad (m) \\
\hline
\Delta, \Delta \xrightarrow{\lambda_0} \Xi, \Xi \quad (k) \\
\hline\hline
\Delta \rightarrow \Xi \\
\vdots \\
\rightarrow
\end{array}\quad \text{(cut for } A\text{)}$$

So m is the height of the upper sequents of I' (the cut for A). Note that the height of the lower sequent of I' is k.

It is obvious that $m = k$ if $k >$ grade of A and $m =$ grade of A otherwise. In either case $k \leq m < l$.

$$h(\Gamma \rightarrow A, \Theta, A \wedge B; P') = h(A \wedge B, \Pi \rightarrow \Lambda; P') = l,$$

since all cut formulas below I in P occur in P' below J_1, all cut formulas below J_1 in P' except A occur in P under I, and grade of $A <$ grade of $A \wedge B \leq l$. Similarly,

$$h(\Gamma \rightarrow \Theta, A \wedge B; P') = h(A \wedge B, \Pi, A \rightarrow \Lambda; P') = l.$$

Let

$$\mu_1 = o(\Gamma \to A, \Theta, A \wedge B; P'), \qquad \nu_1 = o(A \wedge B, \Pi \to \Lambda; P'),$$
$$\lambda_1 = o(\Delta \to A, \Xi; P'), \qquad \mu_2 = o(\Gamma \to \Theta, A \wedge B; P'),$$
$$\nu_2 = o(A \wedge B, \Pi, A \to \Lambda; P'), \qquad \lambda_2 = o(\Delta, A \to \Xi; P'),$$
$$\lambda_0 = o(\Delta, \Delta \to \Xi, \Xi; P').$$

Then $\mu_1 < \mu$, $\nu_1 = \nu$, $\mu_2 = \mu$ and $\nu_2 < \nu$.

Now let

$$J' \quad \frac{S_1' \quad S_2' \quad (k_1)}{S' \qquad (k_2)}$$

be an arbitrary inference between J_1 and $\Delta \to A, \Xi$ and let

$$J \quad \frac{S_1 \quad S_2}{S}$$

be the corresponding inference between I and $\Delta \to \Xi$. Let

$$\alpha_1' = o(S_1'; P'), \qquad \alpha_2' = o(S_2'; P'), \qquad \alpha' = o(S'; P'),$$
$$\alpha_1 = o(S_1; P), \qquad \alpha_2 = (S_2, P), \qquad \alpha = o(S; P),$$
$$k_1 = h(S_1', P') = h(S_2', P'), \qquad k_2 = h(S', P').$$

Then $\alpha = \alpha_1 \# \alpha_2$ if S is not $\Delta \to A, \Xi$, and $\alpha = \omega_{l-k}(\alpha_1 \# \alpha_2)$ if S' is $\Delta \to A, \Xi$. On the other hand $\alpha' = \omega_{k_1-k_2}(\alpha_1' \# {}_2')$.

Starting with $\mu_1 < \mu$ and $\nu_1 = \nu$, it is easily seen by induction on the number of inferences between J_1 and S' that

$$\alpha' < \omega_{l-k_2}(\alpha), \tag{1}$$

if S is not $\Delta \to A, \Xi$. Let $\lambda = \omega_{l-k}(\kappa)$. Then (1) implies that $\lambda_1 < \omega_{l-m}(\kappa)$. Similarly, $\lambda_2 < \omega_{l-m}(\kappa)$. Hence

$$\omega_{m-k}(\lambda_1 \# \lambda_2) < \omega_{l-k}(\kappa),$$

since $l-k=(l-m)+(m-k)$. Therefore $\lambda_0<\lambda$. Finally, from $\lambda_0<\lambda$ it follows that $o(P')<o(P)$.

Case 2. The cut formula of I is of the form $\forall x\, F(x)$. So P has the form:

$$
\begin{array}{llcl}
 & \quad\vdots & & \qquad\vdots \\
I_1 & \dfrac{\Gamma' \to \Theta', F(a)}{\Gamma' \to \Theta', \forall x\, F(x)} & I_2 & \dfrac{F(s), \Pi' \to \Lambda'}{\forall x\, F(x), \Pi' \to \Lambda'} \\
 & \quad\vdots & & \qquad\vdots \\
I & \multicolumn{3}{c}{\dfrac{\Gamma \to \Theta, \forall x\, F(x) \qquad\qquad \forall x\, F(x), \Pi \to \Lambda}{\Gamma, \Pi \to \Theta, \Lambda}} \\
 & \multicolumn{3}{c}{\vdots} \\
 & \multicolumn{3}{c}{\Delta \to \Xi} \\
 & \multicolumn{3}{c}{\vdots} \\
 & \multicolumn{3}{c}{\to}
\end{array}
$$

The definition of $\Delta \to \Xi$ is the same as in case 1. The proof P' is then defined in terms of the following two subproofs P_1 and P_2:

P_1:

$$
\begin{array}{l}
\begin{array}{l}
\Gamma' \to \Theta', F(s) \\ \hline\hline
\Gamma' \to F(s), \Theta' \\ \hline
\Gamma' \to F(s), \Theta', \forall x\, F(x) \\ \hline
\end{array} \\
\qquad\vdots \qquad\qquad\qquad\qquad\qquad \vdots \\
\dfrac{\Gamma \to F(s), \Theta, \forall x\, F(x) \qquad \forall x\, F(x), \Pi \to \Lambda}{\qquad\qquad \Gamma, \Pi \to F(s), \Theta, \Lambda} \\
\qquad\qquad\qquad\qquad\qquad \vdots \\
\qquad\qquad\qquad\qquad
\begin{array}{c}
\Delta \to F(s), \Xi \\ \hline\hline
\Delta \to \Xi, F(s)
\end{array}
\end{array}
$$

P_2:

$$
\begin{array}{r}
\qquad\qquad\qquad\qquad \vdots \\
\begin{array}{r}
F(s), \Pi' \to \Lambda' \\ \hline\hline
\Pi', F(s) \to \Lambda' \\ \hline
\forall x\, F(x), \Pi', F(s) \to \Lambda'
\end{array} \\
\vdots \qquad\qquad\qquad\qquad\qquad\qquad \vdots \\
\dfrac{\Gamma \to \Theta, \forall x\, F(x) \quad \forall x\, F(x), \Pi, F(s) \to \Lambda}{\Gamma, \Pi, F(s) \to \Theta, \Lambda \qquad\qquad\qquad} \\
\vdots \qquad\qquad\qquad\qquad\qquad \\
\begin{array}{c}
\Delta, F(s) \to \Xi \\ \hline\hline
F(s), \Delta \to \Xi
\end{array} \qquad\qquad\qquad
\end{array}
$$

P' is defined to be

$$\dfrac{\dfrac{\overset{P_1}{\vdots}}{\Delta \rightarrow \Xi, F(s)} \qquad \dfrac{\overset{P_2}{\vdots}}{F(s), \Delta \rightarrow \Xi}}{\dfrac{\Delta, \Delta \rightarrow \Xi, \Xi}{\underset{\displaystyle \rightarrow}{\overset{\displaystyle \Delta \rightarrow \Xi}{\vdots}}}}.$$

Note that $o(\Gamma' \rightarrow \Theta', F(s); P') = o(\Gamma' \rightarrow \Theta', F(a); P)$. The argument on ordinals goes through as in case 1.

For the other cases, the proof is similar.

This completes the proof of Lemma 12.8 and hence the consistency proof for **PA** (Theorem 12.1).

REMARK 12.10. We wish to point out the following. One often says that the consistency of **PA** is proved by transfinite induction on the ordinals of proofs, as if we were using a general principle of transfinite induction in order to prove the consistency of mathematical induction.

This is misleading, however. The point is that the consistency proof uses the notion of accessibility of ε_0, as explained in §11, and otherwise strictly finitist method. To re-state the matter from a more formal viewpoint:

The principle of *transfinite induction* on some (definable) well-ordering $<$ of the natural numbers can be expressed (in first-order formal systems) by the schema

$$TI(<, F(x)): \qquad \forall x\,[\forall y\,(y < x \supset F(y)) \supset F(x)] \rightarrow \forall x\, F(x)$$

for arbitrary formulas $F(x)$ of the system considered.

Now Gentzen's consistency proof of **PA** can be formalized in the system of primitive recursive arithmetic, together with the axiom $TI(<, F(x))$, where $<$ is the standard well-ordering of type ε_0 and $F(x)$ is a certain *quantifier-free* formula.

PROBLEM 12.11. We can extend the reduction procedure of Lemma 12.8 to the following situation.

A sequent S (of the language of **PA**) is said to satisfy the property P if:

(1) All sequent-formulas of S are closed;

(2) Each sequent-formula in the succedent of S is either quantifier-free or of the form $\exists y_1, \ldots, \forall y_m\ R(y_1, \ldots, y_m)$, where $R(y_1, \ldots, y_m)$ is quantifier-free;

(3) Each sequent formula in the antecedent of S is either quantifier-free or of the form $\forall y_1, \ldots, \forall y_m\ R(y_1, \ldots, y_m)$, where $R(y_1, \ldots, y_m)$ is quantifier-free.

Show that if a sequent satisfying P is provable in **PA**, then it is provable without an essential cut or ind. [*Hint*: We may assume that there is no free variable which is not used as an eigenvariable in the end-piece of a proof of such a sequent.]

If the end-piece has an explicit logical inference, take the lowermost explicit logical inference I. Without loss of generality, we assume that the proof is of the following form:

$$I \quad \frac{\Gamma \rightarrow \Delta, \exists y_2 \ldots \exists y_m R(t, y_2, \ldots, y_m)}{\Gamma \rightarrow \Delta, \exists y_1 \ldots \exists y_m R(y_1, y_2, \ldots, y_m)}$$

$$\vdots$$

$$\Gamma_0 \rightarrow \Delta_0, \forall y_1 \ldots \forall y_m R(y_1, \ldots, y_m), \Delta_1$$

where $\Gamma_0 \rightarrow \Delta_0, \exists y_1 \ldots \exists y_m R(y_1, \ldots, y_m), \Delta_1$ is the end-sequent of the proof. We can eliminate I by replacing the proof by a proof whose end-sequent is either of the form

$$\Gamma_0 \rightarrow \Delta_0, \exists y_2 \ldots \exists y_m R(t, y_2, \ldots, y_m), \Delta_1$$

or of the form

$$\Gamma_0 \rightarrow \Delta_0, \exists y_1 \ldots \exists y_m R(y_1, \ldots, y_m), \Delta_1, \exists y_2 \ldots \exists y_m R(t, y_2, \ldots, y_m).$$

PROBLEM 12.12. Intuitionistic arithmetic can be formalized as the subsystem of **PA** defined by the condition that in the succedent of every sequent there can be at most one sequent-formula which contains quantifiers. This system may be called **HA** (for Heyting arithmetic). The reduction method for **PA** works for **HA** with a slight modification: roughly, in an essential reduction, if the cut formula of the suitable cut under consideration contains a quantifier then the weakening : right will not be introduced.

Define the reduction for **HA** precisely, thus proving the consistency of **HA** directly (not as a subsystem of **PA**).

PROBLEM 12.13. Let (*) be the property of formulas defined in Theorem 6.14, i.e., a formula satisfies (*) if every $\vee$ and $\exists$ in it is either in the scope of a $\neg$ or in the left scope of a $\supset$. Show that, if each formula in Γ satisfies (*) and all formulas in Γ, A, B and $\exists x\, F(x)$ are closed, then in **HA** (cf. Problem 12.12):

(1) $\Gamma \rightarrow A \vee B$ if and only if $\Gamma \rightarrow A$ or $\Gamma \rightarrow B$,

(2) $\Gamma \rightarrow \exists x\, F(x)$ if and only if for some closed term s, $\Gamma \rightarrow F(s)$.

[*Hint* (B. Scarpellini): By transfinite induction on the ordinal of a proof P of $\Gamma \rightarrow A \vee B$ (for 1) or $\Gamma \rightarrow \exists x\, F(x)$ (for 2), respectively, following the reduction method for the consistency of **PA**. First deal with explicit logical inferences in the end-piece of P.]

REMARK 12.14. As an application of Gentzen's reduction method, one can easily prove the following.

The consistency of arithmetic in which the induction formulas are restricted to those which have at most k quantifiers can be proved by transfinite induction on ω_{k+1}.

The outline of the proof is as follows. Suppose there is a proof of $\rightarrow$ in this system. We shall carry out a reduction of such a proof.

(1) We assume that the induction formulas are in prenex normal form.

(2) A formula A in a proof (in this system) will be temporarily called free if either it has no ancestor which is an induction formula, or it has an induction formula as an ancestor but a logical symbol is introduced in an ancestor of A between any such induction formula and A itself. A cut is called free if both cut formulas are free. Notice that if a formula is not free, then it is in prenex form with at most k quantifiers. Now we can prove the following partial cut-elimination theorem:

If a sequent is provable in our system, then it is provable without free cuts.

(We simply adapt the cut-elimination proof for **LK**.)

Thus we obtain a proof of $\rightarrow$ in which there are no free cuts, and so all the cut formulas, as well as induction formulas, are in prenex form with at most k quantifiers. We assume $k \geq 1$.

(3) Further we can assume, for convenience, that the inference rules are modified in such a way that all formulas in the proof are in prenex form, with at most k quantifiers.

This system is called $\mathbf{PA}_k$.

We must now modify some notions slightly. The grade of a formula A is now defined to be: the number of quantifiers in A, minus 1; the grade of a cut or induction inference is the grade of the cut formula or the induction formula, respectively. The height of a sequent in a proof is defined as before, using the new definition of grade. The ordinals are assigned as before, except that the initial sequents are assigned the ordinal 0 and the propositional inferences as well as quantifier-free cuts are treated in the same manner as the weak inferences, i.e., the ordinals do not change. (In case there are two upper sequents, take the maximum of the two ordinals.) It can easily be seen that the ordinal of a proof (of the kind we are considering) is less than $\omega_k(l)$ for some natural number l.

A boundary inference is defined to be an inference which introduces a quantifier and is a boundary inference in the previous sense. A suitable cut is a cut whose cut formula contains quantifiers and which is suitable in the previous sense. In eliminating initial sequents from the end-piece, one eliminates only those which have quantifiers. The existence of a suitable cut (under certain conditions) can be proved just as before.

(4) In an essential reduction, if the suitable cut is of grade >0, then we can proceed as before (Step 5 in the proof of Lemma 12.8). If its grade is 0, then the cut formula is either of the form $\forall x\, F(x)$ or $\exists x\, F(x)$, where F is quantifier-free. Let us take the first case as an example. Let $F(s)$ be the auxiliary formula of a boundary inference which is an ancestor of the cut formula $\forall x\, F(x)$. s is a closed term, and so either $\rightarrow F(s)$ or $F(s)\rightarrow$ is a mathematical initial sequent (with ordinal 0). Suppose $\rightarrow F(s)$ is a mathematical initial sequent. Consider the proof:

$$
\dfrac{\begin{array}{c}\vdots\\ \Gamma \rightarrow \Theta, \forall x\, F(x)\end{array} \qquad \begin{array}{c}\dfrac{\dfrac{\rightarrow F(s) \qquad \begin{array}{c}\vdots\\ F(s), \Pi' \rightarrow \Lambda'\end{array}}{\Pi' \rightarrow \Lambda'}}{\forall x\, F(x), \Pi' \rightarrow \Lambda'}\\ \vdots\\ \forall x\, F(x), \Pi \rightarrow \Lambda\end{array}}{\begin{array}{c}\Gamma, \Pi \rightarrow \Theta, \Lambda\\ \vdots\\ \rightarrow\end{array}}
$$

(taking $\Gamma, \Pi \rightarrow \Theta, \Lambda$ as the sequent $\Delta \rightarrow \Xi$ shown in Lemma 12.8, Step 5). It is easy to see that the ordinal decreases again.

REMARK 12.15. Here we define an extended notion of primitive recursiveness. Let $<\cdot$ be a primitive recursive well-ordering of natural numbers. The class of $<\cdot$-primitive recursive functions is defined as the class of functions f generated by the following schemata:

(i) $f(a) = a+1$,

(ii) $f(a_1, \ldots, a_n) = 0$,

(iii) $f(a_1, \ldots, a_n) = a_i (1 \leq i \leq n)$,

(iv) $f(a_1, \ldots, a_n) = g(h_1(a_1, \ldots, a_n), \ldots, h_m(a_1, \ldots, a_n))$, where g and $h_i (1 \leq i \leq m)$ are $<\cdot$-primitive recursive.

(v) $f(0, a_2, \ldots, a_n) = g(a_2, \ldots, a_n)$,
$f(a+1, a_2, \ldots, a_n) = h(a, f(a, a_2, \ldots, a_n), a_2, \ldots, a_n)$,
where g and h are $<\cdot$-primitive recursive.

(vi) (Definition by $<\cdot$-recursion.)

$$
f(a_1, \ldots, a_n) = \begin{cases} h(f(\tau(a_1, \ldots, a_n), a_2, \ldots, a_n), a_1, \ldots, a_n) \\ \qquad\qquad\qquad \text{if} \quad \tau(a_1, \ldots, a_n) <\cdot\ a_1, \\ g(a_1, \ldots, a_n) \quad \text{otherwise,} \end{cases}
$$

where g, h and τ are $<\cdot$-primitive recursive.

The idea of (vi) is that $f(a, a_2, \ldots, a_n)$ is defined either outright or in terms of $f(b, a_2, \ldots, a_n)$ for certain $b <\cdot\ a$.

The consistency proof for $\mathbf{PA}_k$ which has just been presented has the following application.

COROLLARY 12.16. *Suppose R is a primitive recursive predicate and there is a proof of* $\rightarrow \exists x\, R(a, x)$ *in* $\mathbf{PA}_k$, *with ordinal* $< \omega_k(l)$ *for some numbers k and l* (as defined just above Definition 12.6). *Then the number-theoretic function f defined by*

$$f(m) = \text{the least } n \text{ such that } R(m, n)$$

is $<\cdot$*-primitive recursive, where* $<\cdot$ *is the initial segment of the standard ordering of* ε_0, *of order type* $\omega_k(l)$.

PROOF. We divide the proof into steps.

(i) Let $P(a)$ be a proof in $\mathbf{PA}_k$ of $\rightarrow \exists x\, R(a, x)$ (where all occurrences of a are indicated). Then for all m, $P(\bar{m})$ is a proof in $\mathbf{PA}_k$ of $\rightarrow \exists x\, R(\bar{m}, x)$ with the same ordinal, and with Gödel number primitive recursive in m. Also note that $\rightarrow \exists x\, R(\bar{m}, x)$ satisfies property P of Problem 12.11.

(ii) We (temporarily) call a proof *reducible* if it is a proof in $\mathbf{PA}_k$, with ordinal $< \omega_k(l)$, containing an essential cut or ind, and with end-sequent satisfying P. If P is reducible, then by applying repeatedly the reduction procedure of Lemma 12.8 (modified for $\mathbf{PA}_k$ as in Remark 12.14), we obtain a proof in $\mathbf{PA}_k$ of the same sequent, without an essential cut or ind. Let r be the function such that if p is a Gödel number of a reducible proof, then $r(p)$ is the Gödel number of the proof obtained by applying this reduction procedure (once), otherwise $r(p) = p$. Clearly r is primitive recursive.

Let O be the reduction such that if p is a Gödel number of a proof in $\mathbf{PA}_k$ with ordinal $< \omega_k(l)$, then $O(p)$ is the Gödel number of its ordinal (and, say $O(p) = 0$ otherwise). Clearly O is primitive recursive. Note also that for all p, $O(r(p)) <\cdot\, O(p) \Leftrightarrow p$ is the Gödel number of a reducible proof.

(iii) Now given a proof P of $\rightarrow \exists x\, R(\bar{m}, x)$ without an essential cut or ind, we can effectively find from P a number n such that $R(m, n)$ holds (and in fact the least such n). This is done in the following way.

First, we may assume that no free variables appear in P. Hence if $\Gamma \rightarrow \Delta$ is a sequent in P, every formula in Γ is a closed atomic formula and every formula in Δ is either $\exists x\, R(\bar{m}, x)$ or a closed atomic formula.

Now consider the following property Q of sequents: Every atomic formula in the antecedent is true and every atomic formula in the succedent is false.

Notice that the end-sequent of P satisfies Q; and if the lower sequent of a cut in P satisfies Q, then so does one upper sequent (since the cut formula is closed and atomic). Now start to construct a thread of sequents in P satisfying Q, working from the bottom upwards: the end-sequent is in the thread, and if the lower sequent of an inference is in the thread, take an upper sequent which satisfies Q. Since no initial sequent of P satisfies Q, this procedure must stop before we reach an initial sequent. The only way for this to happen is in the following case:

$$\frac{\Gamma \rightarrow \Delta, R(\bar{m}, \bar{k})}{\Gamma \rightarrow \Delta, \exists x\, R(\bar{m}, x)}$$

where $R(\bar{m}, \bar{k})$ is true. Finally, take the least $n \leqslant k$ for which $R(m, n)$ holds. Clearly there is a primitive recursive function h such that if P is a proof of $\rightarrow \exists x\, R(\bar{m}, x)$ without an essential cut or ind, then $h(\ulcorner P \urcorner)$ is the number n found as above.

(iv) Now we can define a $<\cdot$-primitive recursive function g such that if P is a proof of $\rightarrow \exists x\, R(\bar{m}, x)$ in $\mathbf{PA}_k$, with ordinal $< \omega_k(l)$, then $g(\ulcorner P \urcorner) =$ the least n such that $R(m, n)$ holds:

$$g(p) = \begin{cases} g(r(p)) & \text{if} \quad O(r(p)) <\cdot O(p), \\ h(p) & \text{otherwise.} \end{cases}$$

Then it is easily seen that g is $<\cdot$-primitive recursive function.

(v) Finally, let $P(a)$ be a proof of $\rightarrow \exists x\, R(a, x)$ in $\mathbf{PA}_k$, with ordinal $< \omega_k(l)$ as stated. Then we define f by:

$$f(m) = g(\ulcorner P(\bar{m}) \urcorner).$$

As a special case of Corollary 12.16 we have: if $\rightarrow \exists x\, R(x, a)$ is provable within the system whose induction formulas have at most one quantifier, then f (defined as above) is primitive recursive (by a theorem of R. Peter that ω^l-primitive recursiveness implies primitive recursiveness for any finite l).

The following corollary is a more precise statement of Corollary 12.16.

COROLLARY 12.17. *In the same situation of Corollary* 12.16 *there exists a function g_0 and primitive recursive functions h, r, s, such that g_0 is defined by*

$$g_0(x) = \begin{cases} g_0(r(x)) & \text{if } O(r(x)) <\cdot O(x) \\ x & \text{otherwise} \end{cases}$$

and

$$f(x) = h(g_0(s(x))),$$

where $<\cdot$ is the initial segment of the standard ordering of ε_0, of order type $\omega_k(l)$.

PROOF. Let g, h, r and P be same as in the proof of Corollary 12.16. In the proof of Corollary 12.16 we used different reduction and ordinal assignment from Gentzen's original reduction and ordinal assignment. In this proof, we can also use Gentzen's original reduction and ordinal assignment. Define s by $s(x) = \ulcorner P(\bar{x}) \urcorner$. Then $f(x) = g(s(x))$. It is easily seen that $g(x) = h(g_0(x))$.

DEFINITION 12.18. A function f is provably recursive in **PA** if there exist a primitive recursive predicate R and a primitive recursive function g such that $\forall x_1 \ldots \forall x_n \exists y R(x_1, \ldots, x_n, y)$ is provable in **PA** and f satisfies

$$f(a_1, \ldots, a_n) = g(\mu y R(a_1, \ldots, a_n, y)),$$

where $\mu y R(a_1, \ldots, a_n, y)$ is the least y such that $R(a_1, \ldots, a_n, y)$. (See also the definition immediately before Problem 13.8.)

Corollary 12.16 gives a characterization of the provably recursive functions in **PA** by $<\cdot$-primitive recursive functions, where $<\cdot$ is some proper initial segment of the standard ordering of ε_0.

S. S. Wainer gave a finer characterization of the provably recursive functions in **PA** by the use of Hardy functions. For the development of the theory of Hardy functions we need a formal treatment of ordinals less than ε_0. In the rest of this section by an ordinal we mean an ordinal less than ε_0 and let $\alpha, \beta, \ldots$ be ordinals.

DEFINITION 12.19. (1) If $\omega^{\alpha_1} + \ldots + \omega^{\alpha_n}$ is the normal form of α (namely $\alpha_1 \geqslant \ldots \geqslant \alpha_n$) and $\omega^{\beta_1} + \ldots + \omega^{\beta_m}$ is the normal form of β and $\alpha_n \geqslant \beta_1$, then we use $\alpha \dot{+} \beta$ to express $\alpha + \beta$. Whenever we use the expression $\alpha \dot{+} \beta$, we assume $\alpha_n \geqslant \beta_1$. We also use $\alpha \dot{+} \beta$ when α is empty. If α is empty, then $\alpha \dot{+} \beta$ is β itself.

(2) If α is a limit ordinal, we define a fixed fundamental sequence of α, $\{\alpha\}(0), \{\alpha\}(1), \{\alpha\}(2), \ldots$ as follows.

(i) If α is of the form $\beta \dot{+} \omega^{\gamma+1}$, then $\{\alpha\}(n) = \beta \dot{+} \omega^{\gamma} \cdot n$.

(ii) If α is of the form $\beta \dot{+} \omega^{\gamma}$ and γ is a limit ordinal, then $\{\alpha\}(n) = \beta \dot{+} \omega^{\{\gamma\}(n)}$.

PROPOSITION 12.20. (1) *If $\alpha > 1$ and $n > 0$, then $\{\omega^{\alpha}\}(n)$ is a limit ordinal.*

(2) *If α is a limit ordinal and x is a positive natural number, then there exist limit ordinals $\alpha = \alpha_1 > \alpha_2 > \ldots > \alpha_k$ such that α_{i+1} $(1 \leqslant i \leqslant k)$ is either $\{\alpha_i\}(0)$ or $\{\alpha_i\}(x)$ and $\alpha_{k+1} = 0$.*

(3) *Let α be a limit ordinal and not of the form $\alpha_0 \dot{+} \omega$. If j and x are positive natural numbers, then there exist limit ordinals*

$$\{\alpha\}(j) = \alpha_1 > \alpha_2 > \ldots > \alpha_k,$$

such that α_{i+1} $(1 \leqslant i \leqslant k)$ is either $\{\alpha_i\}(0)$ or $\{\alpha_i\}(x)$ and $\alpha_{k+1} = \{\alpha\}(j-1)$.

PROOF. (1) This is proved by induction on α. We may assume $\alpha = \omega^\beta$. If $\beta = \beta_0 + 1$ and $\beta_0 \neq 0$, then $\{\omega^{\beta_0+1}\}(n) = \omega^{\beta_0} \cdot n$. If β is a limit ordinal and not of the form $\beta_1 \dot{+} \omega$, then $\{\alpha\}(n) = \omega^{\{\beta\}(n)}$ and $\{\beta\}(n)$ is a limit ordinal by the induction hypothesis. If $\beta = \beta_1 \dot{+} \omega$, then $\{\alpha\}(n) = \omega^{\beta_1 \dot{+} n}$.

(2) This is proved by induction on α. If α is of the form $\alpha_0 \dot{+} \omega$, then the problem is reduced to $\alpha_0 = \{\alpha\}(0)$. If α is not of the form $\alpha_0 \dot{+} \omega$, then by (1) the problem is reduced to $\{\alpha\}(x)$.

(3) This is proved by induction on α. We may assume that $\alpha = \omega^\beta$ and $\beta > 1$. If $\beta = \beta_0 + 1$, then $\{\alpha\}(j) = \omega^{\beta_0} \cdot j$ and $\{\alpha\}(j-1) = \omega^{\beta_0} \cdot (j-1)$. If β_0 is a successor ordinal, then $\{\alpha\}(j-1) = \{\{\alpha\}(j)\}(0)$. If β_0 is a limit ordinal and not of the form $\beta_1 \dot{+} \omega$, then by the induction hypothesis $\{\beta_0\}(j-1)$ is obtained from $\{\beta_0\}(j)$ by successive applications of $\{\ \}(0)$ and $\{\ \}(x)$. Therefore $\{\alpha\}(j-1) = \omega^{\{\beta_0\}(j-1)}$ is obtained from $\{\alpha\}(j) = \omega^{\{\beta_0\}(j)}$ by successive applications of $\{\ \}(0)$ and $\{\ \}(x)$. Finally let $\alpha = \omega^{\beta_1 \dot{+} \omega}$. Then $\{\alpha\}(j) = \omega^{\beta_1 \dot{+} j}$ and $\{\alpha\}(j-1) = \omega^{\beta_1 \dot{+} j-1}$. Then $\{\{\alpha\}(j)\}(x) = \omega^{\beta_1 \dot{+} (j-1)} \cdot x$. If $x = 1$, then $\{\alpha\}(j-1) = \{\{\alpha\}(j)\}(x)$. If $x > 1$, then by (2) $\{\alpha\}(j-1)$ is obtained from $\{\{\alpha\}(j)\}(x)$ by successive applications of $\{\ \}(0)$ and $\{\ \}(x)$.

DEFINITION 12.21. (1) The Hardy function h_α is defined by induction on α as follows.

$$h_0(x) = x$$
$$h_{\beta+1}(x) = h_\beta(x+1)$$
$$h_\alpha(x) = h_{\{\alpha\}(x)}(x) \text{ if } \alpha \text{ is a limit ordinal.}$$

(2) For each unary function f, f^n denotes the nth iterate of f, defined by $f^0(x) = x$, $f^{m+1}(x) = f(f^m(x))$.

(3) A k-ary function g is said to be majorized by a unary function f if there is a number n such that

$$g(x_1, \ldots, x_k) < f(\max(x_1, \ldots, x_k)),$$

whenever $\max(x_1, \ldots, x_k) \geqslant n$.

LEMMA 12.22. *If α is a limit ordinal, then $h_{\alpha \dot{+} \beta}(x) = h_\alpha(h_\beta(x))$.*

PROOF. This is easily proved by induction on β. From this lemma, the following equations can be seen immediately.

$$h_{\omega^0}(x) = h_1(x) = x + 1$$
$$h_{\omega^{\beta+1}}(x) = h^x_{\omega^\beta}(x)$$
$$h_{\omega^\alpha}(x) = h_{\omega^{\{\alpha\}(x)}}(x) \text{ if } \alpha \text{ is a limit ordinal.}$$

LEMMA 12.23. *For every* α,

(1) h_α *is strictly increasing.*

(2) *If* α *is a limit ordinal and* $i < j \leq x$, *then* $h_{\{\alpha\}(i)}(x) \leq h_{\{\alpha\}(j)}(x)$.

(3) *If* $\beta < \alpha$, *then* h_β *is majorized by* h_α.

PROOF. This is proved by induction on α. If $\alpha = 0$, then the lemma is obvious. Therefore we assume that for every $\alpha_0 < \alpha$ the lemma holds. Now we assume (2) and prove (1) and (3). If α is a successor ordinal, then (1) and (3) are obvious by the induction hypothesis. Now let α be a limit ordinal, then (1) is proved as follows:

$$h_\alpha(x) = h_{\{\alpha\}(x)}(x) < h_{\{\alpha\}(x)}(x+1) \leq h_{\{\alpha\}(x+1)}(x+1) = h_\alpha(x+1),$$

where the strict inequality holds by the induction hypothesis and the second inequality holds by (2).

Now we prove (3) from (2) under the hypothesis that α is a limit ordinal. If $\beta < \alpha$, then there exists i such that $\beta < \{\alpha\}(i)$. Then h_β is majorized by $h_{\{\alpha\}(i)}$. If $x > i$, then $h_{\{\alpha\}(i)}(x) \leq h_{\{\alpha\}(x)}(x) = h_\alpha(x)$ by (2).

Now we prove (2). We may assume $i = j - 1$. If α is $\alpha_1 \dot{+} \omega$, then

$$\begin{aligned} h_{\{\alpha\}(j)}(x) &= h_{\alpha_1 + j}(x) = h_{\alpha_1 + (j-1)}(x+1) \\ &> h_{\alpha_1 + (j-1)}(x) = h_{\{\alpha\}(i)}(x). \end{aligned}$$

If α is not of the form $\alpha_1 \dot{+} \omega$, then by Proposition 12.20 $\{\alpha\}(j-1)$ is obtained from $\{\alpha\}(j)$ by successive applications of $\{\ \}(0)$ and $\{\ \}(x)$. We denote this series of ordinals by

$$\{\alpha\}(j) = \alpha_0 > \alpha_1 > \ldots > \alpha_n = \{\alpha\}(i),$$

where α_{k+1} is either $\{\alpha_k\}(0)$ or $\{\alpha_k\}(x)$. It suffices to show $h_{\alpha_{k+1}}(x) \leq h_{\alpha_k}(x)$. If $\alpha_{k+1} = \{\alpha_k\}(0)$, then by the induction hypothesis

$$h_{\alpha_{k+1}}(x) = h_{\{\alpha_k\}(0)}(x) \leq h_{\{\alpha_k\}(x)}(x) = h_{\alpha_k}(x).$$

If $\alpha_{k+1} = \{\alpha_k\}(x)$, then

$$h_{\alpha_{k+1}}(x) = h_{\{\alpha_k\}(x)}(x) = h_{\alpha_k}(x).$$

COROLLARY 12.24. *If* $\alpha > 0$ *and* $x > 0$, *then* $h_\alpha(x) > x$.

PROOF. Is easily done by induction on α.

LEMMA 12.25. *If* α *is a limit ordinal,* $\delta = \omega^{\delta_0}$ *and* $\omega^\delta > \alpha$, *then*

$$\omega^\delta \cdot \{\alpha\}(n) = \{\omega^\delta \cdot \alpha\}(n).$$

PROOF. This is proved by induction on α. Let $\alpha=\omega^{\alpha_1}+\ldots+\omega^{\alpha_n}$ be a normal form. Then $\omega^{\delta}\cdot(\omega^{\alpha_1}+\ldots+\omega^{\alpha_n})=\omega^{\delta+\alpha_1}+\ldots+\omega^{\delta+\alpha_n}$. The lemma is now obvious.

DEFINITION 12.26. The *complexity* $c(\alpha)$ of α is defined as follows.

$$c(0)=0,\qquad c(\omega^{\alpha})=c(\alpha)+1,\qquad c(\alpha\dot{+}\beta)=c(\alpha)+c(\beta).$$

LEMMA 12.27. *Let β be a limit ordinal, $\alpha<\beta$ and $c(\alpha)<n$. Then $\alpha<\{\beta\}(n)$.*

PROOF. Let $\alpha=\delta\dot{+}(\omega^{\alpha_1}\cdot a_1+\ldots+\omega^{\alpha_s}a_s)$ $(\alpha_1>\ldots>\alpha_s)$; $\beta=\delta+(\omega^{\beta_1}\cdot b_1+\ldots+\omega^{\beta_r}\cdot b_r)$ $(\beta_1>\ldots>\beta_r)$ and $\omega^{\beta_1}\cdot b_1>\omega^{\alpha_1}\cdot a_1$. Then $\{\beta\}(n)\geqslant\delta\dot{+}\omega^{\beta_1}(b_1-1)\dot{+}\{\omega^{\beta_1}\}(n)$.

Case (1). $\beta_1>\alpha_1$.

If β_1 is a successor ordinal, then $\{\omega^{\beta_1}\}(n)\geqslant\omega^{\alpha_1}\cdot n$. Since $n>c(\alpha)\geqslant a_1$, $\{\omega^{\beta_1}\}(n)\geqslant\omega^{\alpha_1}\cdot(a_1+1)$. If β_1 is a limit ordinal and $\alpha=\delta$ (i.e., $a_1=0$), then this is obvious since $\{\omega^{\beta_1}\}(n)>0$ by Proposition 12.20. If β_1 is a limit ordinal and $a_1>0$, then $\{\beta_1\}(n)>\alpha_1$ by the induction hypothesis since $n>c(\alpha_1)$. Therefore $\{\omega^{\beta_1}\}(n)=\omega^{\{\beta_1\}(n)}>\omega^{\alpha_1}$.

Case (2). $\beta_1=\alpha_1$ and $b_1>a_1$.

If β_1 $(=\alpha_1)$ is a successor ordinal, then $\omega^{\beta_1}(b_1-1)\geqslant\omega^{\alpha_1}\cdot a_1$ and $\{\omega^{\beta_1}\}(n)>\omega^{\alpha_2}\cdot a_2+\ldots+\omega^{\alpha_s}a_s$ since

$$n>c(\alpha)>c(\omega^{\alpha_2}\cdot a_2+\ldots+\omega^{\alpha_s}\cdot a_s).$$

If β_1 $(=\alpha_1)$ is a limit ordinal and $a_2=a_3=\ldots=a_s=0$, then

$$\begin{aligned}\{\beta\}(n)&\geqslant\delta\dot{+}\omega^{\alpha_1}\cdot a_1+\{\omega^{\alpha_1}\}(n)\\&>\delta\dot{+}\omega^{\alpha_1}\cdot a_1=\alpha.\end{aligned}$$

If β_1 $(=\alpha_1)$ is a limit ordinal and $a_2>0$, then $\{\beta_1\}(n)>\alpha_2$ since $n>c(\alpha_2)$. Therefore $\{\beta\}(n)\geqslant\delta+\omega^{\alpha_1}\cdot a_1+\omega^{\{\alpha_1\}(n)}>\alpha$.

LEMMA 12.28. *Let $\delta=\omega^{\delta_0}$, $\alpha<\beta<\omega^{\delta}$ and $c(\alpha)<x$. Then $h_{\omega^{\delta}\cdot\alpha}(x)<h_{\omega^{\delta}\cdot\beta}(x)$.*

PROOF. This is proved by induction on β. If $\beta=\beta_0+1$, then

$$h_{\omega^{\delta}\cdot\alpha}(x)\leqslant h_{\omega^{\delta}\cdot\beta_0}(x)<h_{\omega^{\delta}\cdot\beta_0}(h_{\omega^{\delta}}(x))=h_{\omega^{\delta}\cdot\beta}(x).$$

If β is a limit ordinal, then by Lemma 12.27 $c(\alpha)<x$ implies $\alpha<\{\beta\}(x)$. Therefore by the induction hypothesis and Lemma 12.25

$$h_{\omega^{\delta}\cdot\alpha}(x)<h_{\omega^{\delta}\cdot\{\beta\}(x)}(x)=h_{\{\omega^{\delta}\cdot\beta\}(x)}(x)=h_{\omega^{\delta}\cdot\beta}(x).$$

LEMMA 12.29. (1) *If* $f(x_1, \ldots, x_m)$ *and* $g_1(x_1, \ldots, x_n), \ldots, g_m(x_1, \ldots, x_n)$ *are majorized by* h_{ω^α}, *then* $f(g_1(x_1, \ldots, x_n), \ldots, g_m(x_1, \ldots, x_n))$ *is majorized by* $h_{\omega^\alpha \cdot 2}$.

(2) *If* $h(x)$ *and* $g(x, y, z)$ *are majorized by* h_{ω^α} *and* $f(x, y)$ *is obtained by a primitive recursion from* h *and* g, *namely*, $f(0, y) = h(y)$ *and* $f(x+1, y) = g(x, y, f(x, y))$, *then* $f(x, y)$ *is majorized by* $h_{\omega^{\alpha+1}+1}$.

PROOF. First $f(x_1, \ldots, x_n)$ is majorized by h_α iff there exists p such that for every $x_1, \ldots, x_n$

$$f(x_1, \ldots, x_n) < \max(h_\alpha(\max(x_1, \ldots, x_n)), p).$$

(1) Let $x = \max(x_1, \ldots, x_n)$ and p be such that for every $y_1, \ldots, y_m$ and $x_1, \ldots, x_n$,

$$\begin{aligned} f(y_1, \ldots, y_m) &\leq \max(h_{\omega^\alpha}(\max(y_1, \ldots, y_n)), p) \\ g_1(x_1, \ldots, x_n) &\leq \max(h_{\omega^\alpha}(x), p) \\ &\vdots \\ g_m(x_1, \ldots, x_n) &\leq \max(h_{\omega^\alpha}(x), p). \end{aligned}$$

Then

$$\begin{aligned} &f(g_1(x_1, \ldots, x_n), \ldots, g_m(x_1, \ldots, x_n)) \\ &\quad \leq \max(h_{\omega^\alpha}(\max(h_{\omega^\alpha}(x), p)), p) \\ &\quad \leq \max(h_{\omega^\alpha} \cdot h_{\omega^\alpha}(x)), p') = \max(h_{\omega^\alpha \cdot 2}(x), p'), \end{aligned}$$

where $p' = h_{\omega^\alpha}(p)$.

(2) Let p satisfy the following inequality for every x, y, and z,

$$\begin{aligned} h(x) &< \max(h_{\omega^\alpha}(x), p) \\ g(x, y, z) &< \max(h_{\omega^\alpha}(\max(x, y, z)), p). \end{aligned}$$

We prove the following inequality by induction on x,

$$f(x, y) < h_{\omega^\alpha}^{x+1}(\max(x, y, p)).$$

If $x = 0$, then the inequality is obvious. The case for $x+1$ is proved as follows.

$$\begin{aligned} f(x+1, y) &< \max(h_{\omega^\alpha}(\max(x, y, h_{\omega^\alpha}^{x+1}(\max(x, y, p)), p), p) \\ &\leq h_{\omega^\alpha} \cdot h_{\omega^\alpha}^{x+1}(\max(x, y, p)) \\ &\leq h_{\omega^\alpha}^{x+2}(\max(x+1, y, p)). \end{aligned}$$

Since $h^{x+1}_{\omega^\alpha} = h_{\omega^\alpha \cdot (x+1)}$, (2) is proved as follows.

$$\begin{aligned} f(x, y) &< h_{\omega^\alpha \cdot (x+1)}(\max(x, y, p)) \\ &\leq h_{\omega^{\alpha+1}}(\max(x, y, p) + 1) \\ &\leq h_{\omega^{\alpha+1}+1}(\max(x, y, p)). \end{aligned}$$

COROLLARY 12.30. *All primitive recursive functions are majorized by* h_{ω^ω}.

DEFINITION 12.31. The Hardy class $\mathscr{H}$ is the smallest class of functions containing 0, all h_α, all projection functions $(I_{n,i}(x_1, \ldots, x_n) = x_i)$, and closed under primitive recursion and substitution.

By Lemma 12.23 and Lemma 12.29, every function in $\mathscr{H}$ is majorized by $h_{\omega_n}(n)$.

From now on, we fix a representation of ordinals less than ε_0 by natural numbers. We assume that the natural number 0 does not represent any ordinal. The following theorem is easily seen from our discussion in §11. However we give here another formal proof which is also given by Gentzen.

THEOREM 12.32. *Let n be a natural number. Then transfinite induction up to* ω_n *is provable in* **PA**.

PROOF. Let $<$ express the order of ordinals in this proof. Then we shall show that

$$(*) \qquad \forall\alpha\,(\forall\beta < \alpha A(\beta) \supset A(\alpha)) \to \forall\alpha \leq \omega_n A(\alpha)$$

Is provable in **PA** for every formula $A(\alpha)$ in **PA**. We show this by induction on n. The case $n = 1$ is obvious since **PA** has mathematical induction. Now we assume that (*) for n is provable in **PA**. Let $A^*(\alpha)$ and φ denote $\forall\beta \leq \alpha A(\beta)$ and $\forall\alpha\,(\forall\beta < \alpha A(\beta) \supset A(\alpha))$ respectively. Then it is easily seen that $\varphi \to \forall\alpha\,(\forall\beta < \alpha A^*(\beta) \supset A^*(\alpha))$. Let $\psi(\alpha)$ denote $\forall\gamma\,(A^*(\gamma) \supset A^*(\gamma + \omega^\alpha))$. Let $\alpha = \alpha_0 + 1$. Then it is easily seen that

$$\forall\gamma < \alpha\psi(\gamma), \quad A^*(\beta + \omega^{\alpha_0} \cdot a) \to A^*(\beta + \omega^{\alpha_0} \cdot (a + 1)).$$

By mathematical induction on a, we have

$$\forall\gamma < \alpha\psi(\gamma), \quad A^*(\beta) \to A^*(\beta + \omega^{\alpha_0} \cdot a).$$

Therefore we have

$$\forall\gamma < \alpha\psi(\gamma), \quad A^*(\beta) \to \forall\gamma < \beta + \omega^\alpha A^*(\gamma)$$

and $\varphi, \forall\gamma < \alpha\psi(\gamma), A^*(\beta) \rightarrow A^*(\beta + \omega^\alpha)$
and $\varphi, \forall\gamma < \alpha\psi(\gamma) \rightarrow \psi(\alpha)$.

It is also easily seen that

$$\mathrm{Lim}(\alpha), \forall\gamma < \alpha\psi(\gamma) \rightarrow \psi(\alpha),$$

where $\mathrm{Lim}(\alpha)$ is a formula expressing "α is a limit ordinal". Hence we have $\forall\gamma < \alpha\psi(\gamma) \rightarrow \psi(\alpha)$. By the induction hypothesis we have $\psi(\omega_n)$ namely $\forall\gamma(A^*(\gamma) \supset A^*(\gamma + \omega^n))$. Therefore we have $A^*(0) \rightarrow A^*(\omega_{n+1})$. Since $\varphi \rightarrow A^*(0)$ is provable, we have $\varphi \rightarrow A^*(\omega_{n+1})$.

COROLLARY 12.33. *For every α, Hardy function h_α is provably recursive in* **PA**.

LEMMA 12.34. *Let f be provably recursive in* **PA**. *Then there exist h_α and a primitive recursive function $u(x)$ such that for every x, $f(x) < h_\alpha(u(x))$.*

PROOF. By Corollary 12.17 and Lemma 12.29 we may assume that $f(x)$ is $g_0(x)$ in Corollary 12.17 namely

$$f(x) = \begin{cases} f(r(x)) & \text{if } O(r(x)) < O(x) < \omega_n \\ x & \text{otherwise,} \end{cases}$$

where $r(x)$ is the Gödel number of the result of Gentzen's reduction of x, if x is the Gödel number of a proof, and $O(x)$ is the ordinal assigned to x. The usual assignment of ordinal satisfies the condition $x > c(O(x))$ if x is the Gödel number of a proof. We assume that this condition is satisfied. We also assume that the ordinal of the proof of $\rightarrow \exists x\, R(a, x)$ is smaller than ω_n. We define an ordinal $|x|$ as follows.

$$|x| = \begin{cases} O(x) & \text{if } O(r(x)) < O(x) < \omega_n \\ x & \text{otherwise.} \end{cases}$$

Since $r(x)$ is primitive recursive, there exists b such that $b \geqslant 2$ and

$$\max(x, \max(r(x), r(r(x)), y) + y) \geqslant h_{\omega_n}(\max(x, y)) + b.$$

We define $u(x)$ as follows.

$$u(x) = \max(x, r(x), b) + b.$$

Then we have

$$\begin{aligned}\max(x, u(r(x))) &= \max(x, \max(r(x), r(r(x)), b) + b)\\ &\leq h_{\omega_n}(\max(x, b)) + b\\ &\leq h_{\omega_n}(\max(x, b) + b)\\ &\leq h_{\omega_n}(u(x)).\end{aligned}$$

By induction on $|x|$, we prove

$$f(x) \leq h_{\omega_n \cdot |x|}(u(x)).$$

If $|x| = 0$, then $f(x) = x \leq u(x) = h_0(u(x))$. So we assume that $|x| > 0$ and the inequality holds for y satisfying $|y| < |x|$. Then we have

$$\begin{aligned}f(x) &\leq \max(x, f(r(x)))\\ &\leq \max(x, h_{\omega_n \cdot |r(x)|}(u(r(x))))\\ &\leq h_{\omega_n \cdot |r(x)|}(\max(x, u(r(x))))\\ &\leq h_{\omega_n \cdot |r(x)|}(h_{\omega_n}(u(x)))\\ &\leq h_{\omega_n \cdot (|r(x)|+1)}(u(x))\\ &\leq h_{\omega_n \cdot |x|}(u(x)).\end{aligned}$$

The last inequality holds since

$$c(|r(x)| + 1) \leq c(|r(x)|) + 1 \leq r(x) < u(x).$$

Since $c(|x|) < x < u(x)$, we have

$$f(x) < h_{\omega_n \cdot \omega_n}(u(x)).$$

THEOREM 12.35 (Wainer). *$\mathcal{H}$ is the class of all provably recursive functions in* **PA**.

PROOF. It suffices to show that the following function f is in $\mathcal{H}$.

$$f(x) = \begin{cases} f(r(x)) & \text{if } 0(r(x)) < 0(x) < \omega_n \\ x & \text{otherwise.} \end{cases}$$

We define primitive recursive functions $t(x)$ and $g(x, y)$ as follows.

$$t(x) = \begin{cases} r(x) & \text{if } 0(r(x)) < 0(x) < \omega_n \\ x & \text{otherwise.} \end{cases}$$

$$q(0, x) = x, \qquad q(y + 1, x) = t(q(y, x)).$$

Then $p(x) = \mu y(q(y+1, x) = q(y, x))$ is provably recursive in **PA**. Therefore there exist β and a primitive recursive function $u(x)$ such that $p(x) < h_\beta(u(x))$. Then $p(x) = \mu y(y < h_\beta(u(x)) \wedge q(y+1, x) = q(y, x))$ and $p(x)$ is in $\mathcal{H}$. Since $f(x) = q(p(x), x)$, $f(x)$ is also in $\mathcal{H}$.

Now we are going to discuss Kirby and Paris' result on Goodstein's Theorem.

DEFINITION 12.36. Let m and n be natural numbers, $n > 1$. We define the *pure base n* representation of m as follows. First write m as the sum of powers of n. For example, if $m = 26$, $n = 2$, write $26 = 2^4 + 2^3 + 2$. Now write each exponent as a sum of powers of n. Repeat with exponents of exponents and so on until the representation stabilizes. For example the pure base 2 representation of 26 is $2^{2^2} + 2^{2+1} + 2$.

We not define the Goodstein number $g_n(m)$ as follows. If $m = 0$ set $g_n(m) = 0$. Otherwise set $g_n(m)$ to be the number produced by replacing every n in the pure base n representation of m by $n+1$ and then subtracting 1. For example $g_2(26) = 3^{3^3} + 3^{3+1} + 3 - 1$. Now define the Goodstein sequence for m by

$$m_0 = m, \qquad m_k = g_{k+1}(m_{k-1}) \quad (k > 1).$$

So, for example,

$$26_0 = 26 = 2^{2^2} + 2^{2+1} + 2,$$
$$26_1 = 3^{3^3} + 3^{3+1} + 3 - 1 = 3^{3^3} + 3^{3+1} + 2.$$
$$26_2 = 4^{4^4} + 4^{4+1} + 2 - 1 = 4^{4^4} + 4^{4+1} + 1.$$
$$26_3 = 5^{5^5} + 5^{5+1} + 1 - 1 = 5^{5^5} + 5^{5+1}.$$

The general Goodstein sequence $m_{n,k}$ for $n \geqslant 2$ is defined as follows.

$$m_{n,0} = m, \qquad m_{n,k} = g_{n+k-1}(m_{n,k-1}) \quad (k > 1).$$

Obviously $m_k = m_{2,k}$.

THEOREM 12.37. (1) (Goodstein) $\forall m \forall n \geqslant 2 \exists k\, m_{n,k} = 0$. *Namely the general Goodstein sequence* $m_{n,0}, m_{n,1}, m_{n,2}, \ldots$ *eventually terminates with* 0.

(2) (Kirby and Paris) *Goodstein's theorem for* $n = 2$ *namely* $\forall m \exists k\, m_k = 0$ *is not provable in* **PA**.

Our proof is due to Cichon. We need several preparation.

DEFINITION 12.38. Let $x < \omega$ and $\alpha < \varepsilon_0$.

(1) $G_x(\alpha)$ is defined as follows.

$$G_x(0)=0, \qquad G_x(\alpha+1)=G_x(\alpha)+1;$$
$$G_x(\alpha)=G_x(\{\alpha\}(x)) \quad \text{for a limit ordinal } \alpha.$$

(2) $P_x(\alpha)$ is defined as follows.

$$P_x(0)=0, \qquad P_x(\alpha+1)=\alpha;$$
$$P_x(\alpha)=P_x(\{\alpha\}(x)) \quad \text{for a limit ordinal } \alpha.$$

LEMMA 12.39. *For $x<\omega$ and $\alpha<\varepsilon_0$, the following equations hold.*
(1) $G_x(\alpha\dot{+}\beta)=G_x(\alpha)+G_x(\beta)$
(2) $G_x(\omega^\alpha)=x^{G_x(\alpha)}$.

PROOF. (1) is easily proved by induction on β. (2) is proved by induction on α. If α is 0 or a limit ordinal, then the equation is obvious. For $\alpha+1$,

$$G_x(\omega^{\alpha+1})=G_x(\omega^\alpha\cdot x)=G_x(\overbrace{\omega^\alpha+\ldots+\omega^\alpha}^{x})$$
$$=\underbrace{x^{G_x(\alpha)}+\ldots+x^{G_x(\alpha)}}_{x}=x^{G_x(\alpha)+1}=x^{G_x(\alpha+1)}.$$

COROLLARY 12.40. *$G_x(\alpha)$ is the result of replacing ω by x in the canonical normal form of α.*

LEMMA 12.41. *For $x<\omega$ and $\alpha<\varepsilon_0$, the following equation holds*

$$G_xP_x(\alpha)=P_xG_x(\alpha).$$

PROOF. Is immediate by induction on α.

Now let $n\geqslant 2$. We are going to form $m_{n,0}, m_{n,1}, m_{n,2},\ldots$ by using G_x and P_x. Let α be obtained from the pure base n expression of m by replacing n by ω. Then $m_{n,0}=m=G_n(\alpha)$.

$$m_{n,1}=G_{n+1}(\alpha)-1=P_{n+1}G_{n+1}(\alpha)=G_{n+1}P_{n+1}(\alpha),$$
$$m_{n,2}=G_{n+2}(P_{n+1}(\alpha))-1$$
$$=P_{n+2}G_{n+2}P_{n+1}(\alpha)$$
$$=G_{n+2}P_{n+2}P_{n+1}(\alpha),$$
$$m_{n,k}=G_{n+k}P_{n+k}P_{n+k-1}\ldots P_{n+1}(\alpha).$$

It is easily seen by induction on α that for $x\neq 0$, $G_x(\alpha)=0$ iff $\alpha=0$.

Therefore, (1) of Theorem 12.37 $\forall m\, \forall n \geq 2\, \exists k\; m_{n,k} = 0$ is implied by $\forall \alpha\, \forall n \geq 2\, \exists k\, P_{n+k} \ldots P_{n+1}(\alpha) = 0$ which is obvious since $\alpha > 0$ then $P_x(\alpha) < \alpha$.

LEMMA 12.42. *If* $0 < \alpha < \varepsilon_0$ *and* $1 \leq n < \omega$, *then* $(\mu x)(P_x P_{x-1} \ldots P_{n+1}(\alpha) = 0) = h_\alpha(n+1) - 1$, *where* $\mu x \varphi(x)$ *is the least* x *satisfying* $\varphi(x)$ *and* h_α *is a Hardy function.*

PROOF. If $\alpha = 1$, then the lemma is obvious. If $\alpha = \beta + 1$, then

$$(\mu x)(P_x P_{x-1} \ldots P_{n+1}(\beta + 1) = 0) = (\mu x)(P_x P_{x-1} \ldots P_{n+2}(\beta) = 0)$$
$$= h_\beta(n+2) - 1 = h_\alpha(n+1) - 1.$$

If α is a limit ordinal, then

$$(\mu x)(P_x P_{x-1} \ldots P_{n+1}(\alpha) = 0) = (\mu x)(P_x \ldots P_{n+2} P_{n+1}(\{\alpha\}(n+1)))$$
$$= h_{\{\alpha\}(n+1)}(n+1) - 1 = h_\alpha(n+1) - 1.$$

Now we prove (2) of Theorem 12.36. We define a_n by $a_0 = 1$, $a_{k+1} = 2^{a_k}$ and b_n by $b_n = a_n + a_{n-1} + \ldots + a_0$. Define $\alpha_n = \omega_n + \omega_{n-1} + \omega_{n-2} + \ldots + 1$. Then $G_2(\alpha_n) = b_n$. Then $\forall m\, \exists k\; m_k = 0$ implies that $h_{\alpha_n}(3)$ as a function of n is provably recursive in **PA**. However $h_{\alpha_n}(3) \geq h_{\omega_n}(n)$. Since $h_{\omega_n}(n)$ is not majorized by any h_α $(\alpha < \varepsilon_0)$, this is a contradiction.

DEFINITION 12.43. Let A be a set and m be a natural number. $A^{[m]}$ is the collection of subset of A of cardinality m. If $F: A^{[m]} \rightarrow X$, a subset B of A is homogeneous for F if F is constant on $B^{[m]}$. We identify a natural number n with the set $\{0, 1, \ldots, n-1\}$.

We state the following Ramsey's theorem without proof.

THEOREM 12.44. *Let* A *be an infinite set and* m *and* n *be natural numbers. Then for every function* $F: A^{[m]} \rightarrow n$, *there exists an infinite subset* B *of* A *which is homogeneous for* F.

DEFINITION 12.45. (1) Let S be a set of natural numbers. S is *large* if S is non-empty, and letting s be its least element, S has at least s elements.

(2) Let a, b, c be positive natural numbers. Then $a \rightarrow (\text{large})^b_c$, if for every map $F: a^{[b]} \rightarrow c$, there is a large homogeneous set for F of cardinality greater than b.

THEOREM 12.46. *Let* a, b, c *range over the natural numbers. Then the following holds*

$$\forall b, c \geq 1\, \exists a \geq 1 (a \rightarrow (\text{large})^b_c).$$

PROOF. Suppose the theorem were false. Then there exist fixed b, $c \geq 1$ such that

$$\forall a \geq 1(a \not\rightarrow (\text{large})^b_c).$$

Let T be the set of all F such that $F: a^{[b]} \rightarrow c$ for some a and there are no large homogeneous set for F of cardinality greater than b. For F_1, F_2 we define $F_1 \leq F_2$ if F_2 is an extension of F_1. Then T is a tree. Then the condition $\forall a \geq 1(a \not\rightarrow (\text{large})^b_c)$ implies that for every a, there exists a branch (a maximal linearly ordered subset) of T with length at least a. By König's lemma, there exists an infinite branch in T. Such a branch produces a function $F: \mathbb{N}^{[b]} \rightarrow C$ such that there are no large homogeneous sets for F of cardinality greater than b. This is a contradiction because of Ramsey's theorem.

Paris and Harrington proved that $\forall b, c \geq 1 \, \exists a \geq 1(a \rightarrow (\text{large})^b_c)$ is not provable in **PA**. Let $\sigma(b, c)$ be the least natural number a such that $a \rightarrow (\text{large})^b_c$. Paris and Harrington proved the following stronger theorem.

THEOREM (Paris, Harrington). *The function $\sigma(n, n)$ majorizes all provably recursive functions in* **PA**.

Ketonen and Solovay investigated upper and lower bounds of $\sigma(n, n)$. As a lower bound, they proved that $\sigma(n, 7)$ majorizes all provably recursive functions in **PA**. J. Quinsey improved the result by showing that $\sigma(n, 3)$ majorizes all provably recursive functions in **PA**. We shall discuss their results. We shall not discuss an upperbound of $\sigma(n, n)$. We simply note that one can get an upperbound of $\sigma(n, n)$ in the form of an ordinal recursive function by the method of §30 if one checks the second order system in which the proof of

$$\forall b, c \geq 1 \, \exists a \geq 1(a \rightarrow (\text{large})^b_c)$$

can be carried out.

First we discuss more on Ramsey's theorem. In his paper: Ramsey's Theorem and Recursion Theory, The Journal of Symbolic Logic, Vol. 37, pp. 268–280, 1972, C. Jockusch proved that Ramsey's Theorem namely Theorem 12.44 is not provable in the second order arithmetic with the arithmetical comprehension axioms and the arithmetical mathematical inductions. It is also easily seen from his paper that if m is a fixed natural number but not a variable, then Ramsey's theorem for this fixed m is provable in the second order arithmetic with the arithmetical comprehension axioms and the arithmetical mathematical inductions. This fact together with the proof of Theorem 12.44 implies that for any fixed natural number $b \geq 1$,

$$\forall c \geq 1 \, \exists a \geq 1(a \rightarrow (\text{large})^b_c)$$

is provable in second order arithmetic with the arithmetical comprehension axioms and the arithmetical mathematical inductions. As is proved in §16, second order arithmetic with the arithmetical comprehension axioms and the arithmetical mathematical inductions is a conservative extension of **PA**. Therefore for every fixed natural number $b \geqslant 1$,

$$\forall c \geqslant 1, \exists a \geqslant 1(a \to (\text{large})_c^b)$$

is provable in **PA**.

DEFINITION 12.47. Let a, b, c, and k be natural numbers. By $a \to (k)_c^b$, we mean the following statement: for every function $f:[a]^b \to c$, there exists a subset A of a such that A is homogeneous for f and the cardinality of A is k.

It is well-known in combinatorics that there exists a primitive recursive function $g(b, c, k)$ such that $g(b, c, k) \to (k)_c^b$.

DEFINITION 12.48. A function $f:\mathbb{N}^n \to \mathbb{N}$ is said to be monotone if for $x_1, \ldots, x_n, y_1, \ldots, y_n \in \mathbb{N}$, the following holds:

$$\text{if } x_1 \leqslant y_1, \ldots, x_n \leqslant y_n, \text{ then } f(x_1, \ldots, x_n) \leqslant f(y_1, \ldots, y_n).$$

Our strategy is expressed by the following simple proposition.

PROPOSITION 12.49. *Let $g(x)$, $p(x)$ and $h(x)$ be monotone functions from $\mathbb{N}$ into $\mathbb{N}$ satisfying the following conditions.*

(1) *Every provably recursive function in* **PA** *is majorized by g.*

(2) *$p(x)$ is provably recursive.*

(3) *For every $x \in \mathbb{N}$, $g(x) \leqslant h(p(x))$.*

Then every provably recursive function in **PA** *is majorized by h.*

PROOF. By replacing $p(x)$ by $\max(p(x), x)$ if necessary, we may assume that $\forall x\,(x \leqslant p(x))$, therefore that $\forall x\,\exists y\,(x \leqslant p(y))$. Let $f(x)$ be a provably recursive function in **PA**. We define $q(x)$ and $q_0(x)$ by the following equations:

$$q(x) = \text{the least } y \leqslant x \text{ such that } x \leqslant p(y)$$

$$q_0(x) = q(x) \dot{-} 1.$$

Since $x \leqslant p(q_0(x)+1)$ we have

$$f(x) \leqslant f(p(q_0(x)+1)).$$

Since $f(p(x+1))$ is provably recursive in **PA** and $p(q_0(x)) < x$ for any large x, we have the following inequality for any large x

$$f(p(q_0(x)+1)) < g(q_0(x)) \leq h(p(q_0(x)) \leq h(x).$$

In the following, we shall find two primitive recursive functions $p(x)$ and $q(x)$ such that $h_{\omega_n}(n) \leq \sigma(p(n), p(n))$ and $\sigma(n, n) \leq \sigma(q(n), 3)$.

For a while, we shall use capital letters to denote subsets of $\mathbb{N}$ and lower case letters to denote elements of $\mathbb{N}$. If X is a set, $|X|$ is the cardinality of X; if $|X| = m$, and we write $X = \{x_0, \ldots, x_{m-1}\}$, it is always tacitly understood that $x_0 < x_1 < \ldots < x_{m-1}$.

DEFINITION 12.50. (1) Let n and c be positive natural numbers. An (n, c)-*algebra* is a map $G: \mathbb{N}^{[n]} \to C$, where C is a finite set of cardinality c. Also n is called the dimension of G and c is called the number of colors.

(2) A finite subset S of $\mathbb{N}$ is suitable for the (n, c)-algebra G if S is large and homogeneous for G and $|S| \geq n+1$.

(3) An algebra G_2 is said to simulate an algebra G_1 if every S suitable for G_2 is suitable for G_1.

(4) Let F_1 and F_2 be (n, c_1)-algebra and (n, c_2)-algebra respectively. An (n, c_1c_2)-algebra F defined by $F(u) = \langle F_1(u), F_2(u)\rangle$ is called the *product algebra* of F_1 and F_2.

PROPOSITION 12.51. (1) *If F is the product algebra of F_1 and F_2, then F simulates F_1 and F_2.*

(2) *Every (n, c_1c_2)-algebra is isomorphic to the product of an (n, c_1)-algebra and an (n, c_2)-algebra.*

(3) *If $c_1 \leq c_2$, any (n, c_1)-algebra can be construed as an (n, c_2)-algebra.*

(4) *Let G be (n, c)-algebra and $S \subseteq \mathbb{N}$. If every subset T of S with $|T| = n+1$ is homogeneous for G, then S is homogeneous for G.*

PROOF. (1), (2), and (3) are obvious.

(4) Let $X, Y \in S^{[n]}$. We prove by induction on $|X \cup Y|$ that $G(X) = G(Y)$. If $|X \cup Y| \leq n+1$, then it is obvious. Let $|X \cup Y| > n+1$, $x \in X - Y$, and $y \in Y - X$. Let $X_0 = (X - \{x\}) \cup \{y\}$. By the induction hypothesis, $G(X) = G(X_0) = G(Y)$.

LEMMA 12.52. (1) *Let $G_i: \mathbb{N}^{[n]} \to C_i$ be an (n, c_i)-algebra $(1 \leq i \leq k)$. If $G: \mathbb{N}^{[n]} \to C_1 \times \ldots \times C_k$ be the product algebra, then G can be simulated by an $(n+1, c_1 + \ldots + c_k + 1)$-algebra.*

(2) *Let $c \leq c_1 c_2 \ldots c_k$. Then any (n, c)-algebra can be simulated by an $(n+1, c_1 + \ldots + c_k + 1)$-algebra.*

(3) *Any $(n, 7)$-algebra can be simulated by an $(n+1, 7)$-algebra.*

PROOF. (1) Define an algebra $G^*: \mathbb{N}^{(n+1]} \to \{0\} \cup \bigcup_{1 \leq i \leq k} \{i\} \times C_i$ as follows. If $X \in \mathbb{N}^{[n+1]}$ is homogeneous for each of the G_i's, then $G^*(X) = 0$.

Otherwise, let i be least such that X is not homogeneous for G_i. Let Y be the first n elements of X. Set $G^*(X) = \langle i, G_i(Y)\rangle$.

We show that G^* simulates G. It suffices to show that S is homogeneous for G if S is suitable for G^*.

Suppose first that the constant value assumed by S is 0. Then by (4) of Proposition 12.49, S is homogeneous for each G_i, and hence for G.

Next suppose that the constant value of G^* is $\langle i, t\rangle$. So each size $n+1$ subset of S is homogeneous for G_j if $1 \leq j < i$, is not homogeneous for G. As a special case the first $n+1$ elements of S is not homogeneous for G. Since S is suitable for G^*, $|S| \geq n+2$. Let $s_0, \ldots, s_{n+1}$ be the first $n+2$ elements of S. Let Y be an n element subset of $\{s_0, \ldots, s_n\}$. Since $G^*(Y \cup \{s_{n+1}\}) = \langle i, t\rangle$, $G_i(Y) = t$. Thus G_i takes the constant value t on the size n subsets of $\{s_0, \ldots, s_n\}$. This is a contradiction.

(2) follows immediately from (1).

(3) This is immediate from the case $k=2$ of (2), since $7 < 3 \cdot 3$ and $3+3+1=7$.

We shall construct many (n, c)-algebras satisfying several properties. As is suggested by Proposition 12.49, the exact values of n and c are not necessary in the end as far as they can be primitive recursively calculated. So we shall skip the calculations of n and c. In this sense, some of the following lemmas are not necessary since what we need is primitive recursive functions. Nevertheless we keep them because of their beauty.

LEMMA 12.53. (1) *Let G be an (n, c)-algebra and d be a natural number ≥ 1. Then there is an $(n, c+d)$-algebra G^* which simulates G and such that any S suitable for G^* has* $\min S \geq d$.

(2) *Let $n \geq 2$. Then there is an $(n, 7)$-algebra G such that if S is suitable for G, then* $\min S \geq 2n+3$.

PROOF. (1) Let $G: \mathbb{N}^{[n]} \to C$. Define $G^*: \mathbb{N}^{[n]} \to (\{0\} \times d) \cup (\{1\} \times C)$ as follows. Let $X \in \mathbb{N}^{[n]}$. If $\min X < d$, set $G^*(X) = \langle 0, \min X\rangle$; otherwise, let $G^*(X) = \langle 1, G(X)\rangle$.

Let $S = \{s_0, \ldots, s_m\}$ be suitable for G^*. Since S is suitable, $m \geq n$. We show first that $s_0 \geq d$. For, otherwise $G^*(\{s_0, \ldots, s_{n-1}\}) = \langle 0, s_0\rangle$. Then $G^*(\{s_1, \ldots, s_n\}) = \langle 0, s_0\rangle$ since S is homogeneous for G^*. So $s_0 = s_1$ which is a contradiction.

Hence G^* must have constant value $\langle 1, z\rangle$ for some z. But then G has constant value z on $S^{[n]}$ and S is suitable for G.

(2) Define a $(1, 4)$-algebra G_1 as follows:

if $0 \leq m < n$,	$G_1(m) = 0$;
if $n \leq m < 2n$,	$G_1(m) = 1$;
if $2n < m < 4n-1$,	$G_1(m) = 2$;
if $4n-1 \leq m$,	$G_1(m) = 3$.

By (3) of Lemma 12.52, let G be an $(n, 7)$-algebra which simulates G_1. Let S be suitable for G. We shall show that $\min S \geqslant 2n+3$.

Let $S = \{s_0, \ldots, s_m\}$. Then $m \geqslant n$ and $m \geqslant s_0 - 1$ since S is suitable for G. Since G simulates G_1, S is homogeneous for G_1. Now $m \geqslant n$ implies $G_1(s_m) \geqslant 1$, so $G_1(s_0) \geqslant 1$. That is $s_0 \geqslant n$. Since $m \geqslant n$, $s_n \geqslant n + s_0 \geqslant 2n$, whence $G_1(s_n) \geqslant 2$. So $G_1(s_0) \geqslant 2$ namely $s_0 \geqslant 2n$. Now $m \geqslant s_0 - 1 \geqslant 2n-1$, whence $s_m \geqslant 4n-1$; so $G_1(s_0) = G_1(s_m) = 3$. That is $s_0 \geqslant 4n-1$. Finally $2n+3 \leqslant 4n-1$ since $n \geqslant 2$.

DEFINITION 12.54. Define functions $E_m : \mathbb{N} \to \mathbb{N}$ as follows:

$$E_0(n) = n \quad \text{and} \quad E_{m+1}(n) = 2^{E_m(n)}.$$

LEMMA 12.55. *Let* $h : \mathbb{N}^{[n]} \to \mathbb{N}$. *Suppose that whenever* $1 \leqslant x_0 < \ldots < x_{n-1}$, $h(x_0, \ldots, x_{n-1}) < E_m(x_0)$. *Then there is an* $(n+m+1, 10^{2m+2})$-*algebra* G *such that for any* S *suitable for* G, *there is a function* $g_s : S \to \mathbb{N}$ *such that for any* $X \in S^{[n]}$, $h(X) = g_s(x_0)$. *We express this last as "on* $S^{[n]}$, h *depends only on the first coordinate".*

PROOF. By induction on m.

Case 1. $m = 0$.

The algebra G will be chosen to simulate a finite number of simpler algebras: G_1 is the $(n, 7)$-algebra provided by (2) of Lemma 12.53 that guarantees $\min S \geqslant 2n+3$.

G_2 is an $(n, 2)$-algebra. $G(X) = 0$ if $h(X) < [\frac{1}{2}x_0]$; otherwise $G_2(X) = 1$, where $[a]$ is Gauss's symbol.

G_3 is an $(n+1, 3)$-algebra. $G_3(X) = 0, 1, 2$ respectively if $h(x_0, x_1, \ldots, x_{n-1}) = , >, < h(x_0, x_2, \ldots, x_n)$ respectively.

G_4 is an $(n+1, 2)$-algebra. $G_4(X) = 0$ if on $X^{[n]}$, h depends only on the first coordinates. Otherwise $G_4(X) = 1$.

Let G be an $(n+1, 100)$-algebra which simulates G_1, G_2, G_3, and G_4.

Let $S = \{s_0, \ldots, s_k\}$ be suitable for G. We shall prove that on $S^{[n]}$, h depends on the first coordinates.

1. Since G simulates each G_i, S is homogeneous for each G_i. By an argument totally similar to the proof of (4) of Proposition 12.51, it suffices to show that G_4 is identically zero on $S^{[n+1]}$. Since S is homogeneous for G_4, it suffices to show that $G_4(s_0, \ldots, s_n) = 0$.

2. Let $W = \{m \mid m < [\frac{1}{2}s_0]\}$ if G_2 has constant value 0 on $S^{[n]}$, i.e., $h(X) < [\frac{1}{2}x_0]$; otherwise let $W = \{m \mid \frac{1}{2}s_0] \leqslant m < s_0\}$. Then W has at most $[\frac{1}{2}s_0] + 1$ elements. Moreover, if $A \in S^{[n]}$ and $\min A = s_0$, then $h(A) \in W$ since S is homogeneous for G_2.

3. We shall show $[\frac{1}{2}s_0] + 1 < s_0 - n$. Since $s_0 - [\frac{1}{2}s_0] = [\frac{1}{2}(s_0+1)]$, it suffices to show $n+1 < [\frac{1}{2}(s_0+1)]$ or equivalently $n+2 \leqslant [\frac{1}{2}(s_0+1)]$. This is again equivalent to $n+2 \leqslant \frac{1}{2}(s_0+1)$, i.e., $2n+3 \leqslant s_0$. This follows from S being suitable for G_1.

4. For $1 \leqslant i \leqslant s_0+1-n$, let $A_i = \{s_0, s_i, \ldots, s_{i+n-2}\}$. Then $A_i \in S^{[n]}$. Let $w_i = h(A_i)$; then $w_i \in W$. By the inequality of 3 and the pigenhole principle, there are i, j with $1 \leqslant i < j \leqslant s_0+1-n$ with $w_i = w_j$.

5. Since G simulates G_3, G_3 takes some constant value on $S^{[n+1]}$. We claim this value is 0. If, for example, the value were 1, the $w_1 > w_2 > \ldots > > w_{s_0+1-n}$, contrary to the conclusion of 4.

6. Since $s_0 \geqslant 2n+3$, s_{2n+2} is defined. We claim that if $Y \in \{s_0, \ldots, s_n\}^{[n]}$ and $\min Y = s_0$, then $h(Y) = h(s_0, s_{n+1}, \ldots, s_{2n-1})$.

To see this, let $\{t_0, \ldots, t_{2n-2}\} = Y \cup \{s_{n+1}, \ldots, s_{2n-1}\}$. Using the fact that G_3 is identically zero on $S^{[n+1]}$, we get

$$\begin{aligned} h(Y) &= h(t_0, t_1, \ldots, t_{n-1}) = h(t_0, t_2, \ldots, t_n) \\ &= h(t_0, t_i, t_{i+1}, \ldots, t_{i+n-2}) = \ldots = h(t_0, t_n, \ldots, t_{2n-2}) \\ &= h(s_0, s_{n+1}, \ldots, s_{2n-1}). \end{aligned}$$

7. Now there is precisely one $Y \in \{s_0, \ldots, s_n\}^{[n]}$ with $\min Y \neq s_0$. Hence 6 implies $G_4(s_0, \ldots, s_n) = 0$. This completes the proof of Case 1.

Case 2. $m = k+1$.
Let $h: \mathbb{N}^{[n]} \to \mathbb{N}$ such that $h(X) < E_m(x_0)$. We define an auxiliary function $g: \mathbb{N}^{[n+1]} \to \mathbb{N}$ such that $g(Y) < E_k(y_0)$.

Let $Y = \{y_0, \ldots, y_n\}$. If $h(\{y_0, \ldots, y_{n-1}\}) = h(\{y_0, y_2, \ldots, y_n\})$, set $g(Y) = 0$. Otherwise, let $g(Y)$ be the largest j at which the binary expansions of the two just stated values of h differ. Clearly $g(Y) < E_k(y_0)$.

As before G will be chosen so as to simulate a finite number of simpler algebras:

G_1 is the $(n, 7)$-algebra provided by (2) of Lemma 12.53 which guarantees $\min S \geqslant 2n+3$.

G_2 is the $(n+m+1, 10^{2m})$-algebra obtained by applying the induction hypothesis to the function g just introduced.

G_3 is an $(n+1, 3)$-algebra. $G_3(y_0, \ldots, y_n) = 0, 1, 2$ respectively if $h(y_0, \ldots, y_{n-1})$ is $=, >, <$ $h(y_0, y_2, \ldots, y_n)$ respectively.

G_4 is an $(n+1, 2)$-algebra. $G_4(Y) = 0$ iff h depends only on its first coordinates on $Y^{[n]}$; otherwise $G_4(Y) = 1$.

G is an $(n+m+1, 10^{2m+2})$-algebra which simulates G_1, G_2, G_3, and G_4.

Let S be suitable for G. We show that the constant value taken by G_3 on $S^{[n+1]}$ is 0. It will then follow, exactly as in Case 1, that the value of $h(X)$ for $X \in S^{[n]}$ depends only on $\min X$.

Suppose then that G_3 takes the constant value 1. (The case when G_3's constant value is 2 is totally analogous.) We shall derive a contradiction.

Let $S = \{s_0, \ldots, s_r\}$. Since G simulates G_1, $r \geqslant n+1$. For $1 \leqslant i \leqslant 3$, let $z_i = h(s_0, s_i, \ldots, s_{n-2+i})$. Since $G_3 = 1$ on $S^{[n+1]}$, $z_1 > z_2 > z_3$.

Let $j_0 = g(s_0, \ldots, s_n)$, $j_1 = g(s_0, s_2, \ldots, s_{n+1})$. Since G simulates G_2, $j_0 = j_1 = j$, say. By the definition of g, the jth digit is the highest order binary digit at which z_1, z_2 differ. Since $z_1 > z_2$, z_2's jth digit is a 0. On the other hand, the jth digit is the highest order binary digit at which z_2, z_3 differ. Since $z_2 > z_3$, z_2's jth digit must be 1. This gives the desired contradiction.

LEMMA 12.56. *Let* $g : \mathbb{N}^{[n]} \to \mathbb{N}$ *satisfying* $g(x_0, \ldots, x_{n-1}) \leq x_0$. *Then there an* $(n+1, 10^4)$-*algebra* G *such that if* S *is suitable for* G:

(1) *On* $S^{[n]}$, g *depends only on its first coordinates.*

(2) *If* $X, Y \in S^{[n]}$ *and* $\min X < \min Y$, *then* $g(X) \leq g(Y)$.

PROOF. G is choosen to simulate a finite number of simpler algebras:

G_1 is an $(n, 2)$-algebra. $G_1(X) = 0$ if $g(X) < \min X$, otherwise $G_1(X) = 1$.

Define $g' : \mathbb{N}^{[n]} \to \mathbb{N}$ as follows: if $g(X) < \min X$, $g'(X) = g(X)$. Otherwise, $g'(X) = 0$. Let G_2 be the $(n+1, 100)$-algebra obtained by applying Lemma 12.55 to g' ($m = 0$ in that lemma).

G_3 is an $(n+1, 2)$-algebra. $G_3(X) = 0$ if $g(x_1, \ldots, x_n) < [\frac{1}{2}x_0]$; otherwise $G_3(X) = 1$.

G_4 is an $(n+1, 2)$-algebra. $G_4(x_0, \ldots, x_n) = 0$ if $g(x_0, \ldots, x_{n-1}) \leq g(x_1, \ldots, x_n)$. Otherwise $G_4(x_0, \ldots, x_n) = 1$.

G_5 is the $(n, 7)$-algebra provided by (2) of Lemma 12.53. It insures that any S suitable for G_5 satisfies $\min S \geq 2n + 3$.

Let G be an $(n+1, 10^4)$-algebra that simulates $G_1, \ldots, G_5$. Let S be suitable for G. We show that on S, g satisfies (1) and (2) of the lemma.

First suppose that on $S^{[n]}$, G_1 takes the constant value 1. Then on $S^{[n]}$, $g(X) = \min X$ and (1) and (2) are clear. So from now on we may assume that on $S^{[n]}$, G_1 takes the constant value 0. Hence $g = g'$ on $S^{[n]}$, so claim (1) is clear since G simulates G_2.

Let $S = \{s_0, \ldots, s_m\}$. For $0 \leq i \leq m - n + 1$, let $z_i = g(s_i, \ldots, s_{i+n-1})$. If G_4 takes the constant value 0 on $S^{[n+1]}$, then $z_0 \leq z_1 \leq z_2 \leq \ldots \leq z_{m-n+1}$, and claim (2) is clear. So suppose, towards a contradiction, that G_4 takes the constant value 1 on $S^{[n+1]}$ so that

$$s_0 > z_0 > z_1 > \ldots > z_{m-n+1}.$$

Let $W = \{i \mid i < [\frac{1}{2}s_0]\}$ if G_3 has constant value 0; otherwise let $W = \{i \mid [\frac{1}{2}s_0] \leq i < s_0\}$. By the definition of G_3, we have $z_i \in W$ for $1 \leq i \leq m - n + 1$. But for $i < j$, $z_i > z_j$; moreover since $m \geq s_0 - 1$ and $s_0 \geq 2n + 3$, we have $|W| < m - n + 1$. (Cf. Lemma 12.55, Case 1, Claim 3.) But this contradicts the pigenhole principle.

In Definition 12.19, we defined $\{\alpha\}(n)$ only when α is a limit ordinal. Now we extend the definition of $\{\alpha\}(n)$ for all ordinals less than ε_0. We assume that the ordinals are $< \varepsilon_0$ in this section without explicit mention.

DEFINITION 12.57. (1) In addition to (2) of Definition 12.19, $\{\alpha\}(n)$ is defined as follows:

$$\{\alpha+1\}(n)=\alpha \quad \text{and} \quad \{0\}(n)=0.$$

(2) Let $\alpha<\beta$. Then $\beta \underset{n}{\to} \alpha$ if for some sequence of ordinals $\gamma_0, \dots, \gamma_r$ we have $\gamma_0=\beta$, $\gamma_{i+1}=\{\gamma_i\}(n)$ for $0 \leq i<r$, and $\gamma_r=\alpha$.

PROPOSITION 12.58. (1) *If* $\alpha>0$, *then* $\alpha \underset{n}{\to} 0$.
(2) *Let* $n \geq 1$. *If* $\alpha_1 \underset{n}{\to} \alpha_2$, *then* $\omega^{\alpha_1} \underset{n}{\to} \omega^{\alpha_2}$.
(3) *If* α *is a limit ordinal*, $i<j<\omega$, *and* $0<n<\omega$, *then* $\{\alpha\}(j) \underset{n}{\to} \{\alpha\}(i)$.
(4) *If* $n>i$ *and* $\alpha \underset{i}{\to} \beta$, *then* $\alpha \underset{n}{\to} \beta$.
(5) *If* $\alpha \underset{n}{\to} \beta$, *then* $h_\beta(n) \leq h_\alpha(n)$.

PROOF. (1) is obvious since $\{\alpha\}(n)<\alpha$ for $\alpha>0$.
(2) We may assume $\alpha_2=\{\alpha_1\}(n)$. If α_1 is a limit ordinal, then $\{\omega^{\alpha_1}\}(n)=\omega^{\{\alpha_1\}(n)}=\omega^{\alpha_2}$. If $\alpha_1=\beta+1$, then $\alpha_2=\beta$ and $\{\omega^{\alpha_1}\}(n)=\omega^{\alpha_2}\cdot n$. We can easily prove $\omega^{\alpha_1} \underset{n}{\to} \omega^{\alpha_2}$ reducing $\omega^{\alpha_2}\cdot(n-1)$ to 0 by successive application of $\{\ \}(n)$.
(3) It suffices to show the case $i=j-1$. Since $\alpha \underset{n}{\to} \beta$ implies $\gamma \dotplus \alpha \underset{n}{\to} \gamma \dotplus \beta$, we may assume that α is of the form ω^β. If $\alpha=\omega$, then $\{\alpha\}(j)=\{\alpha\}(i)+1$ and $\{\{\alpha\}(j)\}(n)=\{\alpha\}(i)$. If $\beta=\beta_0+1$ and $\beta_0 \neq 0$, then $\{\alpha\}(j)=\omega^{\beta_0}(j-1)+\omega^{\beta_0}$ and $\{\alpha\}(i)=\omega^{\beta_0}(j-1)$. Then $\{\alpha\}(j) \underset{n}{\to} \{\alpha\}(i)$ is obvious because of (1). If β is a limit ordinal, then $\{\beta\}(j) \underset{n}{\to} \{\beta\}(i)$. Then $\{\omega^\beta\}(j) \underset{n}{\to} \{\omega^\beta\}(i)$ because of (2).
(4) We may assume $\beta=\{\alpha\}(i)$. Then by (3) $\{\alpha\}(n) \underset{n}{\to} \{\alpha\}(i)=\beta$ and $\alpha \underset{n}{\to} \beta$.
(5) We may assume $\beta=\{\alpha\}(n)$. Then $h_\beta(n) \leq h_\alpha(n)$ is obvious if α is a limit ordinal. It is also obvious if α is a successor ordinal.

DEFINITION 12.59. For $n, x \in \mathbb{N}$, we define

$$T(\omega_n, x)=\{\alpha \mid \omega_n \underset{x}{\to} \alpha\}.$$

PROPOSITION 12.60. (1) $T(\omega_n, x)$ *has cardinality at most*

$$\left. m^{m^{\cdot^{\cdot^{\cdot^{m}}}}} \right\} n \ (n\ m\text{'s},\ m=x+1)$$

(2) *If* $\alpha \in T(\omega_n, x)$ *and* $\alpha=\omega^{\beta_1}\cdot k_1+\dots+\omega^{\beta_l}\cdot k_l$ $(\beta_1<\dots<\beta_l)$ *is its Cantor normal form, then all coefficients* $k_1, \dots, k_l \leq x$.

PROOF. We proceed by induction on n. The case $n=0$ is trivial. For $n=n_0+1$, define $M=\{\dot{\alpha} \mid$ all exponents in the Cantor normal form for α

lie in $T(\omega_n, x)$ and all coefficient in the Cantor normal form for $\alpha \leqslant x\}$. It is easily checked that if $\alpha \in M$, so is $\{\alpha\}(x)$, and that $\{\omega_n\}(x)$ lies in M. Whence $T(\omega_n, x) \subseteq M$. So

$$|T(\omega_n, x)| \leqslant |M| \leqslant (x+1)^{|T(\omega_{n_0}, x)|}.$$

(2) is also clear from $T(\omega_n, x) \subseteq M$.

DEFINITION 12.61. For $m, n \in \mathbb{N}$, $E(m, n)$ is defined by $E(m, 0) = 1$ and $E(m, n+1) = m^{E(m,n)}$. Therefore Proposition 12.60 implies $|T(\omega_n, x)| \leqslant E(m, n)$, where $m = x + 1$.

We need an upperbound for $|T(\omega_n, x)|$ in the form of $E_k(x)$. In their paper, Ketonen and Soloray proved that $|T(\omega_n, x)| \leqslant E_{n-1}(x^6)$. We use the following trivial estimate.

PROPOSITION 12.62. *For* $x \geqslant 1$, $|T(\omega_n, x)| \leqslant E_{2n}(x)$.

PROOF. It suffices to show $E(m, n) \leqslant E_{2n}(x)$, where $m = x + 1$. Obviously $(1+n) \leqslant 2^n$. So $2(n+1) \leqslant 2 \cdot 2^n \leqslant 2^{n+1}$ and $2n \leqslant 2^n$. Therefore $n \cdot 2^n \leqslant 2^{2n} \leqslant 2^{2^n}$. Now we prove $E(m, n) \leqslant E_{2n}(x)$ by induction on n. The case $n = 0$ is obvious. The case $n = k + 1$ is proved as follows.

$$\begin{aligned} \log_2 E(m, k+1) = E(m, k) \log_2 m &\leqslant xE_{2k}(x) \\ &\leqslant E_{2k-1}(x) \cdot E_{2k}(x) \\ &\leqslant 2^{E_{2k}(x)} = E_{2k+1}(x). \end{aligned}$$

DEFINITION 12.63. Let $g : \mathbb{N}^{[n]} \to \varepsilon_0$. Then g is weakly controlled by an algebra G if whenever S is suitable for G. On $S^{[n]}$, g depends only on its first coordinate (i.e., if $X, Y \in S^{[n]}$ and $\min X = \min Y$, then $g(X) = g(Y)$). If g is weakly controlled by G, and S is suitable for G, then we define a function g_s from a subset of S into ε_0 by putting $g_s(x_0) = g(x_0, \ldots, x_{i-1})$.

We say that g is controlled by G if g is weakly controlled by g and if whenever S is suitable for G, and $x_0, x_1 \in S$ are such that $x_0 < x_1$ and $g_s(x_0)$, $g_s(x_1)$ are defined, then $g_s(x_0) \leqslant g_s(x_1)$.

As is remarked before, the exact numbers n and c of (n, c)-algebra is not necessary as far as they are primitive recursively calculated. In the following we simply skip c by saying an adequate n-algebra when c can be primitive recursively calculated.

LEMMA 12.64. *Let* $g : \mathbb{N}^{[n]} \to \varepsilon_0$. *Suppose that* $g(x_0, \ldots, x_{n-1}) \in T(\omega_k, x_0)$ *for all* $x \in \mathbb{N}^{[n]}$. *Then* g *is weakly controlled by an* $n + 2k + 1$*-algebra.*

PROOF. The case $k = 0$ is trivial. So assume $k \geqslant 1$. By Lemma 12.55 and

Proposition 12.62, we can choose an adequate $n+2k+1$-algebra that insures that $g(x_0, \dots, x_{n-1})$ depends only on x_0, provided $x_0 \geqslant 1$. By (1) of Lemma 12.53, we can find an adequate $n+2k+1$-algebra G that simulate the algebra just mentioned and such that if S is suitable for G, $\min S \geqslant 1$. This G weakly controls g.

LEMMA 12.65. *Let* $k \geqslant 1$. *Let* $g : \mathbb{N}^{[n]} \to \varepsilon_0$ *satisfy* $g(x_0, \dots, x_{n-1}) \in T(\omega_k, x_0)$. *Then* g *can be controlled by an adequate* $n+2k+1$-*algebra*.

PROOF. The proof is by induction on k. The case $k=1$ follows from Lemma 12.56. So assume $k \geqslant 2$ and fix g as in the statement of the lemma.

Let G_0 be an adequate $n+2k+1$-algebra which weakly controls g (cf. Lemma 12.64).

Let G_1 be an $(n+1, 2)$-algebra. $G_1(x_0, \dots, x_n) = 0$ if $g(x_0, \dots, x_{n-1}) \leqslant g(x_1, \dots, x_n)$. Otherwise $G_1(x_0, \dots x_n) = 1$.

We define an auxiliary function $h : \mathbb{N}^{[n+1]} \to \varepsilon_0$ as follows. If $g(x_0, \dots, x_{n-1}) \leqslant g(x_1, \dots, x_n)$, $h(x) = 0$. If $g(x_0, \dots, x_{n-1}) > g(x_1, \dots, x_n)$, write $g(x_0, \dots, x_{n-1})$ and $g(x_1, \dots, x_n)$ in the Cantor normal form $\omega^{\alpha_1} n_1 + \dots + \omega^{\alpha_k} n_k$ and $h(x)$ is the largest α_i at which $g(x_0, \dots, x_{n-1})$ differs from $g(x_1, \dots, x_n)$, i.e.,

$$g(x_0, \dots, x_{n-1}) = \omega^{\alpha_1} n_1 + \dots + \omega^{\alpha_{i-1}} n_{i-1} + \omega^{\alpha_i} n_i + \dots$$

and

$$g(x_1, \dots, x_n) = \omega^{\alpha_1} n_1 + \dots + \omega^{\alpha_{i-1}} n_{i-1} + \omega^{\beta_i} m_i + \dots$$

and $\alpha_i > \beta_i$ or $\alpha_i = \beta_i \wedge n_i > m_i$. Let $r = k-1$. Note that $h(x_0, \dots, x_n) \in T(\omega_r, x_0)$. Let G_2 be an adequate $n+2k+1$-algebra that controls h. (This exists by our induction hypothesis.)

Define an auxiliary function $h' : \mathbb{N}^{[n+1]} \to \mathbb{N}$ as follows: $h'(x)$ is the coefficient of $\omega^{h(x)}$ in the Cantor normal form of $g(x_0, \dots, x_{n-1})$. (Set $h'(x) = 0$ if $g(x_0, \dots, x_{n-1}) \leqslant g(x_1, \dots, x_n)$.) Since $g(x_0, \dots, x_{n-1}) \in T(\omega_k, x_0)$, $h'(x) \leqslant x_0$. Let G_3 be an adequate $n+3$-algebra that controls h' (cf. Lemma 12.56).

Let G be an adequate $n+2k+1$-algebra which simulates $G_0, \dots, G_3$. We shall show that G controls g.

Let $S = \{s_0, \dots, s_m\}$ be suitable for G. Since G simulates G_3, $m \geqslant n+3$. Let $S' = \{s_i \mid i \leqslant m-n+1\}$. Let $g_s : S' \to \varepsilon_0$ be given by $g_s(s_i) = g(s_i, \dots, s_{i+n-1})$. Since G simulates G_0, if $X \in S^{[n]}$, $g(X) = g_s(x_0)$.

Since G simulates G_1, we know that G_1 takes a constant value $v \leqslant 1$ on $S^{[n+1]}$. We will be done if $v = 0$. Towards a contradiction, assume $v = 1$, i.e., if $x_0 < x_1$, $x_0, x_1 \in S'$, then $g_s(x_0) > g_s(x_1)$. Moreover there is an ordinal $\eta_s(x_0)$ $(= h(x_0, x_1, \dots))$ with coefficient $m_s(x_0)$ $(= h'(x_0, x_1, \dots))$ such that the Cantor normal forms of $g_s(x_0)$ and $g_s(x_1)$ agree at exponents above

$\eta_s(x_0)$, but not at $\eta_s(x_0)$. (The fact that $\eta_s(x_0)$ does not depend on x_1 comes from G simulating G_2 which weakly controls h. Similarly since G simulates G_3, $m_s(x_0)$ does not depend on x_1.)

Since G simulates G_2 and G_3 which control h and h' respectively, the following is true. If $x_0, x_1 \in S'$ with $x_0 < x_1$, then $\eta_s(x_0) \leqslant \eta_s(x_1)$ and $m_s(x_0) \leqslant m_s(x_1)$.

Since $m \geqslant n+2$, $\{s_0, s_1, s_2\} \subseteq S'$. Now $g_s(s_0)$ and $g_s(s_1)$ agree at exponents above $\eta_s(s_0)$ as do $g_s(s_0)$ and $g_s(s_2)$. Hence $g_s(s_1)$ and $g_s(s_2)$ agree at exponents above $\eta_s(s_0)$; i.e., $\eta_s(s_1) \leqslant \eta_s(s_0)$. Also, by the preceding paragraph, $\eta_s(s_0) \leqslant \eta_s(s_1)$; so $\eta_s(s_0) = \eta_s(s_1)$. But then $g_s(s_0) > g_s(s_1)$ implies $m_s(s_1) < m_s(s_0)$, contradicting the observation of the preceding paragraph that $m_s(s_0) \leqslant m_s(s_1)$.

DEFINITION 12.66. An algebra $G: \mathbb{N}^{[n]} \to C$ *captures* a function $f: \mathbb{N} \to \mathbb{N}$ if whenever S is suitable for G, and $x < y$ are elements of S, then $f(x) \leqslant y$.

THEOREM 12.67. *Let $n \geqslant 1$. Then h_{ω_n} can be captured by an adequate $2n+3$-algebra G.*

PROOF. Let G_0 be a (2,2)-algebra. $G_0(x_0, x_1) = 0$ if $x_1 \geqslant h_{\omega_n}(x_0)$; otherwise $G_0(x_0, x_1) = 1$.

Let G_1 be a (2,5)-algebra that simulates G_0 and such that any S suitable for G_1 has min $S \geqslant 3$.

Define an auxiliary function $f: \mathbb{N}^{[2]} \to \varepsilon_0$ as follows: If there is a $\xi \in T(\omega_n, x_0)$ such that $h_\xi(x_0) \geqslant x_1$, let $f(x_0, x_1)$ be the least such. Otherwise $f(x_0, x_1) = 0$. Let G_2 be an adequate $2n+3$-algebra that controls f.

Let G be an adequate $2n+3$-algebra which simulates G_1 and G_2. We show that G captures h_{ω_n}. Let $S = \{s_0, \ldots, s_m\}$ be suitable for G. Since S is large, $m \geqslant s_0 - 1$. Let v be the constant value taken by G_0 on $S^{[2]}$. We have to show $v = 0$. Towards a contradiction, assume $v = 1$.

Let $\{x_0, x_1\} \in S^{[2]}$. Since $G_0(x_0, x_1) = 1$, $x_1 < h_{\omega_n}(x_0)$. Let $\xi_0 = \{\omega_n\}(x_0)$. Then $x_1 < h_{\xi_0}(x_0)$ and $\xi_0 \in T(\omega_n, x_0)$. It follows that if $\xi_1 = f(x_0, x_1)$, then ξ_1 is the least ordinal in $T(\omega_n, x_0)$ such that $x_1 \leqslant h_{\xi_1}(x_0)$.

Now ξ_1 cannot be a limit ordinal. For then, letting $\xi_2 = \{\xi_1\}(x_0)$, we have $\xi_2 < \xi_1$, $\xi_2 \in T(\omega_n, x_0)$ and $x_1 \leqslant h_{\xi_2}(x_0) = h_{\xi_1}(x_0)$. Also ξ_1 cannot equal 0.

Thus $\xi_1 = f(x_0, x_1)$ has the form $\delta + 1$. Note next that since G simulates G_2 and G_2 controls f, on $S^{[2]}$, f depends only on its first coordinates. Say $f(x_0, x_1) = \delta(x_0) + 1$ (for $\{x_0, x_1\} \in S^{[2]}$). Moreover if $\{x_0, x_1\} \in S^{[2]}$ and $\delta(x_1)$ is defined, then $\delta(x_0) \leqslant \delta(x_1)$. Finally, by the minimal choice of $f(x_0, x_1)$ and the fact that $f(x_0, x_1) \underset{x_0}{\to} \delta(x_0)$, we have $\delta(x_0) \in T(\omega_n, x_0)$ and $x_1 > h_{\delta(x_0)}(x_0)$.

Let $0 \leqslant i < s_0 - 1$. We claim $s_{i+1} > h_{\delta(s_0)}(s_i)$. By the preceding paragraph, $s_{i+1} > h_{\delta(s_i)}(s_i)$. Also $\delta(s_0) \leqslant \delta(s_i)$. If $\delta(s_0) = \delta(s_i)$, our claim is clear. So suppose $\delta(s_i) > \delta(s_0)$. Now $\omega_n \underset{s_0}{\to} \delta(s_0)$. Hence by (4) of Proposition 12.58

$\omega_n \underset{s_i}{\rightarrow} \delta(s_0)$. But then by (5) of Proposition 12.58 $h_{\delta(s_0)}(s_i) \leq h_{\delta(s_i)}(s_i) < s_{i+1}$, establishing our claim.

Now we have

$$h_{\delta(s_0)+1}(s_0) = h_{\delta(s_0)}(s_0+1) \leq h_{\delta(s_0)}(s_1) < s_2.$$

But $\delta(s_0)+1 = f(s_0, s_2)$ and this contradicts the definition of f.

THEOREM 12.68. *Let $n \geq 1$. Then there is an adequate $2n+3$-algebra G such that if S is suitable for G, $\max S > h_{\omega_n}(n)$. Hence there exists a primitive recursive function $p(n)$ such that*

$$h_{\omega_n}(n) < \sigma(2n+3, p(n)).$$

Therefore every provably recursive function in **PA** *is majorized by $\sigma(n, n)$.*

PROOF. Let G_0 be an adequate $2n+3$-algebra that captures h_{ω_n}. Let G_1 be an $(n+2,7)$-algebra that insures that if S is suitable for G_1, $\min S \geq 2n+3$. There exist a primitive recursive function $p(n)$ and an $(2n+3, p(n))$-algebra G that simulates G_0 and G_1. But then, if S is suitable for G,

$$\text{mas } S \geq s_2 \geq s_1 \geq h_{\omega_n}(s_0) \geq h_{\omega_n}(n).$$

Hence if we take the restriction of G to $N^{[2n+3]}$, where $N = h_{\omega_n}(n)$, then no subset of N is suitable for G. Hence $N < \sigma(2n+3, p(n))$. Now the last statement of the lemma follows from Proposition 12.49.

We need a lower bound of m satisfying $m \rightarrow (k+1)^k_2$. J. Quinsey noted in his disertation that $2k+6 \not\rightarrow (k+1)^k_2$. But the following trivial estimate suffices for our purpose.

PROPOSITION 12.69. $2k \not\rightarrow (k+1)^k_2$.

PROOF. Let $M = \{0, 1, 2, \ldots, 2k-1\}$. Define $G: M^{[k]} \rightarrow 2$ as follows. Let $X = \{x_0, \ldots, x_{k-1}\} \in M^{[k]}$. $G(x_0, \ldots, x_{k-1}) = 0$ if $x_0 + \ldots + x_{k-1}$ is even. $G(x_0, \ldots, x_{k-1}) = 1$ otherwise. Suppose that $S \in M^{[k]}$ is homogeneous and $|S| = k+1$. Since there are only k many even numbers and only k many odd numbers in M, there exist an even number x and an odd number y in S. Then

$$G(S - \{x\}) \neq G(S - \{y\})$$

contradicting the homogeneity of S.

THEOREM 12.70. *If* $m \to (e+1)^e_c$ *and* $N \to (\text{large})^m_3$, *then* $N \to (\text{large})^e_c$. *Hence every provably recursive function in* **PA** *is majorized by* $\sigma(n, 3)$.

PROOF. We can suppose that $e, c \geq 2$. By Proposition 12.69, $m \geq 2e+1$. Let $F: N^{[e]} \to c$ be given. Let $k = m - e - 1$ and $t = 2m$. By Proposition 12.69 we can choose $g: t^{[m]} \to 2$ with no homogeneous set of cardinality $m+1$. Define $f: N^{[e+1]} \to 2$ by

$$f(x_0, \ldots, x_e) = \begin{cases} 1 & \text{if } x_0, \ldots, x_e \text{ is homogeneous for } F, \\ 0 & \text{otherwise.} \end{cases}$$

Define $G: N^{[m]} \to 3$ as follows. Let $X = \{x_0, \ldots, x_{m-1}\} \in N^{[m]}$. Let i be the least i such that $x_i \geq t$, if there exists such, and m otherwise. Let

$$G(x_0, \ldots, x_{m-1}) = \begin{cases} g(x_0, \ldots, x_{m-1}) & \text{if } i = m, \\ 2 & \text{if } i = 1 \text{ or } n-1, \\ \frac{1}{2}(1-(-1)^i) & \text{if } 1 < i < m-1, \\ f(x_0 - k, \ldots, x_e - k) & \text{if } i = 0. \end{cases}$$

Let $S \subseteq N$ be suitable for G. Then $|S| \geq \min S$ and $|S| \geq m+1$. We show that $\min S \geq t$. Toward a contradiction suppose $\min S < t$. Let $s_0, \ldots, s_m$ be the first $m+1$ elements of S and $S_0 = \{s_0, \ldots, s_m\}$.

Case 1. $s_m < t$.
In this case $G(X) = g(X)$ for $X \in S_0^{[m]}$. This is a contradiction since S_0 is homogeneous for G and g has no homogeneous set of cardinality $m+1$.

Case 2. $S_i < t \leq S_{i+1}$ for some i with $0 \leq i \leq m-1$.
In this case

$$G(S - \{s_i\}) \neq G(S - \{s_{i+1}\})$$

contradicting the homogeneity of S for G.

Therefore we have $\min S \geq t$ and for every $X \in S^{[m]}$, $G(x_0, \ldots, x_{m-1}) = f(x_0 - k, \ldots, x_e - k)$.

Now let $S = \{s_0, \ldots, s_{a-1}\}$. Then $a \geq t = 2m$. Let $Y = \{s_0, \ldots, s_{m-1}\}$ be the first m elements of S. Define $H: Y^{[e]} \to C$ by

$$H(y_0, \ldots, y_{e-1}) = F(y_0 - k, \ldots, y_{e-1} - k).$$

Since $m \to (e+1)^e_c$, there exists $Z \subseteq Y$ which is homogeneous for H and has the cardinality $e+1$. Now the cardinality of $Z \cup \{s_m, \ldots, s_{m+k-1}\}$ is m and $G(Z \cup \{s_m, \ldots, s_{m+k-1}\}) = 1$. Since S is suitable for G, G has the constant value 1 on $[S]^m$. Since $a \geq s_0$, let $X' = \{s_0 - k, \ldots, s_{s_0 - k - 1} - k\}$.

Then

$$|X'| \geq \min X' \geq 2m - k \geq e + 1.$$

We claim that X' is homogeneous for F. For if $\{x'_0 - k, \ldots, x'_e - k\}$ is any subset of cardinality $e+1$, then the cardinality of $\{x'_0, \ldots, x'_e\} \cup \{s_{s_0-k}, \ldots, s_{s_0-1}\}$ is m and $G(x'_0, \ldots, x'_e, s_{s_0-k}, \ldots, s_{s_0-1}) = 1$. Therefore $\{x'_0, \ldots, x'_e\}$ is homogeneous for F.

Now let $p(e, c)$ be a primitive recursive function such that $p(e, c) \to (e+1)^e_c$. Then we have $\sigma(e, c) \leq \sigma(p(e, c), 3)$ and $\sigma(n, n) \leq \sigma(p(n, n), 3)$. Therefore from Proposition 12.49 follows that every provably recursive function in **PA** is majorized by $\sigma(n, 3)$.

H. Friedman proved that Kruskal's theorem on finite trees is not provable in a certain second order extension of Peano's arithmetic. In the following, we prove a weaker version of Friedman's theorem.

DEFINITION 12.71. (1) A *finite tree* is a finite partially ordered set T satisfying the following conditions:

(i) It has the minimum called its root.

(ii) For every $b \in T$, $\{a \in T; a \leq b\}$ is linearly ordered by $\leq$, where $\leq$ is the order of T.

(2) Let T_1 and T_2 be finite trees. A function $f: T_1 \to T_2$ is an embedding iff f is one-to-one, order-preserving and satisfies the equation

$$f(a \wedge b) = f(a) \wedge f(b) \quad \text{for every } a, b \in T_1,$$

where $a \wedge b$ denotes the greatest lower bound of a and b. We denote $T_1 \leq T_2$ iff there exists an embedding $f: T_1 \to T_2$.

(3) The set of all finite trees is denoted by $\mathcal{T}$. A mapping $o: \mathcal{T} \to \varepsilon_0$ is defined as follows, where ε_0 is the set of all ordinal, less than ε_0 as usual.

If T consists of its root alone, then $o(T) = 0$. If T has some member other than its root, let $T^1, \ldots, T^n$ be all component of $T - \{\text{root}(T)\}$, where root(T) is the root of T. Without loss of generality we assume that $o(T^1), \ldots, o(T^n)$ have been assigned and $o(T^1) \geq o(T^2) \geq \ldots \geq o(T^n)$. Then $o(T)$ is defined by the following equalities

$$o(T) = \begin{cases} \alpha & \text{if } n = 1, \\ \alpha + \beta & \text{if } n = 2, \\ \omega^\alpha & \text{if } n \geq 3, \end{cases}$$

where $\alpha = o(T^1)$ and $\beta = o(T^2)$.

DEFINITION 12.72. We define $\tilde{c}(\alpha)$ to be the number of symbols to represent α in a canonical form, namely, $\tilde{c}(0)=1$, $\tilde{c}(\omega^\alpha)=\tilde{c}(\alpha)+1$ and $\tilde{c}(\alpha \dot{+} \beta)=\tilde{c}(\alpha)+\tilde{c}(\beta)+1$.

LEMMA 12.73. (1) *For every* $\alpha<\varepsilon_0$, *there exists a tree* $T \in \mathscr{T}$ *such that* $o(T)=\alpha$ *and* $|T| \leqslant 3\tilde{c}(\alpha)$.

(2) *Let* $T \in \mathscr{T}$ *and* $c \in T$. *We define* $T^c=\{d \in T \mid d \geqslant c\}$. *For every* $c, d \in T$, $c \leqslant d \to o(T^c) \geqslant o(T^d)$.

(3) *Let* $T_1, T_2 \in \mathscr{T}$ *and* $f: T_1 \to T_2$ *be an embedding. Then for every* $a \in T_1$,

$$o(T_1^a) \leqslant o(T_2^{f(a)}).$$

(4) *Let* $T_1, T_2 \in \mathscr{T}$. *If* $T_1 \leqslant T_2$, *then* $o(T_1) \leqslant o(T_2)$.

PROOF. (1) and (2) are obvious. (3) is proved by induction on the number of elements in T_1^a which is denoted by $|T_1^a|$. If $|T_1^a|=1$, then $o(T_1^a)=0 \leqslant o(T_2^{f(a)})$. Now let $|T_1^a|>1$ and $T_1^{b_1}, \ldots, T_1^{b_n}$ be all the components of $T_1^a-\{a\}$. Let c_i be an immediate successor of $f(a)$ satisfying $f(a) \leqslant c_i \leqslant f(b_i)$. Then c_i is uniquely determined and by the induction hypothesis we have

$$o(T_1^{b_i}) \leqslant o(T_2^{f(b_i)}) \leqslant o(T_2^{c_i}).$$

From this follows $o(T_1^a) \leqslant o(T_2^{f(a)})$ immediately. (4) follows from (3).

Kruskal proved the following theorem on finite trees.

THEOREM (Kruskal). *Let* $\langle T_k \mid k \leqslant \omega\rangle$ *be a sequence of finite trees, then there exist* $i<j<\omega$ *such that* $T_i \leqslant T_j$.

Kruskal's theorem and Lemma 12.73 immediately imply the accessibility of ε_0. However the statements of Kruskal's theorem and the accessibility of ε_0 are of second order. Using Friedman's device, we will make the first order miniature of Kruskal's theorem.

DEFINITION 12.74. By the finite version of Kruskal's theorem we mean the following statement:
For every natural number n, there exists k such that for every sequence $\langle T_0, \ldots, T_k\rangle$ of finite trees satisfying $|T_i| \leqslant n(i+1)$ there exist $i<j \leqslant k$ such that $T_i \leqslant T_j$.

The finite version of Kruskal's theorem denoted by **FKT** is an immediate corollary of Kruskal's theorem itself. In order to see this, suppose that **FKT** were false. Then there exists n such that for every k, there exists $\langle T_0, T_1, \ldots, T_k\rangle$ with $|T_i| \leqslant n \cdot (i+1)$ such that for every $i<j \leqslant$

k, $T_i \not\leq T_j$. We fix one such n and define $\mathcal{K}$ to be the collection of $\langle T_0, T_1, \ldots, T_k \rangle$ satisfying $|T_i| \leq n \cdot (i+1)$ for every $i < k$ and $\forall i \forall j (i < j \leq k \supset T_i \not\leq T_j)$. Then by König's lemma, there exists an infinite sequence $T_0, T_1, T_2, \ldots$ such that $\forall i \forall j (i < j \supset T_i \not\leq T_j)$ which contradicts to Kruskal's theorem.

Remark that **FKT** is of the first order. Our weaker version of Friedman's theorem is the following.

THEOREM 12.75 (H. Friedman). **FKT** *is not provable in* **PA**.

In order to prove the theorem, we need several preparations.

DEFINITION 12.76. (1) A sequence $\langle \beta_i \mid i < \omega \rangle$ from ε_0 is slow iff $\exists_n \forall_i (\tilde{c}(\beta_i) \leq n \cdot (i+1))$.

(2) **SWO**(ε_0) is the statement that for every n there exists k such that for every sequence $\langle \beta_0, \beta_1, \ldots, \beta_k \rangle$ from ε_0,

$$\forall i \leq k (\tilde{c}(\beta_i) \leq n \cdot (i+1)) \supset \neg(\beta_0 > \beta_1 > \ldots > \beta_k).$$

SWO(ε_0) is a first order statement.

LEMMA 12.77. **FKT** $\rightarrow$ **SWO**(ε_0) *is provable in* **PA**.

PROOF. Is immediate from (1) of Lemma 12.73.

DEFINITION 12.78. **PRWO**(ε_0) is the statement that there are no primitive recursive strictly descending sequence from ε_0. **PRWO**(ε_0) is also a first order statement.

LEMMA 12.79. **PRWO**(ε_0) $\rightarrow$ Cons(**PA**) *is provable in* **PA**, *where* Cons(**PA**) *is the consistency of* **PA**.

PROOF. If there exists (a Gödel number of) a proof p to a contradiction in **PA**, then $f(n) = O(r^n(p))$ is a primitive recursive strictly descending sequence from ε_0, where r is the Gentzen's reduction and $O(p)$ is the ordinal assigned to p.

LEMMA 12.80. *Let $f: \mathbb{N} \rightarrow \mathbb{N}$ be primitive recursive. Then there exists a primitive recursive function $g: \mathbb{N}^2 \rightarrow \omega^\omega$ such that*

(1) $g(n, m) > g(n, m+1)$ *if* $m < f(n)$,

(2) $\tilde{c}(g(n, m)) \leq \text{constant}(n + m + 1)$, *where constant means that there exists some constant k such that $\tilde{c}(g(n, m)) \leq k \cdot (n + m + 1)$ for every n and m.*

PROOF. We are going to define $g: \mathbb{N}^2 \rightarrow \omega^k$ for some $k < \omega$, where k depends on f.

Case (1). $f(n) = n+1$.
Define g by

$$g(n, m) = n + 2 \dot{-} m.$$

Case (2). Let $g : \mathbb{N}^2 \to \omega^k$ satisfy the conditions (1) and (2) for f and f' be defined by $f'(n) = f^n(n)$. Then define g' by

$$g'(n, m) = \omega^k \cdot (n - i) + g(f^i(n), m),$$

where $m = f(n) + f^2(n) + \ldots + f^i(n) + j$, $i < n$ and $j < f^{i+1}(n)$. g' satisfies the conditions (1) and (2) for f'. For example,

$$\begin{aligned} \tilde{c}(g'(n, m)) &\leq \text{constant} \cdot n + \text{constant}(f^i(n) + m + 1) \\ &\leq \text{constant}(n + m + 1). \end{aligned}$$

Case (3). For an arbitrary primitive recursive function f, there exists f_k such that $\forall n\, (f(n) \leq f_k(\text{constant} + n))$, where f_k is defined in Grzegorczyk hierarchy namely $f_0(n) = n+1$, $f_{k+1}(n) = f_k^n(n)$. For f_k, there exists $g_k : \mathbb{N}^2 \to \omega^{k+1}$ satisfying (1) and (2) for f_k. We define g by $g(n, m) = g_k(\text{constant} + n, m)$. Then g satisfies the conditions (1) and (2) for f.

LEMMA 12.81. *For a given primitive recursive strictly descending sequence $\langle \beta_n \mid n \in \mathbb{N} \rangle$ from ε_0, one can find a slow primitive recursive strictly descending sequence $\langle \alpha_m \mid m \in \mathbb{N} \rangle$.*

PROOF. Let a primitive recursive function $g : \mathbb{N}^2 \to \omega^\omega$ satisfy the conditions that $g(n, j) > g(n, j+1)$ for every $j < \tilde{c}(\beta_{n+1})$ and $\tilde{c}(g(n, j)) \leq \text{constant}(n + j + 1)$. Define $\alpha_m = \omega^\omega \cdot \beta_n + g(n, j)$, where $m = \tilde{c}(\beta_1) + \tilde{c}(\beta_2) + \ldots + \tilde{c}(\beta_n) + j$, $j < \tilde{c}(\beta_{n+1})$. Then we have

$$\begin{aligned} \tilde{c}(\alpha_m) &\leq \text{constant} \cdot \tilde{c}(\beta_n) + \text{constant}(n + j + 1) \\ &\leq \text{constant}(m + 1). \end{aligned}$$

PROOF OF THEOREM 12.75. Now the theorem follows immediately from Lemmas 12.77, 12.79 and 12.81.

§13. Provable well-orderings

In this section, in order to distinguish between the natural ordering of natural numbers and the order relation on numbers given by the standard ordering of type ε_0, we denote the latter by $\prec$ in this section.

A partial function is a number-theoretic function that may not be defined at all arguments.

DEFINITION 13.1. (1) The class of *partial recursive functions* is the class of partial functions generated by the schemata (i)–(vi) for primitive recursive functions (cf. Definition 10.2), and also the schema:

(vii) $f(x_1, \ldots, x_n) \simeq \mu y[g(x_1, \ldots, x_n, y) = 0]$, where g is partial recursive; the right-hand side means the least y such that $\forall z < y(g(x_1, \ldots, x_n, z)$ is defined and $\neq 0)$ and $g(x_1, \ldots, x_n, y) = 0$, if such a y exists, and undefined otherwise; and $\simeq$ means that the left-hand side is defined if and only if the right-hand side is, in which case they are equal.

(2) A *general recursive* or *total recursive* or *recursive* function is a partial recursive function which is *total*, i.e., defined at all arguments.

(3) A relation on natural numbers, say R, is called *recursive* if there is a recursive function f which assumes values 0 and 1 only such that $R(x_1, \ldots, x_n)$ holds if and only if $f(x_1, \ldots, x_n) = 0$.

(4) A Σ_1^0-formula of the language L is a formula of the form

$$\exists y\,(\bar{f}(x_1, \ldots, x_n, y) = 0),$$

$\bar{f}$ a primitive recursive function symbol. A Π_1^0-formula is similarly of the form $\forall y\,(\bar{f}(x_1, \ldots, x_n, y) = 0$, $\bar{f}$ primitive recursive.

It can be shown that any recursive relation R can be represented in **PA** by a Σ_1^0-formula, i.e., there is a Σ_1^0-formula $\bar{R}(x_1, \ldots, x_n)$ of the language L such that, for all $m_1, \ldots, m_n$:

$$R(m_1, \ldots, m_n) \text{ holds} \leftrightarrow \bar{R}(\bar{m}_1, \ldots, \bar{m}_n) \text{ is } \mathbf{PA}\text{-provable}.$$

Also, any recursive relation can be represented in **PA** by a Π_1^0-formula.

DEFINITION 13.2. Let ε be a new predicate constant. $L(\varepsilon)$ is the language extending L (cf. §12), formed by admitting $\varepsilon(t)$ as an atomic formula for all terms t.

PA(ε) is the system **PA** in the language $L(\varepsilon)$; more precisely, we extend **PA** by admitting as mathematical initial sequents $s = t, \varepsilon(s) \rightarrow \varepsilon(t)$ for all terms s, t and applying the rule ind to all formulas of $L(\varepsilon)$.

DEFINITION 13.3. Let $<\cdot$ be a recursive (infinite) linear ordering of the natural numbers which is actually a well-ordering. (Without loss of generality we may assume that the domain of $<\cdot$ is the set of all natural numbers and the least element with respect to $<\cdot$ is 0.) We use the same symbol $<\cdot$ in order to denote the Σ_1^0-formula in **PA** which represents the ordering $<\cdot$.

Consider the sequent

$$TI(<\cdot): \qquad \forall x\,(\forall y <\cdot\, x(\varepsilon(y)) \supset \varepsilon(x)) \rightarrow \varepsilon(a)$$

(cf. the formula $TI(<, F(x))$ of Remark 12.10). If $TI(<\cdot)$ is provable in $\mathbf{PA}(\varepsilon)$, then we say that $<\cdot$ is a *provable well-ordering* of **PA**.

The following theorem is proved by analyzing Gentzen's proof of the unprovability of the well-ordering of $\prec$ (where $\prec$ was defined at the beginning of this section).

THEOREM 13.4 (Gentzen). *If $<\cdot$ is a provable well-ordering of* **PA**, *then there exists a recursive function which is a $<\cdot - \prec$ order-preserving map into an initial segment of ε_0. That is to say, there is a recursive function f such that $a <\cdot b$ if and only if $f(a) \prec f(b)$, and there is an ordinal $\mu(<\varepsilon_0)$ such that for every a, $f(a) \prec \bar{\mu}$ (where $\bar{\mu}$ is the Gödel number of μ).*

This section is devoted to Gentzen's proof, and the arithmetization of it, which proves Theorem 13.4.

From now on, let $<\cdot$ be a fixed provable well-ordering of **PA**.

13.1) First we define **TJ**-proofs, where **TJ** stands for "transfinite induction". **TJ**-proofs are defined as $\mathbf{PA}(\varepsilon)$-proofs with some modifications:

(1) The initial sequents of a **TJ**-proof are those of $\mathbf{PA}(\varepsilon)$, and the following sequents, called **TJ**-initial sequents:

$$\forall x(x <\cdot t \supset \varepsilon(x)) \rightarrow \varepsilon(t)$$

for arbitrary terms t.

(2) The end-sequent of a **TJ**-proof must be of the form

$$\rightarrow \varepsilon(\bar{m}_1), \ldots, \varepsilon(\bar{m}_n),$$

where $\bar{m}_1, \ldots, \bar{m}_n$ are numerals.

Let $|m|_{<\cdot}$ be the ordinal denoted by m with respect to $<\cdot$, i.e., the order type of the initial segment of $<\cdot$ determined by m. Then the minimum of $|m_1|_{<\cdot}, \ldots, |m_n|_{<\cdot}$ is called the end-number of the **TJ**-proof.

13.2) Since $<\cdot$ is a provable well-ordering of **PA**, the sequent $TI(<\cdot)$ (Definition 13.3) is $\mathbf{PA}(\varepsilon)$-provable, and hence we can obtain in the system formed from $\mathbf{PA}(\varepsilon)$ by adjoining **TJ**-initial sequents, a proof $P(a)$ of $\rightarrow \varepsilon(a)$ (for a free variable a). Note that for each number m, $P(\bar{m})$ is a **TJ**-proof of $\rightarrow \varepsilon(\bar{m})$.

13.3) A **TJ**-proof is called non-critical if one of the reduction steps for **PA** (in the proof of Lemma 12.8) which lower the ordinal (i.e., step 2, 3 or 5) applies to it. Otherwise it is called critical.

13.4) We shall assign ordinals (less than ε_0) to **TJ**-proofs and define a reduction for **TJ**-proofs following the reduction method for **PA** given in the proof of Lemma 12.8: if a **TJ**-proof is critical, then more manipulation is required. The reduction is defined in such a manner that a **TJ**-proof P

with end-number > 0 is reduced to another with the same end-number if P is not critical and with an arbitrary end-number which is smaller than the original one if P is critical. At the same time the ordinal decreases.

13.5) If we can define an ordinal assignment and a reduction method with the properties stated in 13.4), we can prove:

LEMMA 13.5 (Fundamental Lemma). *For any* **TJ**-*proof, its end-number is not greater than its ordinal.*

PROOF. By transfinite induction on the ordinal of the proof. Let P be a **TJ**-proof with ordinal μ and end-number σ. We assume as the induction hypothesis that the lemma is true for any **TJ**-proof whose ordinal is less than μ and show that $\sigma \leqslant \mu$. If P is non-critical then P is reduced to a **TJ**-proof P' with the same end-number σ and an ordinal $\nu < \mu$. By the induction hypothesis $\sigma \leqslant \nu$, and hence $\sigma \leqslant \mu$. Now suppose P is critical. If σ were greater than μ, we could reduce P to a **TJ**-proof whose end-number is μ and whose ordinal is less than μ, contradicting the induction hypothesis.

Now let us proceed to the reduction method for **TJ**-proofs.

13.6) The ordinals are assigned to the sequents of the **TJ**-proofs as in §12; the ordinal of a **TJ**-initial sequent is 7, i.e., $\omega^0 + \ldots + \omega^0$ (7 times). The lower sequent of a term-replacement inference is assigned the same ordinal as the upper sequent. For convenience, the formula in the succedent of a **TJ**-initial sequent will be considered as a principal formula.

13.7) We can follow the reduction steps given for the consistency proof of **PA** up to Step 4 (in the proof of Lemma 12.8), i.e., until we reach a **TJ**-proof P with the following properties p 1–p 4.

p 1. The end-piece of P contains no free variable.

p 2. The end-piece of P contains no induction.

p 3. The end-piece of P contains no logical initial sequent.

p 4. If the end-piece of P contains a weakening I, then any inference below I is a weakening.

REMARK. Since the end-piece of a **TJ**-proof is not empty, the end-sequent S' of the proof obtained from P by eliminating weakenings in the end-piece (in Step 4) may be different from the end-sequent of P. In this case we add weakenings below S' so that the end-sequent becomes the same as the end-sequent of P.

13.8) We can easily show the following. Let P be a **TJ**-proof satisfying p 1–p 4. Then P contains at least one logical inference (which must be implicit) or **TJ**-initial sequent. Therefore the end-piece of P contains a principal formula at the boundary or in a **TJ**-initial sequent.

13.9) Let P be a **TJ**-proof satisfying p 1–p 4. By 13.8), the end-piece of P contains a principal formula either at the boundary or in a **TJ**-initial sequent. We call a formula A in the end-piece of P a principal descendant or a principal **TJ**-descendant, according as A is a descendant of a principal formula at the boundary or a descendant of the principal formula of a **TJ**-initial sequent in the end-piece of P.

Note that a principal **TJ**-descendant in the end-piece of P always occurs in the succedent of a sequent, and has the form $\varepsilon(t)$.

13.10) Let P be a **TJ**-proof satisfying p 1–p 4, and S a sequent in the end-piece of P. If S contains a formula B with a logical symbol, then there exists a formula A in S or in a sequent above S such that A is a principal descendant or a principal **TJ**-descendant.

PROOF. Suppose S contains a formula with a logical symbol. Then S is above the uppermost weakening in the end-piece. The property of sequents, of containing a logical symbol, is preserved upwards, to one of the upper sequents of each inference in the end-piece (but not necessarily beyond a boundary inference), or a **TJ**-initial sequent, when we follow upward the string to which S belongs. Notice that B may not be A, since B may be a descendant of a formula which is "passive" at a boundary inference.

13.11) Let P be a **TJ**-proof satisfying p 1–p 4 and not containing a suitable cut. Then its end-sequent contains a principal **TJ**-descendant.

PROOF. It suffices to prove that the end-sequent of P contains a principal descendant or a principal **TJ**-descendant, since the end-sequent contains no logical symbol. Suppose not. Since the end-piece contains a principal descendant or a principal **TJ**-descendant by 13.8), let us consider the following property (P) of cuts in the end-piece of P: A cut in the end-piece of P is said to have the property (P) if (at least) one of its upper sequents contains such a formula and its lower sequent contains no such formula. Since the end-piece contains such a formula, but the end-sequent does not (by assumption), there must be such a cut. Let I be an uppermost cut with the property (P) in the end-piece of P:

$$I \quad \frac{\Gamma \rightarrow \Delta, D \qquad D, \Pi \rightarrow \Lambda}{\Gamma, \Pi \rightarrow \Delta, \Lambda}.$$

Let S_1 and S_2 be the left and right upper sequents of I, respectively. By our assumption one of the cut formulas is a principal descendant or a principal **TJ**-descendant. First suppose D in S_1 has this property. If D contains a logical symbol, then it is a principal descendant. Then also, S_2 contains a formula with a logical symbol (namely D). Therefore, by 13.10), there is a formula A in S_2 or above it such that A is a principal descendant or a principal **TJ**-descendant. If there is no such formula in S_2, there must be a cut having the property (P) above I, contradicting our choice of I. If such a formula A is in S_2, A must be D itself, which contradicts our assumption that P does not contain a suitable cut. Thus D must be of the form $\varepsilon(t)$. Now suppose S_2 contains a logical symbol. Then there exists a principal descendant or a principal **TJ**-descendant either in S_2 or above it. If it is in S_2, it cannot be D (since D is $\varepsilon(t)$ and is in the left side of a sequent, it cannot be a principal **TJ**-descendant), and so it must also appear in the lower sequent of I, contradicting our assumption that I has the property (P). This means that such a formula is in a sequent above S but not in S itself, contradicting our assumption that I is an uppermost cut with the property (P). Thus S_2 cannot contain a formula with a logical symbol. Since I is an uppermost cut with the property (P), no logical inference at the boundary or **TJ**-initial sequent in the end-piece is above S_2. Therefore the proof down to S_2 is included in the end-piece and no logical initial sequents or **TJ**-initial sequents occur there and it is impossible that S_2 contains $\varepsilon(t)$, and so D cannot be $\varepsilon(t)$. Hence we have shown that D in S_1 cannot be a principal descendant or principal **TJ**-descendant. Next, suppose that the cut formula in S_2 is a principal descendant or principal **TJ**-descendant. As was seen above, D cannot be a principal **TJ**-descendant: D must contain a logical symbol. Hence there is a principal descendant or a principal **TJ**-descendant either in S_1 or in a sequent above S_1. If such a formula is not in S_1, there must be a cut having the property (P) above S_1, which contradicts our assumption about I. Therefore D in S_1, must have that property, since the lower sequent of I cannot contain such a formula. This again contradicts our assumption that P does not contain a suitable cut.

13.12) Now let P be a critical **TJ**-proof to which the reduction of Lemma 12.8 has been applied as far as possible (i.e., up to Step 4). Then P satisfies p 1–p 4 and does not contain a suitable cut (since it is critical). We define the notion of critical reduction. By 13.11), the end-sequent of P contains a principal **TJ**-descendant, $\varepsilon(\bar{m}_i)$, say, the descendant of a principal formula $\varepsilon(r)$ (where the closed term r denotes the number m_i). Let m be any number such that $|m|_{<\cdot}$ is less than the end-number of P. Then $\bar{m} <\cdot\, r$ is a true Σ_1^0-sentence of **PA**, and hence the sequent $\rightarrow \bar{m} <\cdot\, r$ can be derived from a mathematical initial sequent of **PA** (say $\rightarrow F$) by one application of $\exists$: right. So we replace the **TJ**-initial sequent

$$\forall x\,(x <\cdot\, r \supset \varepsilon(x)) \rightarrow \varepsilon(r)$$

in P by an ordinary proof in **PA**(ε):

$$\dfrac{\dfrac{\dfrac{\dfrac{\rightarrow F}{\rightarrow \bar{m} <\cdot\, r} \qquad \varepsilon(\bar{m}) \rightarrow \varepsilon(\bar{m})}{\bar{m} <\cdot\, r \supset \varepsilon(\bar{m}) \rightarrow \varepsilon(\bar{m})}}{\forall x\,(x <\cdot\, r \supset \varepsilon(x)) \rightarrow \varepsilon(\bar{m})}}{\forall x\,(x <\cdot\, r \supset \varepsilon(x)) \rightarrow \varepsilon(\bar{m}), \varepsilon(r).}$$

The ordinal of this proof is 4 and is less than that of a **TJ**-initial sequent (which is 7). By this replacement and some obvious changes, P is transformed into a **TJ**-proof P' whose end-sequent is

$$\rightarrow \varepsilon(\bar{m}),\ \varepsilon(\bar{m}_1), \ldots, \varepsilon(\bar{m}_n),$$

where $\rightarrow \varepsilon(\bar{m}_1), \ldots, \varepsilon(\bar{m}_n)$ is the end-sequent of P, and such that the ordinal of P' is less than that of P and the end-number of P' is $|m|_{<\cdot}$. We shall refer to P' as the proof obtained from P by an application of a critical reduction at m.

Now suppose P is any **TJ**-proof (not necessarily critical), and $|m|_{<\cdot}$ is less than the end-number of P. We shall define what is meant by the proof obtained from P be an application of a critical reduction at m.

If P is critical, the definition is as above. Otherwise, apply a sequence of reductions (as in the proof of Lemma 12.8). At each reduction, the ordinal of the proof *decreases*, so this process must terminate after a finite number of steps with a *critical* proof satisfying p 1–p 4. Now take the proof obtained from *this* proof as above.

13.13) Adjoining the reduction in 13.12) to the previous reductions, and applying the fundamental lemma in 13.5), we obtain the original form of Gentzen's theorem:

THEOREM 13.6. *The order type of* $<\cdot$ *is less than* ε_0.

13.14) Let $P(a)$ be a proof of $\rightarrow \varepsilon(a)$, obtained as described in 13.2. Let us define for each number k a **TJ**-proof P_k by induction on k, where the end-number of P_k is $|k|_{<\cdot}$.

(1) The case where $\forall n < k\ (n <\cdot\ k)$. We define P_k to be the proof $P(\bar{k})$ obtained from $P(a)$ by replacing a by the numeral $\bar{k}$ throughout $P(a)$.

(2) The case where $\exists_n < k(k <\cdot\ n)$. Let

(**) $$n_0 <\cdot \ldots <\cdot\ n_{j-1} <\cdot\ n_j(=k) <\cdot\ n_{j+1} <\cdot \ldots <\cdot\ n_k$$

be the re-ordering of the numbers $\leqslant k$ with respect to $<\cdot$. Then we define P_k to be the proof obtained from $P_{n_{j+1}}$ by applying a critical reduction at k (cf. 13.12)). It is obvious that this definition is recursive.

13.15) We now define a map f, which will turn out to be an order-preserving recursive map as required for Theorem 13.4, by making use of the P_k. Define $f(k)$ by induction on k:

$$f(0) = \omega^{o(P_0)},$$

and for $k > 0$, $f(k) = f(n_{j-1}) + \omega^{o(P_k)}$ where $o(P)$ is (the Gödel number of) the ordinal of P, $+$ is (the primitive recursive function representing) addition of ordinals, ω^a is (the primitive recursive function representing) exponentiation by ω, and n_{j-1} is an in (**) (such a number always existing if $k>0$).

13.16) Let $m_0 <\cdot\, m_1 <\cdot \ldots <\cdot\, m_i$ be the re-ordering of the numbers $< i+1$ with respect to $<\cdot$. Then

$$f(m_{j+1}) = f(m_j) + \omega^{o(P_{m_{j+1}})},$$

where $0 \leq j < i$. This is proved by mathematical induction on i. For $i=0$, this is trivial. Assume it for i. For the case of $i+1$, it is sufficient to show (with $m_0, \ldots, m_i$ as above):

$$f(i+1) = f(m_j) + \omega^{o(P_{i+1})} \tag{1}$$

and

$$f(m_{j+1}) = f(i+1) + \omega^{o(P_{m_{j+1}})}, \tag{2}$$

where $m_j <\cdot\, i+1 <\cdot\, m_{j+1}$. Here (1) holds by definition of f, and (2) follows from (1) and $f(m_{j+1}) = f(m_j) + \omega^{o(P_{m_{j+1}})}$ (by induction hypothesis) and $o(P_{i+1}) < o(P_{m_{j+1}})$ (by definition of P_{i+1}). The second point of Theorem 13.4 is also easily seen if one puts $\mu = \omega^{o(P(a))+1}$. This completes the proof of Theorem 13.4.

To end this section, another result of Gentzen will be stated. The proof is straightforward.

THEOREM 13.7. *Let $<_n$ be the standard well-ordering of ε_0, restricted to ω_n. Then $<_n$ is a provable well-ordering of* **PA**.

Kleene's T-predicate (for unary, i.e., one-argument functions) is a primitive recursive relation T such that for an arbitrary partial recursive function f (of one argument) there exists a number e for which

$$f(x) \simeq U(\mu y\, T(e, x, y))$$

for all x. (U is a fixed primitive recursive function). Such an e is called a Gödel number of f. The definition can be extended to functions of many arguments.

If e is the Gödel number of a unary partial recursive function, then clearly

$$f \text{ is (total) recursive if and only if } \forall x\ \exists y\ T(e, x, y).$$

Further, f is called *provably recursive* (in **PA**) if it has a Gödel number e

such that $\forall x\, \exists y\, T(\bar{e}, x, y)$ is **PA**-provable. Having discussed the Gödel numbering of recursive functions, we can now state a problem which should, in its correct context, actually have been placed in §12. The idea is due to Schütte.

PROBLEM 13.8. Let **PA*** be the system obtained by modifying **PA** as follows. The language is the same as that of **PA**; the initial sequents are those of **PA**; the rules of inference are those of **PA** except cut, $\forall$: right and ind; the constructive ω-rule, which is described below, is added as a new rule of inference:

$$\frac{P_1 \ldots P_i \ldots}{\Gamma \rightarrow \Delta, \forall x\, A(x)} \quad (i < \omega),$$

where P_i is a proof ending with $\Gamma \rightarrow \Delta,\ A(\bar{i})$, and there is a recursive function f such that $f(i) = \ulcorner P_i \urcorner$. Let e be a Gödel number of f. Then the proof ending with $\Gamma \rightarrow \Delta, \forall x\, A(x)$ is assigned the number

$$5^e \cdot 7^{\ulcorner \Gamma \rightarrow \Delta, \forall x A(x) \urcorner}.$$

Show that if a sequent S is **PA**-provable and contains no free variable, then S is provable in **PA***. [*Hint*: We adapt the method of the consistency proof of **PA** as follows. Let P be a (regular) proof in **PA**, with ordinal α (according to the assignment of Definition 12.4). Then assign $\omega^\alpha + m$ to P, where m is the number of free variables in the end-piece of P. The reduction process for the consistency proof goes through almost unchanged, except that if P contains an explicit logical inference and the lowermost such is a $\forall$: right, then replace it by the ω-rule, which is applied at the end of the proof.]

PROBLEM 13.9. Let f be a provably recursive function in **PA**. Then there exists an ordinal μ (less than ε_0) such that f is $<^\mu$-primitive recursive, where $<^\mu$ is the standard ordering of ε_0 restricted to μ. [*Hint*: Let e be a Gödel number of f such that $\forall x\, \exists y\, T(\bar{e}, x, y)$ is **PA**-provable. Then there is a proof, say $P(a)$, of $\exists y\, T(\bar{e}, a, y)$, with free variable a. Let μ be the ordinal assigned to $P(a)$, and let P_m denote $P(\bar{m})$ for each natural number m. By the method of Problem 13.8, P_m can be transformed into a cut-free proof in **PA*** of the same end-sequent. It can be easily shown that the resulting proof does not contain the ω-rule, since $P(\bar{m})$ does not contain any explicit $\forall$: right. The transformation is actually $<^\mu$-primitive recursive. Thus there is a $<^\mu$-primitive recursive function τ such that $\tau(\ulcorner P_m \urcorner)$ is (the Gödel number of) a cut-free proof of $\exists y\, T(\bar{e}, \bar{m}, y)$. By examining this proof, we can find (primitive recursively in its Gödel number) a number n satisfying $T(e, m, n)$. Then n is a $<^\mu$-primitive recursive function of m and $f(m) = U(n)$. Thus f is $<^\mu$-primitive recursive.]

§14. An additional topic

Here we assume again that all the primitive recursive functions are included in the language of **PA** and their defining equations are included as initial sequents.

PROPOSITION 14.1. *Let Φ_n be the set of sentences of* **PA** *which have at most n logical symbols. Then there exists a truth definition for Φ_n in* **PA**, *i.e., a formula $T_n(a)$ of* **PA** *such that for every sentence A of Φ_n*

$$T_n(\overline{\ulcorner A \urcorner}) \equiv A$$

is **PA**-*provable.*

PROOF. T_n is defined by induction on n. We shall present only the induction step, in passing from T_n to T_{n+1}.

A sequence number, say x, is a number which can be decomposed into the form $2^{x_0} \cdot 3^{x_1} \cdot \ldots \cdot p_{n-1}^{x_{n-1}}$, where $x_i = 0$ or 1 for each i, $0 \leq i \leq n$. Let seq(x, n) be a (primitive recursive) predicate which expresses that x is a sequence number of the above form. We call n the length of x. The ith exponent of x, x_i, will be denoted $x(i)$. Let st$(\ulcorner A \urcorner)$ express "A is a sentence", and let ls$(\ulcorner A \urcorner)$ be the number of logical symbols in A. Then T_{n+1} is defined as follows.

$$\begin{aligned}
&T_{n+1}(\ulcorner A \urcorner) \leftrightarrow \\
&\leftrightarrow \mathrm{st}(\ulcorner A \urcorner) \wedge \mathrm{ls}(\ulcorner A \urcorner) \leq n+1 \\
&\quad \wedge \exists x\,[\mathrm{seq}(x, \ulcorner A \urcorner) \wedge \forall i\,(0 \leq i \leq \ulcorner A \urcorner \supset \\
&\qquad (\forall \ulcorner B \urcorner [i = \ulcorner \neg B \urcorner \supset (x(i) = 1 \equiv x(\ulcorner B \urcorner) = 0)] \\
&\qquad \wedge \forall \ulcorner B \urcorner \forall \ulcorner C \urcorner [i = \ulcorner B \wedge C \urcorner \\
&\qquad\qquad\qquad \supset (x(i) = 1 \equiv x(\ulcorner B \urcorner) = 1 \wedge x(\ulcorner C \urcorner) = 1)] \\
&\qquad \wedge \forall \ulcorner \forall y\, B(y) \urcorner [i = \ulcorner \forall y\, B(y) \urcorner \supset (x(i) = 1 \equiv \forall y\, T_n(\ulcorner B(\bar{y}) \urcorner))] \\
&\qquad \wedge \forall \ulcorner \exists y\, B(y) \urcorner [i = \ulcorner \exists y\, B(y) \urcorner \supset (x(i) = 1 \equiv \exists y\, T_n(\ulcorner B(\bar{y}) \urcorner))])) \\
&\qquad \wedge x(\ulcorner A \urcorner) = 1].
\end{aligned}$$

It is easily seen that

$$T_{n+1}(\overline{A(\bar{b}_1, \ldots, \bar{b}_n)}) \equiv A(b_1, \ldots, b_n)$$

is **PA**-provable for every A is Φ_n, where all the free variables of A are among $b_1, \ldots, b_n$.

Let $S: A_1, \ldots, A_m \rightarrow B_1, \ldots, B_l$ be a sequent such that all of $A_1, \ldots, A_m, B_1, \ldots, B_l$ are in Φ_n. Then $T_n(\ulcorner S \urcorner)$ is defined to be

$$\exists i\,(1 \leq i \leq m \wedge \neg T_n(\ulcorner A_i \urcorner)) \vee \exists i\,(1 \leq i \leq l \wedge T_n(\ulcorner B_i \urcorner))).$$

Here of course m and l are primitive recursive functions of $\ulcorner S \urcorner$ and A_i and B_i are determined primitive recursively from $\ulcorner S \urcorner$ and i.

PROPOSITION 14.2. **PA** *cannot be formulated with finitely many axioms; in other words, mathematical induction cannot be expressed by finitely many formulas.*

PROOF (Feferman). First note that

$$\mathbf{PA} \vdash (\vdash \ulcorner S \urcorner \rightarrow \underset{CF}{\vdash} \ulcorner S \urcorner), \tag{1}$$

by formalizing the cut-elimination theorem for **LK** in **PA**.

Next, suppose P is a cut-free proof of a sequent S, and all the formulas in S are in F_n. Then every formula in P is in F_n. Further, if P is in the language of **PA**, then we can prove in **PA** that every numerical instance of S is true; in other words:

$$\mathbf{PA} \vdash \underset{CF}{\vdash} \ulcorner S(b_1, \ldots, b_m) \urcorner \rightarrow \forall x_1 \ldots x_m\, T_n(\ulcorner S(\bar{x}_1, \ldots, \bar{x}_m) \urcorner), \tag{2}$$

where all the free variables of S are among $b_1, \ldots, b_m$. The proof of (2) is by induction on the number of sequents in P.

Now let Γ_0 be any finite set (or rather sequence) of axioms of CA ∪ VJ (Definition 9.5) and let n be the maximum number of logical symbols in any formula of Γ_0. Letting S be $\Gamma_0 \rightarrow \bar{0} = \bar{1}$, we obtain from (2):

$$\mathbf{PA} \vdash \underset{CF}{\vdash} \ulcorner \Gamma_0 \rightarrow \bar{0} = \bar{1} \urcorner \rightarrow T_n(\ulcorner \Gamma_0 \rightarrow \bar{0} = \bar{1} \urcorner). \tag{3}$$

Further (of course):

$$\mathbf{PA} \vdash \neg T_n(\ulcorner \Gamma_0 \rightarrow \bar{0} = \bar{1} \urcorner)$$

and hence, from (1) and (3):

$$\mathbf{PA} \vdash \neg \vdash \ulcorner \Gamma_0 \rightarrow \bar{0} = \bar{1} \urcorner.$$

This sentence, $\neg \vdash \ulcorner \Gamma_0 \rightarrow \bar{0} = \bar{1} \urcorner$, can be taken as expressing the consistency of Γ_0, which, as we see, is provable in **PA**. Hence, by Gödel's second incompleteness theorem (Theorem 10.18), Γ_0 cannot be proof-theoretically equivalent to **PA**.

EXERCISE 14.3. Show that ZF (Zermelo-Fraenkel set theory) cannot be formulated with finitely many axioms; in other words, the axiom of replacement cannot be expressed by finitely many formulas.

PART II

SECOND ORDER AND FINITE ORDER SYSTEMS

PART II

SECOND ORDER AND FINITE ORDER SYSTEMS

Prior to the rigorous development of Part II, we shall first explain the significance of the theory of higher (finite) order systems, and difficulties related to it.

Let us, to begin with, adopt the standpoint of an "infinite mind", which we suppose can examine infinitely many objects one by one. From such a standpoint, the meaning of the first order predicate calculus is quite clear; in other words, the meaning of the quantifiers ($\forall x$ and $\exists x$) is clearly and unambiguously defined. That is, given a structure $\mathcal{D} = \langle D, \phi \rangle$, $\forall x$ means "for every element x of D" and $\exists x$ means "there exists an element of x of D". One could question whether the structure $\mathcal{D}$ is well-defined; nevertheless there is no doubt about the first order predicate calculus itself. Once we begin to consider arbitrary sets over the domain of a structure $\mathcal{D} = \langle D, \phi \rangle$, i.e., subsets of D, however, the situation becomes entirely different. We must then assume that the infinite mind possesses the following capabilities in addition to that mentioned above:

1) It can unite members of D to form arbitrary subsets of D.

2) It can examine each of these subsets.

Let D be a given domain. Consider the situation where the set of subsets of D, the set of subsets of this set, and so on, come under attention. If we assume the above two capabilities of the infinite mind, then it is obvious that the comprehension axioms hold over these sets. As for the axiom of choice, although the meaning is not so clear as for the comprehension axiom, one must accept it also once one has accepted the existence of the infinite mind.

This approach is in its essence the way in which many working mathematicians conceive of sets. It is fair to say that in modern mathematics many of the arguments concerning sets are carried out along these lines, and the higher (finite) order predicate calculus is a formulation of such an approach to sets. It is, therefore, only natural that the investigation of the structure of the higher order predicate calculus (second order in particular) from the proof-theoretic viewpoint should attract our attention, although the finite order predicate calculus is a formulation of only an incomplete portion of the general notion of set. Dealing with this portion of set theory can be justified further when we notice that in practice there are very few theorems which can be obtained only in full set theory.

Furthermore, the basic notions of set theory do not seem to be quite clear, in the following senses.

(1) We do not know whether the totality of ordinals, and hence the totality of sets, is a completed totality or is in the process of creation; in other words, it is not certain whether it is a closed universe or a growing universe, so to speak. If we imagine that it is a growing universe then the meanings of the quantifiers ($\forall x$ and $\exists x$) are not clear.

(2) There are difficulties in justifying the axiom of replacement. This axiom is sometimes justified in the following manner.

(2.1) The ordinals are created endlessly.

(2.2) Suppose α is an ordinal which has already been created and f is a function defined for all $\beta < \alpha$. By virtue of (2.1), there must be an ordinal which is larger than $f(\beta)$ for every $\beta < \alpha$.

Even if we accept (2.1) and (2.2) above, the meaning of the axiom of replacement is still not very clear. The difficulty exists in the fact that the axiom of replacement involves formulas with quantifiers, hence, if the creation of sets is assumed to be endless, it is not clear what these quantifiers mean. Even if we try to interpret these quantifiers in some appropriate manner, we get into a difficulty which stems from the assumption (2.2).

We shall explain this difficulty in some detail. Suppose that (under a certain interpretation) the formula $A(x, y)$ expresses a function $y = f(x)$, and the supremum of the values of f on an ordinal α does not exist at a certain stage. (This is possible if we assume that the ordinals are still in the process of creation.) Suppose further that eventually the supremum of f on α is created. This then means that a new set is added and the meaning of the quantifiers will (in general) change accordingly. Thus $A(x, y)$ may not express the same function f any longer.

In order to illuminate this situation, let us consider a similar situation in the theory of natural numbers. The natural numbers are constructed from 0 by the operation $+1$. In the attempt to prove $\forall x\, \exists y\, (y = x + 1)$ when the numbers $0, 1, \ldots, n$ have been constructed, one finds that this sentence cannot be satisfied for $x = n$; thus one adds $n + 1$ to the set of natural numbers and proves the sentence for n as well as $0, \ldots, n-1$. This does not mean, however, that $\{0, 1, \ldots, n, n+1\}$, the set of natural numbers which has so far been constructed, satisfies the sentence (since the meaning of the quantifier $\forall x$ has changed). This example has been presented as an analogy to the incompleted universe of sets, though in the case of natural numbers we can actually complete the creation of natural numbers and construct a complete universe of natural numbers, namely ω. This is so, because in the case of natural numbers $+1$ is the sole operation for creating new objects and its behavior is quite clear. In the case of set theory the principle of creation is powerful and inexact. Even if we manage to interpret the quantifiers locally, viz. at certain stages of creation (which can be done in a manner analogous to the above example for natural numbers), it is still uncertain whether or not there is a universe in which the axiom of replacement holds. This makes us dubious about the existence of inaccessible cardinals.

The foregoing discussion supplies us with some justification for studying the finite-order predicate calculus. Namely:

1. The world of set theory is more unstable than the finite-order predicate calculus (with the comprehension axiom and the axiom of choice).

2. Most of actual mathematics (at least classical analysis) can be formalized within the finite order predicate calculus.

3. The proof theory of full set theory is just too difficult for a satisfactory investigation yet. So we must do something simpler, which will hopefully point the way to the study of stronger systems.

The above, as they stand, seem rather negative reasons for studying the finite-order predicate calculus. However, by developing the second point above, we can perhaps give a more positive reason: The point is that it is not only unnecessary to embed classical analysis in full set theory, but perhaps even misleading! Historically, full set theory was developed later than classical analysis, and in fact the principles of set theory are unnecessarily powerful for studying this part of mathematics: finite order predicate calculus is more suited to this purpose, since it formalizes (most of) the principles actually involved here. In a similar way perhaps, primitive recursive arithmetic is more appropriate than Peano arithmetic for formalizing finitist mathematics.

(To be more accurate, we should point out that some parts of classical analysis can make use of the (elementary) theory of ordinals; in fact it was this that led Cantor to develop set theory.)

The study of the finite order predicate calculus is still a field of the future, as it is not known to what extent the beautiful metatheorems of the first order predicate calculus hold here. The known results are very few. We shall list the difficulties we have encountered while workong on the finite order predicate calculus.

1. Although the cut-elimination theorem holds for the system with the comprehension axiom, and this in itself is a nice theorem, it has not supplied us with information on the structure of the system (unlike the case for the first order system, with a subformula property for cut-free proofs: see §6).

2. Unlike the first order system, the system with the comprehension axiom and the axiom of choice is not complete, with respect to the semantics of standard structures, where a standard model (or structure) for a finite order language means one in which the second order quantifiers range over *all* the subsets of the given domain, the third order quantifiers range over all subsets of these, and so on. For instance, the set of

sentences

$$\begin{aligned}
&\forall\phi\,\forall x\,(\phi(0)\wedge\forall y\,(\phi(y)\supset\phi(y'))\supset\phi(x)),\\
&\forall x\,\forall y\,\forall\phi\,(x=y\wedge\phi(x)\supset\phi(y)),\\
&\forall x\,\forall y\,(x'=y'\supset x=y),\\
&\forall x\,\neg x'=0,\\
&P(0),\,P(1),\,P(2),\ldots,\\
&\exists x\,\neg P(x)
\end{aligned}$$

is consistent but it has no standard model. (By contrast, completeness for the finite order calculus *does* hold for the semantics of so-called Henkin structures, as will be seen in §21.) Even worse, we have not discovered any nice extension of the finite order predicate calculus, let alone complete extension. We may rephrase it this way: we do not know any natural and meaningful principles for the second order predicate calculus beyond the comprehension axiom and the axiom of choice. This suggests an important new line of investigation: to look for second order systems with some inferences or transcendental character (an infinite inference for example) in order to obtain complete systems.

So far we have discussed the study of the finite order predicate calculus from the standpoint of the "infinite mind". It is certainly just as important to study the finite order predicate calculus from a standpoint which denies the absolute world of an infinite mind. In order to develop such a standpoint, we endeavour to analyze the above systems and establish foundations for them. This latter standpoint may not come to the surface in Part II; it will be in Part III that this standpoint is exploited, with the consistency proofs that are presented there. It is worth noting that though we give the discussion in Part II assuming strong means, such as set theory and the "infinite mind", there are some results here which fit a purely finitist standpoint.

The study of infinitary languages was started in an effort to break the deadlock of second order systems. An infinitary language is stronger than an ordinary first order language, yet the completeness theorem holds for it (hence is relatively weak). As may be expected, the general Henkin quantifier presents the same kind of difficulty as that which lies with the second order language. Its study is therefore still a field of the future.

CHAPTER 3

SECOND ORDER SYSTEMS AND SIMPLE TYPE THEORY

§15. Second order predicate calculus

DEFINITION 15.1. A language for second order predicate calculus (a second order language) is defined by extending a language for first order predicate calculus (Definition 1.1) by adding the following.

5) *Second order variables*:

5.1) Free variables with i argument-places ($i = 0, 1, 2, \ldots$):

$$\alpha_0^i, \alpha_1^i, \ldots, \alpha_j^i, \ldots \qquad (j = 0, 1, 2, \ldots).$$

5.2) Bound variables with i argument-places ($i = 0, 1, 2, \ldots$):

$$\varphi_0^i, \varphi_1^i, \ldots, \varphi_j^i, \ldots \qquad (j = 0, 1, 2, \ldots).$$

We shall call the variables in 2) of Definition 1.1 ($a_0, a_1, \ldots$ and $x_0, x_1, \ldots$) the *first order variables* in order to distinguish them from the second order variables.

Terms are defined as in Definition 1.2.

As in the preceding sections, we use α and φ both as formal and meta-variables; $\alpha, \beta, \gamma, \ldots$ may be used for second order free variables (with or without subscripts) and φ, ψ, χ may be used for second order bound variables. The superscripts i in α_j^i and φ_j^i are mostly omitted.

DEFINITION 15.2. The formulas for a second order language are defined as in Definition 1.3 with the following alteration.

If R^i is a predicate constant or a second order free variable with i argument-places and $t_1, \ldots, t_i$ are terms, then $R^i(t_1, \ldots, t_i)$ is an atomic formula.

In 3) of Definition 1.3 "a is a free variable" and "x is a bound variable" should read "a is a first order free variable" and "x is a first order bound variable", respectively.

We also add the clause:

3′) If A is a formula, α a second order free variable and φ a second order bound variable not occurring in A, which has the same number of argument-places as α, then $\forall\varphi\, A'$ and $\exists\varphi\, A'$ are formulas, where A' is the expression obtained from A by writing φ in place of α at each occurrence of α in A. The outermost logical symbols of $\forall\varphi\, A'$ and $\exists\varphi\, A'$ are $\forall$ and $\exists$, respectively.

The quantifier-free formulas and closed formulas (i.e., sentences) are defined as before.

The replacement of symbols, and the notions of indicated and fully indicated occurrences of certain symbols, are defined as in Definitions 1.4 and 1.6, respectively. Thus from $F(\alpha)$ we obtained $F(R)$ by replacing the indicated occurrences of α by R. Also the notion of alphabetical variant is defined as in Definition 2.15 (where we assume, of course, that bound variables are replaced by other bound variables of the same order and, for second order variables, the same number of argument places).

A sequent is an expression of the form $\Gamma \to \Delta$, where Γ and Δ are finite sequences of formulas of our language.

In the following, we shall assume we have a fixed second order language, which we call L_2.

We shall first define a second order system which does not contain any "comprehension axiom", and is simply **LK** with second order variables. Since this system is basic to second order systems, we shall call it the basic calculus for second order systems and abbreviate it **BC**.

DEFINITION 15.3. The formulas of **BC** are those of L_2 and the sequents of **BC** are those of L_2. The rules of inference of **BC** are defined as those for **LK**: only the following should be added to those in Definition 2.1.

2.5′) Second order $\forall$:

$$\text{left:}\quad \frac{F(R), \Gamma \to \Delta}{\forall \varphi F(\varphi), \Gamma \to \Delta},$$

where R is an arbitrary second order free variable or predicate constant and φ has the same number of argument-places as R.

$$\text{right:}\quad \frac{\Gamma \to \Delta, F(\alpha)}{\Gamma \to \Delta, \forall \varphi F(\varphi)},$$

where α is a second order free variable which is fully indicated in $F(\alpha)$ and does not occur in the lower sequent, and φ is a second order bound variable of the same number of argument-places as α (and does not occur in $F(\alpha)$, of course). Here α is called the *eigenvariable* of the inference.

2.6′) Second order $\exists$:

$$\text{left:}\quad \frac{F(\alpha), \Gamma \to \Delta}{\exists \varphi F(\varphi), \Gamma \to \Delta},$$

where α is a second order free variable which is fully indicated in $F(\alpha)$ and does not occur in the lower sequent, and φ is a second order bound variable of the same number of argument-places as α. Then α is called the eigenvariable of the inference.

$$\text{right:}\quad \frac{\Gamma \to \Delta, F(R)}{\Gamma \to \Delta, \exists \varphi F(\varphi)},$$

where R is an arbitrary second order free variable or predicate constant and φ has the same number of argument-places as R.

The auxiliary and principal formulas of these inferences are defined as for the other cases.

In contrast to 2.5′) and 2.6′), 2.5) and 2.6) will be called "first order $\forall$" and "first order $\exists$", respectively.

DEFINITION 15.4. The proofs of **BC** and the related notions and terminologies are defined as in §2 (cf. Definitions 2.2, 2.3 and 2.8): thus, we can define "a proof ending with S, or of S". "S is provable", "a thread of sequents", the concept of one sequent being "below" or "above" another, etc. The consistency of the system is defined exactly as before (Definition 4.1).

Similarly to Lemma 2.10 we can prove the following.

PROPOSITION 15.5. (1) *Let $P(R)$ be a* **BC**-*proof of a sequent $S(R)$, where R is an arbitrary second order free variable or predicate constant. Let R' be an arbitrary second order free variable or a predicate constant respectively which does not occur in $P(R)$. Assume that R and R' have the same number of argument-places. Then $P(R')$ is a proof of $S(R')$.*

(2) *A proof is called regular if it satisfies the condition that, firstly, all second order eigenvariables are distinct from one another, and, secondly, if a second order α occurs as an eigenvariable in a sequent S of the proof, then α occurs only in sequents above S. If a sequel S is* **BC**-*provable then S is provable with a regular proof.*

From now on we assume that we deal with regular proofs whenever necessary.

DEFINITION 15.6. The concept of "axiom system" is defined as in Definition 4.1; an axiom system $\mathscr{A}$ (of L_2) is a set of sentences (of L_2). Clauses 2)–7) in Definition 4.1 can be adapted to the second order case. Proposition 4.4 is re-stated here.

PROPOSITION 15.7. *Let $\mathscr{A}$ be an axiom system and let* $\mathbf{BC}_{\mathscr{A}}$ *be the system obtained from* **BC** *by adding $\to A$ as initial sequents for all A in $\mathscr{A}$. Then a sequent $\Gamma \to \Delta$ is* $\mathbf{BC}_{\mathscr{A}}$-*provable if and only if for some $A_1, \ldots, A_m$ of $\mathscr{A}$, $A_1, \ldots, A_m, \Gamma \to \Delta$ is* **BC**-*provable.*

When dealing with second order systems it is convenient to work with semi-terms and semi-formulas.

DEFINITION 15.8. (1) Semi-terms are defined as follows. Individual constants and first order variables (free or bound) are semi-terms; if $t_1, \ldots, t_n$ are semi-terms and f is a function constant with n argument-places, then $f(t_1, \ldots, t_n)$ is a semi-term.

(2) Semi-formulas and the free occurrences of bound variables are defined as follows. Let R be a predicate constant or a second order variable (free or bound) with i argument-places, and let $t_1, \ldots, t_i$ be semi-terms. Then $R(t_1, \ldots, t_i)$ is an atomic semi-formula; the bound variables in $t_1, \ldots, t_i$ occur free in $R(t_1, \ldots, t_i)$, and if R is a bound variable, then R occurs free in $R(t_1, \ldots, t_i)$. If B and C are semi-formulas, then so is $B \wedge C$, and the free occurrences of bound variables in $B \wedge C$ are those of B and C. For other propositional connectives, the definition is analogous. If $F(x)$ is a semi-formula in which x is fully indicated, then $\forall x\, F(x)$ is a semi-formula; the free occurrences of bound variables in $\forall x\, F(x)$ are those in $F(x)$ except x. If $F(\varphi)$ is a semi-formula in which φ is fully indicated, then $\forall \varphi\, F(\varphi)$ is a semi-formula and the free occurrences in $\forall \varphi\, F(\varphi)$ are those in $F(\varphi)$ except φ. For $\exists$ the definition is analogous.

It is obvious that terms are semi-terms without bound variables, and formulas are semi-formulas without free occurrences of bound variables.

Now we shall define two important notions of abstracts and substitution.

DEFINITION 15.9. Let $A(b_1, \ldots, b_m)$ be a formula where some occurrences of $b_1, \ldots, b_m$ are indicated. (Some of $b_1, \ldots, b_m$ may not occur in the formula at all.) Let $y_1, \ldots, y_m$ be bound first order variables which do not occur in $A(b_1, \ldots, b_m)$. Then the meta-expression $\{y_1, \ldots, y_m\}A(y_1, \ldots, y_m)$ is called an ***abstract*** of $A(b_1, \ldots, b_m)$.

We should emphasize that this is a meta-expression, i.e., not a formal expression of L_2, and will be used only as an auxiliary aid.

An abstract of the form $\{y_1, \ldots, y_m\}A(y_1, \ldots, y_m)$ is said to have m ***argument-places***. Abstracts are mostly denoted by $V, U, \ldots$. An abstract of the form $\{y_1, \ldots, y_m\}\alpha(y_1, \ldots, y_m)$ is often identified with α. If V denotes the abstract $\{y_1, \ldots, y_m\}A(y_1, \ldots, y_m)$ and $t_1, \ldots, t_m$ are semi-terms, then $V(t_1, \ldots, t_m)$ stands for $A(t_1, \ldots, t_m)$.

DEFINITION 15.10. Substitution of an abstract for a second order free variable in a semi-formula is defined as follows. Let $F(\alpha)$ be a semi-formula where some of the occurrences of α are indicated, and let V be an abstract with the same number of argument-places as α. (In the following we shall not mention the last condition, as the substitution is defined only for α and V which have the same number of argument-places.) We define ***substitution*** of V for α in $F(\alpha)$, denoting the result of $F\binom{\alpha}{V}$ or $F(V)$. In order to simplify the notation, we assume that α and V have one argument-place. One can easily generalize the definition to the case of more than one argument-place. So let V be of the form $\{y\}A(y)$. $F\binom{\alpha}{V}$ is defined by induction on the logical complexity of $F(\alpha)$.

1) (i) $F(\alpha)$ is $\alpha(s)$ and this α is indicated in $F(\alpha)$. Then $F\left(\begin{smallmatrix}\alpha\\ V\end{smallmatrix}\right)$ is $A(s)$. (ii) $F(\alpha)$ is $\alpha(s)$ and this α is not indicated, or $F(\alpha)$ is $\beta(s)$ for some β other than α. Then $F\left(\begin{smallmatrix}\alpha\\ V\end{smallmatrix}\right)$ is $F(\alpha)$ itself.

In the subsequent cases we first replace all the bound variables in F which occur in V by bound variables which do not occur in V in a manner such that each variable is replaced by another of the same order, distinct variables are replaced by distinct ones and a second order variable of i argument-places is replaced by another of i argument-places. Thus we may assume that F does not contain bound variables which occur in V.

2) $F(\alpha)$ is one of $\neg B(\alpha)$, $B(\alpha) \wedge C(\alpha)$, $B(\alpha) \vee C(\alpha)$, and $B(\alpha) \supset C(\alpha)$. Then $F\left(\begin{smallmatrix}\alpha\\ V\end{smallmatrix}\right)$ is, respectively, $\neg B\left(\begin{smallmatrix}\alpha\\ V\end{smallmatrix}\right)$, $B\left(\begin{smallmatrix}\alpha\\ V\end{smallmatrix}\right) \wedge C\left(\begin{smallmatrix}\alpha\\ V\end{smallmatrix}\right)$, $B\left(\begin{smallmatrix}\alpha\\ V\end{smallmatrix}\right) \vee C\left(\begin{smallmatrix}\alpha\\ V\end{smallmatrix}\right)$ and $B\left(\begin{smallmatrix}\alpha\\ V\end{smallmatrix}\right) \supset C\left(\begin{smallmatrix}\alpha\\ V\end{smallmatrix}\right)$.

3) $F(\alpha)$ has one of the forms $\forall x\, G(x)(\alpha)$, $\exists x\, G(x)(\alpha)$, $\forall \varphi\, G(\varphi)(\alpha)$ and $\exists \varphi\, G(\varphi)(\alpha)$. Then $F\left(\begin{smallmatrix}\alpha\\ V\end{smallmatrix}\right)$ is, respectively, $\forall x\,(G(x)\left(\begin{smallmatrix}\alpha\\ V\end{smallmatrix}\right))$, $\exists x\,(G(x)\left(\begin{smallmatrix}\alpha\\ V\end{smallmatrix}\right))$, $\forall \varphi\,(G(\varphi)\left(\begin{smallmatrix}\alpha\\ V\end{smallmatrix}\right))$ and $\exists \varphi\,(G(\varphi)\left(\begin{smallmatrix}\alpha\\ V\end{smallmatrix}\right))$.

It is obvious that $F\left(\begin{smallmatrix}\alpha\\ V\end{smallmatrix}\right)$ is a semi-formula. It is also obvious that if $F(\alpha)$ is a formula then so is $F\left(\begin{smallmatrix}\alpha\\ V\end{smallmatrix}\right)$.

The ambiguity in 2) and 3), viz. the choice of new bound variables, can be eliminated by requiring that these are the first variables in the list of first and second order bound variables which satisfy the conditions. This is not an essential restriction, by virtue of the following.

PROPOSITION 15.11. *Let A and B be two formulas which are alphabetical variants of each other. Then $A \equiv B$ is* **BC**-*provable.*

Thus we shall henceforth deal with any of the alphabetical variants of a given formula.

EXAMPLE 15.12. (1) Let $F(\alpha)$ be $\forall x\, \forall y\, (x = y \supset (\alpha(x) \equiv \alpha(y)))$, where both occurrences of α are indicated, and let V be $\{u\}\ \exists x\,(x + u = 5)$, where it is assumed that 5 is an individual constant, $+$ is a function constant and $=$ is a predicate constant in the language. Since x in $F(\alpha)$ occurs in V, first change it to, say, z: $\forall z\, \forall y\, (z = y \supset (\alpha(z) \equiv \alpha(y)))$. Let us call this formula $F'(\alpha)$. We shall carry out the substitution of V for α in $F'(\alpha)$ step by step.

$$\alpha(z)\begin{pmatrix}\alpha\\ V\end{pmatrix}: \quad \exists x\,(x + z = 5)$$

$$\alpha(y)\begin{pmatrix}\alpha\\ V\end{pmatrix}: \quad \exists x\,(x + y = 5)$$

$$(\alpha(z) \equiv \alpha(y))\begin{pmatrix}\alpha\\ V\end{pmatrix}: \quad \exists x\,(x + z = 5) \equiv \exists x\,(x + y = 5)$$

$$F'\begin{pmatrix}\alpha\\ V\end{pmatrix}: \quad \text{i.e.,} \quad \forall z\, \forall y\, (z = y \supset (\alpha(z) \equiv \alpha(y)))\begin{pmatrix}\alpha\\ V\end{pmatrix}:$$

$$\forall z\, \forall y\, (z = y \supset (\exists x\,(x + z = 5) \equiv \exists x\,(x + y = 5))).$$

This is a familiar formula, in fact an equality axiom. If we did not first replace x by z, the result would be

$$\forall x\, \forall y\, (x = y \supset (\exists x\, (x + x = 5) \equiv \exists x\, (x + y = 5))),$$

which is not even a formula.

This can be generalized to an arbitrary abstract $\{u\}B(u)$ (assuming there is no clash of bound variables), thus obtaining $\forall x\, \forall y\, (x = y \supset (B(x) \equiv B(y)))$, which is an equality axiom. That is to say, the simple schema

$$\forall x\, \forall y\, (x = y \supset (\alpha(x) \equiv \alpha(y)))$$

and substitution produce all the equality axioms.

(2) Let $F(\alpha)$ be $\alpha(0) \wedge \forall x\, (\alpha(x) \supset \alpha(x')) \supset \forall x\, \alpha(x)$, where all occurrences of α are indicated, and let V be $\{u\}B(u)$. Let us assume that x does not occur in V. Then $F\,(^{\alpha}_{V})$ is $B(0) \wedge \forall x\, (B(x) \supset B(x')) \supset \forall x\, B(x)$, which is an induction axiom in arithmetic (in an appropriate language).

(3) Let $F(\alpha)$ be

$$\forall x\, \forall y\, \forall z\, (\alpha(x, y) \wedge \alpha(x, z) \supset y = z) \supset$$
$$\supset \exists v\, \forall y\, (y \in v \equiv \exists x\, (x \in u \wedge \alpha(x, y))),$$

with all occurrences of α indicated, and let V be $\{x^1, y^1\}B(x^1, y^1)$, in the language of set theory. Then $F\,(^{\alpha}_{V})$ is

$$\forall x\, \forall y\, \forall z\, (B(x, y) \wedge B(x, z) \supset y = z) \supset$$
$$\supset \exists v\, \forall y\, (y \in v \equiv \exists x\, (x \in u \wedge B(x, y)),$$

which is an axiom of replacement in ZF set theory. Note that $B(x, y)$ may contain variables other than x and y, including u, but not v (since this is bound in $F(\alpha)$).

We shall return to those examples later.

The following is easily proved by induction on the number of logical symbols in $F(\alpha)$.

PROPOSITION 15.13. *For an arbitrary formula $F(\alpha)$ and arbitrary abstracts U and V, the sequent*

$$\forall x\, (U(x) \equiv V(x)), F(U) \rightarrow F(V),$$

(*where it is assumed that the bound variables are properly taken care of*) *is* **BC**-*provable*.

DEFINITION 15.14. (1) Let $A(b_1, \ldots, b_m, c_1, \ldots, c_n, \beta_1, \ldots, \beta_k)$ be a formula, all of whose free variables are among $b_1, \ldots, b_m$, $c_1, \ldots, c_n, \beta_1, \ldots, \beta_k$ (though not necessarily all of these occur in A), and where all occurrences of these free variables are indicated. Then a sentence of the form

$$(*)\qquad \forall z_1 \dots \forall z_n \forall \psi_1 \dots \forall \psi_k \exists \varphi \forall y_1 \dots \forall y_m (\varphi(y_1, \dots, y_m) \equiv A(y_1, \dots, y_m, z_1, \dots, z_n, \psi_1, \dots, \psi_k))$$

is called a comprehension axiom.

Let V be the abstract

$$\{y_1 \dots y_m\}A(y_1, \dots, y_m, c_1, \dots, c_n, \beta_1, \dots, \beta_k).$$

Then the above comprehension axiom may be written as

$$(**)\qquad \forall z_1, \dots, \forall z_n \forall \psi_1 \dots \forall \psi_k \exists \varphi \forall y_1 \dots \forall y_m(\varphi(y_1, \dots, y_m) \equiv U(y_1, \dots, y_m)),$$

where U is obtained from V by replacing c's and β's by z's and ψ's, respectively.

(2) Let K be an arbitrary set of formulas. (This use of K is only temporary.) A formula which belongs to K is called a K-formula, and if a formula $A(b_1, \dots, b_m)$ is a K-formula, then the abstract $\{y_1 \dots y_m\}A(y_1, \dots, y_m)$ is called a K-abstract. If the formula A in a comprehension axiom (cf. (*)) is a K-formula, then (*) is called a K-comprehension axiom.

(3) A set of formulas K is said to be closed under substitution if for every K-formula or K-abstract $A(\alpha)$ and for every K-abstract V, $A(V)$ again belongs to K.

DEFINITION 15.15. Let K be a set of formulas.

1) The K-system is obtained from **BC** by adding to it all K-comprehension axioms as initial sequents (viz. sequents of the form $\rightarrow A$, where A is a K-comprehension axiom).

2) **KC** is the system obtained from **BC** by adding the following inferences, for arbitrary formulas $F(\alpha)$, and K-abstracts V (where $F(V)$ and $F(\varphi)$ are obtained by replacing the indicated α by, respectively, V and φ):

$$\text{Second order } \forall : \text{left:} \quad \frac{F(V), \Gamma \rightarrow \Delta}{\forall \varphi F(\varphi), \Gamma \rightarrow \Delta};$$

$$\text{Second order } \exists : \text{right:} \quad \frac{\Gamma \rightarrow \Delta, F(V)}{\Gamma \rightarrow \Delta, \exists \varphi F(\varphi)}.$$

The auxiliary and principal formulas of these inferences are defined as usual.

Since a system **KC** has interest only if K is closed under substitution, we shall henceforth assume that K is closed under substitution.

PROPOSITION 15.16. *For an arbitrary set K of formulas (closed under substitution), the K-system is equivalent to* **KC**.

PROOF. It can be easily be shown that the K-comprehension axioms are provable in **KC**, while the lower sequents of second order $\forall$: left and second order $\exists$: right are provable in the K-system from their upper sequents.

Due to the above proposition we shall henceforth only deal with K-comprehension axioms in the form of the system **KC**.

DEFINITION 15.17. If K is the set of all second order formulas, then **KC** is called the second order predicate calculus with full comprehension, and is denoted by $\mathbf{G^1LC}$.

PROPOSITION 15.18. *If the cut-elimination theorem holds for* $\mathbf{G^1LC}$, *then* $\mathbf{G^1LC}$ *is consistent.*

PROOF. The proof is immediate, as for Theorem 6.2.

In fact the cut-elimination theorem does hold for $\mathbf{G^1LC}$, as we will see later (§20). The reason why we put Proposition 15.18 in this form is that the proof of cut-elimination for $\mathbf{G^1LC}$ is non-constructive, and hence, on the basis of our finitist standpoint, we cannot claim the consistency of $\mathbf{G^1LC}$ from that proof.

§16. Some systems of second order predicate calculus

In this section we shall deal with some inessential extensions of the first order predicate calculus.

DEFINITION 16.1. Let $\mathbf{S}_1$ and $\mathbf{S}_2$ be two formal systems which contain **LK**. $\mathbf{S}_2$ is called a conservative extension of $\mathbf{S}_1$ if $\mathbf{S}_1$ is a subsystem of $\mathbf{S}_2$ and for any sequent S of the language of $\mathbf{S}_1$, if S is $\mathbf{S}_2$-provable, then S is $\mathbf{S}_1$-provable.

PROPOSITION 16.2. *The cut-elimination theorem holds for* **BC**.

The proof is exactly as for **LK**, so we shall not repeat the argument.

As consequences of this proposition, consistency, the subformula property, the midsequent property, etc., all hold for **BC**. As another consequence we can claim:

COROLLARY 16.3. **BC** *is a conservative extension of* **LK**.

PROOF. If any inference for a second order quantifier is used in a cut-free proof of **BC**, then that quantifier will occur in all sequents below that inference (as is easily shown by induction on the number of inferences in such a proof).

DEFINITION 16.4. (1) A *first order formula* is one which contains no second order quantifiers (although it may contain second order variables). Such a formula is also called arithmetical if the language is that of second order arithmetic (i.e., **PA**, with second order variables).

A *first order abstract* is one obtained from a first order formula.

(2) K_1 is the set of all first order formulas.

(3) The predicative comprehension axioms are those in which the U in (**) of Definition 15.14 is a first order abstract; in other words, the K_1-comprehension axioms.

THEOREM 16.5 (cut-elimination theorem for the system with predicative comprehension axioms). *If a sequent S is provable in the system* $\mathbf{K_1C}$ (*cf. Definition* 15.15), *then it is provable in* $\mathbf{K_1C}$ *without cut.*

PROOF. The proof for **LK** almost goes through. Here we use triple induction instead of double induction (cf. Proof of Lemma 5.4). Let A be a formula of a second order language. Define a function c by: $c(A) =_{\mathrm{df}}$ the number of second order quantifiers in A. It is easily seen that $c(F(\alpha)) = c(F(V))$ if V is first order. Let $c =_{\mathrm{df}} c(P) =_{\mathrm{df}} c(D)$, where D is the mix formula of P (assuming P has a mix at most as the last inference). Then Lemma 5.4 is proved now by transfinite induction on $\omega^2 \cdot c + \omega \cdot g(P) + \mathrm{rank}(P)$. We may follow the proof in §5 but there are some additional cases here. After 1.5) (i) there, add the cases that D is $\forall \varphi F(\varphi)$ and $\exists \varphi F(\varphi)$. P has the form

$$\dfrac{\dfrac{\Gamma \rightarrow \Delta_0, F(\alpha)}{\Gamma \rightarrow \Delta_0, \forall \varphi F(\varphi)} \qquad \dfrac{F(V), \Pi_0 \rightarrow \Lambda}{\forall \varphi F(\varphi), \Pi_0 \rightarrow \Lambda}}{\Gamma, \Pi_0 \rightarrow \Delta_0, \Lambda} \; (\forall \varphi F(\varphi)),$$

where V is a first order abstract. From the above remark, $c(F(V)) = c(F(\alpha)) = c(\forall \varphi F(\varphi)) - 1$. As α does not occur in Γ, Δ_0 or $F(\varphi)$; $\Gamma \rightarrow \Delta$, $F(V)$ is provable without a mix (cf. 1.5) of proof of Lemma 5.4). Define P' as

$$\dfrac{\Gamma \rightarrow \Delta_0, F(V) \qquad F(V), \Pi_0 \rightarrow \Lambda}{\Gamma, \Pi_0^{\#} \rightarrow \Delta_0^{\#}, \Lambda} \; (F(V)).$$

Since $c(P') = c(F(V)) < c(\forall \varphi F(\varphi)) = c(P)$, the induction hypothesis applies to P'. Thus we can obtain a proof without a mix of $\Gamma, \Pi_0^{\#} \rightarrow \Delta_0^{\#}, \Lambda$, and hence a proof without a mix of $\Gamma, \Pi_0 \rightarrow \Delta_0, \Lambda$.

Finally, after 2.1.3 (ii) of the Proof of Lemma 5.4, add the cases where D is $\forall\varphi\, F(\varphi)$ and $\exists\varphi\, F(\varphi)$.

COROLLARY 16.6. $\mathbf{K_1C}$ *is a conservative extension of* **LK**. *Hence, in particular,* $\mathbf{K_1C}$ *is consistent.*

PROOF. The proof is as for **BC** (cf. Corollary 16.3).

PROPOSITION 16.7. *Let* $\mathbf{LK}^+$ *be the system which is like* **LK** *except that the language includes free second order variables.*

Let $\Gamma \to \Theta$ *be a sequent consisting of first order formulas only and let* $F_i(\beta_i)$ *be a first order formula which has a free second order variable* β_i, $i = 1, 2, \ldots, m$. *Then*

$$\forall\varphi_1\, F_1(\varphi_1), \ldots, \forall\varphi_m\, F_m(\varphi_m), \Gamma \to \Theta \tag{1}$$

is $\mathbf{K_1C}$*-provable if and only if the following is satisfied:*

(*) *For each* $i = 1, 2, \ldots, m$ *there exist first order abstracts* $V_{i,1}, \ldots, V_{i,l_i}$ $(l_i \geq 1)$ *such that*

$$\{\forall z_{1,j}\, F_1(V'_{1,j})\}_{j=1,\ldots,t_1}, \ldots, \{\forall z_{n,j}\, F_n(V'_{n,j})\}_{j=1,\ldots,t_n}, \Gamma \to \Theta \tag{2}$$

is $\mathbf{LK}^+$*-provable;*

Here $\{A_j\}_{j=1,\ldots,m}$ *denotes a sequence of formulas* $A_1, \ldots, A_m$, $\forall z_{i,j}$ *denotes a (possibly empty) sequence of universally quantified first order variables* $\forall z_1 \forall z_2 \ldots \forall z_k$, *where* k *(depending on* i *and* j*) is the number of free first order variables in* $V_{i,j}$, *and* V' *is obtained from* V *by changing the free first order variables in* V *which do not occur in* (1) *to* $z_1, \ldots, z_k$.

PROOF. If: Suppose (*) holds. First we shall prove that for every formula $F(\alpha)$; $\forall\varphi\, F(\varphi) \to \forall z\, F(V')$ is $\mathbf{K_1C}$-provable, if V is first order.

$$\cfrac{\cfrac{\cfrac{F(V) \to F(V)}{\forall\varphi\, F(\varphi) \to F(V)}}{\text{(repeated } \forall : \text{right)}}}{\forall\varphi\, F(\varphi) \to \forall z\, F(V).}$$

Thus we have $\forall\varphi_i\, F_i(\varphi_i) \to \forall z_{i,j}\, F_i(V'_{i,j})$ for $j = 1, 2, \ldots, l_i$ and $i = 1, 2, \ldots, n$. From these and (2), by repeated cuts and contractions, we can construct a $\mathbf{K_1C}$-proof of (1).

Only if: Suppose (1) is $\mathbf{K_1C}$-provable. Then there exists a cut-free proof of (1) in $\mathbf{K_1C}$. Therefore it is sufficient to prove the following proposition.

PROPOSITION 16.8. *Suppose* P *is a cut-free proof in* $\mathbf{K_1C}$ *of a sequent of the form* (1) *above. Then for the end-sequent of* P, (*) *holds.*

Notice that since P is cut-free, all sequents in P have the form (1). The proposition may now be proved by mathematical induction on the number of inferences in P.

PROOF. (1) If P consists of an initial sequent $D \to D$, then D has no second order quantifier. Therefore $D \to D$ itself has the form (2) above.

(2) The induction steps are proved according to the last inference I in P. Notice that the only possible inference in P concerning a second order quantifier is second order $\forall$: left.

2.1) I is second order $\forall$: left. P is of the form

$$\frac{F(V), \Pi \to \Lambda}{\forall \varphi F(\varphi), \Pi \to \Lambda},$$

where V and $F(\varphi)$ are first order. By the induction hypothesis, when $F(V)$, $\Pi \to \Lambda$ is taken for the sequence in (1), there are appropriate abstracts for which a sequent like (2) is provable in $\mathbf{LK}^+$. Denote such a sequent by $F(V), \Pi^* \to \Lambda$. Now add V to the set of abstracts obtained by the induction hypothesis. If V has no first order free variable which does not occur in $\forall \varphi F(\varphi)$, $\Pi \to \Lambda$, then take $F(V)$, $\Pi^* \to \Lambda$ itself for the sequent (2). If V has free variables $b_1, \ldots, b_k$ which do not occur in the above sequent, then replace them by new bound variables $z_1, \ldots, z_k$ and call the result V'. The required sequent is then $\forall z_1, \ldots, \forall z_k F(V')$, $\Pi^* \to \Lambda$.

2.2) I is not a second order $\forall$: left. Such a case is proved trivially from the induction hypothesis.

Then replace free second order variables by $0 = 0$.

Notice that it is the inference contraction : left that results in more than one first order abstract $V_{i,1}, V_{i,2}, \ldots$ being associated with the same formula $F_i(\beta_i)$ in (2).

PROPOSITION 16.9. *If a formula $\exists \varphi F(\varphi)$ is provable in $\mathbf{K}_1\mathbf{C}$, where $F(\varphi)$ does not have second order quantifiers, then there exist first order abstracts $V_1, \ldots, V_n$ such that $\exists z_1 F(V_1') \vee \ldots \vee \exists z_n F(V_n')$ is provable in $\mathbf{LK}^+$.*

This is the dual of Proposition 16.7 and is proved similarly.

PROBLEM 16.10. We define $\mathbf{PA}'$, the predicative (second order) extension of Peano arithmetic, as follows. Let VJ′ and Eq′ be respectively the sentences:

VJ′ $\quad \forall \varphi (\varphi(0) \wedge \forall x (\varphi(x) \supset \varphi(x')) \supset \forall x \varphi(x))$;

Eq′ $\quad \forall \varphi \forall x \forall y (x = y \wedge \varphi(x) \supset \varphi(y))$.

VJ′ is the second order formulation of the principle of mathematical induction and Eq′ is the second order formulation of one of the equality axioms.

PA$'$ is then obtained from $\mathbf{K_1C}$ (in the language of **PA** augmented by second order variables) by adding to it the axioms of $\mathrm{CA} \cup \mathrm{VJ}' \cup \mathrm{Eq}'$ as initial sequents. (CA was defined in Definition 9.2.)

Show that **PA**$'$ is a conservative extension of **PA**. [*Hint*: Let A be a formula of the language of **PA**$'$. then A is **PA**$'$-provable if and only if $\mathrm{CA} \cup \mathrm{VJ}' \cup \mathrm{Eq}' \to A$ is $\mathbf{K_1C}$-provable. Noting that VJ$'$ and Eq$'$ each have one second order $\forall$ in front, apply Proposition 16.7.]

PROBLEM 16.11. Consider ZF (Zermelo-Fraenkel set theory). The language consists of $\in$ (a binary predicate symbol), first order variables and logical symbols ($a = b$ is an abbreviation of $\forall x\,(a \in x \equiv b \in x)$). The axioms of extensionality, pairs, sum, power, regularity and infinity can be stated as single sentences. However, the axiom of replacement is actually an axiom schema, which is formulated as

$$\forall x\,\forall y\,\forall z\,(B(x, y) \land B(x, z) \supset y = z) \supset$$
$$\supset \exists v\,\forall y\,(y \in v \equiv \exists x\,(x \in u \land B(x, y))$$

(cf. Example 15.12, (3)). The basic logical system is **LK**.

On the other hand BG (Barnays-Gödel set theory) is formulated in a second order language. The language is that of ZF augmented by second order variables. The axioms are those of ZF plus an axiom of equality

$$\forall \varphi\,\forall x\,\forall y\,(x = y \supset \varphi(x) \equiv \varphi(y))$$

except that the axiom of replacement is now formulated in a single sentence:

$$\forall \varphi\,(\forall x\,\forall y\,\forall z\,(\varphi(x, y) \land \varphi(x, z) \supset y = z) \supset$$
$$\supset \exists v\,\forall y\,(y \in v \equiv \exists x\,(x \in u \land \varphi(x, y)))).$$

The basic logical system is $\mathbf{K_1C}$.

Show that BG is a conservative extension of ZF. [*Hint*: As for the previous problem.]

PROBLEM 16.12 (Kreisel). Consider the system **PA**$'$ of predicative second order arithmetic defined in Problem 16.10. In order to facilitate the use of some results in recursive function theory, we use the following notation: $\forall f\,A(f)$ (resp. $\exists f\,A(f)$) is an abbreviation for $\forall \varphi$ (φ is a function $\supset A^*(\varphi)$) (resp. $\exists \varphi$ (φ is a function and $A^*(\varphi)$)), where "φ is a function" is expressed by

$$\forall x\,\exists y\,\forall z\,(\varphi(x, z) \equiv y = z)$$

and $A(f)$ and $A^*(\varphi)$ are related in a manner such that subformulas of $A(f)$ of the form $B(f(t))$ are (systematically) replaced in A^* by $\exists y\,(\varphi(t, y) \land B(y))$.

Next we define a Π^1_1 (resp. Σ^1_1) formula as one of the form $\forall \varphi\,A(\varphi)$

(resp. $\exists\varphi\, A(\varphi)$) where A is arithmetical. Any Π^1_1 formula is equivalent (in **PA**′) to one in "Π^1_1-normal form", i.e., $\forall f\, \exists y\, R(\bar{f}y)$, for some R which is primitive recursive (or more strictly, primitive recursive relative to the free second order variables in the formula), where $\bar{f}y$ is (the Gödel number of) the sequence $\langle f(0), \ldots, f(y-1)\rangle$. Similarly, any Σ^1_1 formula can be transformed to Σ^1_1-normal form: $\exists f\, \forall y\, R(\bar{f}y)$, for suitable primitive recursive R. Finally, a Π^1_1 predicate or relation (of k number variables say) is one which can be expressed by a Π^1_1 formula with k free first order variables (and no free second order variables). Similarly for Σ^1_1 predicates or relations.

Now let $<\cdot$ be a Σ^1_1-ordering, i.e., $<\cdot$ is a Σ^1_1 binary relation which is a linear ordering of natural numbers. Let $We(<\cdot)$ express that $<\cdot$ is a well-ordering: $\forall f\, \exists x\, \neg(f(x+1) <\cdot f(x))$. Suppose $We(<\cdot)$ is provable in **PA**′, i.e., $<\cdot$ is a provable well-ordering of **PA**′. Show that the ordinal of $<\cdot$ (the order type of $<\cdot$) is less than ε_0. [*Hint*: First express $a <\cdot b$ in Σ^1_1-normal form: $\exists f\, \forall y\, R(\bar{f}y, ab)$. Now follow the steps listed below.

1) Let $<_n$ be an enumeration of primitive recursive, binary relations (for $n = 0, 1, 2, \ldots$), and let $W(x)$ denote $We(<_x)$. Then $W(x)$ is a (provably) complete Π^1_1-form, viz. there is a primitive recursive function $S(r, x)$ such that for every Π^1_1-predicate $A(x)$, there exists a number r_0 such that $\forall x\, (A(x) \equiv W(S(\bar{r}_0, x)))$ is **PA**′-provable.

2) For any Σ^1_1 predicate $B(x)$ there is a number n such that $B(\bar{n}) \equiv \neg W(\bar{n})$ is **PA**′-provable. (We formalize an argument in recursion theory showing that a Σ^1_1 predicate cannot be Π^1_1-complete.)

3) Let $W_1(x)$ be the formula expressing that there is an embedding of $<_x$ into $<\cdot$, i.e., an order-preserving function from the domain of $<_x$ to the domain of $<\cdot$ (which implies that $<_x$ is a well-ordering, with ordinal less than or equal to that of $<\cdot$). Then $W_1(x)$ is Σ^1_1, and so there is a formula $W^*_1(x)$ in Σ^1_1-normal form such that $\forall x\, (W_1(x) \equiv W^*_1(x))$ is **PA**′-provable.

4) Since $We(<\cdot)$ is provable, $\forall x\, (W^*_1(x) \supset W(x))$ is provable.

5) By 2), there is an n such that $W^*_1(\bar{n}) \equiv \neg W(\bar{n})$ is provable. Hence by 4), $W(\bar{n})$ and $\neg W^*_1(\bar{n})$ are provable.

6) $W(\bar{n})$ being provable in **PA**′ means that the primitive recursive relation $<_n$ is a provable well-ordering of **PA**′, hence of **PA**. By Gentzen's result in the previous chapter, this implies that the ordinal of $<_n$ is less than ε_0.

7) $W(\bar{n})$ and $\neg W^*_1(\bar{n})$ (see 5)) means that $<_n$ is a primitive recursive well-ordering which is not embeddable in $<\cdot$, and hence the ordinal of $<\cdot$ is less than that of $<_n$. This and 6) yield that the order type of $<\cdot$ is less than ε_0.]

§17. The theory of relativization

DEFINITION 17.1. A *system of relativization* consists of a pair of formulas $R^0(a)$ and $R^1(\alpha)$, where $R^0(a)$ and $R^1(\alpha)$ each have the one free variable

a and α, respectively. One or both of $R^0(a)$ and $R^1(\alpha)$ may be missing. A system of relativization is often denoted by r.

For an arbitrary (second order) semi-formula A, A^r (the relativization of A to r) is defined inductively as follows. (Here it is assumed that r consists of two formulas. If one or both of R^0 and R^1 is missing, then the definition should be adjusted accordingly.)

1) If A has no logical symbol, then A^r is A.
2) $(A \wedge B)^r$, $(A \vee B)^r$, $(A \supset B)^r$, $(\neg A)^r$ are, respectively,

$$A^r \wedge B^r,\ A^r \vee B^r,\ A^r \supset B^r,\ \neg A^r.$$

3) $(\forall x\, F(x))^r$ and $(\exists x\, F(x))^r$ are $\forall y\,(R^0(y) \supset F^r(y))$ and $\exists y\,(R^0(y) \wedge F^r(y))$, respectively, where $F^r(y)$ is $(F(y))^r$ and y is a variable which does not occur in $R^0(a)$ or $F^r(a)$.
4) $(\forall \varphi\, F(\varphi))^r$ and $(\exists x\, F(\varphi))^r$ are $\forall \psi\,(R^1(\psi) \supset F^r(\psi))$ and $\exists \psi\,(R^1(\psi) \wedge F^r(\psi))$, respectively, where $F^r(\psi)$ is $(F(\psi))^r$ and ψ is a variable which does not occur in $R^1(\alpha)$ or $F^r(\alpha)$.
5) $(\{y_1, \ldots, y_n\}A(y_1, \ldots, y_n))^r$ is $\{y_1, \ldots, y_n\}(A(y_1, \ldots, y_n))^r$.

LEMMA 17.2. (1) *If A has no quantifiers, then A^r is A.*

(2) *A free variable occurs in A if and only if it occurs in A^r.*

(3) *A bound variable occurs free in A if and only if it occurs free in A^r.*

(4) *A is a formula if and only if A^r is a formula.*

(5) *Let $A(t)$ denote $A(a)\binom{a}{t}$; then $(A(t))^r$ is the same as $A^r(t)$, i.e., $(A(a))^r\binom{a}{t}$ ("the same", that is, up to bound occurrences of bound variables).*

(6) *Let $A(V)$ denote $A(\alpha)\binom{\alpha}{V}$, then $(A(V))^r$ is the same as $A^r(V^r)$, i.e., $(A(\alpha))^r\binom{\alpha}{V^r}$ (again, up to bound occurrences of bound variables).*

PROOF. By mathematical induction on the number of logical symbols in A. (1)–(5) are left to the reader.

(6): Let V be $\{y\}C(y)$. (For the sake of simplicity we assume that V has only one argument place.) If $A(\alpha)$ is $\alpha(t)$, then $(A(V))^r$ is $(C(t))^r$ and $(A(\alpha))^r\binom{\alpha}{V^r}$ is $(\alpha(t))^r\binom{\alpha}{V^r}$, i.e., $C^r(t)$. These are the same by (5).

Suppose $A(\alpha)$ is $\forall x\, F(x, \alpha)$. $(\forall x\, F(x, V))^r$ is $\forall y\,(R^0(y) \supset (F(y, V))^r)$. By the induction hypothesis, this is the same as $\forall y\,(R^0(y) \supset F^r(y, \alpha)\binom{\alpha}{V^r})$, i.e., $(\forall x\, F(x, \alpha))^r\binom{\alpha}{V^r}$.

Suppose $A(\alpha)$ is $\forall \varphi\, F(\varphi, \alpha)$. $(\forall \varphi\, F(\varphi, V))^r$ is $\forall \psi\,(R^1(\psi) \supset (F(\psi, V))^r)$. By the induction hypothesis, this is the same as $\forall \psi\,(R^1(\psi) \supset F^r(\psi, \alpha)\binom{\alpha}{V^r})$, i.e., $(\forall \varphi\, F(\varphi, \alpha))^r\binom{\alpha}{V^r}$.

The other cases are left to the reader.

DEFINITION 17.3. (1) If Γ is a sequence of formulas $A_1, \ldots, A_m$, then Γ^r denotes $A_1^r, \ldots, A_m^r$. For the sake of simplicity, we write both R^0 and R^1 as r. It will be obvious which is meant, since $r(a)$ denotes $R^0(a)$ and $r(\alpha)$ denotes $R^1(\alpha)$.

(2) Suppose a system of relativization r is given. Then Φ is the set of the following formulas.

1) $r(c)$ for every individual constant c (in the language).
2) $\forall y_1 \dots \forall y_m\, (r(y_1) \wedge \dots \wedge r(y_m)) \supset r(f(y_1, \dots, y_m)))$ for every function constant f.
3) $\exists x\, r(x)$.
4) $\exists \varphi\, r(\varphi)$.

EXAMPLE 17.4. The language includes $=$ as a distinguished binary predicate constant. Suppose r consists of R^1 only, where $R^1(\alpha)$ is defined to be

$$\forall x\, \forall y\, (x = y \wedge \alpha(x) \supset \alpha(y)).$$

Let $\mathscr{B}$ be the following axiom system:

$$\forall x\, (x = x),$$

$$\forall x\, \forall y\, (x = y \supset y = x),$$

$$\forall x\, \forall y\, \forall z\, (x = y \wedge y = z \supset x = z),$$

$$\forall x_1 \dots \forall x_m\, \forall y_1 \dots \forall y_m\, (x_1 = y_1 \wedge \dots \wedge x_m = y_m \supset f(x_1, \dots, x_m) = f(y_1, \dots, y_m)) \text{ for every } f,$$

$$\forall x_1 \dots \forall x_m\, \forall y_1 \dots \forall y_m\, (x_1 = y_1 \wedge \dots \wedge x_m = y_m \wedge P(x_1, \dots, x_m) \supset P(y_1, \dots, y_m)) \text{ for every } P.$$

To apply the theory below to this example, we want to check that $\mathscr{B} \rightarrow A$ is provable in the systems considered for every A in Φ. Since R^0 is missing, we only have to consider 4). It is a routine matter to see that this condition is satisfied. (Further, r and $\mathscr{B}$ also satisfy condition 5) in Lemma 17.5.)

LEMMA 17.5. *Let* **S** *be* **KC**, *where K is an arbitrary set of formulas which is closed under substitution (cf. Definitions* 15.14 *and* 15.15*). Suppose the system of relativization r satisfies the condition that for an arbitrary K-abstract V, V′ is also a K-abstract. (This is satisfied if, e.g., K consists of all formulas in the language.) Let* $\mathscr{B}$ *be an axiom system such that* $\mathscr{B} \rightarrow A$ *is* **S**-*provable for every A in* Φ *and*

5) $r(b), \dots, r(\beta), \dots, \mathscr{B} \rightarrow r(V')$ *is* **S**-*provable for every K-abstract V, where b, ... and β, ... are all the free variables which occur in V (and hence in V′: cf. part* (2) *of Lemma* 17.2*).*

Then for an arbitrary sequent $\Gamma \rightarrow \Theta$ *which is* **S**-*provable,*

$$r(a), \dots, r(\alpha), \dots, \mathscr{B}, \Gamma' \rightarrow \Theta'$$

is **S**-*provable, where a, ..., α, ... are all the free variables which occur in* Γ, Θ.

We first prove the following sublemma.

SUBLEMMA 17.6. *If s is a term, then $r(b), \ldots, \mathscr{B} \to r(s)$ is* **S**-*provable, where $b, \ldots$ are all the (free first order) variables in s.*

This is proved by mathematical induction on the number of function constants in s.

PROOF OF LEMMA 17.5. This is proved by mathematical induction on the number of inferences in a proof ending with $\Gamma \to \Theta$.

1) The proof consists of an initial sequent $D \to D$. Then $D^r \to D^r$ is also an initial sequent. Therefore $\mathscr{B}, D^r \to D^r$ is obviously **S**-provable, and hence so is $r(a), \ldots, r(\alpha), \ldots, \mathscr{B}, D^r \to D^r$.

2) The last inference is a first order $\forall$: left:

$$\frac{F(s), \Gamma' \to \Theta}{\forall x\, F(x), \Gamma' \to \Theta}.$$

By the induction hypothesis,

$$r(a), \ldots, r(\alpha), \ldots, \mathscr{B}, F^r(s), \Gamma'^r \to \Theta^r$$

is **S**-provable (cf. part (5) of Lemma 17.2). Also $r(b), \ldots, \mathscr{B} \to r(s)$ by Sublemma 17.6 (where the variables $b, \ldots$ are included among $a, \ldots$). Therefore

$$r(a), \ldots, r(\alpha), \ldots, \mathscr{B}, \forall x\, (r(x) \supset F^r(x)), \Gamma'^r \to \Theta^r.$$

3) The last inference is a first order $\forall$: right:

$$\frac{\Gamma \to \Theta', F(d)}{\Gamma \to \Theta', \forall x\, F(x)}.$$

By the induction hypothesis,

$$r(a), \ldots, r(\alpha), \ldots, \mathscr{B}, \Gamma^r \to \Theta'^r, F^r(d),$$

where d does not occur in the antecedent. So

$$r(a), \ldots, r(\alpha), \ldots, \mathscr{B}, \Gamma^r \to \Theta'^r, \forall x\, (r(x) \supset F^r(x)).$$

4) The last inference is a second order $\forall$: left:

$$\frac{F(V), \Gamma' \to \Theta}{\forall \varphi\, F(\varphi), \Gamma' \to \Theta},$$

where V is a K-abstract. By the induction hypothesis,

$$r(a), \ldots, r(\alpha), \ldots, \mathscr{B}, F^r(V^r), \Gamma'^r \to \Theta^r$$

(cf. part (6) of Lemma 17.2). Hence

$$r(a), \ldots, r(\alpha), \ldots, \mathscr{B} \to r(V^r)$$

by condition 5) (since $a, \ldots, \alpha, \ldots$ include all the free variables in V). So

$$r(a), \ldots, r(\alpha), \ldots, \mathscr{B}, r(V^r) \supset F^r(V^r), \Gamma'' \to \Theta^r.$$

Also by the condition on K, V^r is a K-abstract, so that from the last sequent

$$r(a), \ldots, r(\alpha), \ldots, b, \forall\varphi\, (r(\varphi) \supset F^r(\varphi)), \Gamma'' \to \Theta^r$$

is **S**-provable.

Other cases are treated similarly.

DEFINITION 17.7. (1) An *axiom system* (in this section) is defined as a set of formulas which do not contain any free first order variables.

(2) Let $\mathscr{A}$ be an arbitrary axiom system and let r be a system of relativization. $\mathscr{A}^r$ is the set of formulas A^r for all A in $\mathscr{A}$.

THEOREM 17.8. *Let $\mathscr{A}$ and $\mathscr{B}$ be axiom systems. Suppose that the formal system* **S** *and the axiom system $\mathscr{B}$ satisfy the conditions of Lemma* 17.5, *and, further: for every formula A of $\mathscr{A} \cup \mathscr{B}$, $\mathscr{B} \to A^r$ is* **S***-provable; and for every second order free variable α contained in $\mathscr{A} \cup \mathscr{B}$, $\mathscr{B} \to r(\alpha)$ is* **S***-provable. Then*:

(1) *for any formula B, if $\mathscr{A} \cup \mathscr{B} \to B$ is* **S***-provable, then so is $\mathscr{B} \to B^r$*;

(2) *if $\mathscr{B}$ is consistent (with* **S**), *then so is $\mathscr{A} \cup \mathscr{B}$.*

NOTE. We can express this result (part (1)) by saying that $\mathscr{A} \cup \mathscr{B}$ is interpretable in $\mathscr{B}$ (relative to **S**), or more accurately that r provides an interpretation (or "inner model") of $\mathscr{A} \cup \mathscr{B}$ in $\mathscr{B}$ (relative to **S**).

PROOF. (1) Suppose $\mathscr{A} \cup \mathscr{B} \to B$ (in **S**). Then there are finite sequences Γ and Δ from $\mathscr{A}$ and $\mathscr{B}$, respectively, such that $\Gamma, \Delta \to B$ (in **S**). So by Lemma 17.5,

$$r(\alpha), \ldots, \Gamma^r, \Delta^r \to B^r \text{ (in } \mathbf{S}).$$

(Recall that Γ and Δ contain no free first order variables.) But by our assumption, $\mathscr{B} \to A^r$ and $\mathscr{B} \to r(\alpha)$ are provable for every formula A and second order variable α in $\Gamma \cup \Delta$. Hence $\mathscr{B} \to B^r$ (in **S**).

Part (2) is proved from (1), by taking, for B, say $C \wedge \neg C$ (since $(C \wedge \neg C)^r$ is $C^r \wedge \neg C^r$).

DEFINITION 17.9. For the following theorem, let L′ denote the second order language with constants 0, ′ and =, and let $\mathscr{A}_0$ denote the following axioms (for arithmetic) in this language:

$$\forall x\,(\neg x' = 0),$$
$$\forall x\,\forall y\,(x = y \supset x' = y'),$$
$$\forall x\,\forall y\,(x' = y' \supset x = y),$$
$$\forall x\,(x = x),$$
$$\forall x\,\forall y\,(x = y \supset y = x),$$
$$\forall x\,\forall y\,\forall z\,(x = y \wedge y = z \supset x = z).$$

THEOREM 17.10 (relative consistency of classical analysis). *Consider classical analysis, formalized as* $\mathbf{G^1LC}$ *together with the axiom system* $\mathscr{A}_0 \cup \{\mathrm{Eq}', \mathrm{VJ}'\}$ *in the language* L′. (Eq′ *and* VJ′ *were defined in Problem* 16.10.) *Then*:

(1) *it is interpretable in* $\mathscr{A}_0$ (*relative to* $\mathbf{G^1LC}$);
(2) *it is consistent, assuming cut-elimination for* $\mathbf{G^1LC}$.

PROOF. The interpretation is carried out in two steps.

(i) $\mathscr{A}_0 \cup \{\mathrm{Eq}'\}$ is interpreted in $\mathscr{A}_0$ (relative to $\mathbf{G^1LC}$) by Theorem 17.8, with r defined by: $R^1(\alpha)$ is $\forall x\,\forall y\,(x = y \wedge \alpha(x) \supset \alpha(y))$ (and no R^0). In fact, taking V as $\{u\}(u = 0)$, we can prove in $\mathbf{G^1LC}$ $\mathscr{A}_0 \rightarrow r(V)$, and hence $\mathscr{A}_0 \rightarrow \exists\varphi\, r(\varphi)$. Further, condition 5) of Lemma 17.5 is easily proved by induction on the logical complexity of V; and Eq'^r is provable in $\mathbf{G^1LC}$.

(ii) Next, $\mathscr{A}_0 \cup \{\mathrm{Eq}', \mathrm{VJ}'\}$ is interpreted in $\mathscr{A}_0 \cup \{\mathrm{Eq}'\}$, again by Theorem 17.8, this time with r defined by:

$$R^0(a) \text{ is } \quad \forall\varphi\,(\varphi(0) \wedge \forall y\,(\varphi(y) \supset \varphi(y')) \supset \varphi(a)) \quad (\text{and no } R^1).$$

Now $r(0)$, and hence $\exists x\, r(x)$, are easily proved in $\mathbf{G^1LC}$. Further, for any formula A of $\mathscr{A}_0 \cup \{\mathrm{Eq}'\}$, the sequent $\mathscr{A}_0, \mathrm{Eq}' \rightarrow A^r$ is provable in $\mathbf{G^1LC}$; and so is $\mathscr{A}_0, \mathrm{Eq}' \rightarrow \mathrm{VJ}'^r$.

Thus part (1) is proved. In fact the two steps could be combined, so as to give an interpretation of $\mathscr{A}_0 \cup \{\mathrm{Eq}', \mathrm{VJ}'\}$ directly in $\mathscr{A}_0$, by means of a single system of relativization r. The reader is invited to define such an r.

To prove part (2), we first show that if $\mathscr{A}_0$ is consistent (with $\mathbf{G^1LC}$), then so is $\mathscr{A}_0 \cup \{\mathrm{Eq}', \mathrm{VJ}'\}$. The method is parallel to that for part (1), using Theorem 17.8 part (2) twice.

The argument is completed by showing that $\mathscr{A}_0$ is consistent (with $\mathbf{G^1LC}$), assuming cut-elimination for $\mathbf{G^1LC}$. But this is clear, since a proof of $\mathscr{A}_0 \rightarrow$ in $\mathbf{G^1LC}$ without a cut would in fact be a proof in $\mathbf{LK}$, which is impossible.

§18. Truth definition for first order arithmetic

DEFINITION 18.1. (1) Although we have mentioned second order arithmetic from time to time we shall now formulate it more systematically. The language of the systems of second order arithmetic is that of **PA** (cf. §9) augmented by second order variables. The basic logical system is **BC** (cf. Definition 15.3), and the axioms (i.e., the mathematical initial sequents) are those of **PA** and the generalized equality axioms:

$$s = t,\ A(s) \rightarrow A(t)$$

for arbitrary terms s and t and arbitrary formulas A. The various systems of second order arithmetic are classified according to the forms of the induction and comprehension axioms. They are both introduced as rules of inference rather than axioms, and if both are allowed for all formulas and abstracts, then the system will represent classical analysis. In order to simplify the arguments, we assume only the logical symbols $\neg$, $\wedge$, $\forall$, although other symbols may be used occasionally. We recall that the rule of induction or "ind" has the form

$$\frac{F(a),\ \Gamma \rightarrow \Theta,\ F(a')}{F(0),\ \Gamma \rightarrow \Theta,\ F(s)},$$

where a does not occur in Γ, Θ or $F(0)$, and s is an arbitrary terms. $F(a)$ is called the induction formula, and a is called the eigenvariable, of this inference. The comprehension axiom, or $\forall$: left rule, has the form

$$\frac{F(V),\ \Gamma \rightarrow \Theta}{\forall \varphi F(\varphi),\ \Gamma \rightarrow \Theta},$$

where V and φ have the same number of argument places.

We normally deal with systems where the induction formulas belong to a certain class of formulas which is closed under substitution (cf. Definition 15.14, part (3)) and the abstracts for $\forall$: left also belong to a certain class, closed under substitution. If the induction formula or the abstract for $\forall$: left is restricted to a set K, then the corresponding ind or $\forall$: left is called a K-ind or a K-comprehension axiom, respectively.

(2) Formulas of second order arithmetic which do not contain second order quantifiers are called arithmetical. Also, abstracts are called arithmetical if they are formed from arithmetical formulas.

(3) Let Π_i^1 be the class of formulas of the form $\forall \varphi_1 \exists \varphi_2 \ldots \varphi_i F(\varphi_1, \varphi_2, \ldots, \varphi_i)$, where $\forall \varphi_1 \exists \varphi_2 \ldots \varphi_i$ denotes a string of i alternating quantifiers with second order bound variables starting with $\forall$, and F is arithmetical. The closure of Π_i^1 under substitution will be called Π_i^1 (in wider sense). Σ_i^1 and Σ_i^1 (in wider sense) are defined likewise (with $\exists \varphi_1 \forall \varphi_2 \ldots \varphi_i$ instead of $\forall \varphi_1 \exists \varphi_2 \ldots \varphi_i$). (For $i = 1$, this is essentially the

same as the definition in Problem 16.12, which used function instead of predicate quantification, since predicates or sets can be represented by their characteristic functions.)

The following are straightforward consequences of the definition.

PROPOSITION 18.2. (1) *The Π^1_i-comprehension axiom and the Π^1_i (in wider sense)-comprehension axiom, are equivalent (in* **BC**). *Similarly with the Σ^1_i-comprehension axiom.*

(2) *The Π^1_i-comprehension axiom, and Σ^1_i-comprehension axiom, are equivalent (in* **BC**).

This proposition enables us to identify the Π^1_i-, the Π^1_i (in wider sense)-, the Π^1_i-, and the Σ^1_i (in wider sense)-comprehension axioms. Therefore we shall call them all the Π^1_i-comprehension axiom.

DEFINITION 18.3. We assume a standard Gödel numbering of **PA**, and, if X is a formal object of **PA**, then $\ulcorner X \urcorner$ denotes the Gödel number of X (cf. §9 and §10). We shall list the primitive recursive functions and predicates we need.

ls(a): the number of logical symbols in a.

fl(a): a is a formula of **PA**.

st(a): a is a sentence (i.e., closed formula) of **PA**.

tm(a): a is a term.

ct(a): a is a closed term.

sub$(\ulcorner A \urcorner, \ulcorner a_i \urcorner, \ulcorner t \urcorner)$: the result of substituting t for a_i in A. This may be denoted by $\ulcorner A(t) \urcorner$.

$v(\ulcorner t \urcorner)$: the value of t (if t is closed).

$\nu(a)$: Gödel number of the ath numeral.

Abbreviated notions:

$\forall \ulcorner A \urcorner (\ldots \ulcorner A \urcorner \ldots)$: $\forall x\, (\mathrm{fl}(x) \supset \ldots x \ldots)$,

$\forall \ulcorner A \wedge B \urcorner (\ldots \ulcorner A \wedge B \urcorner \ldots)$: $\forall x\, (\mathrm{fl}(x) \wedge$ "the outermost logical symbol of x is $\wedge$" $\supset \ldots x \ldots)$.

$\forall \ulcorner t \urcorner (\ldots \ulcorner t \urcorner \ldots)$: $\forall x\, (\mathrm{tm}(x) \supset \ldots x \ldots)$.

Also we will write, for terms $t(a_i)$ and formulas $A(a_i)$:

$\ulcorner t(\nu(b)) \urcorner$ for sub$(\ulcorner t(a_i) \urcorner, \ulcorner a_i \urcorner, \nu(b))$,

$\ulcorner A(\nu(b)) \urcorner$ for sub$(\ulcorner A(a_i) \urcorner, \ulcorner a_i \urcorner, \nu(b))$.

In this section, $\mathbf{S}^1$ denotes second order arithmetic with the arithmetical comprehension axiom and ind applied to Π^1_1 (in wider sense) formulas.

PA$'$ is the system of second order arithmetic with the arithmetical ind and the arithmetical comprehension axiom. (This is clearly equivalent to the system of predicative arithmetic, also denoted by **PA**$'$, defined in Problem 16.10.)

In order to avoid too many parentheses and brackets, we use dots for punctuation in a well-known manner; $A . \supset . B \equiv C$, for instance, means $A \supset (B \equiv C)$.

This section is devoted to defining the truth definition for **PA**. The argument is carried out within $\mathbf{S}^1$.

DEFINITION 18.4. $F(\alpha, n)$ stands for the following formula:

$$\forall \ulcorner t_1 \urcorner \forall \ulcorner t_2 \urcorner [\mathrm{ct}(\ulcorner t_1 \urcorner) \wedge \mathrm{ct}(\ulcorner t_2 \urcorner) \supset$$
$$\supset (\alpha(\ulcorner t_1 = t_2 \urcorner) \equiv v(\ulcorner t_1 \urcorner) = v(\ulcorner t_2 \urcorner))]$$
$$\wedge \forall \ulcorner A \urcorner \forall \ulcorner B \urcorner [\mathrm{st}(\ulcorner A \wedge B \urcorner) \wedge \mathrm{ls}(\ulcorner A \wedge B \urcorner) \leqslant n \supset$$
$$\supset 1(\alpha(\ulcorner A \wedge B \urcorner) \equiv \alpha(\ulcorner A \urcorner) \wedge \alpha(\ulcorner B \urcorner))]$$
$$\wedge \forall \ulcorner \neg A \urcorner [\mathrm{st}(\ulcorner \neg A \urcorner) \wedge \mathrm{ls}(\ulcorner \neg A \urcorner) \leqslant n \supset$$
$$\supset (\alpha(\ulcorner \neg A \urcorner) \equiv \neg \alpha(\ulcorner A \urcorner))]$$
$$\wedge \forall \ulcorner \forall x_i\, A(x_i) \urcorner [\mathrm{st}(\ulcorner \forall x_i\, A(x_i) \urcorner) \wedge \mathrm{ls}(\ulcorner \forall x_i\, A(x_i) \urcorner) \leqslant n \supset$$
$$\supset (\alpha(\ulcorner \forall x_i\, A(x_i) \urcorner) \equiv \forall x\, \alpha(\ulcorner A(\nu(x)) \urcorner))].$$

$F(\alpha, n)$ means: α is a truth definition for sentences of complexity $\leqslant n$. In fact, a predicate T_n, satisfying $F(\{y\}T_n(y), \bar{n})$, was defined in §14 for each n (separately). However, we can now go further, and give a truth definition for all sentences, namely:

$$T(a) \text{ abbreviates } \mathrm{st}(a) \wedge \exists \varphi\, (F(\varphi, \mathrm{ls}(a)) \wedge \varphi(a)).$$

(Note: This is not a "truth definition" according to Definition 10.10. We are generalizing the notion of truth definition. For the meaning of this, see Theorem 18.13.)

LEMMA 18.5. (1) *In* **PA**$'$:

$$F(\alpha, n), F(\beta, n), \mathrm{st}(\ulcorner A \urcorner), \mathrm{ls}(\ulcorner A \urcorner) \leqslant n \rightarrow \alpha(\ulcorner A \urcorner) \equiv \beta(\ulcorner A \urcorner).$$

(*This states that any* α *for which* $F(\alpha, n)$ *holds, is unique, at least with regard to the sentences whose complexities are* $\leqslant n$.)

(2) $F(\alpha, n), m \leqslant n \rightarrow F(\alpha, m)$ *in* **PA**$'$.

(3) $F(\alpha, n), E(\beta, \alpha, n) \rightarrow F(\beta, n+1)$ *in* **PA**$'$, *where* $E(\alpha, \beta, n)$ *is an abbreviation of the following*:

$$\forall x\, (\beta(x) \equiv [\mathrm{st}(x) \wedge \mathrm{ls}(x) \leqslant n \wedge \alpha(x)]$$
$$\vee [\mathrm{st}(x) \wedge \mathrm{ls}(x) = n+1$$
$$\wedge \{\exists \ulcorner A \urcorner (x = \ulcorner \neg A \urcorner \wedge \neg \alpha(\ulcorner A \urcorner))$$
$$\vee \exists \ulcorner A \urcorner \exists \ulcorner B \urcorner (x = \ulcorner A \wedge B \urcorner \wedge \alpha(\ulcorner A \urcorner) \wedge \alpha(\ulcorner B \urcorner))$$
$$\vee \exists \ulcorner \forall x_i\, A(x_i) \urcorner (x = \ulcorner \forall x_i\, A(x_i) \urcorner \wedge \forall y\, \alpha(\ulcorner A(\nu(y)) \urcorner))\}]).$$

(*This means that* β *is an extension of* α *to the sentences of complexity* $n+1$.)

(4) $\forall n\, \forall \varphi\, \exists \psi\, E(\psi, \varphi, n)$ *in* **PA**′. (*The existence of an extension.*)

(5) $G(\alpha) \to F(\alpha, 0)$ *in* **PA**′, *where* $G(\alpha)$ *is an abbreviation of*

$$\forall x\, (\alpha(x) \equiv \exists \ulcorner t_1 \urcorner\, \exists \ulcorner t_2 \urcorner\, [\mathrm{ct}(\ulcorner t_1 \urcorner) \wedge \mathrm{ct}(\ulcorner t_2 \urcorner) \wedge (x = \ulcorner t_1 = t_2 \urcorner \wedge v(\ulcorner t_1 \urcorner) = v(\ulcorner t_2 \urcorner))]).$$

(*Truth definition for* $n = 0$.)

PROPOSITION 18.6. $\forall n\, \exists \varphi\, F(\varphi, n)$ *in* $\mathbf{S}^1$.

PROOF. By an application of ind, with induction formula $\exists \varphi\, F(\varphi, a)$, which is Σ^1_1, or Π^1_1 (in wider sense). Use (4), (2) and (3) of Lemma 18.5.

PROPOSITION 18.7. $T(a) \equiv \mathrm{st}(a) \wedge \forall \varphi\, (F(\varphi, \mathrm{ls}(a)) \supset q(a))$ *in* $\mathbf{S}^1$.

PROOF. Use (1) of Lemma 18.5, and Proposition 18.6.

The following propositions (18.8–18.10) assert that T commutes with logical symbols.

PROPOSITION 18.8. $\mathrm{st}(\ulcorner A \urcorner) \to T(\ulcorner \neg A \urcorner) \equiv \neg T(\ulcorner A \urcorner)$ *in* $\mathbf{S}^1$.

PROPOSITION 18.9. $\mathrm{st}(\ulcorner B \urcorner) \wedge \mathrm{st}(\ulcorner C \urcorner) \to T(\ulcorner B \wedge C \urcorner) \equiv T(\ulcorner B \urcorner) \wedge T(\ulcorner C \urcorner)$ *in* $\mathbf{S}^1$.

PROPOSITION 18.10. $\mathrm{st}(\ulcorner \forall x_i\, B(x_i) \urcorner) \to T(\ulcorner \forall x_i\, B(x_i) \urcorner) \equiv \forall x\, T(\ulcorner B(\nu(x)) \urcorner)$ *in* $\mathbf{S}^1$.

These propositions follow from Proposition 18.6 and (1) and (2) of Lemma 18.5. It should be noted that if we assume Proposition 18.6, then the argument goes through in **PA**′.

LEMMA 18.11. (1) $\mathrm{ct}(\ulcorner t_1 \urcorner), \mathrm{ct}(\ulcorner t_2 \urcorner) \to T(\ulcorner t_1 = t_2 \urcorner) \equiv v(\ulcorner t_1 \urcorner) = v(\ulcorner t_2 \urcorner)$ *in* **PA**′.

(2) $v(\ulcorner \nu(b) \urcorner) = b$ *in* **PA**.

(3) *Let* $t(a_1, \dots, a_k)$ *be a term with at most* $a_1, \dots, a_k$ *as free variables. Then*

$$v(\ulcorner t(\nu(b_1), \dots, \nu(b_k)) \urcorner) = t(b_1, \dots, b_k)$$

in **PA**, *where* $b_1, \dots, b_k$ *are arbitrary free variables.*

PROPOSITION 18.12. *Let* $A(a_1, \dots, a_k)$ *be a formula of* **PA** *with at most* $a_1, \dots, a_k$ *as free variables. Then*

$$T(\ulcorner A(\nu(b_1), \dots, \nu(b_k)) \urcorner) \equiv A(b_1, \dots, b_k) \text{ in } \mathbf{S}^1.$$

PROOF. By mathematical induction on the complexity of A. Use (3) of Lemma 18.11, and Propositions 18.8–18.10.

THEOREM 18.13 (property of truth definition). *Let A be a sentence of* **PA**. *Then* $T(\ulcorner A \urcorner) \equiv A$ *is provable in* $\mathbf{S}^1$.

PROOF. This is a special case of Proposition 18.12.

Since we have established this property of T, we can prove the consistency of **PA** in $\mathbf{S}^1$.

DEFINITION 18.14. We recall (cf. §10) that $\mathrm{Prov}_{\mathbf{PA}}(p, a)$ is the proof-predicate for **PA**: p is a proof of a in **PA**. (The subscript **PA** may be omitted.) Also, $\exists p\, \mathrm{Prov}(p, a)$ may be abbreviated to $\mathrm{Pr}(a)$.

LEMMA 18.15.

(1) $\mathrm{ct}(\ulcorner t_1 \urcorner), \mathrm{ct}(\ulcorner t_2 \urcorner), \mathrm{ct}(\ulcorner t(0) \urcorner), v(\ulcorner t_1 \urcorner) = v(\ulcorner t_2 \urcorner) \rightarrow v(\ulcorner t(t_1) \urcorner) = v(\ulcorner t(t_2) \urcorner)$ *in* **PA**.

(2) $\mathrm{ct}(\ulcorner t_1 \urcorner), \mathrm{ct}(\ulcorner t_2 \urcorner), \mathrm{st}(\ulcorner A(0) \urcorner), v(\ulcorner t_1 \urcorner) = v(\ulcorner t_2 \urcorner) \rightarrow$
$\rightarrow T(\ulcorner A(t_1) \urcorner) \equiv T(\ulcorner A(t_2) \urcorner)$ *in* $\mathbf{S}^1$.

PROOF OF (2). By induction on n applied to the following formula:

$$\forall \ulcorner A(a_i) \urcorner [\mathrm{st}(\ulcorner A(0) \urcorner) \wedge \mathrm{ls}(\ulcorner A(0) \urcorner) \leqslant n \,.\!\supset$$
$$\supset\!. \forall \ulcorner t_1 \urcorner \forall \ulcorner t_2 \urcorner (\mathrm{ct}(\ulcorner t_1 \urcorner) \wedge \mathrm{ct}(\ulcorner t_2 \urcorner) \wedge v(\ulcorner t_1 \urcorner) = v(\ulcorner t_2 \urcorner) \supset$$
$$\supset T(\ulcorner A(t_1) \urcorner) \equiv T(\ulcorner A(t_2) \urcorner))].$$

This is Π_1^1 (in wider sense). Use (1) and Propositions 18.8–18.10.

THEOREM 18.16. $\mathrm{st}(a), \mathrm{Pr}(a) \rightarrow T(a)$ *in* $\mathbf{S}^1$.

PROOF. By induction on n applied to the following formula:

$$\forall y\, (i(y) \leqslant n \supset T(u(y))),$$

where, if y is the Gödel number of a proof of $A_1, \ldots, A_m \rightarrow B_1, \ldots, B_p$, then $i(y)$ is the number of inferences of this proof and $u(y)$ denotes the "universal closure" (i.e., a sentence formed by repeated universal quantification) of $(A_1 \wedge \ldots \wedge A_m) \supset (B_1 \vee \ldots \vee B_p)$. Both i and u are primitive recursive functions. The above formula is Π_1^1 (in wider sense). Use Propositions 18.8–18.10 and Lemma 18.15.

THEOREM 18.17. Consis(**PA**) *in* $\mathbf{S}^1$.

PROOF. By Theorems 18.13 and 18.16 applied to the formula $0 = 0'$.

PROBLEM 18.18. Let ZF′ be the second order system which is defined as ZF with the basic logical system $\mathbf{K_1C}$ (cf. Definitions 15.15 and 16.4) and (finite) induction applied to Π_1^1-formulas. Give a truth definition for ZF in ZF′, thus proving the consistency of ZF in ZF′. [*Hint*: Follow the arguments in this section. It is important to notice that $T(\ulcorner A \urcorner)$, where A is an axiom of replacement, is a formula of ZF′ which is again an axiom of replacement.]

§19. The interpretation of a system of second order arithmetic

DEFINITION 19.1. (1) Let $\mathbf{S}^2$ be second order arithmetic with the arithmetical comprehension axiom and full induction, and let $\mathbf{S}^3$ be second order arithmetic with the Π_1^1 (in wider sense)-comprehension axiom and Π_1^1 (in wider sense)-induction. Notice that $\mathbf{S}^2$ and $\mathbf{S}^3$ are extensions of $\mathbf{S}^1$.

(2) We shall assume a standard Gödel numbering of the second order language (of arithmetic). In particular, $\ulcorner \alpha_1 \urcorner, \ulcorner \alpha_2 \urcorner, \ldots, \ulcorner \varphi_1 \urcorner, \ulcorner \varphi_2 \urcorner, \ldots$ denote Gödel numbers of second order variables. We include abstracts among the formal objects. So we need Gödel numbers for them: $\ulcorner\{\urcorner, \ulcorner\}\urcorner, \ulcorner\{x\}A(x)\urcorner$. Actually we only use arithmetical abstracts here.)

Notice however that although abstracts are included among the formal objects for convenience, they do not actually occur in the formulas of $\mathbf{S}^2$ (as explained in §15).

(3) We take over from §18 all the notations for primitive recursive functions and predicates pertaining to first order arithmetic (some of them, like $\mathrm{ls}(a)$, now adapted in an obvious way to the second order language).

Additional primitive recursive functions and predicates needed are:

$\mathrm{fl2}(a)$: a is a first or second order formula (of $\mathbf{S}^2$).

st2(a): a is a sentence, i.e., $\mathrm{fl2}(a)$ and a is closed.

$\mathrm{ab}(a)$: a is an arithmetical abstract.

$\mathrm{cab}(a)$: $\mathrm{ab}(a)$ and a is closed.

$\mathrm{sub}(\ulcorner A \urcorner, \ulcorner \alpha \urcorner, V)$: the result of substituting V for α in A if A is a formula, α a second order variable and V an arithmetical abstract. This may be denoted by $\ulcorner A(\substack{\alpha \\ V}) \urcorner$ or $\ulcorner A(V) \urcorner$.

$\mathrm{q2}(a)$: The number of second order quantifiers in a, if $\mathrm{fl2}(a)$.

We also use the other abbreviations in §18, and T is defined as in Definition 18.4.

DEFINITION 19.2. $F'(\alpha, n)$ stands for:

$$\forall \ulcorner A \urcorner [\mathrm{st2}(\ulcorner A \urcorner) \wedge \mathrm{q2}(\ulcorner A \urcorner) = 0 \,.\supset.\, \alpha(\ulcorner A \urcorner) \equiv T(\ulcorner A \urcorner)]$$

$$\wedge \forall \ulcorner A \urcorner [\mathrm{st2}(\ulcorner A \urcorner) \wedge 0 < \mathrm{q2}(\ulcorner \neg A \urcorner) \leq n \,.\supset.\, \alpha(\ulcorner \neg A \urcorner) \equiv \neg \alpha(\ulcorner A \urcorner)]$$

$$\wedge \forall \ulcorner A \wedge B \urcorner [\mathrm{st}(\ulcorner A \wedge B \urcorner) \wedge 0 < \mathrm{q2}(\ulcorner A \wedge B \urcorner) \leq n \,.\supset$$

$$\supset.\, \alpha(\ulcorner A \wedge B \urcorner) \equiv \alpha(\ulcorner A \urcorner) \wedge \alpha(\ulcorner B \urcorner)]$$

$$\wedge \forall \ulcorner \forall x_i\, A(x_i)\urcorner\, [\mathrm{st2}(\ulcorner \forall x_i\, A(x_i)\urcorner) \wedge 0 < \mathrm{q2}(\ulcorner \forall x_i\, A(x_i)\urcorner) \leqslant n\, .\supset$$
$$\supset .\, \alpha(\ulcorner \forall x_i\, A(x_i)\urcorner) \equiv \forall x\, \alpha(\ulcorner A(\nu(x))\urcorner)]$$
$$\wedge \forall \ulcorner \forall \varphi_i\, A(\varphi_i)\urcorner\, [\mathrm{st2}(\ulcorner \forall \varphi_i\, A(\varphi_i)\urcorner) \wedge 0 < \mathrm{q2}(\ulcorner \forall \varphi_i\, A(\varphi_i)\urcorner) \leqslant n\, .\supset$$
$$\supset .\, \alpha(\ulcorner \forall \varphi_i\, A(\varphi_i)\urcorner) \equiv \forall \ulcorner V \urcorner\, (\mathrm{cab}(\ulcorner V\urcorner) \supset \alpha(\ulcorner A(V)\urcorner))].$$

$F'(\alpha, n)$ means that α is an interpretation for sentences of $\mathbf{S}^2$ of (second order quantifier) complexity $\leqslant n$. We now give an interpretation for all sentences of $\mathbf{S}^2$.

$$\mathrm{I}(a)\colon \exists \psi\, (F'(\psi, \mathrm{q2}(a)) \wedge \psi(a)).$$

The point is that we can give a kind of truth definition for $\mathbf{S}^2$ by interpreting the second order variables as ranging over arithmetical predicates or sets (i.e., sets and relations of natural numbers defined by the closed arithmetical abstracts). This is often expressed by saying that the arithmetical sets form a model of $\mathbf{S}^2$. Further (as Lemma 19.14 essentially says) this can be formalized and proved in $\mathbf{S}^2$.

LEMMA 19.3. (1) $F'(\alpha, n), \mathrm{st}(\ulcorner A\urcorner) \to \alpha(\ulcorner A\urcorner) \equiv T(\ulcorner A\urcorner)$ *in* $\mathbf{PA}'$. (*α coincides with the truth definition T for arithmetical formulas.*)

(2) $\mathrm{st2}(\ulcorner A\urcorner), F'(\alpha, n), F'(\beta, n), \mathrm{q2}(\ulcorner A\urcorner) \leqslant n \to \alpha(\ulcorner A\urcorner) \equiv \beta(\ulcorner A\urcorner)$ *in* $\mathbf{S}^1$.

(3) $F'(\alpha, n), m \leqslant n \to F'(\alpha, m)$ *in* $\mathbf{PA}'$.

PROOF OF (2). By double induction on $(\mathrm{q2}(\ulcorner A\urcorner), \mathrm{ls}(\ulcorner A\urcorner))$ applied to the above sequent, which is Π^1_1 (in wider sense).

DEFINITION 19.4. $E(\alpha, \beta, n, l)$ is defined to be:

$$\forall \ulcorner A\urcorner\, [\mathrm{st2}(\ulcorner A\urcorner) \wedge \mathrm{q2}(\ulcorner A\urcorner) \leqslant n\, .\supset .\, \beta(\ulcorner A\urcorner) \equiv \alpha(\ulcorner A\urcorner)]$$
$$\wedge \forall \ulcorner \neg A\urcorner\, [\mathrm{st2}(\ulcorner \neg A\urcorner) \wedge \mathrm{q2}(\ulcorner \neg A\urcorner) = n' \wedge \mathrm{ls}(\ulcorner \neg A\urcorner) \leqslant l\, .\supset$$
$$\supset .\, \beta(\ulcorner \neg A\urcorner) \equiv \neg \beta(\ulcorner A\urcorner)]$$
$$\wedge \forall \ulcorner A \wedge B\urcorner\, [\mathrm{st2}(\ulcorner A \wedge B\urcorner) \wedge \mathrm{q2}(\ulcorner A \wedge B\urcorner) = n' \wedge \mathrm{ls}(\ulcorner A \wedge B\urcorner) \leqslant l\, .\supset$$
$$\supset .\, \beta(\ulcorner A \wedge B\urcorner) \equiv \beta(\ulcorner A\urcorner) \wedge \beta(\ulcorner B\urcorner)]$$
$$\wedge \forall \ulcorner \forall x_i\, A(x_i)\urcorner [\mathrm{st2}(\ulcorner \forall x_i\, A(x_i)\urcorner) \wedge \mathrm{q2}(\ulcorner \forall x_i\, A(x_i)\urcorner) = n' \wedge \mathrm{ls}(\ulcorner \forall x_i\, A(x_i)\urcorner) \leqslant l\, .\supset$$
$$\supset .\, \beta(\ulcorner \forall x_i\, A(x_i)\urcorner) \equiv \forall x\, \alpha(\ulcorner A(\nu(x))\urcorner)]$$
$$\wedge \forall \ulcorner \forall \varphi_i\, A(\varphi_i)\urcorner\, [\mathrm{st2}(\ulcorner \forall \varphi_i\, A(\varphi_i)\urcorner) \wedge \mathrm{q2}(\ulcorner \forall \varphi_i\, A(\varphi_i)\urcorner) = n'$$
$$\wedge\, \mathrm{ls}(\ulcorner \forall \varphi_i\, A(\varphi_i)\urcorner) \leqslant l\, .\supset$$
$$\supset .\, \beta(\ulcorner \forall \varphi_i\, A(\varphi_i)\urcorner) \equiv \forall \ulcorner V\urcorner\, (\mathrm{cab}(\ulcorner V\urcorner) \supset \beta(\ulcorner A(V)\urcorner)].$$

LEMMA 19.5.

(1) $E(\alpha, \beta, n, l), \mathrm{st2}(\ulcorner A\urcorner), \mathrm{q2}(\ulcorner A\urcorner) \leqslant n \to \beta(\ulcorner A\urcorner) \equiv \alpha(\ulcorner A\urcorner)$ *in* $\mathbf{PA}'$.

(2) $E(\alpha, \beta, n, l)$, $E(\alpha, \gamma, n, k)$, $\mathrm{st2}(\ulcorner A\urcorner)$, $\mathrm{q2}(\ulcorner A\urcorner) \leqslant n'$. $\mathrm{ls}(\ulcorner A\urcorner) \leqslant l, k \rightarrow \beta(\ulcorner A\urcorner) \equiv \gamma(\ulcorner A\urcorner)$ *in* $\mathbf{PA}'$.

(*Uniqueness of extension.*)

(3) $E(\alpha, \alpha, n, 0)$ *in* $\mathbf{PA}'$.

(4) $E(\alpha, \beta, n, l) \rightarrow E(\alpha, \{x\}C(x), n, l')$ *in* $\mathbf{PA}'$, *where* $C(x)$ *stands for*:

$$[(\mathrm{st2}(x) \wedge \mathrm{q2}(x) \leqslant n \,.\vee.\, (\mathrm{q2}(x) = n' \wedge \mathrm{ls}(x) < l')) \wedge \beta(x)]$$

$$.\vee.[\mathrm{st2}(x) \wedge \mathrm{q2}(x) = n' \wedge \mathrm{ls}(x) = l'$$

$$\wedge (\exists \ulcorner A\urcorner (x = \ulcorner \neg A\urcorner \wedge \neg\beta(\ulcorner A\urcorner))$$

$$\vee \exists \ulcorner A \wedge B\urcorner (x = \ulcorner A \wedge B\urcorner \wedge \beta(\ulcorner A\urcorner) \wedge \beta(\ulcorner B\urcorner))$$

$$\vee \exists \ulcorner \forall x_i\, A(x_i)\urcorner (x = \ulcorner \forall x_i\, A(x_i)\urcorner \wedge \forall y\, \beta(\ulcorner A(\nu(y))\urcorner))$$

$$\vee \exists \ulcorner \forall \varphi_i\, A(\varphi_i)\urcorner (x = \ulcorner \forall \varphi_i\, A(\varphi_i)\urcorner \wedge \forall \ulcorner V\urcorner (\mathrm{cab}(\ulcorner V\urcorner) \supset \beta(\ulcorner A(V)\urcorner))))].$$

(*Extension of* β *from* (n, l) *to* (n, l').)

(5) $\forall l \exists \varphi E(\alpha, \varphi, n, l)$ *in* $\mathbf{S}^1$. (*Existence of an extension of* α, *for fixed* n, *for every* l.)

PROOF OF (5). By induction on l applied to $\exists \varphi E(\alpha, \varphi, n, l)$. Use (3) and (4) of this lemma and the arithmetical comprehension axiom.

DEFINITION 19.6. $G(\alpha, n, x)$: $\exists \varphi (E(\alpha, \varphi, n, \mathrm{ls}(x)) \wedge \varphi(x))$. $G(\alpha, n, x)$ shall be abbreviated to $G(x)$.

LEMMA 19.7. (1) $F'(\alpha, n)$, $\mathrm{st}(\ulcorner A\urcorner) \rightarrow G(\ulcorner A\urcorner) \equiv T(\ulcorner A\urcorner)$ *in* $\mathbf{PA}'$.

(2) $E(\alpha, \beta, n, l)$, $\mathrm{st2}(\ulcorner A\urcorner)$, $\mathrm{q2}(\ulcorner A\urcorner) \leqslant n \vee (\mathrm{q2}(\ulcorner A\urcorner) = n' \wedge \mathrm{ls}(\ulcorner A\urcorner) \leqslant l) \rightarrow$
$\rightarrow G(\ulcorner A\urcorner) \equiv \beta(\ulcorner A\urcorner)$ *in* $\mathbf{PA}'$.

(3) $\mathrm{st2}(\ulcorner A\urcorner)$, $F'(\alpha, n)$, $\mathrm{q2}(\ulcorner \neg A\urcorner) \leqslant n' \rightarrow G(\ulcorner A\urcorner) \equiv \neg G(\ulcorner A\urcorner)$ *in* $\mathbf{S}^1$.

(4) $\mathrm{st2}(\ulcorner A \wedge B\urcorner)$, $F'(\alpha, n)$, $\mathrm{q2}(\ulcorner A \wedge B\urcorner) \leqslant n' \rightarrow$
$\rightarrow G(\ulcorner A \wedge B\urcorner) \equiv G(\ulcorner A\urcorner) \wedge G(\ulcorner B\urcorner)$ *in* $\mathbf{S}^1$.

(5) $\mathrm{st2}(\ulcorner \forall x_i\, A(x_i)\urcorner)$, $F'(\alpha, n)$, $\mathrm{q2}(\ulcorner \forall x_i\, A(x_i)\urcorner) \leqslant n' \rightarrow$
$\rightarrow G(\ulcorner \forall x_i\, A(x_i)\urcorner) \equiv \forall x\, G(\ulcorner A(\nu(x))\urcorner)$ *in* $\mathbf{S}^1$.

(6) $\mathrm{st2}(\ulcorner \forall \varphi_i\, A(\varphi_i)\urcorner)$, $F'(\alpha, n)$, $\mathrm{q2}(\ulcorner \forall \varphi_i\, A(\varphi_i)\urcorner) \leqslant n' \rightarrow$
$\rightarrow G(\ulcorner \forall \varphi_i\, A(\varphi_i)\urcorner) \equiv \forall \ulcorner V\urcorner (\mathrm{cab}(\ulcorner V\urcorner) \supset G(\ulcorner A(V)\urcorner))$ *in* $\mathbf{S}^1$.

PROOF. (1) From (1) of Lemma 19.3 and (1) of Lemma of 19.5.

(2) From (2) of Lemma 19.5.

(3) By (2) above and (5) of Lemma 19.5.

(4)–(6) are proved like (3) above. Notice that in (6)

$$\mathrm{q2}(\ulcorner \forall \varphi_i\, A(\varphi_i)\urcorner) \leqslant n' \text{ implies } \mathrm{q2}(\ulcorner A(V)\urcorner) \leqslant n.$$

LEMMA 19.8. (1) $F'(\alpha, n) \rightarrow F'(\{x\}G(x), n')$ *in* $\mathbf{S}^1$.

(2) $\rightarrow F'(\{x\}T(x), 0)$ *in* $\mathbf{PA}'$.

PROOF. By (1) and (3)–(6) of Lemma 19.7, and the definition of F'.

PROPOSITION 19.9. *$\forall n \exists \psi F'(\psi, n)$ in* $\mathbf{S}^3$.

PROOF. By the comprehension axiom applied to $\{x\}G(x)$ and also to $\{x\}T(x)$, and induction applied to $\exists \psi F'(\psi, n)$, which are all Π_1^1 (in wider sense), and Lemma 19.8.

LEMMA 19.10. $F'(\alpha, n)$, st2(${}^\ulcorner A^\urcorner$), q2(${}^\ulcorner A^\urcorner$)$\leq n \to \alpha({}^\ulcorner A^\urcorner) \equiv I({}^\ulcorner A^\urcorner)$ *in* $\mathbf{S}^1$.

PROOF. Use (2) and (3) of Lemma 19.3.

PROPOSITION 19.11. st2(${}^\ulcorner A^\urcorner$), q2(${}^\ulcorner A^\urcorner$)$= 0 \to I({}^\ulcorner A^\urcorner) \equiv T({}^\ulcorner A^\urcorner)$ *in* $\mathbf{S}^3$.

PROOF. By Lemma 19.10, (1) of Lemma 19.3, and (2) of Lemma 19.8.

The following proposition asserts that I commutes with logical connectives. It is proved by Lemma 19.10 and Proposition 19.9: hence in $\mathbf{S}^3$.

PROPOSITION 19.12. (1) st2(${}^\ulcorner A^\urcorner$)$\to I({}^\ulcorner \neg A^\urcorner) \equiv \neg I({}^\ulcorner A^\urcorner)$.
(2) st2(${}^\ulcorner A \wedge B^\urcorner$)$\to I({}^\ulcorner A \wedge B^\urcorner) \equiv I({}^\ulcorner A^\urcorner) \wedge I({}^\ulcorner B^\urcorner)$.
(3) st2(${}^\ulcorner \forall x_i\, A(x_i)^\urcorner$)$\to I({}^\ulcorner \forall x_i\, A(x_i)^\urcorner) \equiv \forall x\, I({}^\ulcorner A(\nu(x))^\urcorner)$.
(4) st2(${}^\ulcorner \forall \varphi_i\, A(\varphi_i)^\urcorner$)$\to I({}^\ulcorner \forall \varphi_i\, A(\varphi_i)^\urcorner) \equiv \forall {}^\ulcorner V^\urcorner$ (cab(${}^\ulcorner V^\urcorner$)$\supset I({}^\ulcorner A(V)^\urcorner)$)).

We can now proceed to the consistency proof of $\mathbf{S}^2$.

DEFINITION 19.13. (1) $i(p)$ is the number of inferences in (the proof with Gödel number) p.
(2) $i(b, a)$ means: "b is a closed instantiation of a", viz. $i(b, a)$ if and only if $a = {}^\ulcorner A(\beta, \ldots, c, \ldots)^\urcorner$ and $b = {}^\ulcorner A(V, \ldots, \nu(n), \ldots)^\urcorner$ for some closed arithmetical abstracts $V, \ldots,$ and numbers $n, \ldots,$ where $\beta, \ldots, c, \ldots$ are all the free variables in A.
(3) $j(b, p)$ means: if p is a Gödel number of proof in $\mathbf{S}^2$ of $A_1, \ldots, A_m \to B_1, \ldots, B_n$, then $i(b, {}^\ulcorner A_1 \wedge \ldots \wedge A_m \supset B_1 \vee \ldots \vee B_n{}^\urcorner)$.
(4) We write Prov_2 for $\text{Prov}_{\mathbf{S}^2}$, and $\text{Pr}_2(a)$ for $\exists x\, \text{Prov}_2(x, a)$.

LEMMA 19.14. $\text{Pr}_2(a) \to \forall x\, (i(x, a) \supset I(x))$ *in* $\mathbf{S}^3$.

PROOF. Define $H(m)$ as:

$$\forall p, x\, [i(p) \leq m \wedge j(x, p) \supset I(x)].$$

The proof is carried out by induction on m applied to $H(m)$, which is Π_1^1 (in wider sense). The argument proceeds according to the "last inference of p". The case where the last inference is an induction causes no problem, since the induction step is then proved by applying induction (on k) to a formula of the form $I({}^\ulcorner A(\nu(k), V, \ldots, r(n), \ldots)^\urcorner)$ (where $A(a, \beta, \ldots, c, \ldots)$ is the induction formula of this inference, with eigenvariable a).

PROPOSITION 19.15. st2(⌜A¹⌝), Pr_2(⌜A¹⌝)$\to I$(⌜A¹⌝) *in* $\mathbf{S}^3$.

PROOF. By Lemma 19.14.

THEOREM 19.16. Consis($\mathbf{S}^2$) *in* $\mathbf{S}^3$.

PROOF. A corollary of Proposition 19.11 and 19.15, and Theorem 8.13.

§20. Simple type theory

In this section we present the higher (finite) order predicate calculus. We shall formulate it in the sequential calculus. It is a simplification of a system called **GLC**, which was defined by the author. Now we restrict ourselves to predicate variables only. Following common practice, the word "type" is used instead of "order" (as in §15), and types start with 0 rather than 1. Thus the individual objects are of type 0.

DEFINITION 20.1. (1) We define *types* inductively as follows: 0 is a type; if $\tau_1, \ldots, \tau_k$ are types ($k \geq 1$), then so is $[\tau_1, \ldots, \tau_k]$; types are only as required by the above.

(2) The symbols of our language are classified as follows:

1) *Constants*:

1.1) individual constants: $c_0, c_1, \ldots$;

1.2) function constants with i argument-places ($i = 1, 2, \ldots$): $f^i_0, f^i_1, \ldots$;

1.3) predicate constants of type $\tau \neq 0$: $R^\tau_0, R^\tau_1, \ldots$.

2) *Variables*:

2.1) free variables: $a^\tau_0, a^\tau_1, \ldots$ of each type τ,

2.2) bound variables: $x^\tau_0, x^\tau_1, \ldots$ of each type τ.

3) *Logical symbols*: $\neg$ (not), $\wedge$ (and), $\vee$ (or), $\supset$ (implies), $\forall$ (for all), $\exists$ (there exists).

4) *Auxiliary symbols*: (,), {, }, [,].

A higher-order language (a language of simple type theory) is given when all the constants are given. A predicate variable means a variable (free or bound) of some type $\neq 0$. We shall use the symbols of the language also as metavariables. Type superscripts are sometimes omitted. We shall, further, take over all the appropriate notational conventions in §1.

Intuitively, variables of type 0 range over individual objects while variables of type $[\tau_1, \ldots, \tau_k]$ range over predicates which are associated with the subsets of $T_1 \times \ldots \times T_k$, where T_i is the set of objects of type τ_i.

DEFINITION 20.2. *Terms* (of given types), *formulas* and outermost *logical symbols* are defined inductively (and simultaneously).

1) Individual constants are terms of type 0.

2) Free variables of type τ are terms of type τ.

3) If f is a function constant with i argument-places and $t_1, \ldots, t_i$ are terms of type 0, then $f(t_1, \ldots, t_i)$ is a term of type 0.

4) Predicate constants of type τ are terms of type τ.

5) If A is a formula, $a_0^{\tau_0}, \ldots, a_k^{\tau_k}$ are distinct free variables of the indicated types, $x_0^{\tau_0}, \ldots, x_k^{\tau_k}$ are distinct bound variables of the indicated types not occurring in A, and A' is the result of simultaneously replacing, in A, a_0 by $x_0, \ldots, a_k$ by x_k, then $\{x_0, \ldots, x_k\} A'$ is a term (called an abstract) of type $[\tau_0, \ldots, \tau_k]$.

6) If α is a predicate constant or a free variable of type $[\tau_1, \ldots, \tau_k]$ and $t_1, \ldots, t_k$ are terms of type $\tau_1, \ldots, \tau_k$, then $\alpha[t_1, \ldots, t_k]$ is a formula, which is called atomic. There is no outermost logical symbol in this case.

7) If A and B are formulas, then $(\neg A)$, $(A \wedge B)$, $(A \vee B)$, $(A \supset B)$ are formulas, and their outermost logical symbols are $\neg$, $\wedge$, $\vee$, $\supset$, respectively.

8) If A is a formula, a^τ is a free variable, x^τ is a bound variable of the same type which does not occur in A, and A' is obtained from A by replacing all occurrences of a^τ by x^τ, then $\forall x^\tau A'$ and $\exists x^\tau A'$ are formulas, and their outermost logical symbols are $\forall$ and $\exists$, respectively.

These formation-rules may result in an excessive number of parentheses: if no ambiguity results, we may omit some of these as we did in the preceding sections.

Notice that here, unlike the preceding sections, abstracts are taken as formal objects. The notion of *alphabetical variant* is defined as before: for two expressions A and B, A is said to be an alphabetical variant of B (and vice versa) if A and B differ only in the names of some bound variables.

DEFINITION 20.3. The *height* of a type τ, $h(\tau)$, is defined inductively as follows:

$$h(0) = 0; \qquad h([\tau_1, \ldots, \tau_k]) = \max(h(\tau_1), \ldots, h(\tau_k)) + 1.$$

By the height $h(t)$ of a term (abstract) t, we mean the height of its type.

The (*logical*) *complexity* of a formula or abstract A is defined to be the total number of logical symbols and pairs of abstraction symbols $\{,\}$ in A.

Substitution of a term t of type τ for a free variable a of type τ in a formula or an abstract A is now defined by double induction on the height of τ and the complexity of A.

1) Basis: the height of τ is 0, i.e., the type τ is 0.

Then $A\binom{a}{t}$ can simply be defined as $(A\ \frac{a}{t})$, in accordance with Definition 1.4, or an alphabetical variant of this.

Let a and b be free variables of type 0 and let t and s be terms of type 0. We can easily prove the following.

(i) If A is a formula (term of type τ) then $A\binom{a}{t}$ is a formula (term of type τ).

(ii) If A is an alphabetical variant of B then $A\binom{a}{t}$ is an alphabetical variant of $B\binom{a}{t}$.

(iii) $A\binom{a}{t}$ contains only those free variables contained in A or t.

(iv) If s does not contain a, then

$$A\binom{a}{t}\binom{b}{s} \quad \text{is} \quad A\binom{b}{s}\binom{a}{t\binom{b}{s}}$$

2) Induction step: suppose $h(t) = n \neq 0$ and for any $m < n$, substitution of a term t of type σ (with $h(\sigma) = m$) for a free variable a of type σ has been defined so as to satisfy the following properties:

(1) If A is a formula (term of type σ), then $A\binom{a}{t}$ is a formula (term of type σ).

(2) If A is an alphabetical variant of B and s is an alphabetical variant of t, then $A\binom{a}{s}$ is an alphabetical variant of $B\binom{a}{t}$.

(3) $A\binom{a}{s}$ contains only free variables contained in A or t.

(4) $A\binom{a}{\{x_1, \ldots, x_k\}a[x_1, \ldots, x_k]}$ is an alphabetical variant of A.

(5) $\{x_1, \ldots, x_k\}a[x_1, \ldots, x_k]\binom{a}{t}$ is an alphabetical variant of t.

(6) If s does not contain a, and the height of b is less than n, then

$$A\binom{a}{t}\binom{b}{s}$$

is an alphabetical variant of

$$A\binom{b}{s}\binom{a}{t\binom{b}{s}}.$$

(7) If A does not contain a, then $A\binom{a}{s}$ is an alphabetical variant of A.

Let a and t be a free variable and a term, respectively, of type τ such that $h(\tau) = n$. If t is a free variable or predicate constant, then $A\binom{a}{t}$ is defined again to be (an alphabetical variant of) $(A\frac{a}{t})$. So suppose t is an abstract. We define $A\binom{a}{t}$ by induction on the complexity of A. Let t be $\{x_1, \ldots, x_k\}U(x_1, \ldots, x_k)$, where x_i is of type τ_i, a is of type $\tau = [\tau_1, \ldots, \tau_k]$ and $\max(h(\tau_1), \ldots, h(\tau_k)) + 1 = n$. First note that for any term s of type 0, $s\binom{a}{t}$ is defined to be s.

2.1) If A is of the form $b[t_1, \ldots, t_k]$, where b is a predicate constant or variable other than a, then

$$A\binom{a}{t} \quad \text{is} \quad b\left[t_1\binom{a}{t}, \ldots, t_k\binom{a}{t}\right].$$

2.2) If A is $a[t_1, \ldots, t_k]$, then

$$A\binom{a}{t} \quad \text{is} \quad U(b_1, \ldots, b_k)\binom{b_1}{t_1\binom{a}{t}}\cdots\binom{b_k}{t_k\binom{a}{t}},$$

where $b_1, \ldots, b_k$ are different from any free variable in A and

$$U(b_1, \ldots, b_k) \quad \text{is} \quad U(x_1, \ldots, x_k)\left(\frac{x_1, \ldots, x_k}{b_1, \ldots, b_k}\right).$$

$h(b_i) < n$ and t_i is "simpler" than A, and hence

$$B\left(\begin{matrix} b_i \\ t_i\begin{pmatrix} a \\ t \end{pmatrix}\end{matrix}\right)$$

has been defined for arbitrary B.

2.3) If A is of the form $\neg B$, $B \wedge C$, $B \vee C$ or $B \supset C$, then $A\binom{a}{t}$ is

$$\neg B\begin{pmatrix} a \\ t \end{pmatrix}, \quad B\begin{pmatrix} a \\ t \end{pmatrix} \wedge C\begin{pmatrix} a \\ t \end{pmatrix}, \quad B\begin{pmatrix} a \\ t \end{pmatrix} \vee C\begin{pmatrix} a \\ t \end{pmatrix} \quad \text{or} \quad B\begin{pmatrix} a \\ t \end{pmatrix} \supset C\begin{pmatrix} a \\ t \end{pmatrix}.$$

2.4) If A is of the form $\forall x F(x)$ or $\exists x F(x)$, then $A\binom{a}{t}$ is $\forall y G(y)$ or $\exists y G(y)$, respectively, where $G(b)$ is $F(b)\binom{a}{t}$, b is different from a and does not occur in A, and y does not occur in $G(b)$.

2.5) If A is of the form $\{x_1, \ldots, x_k\}B(x_1, \ldots, x_k)$, then

$$A\begin{pmatrix} a \\ t \end{pmatrix} \quad \text{is} \quad \{y_1, \ldots, y_k\}C(y_1, \ldots, y_k),$$

where $C(b_1, \ldots, b_k)$ is $B(b_1, \ldots, b_k)\binom{a}{t}$, $b_1, \ldots, b_k$ are different from a and do not occur in A, and none of the y_i's occur in $C(b_1, \ldots, b_k)$.

Then we can prove (1)–(7) for $h(a) = n$. (The proof is omitted.) We often denote $A\binom{a}{t}$ by $A(t)$.

Here and henceforth we use $U, V, \ldots$ with or without type-superscripts, as meta-variables for abstracts. Also, $\alpha, \beta, \ldots$ are often used for free variables instead of $a, b, \ldots$, and $\varphi, \psi, \ldots$ for bound variables instead of $x, y, \ldots$, usually when we are thinking of variables of type $\neq 0$.

EXAMPLE 20.4. Let A be $\forall \varphi \exists x (\alpha[a] \equiv \varphi[x])$, where φ and α are of type $[0]$ and x and a are of type 0. Let V be $\{z\} \forall \varphi \exists x (\varphi[x] \wedge \beta[z])$, where φ and β are of type $[0]$ and x and z are of type 0. V is an abstract of type $[0]$. Consider $A\binom{\alpha}{V}$. The substitution is carried out step by step according to Definition 20.3. Since α is of type $[0]$, we start with clause 2) of the definition and are led repeatedly back to 1). By 2.2) and 1), $\alpha[a]\binom{\alpha}{V}$ is $\forall \varphi \exists x (\varphi[x] \wedge \beta[a])$. Using this, by 2.1) and 2.3), $(\alpha[a] \equiv \gamma[b])\binom{\alpha}{V}$ (for some b, and γ different from a and α, respectively) is $\forall \varphi \exists x (\varphi(x) \wedge \beta[a]) \equiv \gamma[b]$. From this and 2.4), we obtain

$$\forall \psi \exists y (\forall \varphi \exists x (\varphi[x] \wedge \beta[a]) \equiv \psi[y]).$$

EXERCISE 20.5. Let A be

$$\{y\}\alpha^2[\{z\}(\alpha^1[z] \equiv \beta^1[y])],$$

and let V^2 be

$$\{\psi^1\}\forall\varphi^1\,\exists x\,(\psi^1[a]\equiv\varphi^1[x]),$$

where $1=[0]$ and $2=[1]$. Compute $A\left(\begin{smallmatrix}\alpha^2\\ V^2\end{smallmatrix}\right)$.

DEFINITION 20.6. Let V be an abstract of the form

$$\{x_1^{\tau_1}\ldots x_n^{\tau_n}\}A(x_1^{\tau_1},\ldots,x_n^{\tau_n}),$$

and let $V_1,\ldots,V_n$ be terms of types $\tau_1,\ldots,\tau_n$, respectively. Then $V[V_1,\ldots,V_n]$ is defined to be

$$A(a_1^{\tau_1},\ldots,a_n^{\tau_n})\begin{pmatrix}a_1^{\tau_1}\\ V_1\end{pmatrix}\cdots\begin{pmatrix}a_n^{\tau_n}\\ V_n\end{pmatrix}.$$

where $a_1^{\tau_1},\ldots,a_n^{\tau_n}$ are free variables of the indicated types which do not occur in any of $V, V_1,\ldots,V_n$ (and $A(a_1^{\tau_1},\ldots,a_n^{\tau_n})$ is

$$A(x_1^{\tau_1},\ldots,x_n^{\tau_n})\left(\frac{x_1^{\tau_1},\ldots,x_n^{\tau_n}}{a_1^{\tau_1},\ldots,a_n^{\tau_n}}\right).$$

DEFINITION 20.7. The formal system of simple type theory is defined like **G¹LC** in §15. The sequents are, as usual, of the form $\Gamma\rightarrow\Theta$, where Γ and Θ consist of finitely many formulas. The rules of inference are those of **G¹LC** (cf. Definition 15.15) with the following generalization (to higher types). (We take over all relevant notions and terminology from the previous sections.)

$$\forall:\text{left:}\quad\frac{F(V),\,\Gamma\rightarrow\Theta}{\forall\varphi\,F(\varphi),\,\Gamma\rightarrow\Theta},$$

where V is an arbitrary term (of any type) and φ is a bound variable of the same type as V (and if $F(V)$ is $F\left(\begin{smallmatrix}\alpha\\ V\end{smallmatrix}\right)$ then $F(\varphi)$ is $\left(F\frac{\alpha}{\varphi}\right)$).

$$\forall:\text{right:}\quad\frac{\Gamma\rightarrow\Theta,\,F(\alpha)}{\Gamma\rightarrow\Theta,\,\forall\varphi\,F(\varphi)},$$

where α does not occur in the lower sequent and φ is of the same type as α. Here α is called the eigenvariable of the inference. We define $\exists$: left and $\exists$: right similarly.

A sentence of the form

$$\forall y_1,\ldots,y_m\,\exists\varphi\,\forall x_1,\ldots,x_n\,(\varphi(x_1,\ldots,x_n)\equiv A(x_1,\ldots,x_n,y_1,\ldots,y_m)).$$

where $A(a_1,\ldots,a_n,b_1,\ldots,b_m)$ is an arbitrary formula (and $x_1,\ldots,x_n$, $y_1,\ldots,y_m$ are of arbitrary types), is called a comprehension axiom (cf. Definition 15.14).

The following is analogous to Proposition 15.16.

PROPOSITION 20.8. *The comprehension axioms* (*for arbitrary* A) *are provable in simple type theory.*

DEFINITION 20.9. (1) A *semi-formula* is an expression like a formula, except that it may contain free occurrences of bound variables. A semi-term is defined likewise.

(2) The *logical complexity* of a semi-formula or semi-term is defined as for formulas and abstracts (Definition 20.3).

§21. The cut-elimination theorem for simple type theory

In 1954 the author conjectured that the cut-elimination theorem holds for simple type theory as formulated in Definition 20.7. This was known as Takeuti's conjecture and it remained unresolved for many years. W. W. Tait provided support for my conjecture by proving the cut-elimination theorem for second order logic. The full conjecture was then resolved positively by Takahashi, and independently by Prawitz. In this section we will present a proof of the cut-elimination theorem for simple type theory, using the method of Takahashi and Prawitz. We, however, wish to point out that in 1971 J.-Y. Girard made significant improvements on several of the results of this section including the proof of the cut-elimination theorem. (See the *Proceedings of the Second Scandinavian Logic Symposium*, ed., J. E. Fenstad (North Holland, Amsterdam, 1971).) Girard's basic idea was then used by Martin-Löf and Prawitz, independently to produce a variant and somewhat more elegant form of cut-elimination.

Throughout this section we shall deal with variables and abstracts of one argument-place only and restrict the logical symbols to $\neg$, $\land$ and $\forall$ in order to simplify the argument. Thus the types are 0, [0], [[0]], . . . which may be called 0, 1, 2, ... We shall also omit the constants. The results can easily be extended to the general case.

DEFINITION 21.1. An *axiom of extensionality* is a formula of the form

$$\forall x\,(V_1(x) \equiv V_2(x)) \supset \forall \varphi\,(\varphi[V_1] \equiv \varphi[V_2]),$$

where V_1 and V_2 are arbitrary abstracts of the same type. In simple type theory this is equivalent to the following rule of inference (the extensionality rule):

$$\frac{V_1(a), \Gamma \rightarrow \Delta, V_2(a) \qquad V_2(a), \Gamma \rightarrow \Delta, V_1(a)}{\alpha[V_1], \Gamma \rightarrow \Delta, \alpha[V_2]}$$

where a does not occur in the lower sequent.

PROPOSITION 21.2. *The following is an admissible inference in simple type theory augmented by the extensionality rule*:

$$\frac{V_1(a), \Gamma \rightarrow \Delta, V_2(a) \qquad V_2(a), \Gamma \rightarrow \Delta, V_1(a)}{A(V_1), \Gamma \rightarrow \Delta, A(V_2)}$$

where $A(V_1)$ is obtained from an (arbitrary) formula $A(\beta)$ by substitution of V_1 for β and a does not occur in the lower sequent.

PROOF. By mathematical induction on the complexity of A. We shall deal with the case where $A(\beta)$ is of the form $\forall\varphi\, B(\varphi, \beta)$. Assume $V_1(a)$, $\Gamma \rightarrow \Delta$, $V_2(a)$ and $V_2(a)$, $\Gamma \rightarrow \Delta$, $V_1(a)$. By the induction hypothesis, $B(\gamma, V_1)$, $\Gamma \rightarrow \Delta$, $B(\gamma, V_2)$ is provable, where γ is a free variable of appropriate type which does not occur elsewhere in this sequent; hence by introducing $\forall\varphi$ on both sides, $\forall\varphi\, A(\varphi, V_1)$, $\Gamma \rightarrow \Delta$, $\forall\varphi\, A(\varphi, V_2)$ is provable.

THEOREM 21.3 (the cut-elimination theorem for simple type theory with extensionality: Takahashi). *Let* **S** *be simple type theory augmented by the extensionality rule. Then the cut-elimination theorem holds for* **S**.

The proof is obtained by modifying the original Takahashi-Prawitz method. The proof is presented stage by stage, introducing certain notions and notations as needed.

DEFINITION 21.4. (1) A *structure* (for simple type theory) is an ω-sequence of sets, say $\mathcal{S} = (S_0, S_1, \ldots, S_i, \ldots)$, where

1.1) S_0 is a non-empty set,

1.2) S_{i+1} is a subset of $P(S_i)$, the power set of S_i.

(2) An *assignment* ϕ (from $\mathcal{S}$) is a mapping from all (free and bound) variables such that to every variable of type i, ϕ assigns an element of S_i. An *interpretation* $\mathfrak{I}$ is a pair $(\mathcal{S}, \phi)$ consisting of a structure $\mathcal{S}$ and an assignment from $\mathcal{S}$.

(3) Given an interpretation $\mathfrak{I} = (\mathcal{S}, \phi)$, we will define the interpretation (by $\mathfrak{I}$) of semi-formulas and semi-terms. We use the notation $\phi\binom{x}{S}$ in order to express the assignment which agrees with ϕ except at x, where its value is S.

If A is a semi-formula or a semi-term, then its interpretation (by $\mathfrak{I}$) is denoted by $\phi(A)$. It is defined in such a way that for every semi-formula A, exactly one of $\phi(A) = \mathsf{T}$ and $\phi(A) = \mathsf{F}$ holds (where T stands for "truth" and F for "falsehood"), and for a semi-term A of type i, $\phi(A)$ is a subset of S_i. The definition is by induction on the complexity of A. (If A is a free or bound variable then $\phi(A)$ is already defined as the value of ϕ at A.)

3.1) $\phi[\alpha[W]) = \mathsf{T}$ if and only if $\phi(W) \in \phi(\alpha)$.

3.2) $\phi(\forall x\, A(x)) = \mathsf{T}$ (for x of any type) if and only if, for every ϕ' which agrees with ϕ except perhaps at x, $\phi'(A(x)) = \mathsf{T}$.

3.3) $\phi(\{x\}A(x)) = \{S \mid S \in S_i \text{ and } \phi\binom{x}{S}(A(x)) = \mathsf{T}\}$, where x is of type i.

3.4) $\phi(A \wedge B) = \mathsf{T}$ if and only if $\phi(A) = \mathsf{T}$ and $\phi(B) = \mathsf{T}$.

3.5) $\phi(\neg A) = \mathsf{T}$ if and only if $\phi(A) = \mathsf{F}$.

Let $S: A_1, \ldots, A_m \to B_1, \ldots, B_n$ be a sequent. Then

$$\phi(S) = \phi(\neg(A_1 \wedge \ldots \wedge A_m) \vee B_1 \vee \ldots \vee B_n)$$

(where $\vee$ is defined in terms of $\neg$ and $\wedge$).

(4) A structure $\mathcal{S}$ is called a Henkin structure if for every assignment ϕ from $\mathcal{S}$ and every abstract U^i of type i (for $i = 1, 2, \ldots$), $\phi(U^i)$ is a member of S_i.

PROPOSITION 21.5. *Suppose $\mathcal{S}$ is a Henkin structure and ϕ is an assignment from $\mathcal{S}$. If a sequent S is provable in* **S**, *then $\phi(S) = \mathsf{T}$.*

This is proved simply by examining each rule of inference.

DEFINITION 21.6. A semi-valuation with *extensionality* is an assignment v of at most one of the values T and F to formulas, which satisfies the following.

1) If $v(\neg A) = \mathsf{T}$, then $v(A) = \mathsf{F}$; if $v(\neg A) = \mathsf{F}$, then $v(A) = \mathsf{T}$.

2) If $v(A \wedge B) = \mathsf{T}$, then $v(A) = \mathsf{T}$ and $v(B) = \mathsf{T}$; if $v(A \wedge B) = \mathsf{F}$, then $v(A) = \mathsf{F}$ or $v(B) = \mathsf{F}$.

3) If $v(\forall x\, A(x)) = \mathsf{T}$, then $v(A(t)) = \mathsf{T}$ for every term t of the same type as x; if $v(\forall x\, A(x)) = \mathsf{F}$, then there is a free variable a of the same type as x such that $v(A(a)) = \mathsf{F}$.

4) If A is an alphabetical variant of B, then $v(A) = v(B)$.

5) Let α be a free variable of type >1. If $v(\alpha[U_1]) = \mathsf{T}$ and $v(\alpha[U_2]) = \mathsf{F}$, then there is a free variable a of appropriate type such that either $v(U_1[a]) = \mathsf{T}$ and $v(U_2[a]) = \mathsf{F}$ or $v(U_1[a]) = \mathsf{F}$ and $v(U_2[a]) = \mathsf{T}$.

If v is a semi-valuation with extensionality, and S is a sequent

$$A_1, \ldots, A_m \to B_1, \ldots, B_n,$$

then we define

$$v(S) =_{\mathrm{df}} v(\neg(A_1 \wedge \ldots \wedge A_m) \vee B_1 \vee \ldots \vee B_n),$$

if the latter is defined.

PROPOSITION 21.7. *If S is not cut-free provable, then there is a semi-valuation with extensionality, say v, such that $v(S) = \mathsf{F}$.*

PROOF. This is proved like the completeness theorem; construct a reduction tree for S (cf. §8). We shall only outline the definition of the immediate successors (i.e., the sequents written down immediately above those being considered) when "extensionality" and formulas of the form $\forall \varphi\, F(\varphi)$ come under attention. For the former let $\Gamma \to \Delta$ be one of the uppermost sequents in the tree which has been constructed so far. Let

$$\langle \alpha_1[U_{111}], \alpha_1[U_{112}] \rangle, \ldots, \langle \alpha_1[U_{1k_11}], \alpha_1[U_{1k_12}] \rangle, \ldots$$
$$\ldots, \langle \alpha_m[U_{m11}], \alpha_m[U_{m12}] \rangle, \ldots, \langle \alpha_m[U_{mk_m1}], \alpha_m[U_{mk_m2}] \rangle$$

be all the pairs of atomic formulas in Γ, Δ such that $\alpha_i[U_{ij1}]$ occurs in Γ and $\alpha_i[U_{ij2}]$ occurs in Δ for $j = 1, \ldots, k_i$. Let $b_{11}, \ldots, b_{1k_1}, \ldots, b_{m1}, \ldots, b_{mk_m}$ be distinct new variables of appropriate types. Write all sequents of the form $U_{ijl}[b_{ij}],\ \Gamma \to \Delta,\ U_{ijl'}[b_{ij}]$, for $i = 1, \ldots, m$, $j = 1, \ldots, k_i$, $l = 1$ or 2 and $l' = 2$ or 1 (according as $l = 1$ or 2) immediately above $\Gamma \to \Delta$.

For the stages when the higher type quantifiers come to attention, we proceed as follows. We define the notion of available free variable (at a given stage) as in §8, but in such a way that at least one free variable of each type is always available. Now, for a (higher type) $\forall$: left reduction: let $\{\forall \varphi_i\, F_i(\varphi_i)\}_{i=1}^{m}$ be a sequence of all the formulas in the antecedent of a sequent $\Gamma \to \Delta$ which start with higher order quantifiers. Suppose it is the kth stage. For all i, $1 \leq i \leq m$, let $V^i_1, \ldots, V^i_k$ be the first k abstracts in some predetermined list of abstracts of the same type as φ_i. Then the immediate successor of $\Gamma \to \Delta$ is

$$F_1(V^1_1), \ldots, F_1(V^1_k), \ldots, F_m(V^m_1), \ldots, F_m(V^m_k), \Gamma \to \Delta.$$

Next, for a (higher type) $\forall$: right reduction, proceed as before (i.e., case II, 9) in the proof of Lemma 8.3, replacing bound variables by free variables of the same type).

In this way we can complete the prescription for constructing the tree. As in Lemma 8.3, if each branch of this tree is finite, and (hence) ending with a sequent whose antecedent and succedent contain a formula in common, then it is easy to convert this tree to a cut-free proof in **S** of the sequent S. So if S is not cut-free provable, then there is an infinite branch. Take one such infinite branch and define a semi-valuation v as follows: $v(A) = \mathsf{T}$ if A occurs in the antecedent of a sequent in the branch and $= \mathsf{F}$ if it occurs in the succedent. It is not difficult to see that v satisfies the conditions for a semi-valuation. From the definition, $v(S) = \mathsf{F}$. We shall show only that v satisfies conditions 5) and 3) of Definition 21.6.

Suppose $v(\alpha[U_1]) = \mathsf{T}$ and $v(\alpha[U_2]) = \mathsf{F}$. Then $\alpha[U_1]$ occurs in the antecedent and $\alpha[U_2]$ in the succedent of the branch under consideration. From the construction of the tree, it follows that once $\alpha[U_1]$ occurs in the antecedent of a sequent, then it occurs in the antecedents of all the sequents above it, and likewise with $\alpha[U_2]$. Thus there is a sequent in which $\alpha[U_1]$ occurs in the antecedent and $\alpha[U_2]$ occurs in the succedent and to which the "extensionality" stage applies; thus there is a free variable a such that its immediate successor contains $U_1[a]$ in the antecedent and $U_2[a]$ in the succedent. Thus, by the definition of v, $v(U_1[a]) = \mathsf{T}$ and $v((U_2[a]) = \mathsf{F}$.

For the case of a formula $\forall \varphi\, F(\varphi)$, suppose $v(\forall \varphi\, F(\varphi)) = \mathsf{T}$. This means that $\forall \varphi\, F(\varphi)$ occurs in the antecedent of a sequent (and hence of all sequents above it). By the construction of the tree, for every abstract V of the same type as φ, $F(V)$ occurs in the antecedent of some sequent: hence $v(F(V)) = \mathsf{T}$.

DEFINITION 21.8. Given a semi-valuation with extensionality v, we define the *structure* $\mathcal{S} = (S_0, S_1, \ldots)$ induced by v, as follows. The sets $S_0, S_1, \ldots$ and relations $U^{n+1} < S$ for $S \subseteq S_n$ are defined simultaneously.

1) S_0 is the set of all terms of type 0 (in our simplified case these are only free variables). For any terms t_1 and t_2, $t_1 < t_2$ means that t_1 is identical with t_2.

2) Suppose $S_0, \ldots, S_n$ and $<$ for these types have been defined. Suppose $S \subseteq S_n$. Then $U^{n+1} < S$ if and only if for every abstract U_0^n of type n and every S^n which belongs to S_n, if $U_0^n < S^n$ and $v(U^{n+1}[U_0^n]) = \mathsf{T}$, then S^n belongs to S, and if $U_0^n < S^n$ and $v(U^{n+1}[U_0^n]) = \mathsf{F}$, then S^n does not belong to S. S_{n+1} is defined by

$$S_{n+1} =_{\mathrm{df}} \{S \mid S \subseteq S_n \text{ and there exists a } U^{n+1} \text{ such that } U^{n+1} < S\}.$$

From the definition it is obvious that $S_{n+1} \subseteq P(S_n)$.

We can think of $U < S$ as meaning: "U is a possible name for S (under the semi-valuation v)".

For convenience, we use the word "abstract" below to mean (also) a free variable of type 0. Then with any free variable a, we associate an abstract, also written a, namely $\{x\}a[x]$ if a has type >0, and a itself if its type is 0.

PROPOSITION 21.9. *For the structure $\mathcal{S}$ defined as in Definition* 21.8, *it holds that given a free variable of type n, say α, there exists an element of S_n, say S, such that $\alpha < S$ (α denoting an abstract as defined above).*

PROOF. By induction on n.

Basis: $n = 0$. For every variable a of type 0, a belongs to S_0 and $a < a$.

Induction step: let $n > 0$ and suppose the proposition holds for $0, 1, \ldots, n-1$. Let S be the set defined by

$$S =_{\mathrm{df}} \{S^{n-1} \mid S^{n-1} \text{ is in } S_{n-1} \text{ and there exists a } U^{n-1} \text{ such that } U^{n-1} < S^{n-1} \text{ and } v(\alpha[U^{n-1}]) = \mathsf{T}\}.$$

Then, by definition, $S \subseteq S_{n-1}$. We claim that $\alpha < S$. For take arbitrary U^{n-1} and S^{n-1} such that $U^{n-1} < S^{n-1}$. We show the following.

(1) If $v(\alpha[U^{n-1}]) = \mathsf{T}$, then S^{n-1} belongs to S.

(2) If $v(\alpha[U^{n-1}]) = \mathsf{F}$, then S^{n-1} does not belong to S.

(1) is obvious by definition of S. (2) is proved as follows. Suppose not (2): $v(\alpha[U^{n-1}]) = \mathsf{F}$ and S^{n-1} belongs to S. Then there is a W^{n-1} such that $W^{n-1} < S^{n-1}$, $v(\alpha[W^{n-1}]) = \mathsf{T}$.

Case 1. $n = 1$. $W^{n-1} = S^{n-1} = U^{n-1}$, yielding a contradiction.

Case 2. $n > 1$. Since $v(\alpha[U^{n-1}]) = \mathsf{F}$ and $v(\alpha[W^{n-1}]) = \mathsf{T}$, by condition 5) in Definition 21.6, there is an a such that either $v(U^{n-1}[a]) = \mathsf{F}$ and $v(W^{n-1}[a]) = \mathsf{T}$, or $v(U^{n-1}[a]) = \mathsf{T}$ and $v(W^{n-1}[a]) = \mathsf{F}$. By the induction hypothesis, there is an S^{n-2} in S_{n-2} such that $a < S^{n-2}$. If $v(U^{n-1}[a]) = \mathsf{F}$

and $v(W^{n-1}[a]) = \mathsf{T}$, then (since $v(U^{n-1}[a]) = \mathsf{F}$) S^{n-2} does not belong to S^{n-1}, since $U^{n-1} < S^{n-1}$. On the other hand, $v(W^{n-1}[a]) = \mathsf{T}$ implies that S^{n-2} belongs to S^{n-1}, since $W^{n-1} < S^{n-1}$. Thus we have a contradiction.

Similarly, if $v(U^{n-1}[a]) = \mathsf{T}$ and $v(W^{n-1}[a]) = \mathsf{F}$, we obtain a contradiction. So (2) must hold.

From (1) and (2), $\alpha < S$ by definition.

DEFINITION 21.10. We shall extend the relation $<$ to formulas and truth values as follows.

1) $A < \mathsf{T}$ if and only if $v(A) \neq \mathsf{F}$.

2) $A < \mathsf{F}$ if and only if $v(A) \neq \mathsf{T}$.

As immediate consequences of this, the following hold:

(1) if $A < *$ (where $*$ stands for F or T) and $v(A) = \mathsf{T}$, then $* = \mathsf{T}$,

(2) if $A < *$ and $v(A) = \mathsf{F}$, then $* = \mathsf{F}$;

since if $v(A) = \mathsf{T}$, then $v(A) \neq \mathsf{F}$ by the definition of v; so by 1), $A < T$. $*$ cannot be F for this case by virtue of 2), (2) is proved similarly.

PROPOSITION 21.11. *Let $\mathcal{S}$ be the structure which was defined in Definition 21.8, and let ϕ be an assignment from $\mathcal{S}$. Then for any abstract or formula $U(\alpha_1, \dots, \alpha_n)$, where all the free variables in U are among $\alpha_1, \dots, \alpha_n$, and for any abstracts $U_1, \dots, U_n$ (of appropriate types), if $U_i < \phi(\alpha_i)$ for $i = 1, \dots, n$, then $U(U_1, \dots, U_n) < \phi(U(\alpha_1, \dots, \alpha_n))$.*

PROOF. By induction on the complexity of $U(\alpha_1, \dots, \alpha_n)$.

(1) U is α_i. By the hypothesis, $U_i < \phi(\alpha_i)$.

The following are the induction steps.

(2) U is $\alpha_i[W(\alpha_1, \dots, \alpha_n)]$. Let α_i be of type n_i. $U_i < \phi(\alpha_i)$ by hypothesis, which implies (by the definition of $<$) that for every $U_0^{n_i-1}$ and S^{n_i-1} in S_{n_i-1}:

1) *if* $U_0^{n_i-1} < S^{n_i-1}$ and $v(U_i[U_0^{n_i-1}]) = \mathsf{T}$, then $S^{n_i-1} \in \phi(\alpha_i)$,

2) *if* $U_0^{n_i-1} < S^{n_i-1}$ and $v(U_i[U_0^{n_i-1}]) = \mathsf{F}$, then $S^{n_i-1} \notin \phi(\alpha_i)$.

Now take $W(U_1, \dots, U_n)$ as $U_0^{n_i-1}$ and $\phi(W(\alpha_1, \dots, \alpha_n))$ as S^{n_i-1}. By the induction hypothesis, $W(U_1, \dots, U_n) < \phi(W(\alpha_1, \dots, \alpha_n))$, hence the first premiss in 1) and 2) holds. If $v(U_i[W(U_1, \dots, U_n)]) = \mathsf{T}$, then by 1) $\phi(W(\alpha_1, \dots, \alpha_n)) \in \phi(\alpha_i)$, and if it $=$ F, then by 2) $\phi(W(\alpha_1, \dots, \alpha_n)) \notin \phi(\alpha_i)$. The first case implies $U_i[W(U_1, \dots, U_n)] < \mathsf{T}$ by Definition 21.10, and by Definition 21.4, part 3.1), $\phi(\alpha_i[W(\alpha_1, \dots, \alpha_n)]) = \mathsf{T}$; hence $U_i[W(U_1, \dots, U_n)] < \phi(\alpha_i[W(\alpha_i, \dots, \alpha_n)])$. Similarly, the second case implies $U_i[W(U_1, \dots, U_n)] < \mathsf{F} = \phi(\alpha_i[W(\alpha_1, \dots, \alpha_n)])$. (Note that if $v(U(U_1, \dots, U_n))$ is defined then (trivially) $U(U_1, \dots, U_n) < \phi(U(\alpha_1, \dots, \alpha_n))$, by definition of $<$ in Definition 21.10.)

(3) $U(\alpha_1, \dots, \alpha_n)$ is $\forall \varphi A(\varphi, \alpha_1, \dots, \alpha_n)$.

Case 1. $v(\forall \varphi A(\varphi, U_1, \dots, U_n)) = \mathsf{F}$. There is a β such that $v(A(\beta, U_1, \dots, U_n)) = \mathsf{F}$. For this β, take the S which was constructed in

Proposition 21.9 so that $\beta < S$. Let ϕ' be $\phi\binom{\beta}{S}$. Then $\beta < \phi'(\beta)$ and $U_i < \phi'(\alpha_i)$, $1 \le i \le n$. By the induction hypothesis

$$A(\beta, U_1, \ldots, U_n) < \phi'(A(\beta, \alpha_1, \ldots, \alpha_n)),$$

so $\phi'(A(\beta, \alpha_1, \ldots, \alpha_n)) = \mathsf{F}$ by (2) of Definition 21.10, which means that $\phi(\forall\varphi\, A(\varphi, \alpha_1, \ldots, \alpha_n)) = \mathsf{F}$, so $U(U_1, \ldots, U_n) < \phi(U(\alpha_1, \ldots, \alpha_n))$.

Case 2. $v(\forall\varphi\, A(\varphi, U_1, \ldots, U_n)) = \mathsf{T}$. Let β be a new free variable of the same type as φ. For an arbitrary S in S_n, define $\phi' = \phi\binom{\beta}{S}$. There is an abstract V such that $V < S$, so $V < \phi'(\beta)$. By the induction hypothesis,

$$A(V, U_1, \ldots, U_n) < \phi'(A(\beta, \alpha_1, \ldots, \alpha_n)).$$

From the assumption, $v(A(V, U_1, \ldots, U_n)) = \mathsf{T}$, which implies

$$\phi'(A(\beta, \alpha_1, \ldots, \alpha_n)) = \mathsf{T}.$$

This is true for every S in S_n, i.e., for every ϕ' which agrees with ϕ except at β. Thus $\phi(\forall\varphi\, A(\varphi, \alpha_1, \ldots, \alpha_n)) = \mathsf{T}$, hence $U(U_1, \ldots, U_n)) < \phi(U(\alpha_1, \ldots, \alpha_n))$.

(4) $U(\alpha_1, \ldots, \alpha_n)$ is $\{x\}A(x, \alpha_1, \ldots, \alpha_n)$. Let β be a new free variable and

$$Q = \left\{ S \mid \phi'(A(\beta, \alpha_1, \ldots, \alpha_n)) = T, \text{ where } \phi' = \phi\begin{pmatrix}\beta\\ S\end{pmatrix}\right\}$$

$$= \phi(\{x\}\, A(x, \alpha_1, \ldots, \alpha_n)).$$

For arbitrary U_0 and S_0 of appropriate type which satisfy $U_0 < S_0$, consider $A(U_0, U_1, \ldots, U_n)$ and $\phi' = \phi\binom{\beta}{S_0}$; so $U_0 < \phi'(\beta)$. By the induction hypothesis, $A(U_0, U_1, \ldots, U_n) < \phi'(A(\beta, \alpha_1, \ldots, \alpha_n))$. Suppose $v(A(U_0, U_1, \ldots, U_n)) = \mathsf{T}$. Then

$$\phi'(A(\beta, \alpha_1, \ldots, \alpha_n)) = \mathsf{T}$$

by Definition 21.10, so $S_0 \in Q$. If $v(A(U_0, U_1, \ldots, U_n)) = \mathsf{F}$, then

$$\phi'(A(\beta, \alpha_1, \ldots, \alpha_n)) = \mathsf{F},$$

and hence $S_0 \notin Q$. Therefore, by the definition $<$, $U(U_1, \ldots, U_n) < Q$.

Other cases are left to the reader.

PROPOSITION 21.12. *$\mathcal{S}$ (as defined in Definition 21.8) is a Henkin structure.*

PROOF. Let ϕ be an arbitrary assignment from $\mathcal{S}$. We have only to show that for every U, $U' < \phi(U)$ for some U' of the same type as U. Suppose all the free variables in U are among $\alpha_1, \ldots, \alpha_n$. Since $\phi(\alpha_i)$ belongs to S_{n_i}, where n_i is the type of α_i, there exists a U_i of type n_i such that $U_i < \phi(\alpha_i)$. Hence, by Proposition 21.11, $U(U_1, \ldots, U_n) < \phi(U(\alpha_1, \ldots, \alpha_n))$. So take U' to be $U(U_1, \ldots, U_n)$.

PROPOSITION 21.13. *Let $\mathcal{S}$ be the structure we have been dealing with and let ϕ_0 be an assignment from $\mathcal{S}$ which satisfies the following:*

(i) *$\phi_0(a) = a$ if a is a free variable of type 0;*

(ii) *$\phi_0(\alpha)$ = the S which was defined in Proposition 21.9, if α is a free variable of type > 0 (and $\phi_0(x)$, for bound variables x, is arbitrary).*

Let A be any formula. Then $v(A) = \mathsf{T}$ implies $\phi_0(A) = \mathsf{T}$, and $v(A) = \mathsf{F}$ implies $\phi_0(A) = \mathsf{F}$.

PROOF. For ϕ_0 as above, $\alpha < \phi_0(\alpha)$ for every free variable α. Thus, by Proposition 21.11, $A < \phi_0(A)$ (by taking U_i to be α_i). Then the proposition is a consequence of Definition 21.10.

PROPOSITION 21.14. *If a sequent S is not cut-free provable (in* **S**), *then there exists a Henkin structure $\mathcal{S}$ and an assignment from $\mathcal{S}$, say ϕ_0, such that $\phi_0(S) = \mathsf{F}$.*

PROOF. By Proposition 21.7, there is a semi-valuation with extensionality, say v, such that $v(S) = \mathsf{F}$. Let $\mathcal{S}$ be the Henkin structure induced by v (cf. Definition 21.8 and Proposition 21.12). Let ϕ_0 be the assignment from $\mathcal{S}$ defined in Proposition 21.13. Then $v(S) = \mathsf{F}$ implies $\phi_0(S) = \mathsf{F}$, again by Proposition 21.13.

PROOF OF THEOREM 21.3. By Proposition 21.14, if S is not cut-free provable (in **S**), then there is a Henkin structure $\mathcal{S}$ and an assignment ϕ_0 from $\mathcal{S}$ such that $\phi_0(S) = \mathsf{F}$. But this and Proposition 21.5 imply that S is not provable in **S** at all. In other words, if S is provable in **S**, then it is provable without a cut.

REMARK. By Proposition 21.14, we have proved not only cut-elimination for **S**, but also *completeness* of **S** *without the cut rule* (relative to the semantics of Henkin structures). (*Soundness* of **S** follows from Proposition 21.5.)

Next we shall prove the same theorem for the system without the extensionality rule. The method is quite similar to the proof of Theorem 21.3.

THEOREM 21.15 (the cut-elimination theorem for simple type theory without extensionality: Takahashi-Prawitz). *Let $\mathbf{S}^-$ be the system for simple type theory given in §20. Then the cut-elimination theorem holds for $\mathbf{S}^-$.*

PROOF. We follow the proof of Theorem 21.3, pointing out the corresponding items of §21.

DEFINITION 21.16 (cf. Definition 21.4). (1) A structure (for simple type theory without extensionality) is an ω-sequence of sets, say $\mathcal{P} = (P_0, P_1, \ldots, P_i, \ldots)$, with a relation $\in$; where

1.1) P_0 is a non-empty set,

1.2) P_{i+1} is a set of pairs of the form $\langle U^{i+1}, S\rangle$, where U^{i+1} is an abstract of type $i+1$ and S is a subset of P_i. Let $P^{i+1} = \langle U^{i+1}, S\rangle$ be an element of P_{i+1}, and let P^i be an element of P_i.
Then $P^i \in P^{i+1}$ if and only if P^i belongs to S.

(2) An assignment from $\mathcal{P}$ is a map ϕ from variables such that to every variable of type i, ϕ assigns an element of P_i. An interpretation is a pair $\mathfrak{I} = (\mathcal{P}, \phi)$.

(3) For each semi-formula or semi-term A, $\phi(A)$ is defined as in Definition 21.4 except for the following cases: $\phi(\alpha[W]) = \mathsf{T}$ if and only if $\phi(\alpha) \in \phi(W)$;

$$\phi(\{x^n\}A(x^n, x_1, \ldots, x_m)) = \left\langle \{x^n\}A(x^n, U_1, \ldots, U_m), \left\{P^n \,\middle|\, P^n \in P_n \right.\right.$$
$$\left.\left. \wedge\, \phi\binom{x^n}{P^n}(A(x^n, x_1, \ldots, x_m)) = \mathsf{T}\right\}\right\rangle,$$

where $\phi(x_i) = \langle U_i, S_i\rangle$ and all the bound variables occurring free in $\{x\}A(x)$ are among $x_1, \ldots, x_m$.

(4) A structure is called a pre-Henkin structure if for every assignment ϕ from P and every abstract U of type i, $\phi(U)$ belongs to P_i.

REMARK. The reason why we must take pairs $\langle U, S\rangle$ instead of just S in defining P^{i+1} is that $\mathcal{P}$ is a model of the comprehension axioms for which the axiom of extensionality may not hold. Thus, we cannot always identify two objects whenever they have the same extension; in order to distinguish two objects with the same extension, we consider pairs so that the names (of the extension) are explicitly expressed.

PROPOSITION 21.17 (cf. Proposition 21.5). *Suppose $\mathcal{P}$ is a pre-Henkin structure and ϕ is an assignment from $\mathcal{P}$. If a sequent S is provable in* $\mathbf{S}^-$, *then* $\phi(S) = \mathsf{T}$.

DEFINITION 21.18 (cf. Definition 21.6). Semi-valuations are defined as in Definition 21.6, omitting 5).

PROPOSITION 21.19 (cf. Proposition 21.7). *If S is not cut-free provable in* $\mathbf{S}^-$, *then there is a semi-valuation, say v, such that* $v(S) = \mathsf{F}$.

DEFINITION 21.20 (cf. Definition 21.8). Given a semi-valuation v, we define the structure $\mathcal{P}$ induced by v: the sets $P_0, P_1, \ldots$ and relations

$U^i < S$ and $U^i < P^i$ (for abstracts U^i, $P^i \in P_i$ and $S \subseteq P^i$) are defined simultaneously by induction on i.

1) P_0 is the set of all free variables of type 0. $t_1 < t_2$ if t_1 is t_2.

2) Suppose $P_0, \ldots, P_i$ and $<$ for those sets have been defined. Let S be a subset of P_i. $U^{i+1} < S$ is defined to be true if and only if for every U_0^i and every P^i in P_i with $U_0^i < P^i$: $v(U^{i+1}[U_0^i]) = \mathsf{T}$ implies $P' \in S$, and $v(U^{i+1}[U_0^i]) = \mathsf{F}$ implies $P^i \notin S$.

3) $P_{i+1} =_{\text{df}} \{\langle U^{i+1}, S\rangle \mid S \subseteq P_i \text{ and } U^{i+1} < S\}$.

4) Let $P^{i+1} = \langle U^{i+1}, S\rangle$ be an element of P_{i+1} (so $U^{i+1} < S$). Then $U < P^{i+1}$ if and only if U is U^{i+1}.

PROPOSITION 21.21 (cf. Proposition 21.9). *For an arbitrary α of type n there exists a P in P_n such that $\alpha < P$.*

PROOF. There are two cases.

1) $n = 0$. For every a in P_0, $a < a$ by definition.

2) $n > 0$. Define P as

$$P =_{\text{df}} \langle \alpha, S\rangle,$$

where

$$S = \{P^{n-1} \mid P^{n-1} \in P_{n-1} \text{ and there exists a } U^{n-1} \text{ such that } U^{n-1} < P^{n-1} \text{ and } v(\alpha[U^{n-1}]) = \mathsf{T}\}.$$

We have only to show that $\alpha < S$ for this S. In order to prove $\alpha < S$ it suffices to show that for any U^{n-1} and P^{n-1} with $U^{n-1} < P^{n-1}$:

(1) If $v(\alpha[U^{n-1}]) = \mathsf{T}$, then $P^{n-1} \in S$;

(2) if $v(\alpha[U^{n-1}]) = \mathsf{F}$, then $P^{n-1} \notin S$.

The proof of (2) in this case is trivial, since $U^{n-1} < P^{n-1}$ here means that $P^{n-1} = \langle U^{n-1}, \tilde{S}\rangle$ for an appropriate $\tilde{S}$.

DEFINITION 21.22. We can extend $<$ to formulas as in Definition 21.10.

PROPOSITION 21.23 (cf. Proposition 21.11). *Let $\mathcal{P}$ be the structure defined as above. Given an assignment ϕ, define ϕ_1 as follows: If $\phi(\alpha) = \langle U_1, S\rangle$, then $\phi_1(\alpha) = U_1$, and for free variables a of type 0, $\phi_1(a) = \phi(a) = a$. Let U be a formula or abstract whose free variables are $\alpha_1, \ldots, \alpha_n$. If*

$$\phi_1(\alpha_1) = U_1, \ldots, \phi_1(\alpha_n) = U_n,$$

then

$$U(U_1, \ldots, U_n) < \phi(U(\alpha_1, \ldots, \alpha_n)).$$

PROOF. By induction on the complexity of $U(\alpha_1, \ldots, \alpha_n)$. The argument is very much the same as the proof of Proposition 21.11. We shall give one example for the induction step. Suppose U is $\alpha_i[W(\alpha_1, \ldots, \alpha_n)]$. Let α_i be of type n_i and let $\phi(\alpha_i)$ be $\langle U_i, S_i\rangle$ (this is the case for $n_i > 0$, since $\phi_1(\alpha_i) = U_i$; for $n_i = 0$, $\phi(\alpha_i) = \alpha_i = U_i$). Since $U_i < S_i$, for every $U_0^{n_i-1}$ and P^{n_i-1} in P_{n_i-1}:

1) If $U_0^{n_i-1} < P^{n_i-1}$ and $v(U_i[U_0^{n_i}]) = \mathsf{T}$, then $P^{n_i-1} \in S_i$;

2) if $U_0^{n_i-1} < P^{n_i-1}$ and $v(U_i[U_0^{n_i-1}]) = \mathsf{F}$, then $P^{n_i-1} \notin S_i$.

Now take $W(U_1, \ldots, U_n)$ as $U_0^{n_i-1}$ and $\phi(W(\alpha_1, \ldots, \alpha_n))$ as P^{n_i-1}. By the induction hypothesis, $W(U_1, \ldots, U_n) < \phi(W(\alpha_1, \ldots, \alpha_n))$, hence the first premiss in 1) and 2) holds. If $v(U_i[W(U_1, \ldots, U_n)]) = \mathsf{T}$, then by 1), $\phi(W(\alpha_1, \ldots, \alpha_n)) \in \phi(\alpha_i)$ (since $\phi(\alpha_i) = \langle U_i, S_i\rangle$ and $\phi(W(\alpha_1, \ldots, \alpha_n))$ belongs to S_i), so $\phi(\alpha_i[W(\alpha_1, \ldots, \alpha_n)]) = \mathsf{T}$, which implies $U_i[W(U_1, \ldots, U_n)] < \phi(\alpha_i[W(\alpha_1, \ldots, \alpha_n)])$. Similarly, if $v(U_i[W(U_1, \ldots, U_n)]) = \mathsf{F}$, then by 2), $\phi(\alpha_i[W(\alpha_1, \ldots, \alpha_n)]) = \mathsf{F}$, and hence the desired relation holds.

PROPOSITION 21.24 (cf. Proposition 21.12). *$\mathcal{P}$ is pre-Henkin structure.*

PROOF. Let ϕ be an arbitrary assignment from $\mathcal{P}$. We have only to show that if $\phi(U) = \langle V, S\rangle$, then $V < S$. Suppose all the free variables in U are among $\alpha_1, \ldots, \alpha_n$. Let $U_i = \phi(\alpha_i)$. Then $U(U_1, \ldots, U_n) < \phi(U(\alpha_1, \ldots, \alpha_n))$ by Proposition 21.23. By the definition of $<$, this means that $U(U_1, \ldots, U_n)$ is V and hence $V < S$, again by definition of $<$.

PROPOSITION 21.25 (cf. Proposition 21.13). *Let $\mathcal{P}$ be the structure we have been dealing with and let ϕ be an assignment from $\mathcal{P}$ which satisfies the following:*

(i) *$\phi(a) = a$ if a is a free variable of type* 0;

(ii) *$\phi(\alpha) =$ the P which was defined in Proposition* 21.21, *if α is a free variable of type* > 0 *(and $\phi(x)$, for bound variables x, is arbitrary).*

Let A by any formula. Then $v(A) = \mathsf{T}$ implies $\phi(A) = \mathsf{T}$, and $v(A) = \mathsf{F}$ implies $\phi(A) = \mathsf{F}$.

PROOF. For ϕ as above and ϕ_1 as in Proposition 21.23, $\alpha = \phi_1(\alpha)$ for every free variable α. Thus, by Proposition 21.23, $A < \phi(A)$ (by taking U_i to be α_i). Then the proposition is a consequence of Definition 21.22.

PROPOSITION 21.26 (cf. Proposition 21.14). *If a sequent S is not cut-free provable in $\mathbf{S}^-$, then there exists a pre-Henkin structure and an assignment from $\mathcal{P}$, say ϕ, such that $\phi(S) = \mathsf{F}$.*

PROOF. The proof is parallel to Proposition 21.14.

PROOF OF THEOREM 21.15. By Propositions 21.26 and 21.17. Follow the proof of Theorem 21.3.

CHAPTER 4

INFINITARY LOGIC

In this chapter we will deal with a proof-theoretic development of infinitary logic. One reason for our interest in infinitary logic is that it enables us to establish a stronger link between model theory and proof theory. Model theory and proof theory are related to each other in many respects. For example, Craig's theorem, Beth's theorem and Tarski's theorem, stated in Chapter I, can be regarded as theorems of both model theory and proof theory. On the other hand proof theory is somewhat narrower than model theory in the sense that one cannot always express a model-theoretic result in proof-theoretic terms although the converse is usually possible. For example, although there are several proof-theoretic results containing part of the Löwenheim-Skolem theorem, one of the most fundamental theorems in model theory, we do not have a proof-theoretic version of the full theorem in ordinary (finitary) proof theory. However, if we introduce infinitary logic with an appropriate notion of proof, then the Löwenheim-Skolem theorem can be stated syntactically (see Problem 22.20).

Let α be an ordinal number, let f be a mapping from α into $\{\forall, \exists\}$ and let $\boldsymbol{x}_{<\alpha}$ denote the sequence $\{x_\xi\}_{\xi<\alpha}$. Then $\mathbf{Q}^f\boldsymbol{x}_{<\alpha}$ is a quantifier of "arity" α. If all the values of f are $\forall$ or all the values of f are $\exists$, then $\mathbf{Q}^f\boldsymbol{x}_{<\alpha}$ is a *homogeneous* quantifier that we denote by $\forall\boldsymbol{x}_{<\alpha}$ or $\exists\boldsymbol{x}_{<\alpha}$, respectively. A quantifier that is not homogeneous is called *heterogeneous*.

Heterogeneous quantifiers can occur in more general forms (Henkin). Let X and Y be disjoint sets of bound variables and let T be a function that maps Y onto a subset S of $P(X)$. We associate with T, X, Y a quantifier $\mathbf{Q}(T, X, Y)$. For simplicity let $\boldsymbol{x}$ and $\boldsymbol{y}$ be sequences composed of all the elements of X and Y, respectively, ordered by some well-orderings of X and Y. Then for a formula $A(\boldsymbol{a}, \boldsymbol{b})$, $\mathbf{Q}(T, \boldsymbol{x}, \boldsymbol{y})\ A(\boldsymbol{x}, y)$ (denoted $\mathbf{Q}^T\boldsymbol{xy}\,A(\boldsymbol{x}, \boldsymbol{y})$) is a formula having the following meaning. Given any values of the variables $\boldsymbol{x}$ there exist values of the variables $\boldsymbol{y}$ such that (1) for each η, the value of y_η is dependent on the values of those x_ξ's that are in $T(y_\eta)$, (2) for each η, the value of y_η is independent of the values of those x_ξ's that are not in $T(y_\eta)$, and (3) $A(\boldsymbol{x}, \boldsymbol{y})$. In other words $\mathbf{Q}^T\boldsymbol{xy}\,A(\boldsymbol{x}, \boldsymbol{y})$ is equivalent to the second order formula.

$$(\exists f_0, \ldots, f_\eta, \ldots)\,(\forall \boldsymbol{x})\,(A(\boldsymbol{x}, f_0(x_{0_0}, x_{0_1}, \ldots), \ldots, f_\eta(x_{\eta_0}, x_{\eta_1}, \ldots), \ldots)\,,$$

where $x_{\eta_0}, x_{\eta_1}, \ldots$ are the elements of $T(y_\eta)$. For example, if $X = \{x, y\}$,

$Y = \{u, v\}$ and T is defined by

$$T(u) = \{x\}, \qquad T(v) = \{y\} .$$

then we have the formula $\mathbf{Q}(T, X, Y)\, A(X, Y)$ that we denote by

$$\begin{pmatrix} \forall x & \exists u \\ \forall y & \exists v \end{pmatrix} A(x, y; u, v) .$$

It is known (Mostowski) that this quantifier cannot be defined in terms of ordinary quantifiers $\forall$ and $\exists$. Other examples of this kind will be given below.

We shall consider both homogeneous and heterogeneous quantifiers. Were we to restrict ourselves to homogeneous quantifiers, the theory obtained would be more or less like a finitary first order theory, whose nature is well-understood. The situation with regard to heterogeneous quantifiers is more interesting. One of our objectives will be to determine whether logics with heterogeneous quantifiers are like finite first order logics or finite second order logics.

An infinitary logic, with heterogeneous quantifiers $\mathbf{Q}^f \boldsymbol{x}_{<\alpha}$ such that $f(\beta) = \forall$ if β is even and $f(\beta) = \exists$ if β is odd, is of particular interest in connection with the *axiom of determinateness*, an axiom that implies many interesting theorems in set theory. The axiom of determinateness asserts that for each quantifier $\mathbf{Q}^f$, and for every formula ψ, exactly one of the two formulas

$$\mathbf{Q}^f \boldsymbol{x}_{<\alpha}\, \psi(\boldsymbol{x}_{<\alpha}, \boldsymbol{a}_{<\beta}) ,$$

or

$$\mathbf{Q}^{\bar{f}} \boldsymbol{x}_{<\alpha} \neg\psi(\boldsymbol{x}_{<\alpha}, \boldsymbol{a}_{<\beta})$$

is true, where $\bar{f}$ is the dual of f, that is, $\bar{f}(\gamma) = \forall$ if $f(\gamma) = \exists$, and $\bar{f}(\gamma) = \exists$ if $f(\gamma) = \forall$.

Through the axiom of determinateness we can see connections between proof theory and set theory. For example, one of the important properties of rules of inference in **LK** is that they come in symmetrically related pairs. This property was essential in the proof of the cut-elimination theorem of **LK** but apparently cannot be preserved when we introduce heterogeneous quantifiers. So it seems rather hopeless to expect that the cut-elimination theorem holds in infinitary logic with heterogeneous quantifiers. However, in a determinate logic (to be defined in §23, roughly as an infinitary logic in which the axiom of determinateness holds) the rules of inference are symmetric. This offers hope that the cut-elimination theorem might hold in such a logic. It is known, however, that the cut-elimination theorem fails for a determinate logic which has disjunction and conjunction symbols of arity $2^{\aleph_0}$, that is, disjunction symbols and conjunction symbols that operate on sequences of length $2^{\aleph_0}$. Is this also the case for a determinate logic in which disjunction and conjunction are only of arity ω?

There are two approaches to the study of determinate logic, one assuming the axiom of choice, and the other without it. Without the axiom of choice, some proofs need very delicate arguments. Nevertheless we can prove the following without this axiom.

Let M be a transitive model of ZF + DC, that is, Zermelo-Fraenkel set theory augmented by the axiom of dependent choice, and let the power set of ω belong to M. Then the axiom of determinateness, AD, holds in M if and only if the cut-elimination theorem holds for every determinate logic of M, i.e., every determinate logic that is M-definable.

This theorem suggests that there is a close relationship between the cut-elimination theorem and the axiom of determinateness. Furthermore there is a natural reduction in **LK** that provides a basis for proving the cut-elimination theorem. This suggests that by extending the notion of reduction to infinitary languages we may be able to prove the cut-elimination theorem and thereby learn more about the axiom of determinateness. We shall, therefore, generalize the cut rule so that a natural reduction exists for infinitary languages.

The simplest cases of infinitary logic are those systems with propositional connectives of countable arity, but quantifiers only of finite arity. Although these are very interesting logics we will give only one result concerning such systems (cf. Problem 22.21 : Lopez-Escobar). For more information the reader should see: J. Barwise, Infinitary logic and admissible sets, Journal of Symbolic Logic 34 (1969).

An infinitary logic can be regarded as a subsystem of a second order logic simply because one can formulate the truth definition of any significant infinitary system in a reasonable second order system. An example is given as Problem 22.26.

In defining an infinitary language, the basic idea is to determine a set of variables, a set of constants, and formation rules for formulas. There are various ways of defining the formulas of the language:

a) Accept all the formulas that are inductively defined from the constants and the variables.
b) Restrict the "admissible" formulas to some subsets of all the formulas defined as in a), with the provision that the set of admissible formulas must be closed with respect to subformulas.

Unless we state otherwise, the systems we will study are ones in which the formulas are defined as in a). Although it is common practice to set an upper bound on the cardinality of the various sets of language symbols we will not always do so.

By an infinitary language we mean the following:

1) a set of bound variables;
2) a set of free variables;
3) a set of predicate constants each with its own arity, i.e., "number" of argument places;

4) a set of individual constants;
5) a set of logical symbols.

The set of logical symbols consists of the usual unary negation sign $\neg$, and the binary implication sign $\supset$, together with a collection of disjunction, conjunction, universal quantification, and existential quantification signs each with its own arity. However, we will not use different symbols for signs with different arity. We will use only one symbol for disjunction $\bigvee$, one for conjunction $\bigwedge$, one for universal quantification $\forall$, and one for existential quantification $\exists$. We will then rely upon the context to make clear which symbols are "distinct", for example, two $\forall$'s followed by sequences of bound variables are different symbols if the lengths of the sequences are different, i.e., the $\forall$'s in $\forall \boldsymbol{x}_{<\alpha}$ and $\forall \boldsymbol{x}_{<\beta}$ are different if $\alpha \neq \beta$. The same is true of $\exists$, $\bigvee$, and $\bigwedge$. For example, the $\bigwedge$'s in $\bigwedge_{\gamma<\alpha} A_\gamma$ and $\bigwedge_{\gamma<\beta} A_\gamma$ are different if $\alpha \neq \beta$.

In the case of formulas defined by b) the logical symbols of the language are determined by the admissible formulas. That is, a particular symbol $\bigwedge$ is a symbol of the language if it occurs in some admissible formula.

§22. Infinitary logic with homogeneous quantifiers

In this section we shall formulate an infinitary logic with homogeneous quantifiers by extending the Gentzen-style calculi. Although the treatment of languages with function constants is not difficult, we will, for simplicity, consider only languages without function constants.

DEFINITION 22.1. (1) The language L consists of the following:
1) *Logical symbols*:
  $\neg$ (not),
  $\bigwedge$ (conjunction of arity α for certain α's),
  $\bigvee$ (disjunction of arity α for certain α's),
  $\forall$ (universal quantifier of arity α for certain α's),
  $\exists$ (existential quantifier of arity α for certain α's).

We will sometimes write $\bigwedge_\beta$ and $\bigvee_\beta$ for $\bigwedge_{\beta<\alpha}$ and $\bigvee_{\beta<\alpha}$, when the meaning is clear from the context (and especially in the case $\alpha = \omega$).
2) *Auxiliary symbols*: (,) and, (comma).
3) *Constants*:
  3.1) Individual constants; $c_0, c_1, \ldots, c_\xi, \ldots, \xi < \mu$ for some μ.
  3.2) Predicate constants of arity α; $p_0^\alpha, \ldots, p_\xi^\alpha, \ldots, \xi < \gamma$ for some γ and certain α's.
4) *Variables*:
  4.1) Bound variables: $x_0, x_1, \ldots, x_\eta, \ldots, \eta < K_1$.
  4.2) Free variables: $a_0, a_1, \ldots, a_\xi, \ldots, \xi < K_2$.

Here K_1 and K_2 are ordinals but they are not arbitrary. We must have a sufficiently large supply of bound and free variables.

We proceed in the following way. First we fix the number of constants and logical symbols. We then add a sufficiently large supply of bound variables. We then need a very large collection of free variables. Indeed the cardinality of the set of free variables must be the same as the cardinality of the set of all formulas.

Of course the number of free variables we have influences the number of formulas. Nevertheless, in set theory, we can show that if the number of language symbols is fixed, except for the free variables, then for a sufficiently large collection of free variables, the number of free variables will be the same as the number of formulas.

(2) A *term* is either a free variable or an individual constant.

(3) *Formulas* and their outermost logical symbols we define in the following way.

(3.1) If p is a predicate constant with arity α and $\{t_\beta\}_{\beta<\alpha}$, is a sequence of terms, then $p(t_0, \ldots, t_\beta, \ldots)$ is an *atomic formula*. An atomic formula does not have an outermost logical symbol.

(3.2) If A is a formula, then $\neg A$ is a formula and its outermost logical symbol is $\neg$.

(3.3) If $\bigwedge(\bigvee)$ of arity α belongs to our language and $\{A_\xi\}_{\xi<\alpha}$ is a sequence of formulas then $\bigwedge_{\xi<\alpha} A_\xi$ $(\bigvee_{\xi<\alpha} A_\xi)$ is a formula and its outermost logical symbol is $\bigwedge(\bigvee)$.

(3.4) If $\forall$ $(\exists)$ of arity β belongs to our language, if A is a formula, if $\boldsymbol{a}$ is a sequence of free variables of length β, and if $\boldsymbol{x}$ is a sequence of bound variables of length β none of whose terms occur in A, then $(\forall \boldsymbol{x})\, A(\boldsymbol{x})$ $((\exists \boldsymbol{x})\, A(\boldsymbol{x}))$ is a formula whose outermost logical symbol is $\forall$ $(\exists)$, where $A(\boldsymbol{x})$ is the expression obtained from A by writing x's for the corresponding a's at all occurrences of a's in A.

Subformulas are defined as for first order finite languages: If $A = \bigwedge_{\beta<\alpha} A_\beta$ is a formula of L (L-formula), then each A_β is a subformula of A, if $A : \forall \boldsymbol{x}\, A(\boldsymbol{x})$ is an L-formula, then $A(\boldsymbol{s})$ is a subformula of A for an arbitrary sequence of terms $\boldsymbol{s}$.

Of course, L must be defined so that each subformula of an L-formula is an L-formula. Since formulas are defined inductively, properties of formulas are normally proved by transfinite induction on the construction of formulas.

(4) In order to introduce the notion of proof we use auxiliary symbols $\rightarrow$ and — as before. In the following $\Gamma, \Delta, \Pi, \Lambda, \Gamma_0, \Gamma_1, \ldots$ denote sequences of formulas of length $< K^+$, where K is the cardinality of the set of all formulas in L.

$\Gamma \rightarrow \Delta$ is called a *sequent*. Γ and Δ are called the *antecedent* and *succedent* of the sequent, respectively.

The rules of inference of L are as follows:

(4.1) (Weak) structural rule of inference:

$$\frac{\Gamma \to \Delta}{\Gamma' \to \Delta'},$$

where every formula occurring in Γ occurs in Γ', and every formula occurring in Δ occurs in Δ'.

(4.2) Logical rule of inference:

$$\neg: \text{left:} \quad \frac{\Gamma \to \Delta, \{A_\lambda\}_{\lambda<\gamma}}{\{\neg A_\lambda\}_{\lambda<\gamma} \Gamma \to \Delta}$$

for some $\gamma < K^+$.

$$\neg: \text{right:} \quad \frac{\{A_\lambda\}_{\lambda<\gamma}, \Gamma \to \Delta}{\Gamma \to \Delta, \{\neg A_\lambda\}_{\lambda<\gamma}}$$

for some $\gamma < K^+$.

$$\bigwedge: \text{left:} \quad \frac{\{A_{\lambda,\mu}\}_{\mu<\beta_\lambda, \lambda<\gamma}, \Gamma \to \Delta}{\{\bigwedge_{\mu<\beta_\lambda} A_{\lambda,\mu}\}_{\lambda<\gamma}, \Gamma \to \Delta}$$

for some $\gamma < K^+$, where $\bigwedge_{\mu<\beta_\lambda}$ belongs to L for every $\lambda < \gamma$.

$$\bigwedge: \text{right:} \quad \frac{\Gamma \to \Delta, \{A_{\lambda,\mu_\lambda}\}_{\lambda<\gamma} \text{ for all } \{\mu_\lambda\}_{\lambda<\gamma} \text{ such that } \mu_\lambda < \beta_\lambda (\lambda < \gamma)}{\Gamma \to \Delta, \{\bigwedge_{\mu<\beta_\lambda} A_{\lambda,\mu}\}_{\lambda<\gamma}}$$

for some $\gamma < K^+$, where $\bigwedge_{\mu<\beta_\lambda}$ belongs to L for every $\lambda < \gamma$.

$$\bigvee: \text{left:} \quad \frac{\{A_{\lambda,\mu_\lambda}\}_{\lambda<\gamma}, \Gamma \to \Delta \text{ for all } \{\mu_\lambda\}_{\lambda<\gamma} \text{ such that } \mu_\lambda < \beta_\lambda (\lambda < \gamma)}{\{\bigvee_{\mu<\beta_\lambda} A_{\lambda',\mu}\}_{\lambda<\gamma}, \Gamma \to \Delta}$$

for some $\gamma < K^+$, where $\bigvee_{\mu<\beta_\lambda}$ belongs to L for each $\lambda < \gamma$.

$$\bigvee: \text{right:} \quad \frac{\Gamma \to \Delta, \{A_{\lambda,\mu}\}_{\mu<\beta_\lambda, \lambda<\gamma}}{\Gamma \to \Delta, \{\bigvee_{\mu<\beta_\lambda} A_{\lambda,\mu}\}_{\lambda<\gamma}}$$

for some $\gamma < K^+$, where $\bigvee_{\mu<\beta_\lambda}$ belongs to L for each $\lambda < \gamma$.

$$\forall: \text{left:} \quad \frac{\{A_\lambda(\boldsymbol{t}_\lambda)\}_{\lambda<\gamma}, \Gamma \to \Delta}{\{\forall \boldsymbol{x}_\lambda A_\lambda(\boldsymbol{x}_\lambda)\}_{\lambda<\gamma}, \Gamma \to \Delta}$$

for some $\gamma < K^+$, where the $\boldsymbol{t}$'s are sequences of arbitrary terms of appropriate length.

$$\forall: \text{right:} \quad \frac{\Gamma \to \Delta, \{A_\lambda(\boldsymbol{a}_\lambda)\}_{\lambda<\gamma}}{\Gamma \to \Delta, \{\forall \boldsymbol{x}_\lambda A_\lambda(\boldsymbol{x}_\lambda)\}_{\lambda<\gamma}},$$

for some $\gamma < K^+$, where the $\boldsymbol{a}$'s are sequences of distinct free variables of appropriate length. Each variable occurring in the $\boldsymbol{a}$'s is called an eigenvariable of the inference. When an eigenvariable a, of such an inference occurs in $\boldsymbol{a}_\lambda$, then $\forall \boldsymbol{x}_\lambda A_\lambda(\boldsymbol{x}_\lambda)$ is called the principal formula of a and $A_\lambda(\boldsymbol{a}_\lambda)$ is called the auxiliary formula of both a and of the principal formula. The μth variable $a_{\lambda,\mu}$ in $\boldsymbol{a}_\lambda$ is said to be of order μ with respect to the principal formula $\forall \boldsymbol{x}_\lambda A_\lambda(\boldsymbol{x}_\lambda)$.

$$\exists : \text{left:} \quad \frac{\{A_\lambda(\boldsymbol{a}_\lambda)\}_{\lambda<\gamma}, \Gamma \to \Delta}{\{\exists \boldsymbol{x}_\lambda A_\lambda(\boldsymbol{x}_\lambda)\}_{\lambda<\gamma}, \Gamma \to \Delta}$$

for some $\gamma < K^+$, where the $\boldsymbol{a}$'s are sequences of distinct free variables of appropriate length. Each of the $\boldsymbol{a}$'s is called an eigenvariable of the inference. When an eigenvariable a of such an inference occurs in $\boldsymbol{a}_\lambda$, then $\exists \boldsymbol{x}_\lambda A_\lambda(\boldsymbol{x}_\lambda)$ is called the principal formula of the eigenvariable and $A_\lambda(\boldsymbol{a}_\lambda)$ is called the auxiliary formula of a and of the principal formula. The μth variable $a_{\lambda,\mu}$ in $\boldsymbol{a}_\lambda$ is said to be of order μ with respect to the principal formula $\exists \boldsymbol{x}_\lambda A_\lambda(\boldsymbol{x}_\lambda)$.

$$\exists : \text{right:} \quad \frac{\Gamma \to \Delta, \{A_\lambda(\boldsymbol{t}_\lambda)\}_{\lambda<\gamma}}{\Gamma \to \Delta, \{\exists \boldsymbol{x}_\lambda A_\lambda(\boldsymbol{x}_\lambda)\}_{\lambda<\gamma}}$$

for some $\gamma < K^+$, where the $\boldsymbol{t}$'s are sequences of arbitrary terms of appropriate length.

(4.3) Cut rule:

$$\frac{\Gamma \to \Delta, A_1; \Gamma \to \Delta, A_2; \ldots; \Gamma \to \Delta, A_\lambda; \ldots (\lambda < \gamma); \{A_\lambda\}_{\lambda<\gamma}, \Pi \to \Lambda}{\Gamma, \Pi \to \Delta, \Lambda}$$

for some $\gamma < K^+$.

A semi-proof P is a finite or infinite tree of sequents defined as follows: The topmost, or initial, sequents are of the form $D \to D$. Each sequent in P, but one, is an upper sequent of an inference followed by its lower sequent. The exceptional sequent is called the end sequent. A more precise definition of semi-proof is formulated inductively as follows:

1) A sequent of the form $D \to D$ alone is a semi-proof.

2) If each P_α is a semi-proof with end-sequent $\Gamma_\alpha \to \Delta_\alpha$ and

$$\frac{\ldots; \Gamma_\alpha \to \Delta_\alpha; \ldots}{\Gamma \to \Delta}$$

is an inference, then

$$\frac{\ldots; P_\alpha; \ldots}{\Gamma \to \Delta}$$

is a semi-proof.

3) Every semi-proof is obtained by 1) or 2).

Since semi-proofs are defined inductively, one can assign ordinals to sequents in a semi-proof, so that the ordinal assigned to S_1 is smaller than the ordinal assigned to S_2 if S_1 is an ancestor of S_2. Therefore it is also important to note that although a semi-proof may be an infinite figure, that is, the tree form may have infinitely many branches, each string of sequents traced up from the end-sequent or down from an initial sequent through the tree figure will be of finite length.

A semi-proof P is called a proof if P satisfies the following eigenvariable conditions.

(I) If a free variable occurs in two or more places as an eigenvariable, the principal formulas of these eigenvariables must be identical and the order of this eigenvariable with respect to each principal formula is the same in each of the inferences.

(II) For each free variable a in a proof, an ordinal number, $h(a)$ called its height, can be defined so that the height of a free variable occurring in an inference as an eigenvariable is larger than any of the heights of the free variables contained in the principal formula of that eigenvariable.

(III) No variable occurring in an inference as an eigenvariable can occur in the end sequent.

EXAMPLE 22.2. (1) A cut-free proof of the axiom of dependent choice in an infinitary logic with homogeneous quantifiers:

$$\begin{array}{c}
F(a_n, a_{n+1}) \to F(a_n, a_{n+1}) \\ \hline
\{F(a_m, a_{m+1})\}_{m<\omega} \to F(a_n, a_{n+1}) \text{ for each } n<\omega \\ \hline
\{F(a_m, a_{m+1})\}_{m<\omega} \to \bigwedge_{n<\omega} F(a_n, a_{n+1}) \\ \hline
\{F(a_m, a_{m+1})\}_{m<\omega} \to \exists \boldsymbol{x}_{<\omega} \bigwedge_{n<\omega} F_n \\ \hline
\{F(a_m, a_{m+1})\}_{m<\omega} \to \forall x_0\, \exists \boldsymbol{x}_{<\omega} \bigwedge_{n<\omega} F(x_n, x_{n+1}) \\ \hline
\{\exists y\, F(a_m, y)\}_{m<\omega} \to \forall x_0\, \exists \boldsymbol{x}_{<\omega} \bigwedge_{n<\omega} F(x_n, x_{n+1}) \\ \hline
\{\forall x\, \exists y\, F(x, y)\}_{m<\omega} \to \forall x_0\, \exists \boldsymbol{x}_{<\omega} \bigwedge_{n<\omega} F(x_n, x_{n+1}) \\ \hline
\forall x\, \exists y\, F(x, y) \to \forall x_0\, \exists \boldsymbol{x}_{<\omega} \bigwedge_{n<\omega} F(x_n, x_{n+1})
\end{array}$$

Here F_0 is $F(a_0, x_1)$ and F_{i+1} is $F(x_{i+1}, x_{i+2})$ for every $i<\omega$, and $x=(x_1, x_2, \ldots)$. The *heights* are defined by $h(a_m)=m$, $m<\omega$.

(2) A proof of

$$\forall x_0 \ldots (\neg \bigwedge_n x_{n+1} \in x_n), \forall x(\forall y \in x\, A(y) \supset A(x)) \to A(a_0),$$

where $\forall y \in x A(y)$ is an abbreviation of $\forall y\, (y \in x \supset A(y))$:

1) $\forall x\, (\forall y \in x\, A(y) \supset A(x)) \to A(a_0),\ a_1 \in a_0$.

PROOF.

$$\dfrac{\dfrac{A(a_0) \to A(a_0) \qquad \dfrac{\dfrac{\dfrac{a_1 \in a_0 \to a_1 \in a_0}{a_1 \in a_0 \to a_1 \in a_0,\ A(a_1)}}{\to a_1 \in a_0,\ a_1 \in a_0 \supset A(a_1)}}{\to a_1 \in a_0,\ \forall y \in a_0\, A(y)}}{\forall y \in a_0\, A(y) \supset A(a_0) \to A(a_0),\ a_1 \in a_0}}{\forall x\, (\forall y \in x\, A(y) \supset A(x)) \to A(a_0),\ a_1 \in a_0.}$$

2) $\forall x\, (\forall y \in x\, A(y) \supset A(x)) \to A(a_n),\ a_{n+1} \in a_n$.

PROOF. Similar to that of 1).

3) $\forall x\,(\forall y \in x\, A(y) \supset A(x)) \to A(a_{n-k}),\ a_{n+1} \in a_n$ for $k = 0, 1, \ldots, n$.

PROOF. By induction on k we construct a figure ending with the sequent 3). Since the sequent 2) is the case $k = 0$ we need only show how to proceed from k to $k+1$:

$$\dfrac{A(a_{n-(k+1)}) \to A(a_{n-(k+1)}) \qquad \dfrac{\forall x\,(\forall y \in xA(y) \supset A(x)) \to A(a_{n-k}),\ a_{n+1} \in a_n}{\forall x\,(\forall y \in xA(y) \supset A(x)) \to \forall y \in a_{n-(k+1)} A(y),\ a_{n+1} \in a_n}}{\dfrac{\forall y \in a_{n-(k+1)} A(y) \supset A(a_{n-(k+1)}),\ \forall x\,(\forall y \in xA(y) \supset A(x)) \to A(a_{n-(k+1)}),\ a_{n+1} \in a_n}{\forall x\,(\forall y \in xA(y) \supset A(x)) \to A(a_{n-(k+1)}),\ a_{n+1} \in a_n}}$$

4) $\forall x\,(\forall y \in x\, A(y) \supset A(x)) \to A(a_0),\ \bigwedge_n a_{n+1} \in a_n$.

PROOF. From 3) with $k = n$ we have

$$\frac{\forall x\,(\forall y \in x\, A(y) \supset A(x)) \to A(a_0),\ a_{n+1} \in a_n}{\forall x\,(\forall y \in x\, A(y) \supset A(x)) \to A(a_0),\ \bigwedge_n a_{n+1} \in a_n}.$$

From 4) we then conclude

5) $\forall x_0 \ldots (\neg \bigwedge_n x_{n+1} \in x_n),\ \forall x\,(\forall y \in x\, A(y) \supset A(x)) \to A(a_0)$.

In this proof $a_1, a_2, \ldots, a_n, \ldots$ are eigenvariables and $h(a_n) = n$.

(3) Malitz's example. Malitz found a counterexample to the interpolation theorem for homogeneous infinitary languages. His example is the following. Let A and B be two well-ordered sets with the same order type, and let F be a predicate that defines the order preserving map from A one-to-one onto B. That there is exactly one such map is easily proved. If the interpolation theorem held then this order preserving map could be defined in the homogeneous infinitary language without using the predicate F. This, however, is impossible because the length of the defining formula would set an upper bound on the order type of A, but that order type is not bounded. Let $\mathrm{Ln}(=, <)$ be a formula which expresses that $<$ together with $=$ is a linear ordering relation. Let Γ be the following sequence of formulas.

$$\mathrm{Ln}(\overset{1}{=}, \overset{1}{<}),\ \mathrm{Ln}(\overset{2}{=}, \overset{2}{<}).$$

$$\forall x\, \forall y\, \forall u\, \forall v\, (x \overset{1}{=} y \wedge u \overset{2}{=} v \supset (F(x, u) \equiv F(y, v))),$$

$$\forall x\, \forall y\, \forall u\, \forall v\, (x \overset{1}{=} y \wedge u \overset{2}{=} v \supset (G(x, u) \equiv G(y, v))),$$

$$\forall x\, \forall y\, \forall u\, \forall v\, (F(x, u) \wedge F(y, v) \supset (x \overset{1}{<} y \equiv u \overset{2}{<} v) \wedge (x \overset{1}{=} y \equiv u \overset{2}{=} v))$$

$$\forall x\, \forall y\, \forall u\, \forall v\, (G(x, u) \wedge G(y, v) \supset (x \overset{1}{<} y \equiv u \overset{2}{<} v) \wedge (x \overset{1}{=} y \equiv u \overset{2}{=} v))$$

It should be remarked that all the quantifiers in Γ are universal and at the front of a formula. The following sequent is easily proved to be valid.

$$\begin{aligned} &\Gamma, \forall x\, \exists y\, F(x, y), \forall x\, \exists y\, G(x, y) \\ &\quad \forall x\, \exists y\, F(y, x), \forall x\, \exists y\, G(y, x), F(a, b), \\ &\quad \forall x_0 x_1 \ldots \neg \bigwedge_n (x_{n+1} \overset{1}{<} x_n) \quad \rightarrow G(a, b) \end{aligned}$$

We are going to present a cut-free proof of this sequent. Let T be the set of all finite sequences of 1's and 2's. It is understood that the empty sequence is a member of T. We use τ as a variable on T. The set D of free variables is defined as follows.

1) $a \in D$. (a is a^τ, where τ is an empty-sequence.)
2) If $a^\tau \in D$, then $b^{\tau 1}$ and $b^{\tau 2}$ are members of D.
3) If $b^\tau \in D$, then $a^{\tau 1}$ and $a^{\tau 2}$ are members of D.
4) All members of D are obtained by 1), 2) and 3).

The members of D are $a, b^1, b^2, a^{11}, a^{12}, a^{21}, a^{22}, b^{111}, b^{112}, \ldots$. Γ' is a sequence of all the formulas which are obtained from a formula in Γ by deleting all the universal quantifiers and replacing bound variables by the members of D. (From one formula, infinitely many formulas will be obtained. Of course, in one instance of substitution, the same member of D should be substituted for the same bound variable in a formula.) Δ' is a sequence of all the formulas of the form

$$F(a^\tau, b^{\tau 1}), F(a^{\tau 1}, b^\tau), G(a^\tau, b^{\tau 2}), G(a^{\tau 2}, b^\tau), \tau \in T.$$

In the following lemmas, we state several sequents which are provable in the ordinary first order predicate calculus and hence cut-free provable in Gentzen's **LK**.

We define "$\Gamma', \Delta' \rightarrow b^{\tau 11} \overset{2}{=} b^\tau$ is provable" to mean that $\Gamma^*, \Delta^* \rightarrow b^{\tau 11} \overset{2}{=} b^\tau$ is provable for some Γ^* that is a finite subsequence of Γ' and some Δ^* that is a finite subsequence of Δ'.

LEMMA 22.3. *The following are* **LK**-*provable.*

1) $\Gamma', \Delta' \rightarrow b^{\tau 11} = b^\tau$, *where* $b^{\tau 1} = b^{\tau 2}$ *is an abbreviation for* $b^{\tau 1} \overset{2}{=} b^{\tau 2}$. *In the same way,* $a^{\tau 1} = a^{\tau 2}$ *is an abbreviation for* $a^{\tau 1} \overset{1}{=} a^{\tau 2}$.
2) $\Gamma', \Delta' \rightarrow b^{\tau 22} = b^\tau$.
3) $\Gamma', \Delta' \rightarrow a^{\tau 11} = a^\tau$.
4) $\Gamma', \Delta' \rightarrow a^{\tau 22} = a^\tau$.

PROOF. Obviously, Γ', $F(a^{\tau 1}, b^{\tau 11})$, $F(a^{\tau 1}, b^{\tau}) \rightarrow b^{\tau 11} = b^{\tau}$. From this, 1) follows trivially. The proofs of 2), 3) and 4) are similar.

LEMMA 22.4. *The following are* **LK**-*provable*

1) $\Gamma', \Delta', b^{\tau} = b^{\tau 12} \rightarrow a^{\tau 1} = a^{\tau 2}$.
2) $\Gamma', \Delta', a^{\tau} = a^{\tau 12} \rightarrow b^{\tau 1} = b^{\tau 2}$.

PROOF. 1) From $G(a^{\tau 2}, b^{\tau})$, $G(a^{\tau 1}, b^{\tau 12})$, the fourth formula of Γ with $a^{\tau 2}, a^{\tau 1}, b^{\tau 1}, b^{\tau 12}$ as x, y, u, v respectively and from $b^{\tau} = b^{\tau 12}$ it follows that $a^{\tau 1} = a^{\tau 2}$.

LEMMA 22.5. *The following are provable in* **LK**.

1) $\Gamma', \Delta', b^{\tau i 1} = b^{\tau i 2} \rightarrow a^{\tau 1} = a^{\tau 2}$ $(i = 1, 2)$.
2) $\Gamma', \Delta', a^{\tau i 1} = a^{\tau i 2} \rightarrow b^{\tau 1} = b^{\tau 2}$ $(i = 1, 2)$.

PROOF. Under the hypotheses of Γ', and Δ', $b^{\tau 11} = b^{\tau 12} \rightarrow b^{\tau} = b^{\tau 12} \rightarrow a^{\tau 1} = a^{\tau 2}$ (Lemmas 22.3 and 22.4). The other cases are proved similarly.

LEMMA 22.6. *The following is provable in* **LK**.

1) $\Gamma', \Delta', b^{\tau 1} = b^{\tau 2} \rightarrow b^1 = b^2$.

PROOF. By induction on the length of τ, using Lemma 22.5.

LEMMA 22.7. *The following are provable in* **LK**.

1) $\Gamma', \Delta', b^1 = b^2 \rightarrow G(a, b^1)$.
2) $\Gamma', \Delta', b^1 < b^2 \rightarrow a^{12} < a$, *where* $b^{\tau 1} < b^{\tau 2}$ *and* $a^{\tau 1} < a^{\tau 2}$ *are abbreviations for* $b^{\tau 1} \overset{2}{<} b^{\tau 2}$ *and* $a^{\tau 1} \overset{1}{<} a^{\tau 2}$, *respectively*.
3) $\Gamma', \Delta', b^2 < b^1 \rightarrow a^{21} < a$.

PROOF. 1) $\Gamma', G(a, b^2), b^1 = b^2 \rightarrow G(a, b^1)$.

2) $\Gamma', F(a, b^1), b^1 < b^2, G(a, b^2), G(a^{12}, b^1) \rightarrow a^{12} < a$.
3) $\Gamma', F(a, b^1), b^2 < b^1, F(a^{21}, b^2) \rightarrow a^{21} < a$.

LEMMA 22.8. *The following are provable in* **LK**.

1) $\Gamma', \Delta', b^{\tau 1} = b^{\tau 2} \rightarrow G(a, b^1)$.
2) $\Gamma', \Delta', b^{\tau 1} = b^{\tau 2} \rightarrow a^{\tau 12} < a^{\tau}$.
3) $\Gamma', \Delta', b^{\tau 2} < b^{\tau 1} \rightarrow a^{\tau 21} < a^{\tau}$.

PROOF. 1) follows from Lemma 22.6 and 1) of Lemma 22.7. The proofs of 2) and 3) are similar to the proof of Lemma 22.7.

DEFINITION 22.9. (i) $R^0(\tau)$ iff $b^{\tau 1} = b^{\tau 2}$.

(ii) $R^1(\tau)$ iff $b^{\tau 1} < b^{\tau 2}$.

(iii) $R^2(\tau)$ iff $b^{\tau 2} < b^{\tau 1}$.
(iv) $T_0 = \{\tau \in T \mid \text{the length of } \tau \text{ is odd}\}$.

LEMMA 22.10. *The following is cut-free provable for each* $i : T_0 \to \{0, 1, 2\}$

$$\{R^{i_\tau}(\tau)\}_{\tau \in T_0} \Gamma', \Delta' \to \bigwedge_n t_{n+1} \overset{1}{<} t_n, G(a, b^1),$$

where t_n *is a member of* D *whose length is* $2n$.

PROOF. Obvious from Lemma 22.8.

LEMMA 22.11. *The following is cut-free provable.*

$$\Gamma, \Delta', \forall x_0 x_1 \ldots \neg \bigwedge_n (x_{n+1} \overset{1}{<} x_n) \to G(a, b^1).$$

PROOF. This follows from Lemma 22.10, since $\forall x \forall y (x \overset{2}{<} y \vee x \overset{2}{=} y \vee y \overset{2}{<} x)$ is contained in Γ.

THEOREM 22.12. *The following is cut-free provable.*

$$\Gamma, \Delta, \forall x_0 x_1 \ldots \neg \bigwedge_n (x_{n+1} \overset{1}{<} x_n), F(a, b) \to G(a, b),$$

where Δ *consists of* $\forall x \exists y F(x, y)$, $\forall x \exists y F(y, x)$, $\forall x \exists y G(x, y)$ *and* $\forall x \exists y G(y, x)$.

PROOF. Take b to be b^1, and define $h(a^\tau)$ and $h(b^{\tau'})$ to be the length of τ and τ' respectively. The conclusion then follows from Lemma 22.11.

We now introduce a new cut rule, one we will find more convenient in infinitary languages than the old one. As we will prove, the new rule is a generalization of the old one.

DEFINITION 22.13 (the generalized cut rule). Let $\Gamma \to \Delta$ be a sequent and $\mathcal{F}$ be a set of formulas. Let $(\mathcal{F}_1, \mathcal{F}_2)$ denote a partition of $\mathcal{F}$ (i.e., $\mathcal{F}_1 \cup \mathcal{F}_2 = \mathcal{F}$ and $\mathcal{F}_1 \cap \mathcal{F}_2$ is empty). Suppose for an arbitrary partition of $\mathcal{F}$, $(\mathcal{F}_1, \mathcal{F}_2)$, there exists a pair of sets of formulas, say $\Phi \subseteq \mathcal{F}_1$, and $\Psi \subseteq \mathcal{F}_2$, such that there exists a semi-proof of $\Phi, \Gamma \to \Delta, \Psi$. Then the generalized cut rule allows us to infer $\Gamma \to \Delta$. This may be expressed as follows:

$$\text{g.c.} \quad \frac{\Phi, \Gamma \to \Delta, \Psi \text{ for all } (\mathcal{F}_1, \mathcal{F}_2)}{\Gamma \to \Delta}$$

PROPOSITION 22.14. (1) *The usual cut rule is a special case of the g.c. rule.*

(2) *The following is an admissible rule of inference:*

$$\frac{\Gamma \to \Delta}{\tilde{\Gamma} \to \tilde{\Delta}},$$

where $\tilde{\Gamma}$ is obtained from Γ by replacing some of the formulas by alphabetical variants. Similarly with $\tilde{\Delta}$.

(3) *Suppose that some (possibly all) of the upper sequents of a g.c. are obtained by applications of the g.c. rule. Then we can change the proof so that the lower sequent will be obtained by one application of the g.c. rule.*

(4) *In homogeneous systems the g.c. rule is an admissible rule of inference.*

(5) *The g.c. rule can be equivalently expressed as*

$$\frac{\Phi, \Gamma' \to \Delta', \Psi \text{ for all } (\mathscr{F}_1, \mathscr{F}_2)}{\Gamma \to \Delta},$$

where $\Gamma' \subseteq \Gamma$ and $\Delta' \subseteq \Delta$ are determined by $(\mathscr{F}_1, \mathscr{F}_2)$ and Φ and Ψ have the same meaning as before.

PROOF. (1) Consider a cut:

$$\frac{\Gamma \to \Delta, D_\mu \text{ for all } \mu < \lambda; \{D_\mu\}_{\mu<\lambda}, \Pi \to \Lambda}{\Gamma, \Pi \to \Delta, \Lambda}$$

First we obtain $\Gamma, \Pi \to \Delta, \Lambda, D_\mu$ and $\{D_\mu\}, \Gamma, \Pi \to \Delta, \Lambda$ by applications of weakening. Let $\mathscr{F}$ be $\{D_\mu\}_{\mu<\lambda}$ and let $(\mathscr{F}_1, \mathscr{F}_2)$ be a partition of $\mathscr{F}$.

Case 1. $\mathscr{F}_2$ is not empty. Then take Φ to be the empty set and Ψ to be $\{D_\mu\}$, where D_μ is the first formula in $\mathscr{F}_2$.

Case 2. $\mathscr{F}_2$ is empty. Then take Φ to be $\mathscr{F}_1$, which is $\mathscr{F}$, and Ψ to be the empty set.

For any (Φ, Ψ) above, $\Phi, \Gamma, \Pi \to \Delta, \Lambda, \Psi$ is an upper sequent of the cut in consideration. By the g.c. rule

$$\frac{\Phi, \Gamma, \Pi \to \Delta, \Lambda, \Psi, \text{ for all } (\Phi, \Psi) \text{ as above}}{\Gamma, \Pi \to \Delta, \Lambda}.$$

(2) We shall show that if a sequent $\Gamma \to \Delta$ is provable, then another sequent $\tilde{\Gamma} \to \tilde{\Delta}$ can be deduced, where $\tilde{\Gamma}$ is obtained from Γ by simply renaming some of the bound variables. Similarly with $\tilde{\Delta}$. For any formula A, if $\tilde{A}$ is an alphabetical variant of A, then $A \equiv \tilde{A}$ is easily proved. If $\Gamma \to \Delta$ is provable, then $\{A_\lambda \equiv \tilde{A}_\lambda\}_{\lambda<\mu}, \tilde{\Gamma} \to \tilde{\Delta}$ is provable for some A_λ's and $\tilde{A}_\lambda$'s. Using $\to A_\lambda \equiv \tilde{A}_\lambda$ for all $\lambda < \mu$, we obtain $\tilde{\Gamma} \to \tilde{\Delta}$ by the g.c. rule.

(3) Let I be the cut under consideration:

$$\frac{\Phi, \Gamma \to \Delta, \Psi \text{ for all appropriate } (\Phi, \Psi)}{\Gamma \to \Delta}.$$

The proof is by transfinite induction on the complexity of the subproof ending with $\Gamma \to \Delta$. Suppose, as the inductive hypothesis, that there is at most one cut along any string of sequents above I. Let $\{(\Phi_\mu, \Psi_\mu)\}_{\mu<\mu_0}$ be an enumeration of the (Φ, Ψ)'s in I and let $\mathscr{F}$ be the set of cut formulas. Let S_μ denote the sequent $\Phi_\mu, \Gamma \to \Delta, \Psi_\mu$. Let $\{I_\iota^\mu\}_{\iota<\nu_\mu}$ be an enumeration of all the cuts above S_μ and let $\mathscr{F}_\iota^\mu$ be the set of cut formulas of I_ι^μ for each (μ, ι). Let $\{\Phi_\gamma^{\mu,\iota}, \Psi_\gamma^{\mu,\iota})\}_{\gamma<\delta_\iota^\mu}$ be an enumeration of the pairs of formulas which are related to I_ι^μ and hence to $\mathscr{F}_\iota^\mu$.

For each I_ι^μ, consider

$$\frac{\Phi_\gamma^{\mu,\iota}\, \Pi_\iota^\mu \to \Lambda_\iota^\mu\, \Psi_\gamma^{\mu,\iota}}{\Pi_\iota^\mu\, \Phi_\gamma^{\mu,\iota} \to \Psi_\gamma^{\mu,\iota}\, \Lambda_\iota^\mu}$$

for every $\gamma < \delta_\iota^\mu$. For every combination of γ's, i.e., $(\gamma^0, \gamma^1, \ldots, \gamma^\iota, \ldots)$, $\gamma^\iota < \delta_\iota^\mu$, copy the part of the original proof from $\Pi_\iota^\mu \to \Lambda_\iota^\mu$ to S_μ, starting with $\Pi_\iota^\mu, \Phi_\gamma^{\mu,\iota} \to \Psi_\gamma^{\mu,\iota}, \Lambda_\iota^\mu$ obtained as above, in place of $\Pi_\iota^\mu \to \Lambda_\iota^\mu$. Thus we obtain

$$(*) \qquad \Phi_\mu, \Gamma, \{\Phi_{\gamma^\iota}^{\mu,\iota}\}_{\iota<\nu_\mu} \to \{\Psi_{\gamma^\iota}^{\mu,\iota}\}_{\iota<\nu_\mu}, \Delta, \Psi_\mu$$

for every μ and $(\gamma^0, \gamma^1, \ldots, \gamma^\iota, \ldots)$. Call such a sequent $S_\mu(\{\gamma^\iota\}_{\iota<\nu_\mu})$.

Now consider the set of formulas $\mathscr{F}_0 = \mathscr{F} \cup \bigcup_{\mu,\iota} \mathscr{F}_\iota^\mu$ and an arbitrary partition of $\mathscr{F}_0$, say $\mathscr{F}_1$ and $\mathscr{F}_2$. There exist $\mu < \mu_0$ and $\{\gamma^\iota\}_{\iota<\nu_\mu}$ such that $\Phi_\mu \subseteq \mathscr{F}_1 \cap \mathscr{F}$, $\Psi_\mu \subseteq \mathscr{F}_2 \cap \mathscr{F}$, $\Phi_{\gamma^\iota}^{\mu,\iota} \subseteq \mathscr{F}_\iota^\mu \cap \mathscr{F}_1$ and $\Psi_{\gamma^\iota}^{\mu,\iota} \subseteq \mathscr{F}_\iota^\mu \cap \mathscr{F}_2$, for $\mathscr{F}_1$ and $\mathscr{F}_2$ determine partitions of $\mathscr{F}$ and $\mathscr{F}_\iota^\mu$. Define

$$\Phi = \Phi_\mu \cup \bigcup_{\iota<\nu_\mu} \Phi_{\gamma^\iota}^{\mu,\iota}, \qquad \Psi = \Psi_\mu \cup \bigcup_{\iota<\nu_\mu} \Psi_{\gamma^\iota}^{\mu,\iota}.$$

It is obvious that $\Phi \subseteq \mathscr{F}_1$, $\Psi \subseteq \mathscr{F}_2$ and $\Phi, \Gamma \to \Delta, \Psi$ is one of the sequents in (*). There is no cut above it. Since this holds for every partition of $\mathscr{F}_0$, we obtain

$$\text{g.c.} \quad \frac{\Phi, \Gamma \to \Delta, \Psi \text{ for all appropriate } (\Phi, \Psi)}{\Gamma \to \Delta}.$$

(4) As will be proved in Theorem 22.17, this follows easily from the completeness of the homogeneous systems and the fact that the g.c. rule preserves the validity of sequents.

(5) Obvious.

DEFINITION 22.15. Let L be an infinitary language. We define a structure for L, $\langle D, \phi\rangle$, an interpretation $\mathfrak{I} = (\langle D, \phi\rangle, \phi_0)$, the relation that an interpretation $\mathfrak{I}$ satisfies a formula A of L, the validity of a formula, the satisfaction relation for sequents and the validity of a sequent, as in Definition 8.1.

PROPOSITION 22.16 (consistency: Maehara-Takeuti). *Let $\mathscr{A}$ be an arbitrary structure for* L. *Then every provable sequent is valid in $\mathscr{A}$.*

PROOF. For each formula of the form $\exists \boldsymbol{x}\, A(\boldsymbol{x}, \boldsymbol{a})$ in which $\boldsymbol{x}$ is of length α, and $\boldsymbol{a}$ are exactly the free variables in A, we introduce a Skolem function $g_A^\gamma(\boldsymbol{a})$ for each $\gamma < \alpha$, and define the following interpretation of g_A^γ in $\mathscr{A}$:

If $\exists \boldsymbol{x}\, A(\boldsymbol{x}, \boldsymbol{a})$ is satisfied in $\mathscr{A}$ for an assignment ϕ_0, then the values of the g_A^γ's are those satisfying

$$A(\boldsymbol{g}_A^{<\alpha}(\boldsymbol{a}), \boldsymbol{a}).$$

Let 0 be an element of the domain of $\mathscr{A}$. If $\exists \boldsymbol{x}\, A(\boldsymbol{x}, \boldsymbol{a})$ is not satisfied by ϕ_0 in $\mathscr{A}$, then the $g_A^\gamma(\boldsymbol{a})$'s are interpreted to be 0.

Let P be a proof. We well-order all the eigenvariables in P, arranging the well-ordered sequences $a_0, a_1, \ldots, a_\beta, \ldots$, in such a way that $h(a_\beta) \leq h(a_\gamma)$ if $\beta < \gamma$. We define terms t_β by transfinite induction on β. Assuming that $\boldsymbol{t}_{<\beta}$ has been defined, we define t_β in the following way. Let $\forall \boldsymbol{x}\, A(\boldsymbol{x}, \boldsymbol{b})$ (or $\exists \boldsymbol{x}\, A(\boldsymbol{x}, \boldsymbol{b})$) and $A(\boldsymbol{d}, \boldsymbol{b})$ be the principal formula and an auxiliary formula of a_β and let the order of a_β with respect to this principal formula be γ, i.e., let a_β be d_γ. For each b_ν let s_ν be either the already defined t_γ for which b_ν is a_γ, if b_ν is an eigenvariable; or else b_ν itself. Then t_β is defined to be $g_{\neg A}^\gamma(\boldsymbol{s})$ (or $g_A^\gamma(\boldsymbol{s})$). By (I) of the eigenvariable condition, this definition does not depend on the choice of $A(\boldsymbol{a}, \boldsymbol{b})$.

Let P' be the result obtained from P by substituting t_β for a_β for every β. The bottom sequent of P' is the end sequent of P since it contains no eigenvariables.

For an arbitrary assignment of members of D to the free variables, any sequent S in P' is satisfied in $\mathscr{A}$, where the g_A^γ's are interpreted as above. This can be proved by transfinite induction on the complexity of the figure in P above S. As a consequence, the end-sequent of P is valid, since it does not involve eigenvariables. The other cases being obvious, we only consider $\exists$: left and $\forall$: right.

1) $\exists$: left. The corresponding part of P' is

$$\frac{\ldots, A(\boldsymbol{u}, \boldsymbol{s}), \ldots, \Gamma \rightarrow \Delta}{\ldots, \exists \boldsymbol{x}\, A(\boldsymbol{x}, \boldsymbol{s}), \ldots, \Gamma \rightarrow \Delta},$$

where u_γ is $g_A^\gamma(\boldsymbol{s})$. It suffices to show that

$$\exists \boldsymbol{x}\, A(\boldsymbol{x}, \boldsymbol{s}) \rightarrow A(\boldsymbol{u}, \boldsymbol{s}),$$

is satisfied in $\mathscr{A}'$; but this follows from the definition of the g_A^γ's.

2) $\forall$: right. The corresponding part of P' is

$$\frac{\Gamma \rightarrow \Delta, \ldots, A(\boldsymbol{v}, \boldsymbol{s}), \ldots}{\Gamma \rightarrow \Delta, \ldots, \forall \boldsymbol{x}\, A(\boldsymbol{x}, \boldsymbol{s}), \ldots}$$

where v_γ is $g^{\gamma}_{\neg A}(s)$. So it suffices to show that

$$A(v, s) \to \forall x\, A(x, s),$$

is satisfied in $\mathscr{A}$. This follows from

$$\exists x\, \neg A(x, s) \to \neg A(v, s),$$

which follows from the definition of the $g^{\gamma}_{\neg A}$'s.

We shall now prove the completeness theorem in combination with the cut-elimination theorem for an infinitary logic with homogeneous quantifiers. The method is basically the same as that for the proof of the completeness theorem in Chapter I.

As a corollary to the completeness theorem (Theorem 22.17) and Proposition 22.16 we have the cut-elimination theorem: Every provable sequent is provable without the cut rule.

THEOREM 22.17 (Maehara-Takeuti). *Every sequent valid in any non-empty domain is provable without the cut rule.*

PROOF. Let S be an arbitrary sequent and let D_0 be an arbitrary non-empty set containing all free variables and individual constants in S. Let D be the closure of D_0 with respect to all the function symbols g^{γ}_{A} and $\bar{g}^{\gamma}_{A}$ for all formulas A, i.e., let D be generated by all g^{γ}_{A}'s and $\bar{g}^{\gamma}_{A}$'s from D_0. Here $\bar{g}^{\gamma}_{A}$ is $g^{\gamma}_{\neg A}$.

We define the tree $T(S)$ step by step.

Stage 0. We write S.

Stage $n+1$. (1) $n+1 \equiv 1 \pmod 5$. When a sequent $\Pi \to \Lambda$ contains a formula whose outermost logical symbol is $\neg$, we write above $\Pi \to \Lambda$

$$\{D_\mu\}_{\mu<\delta}, \Pi' \to \Lambda' \{C_\lambda\}_{\lambda<\gamma},$$

where $\{\neg C_\lambda\}_{\lambda<\gamma}$ and $\{\neg D_\mu\}_{\mu<\delta}$ are the sequences of all formulas in Π and in Λ, respectively, whose outermost logical symbol is $\neg$, and Π' and Λ' are obtained from Π and Λ respectively by omitting the $\neg C_\lambda$'s and the $\neg D_\mu$'s.

(2) $n+1 \equiv 2 \pmod 5$. When a sequent $\Pi \to \Lambda$ contains a formula whose outermost logical symbol is $\bigwedge$, we write above $\Pi \to \Lambda$

$$\{C_{\mu,\lambda}\}_{\lambda<\gamma_\mu, \mu<\gamma}, \Pi' \to \Lambda', \{D_{\sigma,\rho_\sigma}\}_{\sigma<\delta}$$

for all sequences $\{\rho_\sigma\}_{\sigma<\delta}$ such that $\rho_\sigma < \nu_\sigma$, where $\{\bigwedge_{\lambda<\gamma_\mu} C_{\mu,\lambda}\}_{\mu<\gamma}$ and $\{\bigwedge_{\rho<\nu_\sigma} D_{\sigma,\rho}\}_{\sigma<\delta}$ are the sequences of all formulas in Π and in Λ, respectively, whose outermost logical symbol is $\bigwedge$, and Γ' and Λ' are obtained from Π and Λ, respectively, by omitting the $\bigwedge_{\lambda<\gamma_\mu} C_{\mu,\lambda}$'s and the $\bigwedge_{\rho<\nu_\sigma} D_{\sigma,\rho}$'s.

(3) $n+1 \equiv 3 \pmod 5$. When a sequent $\Pi \to \Lambda$ contains a formula whose outermost logical symbol is $\bigvee$, we write above $\Pi \to \Lambda$.

$$\{C_{\mu,\lambda_\mu}\}_{\mu<\gamma}, \Pi' \to \Lambda', \{D_{\sigma,\rho}\}_{\rho<\nu_\sigma, \sigma<\delta}$$

for all sequences $\{\lambda_\mu\}_{\mu<\gamma}$ such that $\lambda_\mu < \gamma_\mu$, where $\{\bigvee_{\lambda<\gamma_\mu} C_{\mu,\lambda}\}_{\mu<\gamma}$ and

$\{\bigvee_{\rho<\nu_\sigma} D_{\sigma,\rho}\}_{\sigma<\delta}$ are the sequences of all formulas in Π and Λ, respectively, whose outermost logical symbol is $\bigvee$, and Π' and Λ' are obtained from Π and Λ, respectively, by omitting the $\bigvee_{\lambda<\gamma_\mu} C_{\mu,\lambda}$'s and the $\bigvee_{\rho<\delta_\sigma} D_{\sigma,\rho}$'s.

(4) $n+1 \equiv 4 \pmod 5$. When a sequent $\Pi \to \Lambda$ contains a formula whose outermost logical symbol is $\forall$, we write above $\Pi \to \Lambda$

$$\{A_\lambda(\boldsymbol{t}_{\lambda,\mu})\}_{\lambda<\gamma,\mu}\, \Pi' \to \Lambda', \{B_\rho(\boldsymbol{u}_{\rho,\sigma})\}_{\sigma<\delta,\rho},$$

where $\{\forall \boldsymbol{x}_\lambda A_\lambda(\boldsymbol{x}_\lambda)\}_{\lambda<\gamma}$ and $\{\forall \boldsymbol{y}_\rho B_\rho(\boldsymbol{y}_\rho)\}_{\rho<\delta}$ are the sequences of all formulas in Π and Λ, respectively, whose outermost logical symbol is $\forall$, and Π' and Λ' are obtained from Π and Λ respectively by omitting the $\forall \boldsymbol{x}_\lambda A_\lambda(\boldsymbol{x}_\lambda)$'s and the $\forall \boldsymbol{y}_\rho B_\rho(\boldsymbol{y}_\rho)$'s. Furthermore, $\boldsymbol{t}_{\lambda,\mu}$ runs over all sequences of members of D that are the same length as $\boldsymbol{x}_\lambda$.

If $\boldsymbol{u}_{\rho,\sigma}$ is $u_{\rho,\sigma,0}, u_{\rho,\sigma,1}, \ldots, u_{\rho,\sigma,\nu}$, ν being the length of $\boldsymbol{y}_\rho$, then $u_{\rho,\sigma,\xi}$ is $\bar{g}^{\xi}_{A_\rho}(\boldsymbol{v}_\rho)$, where $\xi < \nu$ and $\boldsymbol{v}_\rho$ is the sequence of free variables in $B_\rho(\boldsymbol{y}_\rho)$.

(5) $n+1 \equiv 0 \pmod 5$. When a sequent $\Pi \to \Lambda$ contains a formula whose outermost logical symbol is $\exists$, we write above $\Pi \to \Lambda$

$$\{A_\lambda(\boldsymbol{t}_{\lambda,\mu})\}_{\lambda<\gamma,\mu}, \Pi' \to \Lambda', \{B_\rho(\boldsymbol{u}_{\rho,\sigma})\}_{\rho<\delta,\sigma}$$

where $\{\exists \boldsymbol{x}_\lambda A_\lambda(\boldsymbol{x}_\lambda)\}_{\lambda<\gamma}$ and $\{\exists \boldsymbol{y}_\rho B_\rho(\boldsymbol{y}_\rho)\}_{\rho<\delta}$ are the sequences of all formulas in Π and in Λ, respectively, whose outermost logical symbol is $\exists$, and Π' and Λ' are obtained from Π and Λ, respectively, by omitting the $\exists \boldsymbol{x}_\lambda A_\lambda(\boldsymbol{x}_\lambda)$'s and the $\exists \boldsymbol{y}_\rho B_\rho(\boldsymbol{y}_\rho)$'s. Here $\boldsymbol{u}_{\rho,\sigma}$ runs over all sequences of the same length as that of $\boldsymbol{y}_\rho$, and, if $\boldsymbol{t}_{\lambda,\mu}$ is $t_{\lambda,\mu,0}, t_{\lambda,\mu,1}, \ldots, t_{\lambda,\mu,\zeta}, \ldots$ $(\zeta < \eta)$, η being the length of $\boldsymbol{x}_\lambda$, then $t_{\lambda,\mu,\zeta}$ is $g^{\zeta}_{A_\lambda}(\boldsymbol{s}_\lambda)$ for $\zeta < \eta$, where $\boldsymbol{s}_\lambda$ is the sequence of free variables in $A_\lambda(\boldsymbol{x}_\lambda)$.

Let S_1 and S_2 be sequents in $T(S)$. S_2 is called an immediate ancestor of S_1 if S_2 is one of the sequents written above S_1 by applying one of (1)–(5) to S_1. A branch of $T(S)$ is a sequence $S = S_0, S_1, \ldots$, possibly infinite, such that S_{n+1} is always an immediate ancestor of S_n.

For any sequent $\Gamma \to \Delta$ only one of two cases is possible:

Case 1. In every branch of $T(\Gamma \to \Delta)$ there exists at least one sequent of the form

$$\Gamma_1, D, \Gamma_2 \to \Delta_1, D, \Delta_2.$$

In this case we can obtain a proof of $\Gamma \to \Delta$ without the cut rule by modifying $T(S)$, and regarding the elements of $D - D_0$ as free variables. (The proof is left to the reader).

Case 2. There exists a branch B of $T(\Gamma \to \Delta)$ in which no sequent is of the form

$$\Gamma_1, D, \Gamma_2 \to \Delta_1, D, \Delta_2.$$

In this case we claim that there is an interpretation in which every formula occurring in Γ is true and every formula occurring in Δ is false. In the remainder of this proof we fix such a branch B and consider only the formulas and sequents occurring in B, i.e., "sequent" means "sequent in B".

First observe the following lemmas:

LEMMA 22.18. (1) *If a formula $\neg A$ occurs in the antecedent (succedent) of a sequent, then the formula A occurs in the succedent (antecedent) of a sequent.*

(2) *If a formula $\bigwedge_{\lambda<\beta} A_\lambda$ occurs in the antecedent (succedent) of a sequent, then for every (some) $\lambda < \beta$, A_λ occurs in the antecedent (succedent) of a sequent.*

(3) *If a formula $\bigvee_{\lambda<\beta} A_\lambda$ occurs in the antecedent (succedent) of a sequent, then for some (every) $\lambda < \beta$, A_λ occurs in the antecedent (succedent) of a sequent.*

(4) *If $\forall \boldsymbol{x}\, A(\boldsymbol{x})$ occurs in the antecedent of a sequent, then for every sequence $\boldsymbol{t}$ of elements of D whose length is the same as that of $\boldsymbol{x}$, the formula $A(\boldsymbol{t})$ occurs in the antecedent of a sequent. If $\forall \boldsymbol{x}\, A(\boldsymbol{x})$ occurs in the succedent of a sequent, then the formula $A(\boldsymbol{t})$ occurs in the succedent of a sequent, where t_γ is $\bar{g}^{\gamma}_{A}(\boldsymbol{s})$, $\boldsymbol{s}$ being the sequence of the free variables in $A(\boldsymbol{x})$.*

(5) *If $\exists \boldsymbol{x}\, A(\boldsymbol{x})$ occurs in the antecedent of a sequent, then the formula $A(\boldsymbol{t})$ occurs in the antecedent of a sequent, where t_γ is $g^{\gamma}_{A}(\boldsymbol{s})$, $\boldsymbol{s}$ being the sequence of the free variables in $A(\boldsymbol{x})$. If $\exists \boldsymbol{x}\, A(\boldsymbol{x})$ occurs in the succedent of a sequent, then for an arbitrary sequence $\boldsymbol{t}$ of elements of D whose length is the same as that of $\boldsymbol{x}$, the formula $A(\boldsymbol{t})$ occurs in the succedent of a sequent.*

(6) *If a formula occurs in the antecedent of a sequent, then it does not occur in the succedent of any sequent.*

PROOF. (1)–(5) are obvious from the definition of $T(S)$. (6) can be proved by transfinite induction on the complexity of the formula using (1)–(5).

It is now evident how to define ϕ: For each term $t \in D$ $\phi t = t$. For any predicate constant R, $R(\boldsymbol{t})$ holds in $\langle D, \phi\rangle$ if and only if it occurs in the antecedent of a sequent. This completes the proof of Theorem 22.17.

Note that in the proof of the completeness theorem we need a sequence of new free variables $\boldsymbol{a}$ for every subformula $\exists \boldsymbol{x}\, A(\boldsymbol{x})$ of the end sequent. Moreover for every such sequence $\boldsymbol{a}$ we need another free variable for each instance $A(\boldsymbol{a})$. We then see why we must have a very large supply of free variables available or we must be able to rename the variables that are present.

Briefly we shall consider systems with equality.

DEFINITION 22.19. We define an infinitary logic with homogeneous quantifiers with equality by specifying a binary predicate constant $=$ and adjoining the following rules of inference to those of L:

1) First rules for equality: Let $\Gamma^{(\boldsymbol{a})}$ stand for a sequence of formulas Γ in which some occurrences of $\boldsymbol{a}$ are indicated.

$$\frac{\Gamma^{(\boldsymbol{a})} \to \Delta^{(\boldsymbol{b})}}{\boldsymbol{a} = \boldsymbol{b},\, \Gamma^{(\boldsymbol{b})} \to \Delta^{(\boldsymbol{a})}}; \qquad \frac{\Gamma^{(\boldsymbol{a})} \to \Delta^{(\boldsymbol{a})}}{\boldsymbol{b} = \boldsymbol{a},\, \Gamma^{(\boldsymbol{b})} \to \Delta^{(\boldsymbol{b})}}.$$

Here $\boldsymbol{a} = \boldsymbol{b}$ denotes the sequence $\{a_\lambda = b_\lambda\}_{\lambda<\gamma}$ and $\Gamma^{(\boldsymbol{b})} \to \Delta^{(\boldsymbol{b})}$ denotes the result obtained from $\Gamma^{(\boldsymbol{a})} \to \Delta^{(\boldsymbol{a})}$ by replacing the indicated occurrences of a_λ by b_λ for each $\lambda < \gamma$.

2) Second rule for equality: Let Σ be an arbitrary set of free variables and let $\tilde{\Sigma}$ be the set of all atomic formulas $a = b$ such that a and b belong to Σ. $(\Phi|\Psi)$ is called a *decomposition* of $\tilde{\Sigma}$ if $\Phi \cup \Psi = \tilde{\Sigma}$ and $\Phi \cap \Psi = 0$.

$$\frac{\Phi, \Gamma \to \Delta, \Psi \text{ for all decompositions } (\Phi|\Psi) \text{ of } \tilde{\Sigma}}{\Gamma \to \Delta}.$$

Theorems corresponding to Proposition 22.16 and theorem 22.17 hold for this system. Proofs can be obtained as special cases of the proofs of the corresponding theorems in the next section.

PROBLEM 22.20. Consider a finite, first order language L with K individual constants, where K is an infinite cardinal. The Löwenheim-Skolem theorem is stated as follows: Let $\mathcal{F}$ be a set of L-formulas. If $\mathcal{F}$ has a model then there exists a model of cardinality K. Let L′ be an infinitary homogeneous language which is an extension of L (here L′ has at least K individual constants). It is easily seen that the Löwenheim-Skolem theorem can be stated syntactically as follows: Let $\Gamma \to \Delta$ be a sequent of L, where the lengths of Γ and Δ can be any ordinal less than K^+. If the sequent

$$\exists x_0 \exists x_1 \dots \exists x_\xi \dots \forall y \left(\bigvee_{\xi<K} y = x_\xi\right), \Gamma \to \Delta$$

is provable in the homogeneous system, then so is $\Gamma \to \Delta$, where $=$ is not singled out in L or in L′.

Give a proof-theoretical proof of the Skolem-Löwenheim theorem in the syntactical form.

[*Hint*: 1) Introduce new constants $\{w_\alpha\}_{\alpha<K}$. Let $L_0 = L \cup \{w_\alpha\}_{\alpha<K}$ and consider the closed L_0-formulas of the form $\exists x\, F(x)$. We can define an enumeration (with repetition) of such formulas, $\{\exists x\, F_\alpha(x)\}_{\alpha<K}$, in such a manner that

(i) in $F_\alpha(x)$ no w_γ with $\gamma \geq \alpha$ occurs.

2) Let $\tilde{L} = L' \cup \{w_\alpha\}_{\alpha<K}$ and let $R(a)$ be $\bigvee_{\alpha<K} (a = w_\alpha)$. The relativization of formulas (of $\tilde{L}$) to R, (the relativization of A to R is denoted by A^R), is defined as in §17: $(\exists \boldsymbol{y}\, A(\boldsymbol{y}))^R$ is $\exists \boldsymbol{y} (\bigvee_{\gamma<\delta} R(y_\gamma) \wedge A^R(\boldsymbol{y}))$, where $\boldsymbol{y}$ is $\boldsymbol{y}_{<\delta}$.

3) It is obvious that $R(w_\gamma)$ is provable for every $\gamma < K$; hence $(\exists \boldsymbol{x} \forall y (\bigvee_{\beta<K} y = x_\beta))^R$ is provable. With the same method as in the theory of relativization in §17, we can prove the following:

Let $\Pi \to \Lambda$ be a sequent of L′. Let $\{b_i\}_{i<\beta}$ be the sequent of all free variables in $\Pi \to \Lambda$. If $\Pi \to \Lambda$ is provable in the homogeneous system (with language L′), then

$$\{R(b_i)\}_{i<\beta}, \Pi^R \to \Lambda^R$$

is provable in the homogeneous system with language $\tilde{L}$, where Π^R is obtained from Π by replacing each of its formula, say A, by A^R; similarly with Λ^R.

4) A proof-like figure is called a quasi-proof (of the homogeneous system) if it satisfies all the conditions of the proofs, except (II) and (III) of the eigenvariable conditions.

Besides the condition (i) in 1), we may require furthermore that for the enumeration of $\exists x\, F_\alpha(x)$'s the following holds.

(ii) There is an ω-type subset of $\{w_\alpha\}$, say $\Sigma = \{w_{\nu_0}, \omega_{\nu_1}, \ldots\}$ such that if Γ^* consists of all the $\exists x\, F_\alpha(x) \supset F_\alpha(w_\alpha)$, except those with $w_\alpha \in \Sigma$, then for every closed formula A of L_0 there is a quasi-proof ending with

$$\Gamma^* \to A \equiv A^R.$$

5) Suppose now that

$$\exists x\, \forall y \left(\bigvee_{\alpha<K} y = x_\alpha\right), \Gamma \to \Delta$$

is provable (in the homogeneous system). Then by 3)

$$\{R(b_i)\}_{i<\mu}, \Gamma^R \to \Delta^R$$

is provable, where $\{b_i\}_{i<\mu}$ is the sequence of the free variables in Γ and Δ. Then $\mu \leq \omega$. We may identify b_i with w_{ν_i} in Σ; thus we may assume that

$$\{R(w_{\nu_i})\}_{i<\mu}, \Gamma^R\{w_{\nu_i}\}_{i<\mu} \to \Delta^R\{w_{\nu_i}\}_{i<\mu},$$

is provable where $\Gamma^R\{w_{\nu_i}\}_{i<\mu}$ is obtained from Γ^R by replacing b_i by w_{ν_i}, and similarly with Δ^R.

6) Finally, 3), 4) and 5) imply

$$\Gamma^*, \Gamma\{w_{\nu_i}\} \to \Delta\{w_{\nu_i}\} \quad \text{or} \quad \Gamma^*, \Gamma \to \Delta$$

has a quasi-proof. Recall that if $\exists x\, F(x) \supset F(w)$ belongs to Γ^*, then w does not belong to Σ. Regarding these w's as free variables, we obtain

$$\{\exists y\, (\exists x\, F(x) \supset F(y))\}, \Gamma \to \Delta,$$

where $\exists y\, (\exists x\, F(x) \supset F(y))$ is provable for each F. Therefore, by the cut rule, we have $\Gamma \to \Delta$. Assuming that we have carefully chosen the free variables, we may claim that the eigenvariable conditions are satisfied except for II, on heights. In the quasi-proof of $\{\exists y\, (\exists x\, F(x) \supset F(y))\}$, $\Gamma \to \Delta$, the w where $\exists x\, F(x) \supset F(w)$ is the rth formula in Γ^*, is assigned the height r; each eigenvariable in the quasi-proof in (ii) of 4) is assigned the height K, and any eigenvariable in the proof ending with $\{R(b_i)\}$, $\Gamma^R \to \Delta^R$ is assigned the height K^+.]

If one wishes to study an infinitary logic which is closer to first order logic, he may restrict the quantifiers to those that operate as in the finite case. Lopez-Escobar has defined such a system, called $L_{\omega_1,\omega}$ and proved the completeness and the interpolation theorem for it. The version of these theorems for $L_{\omega_1,\omega}$ is the same as that of **LK**. We shall present the results in the form of a problem.

PROBLEM 22.21 (Lopez-Escobar). The language $L_{\omega_1,\omega}$ is an extension of that of **LK**, and is defined as follows. There are arbitrarily many constants but the arity of each predicate and each function constant is finite. The number of variables is countable. For simplicity, we take only $\neg$, $\bigwedge$ and $\forall$ as logical symbols. The formulas are defined as usual: If A_i, $i<\omega$, is a sequence of formulas, then $\bigwedge_{i<\omega} A_i$ is a formula. Notice that $\forall$ behaves as in the finite case. A sequent consists of at most countably many formulas. The rules of inference, as well as the initial sequents, are those of **LK** except the following:

$$\text{Weak inference:}\quad \frac{\Gamma \rightarrow \Delta}{\Pi \rightarrow \Lambda},$$

where every formula in Γ occurs in Π and every formula in Δ occurs in Λ.

$$\bigwedge\text{: left}\quad \frac{A_i, \Gamma \rightarrow \Delta \text{ for some } i}{\bigwedge_{i<\omega} A_i, \Gamma \rightarrow \Delta}$$

$$\bigwedge\text{: right}\quad \frac{\Gamma \rightarrow \Delta, A_i \text{ for all } i}{\Gamma \rightarrow \Delta, \bigwedge_{i<\omega} A_i}$$

(1) Prove the completeness of the system.

(2) Prove the interpolation theorem for this system; viz. if $A \supset B$ is provable and A and B have at least one predicate symbol in common, then there exists a C of $L_{\omega_1,\omega}$ such that $A \supset C$ and $C \supset B$ are provable.

(3) Show that the following is an admissible rule of inference:

$$\frac{\Gamma \rightarrow \Delta}{\tilde{\Gamma} \rightarrow \tilde{\Delta}}$$

where $\tilde{\Gamma}$ is obtained from Γ by replacing each formula of Γ by one of its alphabetical variants (possibly the formula itself); similarly with $\tilde{\Delta}$.

[*Hint*: (1) Consistency is obvious. For the opposite direction proceed in the following way.

1) Given a sequent of $L_{\omega_1,\omega}$, say S, there are countably many terms which are obtained from the constants which occur in S and all the free variables.

2) Given a sequent S, the S-subformulas are defined as the ordinary subformulas of the formulas of S except for the following case: If $\forall x\, A(x)$ is an S-subformula, then for every term s which satisfies the condition in 1) $A(s)$ is an s-subformula.

3) There are countably many S-subformulas.

4) Given a sequent S, construct a tree $T(S)$. We may assume that there are countably many free variables which do not occur in S. This and the construction of $T(S)$ guarantee that at each step there will still be countably many free variables unused. From 3) we may assume that all the S-subformulas are indexed in ω. We define the tree step by step.

Stage 0. We write the sequent S.

Stage $n+1$. Let $\Gamma \to \Delta$ be a topmost sequent.

Case 1. $n+1 \equiv 1 \pmod 5$. Let $\{\neg A_i\}_{i=1}^m$ and $\{\neg B_j\}_{j=1}^l$ be all the formulas in Γ and Δ, respectively, whose outermost logical symbol is $\neg$ and whose indices in the fixed enumeration of subformulas are $\leq n+1$. Then write $\{B_j\}_{j=1}^l, \Gamma' \to \Delta', \{A_i\}_{i=1}^m$ above $\Gamma \to \Delta$, where Γ' is obtained from Γ by deleting $\{\neg A_i\}_{i=1}^m$ and Δ' is obtained from Δ by deleting $\{\neg B_j\}_{j=1}^l$.

Case 2. $n+1 \equiv 2 \pmod 5$. Let $\{\bigwedge_{i<\omega} A_i^j\}_{j=1}^m$ be all the formulas in Γ whose outermost logical symbol is $\bigwedge$ and whose indices are $\leq n+1$. Then write

$$\{A_i^1\}_{i\leq n}, \ldots, \{A_i^m\}_{i\leq n}, \Gamma \to \Delta,$$

above $\Gamma \to \Delta$.

Case 3. $n+1 \equiv 3 \pmod 5$. Let $\{\bigwedge_{i<\omega} A_i^j\}_{j=1}^m$ be all the formulas in Γ whose outermost logical symbol is $\bigwedge$ and whose indices are $\leq n+1$. Then write

$$\Gamma \to \Delta', \{A_{i_j}^j\}_{j=1}^m$$

for all combinations of $\{i_1, \ldots, i_m\}$ above $\Gamma \to \Delta$.

Case 4. $n+1 \equiv 4 \pmod 5$. Let $\{\forall x_i\, A_i(x_i)\}_{i=1}^m$ be all the formulas in Γ whose outermost logical symbol is $\forall$ and whose indices are $\leq n+1$. Let $A_i(s_i^1), \ldots, A_i(s_i^{n+1})$ be the first $n+1$ formulas in the enumeration that are S-subformulas of $\forall x_i\, A_i(x_i)$. Write

$$\{A_i(s_i^{j_i})\}_{i\leq m}^{j_i\leq n+1}, \Gamma \to \Delta'$$

above $\Gamma \to \Delta$.

Case 5. $n+1 \equiv 0 \pmod 5$. Let $\{\forall x_i\, A_i(x_i)\}_{i=1}^m$ be all the formulas in Δ whose outermost logical symbol is $\forall$ and whose indices are $\leq n+1$. Let $a_{j_1}, \ldots, a_{j_m}$ be the first m free variables which have not occurred so far. Write

$$\Gamma \to \Delta', \{A_i(a_{j_i})\}_{i=1}^m$$

above $\Gamma \to \Delta$.

At any stage, if some formula occurs both in the antecedent and the succedent then stop.

5) Let $T(S)$ be the tree defined in 4).

Case 1. All branches are finite. Then S is provable without the cut rule.

Case 2. There is an infinite branch, say B. Let D be the set of all terms which satisfy the condition in 1). We define the structure with the domain

D and the interpretation of formulas in the usual way. Then the formulas in the antecedent of B are true, while those in the succedent are false. This completes the proof of (1).

(2) From the proof of (1) above, any provable sequent is cut-free provable. Restate the interpolation theorem for sequents. Consider only cut-free proofs and show the described result by induction on the complexities of the proofs. The procedure is exactly the same as the corresponding theorem for **LK**.

(3) Obvious from the completeness.]

PROBLEM 22.22 (corollary to the Lopez-Escobar theorem). Suppose that $\Gamma \to \Delta$ is a provable sequent of $L_{\omega_1,\omega}$ and Γ and Δ are finite sequences. Then there exists a cut-free proof of $\Gamma \to \Delta$ in which every sequent consists of finitely many formulas. (Such a proof may be infinite.)

PROBLEM 22.23. Consider a language consisting of the following:

Predicate symbol: $\in$.

Variables: $x_0, x_1, \ldots, x_\mu, \ldots, \mu \in \mathrm{On}$.

Logical symbols: $\neg, \bigwedge, \forall$.

Formulas are defined as usual. The atomic formulas are of the form $x \in y$. If A is a formula, $\neg A$ is a formula. If A_i, $i < \lambda$, is a sequence of formulas for λ an ordinal, then $\bigwedge_{i<\lambda} A_i$ is a formula. If $A(\boldsymbol{y})$ is a formula, where $\boldsymbol{y}$ is a sequence of variables none of which is in the scope of a quantifier, then $\forall \boldsymbol{y}\, A(\boldsymbol{y})$ is a formula. Show that the truth definition of this language can be developed in a system of second order set theory, i.e., ZF augmented by second order quantifiers and some comprehension axioms.

[*Hint*: The method is similar to the truth definition of **PA** in a second order system. First assign sets to the formal objects of the language. The set assigned to a formal symbol we call the gödelization of that symbol. If A is a formal expression, its gödelization is denoted by $\ulcorner A \urcorner$. For example, $\ulcorner \in \urcorner = \langle 0, 0\rangle$, $\ulcorner x_i \urcorner = \langle 1, i\rangle$, $\ulcorner \neg \urcorner = \langle 3, 0\rangle$, $\ulcorner \bigwedge \urcorner = \langle 5, 0\rangle$, $\ulcorner \forall \urcorner = \langle 7, 0\rangle$, $\ulcorner \boldsymbol{x} \urcorner = \langle 11, \boldsymbol{x}\rangle$, where $\ulcorner x \urcorner$ is the name of a set x, $\ulcorner x \in y \urcorner = \langle \ulcorner \in \urcorner, \ulcorner x \urcorner, \ulcorner y \urcorner\rangle$, $\ulcorner \bigwedge_{i<\lambda} A_i \urcorner = \langle \ulcorner \bigwedge \urcorner, \langle \ulcorner A_0 \urcorner, \ulcorner A_i \urcorner, \ldots\rangle\rangle$, $\ulcorner \forall \boldsymbol{y}\, A(\boldsymbol{y}) \urcorner = \langle \ulcorner \forall \urcorner, \langle \ulcorner y_0 \urcorner, \ulcorner y_1 \urcorner, \ldots\rangle, \ulcorner A(\boldsymbol{y}) \urcorner\rangle$. We can then formally define "A is a closed formula" ($\mathrm{cf}(\ulcorner A \urcorner)$) and "the complexity of a formula A" ($\mathrm{cm}(\ulcorner A \urcorner)$, which is an ordinal). Let α be a second order free variable and let μ be a variable which ranges over ordinals. Define $F(\alpha, \mu)$ as we defined $F(\alpha, n)$ in the case of **PA** to state that α is the truth definition of formulas whose complexities are $\leqslant \mu$. The clause for $\forall \boldsymbol{x}\, A(\boldsymbol{x})$ is expressed as:

$$\forall \ulcorner \forall \boldsymbol{x}\, A(\boldsymbol{x}) \urcorner\, (\mathrm{cf}(\ulcorner \forall \boldsymbol{x}\, A(\boldsymbol{x}) \urcorner) \wedge \mathrm{cm}(\ulcorner \forall \boldsymbol{x}\, A(\boldsymbol{x}) \urcorner) \leqslant \mu \supset$$

$$\supset \forall x\, (x \text{ is a sequence of order type } \lambda, \text{ say } \langle x_0, x_1, \ldots\rangle) \supset$$

$$\supset \alpha(\ulcorner A(x_0, x_1, \ldots) \urcorner).$$

Then define $T(a) \Leftrightarrow_{\mathrm{df}} \mathrm{cf}(a) \wedge \exists \phi\, (F(\phi, \mathrm{cm}(a)) \wedge \phi(a))$. Now prove $T(\ulcorner A \urcorner) \equiv A$ for all closed formulas A.]

REMARK. We can generalize the proposition in Problem 22.23 to the cases where there are predicate constants $p_0, p_1, \ldots$ and where the quantifiers are not homogeneous.

Next we will show that for any homogeneous system there is an equivalent homogeneous system whose eigenvariable conditions are "ordinary" ones, that is, the eigenvariable conditions are conditions on inferences and not on proofs. In order to simplify the argument we take only the logical symbols $\neg$, $\bigvee$, $\exists$, and regard others as a combination of these.

DEFINITION 22.24. The $\forall\exists$-calculus is defined as the homogeneous system with the following alteration: Replace $\exists$: left by

$$\forall\exists \text{ rule:} \frac{\{A_\lambda(\boldsymbol{a}_\lambda, \boldsymbol{b}_\lambda)\}_{\lambda<\alpha}, \Gamma \to \Delta}{\{\forall \boldsymbol{x}_\lambda\, \exists \boldsymbol{y}_\lambda\, A_\lambda(\boldsymbol{x}_\lambda, \boldsymbol{y}_\lambda)\}_{\lambda<\alpha}, \Gamma \to \Delta},$$

where none of the free variables contained in $\boldsymbol{b}_\lambda$'s can occur in the lower sequent.

Each variable in $\boldsymbol{b}_\lambda$ is an eigenvariable. All of the variables of $\boldsymbol{b}_\lambda$ must be distinct and none of them can occur in $A_{\lambda'}(\boldsymbol{a}_{\lambda'}, \boldsymbol{b}_{\lambda'})$ for $\lambda' < \lambda$. There are no other eigenvariable conditions.

Note that $\forall \boldsymbol{x}_\lambda$ can be empty.

PROPOSITION 22.25 (Maehara-Takeuti). *The $\forall\exists$-calculus is equivalent to the homogeneous system (for the same language).*

PROOF. Let P be a proof in the $\forall\exists$-calculus. We may assign a height to every free variable; if b is the μth variable of $\boldsymbol{b}_\lambda$ in the $\forall\exists$-rule, then the height of b is $\sup_a$ (height of a) $+ (1+\mu)$, where a ranges over all the free variables in A_λ other than $\boldsymbol{b}_\lambda$ and by sup we mean the strict supremum. If b is not used as eigenvariable, then the height of b is 0. Transform P as follows. If there is an application of the $\forall\exists$-rule, then replace it by:

$$\cfrac{\cfrac{\cfrac{\cfrac{\{A_\lambda(\boldsymbol{a}_\lambda, \boldsymbol{b}_\lambda)\}_{\lambda<\alpha}, \Gamma \to \Delta}{\{\exists \boldsymbol{y}_x\, A_\lambda(\boldsymbol{a}_\lambda, \boldsymbol{y}_x)\}_{\lambda<\alpha}, \Gamma \to \Delta}}{\Gamma \to \Delta, \{\neg\exists \boldsymbol{y}_x\, A_\lambda(\boldsymbol{a}_\lambda, \boldsymbol{y}_x)\}_{\lambda<\alpha}}}{\Gamma \to \Delta, \{\exists \boldsymbol{x}_\lambda\, \neg\exists \boldsymbol{y}_\lambda\, A_\lambda(\boldsymbol{x}_\lambda, \boldsymbol{y}_x)\}_{\lambda<\alpha}}}{\{\neg\exists \boldsymbol{x}_\lambda\, \neg\exists \boldsymbol{y}_\lambda\, A_\lambda(\boldsymbol{x}_\lambda, \boldsymbol{y}_\lambda)\}_{\lambda<\alpha}, \Gamma \to \Delta.}$$

It can be easily seen that the resulting figure is a proof in the homogeneous system with the same heights as P.

The opposite direction is proved as follows: Let P be a proof in the homogeneous system ending with $\Gamma \to \Delta$. Let $\{\exists \boldsymbol{x}_\lambda\, A_\lambda(\boldsymbol{a}_\lambda, \boldsymbol{x}_\lambda)\}_{\lambda<\alpha}$ be an enumeration of all the principal formulas of the $\exists$: left in P, where $\boldsymbol{a}_\lambda$ is the sequence of all free variables in A_λ. Then eliminate every application of the $\exists$: left as follows: for simplicity we demonstrate a case where there is only one auxiliary formula:

$$\exists : \text{left} \quad \frac{A(a, b), \Pi \to \Delta}{\exists x\, A(a, x), \Pi \to \Lambda}$$

is changed to

$$\frac{\dfrac{A(a, b), \Pi \to \Lambda}{A(a, b), \exists x\, A(a, x), \Pi \to \Lambda} \qquad \neg\exists x\, A(a, x), \exists x\, A(a, x), \Pi \to \Lambda}{\neg\exists x\, A(a, x) \vee A(a, b), \qquad \exists x\, A(a, x), \Pi \to \Lambda}$$

Since no eigenvariables are involved we obtain a proof in the homogeneous system as well as in the ∀∃-calculus. From this proof, which ends with

$$\{\neg\exists x_\lambda\, A_\lambda(a_\lambda, x_\lambda) \vee A_\lambda(a_\lambda, b_\lambda)\}_{\lambda<\alpha}, \Gamma \to \Delta,$$

we obtain, by applying the ∀∃-rule,

$$\{\forall y_\lambda\, \exists z_\lambda\, [\neg\exists x_\lambda A_\lambda(y_\lambda, x_\lambda) \vee A(y_\lambda, z_\lambda)]\}_{\lambda<\alpha}, \Gamma \to \Delta.$$

On the other hand,

$$\forall y_\lambda\, \exists z_\lambda\, [\neg\exists x_\lambda\, A_\lambda(y_\lambda, x_\lambda) \vee A(y_\lambda, z_\lambda)]$$

is provable with the ∀∃-rule. Hence $\Gamma \to \Delta$ is provable in the ∀∃-calculus.

PROBLEM 22.26. The compactness of **LK** (the first order predicate calculus) can be syntactically expressed as follows. Let $\Gamma \to \Delta$ denote a sequent consisting of formulas of **LK** with cardinality $\leq K$, where K is the cardinality of the set of formulas of **LK** (for a given language). For any such sequent that is provable in the homogeneous system (of an appropriate language), there exist finite subsets Γ_0 and Δ_0 of Γ and Δ, respectively, for which $\Gamma_0 \to \Delta_0$ is **LK**-provable.

Prove the compactness of the first order predicate calculus in this syntactic form.

[*Hint*: Let $\Gamma \to \Delta$ be a sequent as above and let P be a proof of $\Gamma \to \Delta$. (*) For each sequent, say $\Pi \to \Lambda$, in P, we can select a finite subsequent $\Pi_0 \to \Lambda_0$, i.e., $\Pi_0 \subseteq \Pi$, $\Lambda_0 \subseteq \Lambda$ and Π_0 and Λ_0 are finite. If $\Pi \to \Lambda$ is the lower sequent of an inference, we can select finitely many upper sequents corresponding to it in such a manner that that part of P which consists of all the selected sequents corresponding to $\Pi_0 \to \Lambda_0$ is a quasi-proof of $\Pi_0 \to \Lambda_0$. In particular, $\Pi \to \Lambda$ can be $\Gamma \to \Delta$; hence there is a finite subsequent $\Gamma_0 \to \Delta_0$, which is provable.

Then applying Proposition 22.25, we can construct an **LK**-proof of $\Gamma_0 \to \Delta_0$. (*) is proved by transfinite induction on the construction of the subproof of P ending with $\Pi \to \Lambda$. For $\bigwedge$: right and $\bigvee$: left, use the generalized König's lemma. (Cf. the proof of Proposition 8.16.)]

PROBLEM 22.27. First we shall define a formal infinitary language in set theoretical terms.

A basic language is an ordered triple $\langle C, P, S\rangle$, where C is a set of individual constants, P is a set of predicate constants, and S is a set of logical symbols. Each element of P is an ordered pair $\langle A, \alpha\rangle$ where α is an ordinal called the arity of $\langle A, \alpha\rangle$. An element of S is either $\neg$ or of the form $\langle \bigwedge, \alpha\rangle$, $\langle \bigvee, \alpha\rangle$, $\langle \forall, \alpha\rangle$ or $\langle \exists, \alpha\rangle$, where α is an ordinal called the arity of $\langle \bigwedge, \alpha\rangle$, $\langle \bigvee, \alpha\rangle$, $\langle \forall, \alpha\rangle$ or $\langle \exists, \alpha\rangle$, respectively. A basic language $\langle C, P, S\rangle$, also satisfies the following conditions.

1) The sets C, P, and S are mutually disjoint.

2) The symbols $\neg$, $\bigwedge$, $\bigvee$, $\forall$, and $\exists$ are different.

A language L is an ordered set $\langle C, P, S, B, F\rangle$, where $\langle C, P, S\rangle$ is a basic language, B and F are a set of bound variables and a set of free variables respectively, and C, P, S, B, and F are mutually disjoint.

Since the terms, formulas, etc. of L are what we commonly, in logic, understand them to be, we skip their formal set theoretical definitions. We however make the following deviations from our previous treatment. We call $\Gamma \to \Delta$ a sequent in L if Γ and Δ are sets of formulas in L. This change is useful when we wish to avoid the use of the axiom of choice as much as possible.

A tree is an ordered pair $\langle T, <\rangle$ satisfying the following conditions.

1. $T \neq \emptyset$.

2. The relation $<$ is a partial ordering on T.

We read $s_1 < s_2$ as "s_1 is below s_2" or "s_2 is above s_1". There is a unique lowermost point s_0 in T, i.e., there is a unique s_0 in T such that

$$\forall s \in T\,(s_0 < s \vee s_0 = s).$$

This lowermost point s_0 is called the end point of the tree. Every point, except the end point, has a unique point below it, i.e.,

$$\forall s \in T\,(s \neq s_0 \supset \exists! t \in T\, \forall u\,(u < s \equiv u = t \vee u < t)).$$

If $s_1 < s_2$ and $\neg \exists s\,(s_1 < s \wedge s < s_2)$, then s_2 is said to be immediately above s_1. If $s_1 < s$, then there exists a unique s_2 such that $s_1 < s_2 \leq s$ and s_2 is immediately above s_1. A topmost point is called an initial point.

3. Any linearly ordered subset of T (with respect to $<$) is finite. A semi-proof P in L is a function f from a tree into a set of sequents in L satisfying the following conditions.

1) If s is an initial point and $f(s)$ is of the form $\Gamma \to \Delta$, then

$$\Gamma \cap \Delta \neq \emptyset.$$

Note that here an initial sequent need not be of the form $D \to D$. This change enables us to prove the completeness theorem with a minimal use of the axiom of choice.

2) Let $\ldots, s_\alpha, \ldots$ be the collection of all points immediately above s.

Then

$$\frac{\ldots, f(s_\alpha), \ldots}{f(s)}$$

is an inference in L.

A proof P in L is an ordered pair $\langle P_0, <\rangle$, where P_0 is a semi-proof in L and $<$ is a well-founded partial ordering on the free variables in P_0 which satisfies our eigenvariable conditions.

A structure for L is defined in the usual manner.

Let M be a transitive set which needs not satisfy the axiom of choice. Let S and A be a structure for L in M and a sentence in L, respectively. Then "S satisfies A in M" denoted by $S \models^{M} A$ is defined as usual except for $\exists$ and $\forall$. Since $\forall$ is defined as $\neg\exists\neg$, we give only the definition for $\exists$.

$$S \models^{M} \exists x_0 x_1 \ldots A(x_0, x_1, \ldots)$$

is defined to be

$$\exists f \in M\,(S \models^{M} A(f(0), f(1), \ldots)).$$

A sentence F is called M-valid, if for every structure S in M, $S \models^{M} F$.

THEOREM A. *Let* L *be a basic language and suppose that* $\forall a \in M$ $(a^\alpha \in M)$ *for every arity* α *in* L. *If an* L*-proof* P *is an element of* M, *then the end-sequent of* P *is* M*-valid.*

[*Hint*: Follow the proof of the validity theorem with the following modification. Define Skolem functions using the axiom of choice. The Skolem functions are possibly outside M but the sequences of terms made by these Skolem functions are members of M by the hypotheses of the theorem, provided their lengths are arities of L. Therefore the proof can be carried out as before.]

THEOREM B. *Let* L *be a basic language and let* λ *be the first regular cardinal greater than all the arities in* L. *We assume* $\lambda > \omega$. *Let* S *be an* L*-sequent. If* M *satisfies the following conditions, then either* S *has a cut free proof in* M *or there exists a counter model of* S *in* M.

1) $L \in M$, $S \in M$ *and* $\lambda \in M$.

2) $\forall a, b \in M(\{a, b\} \in M)$ *and* $\forall a \in M(\bigcup(a) \in M)$, *where* $\bigcup(a)$ *is the union of* a.

3) M *satisfies the axiom of replacement.*

4) $\forall a \in M\, \forall \alpha \in \lambda\, (a^\alpha \in M)$.

In case that L *has* $=$, *the condition* $P(D \times D) \in M$ *is added, where* $P(D \times D)$ *is the power set of* $D \times D$, D *is an adequate set of free variables, and* $D \in M$ *is a consequence of* 1)–4).

[*Hint*: Follow the proof of the completeness theorem in the following manner.

1) Introduce λ-many bound variables in M.

2) Construct all atomic semi-formulas in M.

3) Construct all semi-formulas without free variables in M. Here we use a definition by transfinite induction up to λ. This can be done in M since M satisfies the axiom of replacement.

4) In M, introduce Skolem function letters corresponding to each semi-formula without free variables.

5) Construct the set D of all free variables as the set of all possible function combinations of Skolem function letters in M. Here we again use a definition by transfinite induction up to λ.

6) Construct a reduction tree in M. We define a reduction tree up to the nth step in M by mathematical induction. Then it is easily shown that the whole reduction tree is in M.

7) Construct a countermodel or a cut free proof in M. This can be done as usual.

8) If L has $=$, we need $P(D \times D) \in M$ since all possible partitions of free variables are taken into consideration in the equality axiom.]

Without using $\in$, we can express the axiom of choice in the form

$$\forall x\, \exists y\, A(x, y) \rightarrow \exists f\, \forall x\, A(x, f(x)).$$

Since $\exists f$ occurs in this form, this is a second order form. Since some second order notions can be expressed in an infinitary language, it is natural to ask whether the axiom of choice can be expressed in an infinitary language. Actually, a weak form of the axiom of choice is elegantly expressed in an infinitary language: The axiom of dependent choice can be expressed as

$$\forall x\, \exists y\, A(x, y) \rightarrow \forall x_0 \exists x_1\, x_2 \ldots \bigwedge_{i<\omega} A(x_i, x_{i+1}).$$

COROLLARY. *The axiom of choice is not expressible in an infinitary language.*

[*Hint*: Suppose the axiom of choice can be expressed in an infinitary language L. Then since the axiom of choice is true, there must be a proof P of the axiom. Let α be a large ordinal so that $\mathrm{L} \in R(\alpha)$, $P \in R(\alpha)$ and $\lambda < \alpha$, where $R(\alpha) = \{a \mid \mathrm{rank}(a) < \alpha\}$. This is a contradiction since it is very easy to prove the existence of a transitive set M with the following properties:

1) The axiom of choice is not M-valid.

2) M satisfies the conditions of Theorem A. (For example, take M to be the smallest transitive set satisfying the conditions in Theorem A and $R(\alpha) \in M$.)]

§23. Determinate logic

In this section we will discuss determinate logic with equality (=) as a special case of infinitary logic with heterogeneous quantifiers. In order to simplify the discussion, we will only consider languages that have no individual constants.

DEFINITION 23.1. (1) By a *heterogeneous quantifier of arity* α we mean a symbol $\mathbf{Q}^f$, where f is a map from α into $\{\forall, \exists\}$. For such a map f, the map $\bar{f}$, called the dual of f, is defined in the following way.

(i) The domain of $\bar{f}$ is the same as that of f,

(ii) $\bar{f}(\beta) = \forall$ or $\exists$ according as $f(\beta) = \exists$ or $\forall$ respectively.

If f and g are dual, then $\mathbf{Q}^f$ and $\mathbf{Q}^g$ are called dual quantifiers.

(2) By the language L_D we mean the language obtained from the language in §22 by replacing the quantifiers $\forall$ and $\exists$, of arity α, by heterogeneous quantifiers $\mathbf{Q}^f$ of the same arity.

(3) Let $\mathscr{A}$ be a structure for L_D. We define satisfaction and validity in $\mathscr{A}$ as in Definition 22.15. The structure is said to be *determinate* if for each closed formula A in L_D exactly one of the two formulas

$$\mathbf{Q}^f \boldsymbol{x}_{<\alpha}\, A(\boldsymbol{x})$$

and

$$\mathbf{Q}^{\bar{f}} \boldsymbol{x}_{<\alpha}\, \neg A(\boldsymbol{x})$$

is valid in $\mathscr{A}$.

(4) A logical system **S** with language L_D is called a determinate logic if for every closed formula A in L_D, "A is provable in **S**" is equivalent to "A is valid in every determinate structure".

In this section we will define a logical system **DL** and prove that (i) **DL** is a determinate logic, (ii) if a formula A is provable in **DL** by using heterogeneous quantifier introduction only once at the end of the proof, then A is valid, and (iii) in **DL** the completeness theorem, the cut-elimination theorem, and the interpolation theorem hold in a certain form.

The language of our formal system **DL** is L_D with equality. Consequently, we will develop the theory of determinate logic with equality. First we modify the notion of proof as defined in §22.

DEFINITION 23.2. (1) The rules for = are as in Definition 22.19.

(2) The rules for $\forall$ and $\exists$ in Definition 22.1 are replaced by the following.

$$\mathbf{Q}: \text{left:} \quad \frac{\{A_\lambda(\boldsymbol{a}_\lambda)\}_{\lambda<\gamma},\, \Gamma \to \Delta}{\{\mathbf{Q}^{f_\lambda} \boldsymbol{x}_\lambda\, A_\lambda(\boldsymbol{x}_\lambda)\}_{\lambda<\gamma},\, \Gamma \to \Delta},$$

where $\boldsymbol{a}_\lambda$ denotes a sequence $a_{\lambda,0}, \ldots, a_{\lambda,\alpha}, \ldots$ $(\alpha < \mu_\lambda)$ for some μ_λ. The μth variable of this sequence, $a_{\lambda,\mu}$, we call the variable of $\boldsymbol{a}_\lambda$ of order μ. If

$f_\lambda(\mu) = \exists$, then $a_{\lambda,\mu}$ is called an eigenvariable of the inference

$$\mathbf{Q}: \text{right:}\quad \frac{\Gamma \to \Delta, \{A_\lambda(\boldsymbol{a}_\lambda)\}_{\lambda<\gamma}}{\Gamma \to \Delta, \{\mathbf{Q}^{f_\lambda}\boldsymbol{x}_\lambda\, A_\lambda(\boldsymbol{x}_\lambda)\}_{\lambda<\gamma}}.$$

If $f_\lambda(\mu) = \forall$, then $a_{\lambda,\mu}$ is called an eigenvariable of the inference.

If $a_{\lambda,\mu}$ is an eigenvariable of either inference, $\mathbf{Q}^{f_\lambda}\boldsymbol{x}_\lambda\, A_\lambda(\boldsymbol{x}_\lambda)$ is called a *principal formula* of $a_{\lambda,\mu}$ and also a principal formula of the inference. Furthermore, $A_\lambda(\boldsymbol{a}_\lambda)$ is called an *auxiliary formula* of $\mathbf{Q}^{f_\lambda}\boldsymbol{x}_\lambda\, A_\lambda(\boldsymbol{x}_\lambda)$, of the eigenvariable $a_{\lambda,\mu}$, and of the inference.

If two different variables a and b have the same principal formula then a is said to *precede* b with respect to that principal formula if the order of a is less than the order of b.

(3) Every proof must satisfy the following eigenvariable conditions.

1) If a free variable a occurs in two or more places as an eigenvariable, then for each occurrence a must have the same principal formula and a must have the same order. Moreover, if a occurs in two different auxiliary formulas $A(\boldsymbol{a}_1)$ and $A(\boldsymbol{a}_2)$ as an eigenvariable of order μ then $a_{1,\nu}$ and $a_{2,\nu}$ must be the same variable for all $\nu < \mu$.

2) To each free variable a, we assign an ordinal number $h(a)$, called the height of a, which has the following properties:

2.1) The height, $h(a)$, of an eigenvariable a, is greater than the height, $h(b)$, of every free variable b in the principal formula of the eigenvariable a.

2.2) The height of an eigenvariable a is greater than the height of b if b precedes a with respect to a principal formula of a.

3) No variable occurring in an inference as an eigenvariable may occur in the end sequent.

REMARK. The following weaker modification of the foregoing eigenvariable conditions is enough to assure that a logic is determinate.

Replace the last half of 1) by the following: If $A(\boldsymbol{a})$ is an auxiliary formula of a principal formula $\mathbf{Q}^f\boldsymbol{x}\, A(\boldsymbol{x})$ and a_ν and a_μ are eigenvariables of $\mathbf{Q}^f\boldsymbol{x} A(\boldsymbol{x})$ with $\nu \neq \mu$, then a_ν and a_μ are different. If a occurs in two different auxiliary formulas $A(\boldsymbol{a}_1)$ and $A(\boldsymbol{a}_2)$ as an eigenvariable of a principal formula $\mathbf{Q}^f\boldsymbol{x}\, A(\boldsymbol{x})$ then $a_{1,\nu}$ and $a_{2,\nu}$ are the same for each non-eigenvariable $a_{1,\nu}$ of $\mathbf{Q}^f\boldsymbol{x}\, A(\boldsymbol{x})$ for each ν less than the order of a.

Replace 2.2) by the following: If a is an eigenvariable with principal formula $\mathbf{Q}^f\boldsymbol{x}\, A(\boldsymbol{x})$ then the height of a is greater than the height of b if b precedes a with respect to the principal formula $\mathbf{Q}^f\boldsymbol{x}\, A(\boldsymbol{x})$ but b is not an eigenvariable of this principal formula.

We will use either the original form of 2.2) or the latter version choosing whichever is more convenient for our purposes.

EXAMPLE 23.3. Proof of the axiom of determinateness: Let $\boldsymbol{a}$ be $a_{\lambda<\alpha}$ and $\boldsymbol{b}$ be $b_{\mu<\beta}$.

$$\cfrac{\cfrac{\cfrac{A(\boldsymbol{a}, \boldsymbol{b}) \rightarrow A(\boldsymbol{a}, \boldsymbol{b})}{\rightarrow A(\boldsymbol{a}, \boldsymbol{b}), \neg A(\boldsymbol{a}, \boldsymbol{b})}}{\rightarrow \mathbf{Q}^f\boldsymbol{x}\, A(\boldsymbol{x}, \boldsymbol{b}), \mathbf{Q}^{\bar{f}}\boldsymbol{x} \neg A(\boldsymbol{x}, \boldsymbol{b})}}{\rightarrow \mathbf{Q}^f\boldsymbol{x}\, A(\boldsymbol{x}, \boldsymbol{b}) \vee \mathbf{Q}^{\bar{f}}\boldsymbol{x} \neg A(\boldsymbol{x}, \boldsymbol{b}).}$$

In this proof, $h(a_\lambda) = 1 + \lambda$ and $h(\beta_\mu) = 0$.

THEOREM 23.4 (validity for determinate structures). *Let $\mathscr{A}$ be a determinate structure and $\Gamma \rightarrow \Delta$ be provable in the determinate logic* **DL**. *Then $\Gamma \rightarrow \Delta$ is satisfied in $\mathscr{A}$.*

PROOF. Take an arbitrary formula with a quantifier at the beginning, say

$$\mathbf{Q}^f\boldsymbol{x}\, A(\boldsymbol{x}, \boldsymbol{a}).$$

where $\boldsymbol{a}$ is the sequence of all free variables in this formula and the length of $\boldsymbol{x}$ is α. For each $\gamma < \alpha$, we introduce a Skolem function

$$g_A^{f,\gamma}(x_{\xi_0}, \ldots, x_{\xi_\mu}, \ldots, \boldsymbol{a}) \quad \text{or} \quad \bar{g}_A^{f,\gamma}(x_{\eta_0}, \ldots, x_{\eta_\mu}, \ldots, \boldsymbol{a})$$

according as $f(\gamma) = \exists$ or $f(\gamma) = \forall$, where $\xi_0, \ldots, \xi_\mu, \ldots$ are all the ordinals $< \gamma$ for which $f(\xi) = \forall$ and $\eta_0, \ldots, \eta_\mu, \ldots$ are all the ordinals $< \gamma$ for which $f(\eta) = \exists$. We define the following interpretation of $g_A^{f,\gamma}$ and $\bar{g}_A^{f,\gamma}$ with respect to $\mathscr{A}$.

If $\mathbf{Q}^f\boldsymbol{x}\, A(\boldsymbol{x}, \boldsymbol{a})$ is satisfied in $\mathscr{A}$, then

1) $\forall x_{\xi_0} x_{\xi_1} \ldots A(\tilde{x}_0, \ldots, \boldsymbol{a})$, where $\tilde{x}_\gamma$ is x_{ξ_γ} if $f(\gamma) = \forall$ and $\tilde{x}_\gamma$ is $g_A^{f,\gamma}(x_{\xi_0}, \ldots, \boldsymbol{a})$ if $f(\gamma) = \exists$.

Let D be the universe of $\mathscr{A}$ and 0 be a member of D. Here $\boldsymbol{a}$ is understood to be a sequence of members of D. If $\mathbf{Q}^f\boldsymbol{x}\, A(\boldsymbol{x}, \boldsymbol{a})$ is not satisfied in $\mathscr{A}$, then the $g_A^{f,\gamma}$'s are interpreted to be the constant function 0 in $\mathscr{A}$.

If $\mathbf{Q}^{\bar{f}}\boldsymbol{x} \neg A(\boldsymbol{x}, \boldsymbol{a})$ is satisfied in $\mathscr{A}$, then

2) $\forall x_{\xi_0} x_{\xi_1} \ldots A(\tilde{x}_0, \ldots, \boldsymbol{a})$, where $\tilde{x}_\gamma$ is x_{ξ_γ} if $f(\gamma) = \exists$, and $\tilde{x}_\gamma$ is $\bar{g}_A^{f,\gamma}(x_{\xi_0}, \ldots, \boldsymbol{a})$ if $f(\gamma) = \forall$.

If $\mathbf{Q}^{\bar{f}}\boldsymbol{x} \neg A(\boldsymbol{x}, \boldsymbol{a})$ is not satisfied in $\mathscr{A}$, then the $\bar{g}_A^{f,\gamma}$'s are interpreted to be the constant function 0 in $\mathscr{A}$.

Now let P be a proof in our system. Let

$$a_0, a_1, \ldots, a_\beta, \ldots$$

be a list without repetition, of all the eigenvariables in P with $h(a_\beta) \leq h(a_\gamma)$ if $\beta < \gamma$, h being the height function. By transfinite induction on β we will define terms $t_0, t_1, \ldots, t_\beta, \ldots$ corresponding to the above list of eigenvariables. Assuming that $t_{<\beta}$ have been defined, we define t_β in the following way.

Suppose the principal formula of a_β is $\mathbf{Q}^f\boldsymbol{x}\, A(\boldsymbol{x}, \boldsymbol{b})$ and δ is the order of a_β. Let d be a free variable that precedes a_β with respect to the principal formula. With the variable d we associate a variable u in the following way. If d is not an eigenvariable, let u be d itself. Otherwise, since $h(d) < h(a_\beta)$ by our eigenvariable conditions, d occurs in the above list of eigenvariables as a_γ for some $\gamma < \beta$. By the induction hypothesis t_γ has been defined. Let the u associated with d be this t_γ.

Let b be a free variable in $\boldsymbol{b}$. A variable s associated with b is defined in the same manner as the u associated with d; recall that $h(b) < h(a_\beta)$. It should be noted that these d's and b's are the same for all auxiliary formulas of a_β by virtue of the eigenvariable conditions. Thus t_β can be defined to be $g^{f,\delta}_A(\boldsymbol{u}_1, \boldsymbol{s})$ if the order of a_β is δ and $f(\delta)$ is $\exists$, where $\boldsymbol{u}_1$ is the sequence of the u's corresponding to appropriate d's as defined above. Similarly, t_β is defined to be $\bar{g}^{f,\delta}_A(\boldsymbol{u}_2; \boldsymbol{s})$ if the order of a_β is δ and $f(\delta)$ is $\forall$, where $\boldsymbol{u}_2$ has the same meaning as $\boldsymbol{u}_1$.

Now substitute $t_0, t_1, \ldots, t_\beta, \ldots$ for $a_0, a_1, \ldots, a_\beta, \ldots$ respectively in P. Let P' be the figure thus obtained from P. The end-sequents of P' and P are the same because the end-sequent of P has no eigenvariables. We shall show that every sequent of P' is satisfied in $\mathscr{A}$; this will imply that the end sequent of P is satisfied in $\mathscr{A}$. We have only to show that if the upper sequents of an inference in P' are satisfied in $\mathscr{A}$, then the lower sequent of this inference is also satisfied in $\mathscr{A}$. Since the other cases are obvious, we only consider the inferences on quantifiers.

An introduction of $\mathbf{Q}$: left in P' is of the following form

$$3)\quad \frac{\ldots, A(\boldsymbol{u}, \boldsymbol{s}), \ldots, \Gamma \rightarrow \Delta}{\ldots, \mathbf{Q}^f \boldsymbol{x}\, A(\boldsymbol{x}, \boldsymbol{s}), \ldots, \Gamma \rightarrow \Delta},$$

where u_γ is of the form $g^{f,\gamma}_A(u_{\xi_0}, \ldots, \boldsymbol{s})$ if $f(\gamma) = \exists$.

An introduction of $\mathbf{Q}$: right in P' is of the following form

$$4)\quad \frac{\Gamma \rightarrow \Delta, \ldots, A(\boldsymbol{u}', \boldsymbol{s}), \ldots}{\Gamma \rightarrow \Delta, \mathbf{Q}^f \boldsymbol{x}\, A(\boldsymbol{x}, \boldsymbol{s}), \ldots},$$

where u'_γ is of the form $\bar{g}^{f,\gamma}_A(u'_{\eta_0}, \ldots, s)$, if $f(\gamma) = \forall$.

For 3) we have to show that

5) $\mathbf{Q}^f \boldsymbol{x}\, A(\boldsymbol{x}, \boldsymbol{s}) \rightarrow A(\boldsymbol{u}, \boldsymbol{s})$.

But this is immediate from 1). For 4) we must show that

6) $A(\boldsymbol{u}', \boldsymbol{s}) \rightarrow \mathbf{Q}^f \boldsymbol{x}\, A(\boldsymbol{x}, \boldsymbol{s})$.

Assume that $\neg \mathbf{Q}^f \boldsymbol{x}\, A(\boldsymbol{x}, \boldsymbol{s})$ holds in $\mathscr{A}$. Since $\mathscr{A}$ is determinate, $\mathbf{Q}^{\bar{f}} \boldsymbol{x} \neg A(\boldsymbol{x}, \boldsymbol{s})$ holds in $\mathscr{A}$. Therefore what we have to show is that

$$\mathbf{Q}^{\bar{f}} \boldsymbol{x} \neg A(\boldsymbol{x}, \boldsymbol{s}) \rightarrow \neg A(\boldsymbol{u}', \boldsymbol{s}).$$

But this follows from 2). This completes the proof of Theorem 23.4.

Since in this proof the determinateness of $\mathscr{A}$ was used only for 6) and since the axiom of determinateness always holds for a homogeneous quantifier, we have the following.

PROPOSITION 23.5. *Let P be a proof in our determinate logic in which every quantifier, introduced in a succedent in P, is homogeneous. Then the end-sequent of P is valid.*

Next we shall prove two versions of completeness.

THEOREM 23.6. *Let $\Gamma \to \Delta$ be a sequent. Then either there exists a cut-free proof of $\Gamma \to \Delta$ in our determinate logic or else there exists a structure $\mathcal{A}$ (possibly not determinate) such that every formula in Γ is satisfied in $\mathcal{A}$ and no formula in Δ is satisfied in $\mathcal{A}$.*

PROOF. Let D_0 be an arbitrary non-empty set containing all free variables in Γ and Δ. Let D be the closure of D_0 with respect to all the functions $g_A^{f,\gamma}$ and $\bar{g}_A^{f,\gamma}$ for all formulas A in our language, i.e., D is generated by all $g_A^{f,\gamma}$'s and $\bar{g}_A^{f,\gamma}$'s from D_0. (Actually it is sufficient if D is closed under all the functions $g_A^{f,\gamma}$ and $\bar{g}_A^{f,\gamma}$ for all subformulas A of formulas in Γ and Δ). In this proof, a member of $D - D_0$ is treated as a free variable and a member of D_0 is treated as an individual constant. Let E be the set of all formulas of the form $s = t$, where s and t are members of D. Let $(\Phi|\Psi)$ be an arbitrary decomposition of E and consider the following sequent:

0) $\Phi, \Gamma \to \Delta, \Psi$.

If all the sequents of the form 0) are provable without the cut rule, then $\Gamma \to \Delta$ is also provable without the cut rule.

Let S be $\Gamma \to \Delta$. We shall define a tree $T(S)$ by considering the following eight cases.

1) The lowest sequent is S.

2) Immediate ancesters of S are all the sequents of the form 0).

3) When a sequent $\Pi \to \Lambda$ is

$$\{\neg C_\lambda\}_{\lambda<\gamma}, \Gamma' \to \Delta', \{\neg D_\mu\}_{\mu<\delta},$$

where Γ' and Δ' have no formulas whose outermost logical symbol is $\neg$, and $\Pi \to \Lambda$ is constructed by 2) or 8) (which is to be defined) the immediate ancestor of $\Pi \to \Lambda$ is

$$\{D_\mu\}_{\mu<\delta}, \Gamma' \to \Delta', \{C_\lambda\}_{\lambda<\gamma}.$$

4) When a sequent $\Pi \to \Lambda$ is

$$\left\{\bigvee_{\lambda<\alpha_\mu} C_{\mu,\lambda}\right\}_{\mu<\gamma}, \Gamma' \to \Delta', \left\{\bigvee_{\rho<\beta_\sigma} D_{\sigma,\rho}\right\}_{\sigma<\delta},$$

where Γ' and Δ' have no formulas whose outermost logical symbol is $\bigvee$, and when $\Pi \to \Lambda$ is constructed by 3), the immediate ancestors of $\Pi \to \Lambda$ are

$$\{C_{\lambda_\mu,\mu}\}_{\mu<\gamma}, \Gamma' \to \Delta', \{D_{\sigma,\rho}\}_{\rho<\delta_\sigma, \sigma<\delta}$$

for all sequences $\{\lambda_\mu\}_{\mu<\gamma}$ such that $\lambda_\mu < \alpha_\mu$.

5) When a sequent $\Pi \to \Lambda$ is

$$\Big\{\bigwedge_{\lambda<\alpha_\mu} C_{\mu,\lambda}\Big\}_{\mu<\gamma}, \Gamma' \to \Delta', \Big\{\bigwedge_{\rho<\beta_\sigma} D_{\rho,\sigma}\Big\}_{\sigma<\delta},$$

where Γ' and Δ' have no formulas whose outermost logical symbol is $\bigwedge$, and when $\Pi \to \Lambda$ is constructed by 4), then the immediate ancestors of $\Pi \to \Lambda$ are

$$\{C_{\mu,\lambda}\}_{\lambda<\gamma_\mu,\mu<\gamma}, \Gamma' \to \Delta', \{D_{\rho_\sigma,\sigma}\}_{\sigma<\delta}$$

for all sequences $\{\rho_\sigma\}_{\sigma<\delta}$ such that $\rho_\sigma < \beta_\sigma$.

6) When a sequent $\Pi \to \Lambda$ is

$$\{\mathbf{Q}^{f_\lambda}\boldsymbol{x}_\lambda\, A_\lambda(\boldsymbol{x}_\lambda, \boldsymbol{s}_\lambda)\}_{\lambda<\delta}, \Gamma' \to \Delta',$$

where Γ' has no formulas whose outermost logical symbol is $\mathbf{Q}$, and when $\Pi \to \Lambda$ is constructed by 5), then the immediate ancestor of $\Pi \to \Lambda$ is

$$\{A_\lambda(\boldsymbol{t}_{\lambda,\mu}\boldsymbol{s}_\lambda\}_{\mu,\lambda<\delta}\, \Gamma' \to \Delta'$$

for all $\boldsymbol{t}_{\lambda,\mu}$ satisfying the following:

$$\boldsymbol{t}_{\lambda,\mu} \quad \text{is} \quad \{t_{\lambda,\mu,0}, \ldots, t_{\lambda,\mu,\nu}, \ldots\}_{\nu<\gamma},$$

where γ is the length of $\boldsymbol{x}_\lambda$ and if $\xi_0, \xi_1, \ldots,$ are all the ordinals $<\gamma$ such that $f(\xi) = \forall$ and $\eta_0, \eta_1, \ldots$ are all the ordinals $<\gamma$ such that $f(\xi) = \exists$, then $t_{\lambda,\mu,\xi_0} t_{\lambda,\mu,\xi_1}, \ldots$ is an arbitrary sequence of members of D and

$$t_{\lambda,\mu,\eta} = g_{A_\lambda}^{f,\eta}(t_{\lambda,\mu,\xi_0}, \ldots, \boldsymbol{s}_\lambda),$$

for each $\eta = \eta_0, \eta_1, \ldots$.

7) When a sequent $\Pi \to \Lambda$ is

$$\Gamma' \to \Delta', \{\mathbf{Q}^{f_\lambda}\boldsymbol{x}_\lambda\, A_\lambda(\boldsymbol{x}_\lambda, \boldsymbol{s}_\lambda)\}_{\lambda<\delta},$$

where Δ' has no formulas whose outermost logical symbol is $\mathbf{Q}$, and when $\Pi \to \Lambda$ is constructed by 6), then the immediate ancestor of $\Pi \to \Lambda$ is

$$\Gamma' \to \Delta', \{A_\lambda(\boldsymbol{t}_{\lambda,\mu}, \boldsymbol{s}_\lambda)\}_{\mu,\lambda<\delta}$$

for all $\boldsymbol{t}_{\lambda,\mu}$ satisfying the following:

$$\boldsymbol{t}_{\lambda,\mu} \quad \text{is} \quad \{t_{\lambda,\mu,0}, \ldots, t_{\lambda,\mu,\nu}, \ldots\}_{\nu<\gamma},$$

where γ is the length of $\boldsymbol{x}_\lambda$; if $\xi_0, \xi_1, \ldots,$ are all the ordinals $<\gamma$ such that $f(\xi) = \forall$, and, if $\eta_0, \eta_1, \ldots$ are all the ordinals $<\gamma$ such that $f(\eta) = \exists$, then $t_{\lambda,\mu,\eta_0}, t_{\lambda,\mu,\eta_1}, \ldots$ are arbitrary members of D and $t_{\lambda,\mu,\xi} = \bar{g}_{A_\lambda}^{f,\xi}(t_{\lambda,\mu,\eta_0}, \ldots, \boldsymbol{s}_\lambda)$.

8) When a sequent $\Pi \to \Lambda$ is

$$\{s_\lambda = t_\lambda\}_{\lambda<\beta}, \Gamma' \to \Delta',$$

where Γ' has no formulas of the form $s = t$ and when $\Pi \to \Lambda$ is constructed by 7), then the immediate ancestor of $\Pi \to \Lambda$ is the sequent $\Pi' \to \Lambda'$, where Π' and Λ' are sequences of all the formulas obtained from a formula in Π and Λ, respectively, by arbitrary interchange of s_λ and t_λ ($\lambda < \beta$). (So Π' and Λ' obviously include Π and Λ, respectively.)

This completes the description of $T(S)$.

A branch of $T(S)$ is an infinite sequence $S = S_0, S_1, S_2, \ldots$ such that S_{n+1} is an immediate ancestor of S_n. We have two cases.

Case 1. In every branch of $T(S)$, there exists at least one sequent of the form

$$\Gamma_1, D, \Gamma_2 \rightarrow \Delta_1, D, \Delta_2 \quad \text{or} \quad \Gamma \rightarrow \Delta_1, s = s, \Delta_2.$$

Case 2. There exists at least one branch of $T(S)$, in which there are no sequents of the form

$$\Gamma_1, D, \Gamma_2 \rightarrow \Gamma_1, D, \Delta_2 \quad \text{or} \quad \Gamma \rightarrow \Delta_1, s = s, \Delta_2.$$

For case 1, S is provable without the cut rule. In order to prove this we define the height of the free variables as follows.

(1) If a belongs to D_0, then $h(a) = 0$.

(2) If a is $g_A^{f_A \gamma}(b_0, \ldots, b_\xi, \ldots)$ or $\bar{g}_A^{f_A \gamma}(b_0, \ldots, b_\xi, \ldots)$, then $h(a)$ is the supremum of all $h(b_\xi) + 1$'s.

It is easily shown that $T(S)$ satisfies the conditions 1) and 3) in (3) of Definition 23.2.

In the remainder of this proof we will refer to a figure P as a semi-proof if P satisfies all the conditions of a proof except 4) of Definition 23.2. P is said to be a quasi-proof if P satisfies all the conditions of a proof except 3) in (3) of Definition 23.2.

We now consider the following conditions on P.

(3) P is a cut-free semi-proof.

(4) Every free variable in P occurs in $T(S)$ and every inference on $\mathbf{Q}$ in P occurs in $T(S)$.

(5) The end sequent of P is S.

If P satisfies (3), (4) and (5) then P obviously satisfies 1) and 3) in (3) of Definition 23.2 and therefore P is a cut-free quasi-proof. Now consider the condition C on a sequent S', that S' has a quasi-proof P satisfying (3), (4) and (5). Let S' be in $T(S)$. It is easily seen that if every ancestor of S' satisfies C, then S' satisfies C. Suppose that S is not provable without the cut rule. Then S does not satisfy C. (Recall that the height is defined). Then some ancestor of S, say S_1, does not satisfy C. Continuing this argument, we obtain a sequence $S, S_1, S_2, \ldots$, where S_{n+1} is an ancestor of S_n and does not satisfy C for each n. This contradicts the hypothesis of case 1.

For case 2, we will show that there exists a structure $\mathscr{A}$ in which every formula in Γ is true and every formula in Δ is false. In the rest of this proof, we fix one branch, whose existence is assumed in the hypothesis of case 2, and consider only the formulas and sequents in this branch, that is, throughout this discussion a sequent always means a sequent in this branch. We only have to define an interpretation which makes all the sequents in this branch false with respect to D.

LEMMA 23.7. (1) *If a formula $\neg A$ occurs in the antecedent (succedent) of a sequent, then the formula A occurs in the succedent (antecedent) of a sequent.*

(2) *If a formula $\bigvee_{\lambda<\beta} A_\lambda$ occurs in the antecedent (succedent) of a sequent, then a formula A_λ for some (every) $\lambda < \beta$ occurs in the antecedent (succedent) of a sequent.*

(3) *If a formula $\bigwedge_{\lambda<\beta} A_\lambda$ occurs in the antecedent (succedent) of a sequent, then a formula A_λ for every (some) $\lambda < \beta$ occurs in the antecedent (succedent) of a sequent.*

(4) *If $\mathbf{Q}^f\mathbf{x}\, A(\mathbf{x}, s)$ occurs in the antecedent of a sequent and $\xi_0, \xi_1, \ldots$ are all ordinals such that $f(\xi) = \forall$ and $\eta_0, \eta_1, \ldots$ are all ordinals such that $f(\eta) = \exists$, then for an arbitrary sequence $t_{\xi_0}, t_{\xi_1}, \ldots$ of members of D, the formula $A(\mathbf{t})$ occurs in the antecedent of a sequent, where $t_\eta = g_A^{f,\eta}(t_{\xi_0}, \ldots, \mathbf{s})$ for each $\eta = \eta_0, \eta_1, \ldots$.*

(5) *If $\mathbf{Q}^f\mathbf{x}\, A(\mathbf{x}, \mathbf{s})$ occurs in the succedent of a sequent and $\xi_0, \xi_1, \ldots$ are all ordinals such that $f(\xi) = \forall$ and $\eta_0, \eta_1, \ldots$ are all ordinals such that $f(\eta) = \exists$, then for an arbitrary sequence $t_{\eta_0}, t_{\eta_1}, \ldots$ of members of D, the formula $A(\mathbf{t})$ occurs in the succedent of a sequent, where $t_\xi = \bar{g}_A^{f,\xi}(t_{\eta_0}, \ldots, \mathbf{s})$ for each $\xi = \xi_0, \xi_1, \ldots$.*

PROOF. Obvious.

LEMMA 23.8. *If a formula occurs in the antecedent of a sequent, then it does not occur in the succedent of any sequent.*

PROOF. By transfinite induction on the complexity of formulas using Lemma 23.7.

LEMMA 23.9. (1) *For every member t of D, the formula $t = t$ occurs in the antecedent of a sequent.*

(2) *Let s and t be members of D. If $s = t$ occurs in the antecedent of a sequent, then $t = s$ occurs in the antecedent of a sequent.*

(3) *Let t_1, t_2 and t_3 be members of D. If $t_1 = t_2$ and $t_2 = t_3$ occur in the antecedent of a sequent, then the formula $t_1 = t_3$ occurs in an antecedent of a sequent.*

(4) *Let s_λ, t_λ, $\lambda < \beta$, be members of D. If $A(s_0, \ldots, s_\lambda, \ldots)$ and $\{s_\lambda = t_\lambda\}_{\lambda<\beta}$ occur in the antecedent of a sequent, then $A(u_0, \ldots, u_\lambda, \ldots)$ occurs in the antecedent of a sequent for each sequence $u_0, \ldots, u_\lambda, \ldots$ such that u_λ is s_λ or t_λ.*

PROOF. (1) $t = t$ must be contained in Φ or Ψ in 2) of the tree construction. Since $t = t$ cannot be contained in Ψ because of the hypothesis of case 2, $t = t$ must be contained in Φ.

(2) Let $s = t$ occur in the antecedent of a sequent and $t = s$ occur in the succedent of a sequent, then there is a sequent which contains $s = t$ in the antecedent and $t = s$ in the succedent. By the construction 8) of $T(S)$, there must be a sequent of the form $\Gamma_1 \rightarrow \Delta_1,\ s = s, \Delta_2$. This is a contradiction.

(3) and (4) can be proved similarly.

According to Lemma 23.9, D can be decomposed into equivalence classes by $=$. Let $D_=$ be the set of equivalence classes so obtained; from now on we will denote a class of $D_=$ by a representative of it. We define a structure $\mathscr{A}$ over $D_=$ as follows. Let s be a variable in D. Then the value of s with respect to $\mathscr{A}$ is defined to be the class represented by s. If P is a predicate constant, then $P(t_0, \ldots, t_\lambda, \ldots)$ is defined to be true with respect to $\mathscr{A}$ if $P(t_0, \ldots, t_\lambda, \ldots)$ is in the antecedent of a sequent and is defined to be false with respect to $\mathscr{A}$ otherwise. By transfinite induction on the complexity of A, we shall prove that A is true with respect to $\mathscr{A}$ if A is in the antecedent of a sequent and A is false with respect to $\mathscr{A}$ if A is in the succedent of a sequent. Since the other cases are easy, we only consider the cases where A is $\mathbf{Q}^f \boldsymbol{x}\, A(\boldsymbol{x}, \boldsymbol{s})$.

Case 1. $\mathbf{Q}^f \boldsymbol{x}\, A(\boldsymbol{x}, \boldsymbol{s})$ occurs in the antecedent of a sequent. In this case, it follows from the induction hypothesis and 6) of the construction of $T(S)$, that $A(\boldsymbol{t}, \boldsymbol{s})$ is true with respect to $\mathscr{A}$ for every $\boldsymbol{t}$ satisfying the following condition. If $\xi_0, \xi_1, \ldots$ are all the ordinals such that $f(\xi) = \forall$ and $\eta_0, \eta_1, \ldots$ are all the ordinals such that $f(\eta) = \exists$, then $t_\eta = g^{f,\eta}_{A}(t_{\xi_0}, \ldots, \boldsymbol{s})$ for every η. This implies that $\mathbf{Q}^f \boldsymbol{x}\, A(\boldsymbol{x}, \boldsymbol{s})$ is true with respect to $\mathscr{A}$.

Case 2. $\mathbf{Q}^f \boldsymbol{x}\, A(\boldsymbol{x}, \boldsymbol{s})$ is in the succedent of a sequent. In this case, it follows from the induction hypothesis and 7) of the construction of $T(S)$, that $A(\boldsymbol{t}, \boldsymbol{s})$ is false with respect to $\mathscr{A}$ for every $\boldsymbol{t}$ satisfying the following condition. If $\xi_0, \xi_1, \ldots$ are all the ordinals such that $f(\xi) = \forall$ and $\eta_0, \eta_1, \ldots$ are all the ordinals such that $f(\eta) = \exists$, then $t_\xi = \bar{g}^{f,\xi}_{A}(t_{\eta_0}, \ldots, \boldsymbol{s})$. This implies that $\neg A(\boldsymbol{t}, \boldsymbol{s})$ is true with respect to $\mathscr{A}$ for every such $\boldsymbol{t}$. Therefore $\mathbf{Q}^{\bar{f}} \boldsymbol{x} \neg A(\boldsymbol{x}, \boldsymbol{s})$ is true with respect to $\mathscr{A}$. Since $\mathbf{Q}^{\bar{f}} \boldsymbol{x} \neg A(\boldsymbol{x}, \boldsymbol{s}) \rightarrow \neg \mathbf{Q}^f \boldsymbol{x}\, A(\boldsymbol{x}, \boldsymbol{s})$ is satisfied in all the structures, $\mathbf{Q}^f \boldsymbol{x}\, A(\boldsymbol{x}, \boldsymbol{s})$ is false with respect to $\mathscr{A}$.

This completes the proof of our first version of completeness.

Before we proceed to the second version of completeness, we shall first prove the following.

PROPOSITION 23.10. *Let D and D_0 be the same as in the proof of Theorem 23.6. Γ_0 is defined to be the sequence consisting of all formulas of the form*

$$\mathbf{Q}^f \boldsymbol{x}\, A(\boldsymbol{x}, \boldsymbol{s}) \vee \mathbf{Q}^{\bar{f}} \boldsymbol{x} \neg A(\boldsymbol{x}, \boldsymbol{s})$$

where $A(\boldsymbol{x}, \boldsymbol{s})$ is an arbitrary formula in our language and $\boldsymbol{s}$ is an arbitrary sequence of members of D. Without loss of generality, we may assume that no member of D_0 is ever used as an eigenvariable in any quasi-proof. Now let $\Gamma \rightarrow \Delta$ be a sequent of the original language and let $\tilde{\Gamma}$ be Γ_0, Γ. Then either there is a cut-free quasi-proof whose end-sequent is $\tilde{\Gamma} \rightarrow \Delta$ or else there exists a determinate structure $\mathscr{A}$ such that every formula in $\tilde{\Gamma}$ is satisfied in $\mathscr{A}$ and no formula in Δ is satisfied in $\mathscr{A}$.

PROOF. This is proved similarly to the proof of Theorem 23.6 by replacing "proof" and "Γ" by "quasi-proof" and "$\tilde{\Gamma}$", respectively. Since $\tilde{\Gamma}$ includes Γ_0, it is easily shown that $\mathscr{A}$ is determinate.

THEOREM 23.11. *Let $\Gamma \rightarrow \Delta$ be a sequent. Then either $\Gamma \rightarrow \Delta$ is provable in our determinate logic or there exists a determinate structure $\mathscr{A}$ such that every formula in Γ is satisfied in $\mathscr{A}$ and no formula in Δ is satisfied in $\mathscr{A}$.*

PROOF. Since every formula in Γ_0 is provable in our determinate logic, (cf. Example 23.3) $\Gamma \rightarrow \Delta$ is obtained from $\tilde{\Gamma} \rightarrow \Delta$ by the cut rule as follows.

$$\frac{\rightarrow B_0, \ldots \qquad \rightarrow B_\beta, \ldots \qquad B_0, \ldots, B_\beta, \ldots, \Gamma \rightarrow \Delta}{\Gamma \rightarrow \Delta},$$

where $\{B_0, \ldots, B_\beta, \ldots\}$ is Γ_0. Thus, if $\tilde{\Gamma} \rightarrow \Delta$ has a quasi-proof, then from this quasi-proof we can obtain a proof of $\Gamma \rightarrow \Delta$, since $\Gamma \rightarrow \Delta$ is a sequent of the original language. Otherwise Proposition 23.10 guarantees that there is a determinate structure $\mathscr{A}$ in which every formula of $\tilde{\Gamma}$, and hence every formula of Γ, is satisfied, while no formula of Δ is satisfied.

REMARK. We cannot improve Theorem 23.11 by replacing "provable" by "provable without the cut rule". This is clear from the following example by Gale and Stewart. Let α_0 be the cardinal number of 2^ω, the set of functions from ω to 2. Let $f \in 2^\omega$. Then $\psi(f)$ is defined to be $a_0 = i_0, a_1 = i_1, \ldots,$ where $i_k = 0$ or 1 according as $f(k) = 0$ or 1. The formula $\psi(f)$ implicitly defines the function f. If $A \subseteq 2^\omega$, then A is implicitly defined by the formula $\bigvee_{f \in A} \psi(f)$ where $\bigvee_{f \in A}$ is defined in terms of $\bigvee_{\alpha_0}$. It can be shown that there exists a set $A \subseteq 2^\omega$ such that the axiom of determinateness fails for the game defined by A. (The proof is given below.) If a formula ψ implicitly defines A, then

$$\forall x\,(x = 0 \vee x = 1) \rightarrow 0 = 1,$$

$$\neg(\forall x_0\, \exists x_1\, \forall x_2 \ldots \psi(x_0, x_1, \ldots) \vee \exists x_0\, \forall x_1\, \exists x_2 \ldots \neg\psi(x_0, x_1, \ldots))$$

is provable in our determinate logic, where ψ is constructed from $0, 1, =, \bigwedge_{\alpha_0}$ and $\bigvee_{\alpha_0}$. This means that $\forall x\,(x = 0 \vee x = 1) \rightarrow 0 = 1$ is provable in our determinate logic if our language has $\bigvee_{\alpha_0}$, since the negation of the last formula is an instance of the axiom of determinateness. However, this is not provable without the cut rule even if our language has $\bigvee_{\alpha_0}$.

The proof of the existence of A goes as follows. We shall show that there is a subset A of 2^ω for which there is no winning strategy.

DEFINITION 23.12. (1) For any subset of 2^ω, say A, $G(A)$, a game for A, is defined as follows: A first player I and a second player II alternately chooses a 0 or 1; thus

I: $x_0 x_2 x_4 \ldots x_{2i} \ldots$

II: $x_1 x_3 x_5 \ldots x_{2i+1} \ldots$

for $i < \omega$.

(2) the sequence $\langle x_0, x_1, x_2, \ldots \rangle$ generated in this manner, called a play of the game, determines the winner, that is, if $\langle x_0, x_1, x_2, \ldots \rangle \in A$, then I wins, otherwise II wins.

(3) A sequence $\langle x_0, f_2, f_4, \ldots, f_{2i}, \ldots \rangle$ $i < \omega$ is called a strategy for I if $x_0 \in 2$ and f_{2i} is a function from all i-tuples of 0's and 1's to 2.

(4) Let $\sigma = \langle x_0, f_2, f_4, \ldots \rangle$ be a strategy for I and let $x = \langle x_1, x_3, \ldots, x_{2i+1}, \ldots \rangle$ be a function from odd numbers to 2. Then $\sigma(x)$ is defined by

$$\sigma(x) = \langle x_0, x_1, f_2(x_1), x_3, f_4(x_1, x_3), \ldots \rangle.$$

(5) A strategy σ for I is called a winning strategy for I if

$$\forall x \in 2^{\{2i+1 \mid i<\omega\}}\ \sigma(x) \in A.$$

(6) A sequence $\langle f_1, f_3, \ldots, f_{2i+1}, \ldots \rangle$ is called a strategy for II if f_{2i+1} is a function from all $(i+1)$-tuples of 0's and 1's for every $i < \omega$.

(7) Let $\tau = \langle f_1, f_3, \ldots \rangle$ be a strategy for II and let $x = \langle x_0, x_2, \ldots, x_{2i}, \ldots \rangle$ be a function from even numbers to 2. Then $\tau(x)$ is defined by

$$\tau(x) = \langle x_0, f_1(x_0), x_2, f_3(x_0, x_2), \ldots \rangle.$$

(8) A strategy τ for II is called a winning strategy for II if $\forall x \in 2^{\{2i \mid i<\omega\}}\ \tau(x) \notin A$.

THEOREM 23.13 (Gale-Stewart). *In* ZF, *we can show that if* 2^ω *is well-ordered, then there exists a subset of* 2^ω, *say* A, *for which neither* I *nor* II *has a winning strategy in the game* G(S).

PROOF. It is easy to see that

1. The cardinality of all strategies for I is α_0, and
2. The cardinality of all strategies for II is α_0.
3. If σ is a strategy for I, then the cardinality of the set $\{(\sigma, \tau) \mid \tau$ is a strategy for II$\}$ is α_0.
4. If τ is a strategy for II, then the cardinality of $\{(\sigma, \tau) \mid \sigma$ is a strategy for I$\}$ is α_0.

Let $\sigma_0, \sigma_1, \dots, \sigma_\alpha, \dots, \alpha < \alpha_0$ and $\tau_0, \tau_1, \dots, \tau_\alpha, \dots, \alpha < \alpha_0$, be enumerations of all strategies for I and II, respectively. By transfinite induction we define plays

$$x^\alpha = \langle x_0^\alpha, x_1^\alpha, \dots, x_i^\alpha, \dots \rangle, \qquad y^\alpha = \langle y_0^\alpha, y_1^\alpha, \dots, y_i^\alpha, \dots \rangle,$$

where $x_i^\alpha, y_i^\alpha = 0$ or 1:

(1) $x^0 = (\sigma_0, \tau_0)$.

(2) $y^0 = (\sigma_0, \tau_\beta)$, where β is the smallest ordinal such that $(\sigma_0, \tau_\beta) \neq x^0$.

(3) $S_\alpha = \{x^\beta \mid \beta < \alpha\}$ and $T_\alpha = \{y^\beta \mid \beta < \alpha\}$.

(4) $x^\alpha = (\sigma_\beta, \tau_\beta)$, where β is the smallest ordinal such that $(\sigma_\beta, \tau_\alpha), \notin S_\alpha \cup T_\alpha$.

(5) $y^\alpha = (\sigma_\alpha, \tau_\beta)$, where β is the smallest ordinal such that $(\sigma_\alpha, \tau_\beta) \notin S_\alpha \cup T_\alpha$ and $(\sigma_\alpha, \tau_\beta) \neq x^\alpha$. It is obvious that if $\alpha < \alpha_0$, then $S_\alpha \cap T_\alpha = \emptyset$, $\bar{\bar{S}}_\alpha < \alpha_0$, and $\bar{\bar{T}}_\alpha < \alpha_0$.

(6) $A = \bigcup_{\alpha < \alpha_0} S_\alpha$.

We claim that for this A neither I nor II has a winning strategy. Suppose that I has a winning strategy, say σ_α. Let β be the smallest ordinal such that

$$(\sigma_\alpha, \tau_\beta) \notin S_\alpha \cup T_\alpha \wedge (\sigma_\alpha, \tau_\beta) \neq x^\alpha.$$

Then $(\sigma_\alpha, \tau_\beta) = y^\alpha \notin A$, which means that II has a winning strategy, yielding a contradiction.

In order to prove the interpolation theorem, we need the following proof-theoretic notion.

DEFINITION 23.14. Let P be a cut-free semi-proof and let I be an inference in P. Let A be a formula in an upper sequent of I and B be a formula in the lower sequent of I. B is said to be the immediate successor of A if the following is satisfied.

Case 1. If I is a structural inference

$$\frac{\Gamma \to \Delta}{\Pi \to \Lambda},$$

and A is a formula of $\Gamma(\Delta)$, then B is the first formula in $\Pi(\Lambda)$ which is identical with A.

Case 2. If I is a logical inference

$$\frac{\Pi, \Gamma \to \Delta, \Lambda}{\Pi', \Gamma \to \Delta, \Lambda'},$$

where I applies to the formulas of Π and Λ, and A is in $\Gamma(\Lambda)$, then B is the corresponding formula in $\Gamma(\Lambda)$.

Case 3. If I is a logical inference and A is an auxiliary formula of I, then B is the corresponding principal formula.

Case 4. If I is the first equality rule (cf. Definition 22.19) and A is a formula in $\Gamma^{(a)}$ $(\Delta^{(a)})$ then B is the corresponding formula in $\Gamma^{(b)}$ $(\Delta^{(b)})$.

Case 5. If I is the second equality rule, (cf. Definition 22.19) and A is a formula in $\Gamma(\Delta)$, then B is the corresponding formula in $\Gamma(\Delta)$.

Our interpolation theorem is then stated in the following form.

THEOREM 23.15 (an interpolation theorem for homogeneous languages). *If a sequent $\Gamma_1, \Gamma_2 \to \Delta_1, \Delta_2$ is valid and has no heterogeneous quantifiers, then there exists a formula C such that both the sequents*

$$\Gamma_1 \to \Delta_1, C \quad \textit{and} \quad C, \Gamma_2 \to \Delta_2$$

are valid and every free variable or predicate constant in C, except $=$, occurs in both Γ_1, Δ_1 and Γ_2, Δ_2. (C may have heterogeneous quantifiers and also logical connectives or quantifiers that are longer than the logical symbols in the original language).

PROOF. The proof will be divided into several parts.

1. First we shall introduce two auxiliary systems.

DEFINITION 23.16. A proof P in our determinate logic is said to satisfy condition (Q) if every inference I in P of $\mathbf{Q}$: right is either homogeneous or is of the form

(Q)
$$\frac{\Gamma \to \Delta, A(\boldsymbol{d})}{\Gamma \to \Delta, Q^f \boldsymbol{x}\, A(\boldsymbol{x})},$$

where no eigenvariable in P used above $\Gamma \to \Delta, \mathbf{Q}^f \boldsymbol{x}\, A(\boldsymbol{x})$ occurs in $\Gamma \to \Delta$, $\mathbf{Q}^f \boldsymbol{x}\, A(\boldsymbol{x})$.

PROPOSITION 23.17. *If a sequent S is provable with a proof which satisfies* (Q), *then S is valid.*

PROOF. Define $g^{f,\gamma}_{A}$ and $\bar{g}^{f,\gamma}_{A}$ as in the proof of Theorem 23.4 except that $\bar{g}^{f,\gamma}_{A}$ is defined only for homogeneous f. Then define substitution also as in the proof of Theorem 23.4 except that all eigenvariables in the inference of (Q) remain unsubstituted. Then P will be transformed into P'. What we have to show is that every sequent S' in P' is satisfied in $\mathscr{A}$. This is shown by transfinite induction on the complexity of the semi-proof of S. We can repeat the proof of Theorem 23.4 except in the following case. S is inferred by the inference I:

$$\frac{\Gamma \to \Delta, A(\boldsymbol{d}, \boldsymbol{b})}{\Gamma \to \Delta, \mathbf{Q}^f \boldsymbol{x}\, A(\boldsymbol{x}, \boldsymbol{b})},$$

where $\mathbf{Q}^f$ is not homogeneous. In order to illustrate the proof, we assume that $\mathbf{Q}^f \boldsymbol{x}$ is $\forall x_0 \exists x_1 \forall x_2 \exists x_3 \ldots$ and $\boldsymbol{d}$ is $d_0, d_1, d_2, \ldots$. Since I satisfies (Q) and $h(d_0) < h(d_1) < h(d_2) < \ldots$, $(\Gamma \to \Delta, A(\boldsymbol{d}, \boldsymbol{b}))'$ is of the form

$$(*) \qquad \Gamma' \to \Delta', A(d_0, t_1(d_0, \boldsymbol{s}), d_2, t_3(d_0, d_2, \boldsymbol{s}), \ldots, \boldsymbol{s}).$$

It follows from the induction hypothesis that (*) is satisfied in $\mathscr{A}$ for every sequence $d_0, d_2, d_4, \ldots$ of members of $\mathscr{A}$. Therefore $\Gamma' \to \Delta', \mathbf{Q}^f \boldsymbol{x}\, A(\boldsymbol{x}, \boldsymbol{s})$ is satisfied in $\mathscr{A}$.

Next we shall consider another logical system, the restricted homogeneous system **RHS**.

DEFINITION 23.18. A figure P is said to be a proof in **RHS** if P satisfies the following conditions:

1) All quantifiers in P are $\exists$.

2) P satisfies all conditions of a proof of determinate logic except (3) of Definition 23.2.

3) Every inference in P on the introduction of $\mathbf{Q}$ in the antecedent is of the following form

$$\frac{\{A_\lambda(\boldsymbol{a}_\lambda)\}_{\lambda<\gamma}, \Gamma \to \Delta}{\{\exists \boldsymbol{x}_\lambda\, A_\lambda(\boldsymbol{x}_\lambda)\}_{\lambda<\gamma}, \Gamma \to \Delta},$$

where no variable in $\boldsymbol{a}_\lambda$ occurs in the lower sequent.

Then we have the following proposition.

PROPOSITION 23.19. *If $\Gamma \to \Delta$ is provable in* **RHS** *and height (see Definition 23.2) is defined for all free variables in $\Gamma \to \Delta$, then there exists a proof P' in* **RHS** *ending with $\Gamma \to \Delta$ for which the heights are defined in such a way that the free variables in $\Gamma \to \Delta$ have the same heights as the original ones.*

PROOF. We may assume that the same eigenvariable is never used in two different places. (Otherwise, we can reletter some eigenvariables.) Then it is easy to define heights of free variables from the bottom.

2. Next we prove the following lemma.

LEMMA 23.20. *Let P be a cut-free proof of $\Gamma_1, \Gamma_2 \to \Delta_1, \Delta_2$ in the homogeneous system (see Definition 22.1), a proof satisfying the following conditions:*

(1) *Every quantifier in P is $\exists$.*

(2) *Every* $\mathbf{Q}$*-introduction inference in P is a $\exists$-introduction inference in the succedent.*

Then there exist cut-free proofs P_1 and P_2 in **RHS** *and a formula C satisfying the following conditions.*

(2.1) *The end-sequent of* P_1 *is* $C, \Gamma_1 \to \Delta_1$ *and the end-sequent of* P_2 *is* $\Gamma_2 \to \Delta_2, C$.

(2.2) *Every free variable or predicate constant in* C, *except* $=$, *occurs in both* Γ_1, Δ_1 *and* Γ_2, Δ_2.

PROOF. The proof is by transfinite induction on the complexity of P.

Case 1. P consists of a single initial sequent. The theorem is obvious.

Case 2. The last inference of P is of the form

$$\frac{\Gamma_1, \Gamma_2 \to \Delta'_1, \{A_\lambda(\boldsymbol{a}_\lambda)\}_{\lambda<\beta_1}, \Delta'_2, \{B_\mu(\boldsymbol{b}_\mu)\}_{\mu<\beta_2}}{\Gamma_1, \Gamma_2 \to \Delta'_1, \{\exists \boldsymbol{x}_\lambda\, A_\lambda(\boldsymbol{x}_\lambda)\}_{\lambda<\beta_1}, \Delta'_2, \{\exists \boldsymbol{y}_\mu\, B_\mu(\boldsymbol{y}_\mu)\}_{\mu<\beta_2}},$$

where Δ_1 is $\Delta'_1, \{\exists \boldsymbol{x}_\lambda\, A_\lambda(\boldsymbol{x}_\lambda)\}_{\lambda<\beta_1}$ and Δ_2 is $\Delta'_2, \{\exists \boldsymbol{y}_\mu\, B_\mu(\boldsymbol{y}_\mu)\}_{\mu<\beta_2}$.

By the induction hypothesis, there exists a $C'(\boldsymbol{a}, \boldsymbol{b})$ for which

$$C'(\boldsymbol{a}, \boldsymbol{b}), \Gamma_1 \to \Delta'_1, \{A_\lambda(\boldsymbol{a}_\lambda)\}_{\lambda<\beta_1}$$

and

$$\Gamma_2 \to \Delta'_2, \{B_\mu(\boldsymbol{b}_\mu)\}_{\mu<\beta_2}, C'(\boldsymbol{a}, \boldsymbol{b})$$

are provable in **RHS**.

Moreover, every free variable and predicate constant in $C'(\boldsymbol{a}, \boldsymbol{b})$ is either $=$ or contained in both $\Gamma_1, \Delta'_1, \{A_\lambda(\boldsymbol{a}_\lambda)\}_{\lambda<\beta_1}$ and $\Gamma_2, \Delta'_2, \{B_\mu(\boldsymbol{b}_\mu)\}_{\mu<\beta_2}$. Here $\boldsymbol{a}$ is a sequence of all the variables in $C'(\boldsymbol{a}, \boldsymbol{b})$ which are not in Γ_1, Δ_1 and $\boldsymbol{b}$ is a sequence of all the variables in $C'(\boldsymbol{a}, \boldsymbol{b})$ which are not in Γ_2, Δ_2. Then the required formula C is $\exists \boldsymbol{x}\, \forall \boldsymbol{y}\, C'(\boldsymbol{x}, \boldsymbol{y})$, where $\forall$ is considered as an abbreviation of $\neg\exists\neg$.

Case 3. The last inference of P is of the form

$$\frac{\Gamma_1'^{(\boldsymbol{a})}, \Gamma_2'^{(\boldsymbol{a})} \to \Delta_1^{(\boldsymbol{a})}, \Delta_2^{(\boldsymbol{a})}}{\boldsymbol{a}_1 = \boldsymbol{b}_1, \boldsymbol{a}_2 = \boldsymbol{b}_2, \Gamma_1'^{(\boldsymbol{b})}, \Gamma_2'^{(\boldsymbol{b})} \to \Delta_1^{(\boldsymbol{b})}, \Delta_2^{(\boldsymbol{b})}}$$

where Γ_1 is $\boldsymbol{a}_1 = \boldsymbol{b}_1, \Gamma_1'^{(\boldsymbol{b})}$ and Γ_2 is $\boldsymbol{a}_2 = \boldsymbol{b}_2, \Gamma_2'^{(\boldsymbol{b})}$. This can be divided into two steps; first, the substitution of $\boldsymbol{a}_1$ for $\boldsymbol{b}_1$; then the substitution of $\boldsymbol{a}_2$ for $\boldsymbol{b}_2$. So we may assume that $\boldsymbol{a}_1 = \boldsymbol{b}_1$ is empty. By the induction hypothesis, there exists a formula $C'(\boldsymbol{a}, \boldsymbol{b})$ which satisfies the lemma for $\Gamma_1'^{(\boldsymbol{a})}, \Gamma_2'^{(\boldsymbol{a})} \to \Delta_1^{(\boldsymbol{a})}, \Delta_2^{(\boldsymbol{a})}$, where $\boldsymbol{a}$ is a sequence consisting of all variables in $C'(\boldsymbol{a}, \boldsymbol{b})$ which are not in Γ_1, Δ_1 and $\boldsymbol{b}$ is a sequence of all the variables in $C'(\boldsymbol{a}, \boldsymbol{b})$ which are not in Γ_2, Δ_2. If there exists a unique i such that $a_{2,\mu}$ is the ith variable of $\boldsymbol{a}$ then we define $\breve{a}_{2,\mu}$ to be the ith variable in $\boldsymbol{x}$. Otherwise we define $\breve{a}_{2,\mu}$ to be $a_{2,\mu}$. Then take C to be $\exists \boldsymbol{x}\, \forall \boldsymbol{y}\, (\bigwedge_\mu \breve{a}_{2,\mu} = \breve{b}_{2,\mu} \wedge C'(\boldsymbol{x}, \boldsymbol{y}))$.

Case 4. The last inference of P is of the form

$$\frac{\Phi, \Gamma_1, \Gamma_2 \to \Delta_1, \Delta_2, \Psi}{\Gamma_1, \Gamma_2 \to \Delta_1, \Delta_2} \quad \text{for all } (\Phi | \Psi).$$

By the induction hypothesis, there exist formulas $C_{(\Phi|\Psi)}$ such that $C_{(\Phi|\Psi)}$,

$\Gamma_1 \to \Delta_1$ and $\Phi, \Gamma_2 \to \Delta_2, \Psi, C_{(\Phi|\Psi)}$ are provable in **RHS**. So

$$\bigvee_{(\Phi|\Psi)} C_{(\Phi|\Psi)}, \Gamma_1 \to \Delta_1$$

and

$$\Gamma_2 \to \Delta_2 \bigvee_{(\bar{\Phi}|\Psi)}, C_{(\Phi|\Psi)}$$

are provable in **RHS**. Let $\boldsymbol{a}$ be a sequence of all the free variables in $\bigvee_{(\Phi|\Psi)} C_{(\Phi|\Psi)}$ which do not appear in Γ_2, Δ_2 and let $\boldsymbol{b}$ be those which do not occur in Γ_1, Δ_1. We rewrite $\bigvee_{(\Phi|\Psi)} C_{(\Phi|\Psi)}$ as $C'(\boldsymbol{a}, \boldsymbol{b})$. Then take C to be $\forall \boldsymbol{x} \exists \boldsymbol{y} C'(\boldsymbol{x}, \boldsymbol{y})$.

Other cases. The proof is similar to the one above.

REMARK. In Lemma 23.20, note that (1) is not an essential restriction on P because $\forall$ can be expressed by $\neg$ and $\exists$. Note also that any sub-proof of P, i.e., any part of P consisting of all sequents above and including a given sequent, is a proof in **RHS** because there are no eigenvariables in P.

3. Let $\Gamma_1, \Gamma_2 \to \Delta_1, \Delta_2$ be as in the statement of the theorem. There exists a cut-free proof P of $\Gamma_1, \Gamma_2 \to \Delta_1, \Delta_2$ in the homogeneous system. By the proof of completeness theorem, we may assume that P satisfies the following condition.

3.1. If a variable occurs in two different auxiliary formulas as an eigenvariable, then these two formulas are the same.

Moreover, without loss of generality we may assume the following for P.

3.2. Every quantifier in P is $\exists$.

3.3. The height of a free variable in $\Gamma_1, \Gamma_2 \to \Delta_1, \Delta_2$ is less than the height of any eigenvariable in P.

3.4. The heights of two different variables in P are different.

Let $\Gamma'_1 \to \Delta'_1$ be a sequent in P. Let $\Phi(\Gamma'_1, \Delta'_1)$ be the sequence $A_0, A_1, \ldots, A_\mu, \ldots$ of all A_μ's such that A_μ is of the form $\neg\exists \boldsymbol{x} A(\boldsymbol{x}) \vee A(\boldsymbol{a})$, where $\exists \boldsymbol{x} A(\boldsymbol{x})$ is a principal formula of a $\exists$: left above $\Gamma'_1 \to \Delta'_1$ and $A(\boldsymbol{a})$ is its auxiliary formula. Replacing $\Gamma'_1 \to \Delta'_1$ by $\Phi(\Gamma'_1, \Delta'_1), \Gamma'_1 \to \Delta'_1$ and inserting some appropriate structural inferences, we obtain a new figure P' satisfying the following conditions:

1) P' satisfies (1) and (2) of Lemma 23.20.

2) The end-sequent of P' is of the form (cf. the proof of Proposition 22.25)

$$\{\neg\exists \boldsymbol{x}_\lambda A_\lambda(\boldsymbol{x}_\lambda, \boldsymbol{c}_\lambda) \vee A_\lambda(\boldsymbol{a}_\lambda, \boldsymbol{c}_\lambda)\}, \Gamma_1,$$

$$\{\neg\exists \boldsymbol{y}_\mu B_\mu(\boldsymbol{y}_\mu, \boldsymbol{d}_\mu) \vee B_\mu(\boldsymbol{b}_\mu, \boldsymbol{d}_\mu)\}, \Gamma_2 \to \Delta_1, \Delta_2.$$

3) The height of any $c_{\lambda,\alpha}$ is less than the height of any $a_{\lambda,\beta}$. The height of any $d_{\mu,\alpha}$ is less than the height of any $b_{\mu,\beta}$.

4) Every free variable or predicate constant, except $=$, in $\exists \boldsymbol{z}_\lambda \exists \boldsymbol{x}_\lambda A_\lambda(\boldsymbol{x}_\lambda, \boldsymbol{z}_\lambda)$ occurs in Γ_1, Δ_1 and every free variable or predicate constant, except $=$, occurring in $\exists \boldsymbol{z}_\mu \exists \boldsymbol{x}_\mu B_\mu(\boldsymbol{y}_\mu, \boldsymbol{z}_\mu)$ occurs in Γ_2, Δ_2.

5) Any $a_{\lambda,\alpha}$ and $b_{\mu,\beta}$ are different. (Otherwise we can modify P' so that P' satisfies 5) because P satisfies 3.1.)

Applying Lemma 23.20, there is a formula $C(\boldsymbol{a})$ such that

(a) $$C(\boldsymbol{a}), \{\neg\exists \boldsymbol{x}_\lambda\, A_\lambda(\boldsymbol{x}_\lambda, \boldsymbol{c}_\lambda) \vee A_\lambda(\boldsymbol{a}_\lambda, \boldsymbol{c}_\lambda)\}, \Gamma_1 \to \Delta_1$$

and

$$\{\neg\exists \boldsymbol{y}_\mu\, B_\mu(\boldsymbol{y}_\mu, \boldsymbol{d}_\mu) \vee B_\mu(\boldsymbol{b}_\mu, \boldsymbol{d}_\mu)\}, \Gamma_2 \to \Delta_2, C(\boldsymbol{a})$$

are provable in **RHS**. Let Q_1 and Q_2 be proofs of these sequents in **RHS**.

(b) Every free variable or predicate constant, except $=$, occurring in $C(\boldsymbol{a})$ is in both $\{A_\lambda(\boldsymbol{a}_\lambda, \boldsymbol{c}_\lambda)\}, \Gamma_1, \Delta_1$ and $\{B_\mu(\boldsymbol{b}_\mu, \boldsymbol{d}_\mu)\}, \Gamma_2, \Delta_2$.

(c) $\boldsymbol{a}$ is the sequence of all the free variables in $C(\boldsymbol{a})$ which are not in both Γ_1, Δ_1 and Γ_2, Δ_2 and well-ordered according to heights.

4. Then consider the following figure

$$\begin{array}{r}
Q_1 \\
C(\boldsymbol{a}), (\neg\exists \boldsymbol{x}_\lambda\, A_\lambda(\boldsymbol{x}_\lambda, \boldsymbol{c}_\lambda) \vee A_\lambda(\boldsymbol{a}_\lambda, \boldsymbol{c}_\lambda)), \Gamma_1 \to \Delta_1 \\ \hline
C(\boldsymbol{a}), \{\exists \boldsymbol{x}'_\lambda\, (\neg\exists \boldsymbol{x}_\lambda\, A_\lambda(\boldsymbol{x}_\lambda, \boldsymbol{c}_\lambda) \vee A_\lambda(\boldsymbol{x}'_\lambda, \boldsymbol{c}_\lambda))\}, \Gamma_1 \to \Delta_1 \\ \hline
C(\boldsymbol{a}), \{\forall \boldsymbol{z}_\lambda\, \exists \boldsymbol{x}'_\lambda\, (\neg\exists \boldsymbol{x}_\lambda\, A_\lambda(\boldsymbol{x}_\lambda, \boldsymbol{z}_\lambda) \vee A_\lambda(\boldsymbol{x}'_\lambda, \boldsymbol{z}_\lambda))\}, \Gamma_1 \to \Delta_1 \\ \hline
\mathbf{Q}^f \boldsymbol{x}\, C(\boldsymbol{x}), \{\forall \boldsymbol{z}_\lambda\, \exists \boldsymbol{x}_\lambda\, (\neg\exists \boldsymbol{x}_\lambda\, A_\lambda(\boldsymbol{x}_\lambda, \boldsymbol{z}_\lambda) \vee A_\lambda(\boldsymbol{x}'_\lambda, \boldsymbol{z}_\lambda))\}, \Gamma_1 \to \Delta_1
\end{array}$$

where f is defined as follows.

(e) If a_α is $b_{\mu,\gamma}$ for some γ, then $f(\alpha) = \exists$.

(f) If a_α is $a_{\lambda,\gamma}$, for some γ, then $f(\alpha) = \forall$.

(g) If a_α is contained in Γ_1, Δ_1 but not in Γ_2, Δ_2, then $f(\alpha) = \forall$.

(h) If a_α is contained in Γ_2, Δ_2, but not in Γ_1, Δ_1, then $f(\alpha) = \exists$.

(i) If (e)–(h) are not the case, then $f(\alpha) = \exists$.

The heights for the free variables in $\boldsymbol{a}_\lambda$, $\boldsymbol{c}_\lambda$, $C(\boldsymbol{a})$, Γ_1, Δ_1 are defined to be the heights in P. The heights of all other variables in Q_1 can be so defined, according to Proposition 23.19, that the whole figure will beceome a proof in determinate logic. This means that $\mathbf{Q}^f \boldsymbol{x}\, C(\boldsymbol{x}), \Gamma_1 \to \Delta_1$ is valid. The validity of $\Gamma_2 \to \Delta_2, \mathbf{Q}^F \boldsymbol{x}\, C(\boldsymbol{x})$ is also easily seen from the following proof which satisfies (**Q**) (cf. Definition 23.14).

$$\begin{array}{r}
Q_2 \\
\{\neg\exists \boldsymbol{y}_\mu\, B_\mu(\boldsymbol{y}_\mu, \boldsymbol{d}_\mu) \vee B_\mu(\boldsymbol{b}_\mu, \boldsymbol{d}_\mu)\}, \Gamma_2 \to \Delta_2, C(\boldsymbol{a}) \\ \hline
\{\exists \boldsymbol{y}'_\mu\, (\neg\exists \boldsymbol{y}_\mu\, B_\mu(\boldsymbol{y}_\mu, \boldsymbol{d}_\mu) \vee B_\mu(\boldsymbol{y}'_\mu, \boldsymbol{d}_\mu))\}, \Gamma_2 \to \Delta_2, C(\boldsymbol{a}) \\ \hline
\{\forall \boldsymbol{z}_\mu\, \exists \boldsymbol{y}'_\mu\, (\neg\exists \boldsymbol{y}_\mu\, B_\mu(\boldsymbol{y}_\mu, \boldsymbol{z}_\mu) \vee B_\mu(\boldsymbol{y}'_\mu, \boldsymbol{z}_\mu))\}, \Gamma_2 \to \Delta_2, C(\boldsymbol{a}) \\ \hline
\{\forall \boldsymbol{z}_\mu\, \exists \boldsymbol{y}'_\mu\, (\neg\exists \boldsymbol{y}_\mu\, B_\mu(\boldsymbol{y}_\mu, \boldsymbol{z}_\mu) \vee B_\mu(\boldsymbol{y}_\mu, \boldsymbol{z}_\mu))\}, \Gamma_2 \to \Delta_2\, \mathbf{Q}^f \boldsymbol{x}\, C(\boldsymbol{x}).
\end{array}$$

This completes the proof of Theorem 23.15.

Using the same method we can prove the following theorem.

THEOREM 23.21 (cf. Theorem 23.15). *If every quantifier in* $\Gamma_1, \Gamma_2 \to$

Δ_1, Δ_2 is homogeneous, if $\Gamma_1, \Gamma_2 \to \Delta_1, \Delta_2$ is valid and does not contain $=$, and if Γ_1, Δ_1 and Γ_2, Δ_2 have at least one predicate constant in common, then there exists a formula C such that both $C, \Gamma_1 \to \Delta_1$ and $\Gamma_2 \to \Delta_2, C$ are valid and every free variable or predicate constant in C is contained in both Γ_1, Δ_1 and Γ_2, Δ_2.

REMARK 23.22. In Theorems 23.15 and 23.21, we may add the condition that C contains only one heterogeneous quantifier in the front of C.

REMARK 23.23. For Malitz's example (cf. §22) we can construct an isomorphism between $\overset{1}{<}$ and $\overset{2}{<}$ by the following formula.

$$\forall x_1 \exists y_1 \forall x_2 \exists y_2 \ldots \Big(\bigwedge_i x_i \overset{1}{<} a \to \bigwedge_i y_i \overset{2}{<} b \wedge \bigwedge_i \bigwedge_j (x_i \overset{1}{<} x_j \leftrightarrow y_i \overset{2}{<} y_j) \wedge (x_i \overset{1}{=} x_j \leftrightarrow y_i \overset{2}{=} y_j)\Big)$$

$$\wedge \forall y_1 \exists x_1 \forall y_2 \exists x_2 \ldots \Big(\bigwedge_i y_i \overset{2}{<} b \to \bigwedge_i x_i \overset{1}{<} a \wedge \bigwedge_i \bigwedge_j (x_i \overset{1}{<} x_j \leftrightarrow y_i \overset{2}{<} y_j) \wedge (x_i \overset{1}{=} x_j \leftrightarrow y_i \overset{2}{=} y_j)\Big).$$

The order type of a in $(\overset{1}{=}, \overset{1}{<})$ is denoted by $|a|_1$ and the order type of b in $(\overset{2}{=}, \overset{2}{<})$ is denoted by $|b|_2$. Then $|a|_1 \leqslant |b|_2$ is equivalent to

$$(*) \quad \forall x_1 \exists y_1 \forall x_2 \exists y_2 \ldots \Big(\bigwedge_i x_i \overset{1}{<} a \to \bigwedge_i y_i \overset{2}{<} b \wedge \bigwedge_i \bigwedge_j (x_i \overset{1}{<} x_j \leftrightarrow y_i \overset{2}{<} y_j) \wedge (x_i \overset{1}{=} x_j \leftrightarrow y_i \overset{2}{=} y_j)\Big).$$

This is easily shown by transfinite induction on $|a|_1$, as follows.

Let the formula (*) be denoted by $A(a, b)$. Suppose that for each $c \overset{1}{<} a$ and each d, $A(c, d)$ is equivalent to $|c|_1 \leqslant |d|_2$. Suppose also that $A(a, b)$ holds. Then for each $a_1 \overset{1}{<} a$, there exists a b_1 such that $A(a_1, b_1)$ holds because $\bigwedge_{i \geqslant 2} (x_1 \overset{1}{<} a_i)$ implies $\bigwedge_{i \geqslant 2} (x_i \overset{1}{<} a)$ and hence for $x_2, y_2, \ldots$ selected for (a_1, b_1) in $A(a, b)$ we have $\bigwedge_{i \geqslant 2} y_i \overset{2}{<} b_1$. Then making the appropriate substitutions into $A(a, b)$ we obtain $A(a_1, b_1)$. Since $a_1 \overset{1}{<} a$ we have, by the induction hypothesis $|a_1|_1 \leqslant |b_1|_2 < |b|_2$. Therefore $|a|_1 \leqslant |b|_2$.

The converse is obvious.

The axiom of determinateness, AD, is a very powerful axiom that has numerous interesting and important applications. Augmented by the axiom of dependent choice

$$\text{DC} \qquad \forall x\, \exists y\, R(x, y) \rightarrow \forall x_0\, \exists x_1\, \exists x_2 \ldots \bigwedge_i R(x_i, x_{i+1})$$

the AD has even more implications for mathematics.

Unlike the axiom of choice, AC, which also has important implications for mathematics, the status of the AD is as yet unsettled. We do not know whether the AD is consistent with set theory. Neither do we know whether the AD and DC are consistent. We do know that although the AD implies the axiom of countable choice it is incompatible with the AC.

If it should develop that the AD is inconsistent with set theory we would, of course, cease to be interested in it. But even a proof of consistency would not be sufficient for our purposes, for in order to reap the benefits of the AD for mathematics we must have a transitive model of ZF + AD that contains the power set of ω, $P(\omega)$, as an element. Indeed we would like to have a transitive model of ZF + AD + DC that contains $P(\omega)$ as an element.

Concerning the existence of such models we know the following. Let $\mathrm{L}_\beta(P(\omega))$ be the set obtained from $P(\omega)$ by a β-fold transfinite iteration of Gödel's eight fundamental operations and let α be the smallest β such that $\mathrm{L}_\beta(P(\omega))$ is a model of ZF. Then we know that if there exists a transitive model of ZF + AD that contains $P(\omega)$, $\mathrm{L}_\alpha(P(\omega))$ is a model of the AD. But we also know that $\mathrm{L}_\alpha(P(\omega))$ satisfies DC.

There are then three possibilities:

1) The AD is inconsistent with set theory.

2) The AD is consistent with set theory but no transitive model of ZF + AD exists that contains $P(\omega)$.

3) $\mathrm{L}_\alpha(P(\omega))$ is a model of the AD.

If alternative 1) or 2) should be the case, we would have no further interest in the AD. Our hopes center around alternative 3) which we conjecture to be true. We are, however, unable to prove that $\mathrm{L}_\alpha(P(\omega))$ is the model we conjecture it to be. Moreover, at the present time no one appears to have a method that might resolve the question. In view of the implications of this conjecture for mathematics it is important that a thorough study of the AD be made. As a contribution to this study we will prove a relation between the AD and the cut-elimination theorem.

Let M be a transitive model of ZF + DC that contains $P(\omega)$ as an element. M may be a set or a proper class. Although we cannot assume the AC in M we will assume it in V, the universe of all sets. Using the AC in V we will prove that the AD holds in a model M if and only if the cut-elimination theorem holds in M-definable determinate logic. For the proof we need the following definitions.

DEFINITION 23.24. A set A is *at most the continuum* in M iff $A \in M$, $A \neq \emptyset$ and there is a function f in M such that f maps $P(\omega)$ onto A.

Clearly, if A and B are nonempty sets in M and $B \subseteq A$, then B is at most the continuum in M if A is at most the continuum in M.

Since M is a model that contains $P(\omega)$ as an element it follows that for any language L, having not more than $\aleph_1$ symbols, we can assign to each symbol and to each formula of L a gödelization in M. This enables us to identify collections of language symbols and formulas with sets in M. Since we know that these identifications can be made we will follow the convention of speaking simply of sets of language symbols as being in M. With this convention in mind we define M-definable determinate logic.

DEFINITION 23.25. A language L for *M-definable determinate logic* consists of the following:

1) *Free variables*: A free variable a_s for each s in $P(\omega)$.
2) *Bound variables*: $x_0, x_1, \ldots, x_\alpha, \ldots, \alpha < \omega_1$.
3) *Individual constants*: A set of individual constants that is at most the continuum in M and which contains $0, 1, 2, \ldots$.
4) *Predicate constants*: A set of predicate constants that is at most the continuum in M. The arity of each predicate constant is at most ω.
5) *Logical symbols*:
   $=$ (equality),
   $\neg$ (not),
   $\bigwedge$ (conjunctions of arity α for $\alpha \leqslant \omega$),
   $\bigvee$ (disjunctions of arity α for $\alpha \leqslant \omega$),
   $\mathbf{Q}^f$ (heterogeneous quantifiers of arity α for $1 \leqslant \alpha \leqslant \omega$).

Note that the set of free variables is at most the continuum in M. Furthermore, since $P(\omega)$ is in M, $\omega_1 = \omega_1^M$, that is, ω_1 is M-absolute.

The formulas of L we define in the following way:

Let R be a predicate constant or $=$ and let the arity of R be α. Let $\{t_i\}_{i<\alpha}$ be a sequence of terms. Then $R(t_0, t_1, \ldots)$ is an atomic formula.

Let A be a formula. Then $\neg A$ is a formula.

Let $\{A_i\}_{i<\alpha}$ be a sequence of formulas. Then $\bigwedge_{i<\alpha} A_i$ and $\bigvee_{i<\alpha} A_i$ are formulas.

Let $A(\boldsymbol{a})$ be a formula where $\boldsymbol{a}$ is the sequence of free variables $\{a_i\}_{i<\alpha}$, with $\alpha \leqslant \omega$. Let $\boldsymbol{x}$ be the sequence of bound variables $\{x_i\}_{i<\alpha}$ and let f map α into $\{\forall, \exists\}$. Then $\mathbf{Q}^f\boldsymbol{x}\, A(\boldsymbol{x})$ is a formula, where $A(\boldsymbol{x})$ is obtained from $A(\boldsymbol{a})$ by replacing some occurrences of a_i by x_i for each i.

Let Γ and Δ be sets of formulas of at most the continuum. Then $\Gamma \rightarrow \Delta$ is called a sequent.

Notice that we cannot assume the well-ordering of Γ and Δ, since the axiom of choice is not assumed in M.

COROLLARY 23.26. (1) *If a language* L *is fixed, then the set of* L*-formulas is at most the continuum.*

(2) *Given an* L*-formula, the number of variables, constants and logical symbols which occur in it is at most countable.*

When we consider a set of formulas $\{A_\lambda\}_\lambda$ or a set of free variables, we must remember that they are just sets; they may not be well-ordered.

LEMMA 23.27. (1) *Consider a tree of length ω which has ω branches extending from each node. We may identify this with ω^ω, which is at most the continuum.*

(2) *Let α be an ordinal which is at most the continuum. Consider a tree of length ω which has α branches extending from each node. We may identify this with α^ω, which is at most the continuum.*

DEFINITION 23.28. The notion of proof and the rules of inference for M-definable determinate logic are defined as follows.

1) The initial sequents are the logical initial sequents; $\rightarrow t = t$, where t is an arbitrary term; those sequents of the form $i = j \rightarrow$, where $i \neq j$ and $i, j < \omega$; and those sequents of the form $\rightarrow t = 0, t = 1, t = 2, \ldots$, where t is an arbitrary term.

2) The rules of inference are those rules of determinate logic, which we have already presented. One should keep in mind that in the sequents the formulas form sets that are not necessarily well-ordered. As an example, $\bigvee$: right looks like this:

$$\frac{\Gamma \rightarrow \Delta, \{A_{\lambda, i_\lambda}\}_{i_\lambda < \alpha_\lambda}}{\Gamma \rightarrow \Delta, \{\bigvee_{i < \alpha_\lambda} A_{\lambda, i}\}}$$

where $\alpha_\lambda \leq \omega$ and λ ranges over a set of at most the continuum.

3) A proof in M, say P, is a member of M which is a proof in the ordinary sense except that the notion of height must be replaced by a relation $<$:

3.1) Suppose a is an eigenvariable in P, and b is a free variable which occurs in its principal formula. Then $b < a$.

3.2) Suppose $\boldsymbol{a}$ is an eigenvariable, $A(\boldsymbol{a})$ is its auxiliary formula, $\mathbf{Q}^f \boldsymbol{x}\, A(\boldsymbol{x})$ is its principal formula, and suppose $\boldsymbol{a}$ corresponds to $f(i)$. If b also occurs in $\boldsymbol{a}$ and b corresponds to $f(j)$, where $j < i$, then $b < a$.

The eigenvariable condition is simply that $<$ is well-founded. If $b < a$, let us say that a depends on b.

REMARK. "$<$ is well-founded" is M-absolute because a countable subset of free variables (in V) is a countable subset of $P(\omega)$ and $P(\omega)$ belongs to M, hence this set of free variables is a countable subset of M.

Before we get into the next argument, we should remark that we may and will restrict the indices of free variables, the s in a_s, where $s \in P(\omega)$, to subsets of even numbers. This way, we will be free to introduce new free variables.

LEMMA 23.29. *Consider a countable set of free variables, say* $\tilde{A} =$

$\{a_{s_1}, a_{s_2}, \ldots\}$ where $\tilde{A}$ belongs to M and $\ldots < a_{s_2} < a_{s_1}$ (in V). Define $R(a, b)$ by

$$R(a, b) \Leftrightarrow_{\mathrm{df}} a \in \tilde{A} \supset (b \in \tilde{A} \wedge b < a).$$

Then $\forall x_0 \exists x_1 x_2 \ldots \bigwedge_i R(x_i, x_{i+1})$ in M.

PROOF. It is easily seen that $\forall x \exists y R(x, y)$ in M; hence by DC the desired formula is obtained.

DEFINITION 23.30. A quantifier $\mathbf{Q}$ in a formula A is said to be essentially succedent in A if it is in the scope of an even number of $\neg$'s. A sequent $\Gamma \rightarrow \Delta$ is said to be succedent-homogeneous if every quantifier in a formula of Δ which is essentially succedent is homogeneous and every quantifier in a formula of Γ which is not essentially succedent is homogeneous.

For the following proposition, we assume that $0, 1, 2, \ldots$ are the only individual constants in L. This simplifies the discussion.

PROPOSITION 23.31. (1) *If the* AD *holds in M, then all the provable sequents of M-definable determinate logic are M-valid, that is, valid in every M-definable structure.*

(2) *If a sequent is provable with a proof in which all the sequents are succedent-homogeneous, then it is M-valid.*

Note that in (2) the AD is not assumed for M.

PROOF. Suppose P is a proof for $\Gamma \rightarrow \Delta$. Let $<$ be the well-founded relation defined for free variables of P. We can assign ordinals to these free variables in such a manner that if $a < b$, then the ordinal of a is less than the ordinal of b. Start with those variables that do not depend on any other variables and assign them the value 0. Next, assign the ordinal 1 to those variables that depend only on variables whose ordinals are 0. Continuing in this way we will assign ordinals to all variables in P for the following reason. Suppose there are variables in P which are not assigned ordinals by this process. Let $\tilde{A}$ be the set of those free variables. Then by Lemma 23.29,

$$\forall x_0 \exists x_1 x_2 \ldots \bigwedge_i (x_i \in \tilde{A} \supset (x_{i+1} \in \tilde{A} \wedge x_{i+1} < x_i)).$$

Since $\tilde{A}$ is not empty, this means there is an infinite sequence $\{a_i\}_i$ from $\tilde{A}$ such that $a_{i+1} < a_i$, contradicting the well-foundedness of $<$.

The ordinals assigned to the free variables of P as above will be called heights. It is easy to see that they satisfy the conditions of heights in the previous sense.

Consider an M-definable structure $\mathscr{A}$. Notice that the natural numbers of $\mathscr{A}$ are not necessarily the natural numbers in the absolute sense. They are, however, in one-to-one correspondence with the actual natural numbers. Therefore we may assume, without loss of generality, that the universe of $\mathscr{A}$ is ω and the constants $0, 1, 2, \dots$ in the language are interpreted in the obvious way; thus $\mathscr{A} = \langle \omega, 0, 1, 2, \dots \rangle$.

Then consider all the formulas and subformulas in P and their Skolem functions, $g^{f_\gamma}_{A}$ and $\bar{g}^{f_\gamma}_{A}$, defined as before. Let $\mathbf{Q}^f \boldsymbol{x}\, A(\boldsymbol{x}, \boldsymbol{a})$ be a formula and suppose $\boldsymbol{a}$ exhausts all the free variables in this formula. If $g^{f_\gamma}_{A}$ is regarded as a function of $\boldsymbol{x}$ or of some variables of $\boldsymbol{x}$, which $\boldsymbol{a}$ is held fixed, then such a function is a member of M. If $g^{f_\gamma}_{A}$ is regarded as a function of $\boldsymbol{a}$ as well as some variables of $\boldsymbol{x}$, it is not guaranteed that the function belongs to M. In spite of this, we can carry out the subsequent argument entirely in M, for once the values of $\boldsymbol{a}$ are assigned, $g^{f_\gamma}_{A}$ is an element of M, and $g^{f_\gamma}_{A}$ occurs only in this context. What we will do is to construct such functions and substitute them for eigenvariables. The resulting figure P' may not be an element of M, but each formula in P' becomes a formula of M once those functions are computed.

The process of obtaining P' and determining the interpretation of the $g^{f_\gamma}_{A}$'s and $\bar{g}^{f_\gamma}_{A}$'s parallels the proof of Theorem 23.4 for (1), and the proof of Proposition 23.5 for (2). In a similar manner we can show that for an arbitrary sequent in P', say $\Gamma \rightarrow \Delta$, either there exists a formula of Γ which is false in $\mathscr{A}$ or there is a formula of Δ which is true in $\mathscr{A}$. For 6) in the proof of Theorem 23.4 we need the determinateness of $\mathscr{A}$. That $\mathscr{A}$ is a determinate structure is a consequence of the fact that the AD holds in M. Since substitution for eigenvariables does not change the end-sequent, this means that the end-sequent is true in $\mathscr{A}$.

DEFINITION 23.32. A generalized cut is called inessential if all its cut formulas are equalities, i.e., of the form $t_1 = t_2$; it is called essential otherwise.

Throughout the remainder of this chapter we call a proof cut-free if it has only inessential generalized cuts.

PROPOSITION 23.33. (1) *If the* AD *holds in* M, *then all the valid* M-*sequents are cut-free provable.*

(2) *If* $\Gamma \rightarrow \Delta$ *is succedent-homogeneous and valid, then* $\Gamma \rightarrow \Delta$ *is cut-free provable.*

PROOF. We follow the proof of Theorem 23.6. As was mentioned before, we may assume that the indices of the free variables are subsets of even numbers; this way we can introduce new free variables when necessary. Let

$$\tilde{\Gamma} = \Gamma \cup \{\forall x\, (x = 0 \vee x = 1 \vee \dots)\}$$

and let $\mathbf{Q}^f \boldsymbol{x}\, A(\boldsymbol{x}, \boldsymbol{a})$ be an arbitrary formula in $\tilde{\Gamma} \rightarrow \Delta$. For each such formula we introduce a function symbol $g^{f_A\gamma}$ (interpreted as a Skolem function) for each x_γ in $\boldsymbol{x}$ if $\mathbf{Q}^f\boldsymbol{x}\, A(\boldsymbol{x}, \boldsymbol{a})$ is essentially antecedent, and we introduce $\bar{g}^{f_A\gamma}$ if it is essentially succedent. Let D be the set of terms which are generated from the individual constants and free variables by these Skolem functions. By Lemma 23.27, the set of those subformulas is at most the continuum, hence D is at most the continuum, because the individual constants and free variables form a set, D_0, of at most the continuum, and each stage of applying the Skolem functions increases the set by at most the continuum and we need to repeat the application of Skolem functions ω_1 times, more precisely α times for all $\alpha < \omega_1$.

We regard the terms in $D - D_0$ as free variables and identify them with the free variables which have been saved. (So it may happen that more than one such free variable corresponds to one term in $D - D_0$.) A natural partial ordering $\prec$ can be defined for the free variables from $D - D_0$. If s occurs in t, then $s \prec t$. It can be easily shown that $\prec$ is a well-founded relation and $\prec$ is a member of M.

We are now prepared for the completeness proof of Theorem 23.6. In this proof we use appropriate terms from $D - D_0$ in the reduction of quantifiers.

Since all the terms in D are members of M, it is obvious that the sequents thus obtained (in forming a tree) are members of M, and that they consist of at most the continuum of formulas. For example, in part 6) in the proof of Theorem 23.6 the possibility for $\boldsymbol{t}_{\lambda,\mu}$ is at most $(2^\omega)^\omega = 2^\omega$, which is at most the continuum.

Case 1. Every branch of $T(\tilde{\Gamma} \to \Delta)$ has a sequent of the form

$$\ldots D \ldots \to \ldots D \ldots \quad \text{or} \quad \ldots \to \ldots s = s \ldots .$$

As before, consider the condition C. In order to show that $\tilde{\Gamma} \to \Delta$ satisfies C, we take the following step. Let $S', S'', \ldots$ denote sequents in $T(\tilde{\Gamma} \to \Delta)$ and define $R(S', S'')$ by

$$\begin{aligned} R(S', S'') \Leftrightarrow_{\mathrm{df}} (S' \in T(\tilde{\Gamma} \to \Delta) \wedge (S' \text{ does not satisfy } C) \supset \\ \supset S'' \in T(\tilde{\Gamma} \to \Delta) \wedge (S'' \text{ does not satisfy } C) \\ \wedge (S'' \text{ is an immediate ancestor of } S'). \end{aligned}$$

If we assume that $\tilde{\Gamma} \to \Delta$ does not satisfy C, then $\forall S'\, \exists S''\, R(S', S'')$ is true in M', hence by DC,

$$\forall S_0\, \exists S_1\, S_2 \ldots \bigwedge_i R(S_i, S_{i+1}).$$

Letting S_0 be $\Gamma \to \Delta$, we conclude that there is an infinite branch which does not satisfy C, contradicting the assumption of case 1.

Case 2. In this case we can construct a counterexample for $\tilde{\Gamma} \to \Delta$ in the same manner as before. Recall that the domain D belongs to M.

Consequently, the fact that a formula occurs in the antecedent or in the succedent can be expressed in M.

In proving (1), we need to show that $\mathbf{Q}^{\bar{f}}\mathbf{x}\neg A(\mathbf{x}, \mathbf{s}) \to \neg\mathbf{Q}^{f}\mathbf{x}\, A(\mathbf{x}, \mathbf{s})$ is true in any M-structure, (cf. 2 in the proof of Theorem 23.6). This holds since the AD holds in M. In proving (2), this case does not arise, since the given sequent is succedent-homogeneous.

We now have a cut-free proof of

$$\forall x\,(x=0 \vee x=1 \vee \ldots), \Gamma \to \Delta$$

and since

$$\dfrac{\dfrac{\to a=0,\ a=1, \ldots}{\to a=0 \vee a=1 \vee \ldots}}{\to \forall x\,(x=0 \vee x=1 \vee \ldots)}$$

we have, from the cut rule

$$\Gamma \to \Delta.$$

This last cut however is easily eliminated.

THEOREM 23.34. *The* AD *holds in* M *if and only if the cut-elimination theorem holds for* M*-definable determinate logic.*

PROOF. 1. Suppose that the cut-elimination theorem holds for M-definable determinate logic but the AD does not hold in M, namely, there is a set of M, say A, such that $A \subset \omega^\omega$ and the AD fails for A. Use A as a predicate symbol and let $i_0, i_1, \ldots$ be individual constants corresponding to $0, 1, \ldots$. Then consider two sets of atomic sentences:

$$\Gamma_0 = \{A(i_0, i_1, \ldots) \mid A(i_0, i_1, \ldots) \text{ is true when } A \text{ and } i_0, i_1, \ldots \text{ are interpreted as above}\}$$

and

$$\Delta_0 = \{A(i_0, i_1, \ldots) \mid A(i_0, i_1, \ldots) \text{ is false}\}.$$

We claim that

$$(1) \qquad \forall x_0 \exists y_0 \forall x_1 \exists y_1 \ldots A(x_0, y_0, x_1, y_1, \ldots), \Gamma_0 \to \Delta_0$$

and

$$(2) \qquad \exists x_0 \forall y_0 \exists x_1 \forall y_1 \ldots \neg A(x_0, y_0, x_1, y_1, \ldots), \Gamma_0 \to \Delta_0$$

are both valid. Then, since both are homogeneous-succedent, 2) of Proposition 23.33 implies that those sequents are cut-free provable.

Case 1. A is interpreted differently from the given set A. Then, since $\Gamma_0 \cup \Delta_0$ exhausts all the possibilities for $A(i_0, i_1, \ldots)$, there is at least one $(i_0, i_1, \ldots)$ such that either $A(i_0, i_1, \ldots)$ is in Γ_0 and it is false or

$A(i_0, i_1, \ldots)$ is in Δ_0 and it is true. Hence $\Gamma_0 \to \Delta_0$ is true, which implies that both sequents are true.

Case 2. A is interpreted as the given set A. Then neither

$$\forall x_0 \exists y_0 \forall x_1 \exists y_1 \ldots A(x_0, y_0, x_1, y_1, \ldots)$$

nor

$$\exists x_0 \forall y_0 \exists x_1 \forall y_1 \ldots \neg A(x_0, y_0, x_1, y_1, \ldots)$$

is true; hence both sequents are true.

From (1) and (2) we obtain

$$\forall x_0 \exists y_0 \forall x_1 \exists y_1 \ldots A(x_0, y_0, x_1, y_1, \ldots)$$
$$\vee \exists x_1 \forall y_0 \exists x_1 \forall y_1 \ldots \neg A(x_0, y_0, x_1, y_1, \ldots), \Gamma_0 \to \Delta_0$$

is provable in M-definable determinate logic. On the other hand

$$\forall x_0 \exists y_0 \forall x_1 \exists y_1 \ldots A(x_0, y_0, x_1, y_1, \ldots)$$
$$\vee \exists x_0 \forall y_1 \exists x_1 \forall y_1 \ldots \neg A(x_0, y_0, x_1, y_1, \ldots)$$

is provable in the same system. Hence by the cut rule $\Gamma_0 \to \Delta_0$ is provable; but this is impossible. Therefore the AD must hold in M.

2. If AD holds in M, then by (1) of Proposition 23.33 together with 1) of Proposition 23.31, the cut-elimination theorem holds.

This completes the proof of the theorem.

Next we point out a relation between the cut-elimination theorem and the infinitary propositional calculus **IPC** which is the quantifier-free part of infinitary logic. **IPC** is common to determinate logic and ordinary infinitary logic. Consequently, provable sequents in **IPC** are valid.

Let Γ_0 be a set of quantifier-free sentences. It is well known that if Γ_0 is consistent (with **IPC**), then Γ_0 has a model.

PROPOSITION 23.35. *Let M be as above. Then the following two conditions are equivalent.*

(1) *The cut-elimination theorem holds in M-definable determinate logic.*

(2) *Let Γ_0 be a set of quantifier-free sentences and suppose Γ_0 belongs to M. If Γ_0 is consistent with* **IPC**, *then Γ_0 is consistent with M-definable determinate logic.*

PROOF. Obviously (1) implies (2). Suppose the cut-elimination theorem does not hold. Then by Theorem 23.34 there exists a counterexample for the AD in M, say $A \subseteq \omega^\omega$. In the proof of the theorem, the set of formulas $\Gamma_0 \cup \neg\Delta_0$, where $\neg\Delta_0$ consists of all the formulas of the form $\neg B$ for B in Δ_0, is consistent with **IPC**, since an A as above can be a model. On the other hand, $\Gamma_0 \to \Delta_0$ is provable in M-definable determinate logic, hence $\Gamma_0 \cup \neg\Delta_0$ is not consistent with determinate logic.

PROBLEM 23.36. Let us again assume the AC and assume that the antecedents and succedents of sequents are well-ordered. Consider the following language.

1. Individual constants: $0, 1, 2, \ldots$.
2. Bound variables: $x_0, x_1, \ldots x_\alpha, \ldots$ $(\alpha < \omega_1)$.
3. Free variables: $a_0, a_1, \ldots, a_\alpha, \ldots$ $(\alpha < 2\ {}^{0})$.
4. Predicate constants: $=$ and $R_1, R_2, R_3, \ldots$, where R_i has i argument places.
5. Logical symbols: $\neg, \bigwedge, \bigvee, \forall, \exists, \mathbf{Q}_1, \mathbf{Q}_2$.

By a term, we mean an individual constant or a free variable.

Formulas and their orders are defined simultaneously as follows. The order of a formula is a natural number.

1. If t_i is a term for $i < n$, then $t_0 = t_1$ and $R_n(t_0, \ldots, t_{n-1})$ are atomic formulas and an order of an atomic formula is zero.
2. If A is a formula, so is $\neg A$. The order of $\neg A$ is the same with the order of A.
3. If A_i is a formula for each $i < \omega$ and the maximum of orders of A_i $(i < \omega)$ exists, then $\bigwedge_{i<\omega} A_i$ and $\bigvee_{i<\omega} A_i$ are formulas and the order of these formulas are the same with the maximum of orders of A_i $(i < \omega)$.
4. If $A(a_0, a_1, \ldots, a_i, \ldots)$ $(i < \omega)$ is a formula and $x_0, x_1, \ldots, x_i, \ldots$ $(i < \omega)$ are distinct bound variables not occurring in $A(a_0, a_1, \ldots)$, then

$$\forall x_0 x_1 \ldots A(x_0, x_1, \ldots) \quad \text{and} \quad \exists x_0 x_1 \ldots A(x_0, x_1, \ldots)$$

are formulas and the order of these formulas is $n+1$, where n is the order of $A(a_0, a_1, \ldots)$.

5. If $A(a_0, a_1, \ldots, a_i, \ldots)$ $(i < \omega)$ is a formula without any occurrence of $\mathbf{Q}_1$ or $\mathbf{Q}_2$ and $x_0, x_1, \ldots, x_i, \ldots$ $(i < \omega)$ are distinct bound variables not occurring in $A(a_0, a_1, \ldots)$, then $\mathbf{Q}_i x_0 x_1 \ldots A(x_0, x_1, \ldots)$ $(i = 1, 2)$ are formulas and the order of these formulas is $n+1$, where n is the order of $A(a_0, a_1, \ldots)$. $\mathbf{Q}_1 x_0 x_1 \ldots$ and $\mathbf{Q}_2 x_0, x_1 \ldots$ are also denoted by $\exists x_0 \forall x_1 \exists x_2 \ldots$ and $\forall x_0 \exists x_1 \forall x_2 \ldots$, respectively.

A sequent is of the form

$$A_0, A_1, \ldots, A_\alpha, \ldots \to B_0, B_1, \ldots, B_\beta, \ldots,$$

where A_α's and B_β's are formulas and α and β range over ordinals less than α_0 and β_0, respectively, where α_0 and β_0 are some countable ordinals.

Inference rules are the same as in §22, except that the length of a sequent is restricted to be countable. Of course, $\forall$, $\exists$, $\mathbf{Q}_1$, $\mathbf{Q}_2$ should be exprèssed by an adequate form $\mathbf{Q}^f$.

Non-logical initial sequents are of the following form.

1. $\to t = t$, where t is an arbitrary term.
2. $i = j \to$, where i and j are distinct individual constants.
3. $\to t = 0, t = 1, t = 2, \ldots$, where t is an arbitrary term.

Now prove the following theorem.

THEOREM 23.37. *The projectile determinacy holds if and only if every provable sequent in the system defined above is provable without essential cuts.*

In his paper: Borel Determinacy, Annals of Mathematics (1975) 363–371, D. A. Martin proved Borel determinacy. In the following we consider the proof-theoretical meaning of Borel determinacy. First we define a logical system which is suitable for Borel determinacy.

DEFINITION 23.38. A Borel determinate logic denoted by **BDL** is obtained from **DL** by changing the language as follows.

(1) The heterogeneous quantifiers are restricted to the heterogeneous quantifiers of arity ω. Namely when we use $\mathbf{Q}^f$ in **BDL**, f is a map from ω into $\{\forall, \exists\}$.

(2) The conjunctions $\bigwedge$ and the disjunctions $\bigvee$ are also limited to arity at most ω.

(3) The arities of all predicate constants are finite.

(4) If $\mathbf{Q}^f$ is heterogeneous, then $\mathbf{Q}^f \boldsymbol{x}\, A(\boldsymbol{x})$ forms a formula in **BDL** only when $A(\boldsymbol{x})$ does not have any quantifiers. (If $\mathbf{Q}^f$ is homogeneous, there are no such restriction.)

(5) Individual constants $0, 1, 2, \ldots$ and the following initial sequents are introduced.

$$t = 0,\ t = 1,\ t = 2, \ldots,$$

where t is an arbitrary term.

Then Borel determinacy can be expressed by the following proposition.

PROPOSITION 23.39. *Let $\Gamma \rightarrow \Delta$ be a sequent in* **BDL**.

(1) *If $\Gamma \rightarrow \Delta$ is provable in* **BDL**, *then it is valid in any structure.*

(2) *Either there exists a cut-free proof of $\Gamma \rightarrow \Delta$ in* **BDL** *or else there exists a structure $\mathscr{A}$ such that every formula in Γ is satisfied in $\mathscr{A}$ and no formula in Δ is satisfied in $\mathscr{A}$.*

(3) *If $\Gamma \rightarrow \Delta$ is provable in* **BDL**, *then there exists a cut-free proof of $\Gamma \rightarrow \Delta$. Namely the cut-elimination theorem holds in* **BDL**.

PROOF. (1) is nothing but Borel determinacy. (2) is immediate from Theorem 23.6 and (3) is the consequence of (1) and (2).

A direct proof of the cut-elimination theorem of **BDL** is still open. This is a very interesting problem since the cut-elimination theorem of **BDL** implies Borel determinacy by the method of Theorem 23.34. If one finds such a direct proof, it will be very suggestive to the projective determinacy because of Theorem 23.37.

§24. A general theory of heterogeneous quantifiers

The problem of the completeness of logical systems is an interesting and important one. While much is known, open questions still exist. We know, for example, that first order logic is complete and second order logic is incomplete. For infinitary languages we know that homogeneous systems are complete but whether heterogeneous systems are complete is an open question.

Incompleteness is an inherent weakness in any logical system. In second and higher order systems we can partially compensate for this weakness by a heavy dependence on comprehension axioms and the axiom of choice. In the infinitary languages however, we do not have the comprehension axioms and indeed in determinate logic we are even denied the axiom of choice. This raises the very practical question of whether there exist useful alternatives to the comprehension axioms and the axiom of choice for infinitary languages. In this section we will explore such alternatives. In order to do this we will develop a very general theory of heterogeneous quantifiers, a theory that encompasses the quantifiers $\mathbf{Q}^f$ of determinate logic as a special form (well ordered) of heterogeneous quantifiers.

The system we will present is a very useful one. How to extend it to a complete system is an open question. But before we take up the definition we would like to point out a few things about the system. For one thing, in the right and left quantifier introduction rules we do not have the duality that exists in finite languages and in determinate logic. Although we will assume that the formal objects of our system are well ordered, that assumption is only for convenience and is not essential for the theory. For further simplification we will always omit the individual and function constants unless otherwise stated. Finally we point out that we will not specify the number of bound and free variables. It is understood that we first adjoin a sufficient supply of bound variables. Having fixed the number of bound variables we then adjoin a sufficient supply of free variables. For an explanation of what constitutes a "sufficient supply" we refer the reader to the discussion after Definition 22.1.

DEFINITION 24.1. (1) By a *language with heterogeneous quantifiers*, $L(J)$, where J is a set of mapping such that each T in J is a mapping from β to $P(\alpha)$ for some α and β, we mean the following collection of symbols.

1) Variables:
   1.1) Free variables: $a_0, a_1, \ldots, a_\mu, \ldots.$
   1.2) Bound variables: $x_0, x_1, \ldots, x_\nu, \ldots.$
2) Predicate constants of arity α for certain α's:

$$P_0^\alpha, P_1^\alpha, \ldots, P_\xi^\alpha, \ldots.$$

3) Logical symbols:
   $\neg$ (not),

$\supset$ (implication),
$\bigvee$ (disjunctions of arity α for certain α's),
$\bigwedge$ (conjunctions of arity α for certain α's),
$\forall$ (universal quantifiers of arity α for certain α's),
$\exists$ (existential quantifiers of arity α for certain α's).
4) Heterogeneous quantifiers: We have a quantifier $\mathbf{Q}^T$ for each T in the set J.
5) Auxiliary symbols: (,).

(2) The formulas of L(J) are defined in the usual way with the following modifications.

(2.1) If $\bigvee$ ($\bigwedge$) of arity α belongs to L(J) and A_μ, $\mu < \alpha$ is a sequence of formulas then $\bigvee_{\mu<\alpha} A_\mu$ ($\bigwedge_{\mu<\alpha} A_\mu$) is a formula.

(2.2) Let T be a function in J. Then for some α and β, T maps β into $P(\alpha)$. Let $A(\boldsymbol{a}, \boldsymbol{b})$ be a formula, where $\boldsymbol{a}$ and $\boldsymbol{b}$ are sequences of free variables of length α and β, respectively. We assume that some (possibly none) of the occurrences of $\boldsymbol{a}$ and $\boldsymbol{b}$ in A are indicated. Let $\boldsymbol{x}$ and $\boldsymbol{y}$ be sequences of bound variables, of length α and β respectively which do not occur in $A(\boldsymbol{a}, \boldsymbol{b})$. Then $\mathbf{Q}^T(\boldsymbol{x}; \boldsymbol{y})\, A(\boldsymbol{x}, \boldsymbol{y})$ is a formula.

(3) As before we assume that the collection of formulas of the language L(J) is closed under subformulas.

(4) Let K be the cardinality of the formulas of the language. A sequent $\Gamma \to \Delta$ is defined as usual, where the lengths of Γ and Δ are less than K^+.

EXAMPLE 24.2. Consider a language L(J) where there are countably many free variables (arranged in the order type of ω), the logical symbols are of finite arity and J is arbitrary. This means that the propositional connectives are defined as for the usual, finite languages and there are ω-many free variables. We shall assume that there are ω-many bound variables and a single predicate constant $=$. Let T be a function from 2 to $P(2)$ such that $T(0) = \{0\}$ and $T(1) = \{1\}$. Then

$$(a_0 = a_1 \equiv b_0 = b_1) \wedge b_0 \neq c$$

is a formula of the language which we denote by $A(a_0, a_1, b_0, b_1)$, where all the occurrences of a_0, a_1, b_0, b_1 are supposed to be indicated. Then

$$\mathbf{Q}^T(x_0, x_1; y_0, y_1)\, A(x_0, x_1, y_0, y_1)$$

is a formula. We are going to define a system in which this formula will have the meaning that for every x_0 there exists a y_0, depending on x_0 only, and for every x_1 there exists a y_1, depending on x_1 only, such that $A(x_0, x_1, y_0, y_1)$ holds. Such quantifiers, called dependent quantifiers, were first proposed by Henkin.

DEFINITION 24.3. The rules of inference of our intended system are those of Definition 22.1 with some alterations. We shall remark only on the crucial changes.

1) The $\bigwedge$: left, $\bigwedge$: right, $\bigvee$: left and $\bigvee$: right rules in (4.2) of Definition 22.1 are admitted only for conjunctions and disjunctions that belong to the language L(J), that is, only for values β_λ that are arities of conjunctions and disjunctions that belong to L(J).

2) The rules for quantifiers are quite different here.

2.1) $\mathbf{Q}$: left:

$$\frac{\{A_\lambda(\boldsymbol{a}_\lambda, \boldsymbol{b}_\lambda)\}_{\lambda<\gamma}, \Gamma \to \Delta}{\{\mathbf{Q}^{T_\lambda}(\boldsymbol{x}_\lambda\,;\, \boldsymbol{y}_\lambda)\, A_\lambda(\boldsymbol{x}_\lambda, \boldsymbol{y}_\lambda)\}_{\lambda<\gamma}, \Gamma \to \Delta},$$

where the variables of $\boldsymbol{b}_\lambda$ do not occur in $A_\lambda(\boldsymbol{x}_\lambda, \boldsymbol{y}_\lambda)$. If $\boldsymbol{a}_\lambda$ and $\boldsymbol{b}_\lambda$ are types α_λ and β_λ respectively, then T_λ is a function in J from β_λ to $P(\alpha_\lambda)$.

Any variable of $\boldsymbol{b}_\lambda$, say b, is called an eigenvariable of the inference, $A_\lambda(\boldsymbol{a}_\lambda, \boldsymbol{b}_\lambda)$ is called an auxiliary formula of b, and $\mathbf{Q}^{T_\lambda}(\boldsymbol{x}_\lambda; \boldsymbol{y}_\lambda)\, A_\lambda(\boldsymbol{x}_\lambda, \boldsymbol{y}_\lambda)$ is called the principal formula of b.

2.2) $\mathbf{Q}$: right:

$$\frac{\Gamma \to \Delta, \{A_\lambda(\boldsymbol{a}_\lambda, \boldsymbol{b}_\lambda)\}_{\lambda<\gamma}}{\Gamma \to \Delta, \{\mathbf{Q}^{T_\lambda}(\boldsymbol{x}_\lambda; \boldsymbol{y}_\lambda)\, A_\lambda(\boldsymbol{x}_\lambda, \boldsymbol{y}_\lambda)\}_{\lambda<\gamma}},$$

where if $\boldsymbol{a}_\lambda$ and $\boldsymbol{b}_\lambda$ are of types α_λ and β_λ respectively, then T_λ is a function in J from β_λ to $P(\alpha_\lambda)$.

Every variable of $\boldsymbol{a}_\lambda$, say a, is called an eigenvariable of the inference, $A_\lambda(\boldsymbol{a}_\lambda, \boldsymbol{b}_\lambda)$ is called an auxiliary formula of a, and $\mathbf{Q}^{T_\lambda}(\boldsymbol{x}_\lambda; \boldsymbol{y}_\lambda)\, A_\lambda(\boldsymbol{x}_\lambda, \boldsymbol{y}_\lambda)$ is called the principal formula of a.

3) The cut rule is replaced by the generalized cut rule (g.c.).

DEFINITION 24.4. A proof P in the system is defined in the usual way, as a tree consisting of sequents, where the following eigenvariable conditions must hold in P.

1) If a free variable b is used as an eigenvariable in more than one place, then the principal formulas of b must be the same.

2) Suppose $A(\boldsymbol{a}_1, \boldsymbol{b}_1)$ and $A(\boldsymbol{a}_2, \boldsymbol{b}_2)$ are auxiliary formulas of applications of $\mathbf{Q}$: left in which b is an eigenvariable.

2.1) If b occurs as the αth variable of $\boldsymbol{b}_1$ then b occurs as the αth variable of $\boldsymbol{b}_2$.

2.2) Let $a_{1,\lambda}$ and $a_{2,\lambda}$ denote the λth variable in $\boldsymbol{a}_1$ and $\boldsymbol{a}_2$, respectively. Suppose b is the αth variable in $\boldsymbol{b}_1$ and $\boldsymbol{b}_2$. Then for any λ in $T(\alpha)$, $a_{1,\lambda}$ and $a_{2,\lambda}$ are the same.

Let a be a free variable which is either an $a_{1,\lambda}$ or $a_{2,\lambda}$ where $\lambda \in T(\alpha)$, or it is a free variable in the principal formula of b. Then we say that b depends on a, and we write $a < b$.

REMARK 24.5. The above conditions do not imply that, in the notation of 2), the sequences b_1 and b_2 are identical. It is not guaranteed either that $\boldsymbol{a}_1$ and $\boldsymbol{a}_2$ are the same. Even if $\boldsymbol{b}_1$ and $\boldsymbol{b}_2$ happened to be the same, $\boldsymbol{a}_1$ and $\boldsymbol{a}_2$ may not be the same. For example, it is possible to have different $a_{1,\lambda}$ and

$a_{2,\lambda}$ if none of the variables of $\boldsymbol{b}_1$ depend on $a_{1,\lambda}$ and none of the variables of $\boldsymbol{b}_2$ depend on $a_{2,\lambda}$.

3) All the auxiliary formulas of an eigenvariable of the $\mathbf{Q}$: right rule are identical.

For this case the dependence is defined only between the eigenvariables and the free variables in the principal formula: Let a be an eigenvariable in $\boldsymbol{a}$ and let c be a free variable in $A(\boldsymbol{x}, \boldsymbol{y})$. Then a depends on c, i.e., $c < a$.

4) No eigenvariable occurs in the end-sequent of P.

5) We shall relate a and b by $a < b$ if there is a finite sequence of free variables $a_0, \ldots, a_n$, where $a_0 = a$, $a_n = b$, and $a_i < a_{i+1}$ for $0 \leqslant i \leqslant n-1$ in the sense of 2) and 3). Then $<$ is a partial well-ordering; that is there is no cycle $a_1 < a_2 < \ldots < a_1$ and there is no infinite decreasing sequence of variables $\{a_i\}_{i<\omega}$ such that $a_{i+1} < a_i$.

6) We define $a_1 \leqslant a_2$ to mean that $a_1 < a_2$ or a_1 and a_2 are the same. There exists a well ordering of all the auxiliary formulas of occurrences of the $\mathbf{Q}$: right rule in the proof, say $\{A_\xi(\boldsymbol{a}_\xi, \boldsymbol{b}_\xi)\}_\xi$, which satisfies the following, where the same $A_\xi(\boldsymbol{a}_\xi, \boldsymbol{b}_\xi)$ may appear in several different places in P.

Let $\mathbf{Q}^{T_\xi}(\boldsymbol{x}_\xi : \boldsymbol{y}_\xi)\, A_\xi(\boldsymbol{x}_\xi, \boldsymbol{y}_\xi)$ be the principal formula of an eigenvariable and let $A_\xi(\boldsymbol{a}_\xi, \boldsymbol{b}_\xi)$ be the corresponding auxiliary formula.

6.0) If $b_{0,\sigma}$ is the σth variable of $\boldsymbol{b}_0$, then

$$\{e \mid e \leqslant b_{0,\sigma} \text{ and } e \text{ is an eigenvariable of a } \mathbf{Q} : \text{right}\}$$

is a subset of

$$\{a_{0,\lambda} \mid a_{0,\lambda} \text{ is the } \lambda\text{th variable of } \boldsymbol{a}_0 \text{ and } \lambda \in T_0(\sigma)\}$$

and if c is a free variable in A_0 which is neither in $\boldsymbol{a}_0$ nor $\boldsymbol{b}_0$, then for no eigenvariable e of a $\mathbf{Q}$: right, is $e \leqslant c$.

6.ξ) If $b_{\xi,\alpha}$ is the αth variable of $\boldsymbol{b}_\xi$, then

$$\{e \mid e \leqslant b_{\xi,\alpha} \text{ and } e \text{ is an eigenvariable of a } \mathbf{Q} : \text{right}\}$$

is a subset of

$$\bigcup_{\eta<\xi} \boldsymbol{a}_\eta \cup \{a_{\xi,\lambda} \mid \lambda \in T_\xi(\alpha)\}.$$

Also, if c is a free variable in A_ξ which does not belong to $\boldsymbol{a}_\xi$, or $\boldsymbol{b}_\xi$, then

$$\{e \mid e \leqslant c \text{ and } e \text{ is an eigenvariable of a } \mathbf{Q} : \text{right}\}$$

is a subset of $\bigcup_{\eta<\xi} \boldsymbol{a}_\eta$.

NOTATIONAL CONVENTION. For quantifiers that are relatively simple, we shall use more intuitive notation. For example

$$\begin{pmatrix} \forall x & \exists u \\ \forall y & \exists w \end{pmatrix} A(x, y, u, w)$$

can express $\mathbf{Q}^T(xy;\, uw)\, A(x, y, u, w)$, where $T(0) = \{0\}$ and $T(1) = \{1\}$. $\mathbf{Q}^T(x_0x_1 \ldots;)$ can be expressed with the usual notation $\forall x_0\, x_1 \ldots$ and $\mathbf{Q}^T(;\, x_0x_1 \ldots)$ with $\exists x_0\, x_1 \ldots$.

EXAMPLE 24.6. Examples of proofs with heterogeneous quantifiers.

$$1) \quad \begin{pmatrix} \forall x & \exists u \\ \forall y & \exists w \end{pmatrix} ((x = y \equiv u = w) \wedge u \neq a),\, \Gamma_0 \rightarrow \exists z_0 z_1 \ldots \bigwedge_{i \neq j} (z_i \neq z_j)$$

is provable, where Γ_0 is the equality axioms $\forall x\, (x = x)$, $\forall x \forall y\, (x = y \supset y = x)$, $\forall x \forall y \forall z\, (x = y \wedge y = z \supset x = z)$.

PROOF.

$$1) \quad \{(c_{k+j} = c_k \equiv c_{k+j+1} = a_k) \wedge c_{k+j+1} \neq c_0\}_{k,j<\omega},\, \Gamma_0 \rightarrow \bigwedge_{i \neq j} (c_i \neq c_j)$$

is obvious. From (1), by a $\mathbf{Q}$: right, where there are no eigenvariables involved,

(2)

$$\{(c_{h+j} = c_h \equiv c_{h+j+1} = a_h) \wedge c_{h+j+1} \neq c_0\}_{h,j<\omega},\, \Gamma_0 \rightarrow \exists z_0 z_1 \ldots \bigwedge_{i \neq j} (z_i \neq z_j).$$

Consider $(c_{k+j} = c_k \equiv c_{k+j+1} = a_k) \wedge c_{k+j+1} \neq c_0$, or $A(c_{k+j}, c_{k+j+1}, a_k)$, leaving out c_0. Define $T(0) = \{0\}$ and $T(1) = \{1\}$. Then $\mathbf{Q}^T(xy;\, uw)\, A(x, y, u, w)$ means

$$\begin{pmatrix} \forall x & \exists u \\ \forall y & \exists w \end{pmatrix} ((x = y \equiv u = w) \wedge u \neq c_0).$$

This applies to all pairs (k, j); hence by a $\mathbf{Q}^T$: left applied to all the formulas in the antecedent of (2), followed by contraction, we obtain

$$\mathbf{Q}^T(xy;\, uw)\, A(x, y, u, w),\, \Gamma_0 \rightarrow \exists z_0 z_1 \ldots \bigwedge_{i \neq j} (z_i \neq z_j).$$

By renaming c_0 as a, we obtain the required sequent. In order to see that we have given a proper proof, we examine the conditions in Definition 24.4. Since all the principal formulas are the same, 1) is obvious. Also 2.1) is obvious, since the c's are the first and the a's are the second eigenvariables in any auxiliary formulas. For 2.2), let $b = c_{k+j+1}$. Then $T(0) = \{0\}$ and the 0th variable in (c_{k+j}, c_k), i.e., c_{k+j} is uniquely determined by b. Similarly with a_k. Since there is no eigenvariable for a $\mathbf{Q}$: right, we do not have to worry about 3). Since the eigenvariables are c_{k+j+1} and a_k and not c_0 4) is obvious. As for 5), $c_{k+j} < c_{k+j+1}$, $c_k < a_k$, $c_0 < c_{k+j+1}$, $c_0 < a_k$ exhaust all the dependence relations. It is then easily seen that $<$ is a partial well-ordering. Clearly 6) is irrelevant.

2) $$\forall x\, \exists y\, A(x, y),$$
$$\forall x\, \forall y\, (A(x, y) \supset y \neq c_0),$$
$$\forall x\, \forall y\, \forall u\, \forall v\, (A(x, u) \wedge A(y, v) \supset (x = y \equiv u = v))$$
$$\rightarrow \begin{pmatrix} \forall x & \exists u \\ \forall y & \exists v \end{pmatrix} ((x = y \equiv u = v) \wedge u \neq c_0\,.$$

PROOF.

$$A(a, c),\, A(b, d),\, A(a, c) \supset c \neq c_0,$$
$$A(a, c) \wedge A(b, d) \supset (a = b \equiv c = d) \rightarrow$$
$$\rightarrow (a = b \equiv c = d) \wedge c \neq c_0$$

is obvious. We can introduce $\forall$'s to all the variables in the antecedent except c_0. Let $A(a, b, c, d)$ denote $(a = b \equiv c = d) \wedge c \neq c_0$. Let $T(0) = \{0\}$ and $T(1) = \{1\}$. Then $\mathbf{Q}^T(xy : uv)\, A(x, y, u, v)$ is the formula in the succedent.

Furthermore, 5) of the eigenvariable conditions is obviously satisfied. Since there is only one auxiliary formula of a $\mathbf{Q}$: right introduction, 6) is also easy to see.

DEFINITION 24.7. Let $\mathscr{A}$ be a structure for our language. Let Φ be an assignment from $\mathscr{A}$. The relation that a formula A is satisfied in $\mathscr{A}$ by Φ is defined as usual. $\mathbf{Q}^T(\boldsymbol{x}; \boldsymbol{y})\, A(\boldsymbol{x}, \boldsymbol{y})$ is satisfied if and only if the following holds. Let $\boldsymbol{x}$ and $\boldsymbol{y}$ be of lengths α and β and let $\boldsymbol{a}$ and $\boldsymbol{b}$ be new free variables corresponding to $\boldsymbol{x}$ and $\boldsymbol{y}$. There exists a sequence of functions $\boldsymbol{f}$ corresponding to $\boldsymbol{b}$ such that for every sequence $\boldsymbol{d}$ of elements of $\mathscr{A}$ of length α, if

$$\Phi' = \begin{pmatrix} \boldsymbol{a} & \boldsymbol{b} \\ \boldsymbol{d} & \boldsymbol{f}(\bar{\boldsymbol{d}}) \end{pmatrix},$$

then $A(\boldsymbol{a}, \boldsymbol{b})$ is satisfied in $\mathscr{A}$ by Φ', where $f(\bar{\boldsymbol{d}})$ is a sequence of terms such that if

$$T(\gamma) = \{\xi_0, \xi_1, \ldots, \xi_i, \ldots\},$$

then the γth expression is $f_\gamma(d_{\xi_0}, d_{\xi_1}, \ldots, d_{\xi_i}, \ldots)$.

THEOREM 24.8 (validity for heterogeneous quantifiers). *Every theorem of our system of heterogeneous quantifiers is valid.*

PROOF. The proof is similar to the proof of Theorem 23.4. Given a proof P in the system, a structure $\mathscr{A}$ and an assignment from $\mathscr{A}$ we first take an arbitrary formula in the antecedent of a sequent with a quantifier at the beginning, say

$$\mathbf{Q}^T(\boldsymbol{x}; \boldsymbol{y})\, A(\boldsymbol{x}, \boldsymbol{y}, \boldsymbol{c}),$$

where $\boldsymbol{c}$ is the sequence of all the free variables in this formula and the lengths of $\boldsymbol{x}$ and $\boldsymbol{y}$ are α and β respectively. For each $\gamma < \beta$ we introduce the Skolem function

$$g_A^{T,\gamma}(x_{\xi_0}, x_{\xi_1}, \ldots, \boldsymbol{c}),$$

where $x_{\xi_0}, x_{\xi_1}, \ldots$ are all the variables of $\boldsymbol{x}$ such that $\xi_i \in T(\gamma)$. $g_A^{T,\gamma}$ is interpreted as follows with respect to $\mathcal{A}$. If $\mathbf{Q}^T(\boldsymbol{x}; \boldsymbol{y})\, A(\boldsymbol{x}, \boldsymbol{y}, \boldsymbol{c})$ is satisfied in $\mathcal{A}$, then the $g_A^{T,\gamma}$'s are the functions satisfying

$$(+) \qquad \forall \boldsymbol{x}\, A(\boldsymbol{x}, \boldsymbol{y}', \boldsymbol{c}),$$

where the γth expression of $\boldsymbol{y}'$ is $g_A^{T,\gamma}(x_{\xi_0}, x_{\xi_1}, \ldots, \boldsymbol{c})$. If $\mathbf{Q}^T(\boldsymbol{x}; \boldsymbol{y})\, A(\boldsymbol{x}, \boldsymbol{y})$ is not satisfied in $\mathcal{A}$, then the $g_A^{T,\gamma}$'s are interpreted to be constant functions for a distinguished element of the universe of $\mathcal{A}$.

Well-order all the eigenvariables in P for $\mathbf{Q}$: left introductions in such a way that if $a < b$ (b depends on a), then a precedes b in the ordering:

$$b_0, b_1, \ldots, b_\delta, \ldots.$$

We shall define terms $t_0, t_1, \ldots, t_\delta, \ldots$ by transfinite induction on β. Assuming that $\boldsymbol{t}_{<\delta}$ have been defined, we show how to define t_δ.

Suppose the principal formula of b_δ is $\mathbf{Q}^T(\boldsymbol{x}; \boldsymbol{y})\, A(\boldsymbol{x}, \boldsymbol{y}, \boldsymbol{c})$ and let d be a variable in an auxiliary formula of b_δ which corresponds to a variable in $\boldsymbol{x}$ with $d < b$. If d is not used as an eigenvariable of any $\mathbf{Q}$: left, then define u corresponding to d to be d itself; otherwise d occurs in the above list of eigenvariables, hence is a b_κ, $\kappa < \delta$, since $d < b$. Therefore t_κ has been defined and we take u to be this t_κ. Let c be a free variable in $\boldsymbol{c}$. A term s corresponding to c is defined as the u corresponding to d; recall that $c < b$ by the eigenvariable condition. It should be noticed that those d's and c's are the same for all auxiliary formulas of b_δ by virtue of the eigenvariable condition. Thus t_δ can be defined to be $g_A^{T,\lambda}(\boldsymbol{u}, \boldsymbol{s})$. From the definition, a free variable in t_δ, say d, satisfies $d < b_\delta$. Now substitute $t_0, t_1, \ldots, t_\delta, \ldots$ for $b_0, b_1, \ldots, b_\delta, \ldots$, respectively, in P. Let P' be the figure thus obtained from P.

Let $\{A_\xi(\boldsymbol{a}_\xi, \boldsymbol{b}_\xi, \boldsymbol{c}_\xi)\}_\xi$ be the well-ordering of the auxiliary formulas of the $\mathbf{Q}$: right in P satisfying the conditions in 6) of Definition 24.3, where $\boldsymbol{c}_\xi$ is the sequence of all the free variables in A_ξ which do not occur in $\boldsymbol{a}_\xi$ or $\boldsymbol{b}_\xi$.

We shall define substitutions of terms for eigenvariables of the $\mathbf{Q}$: right introductions in P' in the following manner. Suppose the substitution has been completed for the ξth stage, giving us a figure P_ξ.

Applying this substitution to $\{A_\xi(\boldsymbol{a}_\xi, \boldsymbol{b}_\xi, \boldsymbol{c}_\xi)\}_\xi$ we obtain formulas in P' that we will denote by $\{A_\xi(\boldsymbol{a}_\xi, \boldsymbol{b}'_\xi, \boldsymbol{c}'_\xi)\}_\xi$. Here $\boldsymbol{b}'_\xi$ and $\boldsymbol{c}'_\xi$ are terms that may contain many free variables. Because of (6) of Definition 24.3, $b'_{\xi,\sigma}$ and $c'_{\xi,\sigma}$ satisfy the following condition.

(*) The eigenvariables of the $\mathbf{Q}$: right that are contained in $b'_{\xi,\sigma}$ form a subset of

$$\bigcup_{\eta<\xi} \boldsymbol{a}_\eta \cup \{a_{\xi,\lambda_0}, a_{\xi,\lambda_1}, \ldots\}$$

where

$$\{\lambda_0, \lambda_1, \ldots\} = T_\xi(\sigma)$$

and any eigenvariable of a $\mathbf{Q}$: right that is contained in $c'_{\xi,\sigma}$ is an eigenvariable of an $\boldsymbol{a}_\eta$ for some $\eta < \xi$.

Next we define, by transfinite induction on ξ, substitutions of members of $\mathscr{A}$ for $\boldsymbol{a}_\xi$.

Suppose we have completed the definition of substitutions for eigenvariables of $\boldsymbol{a}_\eta$, $\eta < \xi$. Consider the stage ξ. We shall define $a^*_{\zeta,\gamma}$ corresponding to each $a_{\zeta,\gamma}$ in $\boldsymbol{a}_\xi$.

Case 1. $\mathbf{Q}^{T_\xi}(\boldsymbol{x}_\xi; \boldsymbol{y}_\xi)\, \tilde{A}_\xi(\boldsymbol{x}_\xi, \boldsymbol{y}_\xi)$ is true, where $\tilde{A}_\xi(\boldsymbol{x}_\xi, \boldsymbol{y}_\xi)$ is obtained from $A_\xi(\boldsymbol{x}_\xi, \boldsymbol{y}_\xi)$ by substitutions up to the ξth stage. Let k_0 be a distinguished element of the given structure $\mathscr{A}$. Then $a^*_{\zeta,\gamma}$ is interpreted to be k_0.

Case 2. $\mathbf{Q}^{T_\xi}(\boldsymbol{x}_\xi; \boldsymbol{y}_\xi)\, \tilde{A}_\xi(\boldsymbol{x}_\xi, \boldsymbol{y}_\xi)$ is false. Henceforth we omit the subscript ξ unless needed.

By hypothesis, $\neg\mathbf{Q}^T(\boldsymbol{x}; \boldsymbol{y})\, \tilde{A}(\boldsymbol{x}, \boldsymbol{y})$ is true, or $\exists \boldsymbol{x} \neg\tilde{A}(\boldsymbol{x}, \boldsymbol{f}(\boldsymbol{x}))$ is true no matter what the interpretation of $\boldsymbol{f}$ is, where $\boldsymbol{f}(\boldsymbol{x})$ stands for the sequence

$$f_0(\boldsymbol{x}_0), f_1(\boldsymbol{x}_1), \ldots, f_\sigma(\boldsymbol{x}_\delta), \ldots (\sigma < \beta),$$

$\boldsymbol{x}_\sigma$ being $x_{\sigma_0}, x_{\sigma_1}, \ldots, x_{\sigma_i}, \ldots$, with $T(\sigma) = \{\sigma_0, \sigma_1, \ldots, \sigma_i, \ldots\}$. $A(\boldsymbol{a}, \boldsymbol{b}, \boldsymbol{c})$ has become $A(\boldsymbol{a}, \boldsymbol{u}, \boldsymbol{v})$ where u_σ is a term that may contain unsubstituted a's as free variables. By (*), the free variables in u_σ, form a subset of $\boldsymbol{a}_\sigma = \{a_{\sigma_0}, a_{\sigma_1}, \ldots, a_{\sigma_i}, \ldots\}$, where

$$\{\sigma_0, \sigma_1, \ldots, \sigma_i, \ldots\} = T(\sigma).$$

Therefore, we may put $f_\sigma(\boldsymbol{a}_\sigma) = u_\sigma$. Since for this interpretation of the f_σ's, $\exists \boldsymbol{x} \neg\tilde{A}(\boldsymbol{x}, f(\boldsymbol{x}))$ is true, there are values of $\boldsymbol{a}$, say $\boldsymbol{a}^*$, for which $\neg\tilde{A}(\boldsymbol{a}^*, f(\boldsymbol{a}^*))$. When these substitutions are completed, all the eigenvariables are replaced by the g_A's and the $\boldsymbol{a}^*$'s. We shall call the resulting figure P^*.

Now we can show that every sequent in P^* is true. Since the end-sequent does not contain any eigenvariable, this implies that the end-sequent of P is true.

For the proof that every sequent in P^* is true in $\mathscr{A}$, under the given assignment, we shall deal with three crucial cases only.

1) g.c.:

$$\frac{\Phi, \Gamma \to \Delta, \Psi \text{ for all appropriate } (\Phi, \Psi)}{\Gamma \to \Delta},$$

where $\mathscr{F}$ is the set of cut formulas.

Let $\mathscr{F}_1$ be the set of all formulas of $\mathscr{F}$ which are true and let $\mathscr{F}_2$ be the rest of the formulas. If for $\Phi \subseteq \mathscr{F}_1$ and $\Psi \subseteq \mathscr{F}_2$, $\Phi, \Gamma \to \Delta, \Psi$ is provable, then by the induction hypothesis this sequent is true, where all the formulas in Φ are true and those in Ψ are false. Therefore $\Gamma \to \Delta$ must be true.

2) **Q** : left:

$$\frac{\{\tilde{A}'(\boldsymbol{w}, \boldsymbol{u})\}, \Gamma \to \Delta}{\{\mathbf{Q}^T(\boldsymbol{x}; \boldsymbol{y})\, A'(\boldsymbol{x}, \boldsymbol{y})\}, \Gamma \to \Delta}.$$

It suffices to show that if $\mathbf{Q}^T(\boldsymbol{x}; \boldsymbol{y})\, A'(\boldsymbol{x}, \boldsymbol{y})$ is true then so is $\tilde{A}'(\boldsymbol{w}, \boldsymbol{u})$. However, this is obvious since the $g_A^{T,\gamma}$'s are so chosen (cf. (+) in the definition of $g_A^{T,\gamma}$).

3) **Q** : right:

$$\frac{\Gamma \to \Delta, \{A(\boldsymbol{w}, \boldsymbol{u})\}}{\Gamma \to \Delta, \{\mathbf{Q}^T(\boldsymbol{x}; \boldsymbol{y})\, A(\boldsymbol{x}, \boldsymbol{y})\}}.$$

It suffices to show that if $\mathbf{Q}^T(\boldsymbol{x}; y)\, A(\boldsymbol{x}, \boldsymbol{y})$ is false, then so is $A(\boldsymbol{w}, \boldsymbol{u})$, but this too is obvious, since the a^*_γ's are so chosen.

Since a homogeneous system is a subsystem of a system which satisfies (Q), Proposition 22.14 implies that a homogeneous system is a subsystem of a heterogeneous system.

Heterogeneous quantifiers, even the finite ones, are stronger than homogeneous quantifiers.

PROPOSITION 24.9. *The heterogeneous quantifier*

$$\begin{pmatrix} \forall x & \exists u \\ \forall y & \exists v \end{pmatrix}$$

(*cf. Example* 24.6) *cannot be expressed by* (*finite*) *first order quantifiers.*

PROOF. Consider a formula of the form

$$\begin{pmatrix} \forall x & \exists u \\ \forall y & \exists v \end{pmatrix} ((x = y \equiv u = v) \wedge A(x, y, u, v)), \tag{1}$$

in which

$$\begin{pmatrix} \forall x & \exists u \\ \forall y & \exists v \end{pmatrix}$$

is the only quantifier that occurs. In a second order system with function quantifiers and the axiom of choice, (1) is equivalent to the following formulas:

$$\exists f\, \exists g\, \forall x\, \forall y\, ((x = y \equiv f(x) = g(y)) \wedge A(x, y, f(x), g(y)));$$

$$\exists f\, \exists g\, (\forall x\, \forall y\, (x = y \equiv f(x) = g(y)) \wedge \forall x\, \forall y\, A(x, y, f(x), g(y)));$$

$$\exists f\, \forall x\, \forall y\, A(x, y, f(x), f(y)).$$

Define $A(x, y, f(x), f(y))$ to be

$$y = x + 1 \supset f(y) < f(x).$$

Then $\exists f \forall x \forall y A(x, y, f(x), f(y))$ expresses "$<$ is not well-founded". Although the set of natural numbers is well-founded its non-standard enlargement is not well-founded. Since both of them satisfy the same first order sentences, we conclude that (1) cannot be expressed in terms of homogeneous, first order quantifiers.

Proposition 24.9 explains why we have to place different eigenvariable conditions according as a quantifier is introduced in the antecedent or the succedent. If we were to use the same conditions on eigenvariables in the succedent as those in the antecedent, we would have

$$\neg\begin{pmatrix}\forall x & \exists u\\ \forall y & \exists v\end{pmatrix} A(x, y, u, v) \leftrightarrow \begin{pmatrix}\exists x & \forall u\\ \exists y & \forall v\end{pmatrix} \neg A(x, y, u, v).$$

On the other hand,

$$\begin{pmatrix}\exists x & \forall u\\ \exists y & \forall v\end{pmatrix} \neg A(x, y, u, v) \leftrightarrow \exists x \exists y \forall u \forall v \neg A(x, u, y, v).$$

So it would follow that

$$\begin{pmatrix}\forall x & \exists u\\ \forall y & \exists v\end{pmatrix} A(x, y, u, v) \leftrightarrow \forall x \forall y \exists u \exists v A(x, y, u, v),$$

contradicting Proposition 24.9.

EXAMPLE 24.10. Let $S(a_1, b_1, \ldots, a_n, b_n, \ldots)$ be a formula with free variables $a_1, b_1, \ldots, a_n, b_n, \ldots$ and let $S_n(a_1, b_1, \ldots, a_n, b_n)$ be short for

$$\exists x_{n+1} \forall y_{n+1} \ldots S(a_1, b_1, \ldots, a_n, b_n, x_{n+1}, y_{n+1}, \ldots).$$

Then

$$\exists u_1 \forall v_1 \ldots (\bigvee_n S_n(u_1, v_1, \ldots, u_n, v_n)) \rightarrow$$
$$\rightarrow \exists x_1 \forall y_1 \ldots \exists x_n \forall y_n \ldots S(x_1, y_1, \ldots, x_n, y_n, \ldots).$$

First of all, since

$$S_{n-1}(a_1, b_1, \ldots, a_{n-1}, b_{n-1}) \equiv \exists x_n \forall y_n S_n(a_1, \ldots, x_n, y_n)$$

is easily proved, we can identify

$$S_{n-1}(a_1, \ldots, b_{n-1}) \quad \text{and} \quad \exists x_n \forall y_n S_n(a_1, \ldots, b_{n-1}, x_n, y_n)$$

by using cuts.

For every n, we first consider a figure P_n ending with

$$S_n(a_1, b_1, \ldots, a_n, b_n) \rightarrow S_0.$$

We shall demonstrate P_3 as an example:

$$\cfrac{\cfrac{\cfrac{S_3(a_1, b_1, a_2, b_2, \ldots) \to S_3(a_1, b_1, a_2, b_2, a_3, b_3)}{S_3(\ldots) \to \exists x_3 \forall y_3\, S_3(a_1, b_1, a_2, b_2, x_3, y_3)}}{S_3(\ldots) \to \exists x_2 \forall y_2 \exists x_3 \forall y_3\, S_3(a_1, b_1, x_2, y_2, x_3, y_3)}}{S_3(\ldots) \to \exists x_1 \forall y_1 \exists x_2 \forall y_2 \exists x_3 \forall y_3\, S_3(x_1, y_1, x_2, y_2, x_3, y_3).}$$

It is important to note that we do not introduce $S_3(\ldots) \to S_0$ in one step. Now,

$$\cfrac{\cfrac{P_n,\ n < \omega}{\bigvee_n S_n(a_1, b_1, \ldots, a_n, b_n) \to S_0}}{\exists u_1 \forall v_1 \ldots (\bigvee_n S_n(u_1, v_1, \ldots, u_n, v_n)) \to S_0.}$$

It is easy to see that the eigenvariable conditions are satisfied: The P_n's are carefully constructed so that the auxiliary formulas of an eigenvariable b_i are all identical. $\{S_n(a_1, b_1, \ldots, a_n, b_n)\}_{n>0}$ is the required enumeration of the auxiliary formulas of the $\mathbf{Q}$: right introductions.

EXAMPLE 24.11. In order to state the next example we need some auxiliary definitions. Consider a quantifier of the form $\mathbf{Q}^T(\boldsymbol{x}; \boldsymbol{y})$, where the lengths of $\boldsymbol{x}$ and $\boldsymbol{y}$ are α and β, respectively. Suppose α and β can each be decomposed into two sets $\tilde{\alpha}_1, \tilde{\alpha}_2$ and $\tilde{\beta}_1, \tilde{\beta}_2$ respectively, i.e., $\alpha = \tilde{\alpha}_1 \cup \tilde{\alpha}_2$ and $\beta = \tilde{\beta}_1 \cup \tilde{\beta}_2$ where $\tilde{\alpha}_1 \cap \tilde{\alpha}_2 = \emptyset$ and $\tilde{\beta}_1 \cap \tilde{\beta}_2 = \emptyset$, and in addition,

(1) $T(\gamma) \subseteq \tilde{\alpha}_1$ if $\gamma \in \tilde{\beta}_1$,

(2) $T(\gamma) \subseteq \tilde{\alpha}_2$ if $\gamma \in \tilde{\beta}_2$.

If we well-order $\tilde{\alpha}_1, \tilde{\alpha}_2, \tilde{\beta}_1, \tilde{\beta}_2$ each and restrict T to $\tilde{\beta}_1$ and $\tilde{\beta}_2$ respectively, then we obtain T_1 and T_2 such that

$$\forall \gamma \in \tilde{\beta}_1 \,(T_1(\gamma) = T(\gamma) \subseteq \tilde{\alpha}_1) \qquad \forall \gamma \in \tilde{\beta}_2 \,(T_2(\gamma) = T(\gamma) \subseteq \tilde{\alpha}_2).$$

Suppose $A(\boldsymbol{x}, \boldsymbol{y})$ can be expressed as $A'(\boldsymbol{x}_1, \boldsymbol{x}_2, \boldsymbol{y}_1, \boldsymbol{y}_2)$, where $(\boldsymbol{x}_1, \boldsymbol{x}_2)$ is the partition of $\boldsymbol{x}$ determined by $(\tilde{\alpha}_1, \tilde{\alpha}_2)$; and similarly with $(\boldsymbol{y}_1, \boldsymbol{y}_2)$.

We now show that under those circumstances

$$\mathbf{Q}^T(\boldsymbol{x}; \boldsymbol{y})\, A(\boldsymbol{x}, \boldsymbol{y}) \leftrightarrow \mathbf{Q}^{T_1}(\boldsymbol{x}_1; \boldsymbol{y}_1)\, \mathbf{Q}^{T_2}(\boldsymbol{x}_2; \boldsymbol{y}_2)\, A'(\boldsymbol{x}_1, \boldsymbol{x}_2, \boldsymbol{y}_1, \boldsymbol{y}_2)$$

is provable.

Suppose that $A(\boldsymbol{a}, \boldsymbol{b})$ can be written as $A'(\boldsymbol{a}_1, \boldsymbol{a}_2, \boldsymbol{b}_1, \boldsymbol{b}_2)$ corresponding to $\tilde{\alpha}_1, \tilde{\alpha}_2, \tilde{\beta}_1, \tilde{\beta}_2$.

1)
$$\cfrac{\cfrac{\cfrac{A'(\boldsymbol{a}_1, \boldsymbol{a}_2, \boldsymbol{b}_1, \boldsymbol{b}_2) \to A(\boldsymbol{a}, \boldsymbol{b})}{\mathbf{Q}^{T_2}(\boldsymbol{x}_2; \boldsymbol{y}_2)\, A'(\boldsymbol{a}_1, \boldsymbol{x}_2, \boldsymbol{b}_1, \boldsymbol{y}_2) \to A(\boldsymbol{a}, \boldsymbol{b})}}{\mathbf{Q}^{T_1}(\boldsymbol{x}_1; \boldsymbol{y}_1)\, \mathbf{Q}^{T_2}(\boldsymbol{x}_2; \boldsymbol{y}_2)\, A'(\boldsymbol{x}_1, \boldsymbol{x}_2, \boldsymbol{y}_1, \boldsymbol{y}_2) \to A(\boldsymbol{a}, \boldsymbol{b})}}{\mathbf{Q}^{T_1}(\boldsymbol{x}_1; \boldsymbol{y}_1)\, \mathbf{Q}^{T_2}(\boldsymbol{x}_2; \boldsymbol{y}_2)\, A'(\boldsymbol{x}_1, \boldsymbol{x}_2, \boldsymbol{y}_1, \boldsymbol{y}_2) \to \mathbf{Q}^T(\boldsymbol{x}; \boldsymbol{y})\, A(\boldsymbol{x}, \boldsymbol{y})}$$

In introducing $\mathbf{Q}^{T_2}$, the variables of $\boldsymbol{b}_2$ depend on all the variables of $\boldsymbol{a}_1$ as well as those of $\boldsymbol{b}_1$ and some of the variables of $\boldsymbol{a}_2$ (determined by T_2).

(2)
$$\cfrac{\cfrac{\cfrac{A(\boldsymbol{c}, \boldsymbol{d}) \to A'(\boldsymbol{c}_1, \boldsymbol{c}_2, \boldsymbol{d}_1, \boldsymbol{d}_2)}{A(\boldsymbol{c}, \boldsymbol{d}) \to \mathbf{Q}^{T_2}(\boldsymbol{x}_2; \boldsymbol{y}_2)\, A'(\boldsymbol{c}_1, \boldsymbol{x}_2, \boldsymbol{d}_1, \boldsymbol{y}_2)}}{A(\boldsymbol{c}, \boldsymbol{d}) \to \mathbf{Q}^{T_1}(\boldsymbol{x}_1; \boldsymbol{y}_1)\, \mathbf{Q}^{T_2}(\boldsymbol{x}_2; \boldsymbol{y}_2)\, A'(\boldsymbol{x}_1, \boldsymbol{x}_2, \boldsymbol{y}_1, \boldsymbol{y}_2)}}{\mathbf{Q}^{T}(\boldsymbol{x}; \boldsymbol{y})\, A(\boldsymbol{x}, \boldsymbol{y}) \to \mathbf{Q}^{T_1}(\boldsymbol{x}_1; \boldsymbol{y}_1)\, \mathbf{Q}^{T_2}(\boldsymbol{x}_2; \boldsymbol{y}_2)\, A'(\boldsymbol{x}_1, \boldsymbol{x}_2, \boldsymbol{y}_1, \boldsymbol{y}_2)}$$

From the partition of the variables, it is evident that the variables of $\boldsymbol{d}_1$ do not depend on the variables of $\boldsymbol{c}_2$, hence a cycle can be avoided.

It is easy to see that 6) of the eigenvariable conditions (Definition 24.4) is satisfied. The other conditions are obvious.

EXERCISE 24.12. Show that the following is provable

$$\exists x_1\, P_1(x_1), \forall x_1\, \forall y_1\, \exists x_2\, (P_1(x_1) \supset P_2(x_1, y_1, x_2)), \ldots,$$
$$\forall x_1 \ldots y_{n-1}\, \exists x_n\, (P(x_1) \wedge P_2(x_1, y_1, x_2) \wedge \ldots \wedge P_{n-1}(x_1, y_1, \ldots, x_{n-1}) \supset$$
$$\supset P_n(x_1, \ldots, y_{n-1}, x_n)), \ldots,$$
$$\forall x_1\, y_1 \ldots (\bigwedge_n P_n(x_1, \ldots, x_n) \supset S(x_1, y_1, \ldots)) \to$$
$$\to \exists x_1\, \forall y_1\, \exists x_2\, \forall y_2 \ldots S(x_1, y_1, x_2, y_2, \ldots).$$

[*Hint*: The sequent

$$P_1(a_1), P_1(a_1) \supset P_2(a_1, b_1, a_2), \ldots,$$
$$\bigwedge_n P_n(a_1, \ldots, b_{n-1}, a_n) \supset S(a_1, b_1, \ldots) \to S(a_1, b_1, \ldots)$$

can easily be seen to be provable. From this the desired sequent follows.]

We can improve our system of heterogeneous quantifiers as follows.

DEFINITION 24.13 (cf. Definitions 24.3 and 24.4). (1) Add to 2) of Definition 24.4 the following. If $A(\boldsymbol{a}_1)$ and $A(\boldsymbol{a}_2)$ are auxiliary formulas of an eigenvariable a of an application of a homogeneous $\mathbf{Q}$: right and a is the αth variable of $\boldsymbol{a}_1$, then a is the αth variable of $\boldsymbol{a}_2$.

To the definition of $<$, the following is added. If a is an eigenvariable of a homogeneous $\mathbf{Q}$: right and b is a free variable which occurs in the principal formula of a, then $b < a$, i.e., a depends on b.

(2) Part 3) of Definition 24.4 should read: All the auxiliary formulas of an eigenvariable of a heterogeneous $\mathbf{Q}$: right are identical.

(3) In 6) of Definition 24.4, read "heterogeneous $\mathbf{Q}$: right" in place of "$\mathbf{Q}$: right".

(4) Suppose $\mathbf{Q}^T(\boldsymbol{x}; \boldsymbol{y})$ can be split into $\mathbf{Q}^{T_1}(\boldsymbol{x}_1; \boldsymbol{y}_1)$ and $\mathbf{Q}^{T_2}(\boldsymbol{x}_2; \boldsymbol{y}_2)$ in the sense of Example 24.11. We shall abbreviate those quantifiers as $\mathbf{Q}$, $\mathbf{Q}_1$ and $\mathbf{Q}_2$.

Then introduce a new rule of inference:

$$\frac{\Gamma_1, \{\mathbf{Q}_1^\alpha \mathbf{Q}_2^\alpha\, A_\alpha(\boldsymbol{x}^\alpha, \boldsymbol{y}^\alpha)\}_\alpha, \Gamma_2 \to \Delta_1, \{\mathbf{Q}_1^\beta \mathbf{Q}_2^\beta\, B_\beta(\boldsymbol{u}^\beta, \boldsymbol{v}^\beta)\}_\beta, \Delta_2}{\Gamma_1, \{\mathbf{Q}^\alpha(\boldsymbol{x}^\alpha; \boldsymbol{y}^\alpha)\, A(\boldsymbol{x}^\alpha, \boldsymbol{y}^\alpha)\}_\alpha, \Gamma_2 \to \Delta_1, \{\mathbf{Q}^\beta(\boldsymbol{u}^\beta; \boldsymbol{v}^\beta)\, B(\boldsymbol{u}^\beta, \boldsymbol{v}^\beta)\}_\beta, \Delta_2}.$$

This new system has the advantage that some proofs become much simpler, even cut-free.

EXAMPLE 24.14. A game is called an open game if the winner is determined after a finite number of steps. Let $Y(a_1, b_1, a_2, b_2, \ldots)$ denote the game in which the players I and II choose the terms of the sequences $a_1, a_2, \ldots$ and $b_1, b_2, \ldots$ alternately and the winner of the game is determined. Then a game is open if

(1)

$$Y(a_1, b_1, \ldots) \to \bigvee_n (\forall x_{n+1} y_{n+1} \ldots)\, Y(a_1, b_1, \ldots, a_n, b_n, x_{n+1}, y_{n+1}, \ldots).$$

In our new system, there is a simple proof of the fact that an open game is determinate, i.e., (1) implies

$$\forall x_1 \exists y_1 \forall x_2 \exists y_2 \ldots Y(x_1, y_1, \ldots) \vee \exists x_1 \forall y_1 \exists x_2 \forall y_2 \ldots \neg Y(x_1, y_1, \ldots).$$

(2) $$\forall x_{n+1} y_{n+1} \ldots Y(a_1, \ldots, b_n, x_{n+1}, y_{n+1}, \ldots) \to$$
$$\to \forall x_1 \exists y_1 \ldots \forall x_n \exists y_n \ldots Y(x_1, y_1, \ldots)$$

for each n:

$$\cfrac{\cfrac{\cfrac{\cfrac{\cfrac{Y(a_1, \ldots, b_n, a_{n+1}^n, b_{n+1}^n, \ldots) \to Y(a_1, \ldots, b_n, a_{n+1}^n, \ldots)}{\forall x_{n+1} y_{n+1} \ldots Y(a_1, \ldots, b_n, x_{n+1}, y_{n+1}, \ldots) \to \forall x_{n+1} \exists y_{n+1} \ldots Y(a_1, \ldots, b_n, b_n, x_{n+1}, y_{n+1}, \ldots)}}{\forall x_{n+1} y_{n+1} \ldots Y(a_1, \ldots, b_n, x_{n+1}, \ldots) \to \exists y_n (\forall x_{n+1} \exists y_{n+1} \ldots)\, Y(a_1, \ldots, a_n, y_n, x_{n+1}, \ldots)}}{\forall x_{n+1} y_{n+1} \ldots Y(a_1, \ldots, b_n, x_{n+1}, \ldots) \to \exists y_n \forall x_{n+1} \exists y_{n+1} \ldots Y(a_1, \ldots, a_n, y_n, x_{n+1}, \ldots)}}{\forall x_{n+1} y_{n+1} \ldots Y(a_1, \ldots, b_n, x_{n+1}, \ldots) \to \forall x_n (\exists y_n \forall x_{n+1} \exists y_{n+1} \ldots)\, Y(a_1, \ldots, b_{n-1}, x_n, y_n \ldots)}}{\forall x_{n+1} y_{n+1} \ldots Y(a_1, \ldots, b_n, x_{n+1}, \ldots) \to \forall x_n \exists y_n \forall x_{n+1} \exists y_{n+1} \ldots Y(a_1, \ldots, b_{n-1}, x_n, y_n, \ldots)}$$

$$\vdots$$

$$\forall x_{n+1} y_{n+1} \ldots Y(a_1, \ldots, b_n, x_{n+1}, \ldots) \to \forall x_1 \exists y_1 \ldots \forall x_n \exists y_n \ldots Y(x_1, \ldots, x_n, y_n, \ldots).$$

Notice that our new rule of inference is applied repeatedly. From (2), by $\bigvee$: left,

(3)

$$\bigvee_n \forall x_{n+1} y_{n+1} \ldots Y(a_1, \ldots, b_n, x_{n+1}, y_{n+1}, \ldots) \to \forall x_1 \exists y_1 \ldots Y(x_1, y_1, \ldots).$$

Hence,

$$\frac{\begin{array}{r}Y(a_1, b_1, \ldots) \supset \bigvee_n \forall x_{n+1} y_{n+1} \ldots Y(a_1, \ldots, b_n, x_{n+1}, y_{n+1}, \ldots) \rightarrow \\ \rightarrow \forall x_1 \exists y_1 \ldots Y(x_1, y_1, \ldots), \neg Y(a_1, b_1, \ldots)\end{array}}{\begin{array}{r}\forall v_1 w_1 \ldots (Y(v_1, w_1, \ldots) \supset \bigvee_n (\forall x_{n+1} y_{n+1} \ldots Y(u_1, w_1, \ldots, x_{n+1}, y_{n+1}, \ldots))) \rightarrow \\ \rightarrow \forall x_1 \exists y_1 \ldots Y(x_1, y_1, \ldots) \vee \exists x_1 \forall y_1 \ldots \neg Y(x_1, y_1, \ldots).\end{array}}$$

In this proof, the auxiliary formulas of the heterogeneous $\mathbf{Q}$: right are

$$Y(a_1, \ldots, b_n, a^n_{n+1}, \ldots), \qquad \neg Y(a_1, \ldots, b_n, a_{n+1}, \ldots)$$

with eigenvariables $a^n_{n+1}, a^n_{n+2}, \ldots$ and $b_1, b_2, \ldots$, respectively. The eigenvariables of homogeneous quantifications are $a_1, a_2, \ldots$. Enumerate the auxiliary formulas of applications of $\mathbf{Q}$: right introductions in the order of

$$\neg Y(a_1, b_1, \ldots, a_n, b_n, \ldots), \{Y(a_1, \ldots, b_n, a^n_{n+1}, \ldots)\}_n.$$

Then one can check the eigenvariable conditions (cf. Definition 24.4). We shall examine only condition 6). Here A_0 is $\neg Y(a_1, b_1, \ldots)$, each a_i satisfies the condition, and in 6.0) a_i is used as an eigenvariable of a homogeneous inference.

If e is an eigenvariable of a heterogeneous $\mathbf{Q}$: right and $e < a_i$, then e is one of $b_1, \ldots, b_{i-1}$. If $\exists x_1 \forall y_1 \ldots \exists x_i \forall y_i \ldots$ is denoted by $\mathbf{Q}^{T_0}(\boldsymbol{y}; \boldsymbol{x})$, then b_j $(1 \leq j \leq i-1)$ is the jth eigenvariable of the $\mathbf{Q}$: right applied to A_0 and $j \in T_0(i)$. Consider 6.n), where $n \geq 1$. Let A_n be $Y(a_1, \ldots, b_n, a^n_{n+1}, \ldots)$. Suppose d is one of

$$a_1, b_1, \ldots, a_n, b_n, b^n_{n+1}, b^n_{n+2}, \ldots.$$

If d is used as an eigenvariable of a heterogeneous $\mathbf{Q}$: right then it is one of $b_1, \ldots, b_n$, its auxiliary formula is A_0 and $0 < n$. Since b^n_j, $j = n+1, n+2, \ldots$ is not used as an eigenvariable, a relation $e < b^n_j$ never happens. Next, let c be one of the variables $a_1, b_1, \ldots, a_n, b_n$. If c is an a_i, then the first possibility of an eigenvariable e of a heterogeneous $\mathbf{Q}$: right, such that $e < a_i$ is one of the variables $b_1, \ldots, b_{i-1}$. These are eigenvariables of A_0. Since the $b_1, \ldots, b_{i-1}$ do not depend on any variable, this completes the discussion of eigenvariables.

EXAMPLE 24.15. Let us consider another formulation of determinateness of an open game. An open game is expressed as $\bigvee_m Y_m(a_1, b_1, \ldots, a_m, b_m)$. Let us prove that this is determinate. Let

$Y(a_1, \ldots, b_m, a^m_{m+1}, \ldots, b^m_m)$ be $Y_k(a_1, \ldots, b_k)$ if $1 \leqslant k \leqslant m$

$$\begin{array}{l}
Y_m(a_1, \ldots, b_m) \to Y_m(a_1, \ldots, b_m) \\ \hline
Y_m(a_1, \ldots, b_m) \to \bigvee_k Y_k(a_1, \ldots, b_m, a^m_{m+1}, \ldots, b^m_k) \\ \hline
Y_m(a_1, \ldots, b_m) \to \forall x_{m+1} \exists y_{m+1} \ldots \bigvee_k Y_k(a_1, \ldots, b_m, x_{m+1}, \ldots, y_k) \\ \hline
Y_m(a_1, \ldots, b_m) \to \exists y_m (\forall x_{m+1} \exists y_{m+1} \ldots) \bigvee_k Y_k(a_1, \ldots, y_m, x_{m+1}, \ldots, y_k) \\ \hline
Y_m(a_1, \ldots, b_m) \to \exists y_m \forall x_{m+1} \exists y_{m+1} \ldots \bigvee_k Y_k(a_1, \ldots, y_m, x_{m+1}, \ldots, y_k) \\ \hline
Y_m(a_1, \ldots, b_m) \to \forall x_m (\exists y_m \forall x_{m+1} \exists y_{m+1} \ldots \bigvee_k Y_k(a_1, \ldots, x_m, y_m, x_{m+1}, \ldots, y_l \\ \hline
Y_m(a_1, \ldots, b_m) \to \forall x_m \exists y_m \forall x_{m+1} \exists y_{m+1} \ldots \bigvee_k Y_k(a_1, \ldots, x_m, y_m, x_{m+1}, \ldots, y \\ \hline
\end{array}$$

$$\begin{array}{c}
Y_m(a_1, \ldots, b_m) \to \forall x_1 \exists y_1 \ldots \bigvee_k Y_k(x_1, \ldots, y_k) \\ \hline
\bigvee_m Y_m(a_1, \ldots, b_m) \to \forall x_1 \exists y_1 \ldots \bigvee_k Y_k(x_1, y_1, \ldots, y_k) \\ \hline
\to \forall x_1 \exists y_1 \ldots \bigvee_k Y_k(x_1, y_1, \ldots, y_k), \neg \bigvee_m Y_m(a_1, b_1, \ldots, b_m) \\ \hline
\to \forall x_1 \exists y_1 \ldots \bigvee_m Y_m(x_1, y_1, \ldots, y_m), \exists x_1 \forall y_1 \ldots \neg \bigvee_m Y_m(x_1, y_1, \ldots, y_m)
\end{array}$$

The auxiliary formulas of applications of the heterogeneous **Q** : right can be listed in the order of

$$\neg \bigvee_m Y_m(a_1, b_1, \ldots, a_m, b_m), \left\{ \bigvee_n Y_n(a_1, \ldots, b_m, a^m_{m+1}, \ldots, b^m_n) \right\}_m$$

with eigenvariables $b_1, b_2, \ldots,$ for the first formula, and

$$a^m_{m+1}, a^m_{m+2}, \ldots \quad \text{for} \quad \bigvee_k Y_k(a_1, \ldots, b_m, a^m_{m+1}, \ldots).$$

The reader should go over the eigenvariable conditions,

EXAMPLE 24.16.

$$\begin{array}{c}
A(a_1, b_1, a_2, b_2, \ldots) \to A(a_1, b_1, a_2, b_2, \ldots) \\ \hline
\exists x_2 \forall y_2 \ldots A(a_1, b_1, x_2, y_2, \ldots) \to \exists x_1 \forall y_1 \ldots A(x_1, y_1, \ldots) \\ \hline
\neg \exists x_1 \forall y_1 \ldots A(x_1, y_1, \ldots) \to \neg \exists x_2 \forall y_2 \ldots A(a_1, b_1, x_2, y_2, \ldots) \\ \hline
\neg \exists x_1 \forall y_1 \ldots A(x_1, y_1, \ldots) \to \exists y_1 \neg \exists x_2 \forall y_2 \ldots A(a_1, y_1, x_2, y_2, \ldots) \\ \hline
\neg \exists x_1 \forall y_1 \ldots A(x_1, y_1, \ldots) \to \forall x_1 \exists y_1 \neg \exists x_2 \forall y_2 \ldots A(x_1, y_1, \ldots)
\end{array}$$

REMARK. There are examples which are cut-free provable in our new system, but which require the cut rule in the old system. For example, consider

$$\to (\forall x \, \exists y) \, A(x, y), (\exists x \, \forall y) \neg A(x, y).$$

Here $(\forall x \, \exists y)$ and $(\exists x \, \forall y)$ are regarded as heterogeneous, while $\forall x \, \exists y$ and $\exists x \, \forall y$ are considered two applications of homogeneous quantifiers.

(1) This is provable with a cut in the old system:

$$\dfrac{\dfrac{\dfrac{A(a,b)\to A(a,b)}{\forall x\,\exists y\,A(x,y)\to(\forall x\,\exists y)\,A(x,y)}}{\to\forall x\,\exists y\,A(x,y)\vee\exists x\,\forall y\neg A(x,y)}\qquad\dfrac{\dfrac{\neg A(c,d)\to\neg A(c,d)}{\exists x\,\forall y\neg A(x,y)\to(\exists x\,\forall y)\neg A(x,y)}}{\forall x\,\exists y\,A(x,y)\vee\exists x\,\forall y\neg A(x,y)\to(\forall x\,\exists y)\,A(x,y),(\exists x\,\forall y)\neg A(x,y)}}{\to(\forall x\,\exists y)\,A(x,y),(\exists x\,\forall y)\neg A(x,y).}$$

(2) This cannot be proved without the cut rule in the old system, for if we assume a cut-free proof, then

$$\to\ldots A(a_\alpha, b_\alpha),\ldots\neg A(c_\beta, d_\beta)\ldots$$

is provable for some free variables. Then for some α and β, a_α is c_β and b_α is d_β. Therefore in

$$\to\ldots A(a,b),\ldots,\neg A(a,b),\ldots$$

both a and b are eigenvariables of a heterogeneous $\mathbf{Q}$: right. But this implies that $A(a,b)$ and $\neg A(a,b)$ cannot be ordered.

(3) This sequent is provable in our new system without the cut rule:

$$\dfrac{\dfrac{\vdots\\ \to A(a,b),\ \neg A(a,b)}{\to\forall x\,\exists y\,A(x,y),\exists x\,\forall y\neg A(x,y)}}{\to(\forall x\,\exists y)\,A(x,y),(\exists x\,\forall y)\neg A(x,y).}$$

PROPOSITION 24.17. *A proof in determinate logic that satisfies the condition (Q) in Definition* 23.16 *is a proof in the (extended) heterogeneous system.*

PROOF. Suppose a proof P in a given language, say L, of determinate logic satisfies (Q). Define a language L(J) by admitting logical symbols of precisely the same arities as those of L and defining J as follows. Let $\mathbf{Q}^f\mathbf{z}$ be a quantifier of the determinate logic.

Let $\mathbf{x}$ be the sequence of all variables of $\mathbf{z}$ for which f assumes the value $\forall$ and let $\mathbf{y}$ be the sequence of all the variables of $\mathbf{z}$ for which f assures the value $\exists$. Let y be the βth variable of $\mathbf{y}$. Then $\alpha\in T(\beta)$ if and only if the αth variable of $\mathbf{x}$ precedes y in $\mathbf{z}$. Such a T belongs to J. Therefore we can translate $\mathbf{Q}^f\mathbf{z}$ into $\mathbf{Q}^T(\mathbf{x};\mathbf{y})$. Thus the formulas of P are regarded as those of the language L(J).

By renaming variables in P (if necessary), we can assume the following condition because of (Q).

(*) If the inference

$$\frac{\Gamma\to\Delta, A(\mathbf{a},\mathbf{b})}{\Gamma\to\Delta,\mathbf{Q}^T(\mathbf{x},\mathbf{y})\,A(\mathbf{x},\mathbf{y})}$$

is a heterogeneous $\mathbf{Q}$: right in P, then no eigenvariables in P used above $\Gamma \to \Delta, \mathbf{Q}^T(\boldsymbol{x}, \boldsymbol{y})\, A(\boldsymbol{x}, \boldsymbol{y})$ occur below $\Gamma \to \Delta, \mathbf{Q}^T(\boldsymbol{x}, \boldsymbol{y})\, A(\boldsymbol{x}, \boldsymbol{y})$.

In order to see that P is a proof in this heterogeneous system, it suffices to examine the eigenvariable conditions in Definitions 24.4 and 24.14. There 1), 2) and 4) are exactly conditions on P. By virtue of (Q), 3) is satisfied. Suppose $c < a$ holds in P. Then the height of c is less than the height of a; hence 5) is obvious.

We can define an enumeration of the auxiliary formulas of the heterogeneous $\mathbf{Q}$: right in P in such a manner that it satisfies the eigenvariable condition 6). Let J_1 and J_2 be heterogeneous $\mathbf{Q}$: right in P, with J_1 above J_2 and with J_1 and J_2 having the form

$$J_1 \quad \frac{\Gamma_1 \to \Delta_1, A(\boldsymbol{a}, \boldsymbol{b})}{\Gamma_1 \to \Delta_1, \mathbf{Q}^{T_1}(\boldsymbol{x}, \boldsymbol{y})\, A(\boldsymbol{x}, \boldsymbol{y})}$$

$$J_2 \quad \frac{\Gamma_2 \to \Delta_2, B(\mathbf{c}, \boldsymbol{d}, \boldsymbol{e})}{\Gamma_2 \to \Delta_2, \mathbf{Q}^{T_2}(\boldsymbol{x}', \boldsymbol{y}')\, B(\boldsymbol{x}', \boldsymbol{y}', \boldsymbol{e})}$$

where $\boldsymbol{e}$ is the sequence of variables that are neither in $\boldsymbol{c}$ nor in $\boldsymbol{d}$. Suppose that d is in $\boldsymbol{d}$ (or $\boldsymbol{e}$), a' is an eigenvariable in P, and $a' < d$ (or $a' < e$). Since (*) implies that d (or e) is not an eigenvariable above $\Gamma_1 \to \Delta_1, \mathbf{Q}^{T_1}(\boldsymbol{x}, \boldsymbol{y})\, A(\boldsymbol{x}, \boldsymbol{y})$, a' cannot be in $\boldsymbol{a}$. An appropriate enumeration of the auxiliary formulas in P is obtained if we enumerate them from the bottom.

We shall now investigate the interpolation theorem for subsystems of heterogeneous systems.

DEFINITION 24.18. Let q be an arbitrary symbol and let A be a formula. We say that an occurrence of q is *positive* or *negative* according as q lies in the scope of an even or odd number of $\neg$'s. $A \supset B$ is understood to be $\neg A \vee B$. We say that q is positive in $\Gamma \to \Delta$ if it is positive in Δ or negative in Γ; and that q is negative in $\Gamma \to \Delta$ if it is negative in Δ or positive in Γ.

A sequent $\Gamma \to \Delta$ is said to be negative if all the heterogeneous quantifiers in it are negative.

PROPOSITION 24.19. (1) *In our old system, every negative sequent is either cut-free provable or has a counter-model.*

(2) *The interpolation theorem holds for negative sequents. Suppose $\Gamma \to \Delta$ is a valid negative sequent and $[\{\Gamma_1; \Delta_1\}, \{\Gamma_2; \Delta_2\}]$ is a decomposition of $\Gamma \to \Delta$ such that $\{\Gamma_1; \Delta_1\}$ and $\{\Gamma_2; \Delta_2\}$ have at least one predicate symbol in common. Then there exists a formula C (not necessarily negative) such that $\Gamma_1 \to \Delta_1, C$ and $C, \Gamma_2 \to \Delta_2$ are valid, and all the predicate symbols and free variables in C occur both in $\{\Gamma_1; \Delta_1\}$ and $\{\Gamma_2; \Delta_2\}$.*

PROOF. (1) This can be proved similarly to other completeness proofs, by constructing a tree for a given sequent. Notice that in a stage which concerns heterogeneous quantifiers the reduction is done in the antecedent only.

(2) For technical reasons, we assume that all the homogeneous quantifiers are $\exists$. This restriction is not essential. Following Definition 23.18, we say that a figure P is a proof in **RHS′** if P satisfies the following:

(i) P satisfies all the conditions of a proof in our system except the eigenvariable conditions.

(ii) The only inferences which introduce quantifiers are $\exists$: left introductions,

$$\frac{\{A_\lambda(\boldsymbol{a}_\lambda)\}_{\lambda<\gamma}, \Gamma \rightarrow \Delta}{\{\exists \boldsymbol{x}_\lambda\, A_\lambda(\boldsymbol{x}_\lambda)\}_{\lambda<\gamma}, \Gamma \rightarrow \Delta},$$

where no variable in $\boldsymbol{a}_\lambda$ occurs in the lower sequent.

Notice that P may contain heterogeneous quantifiers, but they are introduced either by initial sequents or weakenings. Then in a manner similar to the proof of Proposition 23.19, we can prove the following: Let P be a proof in **RHS′** ending with $\Gamma \rightarrow \Delta$, and suppose a well-founded relation $<_0$ is defined for the free variables in $\Gamma \rightarrow \Delta$. Then $<_0$ can be extended to a dependence relation for eigenvariables in P.

LEMMA 24.20. *Let P be a cut-free proof of $\Gamma_1, \Gamma_2 \rightarrow \Delta_1, \Delta_2$ in our system in which every homogeneous quantifier is $\exists$, no rule of heterogeneous quantifiers applies, and every inference of the introduction of a quantifier is an inference of the introduction of $\exists$ in the succedent. Suppose also that $\{\Gamma_1; \Delta_1\}$ and $\{\Gamma_2; \Delta_2\}$ have a predicate symbol in common. Then there exist cut-free proofs P_1 and P_2 in* **RHS′** *and a formula C such that the end-sequent of P_1 is*

$$C, \Gamma_1 \rightarrow \Delta_1,$$

the end-sequent of P_2 is

$$\Gamma_2 \rightarrow \Delta_2, C,$$

and every free variable and predicate symbol in C is common to $\{\Gamma_1; \Delta_1\}$ and $\{\Gamma_2, \Delta_2\}$.

PROOF. Similar to that of Lemma 23.20.

We now retrun to the proof of Proposition 24.19 (2) which we will do in a manner similar to 3, of the proof of Theorem 23.15.

Consider a cut-free proof, P_1 of $\Gamma_1, \Gamma_2 \rightarrow \Delta_1, \Delta_2$. The eigenvariable conditions for such a proof can be expressed in terms of a well-founded relation $<$. Fix such a $<$ throughout. We may assume that every homogeneous quantifier in P is $\exists$, that heterogeneous quantifiers are introduced in the antecedent only, and that free variables which occur in

$\Gamma_1, \Gamma_2 \to \Delta_1, \Delta_2$ do not depend on any eigenvariables in P. Let $\{\mathbf{Q}^T(\boldsymbol{x}; \boldsymbol{y})\, A(\boldsymbol{x}, \boldsymbol{y}, \boldsymbol{d})\}$ be an enumeration of all principal formulas of the $\mathbf{Q}$: left introductions (heterogeneous $\mathbf{Q}$ or homogeneous $\exists$: left) in the given proof whose descendents are in Γ_1 or Δ_1. We define $\{\mathbf{Q}^T(\boldsymbol{x}; \boldsymbol{y})\, B(\boldsymbol{x}, \boldsymbol{y}, \boldsymbol{d}')\}$ similarly for Γ_2 and Δ_2. Then we can construct a proof P' of

$$\{\neg \mathbf{Q}^T(\boldsymbol{x}; \boldsymbol{y})\, A(\boldsymbol{x}, \boldsymbol{y}, \boldsymbol{d}) \vee A(\boldsymbol{c}, \boldsymbol{a}, \boldsymbol{d})\}, \{\neg \mathbf{Q}^T(\boldsymbol{x}; \boldsymbol{y})\, B(\boldsymbol{x}, \boldsymbol{y}, \boldsymbol{d}') \vee B(\boldsymbol{e}, \boldsymbol{b}, \boldsymbol{d}')\},$$
$$\Gamma_1, \Gamma_2 \to \Delta_1, \Delta_2,$$

such that every homogeneous quantifier in P' is $\exists$; such that the only inference which introduces a quantifier is $\exists$: right; such that $c, d < a$ for any a in $\boldsymbol{a}$, c in $\boldsymbol{c}$ and d in $\boldsymbol{d}$; such that $e, d' < b$ for any b in $\boldsymbol{b}$, e in $\boldsymbol{e}$ and d' in $\boldsymbol{d}'$; such that every free variable in $A(\boldsymbol{x}, \boldsymbol{y}, \boldsymbol{z})$ occurs in Γ_1 or Δ_1; such that every free variable in $B(\boldsymbol{x}, \boldsymbol{y}, \boldsymbol{z})$ occurs in Γ_2 or Δ_2; and such that all variables in $\boldsymbol{a}$ and $\boldsymbol{b}$ are different. (We have used ambiguous notation such as the same letters $\boldsymbol{x}$ and $\boldsymbol{y}$ for different quantifiers. The meaning of the above expressions should, however, be obvious.)

Then, by Lemma 24.20, there exists a formula C such that

$$C, \{\neg \mathbf{Q}^T(\boldsymbol{x}; \boldsymbol{y})\, A(\boldsymbol{x}, \boldsymbol{y}, \boldsymbol{d}) \vee A(\boldsymbol{c}, \boldsymbol{a}, \boldsymbol{d})\}, \Gamma_1 \to \Delta_1$$

and

$$\{\neg \mathbf{Q}^T(\boldsymbol{x}; \boldsymbol{y})\, B(\boldsymbol{x}, \boldsymbol{y}, \boldsymbol{d}) \vee B(\boldsymbol{c}, \boldsymbol{b}, \boldsymbol{d}), \Gamma_2 \to \Delta_2, C$$

are provable in **RHS**′ and C satisfies some appropriate conditions. Let $\boldsymbol{f}$ be a sequence of all free variables in C which are not common to $\{\Gamma_1; \Delta_1\}$ and $\{\Gamma_2; \Delta_2\}$. We assume that the variables in $\boldsymbol{f}$ are arranged so that if f_1 precedes f_2, then it is not the case that $f_2 < f_1$. ($<$ is the relation which is defined for the original proof P.) Let P_1 and P_2 be proofs in **RHS**′ of the above two sequents and let $<_1$ and $<_2$ be the dependence relations for the free variables in the end sequents of P_1 and P_2, respectively, which are induced from $<$ (for P). Then, we can extend these relations to all the free variables. Let us denote these extended relations by $<_{P_1}$ and $<_{P_2}$.

Next, consider the following quasi-proofs, P_1' and P_2':

$$\begin{array}{r} P_1 \\ C(\boldsymbol{f}), \{\forall \boldsymbol{z}\, \mathbf{Q}^T(\boldsymbol{u}; \boldsymbol{v})\, (\neg \mathbf{Q}^T(\boldsymbol{x}; \boldsymbol{y})\, A(\boldsymbol{x}, \boldsymbol{y}, \boldsymbol{z}) \vee A(\boldsymbol{u}, \boldsymbol{v}, \boldsymbol{z}))\}, \Gamma_1 \to \Delta_1 \end{array}$$

and

$$\begin{array}{r} P_2 \\ \{\forall \boldsymbol{z}\, \mathbf{Q}^T(\boldsymbol{u}; \boldsymbol{v})\, (\neg \mathbf{Q}^T(\boldsymbol{x}; \boldsymbol{y})\, B(\boldsymbol{x}, \boldsymbol{y}, \boldsymbol{z}) \vee B(\boldsymbol{u}, \boldsymbol{v}, \boldsymbol{z}))\}, \Gamma_2 \to \Delta_2, C(\boldsymbol{f}). \end{array}$$

There are three kinds of variables in $\boldsymbol{f}$, those of $\boldsymbol{a}$, those of $\boldsymbol{b}$, and the rest. The first ones, denoted by $\boldsymbol{f}_1$, are eigenvariables in P_1'; the second ones, denoted by $\boldsymbol{f}_2$, are eigenvariables in P_2'; and the third ones are denoted by f_3. Define T', a function from f_2 to subsets of f_1, so that it satisfies the dependence relation $<$. Note that

$$\forall z\, (\mathbf{Q}^T(\boldsymbol{u}; \boldsymbol{v})\, (\neg\mathbf{Q}^T(\boldsymbol{x}; \boldsymbol{y})\, A(\boldsymbol{x}, \boldsymbol{y}, z) \vee A(\boldsymbol{u}, \boldsymbol{v}, z))$$

is provable. Therefore, we obtain

$$\begin{array}{c} P'_1 \\ \vdots \\ \forall \boldsymbol{w}\, \mathbf{Q}^{T'}(\boldsymbol{w}_1; \boldsymbol{w}_2)\, C(\boldsymbol{w}_1, \boldsymbol{w}_2, \boldsymbol{w}), \Gamma_1 \to \Delta_1 \end{array}$$

where $\boldsymbol{w}_1$, $\boldsymbol{w}_2$, and $\boldsymbol{w}$ replace f_1, f_2 and f_3, respectively. Similarly, we obtain

$$\begin{array}{c} P'_2 \\ \vdots \\ \Gamma_2 \to \Delta_2, \forall \boldsymbol{w}\, \mathbf{Q}^{T'}(\boldsymbol{w}_1; \boldsymbol{w}_2)\, C(\boldsymbol{w}_1, \boldsymbol{w}_2, \boldsymbol{w}). \end{array}$$

Since we may assume that the eigenvariables of the proofs of

$$\forall z\, \mathbf{Q}^T(\boldsymbol{u}; \boldsymbol{v})\, (\neg\mathbf{Q}^T(\boldsymbol{x}; \boldsymbol{y})\, A(\boldsymbol{x}, \boldsymbol{y}, z) \vee A(\boldsymbol{u}, \boldsymbol{v}, z))$$

are different from those of P'_1, we naturally extend $<_{P_1}$ to the entire proof of

$$\forall \boldsymbol{w}\, \mathbf{Q}^{T'}(\boldsymbol{w}_1; \boldsymbol{w}_2)\, C(\boldsymbol{w}_1, \boldsymbol{w}_2, \boldsymbol{w}), \Gamma_1 \to \Delta_1.$$

It is easily shown that it satisfies the dependence relation; indeed, T' was so chosen. In a similar manner we extend $<_{P_2}$.

Even with our new inference, the cut-elimination theorem for the entire system can hardly be expected to hold. Defining a complete system for a heterogeneous language is a difficult problem. It is even more difficult if we wish to define a cut-free complete system. We can see this clearly from the following example.

EXAMPLE 24.21. Consider

$$\{c_i \neq c_j\}_{i \neq j, i, j < w} \to \begin{pmatrix} \forall a\, \exists u \\ \forall b\, \exists v \end{pmatrix} (a = b \equiv u = v \wedge u \neq c_0).$$

This sequent is provable in our system:

$$\{c_i \neq c_j\}_{i \neq j}, \bigwedge_i (a = c_i \supset u = c_{i+1}) \wedge (\bigwedge_i a \neq c_i \supset u = a),$$

$$\bigwedge_i (b = c_i \supset v = c_{i+1}) \wedge (\bigwedge_i b \neq c_i \supset v = b) \to (a = b \equiv u = v) \wedge u \neq c_0$$

can be easily proved. It then follows that

$$\forall a\, \exists u\, (\bigwedge_i (a = c_i \supset u = c_{i+1}) \wedge (\bigwedge_i a \neq c_i \supset u = a)),$$

$$\forall b\, \exists v\, (\bigwedge_i (b = c_i \supset v = c_{i+1}) \wedge (\bigwedge_i b \neq c_i \supset v = b)),$$

$$\{c_i \neq c_j\}_{i \neq j} \to \begin{pmatrix} \forall a\ \exists u \\ \forall b\ \exists v \end{pmatrix} ((a = b \equiv u = v) \wedge u \neq c_0).$$

On the other hand

$$\{c_i \neq c_j\}_{i \neq j} \to \forall a\ \exists u\ (\bigwedge_i (a = c_i \supset u = c_{i+1}) \wedge (\bigwedge_i a \neq c_i \supset u = a))$$

and

$$\{c_i \neq c_j\}_{i \neq j} \to \forall b\ \exists v\ (\bigwedge_i (b = c_i \supset v = c_{i+1}) \wedge (\bigwedge_i b \neq c_i \supset v = b))$$

are provable in our system. Therefore by the cut rule

$$\{c_i \neq c_j\}_{i \neq j} \to \begin{pmatrix} \forall a\ \exists u \\ \forall b\ \exists v \end{pmatrix} ((a = b \equiv u = v) \wedge u \neq c_0)$$

is provable.

If we apply Gentzen-type reduction, which was primarily defined for finite languages, to the proof given above, which contains a cut, then we will obtain a "proof-like" figure

$$\frac{\begin{array}{l} \{c_i \neq c_j\}_{i \neq j} \to \{(a = b \equiv c_{i+1} = c_{j+1}) \wedge c_{i+1} \neq c_0\}_{i,j}, \\ \qquad \{(a = b \equiv a = c_{j+1}) \wedge a \neq c_0\}_j, \\ \qquad \{(a = b \equiv c_{i+1} = b) \wedge c_{i+1} \neq c_0\}_i, \\ \qquad (a = b \equiv a = b) \wedge a \neq c_0 \end{array}}{\{c_i \neq c_j\}_{i \neq j} \to \begin{pmatrix} \forall a\ \exists u \\ \forall b\ \exists v \end{pmatrix} ((a = b \equiv u = v) \wedge u \neq c_0).}$$

It is obvious that in this figure the auxiliary formulas of the eigenvariables a and b are not unique. Therefore it cannot be a proof in our sense.

From this figure, we see that there is little hope of expanding our system so that this figure will fit into it and hence little hope of establishing a complete cut-free system.

Although our system is far from being complete, a weak completeness can be proved. For the proof, we employ a more general formulation of the generalized cut rule that we will call the strong generalized cut rule (s.g.c.): Let $\mathscr{F}$ be a non-empty set of formulas. Suppose, for an arbitrary decomposition of $\mathscr{F}$, say $(\mathscr{F}_1, \mathscr{F}_2)$, there are subsets of $\mathscr{F}_1$ and $\mathscr{F}_2$, say Φ and Ψ, respectively, and subsets of Γ and Δ, say Γ' and Δ', respectively, such that $\Phi, \Gamma' \to \Delta', \Psi$. Then $\Gamma \to \Delta$ can be inferred. We can allow the case where $\Gamma'(\Delta')$ has some repetition of some formulas of $\Gamma(\Delta)$.

It is obvious that the s.g.c. rule is a generalization of the g.c. rule that involves two types of inferences, inferences that are essential cuts and inferences that are basically weakenings.

PROPOSITION 24.22. *Consider a language with heterogeneous quantifiers* L *that contains individual constants* $c_0, c_1, \ldots, c_\alpha, \ldots, \alpha < K$, *and contains enough logical symbols* $\wedge, \vee, \forall$, *and* $\exists$ *in the following sense. If* $\boldsymbol{Q}^T(\boldsymbol{x}, \boldsymbol{y})$ *is in* L *and* α *is the arity of* $\boldsymbol{x}$ *or* $\boldsymbol{y}$, *then* L *has* $\bigwedge$ *and* $\bigvee$ *of arity of the cardinal of* K^α *and also* $\forall$ *and* $\exists$ *of arity* α. *Then this system, augmented by the axioms*

$$\rightarrow t = c_0, t = c_1, \ldots, t = c_\alpha, \ldots, (\alpha < K)$$

for arbitrary terms t, is complete.

PROOF. Let θ be the length of J and let τ denote the set $\{c_0, c_1, \ldots, c_\alpha, \ldots\}$. Consider a formula of L of the form $\boldsymbol{Q}^T(\boldsymbol{x}; \boldsymbol{y})\, A(\boldsymbol{x}, \boldsymbol{y})$. Let α be an ordinal and suppose that $T(\alpha)$ is of type β_α, according to the natural ordering of ordinals. Let f^α be a function from τ^{β_α}, the cartesian product of τ, to τ; let $\boldsymbol{m}, \boldsymbol{n}$, etc., be sequences (of appropriate type) of elements of τ. Suppose $\boldsymbol{m}_\alpha$, a sequence $m_{\gamma_0}, m_{\gamma_1}, \ldots$, where $T(\alpha) = \{\gamma_0, \gamma_1, \ldots\}$, is given for each $\alpha < \theta$. Then $\boldsymbol{f}(\boldsymbol{m})$ will denote the sequence of sequences $f^0(\boldsymbol{m}_0), f^1(\boldsymbol{m}_1), \ldots, f^\alpha(\boldsymbol{m}_\alpha), \ldots$. Finally, let L′ be the extended language in which $\bigvee_{\boldsymbol{f}}$ and $\bigwedge_{\boldsymbol{m}}$ are allowed, where in $\bigvee_{\boldsymbol{f}}$, $\boldsymbol{f}$ ranges over all sequences of functions defined as above and in $\bigwedge_{\boldsymbol{m}}$, $\boldsymbol{m}$ ranges over all the sequences of elements of τ defined as above. When those symbols are involved, provability means provability in the system with language L′. Note that L′ is an extension of L.

Under these conventions we shall first show that

$$\text{(1)} \qquad \boldsymbol{Q}^T(\boldsymbol{x}; \boldsymbol{y})\, A(\boldsymbol{x}, \boldsymbol{y}) \rightarrow \bigvee_{\boldsymbol{f}} \bigwedge_{\boldsymbol{m}} A(\boldsymbol{m}, \boldsymbol{f}(\boldsymbol{m}))$$

is provable. Let us abbreviate $\bigwedge_\beta\ (a_\beta = m_\beta)$ to $\boldsymbol{a} = \boldsymbol{m}$. Let $\boldsymbol{g}$ be an arbitrary, but fixed sequent of functions, defined as $\boldsymbol{f}$ above, and let $\boldsymbol{m_f}$ denote an arbitary, but fixed sequence of elements of τ, of appropriate length, chosen for $\boldsymbol{f}$. For each $\boldsymbol{g}$ and $\{\boldsymbol{m_f}\}_{\boldsymbol{f}}$, let

$$\text{(2)} \qquad \ldots, A(\boldsymbol{n}, \boldsymbol{g}(\boldsymbol{n})), \ldots \rightarrow \ldots, A(\boldsymbol{m_f}, \boldsymbol{f}(\boldsymbol{m_f})), \ldots$$

be a sequent, where $\boldsymbol{g}$ and $\{\boldsymbol{m_f}\}_{\boldsymbol{f}}$ are fixed, $\boldsymbol{n}$ and $\boldsymbol{f}$ range over all possibilities. There is an instance that $\boldsymbol{f} = \boldsymbol{g}$ and $\boldsymbol{n} = \boldsymbol{m_f}$. Therefore (2) is provable.

Now we assume that L has an adequate number of free variables so we can carry out the subsequent argument. For each $\alpha < \theta$, and for each $\boldsymbol{n}$, choose a free variable $a_{\alpha, \boldsymbol{n}}$ of L. Then

$$\text{(3)} \qquad \rightarrow a_{\alpha, \boldsymbol{n}} = c_0, a_{\alpha, \boldsymbol{n}} = c_1, \ldots$$

is an axiom. We assume that we can choose different variables for different $(\alpha, \boldsymbol{n})$. For each c, there is a $\boldsymbol{g}$ such that $g^\alpha(\boldsymbol{n}_\alpha) = c$. Therefore, for an arbitrary choice of $(c_{\sigma_0}, c_{\sigma_1}, \ldots)$ a sequence of constants of length θ, (2) for such a $\boldsymbol{g}$ implies

(4)

$$\ldots, a_{0,\boldsymbol{n}} = c_{\sigma_0}, a_{1,\boldsymbol{n}} = c_{\sigma_1}, \ldots, A(\boldsymbol{n}, \boldsymbol{a_n}), \ldots \to \ldots, A(\boldsymbol{m_f}, \boldsymbol{f}(\boldsymbol{m_f})), \ldots,$$

where $\boldsymbol{a_n}$ is the sequence $a_{0,\boldsymbol{n}}, a_{1,\boldsymbol{n}}, \ldots$. Then by an application of the strong generalized cut rule to (3) and (4),

(5) $$\ldots, A(\boldsymbol{n}, \boldsymbol{a_n}), \ldots \to \ldots, A(\boldsymbol{m_f}, \boldsymbol{f}(\boldsymbol{m_f})), \ldots$$

for each $\{\boldsymbol{m_f}\}_{\boldsymbol{f}}$. Therefore from (5)

$$\ldots, A(\boldsymbol{n}, \boldsymbol{a_n}), \ldots \to \ldots, \bigwedge_{\boldsymbol{m}} A(\boldsymbol{m}, \boldsymbol{f}(\boldsymbol{m})), \ldots$$

or

(6) $$\ldots, A(\boldsymbol{n}, \boldsymbol{a_n}), \ldots \to \bigvee_{\boldsymbol{f}} \bigwedge_{\boldsymbol{m}} A(\boldsymbol{m}, \boldsymbol{f}(\boldsymbol{m})).$$

Finally, introducing quantifiers in (6), we obtain

$$\mathbf{Q}^T(\boldsymbol{x}; \boldsymbol{y})\, A(\boldsymbol{x}, \boldsymbol{y}) \to \bigvee_{\boldsymbol{f}} \bigwedge_{\boldsymbol{m}} A(\boldsymbol{m}, \boldsymbol{f}(\boldsymbol{m})),$$

which is (1). Since we have chosen distinct free variables for different $(\alpha, \boldsymbol{n})$, it is obvious that the eigenvariable conditions are satisfied.

In the proof of (1), the strong generalized cut rule is applied only to atomic formulas.

Next, we want to prove the converse of (1) in the form:

(7)

$$\forall \boldsymbol{x}\, \exists \boldsymbol{y} \bigwedge_{\boldsymbol{n}} (\boldsymbol{x} = \boldsymbol{n} \supset \boldsymbol{y} = \boldsymbol{f}(\boldsymbol{n})), \ldots, \bigvee_{\boldsymbol{f}'} \bigwedge_{\boldsymbol{m}} A(\boldsymbol{m}, \boldsymbol{f}'(\boldsymbol{m})) \to \mathbf{Q}^T(\boldsymbol{x}; \boldsymbol{y})\, A(\boldsymbol{x}; \boldsymbol{y}),$$

where $\boldsymbol{f}$ ranges over all possibilities.

First, we have

$$\boldsymbol{a} = \boldsymbol{n}, \boldsymbol{a} = \boldsymbol{n} \supset \boldsymbol{b} = \boldsymbol{f}(\boldsymbol{n}), A(\boldsymbol{n}, \boldsymbol{f}(\boldsymbol{n})) \to A(\boldsymbol{a}, \boldsymbol{b})$$

for every $\boldsymbol{f}$ and $\boldsymbol{n}$, and

$$\to a_\alpha = c_0, a_\alpha = c_1, \ldots.$$

Therefore, by the strong generalized cut rule,

$$\bigwedge_{\boldsymbol{n}} (\boldsymbol{a} = \boldsymbol{n} \supset \boldsymbol{b} = \boldsymbol{f}(\boldsymbol{n})), \bigwedge_{\boldsymbol{m}} A(\boldsymbol{m}, \boldsymbol{f}(\boldsymbol{m})) \to A(\boldsymbol{a}, \boldsymbol{b}).$$

From this we obtain

$$\forall \boldsymbol{x}\, \exists \boldsymbol{y} \bigwedge_{\boldsymbol{n}} (\boldsymbol{x} = \boldsymbol{n} \supset \boldsymbol{y} = \boldsymbol{f}(\boldsymbol{n})), \bigwedge_{\boldsymbol{m}} A(\boldsymbol{m}, \boldsymbol{f}(\boldsymbol{m})) \to A(\boldsymbol{a}, \boldsymbol{b}),$$

and

$$(8)\quad \forall \boldsymbol{x}\, \exists \boldsymbol{y} \bigwedge_{\boldsymbol{n}} (\boldsymbol{x} = \boldsymbol{n} \supset \boldsymbol{y} = \boldsymbol{f}(\boldsymbol{n})), \bigwedge_{\boldsymbol{m}} A(\boldsymbol{m}, \boldsymbol{f}(\boldsymbol{m})) \rightarrow \mathbf{Q}^T(\boldsymbol{x}; \boldsymbol{y})\, A(\boldsymbol{x}, \boldsymbol{y}).$$

Since (8) holds for every $\boldsymbol{f}$, it follows that

(9)

$$\forall \boldsymbol{x}\, \exists \boldsymbol{y} \bigwedge_{\boldsymbol{n}} (\boldsymbol{x} = \boldsymbol{n} \supset \boldsymbol{y} = \boldsymbol{f}(\boldsymbol{n})), \ldots, \bigvee_{\boldsymbol{f}'} \bigwedge_{\boldsymbol{m}} A(\boldsymbol{m}, \boldsymbol{f}'(\boldsymbol{m})) \rightarrow \mathbf{Q}^T(\boldsymbol{x}; \boldsymbol{y})\, A(\boldsymbol{x}, \boldsymbol{y}),$$

which is (7).

On the other hand

$$(10)\qquad \rightarrow \forall \boldsymbol{x}\, \exists \boldsymbol{y} \bigwedge_{\boldsymbol{n}} (\boldsymbol{x} = \boldsymbol{n} \supset \boldsymbol{y} = \boldsymbol{f}(\boldsymbol{n}))$$

is provable for every $\boldsymbol{f}$ in the following way. For an arbitrary $\boldsymbol{n}$ and $\boldsymbol{m}$,

$$\boldsymbol{a} = \boldsymbol{n} \rightarrow \boldsymbol{a} = \boldsymbol{m} \supset \boldsymbol{f}(\boldsymbol{n}) = \boldsymbol{f}(\boldsymbol{m}),$$

hence

$$\boldsymbol{a} = \boldsymbol{n} \rightarrow \exists \boldsymbol{y} \bigwedge_{\boldsymbol{m}} (\boldsymbol{a} = \boldsymbol{m} \supset \boldsymbol{y} = \boldsymbol{f}(\boldsymbol{m})).$$

Also,

$$\rightarrow a_\alpha = c_0, a_\alpha = c_1, \ldots \text{ for each } \alpha.$$

Therefore by the strong generalized cut rule,

$$\rightarrow \exists \boldsymbol{y} \bigwedge_{\boldsymbol{m}} (\boldsymbol{a} = \boldsymbol{m} \supset \boldsymbol{y} = \boldsymbol{f}(\boldsymbol{m})).$$

From this, (10) follows.

The eigenvariable conditions in the proofs of (1) and (7) can be easily examined.

We can extend the above method. Let A be a formula of the form $\mathbf{Q}^T(\boldsymbol{x}; \boldsymbol{y})\, A(x, y)$. Then the set of formulas

$$\{\forall \boldsymbol{x}\, \exists \boldsymbol{y} \bigwedge_{\boldsymbol{n}} (\boldsymbol{x} = \boldsymbol{n} \supset \boldsymbol{y} = \boldsymbol{f}(\boldsymbol{n}))\}_{\boldsymbol{f}}$$

in (7) corresponding to A will be denoted by $\Phi(A)$. Note that $\Phi(A)$ actually depends only on the length of $\boldsymbol{x}$; therefore it does not matter what free variables A may contain.

LEMMA 24.23. *Let $B_1, B_2, \ldots$ be all the subformulas of A of the above form. Then there is a quantifier free formula $\tilde{A}$ (in the extended language) such that $\tilde{A}$ contains exactly the same free variables as A and*

$$(11)\quad \Phi(B_1), \Phi(B_2), \ldots, A \rightarrow \tilde{A}, \qquad \Phi(B_1), \Phi(B_2), \ldots, \tilde{A} \rightarrow A,$$

are provable without essential cuts.

PROOF (by transfinite induction on the complexity of A). Since other cases are obvious, we shall work on the case where A is of the form $\mathbf{Q}^T(\boldsymbol{x}; \boldsymbol{y})\, A(\boldsymbol{x}, \boldsymbol{y})$. To prove the first sequent, we proceed as follows. Consider $A(\boldsymbol{d}, \boldsymbol{e})$ where $\boldsymbol{d}$ and $\boldsymbol{e}$ are new free variables. Then, by the induction hypothesis, there is a quantifier free formula $\tilde{A}(\boldsymbol{d}, \boldsymbol{e})$ such that

$$\Phi(B_1), \ldots, A(\boldsymbol{d}, \boldsymbol{e}) \rightarrow \tilde{A}(\boldsymbol{d}, \boldsymbol{e})$$

is provable without essential cuts. Therefore with the same reasoning as for (2),

$$\Phi(B_1), \ldots, A(\boldsymbol{n}, \boldsymbol{g}(\boldsymbol{n})), \ldots \rightarrow \ldots, \tilde{A}(\boldsymbol{m_f}, \boldsymbol{f}(\boldsymbol{m_f})), \ldots$$

is provable without essential cuts (cf. (2) above). Therefore the proof of the first sequent goes in the same way as in (1).

In proving the second sequent, start with

$$\Phi(B_1), \ldots, \tilde{A}(\boldsymbol{d}, \boldsymbol{e}) \rightarrow A(\boldsymbol{d}, \boldsymbol{e}).$$

From this we obtain

$$\Phi(B_1), \ldots, \boldsymbol{a} = \boldsymbol{n}, \boldsymbol{a} = \boldsymbol{n} \supset \boldsymbol{b} = \boldsymbol{f}(\boldsymbol{n}), \tilde{A}(\boldsymbol{n}, \boldsymbol{f}(\boldsymbol{n})) \rightarrow A(\boldsymbol{a}, \boldsymbol{b}),$$

without cuts. Then, by following the proof of (7), we obtain a cut-free proof of

$$\Phi(A), \Phi(B_1), \ldots, \tilde{A} \rightarrow A.$$

Now consider an arbitrary, valid sequent of L, say

$$A_0, A_1, \ldots \rightarrow B_0, B_1, \ldots,$$

and attempt to prove it. Since our system with language L′ is consistent, the validity of the given sequent and (11) imply that

$$\Phi, \tilde{A}_0, \tilde{A}_1, \ldots \rightarrow \tilde{B}_0, \tilde{B}_1, \ldots \tag{12}$$

is valid, where Φ consists of the formulas of the form

$$\forall \boldsymbol{x}\, \exists \boldsymbol{y} \bigwedge_{\boldsymbol{n}} (\boldsymbol{x} = \boldsymbol{n} \supset \boldsymbol{y} = \boldsymbol{f}(\boldsymbol{n})).$$

Therefore (12) is provable (without essential cuts) in the homogeneous part of our extended system.

On the other hand, Φ, $A_\alpha \rightarrow \tilde{A}_\alpha$ and Φ, $\tilde{B}_\beta \rightarrow B_\beta$ are provable in the system with language L′ without essential cuts. Now by the strong generalized cut rule with the cut formulas

$$\tilde{A}_0, \tilde{A}_1, \ldots, \tilde{B}_0, \tilde{B}_1, \ldots \tag{13}$$

we obtain the sequent

$$\Phi, A_0, A_1, \ldots \rightarrow B_0, B_1, \ldots.$$

Then using the generalized cut again, we obtain the given sequent

$$A_0, A_1, \ldots \rightarrow B_0, B_1, \ldots .$$

Therefore, what remains to be shown is that this essential cut in the proof of the given sequent can be eliminated (in the language of L′). If so, then the given sequent of the language L can be proved in the original system with language L and without essential cuts.

In proving the cut-elimination theorem, we shall make use of the proof given above.

The proof is carried out with a generalized Gentzen's reduction method. In order to simplify the discussion, we assume that the language L has no function symbols. It will also be assumed that the initial sequents consist of atomic formulas only. We shall prove the cut-elimination theorem in the following form.

(*) Let P be a proof (in L′) such that

(i) Along each branch of the sequents there is at most one essential s.g.c.,

(ii) the ancestors of cut formulas have no quantifiers.

Let $S : \Gamma \rightarrow \Delta$ be the lower sequent of a s.g.c. We then list some consequences of a reduction.

(iii) S as well as the descendents of S remain unchanged.

(iv) Either the s.g.c. by which S is obtained in P is eliminated, or sequents of the form $\ldots D \ldots \rightarrow \ldots D \ldots$ are eliminated, or some introductions of logical symbols above S are eliminated, or the essential s.g.c. is pushed up one step.

Let $\mathscr{F}$ be the set of cut formulas of the s.g.c. above S. The reduction is defined according to the stage number k (mod 10).

$k \equiv 0$ (mod 10). Look for a sequent of the form

$$\ldots, D, \ldots \rightarrow \ldots, D, \ldots \quad \text{or} \quad \ldots \rightarrow \ldots, a = a, \ldots$$

among the upper sequents of S. Suppose there is one. In the subsequent treatment, the cases with earlier numbers have priority over those with later numbers.

Case 1. For some D as above, D occurs both in Γ and Δ. Change the figure above S to

$$\frac{D \rightarrow D}{\Gamma \rightarrow \Delta} \quad \text{(no cuts)}.$$

Case 2. An equality $a = a$ occurs in Δ. Change the figure above S to

$$\frac{\rightarrow a = a}{\ldots \rightarrow \ldots, a = a, \ldots}.$$

Case 3. For any such D, D does not occur in either Γ or Δ. This case cannot happen, since then D must belong to both $\mathscr{F}_1$ and $\mathscr{F}_2$ for some partition $(\mathscr{F}_1, \mathscr{F}_2)$ of $\mathscr{F}$.

Case 4. D occurs in Γ but not in Δ. Then the D in the antecedent is not a cut formula but the D in the succedent is a cut formula. Eliminate all the upper sequents which contain D in the succedent. If D occurs in the antecedent in one of the remaining sequents (and such a sequent exists), then regard it as a formula in Γ. Do the same to all such D's. Thus the set of cut formulas will be $\mathscr{F}$—{all such D's} and Case 4 can be eliminated.

Case 5. D occurs in Δ but not Γ. The reduction is defined as for Case 4.

$k \equiv 1 \pmod{10}$. There is an upper sequent of the form

$$\ldots \rightarrow \ldots, t = c_0, \ldots, t = c_1, \ldots, t = c_\alpha, \ldots \quad (\alpha < K).$$

If all the $t = c_\alpha$ occur in Δ, then $\Gamma \rightarrow \Delta$ can be obtained from

$$\frac{\rightarrow t = c_0, t = c_1, \ldots, t = c_\alpha, \ldots}{\Gamma \rightarrow \Delta}$$

and hence without cuts. Otherwise consider the following. Let G be the set of all formulas of the form $t = c_\alpha$ which are cut formulas and let H be the rest of the cut formulas. Let $(\mathscr{F}_1, \mathscr{F}_2)$ be a partition of $\mathscr{F}$. Then there is a $\Phi_1 \subseteq \mathscr{F}_1 \cap G$, a $\Psi_1 \subseteq \mathscr{F}_1 \cap H$, a $\Phi_2 \subseteq \mathscr{F}_2 \cap G$ and a $\Psi_2 \subseteq \mathscr{F}_2 \cap H$ such that

$$\Phi_1, \Psi_1, \Gamma' \rightarrow \Delta', \Phi_2, \Psi_2$$

is an upper sequent of S.

Consider all the partitions of $\mathscr{F}$ which leave $G_1 = \mathscr{F}_1 \cap G$ and $G_2 = \mathscr{F}_2 \cap G$ fixed. Then for each such partition there are Ψ_1 and Ψ_2 such that

$$G_1, \Psi_1, \Gamma' \rightarrow \Delta', G_2, \Psi_2$$

is provable without cuts, without increasing the number of inferences. Hence, by the s.g.c. rule applied to H, we obtain

$$G_1, \Gamma \rightarrow \Delta, G_2.$$

This is true for all possible partitions of G; hence by the s.g.c. rule applied to G, we obtain $\Gamma \rightarrow \Delta$. Here the last s.g.c. is regarded as a weak inference. Thus this case can be eliminated.

$k \equiv 2$. Suppose there are cut formulas whose outermost logical symbols are $\neg$. (If there is no such symbol, pass on to the next stage.) Take an arbitrary upper sequent S_0:

$$S_0\colon \ldots, \neg A, \ldots, \Gamma_0 \rightarrow \Delta_0, \ldots, \neg B, \ldots,$$

where $\neg A, \neg B, \ldots$ are cut formulas and $\Gamma_0 \subseteq \Gamma$ and $\Delta_0 \subseteq \Delta$. Let $\mathbf{Q}_0$ be the sub-proof of P above and including S_0. Then, by changing $\mathbf{Q}_0$ slightly, we obtain a proof of

$$S_0'\colon \ldots, B, \ldots, \Gamma_0 \rightarrow \Delta_0, \ldots, A, \ldots$$

without increasing the number of inferences. Recall that there is no essential cut in $\mathbf{Q}_0$. Replace each formula of $\mathscr{F}$ whose outermost logical symbol is $\neg$, say $\neg A$, by A, thus obtaining $\mathscr{F}'$. Then an arbitrary partition of $\mathscr{F}$, say $(\mathscr{F}'_1, \mathscr{F}'_2)$, induces a partition of $\mathscr{F}$, say $(\mathscr{F}_1, \mathscr{F}_2)$, in a natural manner. If S_0, as above, is an upper sequent of S corresponding to $(\mathscr{F}_1, \mathscr{F}_2)$, then S'_0 corresponds to $(\mathscr{F}'_1, \mathscr{F}'_2)$. Thus the assumption for the s.g.c. rule applied to $\mathscr{F}'$ is satisfied. By a s.g.c. we obtain S. In this case some inferences which introduce $\neg$ are eliminated.

$k \equiv 3$. Consider all the formulas in $\mathscr{F}$ whose outermost logical symbols are $\bigwedge$. Let S_0 be an upper sequent of S;

$$\ldots, \bigwedge_i C_i, \ldots, \Gamma_0 \rightarrow \Delta_0, \ldots, \bigwedge_j A_j, \ldots, \bigwedge_h B_h, \ldots,$$

where $\Gamma_0 \subseteq \Gamma$ and $\Delta_0 \subseteq \Delta$ and $\bigwedge_i C_i$ etc., are only some of the cut formulas whose outermost logical symbols are $\bigwedge$. Let $\mathbf{Q}_0$ be the figure which is above and including S_0. Then, changing $\mathbf{Q}_0$ in an obvious manner, we can construct a quasi-proof of the sequent

$$\ldots, \{C_i\}_i, \ldots, \Gamma_0 \rightarrow \Delta_0, \ldots, A_j, \ldots, B_h, \ldots$$

for every combination of $(\ldots, j, \ldots, h, \ldots)$. Let $\mathscr{F}'$ be the set of formulas obtained from $\mathscr{F}$ by replacing all the formulas of the form $\bigwedge_i C_i$ by

$$\{C_0, C_1, \ldots, C_i, \ldots\}.$$

Let $(\mathscr{F}'_1, \mathscr{F}'_2)$ be a partition of $\mathscr{F}$ for which $\bigwedge_i C_i$ belongs to $\mathscr{F}_2$ if $\bigwedge_i C_i$ belongs to $\mathscr{F}$ and there is one C_i which belongs to $\mathscr{F}'_2$; $\bigwedge_i C_i$ belongs to $\mathscr{F}_1$ if all $C_0, C_1, \ldots, C_i, \ldots$ belong to $\mathscr{F}'_1$; and all other formulas belong to $\mathscr{F}_1(\mathscr{F}_2)$ if and only if they belong to $\mathscr{F}'_1(\mathscr{F}'_2)$. There is an upper sequent of S corresponding to $(\mathscr{F}_1, \mathscr{F}_2)$:

$$\ldots, \bigwedge_i C_i, \ldots, \Gamma_0 \rightarrow \Delta_0, \ldots, \bigwedge_j A_j, \ldots, \bigwedge_h B_h, \ldots.$$

Then, as was shown above, we can change this to

$$\ldots, \{C_i\}_i, \ldots, \Gamma_0 \rightarrow \Delta_0, \ldots, A_j, \ldots, B_h, \ldots$$

for every combination of $\ldots, j, \ldots, h, \ldots$. Since $\bigwedge_i C_i$ belongs to $\mathscr{F}_1$ only if all C_i's belong to $\mathscr{F}'_1$,

$$\ldots, \{C_i\}_i, \ldots \subseteq \mathscr{F}'_1 \quad \text{and} \quad \ldots, A_j, \ldots, B_h, \ldots, \subseteq \mathscr{F}'_2.$$

This argument goes through for any partition of $\mathscr{F}'$. Therefore the s.g.c. rule can apply to $\mathscr{F}'$.

$k \equiv 4$. The reduction for $\bigvee$ can be defined likewise.

$k \equiv 5$. Suppose $\Phi, \Gamma_0 \rightarrow \Delta_0, \Psi$ is an upper sequent of S. Let $\neg A, \ldots$ be all the formulas in Γ_0 whose outermost logical symbols are $\neg$, and let $\tilde{\Gamma}$ be the rest of Γ_0. $\neg B, \ldots$ and $\tilde{\Delta}$ are defined likewise. Then we can construct a quasi-proof of

$$\Phi, \tilde{\Gamma}, B, \ldots \rightarrow \tilde{\Delta}, A, \ldots, \Psi$$

without increasing the number of inferences. By the s.g.c. rule applied to the same cut formulas, we obtain $\Gamma' \rightarrow \Delta'$, where Γ' is obtained from Γ by replacing some $\neg A$'s by B's, and Δ' is defined likewise. So we obtain

$$\frac{\Gamma' \rightarrow \Delta'}{\overline{\Gamma \rightarrow \Delta}}$$

where ══ stands for two inferences, a $\neg$: left and a $\neg$: right. In this case some logical inferences (introduction of $\neg$) are eliminated from above the s.g.c.

$k \equiv 6$. Let $\Phi, \Gamma_0 \rightarrow \Delta_0, \Psi$ be an upper sequent of S. Consider all the formulas in Δ_0 whose outermost logical symbols are $\bigwedge$, say $\ldots \bigwedge_i A_i, \ldots, \bigwedge_j B_j, \ldots$ and let $\tilde{\Delta}_0$ denote the remaining formulas in Δ_0. We can construct a quasi-proof of

$$\Phi, \Gamma_0 \rightarrow \tilde{\Delta}_0, \ldots, A_i, \ldots, B_j, \ldots, \Psi$$

for every combination of $(\ldots, i, \ldots, j, \ldots)$. Hence by the s.g.c. rule we obtain

$$\Gamma' \rightarrow \tilde{\Delta}, \ldots, A_i, \ldots, B_j, \ldots,$$

from which we can infer $\Gamma \rightarrow \Delta$.

$k \equiv 7$. Consider the formulas in Γ_0 whose outermost logical symbol is $\bigwedge$.

$k \equiv 8$. Let $\Phi, \Gamma_0 \rightarrow \Delta_0, \Psi$ be an upper sequent of S. Let $\mathbf{Q}^T(\boldsymbol{x}; \boldsymbol{y})\, A(\boldsymbol{x}, \boldsymbol{y}), \ldots$ be all the formulas in Γ_0 whose outermost logical symbol is $\mathbf{Q}^T$, and let $\tilde{\Gamma}_0$ be the rest of Γ_0. Let $\mathbf{Q}^{T'}(\boldsymbol{u}; \boldsymbol{v})\, B(\boldsymbol{u}, \boldsymbol{v}), \ldots$ be all the formulas in Δ_0 whose outermost logical symbol is $\mathbf{Q}^{T'}$, and let $\tilde{\Delta}_0$ be the rest of Δ_0. Then we can construct a quasi-proof of

$$\Phi, \tilde{\Gamma}_0, \{A(\boldsymbol{s}, \boldsymbol{a})\}_{\boldsymbol{a}, \boldsymbol{s}}, \ldots \rightarrow \{B(\boldsymbol{b}_i, \boldsymbol{t}_i)\}_i, \ldots, \tilde{\Delta}_0, \Psi$$

for some $\boldsymbol{a}$, $\boldsymbol{s}$ and i's. Applying the s.g.c. rule we obtain

$$\tilde{\Gamma}, \{A(\boldsymbol{s}, \boldsymbol{a})\}_{\boldsymbol{a}, \boldsymbol{s}}, \ldots \rightarrow \{B(\boldsymbol{b}_i, \boldsymbol{t}_i)\}_i, \ldots, \tilde{\Delta}.$$

Introducing quantifiers to both sides, we can infer $\Gamma \rightarrow \Delta$. Since the eigenvariable conditions are defined for an entire proof, it is obvious that those conditions, which are satisfied in P, are transferred to the new figure.

This completes the description of the reduction.

Take an arbitrary inference which is not one of the weak inferences, i.e., which is an introduction of a logical symbol. The principal formulas of such an inference are either cut formulas or formulas in $\Gamma \rightarrow \Delta$. Let $\mathscr{L}$ be a string of sequents to which the lower sequent of this inference belong.

Since the number of inferences along a string is finite, and the reduction process reduces the number of logical inferences above the s.g.c., the inference under consideration will eventually be either eliminated or carried down under the s.g.c. Hence along any string of sequents, there will eventually be no s.g.c. within a finite number of stages. Therefore we will obtain a cut-free proof of the given sequent.

PART III

CONSISTENCY PROBLEMS

PART III

CONSISTENCY PROBLEMS

We next take up the study of consistency proofs and their applications. For this study we will develop a theory of ordinal diagrams which, as we will see, will play a major role in our investigation. The theory of ordinal diagrams that we will develop is slightly different from the original development in the author's published papers. Nevertheless the two formulations are essentially the same as far as their important properties and their strengths are concerned. The modifications made here are for convenience in our study.

We think it important that consistency problems be formulated in an appropriate philosophical framework. What framework is appropriate is a question we will discuss in the next chapter. In recent years there has been a tendency to push the foundations of mathematics in the direction of what might be called "quasi" foundations. We think this trend in foundations is an unfortunate one. In Chapter 5 we will explain what we mean by quasi foundations and why we oppose the movement of foundational studies in this direction. For an account of a variety of views on consistency problems the reader should consult the articles on proof theory in: *Intuitionism and proof theory*, Eds. Kino, Myhill and Vesley (North-Holland, Amsterdam, 1970).

CHAPTER 5

CONSISTENCY PROOFS

§25. Introduction

This chapter is devoted to the consistency problems of systems of second order arithmetic. Before we take up these problems there are two points we would like to call to the reader's attention.

1. Mathematicians have an extremely good intuition about the world of the natural numbers as conceived by an infinite mind. Consequently, consistency for the natural numbers is not a particularly important question. In contrast, we can conceive of the world of sets only through our imagination and our mathematical experience. Consequently, the problem of the consistency of the comprehension axioms is a serious and important foundational question.

2. Mathematicians use the term "consistency" as a sort of foundational watchword. The first implication of the term is that no contradiction is derived. Of course this is the most important assurance for our imaginary world of the infinite mind. But sometimes we would like to know more. For example, the fact that no contradiction arises does not explain what it means to say that a theorem is provable from the comprehension axiom. Nonconstructive proofs provide no insight into this important question. On the other hand a constructive proof strengthens our intuition and adds meaning to the theorem. In particular, a constructive proof of the cut-elimination theorem would give us greater confidence in the comprehension axioms and hence strengthen our convictions about our imaginary world of sets.

In this chapter we will be interested in second order arithmetic. Here our comprehension axioms are

$$\forall:\text{left}\quad \frac{F(V),\Gamma\to\Delta}{\forall\phi\, F(\phi),\Gamma\to\Delta},\qquad \exists:\text{right}\quad \frac{\Gamma\to\Delta, F(V)}{\Gamma\to\Delta,\exists\phi\, F(\phi)}.$$

In any cut free proof our comprehension axioms mean roughly that we can introduce a new variable ϕ to express a given abstract V, and that we later use an abbreviation $\forall\phi\, F(\phi)$ or $\exists\phi\, F(\phi)$ in place of

$$F(\phi_1)\wedge\ldots\wedge F(\phi_n)\quad\text{or}\quad F(\phi_1)\vee\ldots\vee F(\phi_n).$$

We will discuss this at greater length later. We point out that a similar interpretation holds in systems of higher order although there the situation is more complicated.

We begin our study with a theory of modulations, a theory that provides our basic argument against "practical" foundations. Given a formula A, we will define left modulations of A and right modulations of A. The definition is according to the outermost logical symbol of A. We assume these symbols to be $\neg, \wedge, \vee$, and $\forall$. For each case a left modulation of A is defined as a formula of the form $A_1 \wedge \ldots \wedge A_m$, where $A_1, \ldots, A_m$ are some left atomic modulations of A, while a right modulation of A is defined as a formula of the form $B_1 \vee \ldots \vee B_n$, where $B_1, \ldots, B_n$ are some right atomic modulations of A. It is also required that if A' is a modulation of A, then every free variable in A' occurs in A.

DEFINITION 25.1. (1) If A is an atomic formula, then the left and the right (atomic) modulations of A are A itself.

(2) If A is of the form $\neg B$, a left atomic modulation of A is of the form $\neg B'$, where B' is a right modulation of B; a right atomic modulation of A is of the form $\neg B''$, where B'' is a left modulation of B.

(3) If A is of the form $B \wedge C$, a left atomic modulation of A is of the form $B' \wedge C'$, where B' and C' are left modulations of B and C, respectively. A right atomic modulation of A is of the form $B'' \wedge C''$, where B'' and C'' are right modulations of B and C respectively. For $B \vee C$, the definition is similar.

(4) If A is of the form $\forall x\, F(x)$, and t is an arbitrary term such that no free variables in t occur in $\forall x\, F(x)$, then $\forall x_1 \ldots \forall x_n\, G$ is an atomic left modulation of A, where G is a left modulation of $F(t)$ and $\forall x_1 \ldots \forall x_n$ bind all free variables in t. A right atomic modulation of $\forall x\, F(x)$ is of the form $\forall x\, G(x)$, where a does not occur in $F(x)$ and $G(a)$ is a right modulation of $F(a)$.

(5) If A is of the form $\forall \phi\, F(\phi)$, and V is an arbitrary abstract such that no free variables in V occur in $\forall \phi\, F(\phi)$, then $\forall \phi_1 \ldots \forall \phi_m \forall x_1 \ldots \forall x_n\, G$ is a left atomic modulation of A, where G is a left modulation of $F(V)$ and $\forall \phi_1 \ldots \forall \phi_m \forall x_1 \ldots \forall x_n$ bind all free variables in V. A right atomic modulation of $\forall \phi\, F(\phi)$ is of the form $\forall \phi\, F'(\phi)$, where α does not occur in $F(\phi)$ and $F'(\alpha)$ is a right modulation of $F(\alpha)$.

PROPOSITION 25.2. *Let α and β be either free variables or constants. If $A'(\alpha)$ is a modulation (right or left) of $A(\alpha)$, then $A'(\beta)$ is also a modulation (right or left) of $A(\beta)$.*

PROPOSITION 25.3. *If A' is a left modulation of A, then $A \rightarrow A'$ is provable. If A'' is a right modulation of A, then $A'' \rightarrow A$ is provable.*

PROOF. Since other cases are easy, we shall consider only the first part for the crucial case, (5): i.e., A is of the form $\forall \phi\, F(\phi)$. A' is of the form

$A'_1 \wedge \ldots \wedge A'_n$, where A'_i is an atomic left modulation of A. It suffices to show $A \to A'_i$ for each i. Let A'_i be of the form $\forall \phi_1 \ldots \forall \phi_l \forall x_1 \ldots \forall x_m G$, where G is a left modulation of $F(V)$. Suppose $F(V) \to G$ is provable. Then

$$\cfrac{\cfrac{F(V) \to G}{\forall \phi F(\phi) \to G}}{\forall \phi F(\phi) \to \forall \phi_1 \ldots \forall \phi_l \forall x_1 \ldots \forall x_m G.}$$

The eigenvariable condition is satisfied.

DEFINITION 25.4. (1) A sequent $A'_1, \ldots, A'_n \to B'_1, \ldots, B'_m$ is called a modulation of $A_1, \ldots, A_n \to B_1, \ldots, B_m$ if A'_i is a left modulation of A_i for each i and B'_j is a right modulation of B_j for each j.

(2) Let P be a cut free proof. For each sequent $\Pi \to \Lambda$ in P, we define a modulation $\Pi' \to \Lambda'$ of $\Pi \to \Lambda$ by induction on the number of inferences above $\Pi \to \Lambda$. $\Pi' \to \Lambda'$ is called the P-modulation of $\Pi \to \Lambda$ and is so defined that if $\Pi \to \Lambda$ is $A_1, \ldots, A_m \to B_1, \ldots, B_n$, then $\Pi' \to \Lambda'$ is $A'_1, \ldots, A'_m \to B'_1, \ldots, B'_n$, where A'_i is a left modulation of A_i and B_j is a right modulation of B_j. We shall give explicit definitions only for some exemplary cases.

1) If $\Pi \to \Lambda$ is an initial sequent in P, then $\Pi' \to \Lambda'$ is $\Pi \to \Lambda$ itself.

2) The last inference is an interchange : left.

$$\frac{\Gamma, C, D, \Xi \to \Delta}{\Gamma, D, C, \Xi \to \Delta}.$$

If $\Gamma', C', D', \Xi' \to \Delta'$ is the P-modulation of $\Gamma, C, D, \Xi \to \Delta$, then $\Gamma', D', C', \Xi' \to \Lambda'$ is the P-modulation of $\Gamma, D, C, \Xi \to \Lambda$.

(3) The last inference is a weakening : left.

$$\frac{\Gamma \to \Delta}{C, \Gamma \to \Delta}.$$

If $\Gamma' \to \Delta'$ is the P-modulation of $\Gamma \to \Delta$, then $C, \Gamma' \to \Delta'$ is the P-modulation of $C, \Gamma \to \Delta$.

4) The last inference is a contraction : left.

$$\frac{C, C, \Gamma \to \Delta}{C, \Gamma \to \Delta}.$$

If $C_1, C_2, \Gamma' \to \Delta'$ is the P-modulation of $C, C, \Gamma \to \Delta$, then $C_1 \wedge C_2, \Gamma' \to \Delta'$ is the P-modulation of $C, \Gamma \to \Delta$.

The last inference is a contraction : right.

$$\frac{\Gamma \to \Delta, C, C}{\Gamma \to \Delta, C}.$$

If $\Gamma' \to \Delta', C_1, C_2$ is the P-modulation $\Gamma \to \Delta, C, C$ then $\Gamma' \to \Delta', C_1 \vee C_2$ is the P-modulation of $\Gamma \to \Delta, C$.

5) The last inference is a second order $\forall$: left.

$$\frac{F(V), \Gamma \rightarrow \Delta}{\forall \phi\, F(\phi), \Gamma \rightarrow \Delta}.$$

If $G, \Gamma' \rightarrow \Delta'$ is the P-modulation of $F(V), \Gamma \rightarrow \Delta$, then

$$\forall \phi_1 \ldots \forall \phi_n \forall x_1 \ldots \forall x_m\, G, \Gamma' \rightarrow \Delta'$$

is the P-modulation of $\forall \phi\, F(\phi), \Gamma \rightarrow \Delta$, where $\forall \phi_1 \ldots \forall \phi_n \forall x_1 \ldots \forall x_m$ binds all free variables in V. More precisely, let $V = V(\alpha_1, \ldots, \alpha_n, a_1, \ldots, a_m)$, where all the free variables are indicated, and let $\beta_1, \ldots, \beta_n, b_1, \ldots, b_m$ be free variables which are not contained in $F(V)$. Let $G'(\beta_1, \ldots, \beta_n, b_1, \ldots, b_m)$ be a left modulation of $F(V(\beta_1, \ldots, \beta_n, b_1, \ldots, b_m))$ so that $G'(\alpha_1, \ldots, \alpha_n, a_1, \ldots, a_m)$ is G. By $\forall \phi_1 \ldots \forall \phi_n \forall x_1 \ldots \forall x_m\, G$ we mean

$$\forall \phi_1 \ldots \forall \phi_n \forall x_1 \ldots \forall x_m\, G'(\phi_1, \ldots, \phi_n, x_1, \ldots, x_m).$$

6) The last inference is a second order $\forall$: right.

$$\frac{\Gamma \rightarrow \Delta, F(\alpha)}{\Gamma \rightarrow \Delta, \forall \phi\, F(\phi)}.$$

Suppose $\Gamma' \rightarrow \Delta', F'(\alpha)$ is the P-modulation of the upper sequent. Then $\Gamma' \rightarrow \Delta', \forall \phi\, F'(\phi)$ is the P-modulation of the lower sequent.

PROPOSITION 25.5. *Let P be a cut-free proof of $\Gamma \rightarrow \Delta$, and let $\Gamma' \rightarrow \Delta'$ be the P-modulation of $\Gamma \rightarrow \Delta$. Then there exists a cut-free predicative proof of $\Gamma' \rightarrow \Delta'$.*

PROOF. This is an immediate corollary of (2) of Definition 25.4. That is, in Definition 25.4, if S_2 is a lower sequent of S_1 (and S) then S_2', the P-modulation of S_2, can be derived from S_1' (and S') without a cut and without impredicative comprehension axioms. The latter fact is obvious from the definition of P-modulation in Case 5) of Definition 25.4. From Proposition 1 it is obvious that the eigenvariable condition is satisfied for any application of $\forall$: right.

Our theory of modulations, and in particular Proposition 25.5, makes clear one important reason for our interest in cut-free proofs, namely, a cut-free proof of a theorem enables us to give to that theorem an interpretation that avoids Russell's vicious circle. Of course, each theorem has as many vicious-circle-free interpretations as it has cut-free proofs, and each different interpretation of a theorem determines a different interpretation of the theory.

If we had a uniform "method" that transformed every formal proof into a cut-free proof, then that "method" would itself determine an interpretation of mathematics that avoids Russell's vicious circle. Consequently, producing such a method of transformation should be a matter of high priority.

We would, of course, prefer an interpretation that is as close to the original (natural) meaning as possible. Consequently, the cut-elimination theorem should be proved by an elimination procedure that preserves as much as possible of the meaning of the original proofs.

It is on precisely this issue that we oppose the recent trend of foundational studies in the direction of what we have chosen to call "quasi" foundations. Let us illustrate our point with an example.

One may view analysis in different ways. One view is that analysis is a theory. An alternate view is that analysis is not a system of axioms but a collection of results. It is this alternate view that gives rise to the problem of quasi foundations.

The job of quasi foundations is to develop a kind of quasi-analysis; to find a collection of theorems, that are similar to a given collection of theorems of analysis, but which are in fact weaker results than the given theorems.

There are two points we wish to make that explain why we do not view "quasi" foundations as the proper direction for foundational studies to take. First of all, when one develops a collection of results that are "similar" to a given set of theorems of analysis it is not at all clear what metamathematical conclusions can be drawn or what kind of theory has been founded. But, second, and of greater importance, our theory of modulations shows that in principle the problems of quasi foundations are solved: If we wish a "quasi foundation" for a certain collection of results we simply start with predicative comprehension axioms and develop the modulations of the results that we wish. Of course the results we obtain will not be classical theorems but modulations. However, as we have shown, modulations are stronger results than the theorems of classical mathematics.

We may, therefore, regard the predicative comprehension axioms as what is really important. The theorems of classical mathematics are simply approximations to stronger theorems. Consequently, so long as our concern is with practical foundations we need not prove the cut-elimination theorem; we only need to justify the predicative comprehension axioms. Having done this, our task is to construct sufficient and appropriate modulations for our purposes.

We, however, advocate a different viewpoint with respect to which proving the cut-elimination theorem constructively is a matter of paramount importance. Indeed, only through a constructive proof of cut-elimination can the theory of modulations become truly a significant theory in a world of predicative mathematics. The proof of the cut-elimination theorem as presented in Chapter 3 is set-theoretical, and therefore useless for our purpose.

§26. Ordinal diagrams

In this section we will develop a theory of ordinal diagrams which, as we pointed out earlier, will play a fundamental role in our study of consistency problems. For each pair of nonempty, well-ordered sets I and A we will define the set of ordinal diagrams based on I and A. At the same time we will also define the notion of connected ordinal diagrams.

DEFINITION 26.1. Let I and A be nonempty, well-ordered sets with 0 the smallest element in I. The system of *ordinal diagrams*, based on I and A, we define inductively in the following way.

1) 0 is a connected ordinal diagram.

2) Let i be an element of I, a an element of A, and let α be an ordinal diagram. Then (i, a, α) is a *connected* ordinal diagram.

3) Let $n \geqslant 2$ and let $\alpha_1, \ldots, \alpha_n$ be connected ordinal diagrams. Then $\alpha_1 \# \alpha_2 \# \ldots \# \alpha_n$ is a *nonconnected* ordinal diagram.

For convenience, in discussing ordinal diagrams we will use i, j, k etc., as variables on I; a, b, c, etc., as variables on A; and α, β, γ, etc., as variables on ordinal diagrams. Hereafter "ordinal diagram" will mean "ordinal diagram based on I and A".

Definition 26.1 defines an inductive procedure for constructing ordinal diagrams. If the ordinal diagram γ enters the construction of α we call γ a sub-ordinal diagram of α. But from the definition it is clear that an ordinal diagram γ could enter the construction of α at several different stages of the construction. In the work to follow it is sometimes important to identify a specific occurrence of a sub-ordinal diagram. For this purpose we will use the notation $\bar{\gamma}$, that is, the notation $\bar{\gamma}$ is to indicate that we are talking about a specific occurrence of γ and not just about the ordinal diagram γ itself. Thus $\bar{\gamma} = \bar{\beta}$ means not only that $\gamma = \beta$ but also that $\bar{\gamma}$ and $\bar{\beta}$ denote the same occurrence of this ordinal diagram.

DEFINITION 26.2. (1) Each connected ordinal diagram $\alpha_1, \ldots, \alpha_n$ is a component of the nonconnected ordinal diagram $\alpha_1 \# \ldots \# \alpha_n$.

(2) Each connected ordinal diagram α has only one component, namely itself.

DEFINITION 26.3. Let $l(\alpha)$ be the total number of ( )'s and #'s in α. Then

$$l(\alpha, \beta, \ldots, \gamma) =_{\text{df}} l(\alpha) + l(\beta) + \ldots + l(\gamma).$$

DEFINITION 26.4. *Equality* for ordinal diagrams we define inductively on $l(\alpha, \beta)$ in the following way.

(1) $0 = 0$.

(2) Let α be of the form (i, a, γ). Then $\alpha = \beta$ if and only if β is of the form (i, a, δ), where $\gamma = \delta$.

(3) Let α be $\alpha_1 \# \ldots \# \alpha_m$ and let β be $\beta_1 \# \ldots \# \beta_n$, where $\alpha_1, \ldots, \alpha_m$ and $\beta_1, \ldots, \beta_n$ are connected. Then $\alpha = \beta$ if and only if $m = n$ and there is a permutation of $\{1, 2, \ldots, m\}$, say $\{j_1, \ldots, j_m\}$, such that

$$\alpha_1 = \beta_{j_1}, \alpha_2 = \beta_{j_2}, \ldots, \alpha_m = \beta_{j_m}.$$

It can be easily proved that $=$ is an equivalence relation. Note also that $\alpha = 0$ or $0 = \alpha$ if and only if α is 0.

DEFINITION 26.5. (1) Consider an occurrence $\overline{(i, a, \gamma)}$ in α. If $\bar{\beta}$ is an occurrence of β in γ, then the occurrence of i and of a in $\overline{(i, a, \gamma)}$ are said to be connected to $\bar{\beta}$ in α. We also say that the occurrence of the pair (i, a) in $\overline{(i, a, \gamma)}$ is connected to $\bar{\beta}$.

(2) Let $\bar{\beta}$ be an occurrence of β in α and let j be an element of I. If for every element i of I that occurs in α and is connected to $\bar{\beta}$ we have that $i \geqslant j$, then $\bar{\beta}$ is said to be j-active in α.

(3) A connected, j-active occurrence of a sub-ordinal diagram of α is called a j-subsection of α.

(4) Let $\overline{(i, a, \gamma)}$ be a j-subsection of α for some $j > i$. Then the occurrence $\bar{\gamma}$ in (i, a, γ) is called an i-section of α. If there is an i-section of α, then we say that i is an index of α.

Note that an i-section of α is a special case of an i-subsection. An i-section of α is an occurrence of a proper sub-ordinal diagram of α, but an i-subsection of α may be α itself.

For certain purposes ahead we introduce a special symbol ∞ that we adjoin to I and regard as the maximal element of the extended set.

DEFINITION 26.6. (1) $\tilde{I} = I \cup \{\infty\}$. The *ordering* of $\tilde{I}$ is that of I with ∞ the maximal element of $\tilde{I}$.

(2) If $i \in \tilde{I}$ and $j \in I$, then i is called a super index of j with respect to $\alpha, \beta, \ldots, \gamma$, if either i is ∞ or $i > j$ and i is an index of at least one of $\alpha, \beta, \ldots, \gamma$.

(3) $j_0(j, \alpha, \beta, \ldots, \gamma)$ is the least super index of j with respect to $\alpha, \beta, \ldots, \gamma$.

(4) $\iota(j, \alpha, \beta, \ldots, \gamma)$ denotes the number of super indices of j with respect to $\alpha, \beta, \ldots, \gamma$ when j is an element of I; it is defined to be 0 if j is ∞.

(5) The outermost index of the ordinal diagram (i, a, α) is i.

(6) A pair (i, a), where $i \in I$ and $a \in A$ is called a *value*. The set of values is ordered lexicographically.

(7) The value (i, a) is the outermost value of (i, a, α).

Note that a value is not an occurrence.

For each i in $\tilde{I}$ we now define an ordering $<_i$ of ordinal diagrams (based on I and A). The definition is by transfinite induction on $\omega \cdot l(\alpha, \beta) + \iota(i, \alpha, \beta)$.

DEFINITION 26.7. (1) $0 <_i \beta$ if $\beta \neq 0$.

(2) If α is $\alpha_1 \# \ldots \# \alpha_m$ and β is $\beta_1 \# \ldots \# \beta_n$, where $m+n>2$, then $\alpha <_i \beta$ if one of the following conditions hold.

i) There exists a q such that $1 \leq q \leq n$ and $\alpha_p <_i \beta_q$ for all p with $1 \leq p \leq m$.

ii) $m=1$, $n>1$ and $\alpha_1 = \beta_q$ for some q with $1 \leq q \leq n$.

iii) $m>1$, $n>1$ and there exists a q and a p such that $1 \leq q \leq m$, $1 \leq p \leq n$, $\alpha_q = \beta_p$ and

$$\alpha_1 \# \ldots \# \alpha_{q-1} \# \alpha_{q+1} \# \ldots \# \alpha_m <_i \beta_1 \# \ldots \# \beta_{p-1} \# \beta_{p+1} \# \ldots \# \beta_n.$$

(3) If α and β are connected, if $i \neq \infty$ and if $j = j_0(i, \alpha, \beta)$, (cf. (2) of Definition 26.6) then $\alpha <_i \beta$ if one of the following holds.

i) There exists an i-section $\bar{\sigma}$ of β such that $\alpha \leq_i \sigma$, i.e., $\alpha <_i \sigma$ or $\alpha = \sigma$.

ii) $\alpha <_j \beta$ and for every i-section $\bar{\delta}$ of α, $\delta <_i \beta$.

(4) If $\alpha = (j, a, \gamma)$ and $\beta = (k, b, \delta)$, then $\alpha <_\infty \beta$ if

i) $j < k$ (in I) or

ii) $j = k$ and $a < b$ (in A) or

iii) $j = k$, $a = b$, and $\gamma <_j \delta$.

The ordering $<_\infty$ is slightly different from the original version in which a and b were compared first.

PROPOSITION 26.8. *For each i in $\bar{I}$ the definition of $<_i$ is sound and $<_i$ is a linear ordering of the ordinal diagrams based on I and A.*

PROOF. By induction on $\omega \cdot l(\alpha, \beta, \gamma) + \iota(i, \alpha, \beta, \gamma)$ and $\omega \cdot l(\alpha, \beta) + \iota(i, \alpha, \beta)$, respectively, we can prove simultaneously that

I if $\alpha <_i \beta$ and $\beta <_i \gamma$, then $\alpha <_i \gamma$,

II if $\alpha = \beta$ and $\beta <_i \gamma$, then $\alpha <_i \gamma$ and if $\alpha <_i \beta$ and $\beta = \gamma$ then $\alpha <_i \gamma$,

III exactly one of $\alpha <_i \beta$, $\alpha = \beta$ and $\beta <_i \alpha$ holds.

If $l(\alpha, \beta, \gamma) = 0$ or $l(\alpha, \beta) = 0$, respectively, then $\alpha = \beta = \gamma = 0$ and hence I, II and III hold trivially. For $l(\alpha, \beta, \gamma) > 0$ and $l(\alpha, \beta) > 0$ we will present proofs only for some exemplary cases of I and III. We assume that α, β and γ are (j, a, α'), (k, b, β') and (m, c, γ'), respectively.

(1) $\iota(i, \alpha, \beta, \gamma) = \iota(i, \alpha, \beta) = 0$ and hence $i = \infty$.

I. If $\alpha <_\infty \beta$ and $\beta <_\infty \gamma$, it follows from the definition of $<_\infty$ that $j \leq m$ and $a \leq c$. If $j < m$ or $a < c$, then $\alpha <_\infty \beta$. If $j = m$ and $a = c$, then $j = k = m$ and $a = b = c$; hence $\alpha' <_j \beta'$ and $\beta' <_j \gamma'$. Then by the induction hypothesis $\alpha' <_j \gamma'$, hence $\alpha <_\infty \gamma$.

III. If $\alpha \neq \beta$, then $(j, a) \neq (k, b)$ or $(j, a) = (k, b)$ and $\alpha' \neq \beta'$. If $(j, a) \neq (k, b)$, then $\alpha <_\infty \beta$ or $\beta <_\infty \alpha$ by definition of $<_\infty$. If $(j, a) = (k, b)$, then by the induction hypothesis $\alpha' <_j \beta'$ or $\beta' <_j \alpha'$ and hence $\alpha <_\infty \beta$ or $\beta <_\infty \alpha$ as the case may be

(2) $\iota(i. \alpha, \beta, \gamma) > 0$ and $\iota(i, \alpha, \beta) > 0$, respectively.

I. We consider only one case: There exists an i-section of β, say $\bar{\sigma}$, such that $\alpha \leqslant_i \sigma$, and $\beta <_{i_0} \gamma$, where $i_0 = j_0(i, \beta, \gamma)$ and for each i-section of β, say $\bar{\delta}$, $\delta <_i \gamma$. Then $\alpha \leqslant_i \sigma$, $\sigma <_i \gamma$ and $l(\alpha, \sigma, \gamma) < l(\alpha, \beta, \gamma)$; hence by the induction hypothesis $\alpha <_i \gamma$.

III. Suppose $\alpha \not\leqslant_i \beta$, i.e., $\alpha \neq \beta$ and it is not the case that $\alpha <_i \beta$. Then

(*) for every i-section of β, say $\bar{\sigma}$, $\alpha \not\leqslant_i \sigma$,

hence by the induction hypothesis $\sigma <_i \alpha$.

Let $i_0 = j_0(i, \alpha, \beta)$. If $\beta <_{i_0} \alpha$, then by (*) and Definition 26.7 $\beta <_i \alpha$. If $\beta \not<_{i_0} \alpha$ then since $\iota(i_0, \alpha, \beta) < \iota(i, \alpha, \beta)$ it follows from the induction hypothesis that $\alpha <_{i_0} \beta$. If for every i-section of α, say $\bar{\delta}$, $\delta <_i \beta$, then $\alpha <_i \beta$. But this contradicts our initial assumption. Therefore there is an i-section of α, say $\bar{\delta}$, such that $\beta \leqslant_i \delta$. But then by Definition 26.7 $\beta <_i \alpha$.

Proposition 26.9. *If $\bar{\sigma}$ is an i-section of α, then $\sigma <_i \alpha$.*

Proof. If σ is connected, then by (3)i) of Definition 26.7 applied to σ and the component of α in which $\bar{\sigma}$ is an i-section, $\sigma <_i \alpha$. If σ is not connected, then for each component of σ, say δ, $\delta <_i \sigma$ by (2)ii) of Definition 26.7 and hence by (3)i) of Definition 26.7, $\delta <_i \alpha$. Then from (2)i) it follows that $\sigma <_i \alpha$.

Proposition 26.10. *Let α be a connected ordinal diagram and let $\bar{\beta}$ be a proper i-subsection of α. Then $\beta <_j \alpha$ for every $j \leqslant i$.*

Proof. The proof is by induction on $\omega \cdot l(\alpha, \beta) + \iota(j, \alpha, \beta)$ for each $j \leqslant i$.

1. If $\beta = 0$, then $\beta <_j \alpha$ for all j.

2. Suppose $\alpha = (k, b, \gamma)$ and β is a component of γ. Then $i \leqslant k$ and since γ occurs as a k-section of α we have by Proposition 26.9 that $\beta \leqslant_k \gamma <_k \alpha$. Consequently, $\beta <_k \alpha$. If $j < k$, then every j-section of β is a j-section of α. But this implies that for every $\bar{\sigma}$, a j-section of β, $\sigma <_j \alpha$, and hence $\beta <_j \alpha$ (by induction on $\iota(j, \alpha, \beta)$).

3. Suppose $\alpha = (k, b, \gamma)$ and β is not a component of γ. Then $k \geqslant i$ and there is a component of γ, say δ, such that β occurs as an i-subsection of δ. Therefore by the induction hypothesis $\beta <_j \delta$ for all $j \leqslant i$. But $\delta \leqslant_j \gamma$ by definition, and $\gamma <_j \alpha$ for every $j \leqslant k$; therefore, $j \leqslant i$ by 2. above and Definition 26.7(2). Consequently, $\beta <_j \alpha$ if $j \leqslant i$.

Proposition 26.11. *If β has an i-active occurrence as a proper subordinal diagram of α, then $\beta <_j \alpha$ for every $j \leqslant i$.*

Proof. Apply Proposition 26.10 to each component of such an occurrence of β.

Definition 26.12. Let α be an ordinal diagram and let i be an element of

$\tilde{I}$. Then $[\alpha]_i = [\alpha_1, \ldots, \alpha_m]_i$ will mean that $\alpha_1, \alpha_2, \ldots, \alpha_m$ are the components of α and

$$\alpha_1 \geqslant_i \alpha_2 \geqslant_i \ldots \geqslant_i \alpha_m.$$

DEFINITION 26.13. $O(I, A)$ will denote the structure consisting of the set of ordinal diagrams based on I and A and the orderings $<_i$ for all i in $\tilde{I}$.

We will follow the usual convention of using $O(I, A)$ to denote the universe of this structure, that is, the set of ordinal diagrams based on I and A.

Now let I and A be a primitive recursive ordering of the natural numbers and accessibility assume that proofs of I and A be given. We shall present an accessibility proof of $O(I, A)$ for $<_i$.

THEOREM 26.14. *For each i in $\tilde{I}$, $O(I, A)$ is accessible with respect to $<_i$.*

For the proof of the theorem, we shall make several definitions and lemmas.

DEFINITION 26.15. Let α be of the form $\alpha_1 \# \alpha_2 \# \ldots \# \alpha_m$ and β be of the form $\beta_1 \# \beta_2 \# \ldots \# \beta_n$. Then $\alpha \# \beta$ will denote the ordinal diagram

$$\alpha_1 \# \alpha_2 \# \ldots \# \alpha_m \# \beta_1 \# \beta_2 \# \ldots \# \beta_n.$$

DEFINITION 26.16. (1) Let B be a subset of $O(I, A)$ and let i be an element of $\tilde{I}$. An ordinal diagram α is $<_i$-accessible in B if α is an element of B and, with respect to $<_i$, there is no infinite (strictly) decreasing sequence of elements of B starting with α.

(2) α is $<_i$-accessible if α is $<_i$-accessible in $O(I, A)$.

DEFINITION 26.17. We define F_i a subset of $O(I, A)$ for every i in I.

(1) $F_0 = O(I, A)$.

(2) $F_{i+1} = \{\alpha \in F_i \mid \text{For every } \bar{\sigma} \text{ an } i\text{-section of } \alpha, \sigma \text{ is } <_i\text{-accessible in } F_i\}$, where $i+1$ denotes the successor of i in I.

(3) $F_i = \bigcap_{j<i} F_j$ if i is a limit element of I.

From the definition it is obvious that if $\alpha \in F_i$, then $\alpha \in F_j$ for all $j \leqslant i$.

The following lemma can be proved in a much more general case but we present it in a very special case.

LEMMA 26.18. *Let C be the set of all connected ordinal diagrams. Let $\alpha_1, \ldots, \alpha_n \in C$ be $<_i$-accessible in C. Then $\alpha_1 \# \ldots \# \alpha_n$ is $<_i$-accessible in all ordinal diagrams.*

PROOF. Let $\{\beta_n\}$ be a $<_i$-decreasing sequence starting with $\alpha_1 \# \ldots \# \alpha_n$. Let $\beta_n = [\gamma_1^n, \ldots, \gamma_m^n]_i$. Obviously all γ_j^n are $<_i$-accessible in C. Let us first check $\{\gamma_1^n\}$. Since γ_1^1 is $<_i$-accessible in C, it suffices to show that either $\{\beta_n\}$ is terminated in a finite number of steps or there will be a $\gamma_1^{m_1} <_i \gamma_1^{n_1}$ $(m_1 > n_1)$ after any stage n_1. Suppose $\gamma_1^{n_1} = \gamma_1^{n_1+1} = \ldots$ continue. Then check $\gamma_2^{n_1}, \gamma_2^{n_1+1}, \ldots$. In the same way as in the case of γ_1^n, it suffices to show that either $\{\beta_n\}$ terminates in a finite number of steps or there will be a $\gamma_2^{m_2} <_i \gamma_2^{n_2}$ $(m_2 > n_2)$ after any stage n_2. Suppose $\gamma_2^{n_2} = \gamma_2^{n_2+1} = \ldots$ continue. Then check $\gamma_3^{n_2}, \gamma_3^{n_2+1}, \ldots$. In this way we have a sequence $\gamma_1^{n_1} \geqslant \gamma_2^{n_2} \geqslant \gamma_3^{n_3} \geqslant \ldots$, where $n_1 \leqslant n_2 \leqslant n_3 \leqslant \ldots$. There is a possibility that $\gamma_1^{n_1} = \gamma_2^{n_2} = \gamma_3^{n_3} = \ldots$ for a while. However, $\gamma_1^{n_1} {}_i\!> \gamma_{m+1}^{n_{m+1}}$, where m is the length of β_{n_1}, i.e., $\beta_{n_1} = [\gamma_1^{n_1}, \ldots, \gamma_m^{n_1}]_i$. In this way, one can find a $<_i$-decreasing sequence $\gamma_1^{n_1}, \gamma_{m+1}^{n_{m+1}}, \ldots$. Since $\gamma_1^{n_1}$ is $<_i$-accessible in C, the whole procedure terminates in a finite number of steps.

COROLLARY 26.19. *Let* $[\alpha]_i = [\alpha_1, \alpha_2, \ldots, \alpha_m]_i$. *Then* α *is* $<_i$*-accessible if and only if each* $\alpha_1, \alpha_2, \ldots, \alpha_m$ *is* $<_i$*-accessible in* C.

LEMMA 26.20. *Every ordinal diagram in* F_∞ *is* $<_\infty$*-accessible in* F_∞.

PROOF. Let $\alpha_1 {}_\infty\!> \alpha_2 {}_\infty\!> \alpha_3 \ldots$ be in F_∞. We would like to show that the sequence terminates in a finite number of steps. As in Lemma 26.18, this is reduced to the case where $\alpha_1, \alpha_2, \ldots$ are connected. Since for $<_\infty$, ordinal diagrams are first compared by their outermost value (i, a) and every decreasing sequence of (i, a) terminates in a finite number of steps, it suffices to show that the decreasing sequence of ordinal diagrams for $<_\infty$, with the same outermost value (i, a), terminates in a finite number of steps. Let α_m be (i, a, δ_m). Then δ_m is $<_i$-accessible in F_i and the comparison of α_m for $<_\infty$ is the same as the comparison of δ_m for $<_i$. Therefore it terminates in a finite number of steps.

In the following let $i \in I$ or $i = \infty$. If j is the maximal element of I, then $j+1$ denotes ∞ and ∞ is said to be a successor. If ∞ is not a successor, it is said to be a limit.

LEMMA 26.21. *Let* α *be in* F_{i+1}. *If* α *is* $<_{i+1}$*-accessible in* F_{i+1}, *then* α *is* $<_i$*-accessible in* F_i.

PROOF. Let β be in F_i with $\beta <_i \alpha$. We would like to show that β is $<_i$-accessible in F_i. We assume that β is connected since the method of Lemma 26.18 also works here. Let $c(\beta)$ be the maximal length of all sequence $\beta = \gamma_0, \gamma_1, \ldots, \gamma_m$, where γ_{j+1} is an i-section of γ_j. We prove the accessibility of β by mathematical induction on $c(\beta)$, that is, we reduce our construction showing the accessibility of β to a construction on an ordinal diagram with smaller $c(\beta)$.

Let $c(\beta) = 0$, that is, let β have no i-section.

Case 1. $\alpha <_{i+1} \beta$.
In this case there exists an i-section of α, say σ, such that $\beta \leqslant_i \sigma$. Then σ is $<_i$-accessible in F_i. Therefore β is $<_i$-accessible in F_i.

Case 2. $\beta <_{i+1} \alpha$.
In this case β is in F_{i+1} and $\beta <_{i+1} \alpha$. Since α is $<_{i+1}$-accessible in F_{i+1}, this procedure terminates in a finite number of steps.

Now let $c(\beta) > 0$.

Case 1. $\alpha <_{i+1} \beta$.
In this case, there exists an i-section of α, say σ, such that $\beta \leqslant_i \sigma$. Then σ is $<_i$-accessible in F_i. Therefore β is $<_i$-accessible in F_i.

Case 2. $\beta <_{i+1} \alpha$.
By the induction hypothesis, every i-section of β is $<_i$-accessible in F_i. Therefore β is in F_{i+1} and $\beta <_{i+1} \alpha$. Since α is $<_{i+1}$-accessible in F_{i+1}, this procedure terminates in a finite number of steps.

REMARK. What we have done can be restated as follows. First we consider a $<_i$-decreasing sequence in F_i, say $\beta_1 {}_i\!> \beta_2 {}_i\!> \beta_3 {}_i\!> \dots$ such that $\beta_1 <_i \alpha$. If $\alpha <_{i+1} \beta_1$, then we can immediately prove the lemma. If $\beta_1 <_{i+1} \alpha$, then check the i-sections of β_1. If every i-section β_1' of β_1 satisfies $\alpha <_{i+1} \beta_1'$, then β_1 is in F_{i+1} and we can continue, that is, we can discuss β_2 and so on. However if some i-section β_1' of β_1 satisfies $\beta_1' <_{i+1} \alpha$, then we switch the problem to β_1', that is, we consider a $<_i$-decreasing sequence in F_i, $\beta_1' {}_i\!> \beta_2' {}_i\!> \beta_3' {}_i\!> \dots$ and discuss the matter.

LEMMA 26.22. *Let α be $<_i$-accessible in F_i and $j < i$. Then α is $<_j$-accessible in F_j.*

PROOF. By Lemma 26.21, we may assume that i is a limit. Let i_0 be the greatest index of α which is less than i. If we prove the lemma for every j with $i_0 < j < i$, then by Lemma 26.21 we have the lemma for $j = i_0$. Therefore we get the whole lemma in a finite number of steps. Let β be in F_j and $\beta <_j \alpha$. Let $d(\beta)$ be the maximal length of all the sequences $\beta = \gamma_0, \gamma_1, \dots, \gamma_m$, where γ_{k+1} is a proper j-subsection of γ_k. We prove the accessibility of β by mathematical induction on $d(\beta)$. Now let $j \leqslant k < i$ and let γ be a k-section of β. Then by the induction hypothesis, γ is $<_k$-accessible in F_k. This is true for every such k. So $\beta \in F_j$ implies $\beta \in F_i$. Now first notice that $\alpha <_i \beta$ never happens since α does not have any k-section such that $j \leqslant k < i$. Therefore $\beta <_i \alpha$. This type of case terminates in a finite number of steps since α is $<_i$-accessible in F_i.

LEMMA 26.23. *Every ordinal diagram of F_j is $<_j$-accessible in F_j.*

PROOF. Let α be an element of F_j.

Case 1. I has no maximal element. Let i be an element of I which is greater than all the elements of I occurring in α. Then $(i, 0, 0) \in F_\infty$ and $\alpha <_j (i, 0, 0)$. By Lemma 26.22, $(i, 0, 0)$ is $<_j$-accessible in F_j, hence so is α.

Case 2. I has a maximal element but A does not. Let i be the greatest element of I occurring in α and let a be an element of A which is greater than any element of A that occurs in α. Then $(i, a, 0) \in F_\infty$ and $\alpha <_j (i, a, 0)$. By Lemma 26.22 $(i, a, 0)$ is $<_j$-accessible, hence so is α.

Case 3. Both I and A have maximal elements. Let i and a be the greatest elements in I and A, respectively. Then there is a β of the form

$$(i, a, (i, a, \dots, (i, a, 0) \dots))$$

such that $\alpha <_j \beta$. If we can show that $\beta \in F_\infty$, then it will follow from Lemma 26.22 that β is $<_j$-accessible in F_j. From this in turn it follows that α is $<_j$-accessible in F_j and this will complete the proof.

If $\beta = (i, a, 0)$, then obviously $\beta \in F_\infty$. Suppose

$$\beta_0 = (i, a, \dots (i, a, 0) \dots) \in F_\infty.$$

Then by Lemma 26.22, β_0 is $<_i$-accessible in F_i. Therefore, by definition, $\beta = (i, a, \beta_0) \in F_\infty$.

As a special case of Lemma 26.23 we have the following.

THEOREM 26.24. *Every ordinal diagram is $<_0$-accessible.*

Since we use the accessibility of ordinal diagrams for consistency proofs, it is important to pursue the constructive content of our accessibility proof. For this purpose we discuss more on the notion involved. Our definition of F_i is made by induction on i. However if we take an ordinal diagram α, then the indexes of α are finite. Therefore we can make the definition of F_i in the following way.

DEFINITION 26.25. (1) Every ordinal diagram is an ordinal diagram in F_0.

(2) Let $i > 0$ and α be an ordinal diagram. Let $\beta_1, \dots, \beta_n$ be all sections of α whose indexes i_k $(1 \leq k \leq n)$ are less than i. Then (α, M) is said to be in F_i if M is a method for showing that β_k is $<_{i_k}$-accessible in F_{i_k} for every k $(1 \leq k \leq n)$. In particular an ordinal diagram α in F_i, together with a method for showing that every i-section of α is $<_i$-accessible in F_i, is an ordinal diagram in F_{i+1}.

(3) Let α be an ordinal diagram, $\beta_1, \dots, \beta_n$ be all sections of α, and $i_1, \dots, i_n$ be their indexes, respectively. Then (α, M) is said to be in F_∞ if M is a method for showing that β_k is $<_{i_k}$-accessible in F_{i_k} for every k $(1 \leq k \leq n)$.

This definition of F_i together with the construction of fundamental sequences of ordinal diagrams make our proof of accessibility closer to the accessibility proof in §11. The construction of the fundamental sequences

are done in the following paper:
G. Takeuti and M. Yasugi: Fundamental Sequences of Ordinal Diagrams, Comment. Math. Univ. St. Pauli 25, 1976, pp. 1–80.
For further discussion of the constructive content of the accessibility of ordinal diagrams, see the papers in Postscript.

In the rest of the section, we shall make some preparation called the theory of approximations for the paper on the fundamental sequences mentioned above. The reader is advised to skip the rest of the section unless he is very much interested in the matter.

Now we shall explain the theory of approximations. That is, given an element j of I and a connected ordinal diagram α, we are to define the (n, k)th j-approximation of α for $n, k = 0, 1, 2, \ldots$, and see that they present good criteria for the comparison of two ordinal diagrams with respect to j.

We shall now define j-valuations and j-approximations of α for every $j \in I$ and every connected nonzero ordinal diagram α.

DEFINITION 26.25. (1) When (i, a) is the outermost value of a connected ordinal diagram, i is its *outermost index*.

(2) Let $\bar{\gamma}$ be a j-subsection of α where the outermost index of γ is $< j$. Then we say that $\bar{\gamma}$ is a *j-kernel* of α. We include $\bar{0}$ as a j-kernel.

DEFINITION 26.26. Let $v_0(j, \alpha)$ be the maximum of the outermost values of the ordinal diagrams represented by the j-subsections of α. Then $v_0(j, \alpha)$ is called the 0th *j-valuation* of α.

Note that every non-zero connected ordinal diagram has a 0th j-valuation.

PROPOSITION 26.27. *Let β and α be non-zero connected ordinal diagrams. If $v_0(j, \beta) < v_0(j, \alpha)$ then $\beta <_j \alpha$.*

PROOF. Let $v_0(j, \alpha) = (i, a)$. For any j-subsection of α, say $\bar{\delta}$, $\delta \leqslant_j \alpha$ (cf. Proposition 26.10). Therefore it suffices to show that $\beta <_j (i, a, \gamma)$ for any j-subsection of α of the form $\overline{(i, a, \gamma)}$ (i.e., the outermost value of the ordinal diagram represented by it is (i, a)). We shall, however, show that (*) $\eta <_m (i, a, \gamma)$ for all $m \geqslant j$ and each $\bar{\eta}$ a j-subsection of β.

Then as a special case we have $\beta <_j (i, a, \gamma)$.

The proof of (*) is by induction on $l(\eta)$. Note that $v_0(j, \eta) \leqslant v_0(j, \beta) < (i, a)$.

(1) $\bar{\eta}$ is a j-kernel of β. Either η is 0 or of the form (k, b, η') where $k < j$. Then obviously $\eta <_m (i, a, \gamma)$ for every $m \geqslant j$.

(2) η is of the form (k, b, η'), where $k \geqslant j$. Since $(k, b) < (i, a)$, $\eta <_m (i, a, \gamma)$ if $m > k$. Let $j \leqslant m \leqslant k$ and let $\bar{\delta}$ be an m-section of $\bar{\eta}$. Let $\bar{\delta}_0$ be any component of $\bar{\delta}$. Then $\bar{\delta}_0$ is a j-subsection of β; hence by the

induction hypothesis $\delta_0 <_m (i, a, \gamma)$. This implies that $\delta <_m (i, a, \gamma)$ so by induction on $\iota(m, \eta)$, $\eta <_m (i, a, \gamma)$ for all m for which $j \leqslant m \leqslant k$.

Definition 26.28. (1) Let α be a connected ordinal diagram and let $(i, a) = v_0(j, \alpha)$. Consider any j-subsection of α whose outermost value is (i, a), say (i, a, γ). Let $\mathrm{apr}(0, j, \alpha)$ be the greatest such (i, a, γ) with respect to $<_i$. Any j-active occurrence of $\mathrm{apr}(0, j, \alpha)$, say $\overline{\mathrm{apr}(0, j, \alpha)}$, is called a 0th *j-approximation* of α.

We shall use α_0 as an abbreviation for $\mathrm{apr}(0, j, \alpha)$. There may, of course, be many occurrences of α_0 that are not 0th j-approximations of α. However, we are not interested in such occurrences. Therefore, for notational convenience we will hereafter use the symbols $\overline{\alpha_0}$ and $\overline{\mathrm{apr}(0, j, \alpha)}$ only for occurrences that are 0th j-approximations of α.

(2) If a j-subsection of α, say $\bar{\eta}$, does not contain an $\overline{\alpha_0}$, a j-active occurrence of $\mathrm{apr}(0, j, \alpha)$, and is not contained by any $\overline{\alpha_0}$, then we say that $\bar{\eta}$ j-omits $\overline{\alpha_0}$. When j is understood we will say simply that $\bar{\eta}$ omits $\overline{\alpha_0}$.

Lemma 26.29. *(1) $\overline{\mathrm{apr}(0, j, \alpha)}$ is a j-subsection of α.*

(2) Let $\overline{\alpha'} = (i, a, \delta)$ be a j-subsection of α whose outermost value is (i, a) and which is different from α_0. Then $\alpha' <_i \alpha_0$, and hence $\delta <_i \gamma$ (where $\alpha_0 = (i, a, \gamma)$). This implies that $\alpha' <_m \alpha_0$ for all $m \geqslant i$.

(3) If $\bar{\eta}$ is a j-kernel of α which is not $\overline{\alpha_0}$, then $\eta <_m \alpha_0$ for all $m \geqslant j$.

(4) If $\bar{\eta}$, a j-subsection of α, j-omits $\overline{\alpha_0}$, then $\eta <_m \alpha_0$ for all $m \geqslant j$.

(5) Let $(i, a) = v_0(j, \alpha)$ and $\bar{\delta}_1, \bar{\delta}_2, \ldots, \bar{\delta}_m$ be all the j-subsections of α. Then

$$\alpha_0 = \mathrm{apr}(0, j, \alpha) = \max_{<_{i+1}}(\delta_1, \delta_2, \ldots, \delta_m).$$

Proof. (3) Let $\eta = (k, b, \eta')$ and $k < j$. If $j \leqslant i$, then evidently $\eta <_m \alpha_0$ for all $m \geqslant j$. Suppose $j > i$. Then $\alpha = \alpha_0$ and there is no j-kernel except α_0.

(4) (i) $\bar{\eta}$ is a j-kernel of α. Then $\eta <_m \alpha$ for all $m \geqslant j$ by (3).

(ii) Let $\eta = (k, b, \eta')$, where $k \geqslant j$. If $(k, b) = (i, a)$, then by (2) $\eta <_m \alpha_0$ for all $m \geqslant i$. If $j \geqslant i$ this will do. Suppose $j < i$. Let $j \leqslant m < i$ and let $\bar{\delta}$ be a component of an m-section of η. Then $\bar{\delta}$ is a j-subsection of α which j-omits $\bar{\alpha}_0$. Therefore by the induction hypothesis $\delta <_m \alpha_0$ for all $m \geqslant j$. By induction on $\iota(m, \eta)$ we can then prove that $\eta <_m \alpha_0$. If $(k, b) < (i, a)$, then $\eta <_m \alpha$, for all $m > k$. Let $j \leqslant m \leqslant k$, and let $\bar{\delta}$ be a component of an m-section of η. Then by the induction hypothesis $\delta <_m \alpha_0$. From this in turn it follows that $\eta <_m \alpha_0$.

Proposition 26.30. *Let α and β be connected ordinal diagrams, where $v_0(j, \alpha) = v_0(j, \beta) = (i, a)$ and* $\mathrm{apr}(0, j, \beta) <_i \mathrm{apr}(0, j, \alpha)$. *Then $\beta <_j \alpha$.*

Proof. Since $\bar{\alpha}_0$ is a j-subsection of α, it suffices to show that $\beta <_j \alpha_0$, for then $\beta <_j \alpha_0 \leqslant_j \alpha$. We can easily show that

(1) If $\bar{\eta}$ is a j-subsection of β, then $v_0(j, \eta) \leqslant (i, a)$.

Furthermore, if $v_0(j, \eta) = (i, a)$, then $\mathrm{apr}(0, j, \eta) <_i \alpha_0$, and

(2) for any two ordinal diagrams of the form (i, a, δ) and (i, a, γ), $(i, a, \delta) <_i (i, a, \gamma)$ implies $(i, a, \delta) <_m (i, a, \gamma)$ for all $m \geqslant i$.

Using (1) and (2) we shall prove by induction on $l(\eta)$

(3) for any j-subsection of β, say $\bar{\eta}$, $\eta <_m \alpha_0$ for all $m \geqslant j$.

As a special case of (3) we have $\beta <_j \alpha_0$.

1) $\bar{\eta}$ is a j-kernel of β. Then η is 0, or η is of the form (k, b, η'), where $k < j$. Then $(k, b) \leqslant (i, a)$. If $(k, b) < (i, a)$, then $\eta <_m \alpha_0$ for all $m > k$, hence for all $m \geqslant j$. If $(k, b) = (i, a)$, then $\eta \leqslant_i \beta_0 <_i \alpha_0$, hence by (2) above, $\eta <_m \alpha_0$ for all $m \geqslant i$. Since $k = i$ and $k < j$, we have $j > i$. So $\eta <_m \alpha_0$ for all $m \geqslant j$.

2) η is of the form (k, b, η'), where $k \geqslant j$ and $(k, b) < (i, a)$. It is obvious that $\eta <_m \alpha_0$ if $m > k$. Let $j \leqslant m \leqslant k$ and let $\bar{\delta}$ be a component of an m-section of η. Then by the induction hypothesis $\delta <_m \alpha_0$ for all $m \geqslant j$. Therefore, by induction on $\iota(m, \eta)$, $\eta <_m \alpha_0$ for all m such that $j \leqslant m \leqslant k$.

3) η is of the form (i, a, η'). Since $\eta \leqslant_i \beta_0 <_i \alpha_0$ it follows from (2) that $\eta <_m \alpha_0$ for all $m \geqslant i$. If $j \geqslant i$ this will do. Suppose $j < i$. Consider any m such that $j \leqslant m < i$ and suppose $\bar{\delta}$ is a component of an m-section of η. Then by the induction hypothesis $\delta <_m \alpha_0$ for all $m \geqslant j$. From this it follows that $\eta <_m \alpha_0$ for all such m.

PROPOSITION 26.31. *Let $\bar{\eta}$ be a j-subsection of α that contains an $\bar{\alpha}_0$. Suppose in addition that for each $\bar{\alpha}_0$ in $\bar{\eta}$ there is an occurrence of an element of I that is less than i and connected to $\bar{\alpha}_0$. Let $\bar{\alpha}_0^1, \alpha_0^2, \ldots, \bar{\alpha}_0^m$ be all such occurrences of α_0 in $\bar{\eta}$ and let q_k be the least such element of I for $\bar{\alpha}_0^k$ as described above. Define $q = q(\eta) = \max(q_1, \ldots, q_m)$. Then $\eta <_p \alpha_0$ for every p such that $q < p \leqslant i$.*

PROOF. First note that $q \geqslant j$. Since $\bar{\eta}$ is a j-subsection of α it can be easily shown that $v_0(p, \eta) \leqslant v_0(j, \alpha)$ $(= (i, a))$ for every $p \geqslant j$, and in particular for a p such that $q < p \leqslant i$. Also $v_0(p, \alpha_0) = (i, a)$ and $\mathrm{apr}(0, p, \alpha_0) = \alpha_0$. So if $v_0(p, \eta) < (i, a)$, then $\eta <_p \alpha_0$ by Proposition 26.43. Suppose $v_0(p, \eta) = (i, a)$. Then $\mathrm{apr}(0, p, \eta) <_i \alpha_0 = \mathrm{apr}(0, p, \alpha_0)$, for $\bar{\alpha}_0$ is not p-active in $\bar{\eta}$ (cf. (2) of Lemma 26.29). Therefore $\eta <_p \alpha_0$.

DEFINITION 26.32. Let $\bar{\gamma}$ be a j-subsection of α for which there is an $\bar{\alpha}_0$ in γ as a j-subsection of γ and such that i is the only element of I that occurs in γ and is connected to $\bar{\alpha}_0$. (Namely, such an $\bar{\alpha}_0$ is i-active in $\bar{\gamma}$.) Let $\mathrm{apr}(1, j, \alpha)$ be the greatest with respect to $<_i$ of such γ. Any such occurrence of $\mathrm{apr}(1, j, \alpha)$ is called a *first j-approximation* of α.

We will use the symbol α_1 as an abbreviation for $\mathrm{apr}(1, j, \alpha)$. Hereafter we will use the symbols $\bar{\alpha}_1$ and $\mathrm{apr}(1, j, \alpha)$ only for occurrences that are first j-approximations of α.

Note that according to the definition, $\alpha_0 = \alpha_1$ is possible.

LEMMA 26.33. *Let* $v_0(j, \alpha) = (i, a)$, $\alpha_0 = \text{apr}(0, j, \alpha)$ *and* $\alpha_1 = \text{apr}(1, j, \alpha)$.
(1) *If* $\bar{\alpha}_1$ *properly contains an* $\bar{\alpha}_0$, *then* $j \leqslant i$ *and* $\alpha_0 <_k \alpha_1$ *for every* $k \leqslant i$.
(2) *If* (i, b, δ) *is a sub-ordinal diagram of* α_1 *such that* δ *contains an* $\bar{\alpha}_0$ *i-active, then* $b < a$.
(3) $\bar{\alpha}_1$ *is "maximal" in the sense that if* $\overline{(k, c, \gamma)}$ *is a j-subsection of* α, *where* $\bar{\alpha}_1$ *is a component of* $\bar{\gamma}$, *then* $k < i$.

DEFINITION 26.34. Let $\bar{\eta}$ be a j-subsection of α. If $\bar{\eta}$ neither contains an $\bar{\alpha}_1$ nor is contained by an $\bar{\alpha}_1$ and is not properly contained by any $\bar{\alpha}_0$, then $\bar{\eta}$ is said to *j-omit* $\bar{\alpha}_1$. When j is understood we will say simply that $\bar{\eta}$ omits $\bar{\alpha}_1$.

PROPOSITION 26.35. *If a j-subsection of* α, *say* $\bar{\eta}$, *omits* $\bar{\alpha}_1$, *then* $\eta <_k \alpha_1$ *for all k such that* $j \leqslant k \leqslant i$.

PROOF. (By induction on $l(\eta)$.)
1) $\bar{\eta}$ omits $\bar{\alpha}_0$. Then $\eta <_k \alpha_0$ for all $k \geqslant j$ by (4) of Lemma 26.29. If $j \leqslant k \leqslant i$, then $\alpha_0 <_k \alpha_1$ (see (1) of Lemma 26.33), hence $\eta <_k \alpha_1$.
2) $\eta = \alpha_0 <_k \alpha$ if $j \leqslant k \leqslant i$.
3) $\bar{\eta}$ contains $\bar{\alpha}_0$ and i is the only element of I that occurs in η and is connected to $\bar{\alpha}_0$. Then $\eta <_i \alpha_1$ by definition of α_1. Let $j \leqslant k < i$ and let $\bar{\delta}$ be a component of a k-section of $\bar{\eta}$. If $\bar{\delta}$ omits $\bar{\alpha}_0$ or contains $\bar{\alpha}_0$, then $\bar{\delta}$ omits $\bar{\alpha}_1$, so by the induction hypothesis $\delta <_k \alpha_1$, if $\bar{\delta}$ is contained in $\bar{\alpha}_0$, then $\bar{\delta}$ is a k-subsection of α_0, hence $\delta <_k \alpha_0 <_k \alpha_1$. So $\delta <_k \alpha_1$ in any case, and by induction on $\iota(k, \eta)$, $\eta <_k \alpha_1$.
4) $\bar{\eta}$ contains $\bar{\alpha}_0$ and for every occurrence of $\bar{\alpha}_0$ there is an element of I that has an occurrence in $\bar{\eta}$ connected to $\bar{\alpha}_0$ and which is less than i. By Proposition 26.31, $\eta <_i \alpha_0 <_i \alpha_1$. For a k satisfying $j \leqslant k < i$, refer to 3) above.

PROPOSITION 26.36. *Let* α *and* β *be connected ordinal diagrams and* $v_0(j, \alpha) = v_0(j, \beta) = (i, a)$. *Let* $\alpha_0 = \text{apr}(0, j, \alpha) = \text{apr}(0, j, \beta) = \beta_0$, $\alpha_1 = \text{apr}(1, j, \alpha)$ *and* $\beta_1 = \text{apr}(1, j, \beta)$. *If* $\beta_1 <_i \alpha_1$, *then* $\beta <_j \alpha$.

PROOF. In order that $\beta_1 <_i \alpha_1$ under the assumption, $\bar{\alpha}_1$ must properly contain $\bar{\alpha}_0$. Therefore $j \leqslant i$. We shall show that for any j-subsection of β, say $\bar{\eta}$, which either contains $\bar{\beta}_0$ or omits $\bar{\beta}_0$.
(*) $\eta <_k \alpha_1$ for all k such that $j \leqslant k \leqslant i$.
As a special case we have $\beta <_j \alpha_1 \leqslant_j \alpha$. The proof of (*) is by induction on $l(\eta)$.
1) $\eta = \beta_0$ or $\bar{\eta}$ omits $\bar{\beta}_0$. Then $\eta \leqslant_k \beta_0 = \alpha_0 <_k \alpha_1$ if $j \leqslant k \leqslant i$.
2) $\bar{\eta}$ properly contains $\bar{\beta}_0$ and there is a $\bar{\beta}_0$ such that the only element of I which has an occurrence in $\bar{\eta}$ connected to $\bar{\beta}_0$ is i. Then by definition of

β_1, $\eta \leq_i \beta_1 <_i \alpha_1$. Let $j \leq k < i$ and let $\bar{\delta}$ be a component of a k-section of $\bar{\eta}$. Then $\bar{\delta}$ either omits $\bar{\beta}_0$ in which case $\delta <_k \alpha_1$ by 1) above; or $\bar{\delta}$ contains $\bar{\beta}_0$ and hence $\delta <_k \alpha_1$ by the induction hypothesis; or $\bar{\delta}$ is a k-subsection of $\bar{\beta}_0$ and hence there is a k-active occurrence of δ in $\bar{\alpha}_1$, so $\delta <_k \alpha_1$. In any case $\delta <_k \alpha_1$. So by induction on $\iota(k, \eta)$, $\eta <_k \alpha_1$ can be proved.

DEFINITION 26.37. Let $v_0(j, \alpha) = (i, a)$, $\operatorname{apr}(0, j, \alpha) = \alpha_0$ and $\operatorname{apr}(1, j, \alpha) = \alpha_1$. Define, as a matter of notational convenience $i_0 = i_1 = i$.

Suppose we have defined some pairs of j-subsections of α and elements of I which have occurrences in α, say $(\alpha_0, i_0), (\alpha_1, i_1), \ldots, (\alpha_n, i_n)$, which satisfy the following conditions:

(*) i) For every m, $1 \leq m < n$, $j \leq i_{m+1} < i_m$.

ii) For each m, $m \geq 1$, i_{m+1} is the maximum k for which there is a j-subsection of α, of the form $\overline{(k, b, \gamma)}$ such that $\bar{\alpha}_m$ is a component of $\bar{\gamma}$.

iii) Let $\bar{\eta}$ denote any j-subsection of α such that $\bar{\eta}$ contains $\bar{\alpha}_m$ and all the elements of I that occur in $\bar{\eta}$ and are connected to $\bar{\alpha}_m$ are $\geq i_{m+1}$. Then α_{m+1} is the maximum, with respect to $<_{i_{m+1}}$, among those η and $\bar{\alpha}_{m+1}$ denotes such an occurrence of α_{m+1}.

Now we define (α_{n+1}, i_{n+1}) as follows, provided that $\bar{\alpha}_n$ is not α. We define i_{n+1} as i_{m+1} in ii) of (*) reading n in place of m and we define $\bar{\alpha}_{n+1}$ as $\bar{\alpha}_{m+1}$ in iii) of (*) reading n in place of m.

We call $\bar{\alpha}_n$ an nth j-approximation of α and denote it by $\overline{\operatorname{apr}(n, j, \alpha)}$, i.e., $\alpha_n = \operatorname{apr}(n, j, \alpha)$. Define v_n by $v_n(j, \alpha) = i_n$, $n = 1, 2, \ldots$.

If $\bar{\alpha}_n = \alpha$, then $\bar{\alpha}_{n+1}$ needs not be defined. We may, however, use the expression $v_{n+1}(j, \beta) < v_{n+1}(j, \alpha)$ to mean that $v_{n+1}(j, \beta)$ is not defined while $v_{n+1}(j, \alpha)$ is.

COROLLARY 26.38. (1) *Let (p, e, ξ) be a j-subsection of α in which $\bar{\alpha}_n$ is a component of $\bar{\xi}$. Then $p < i_n$.*

(2) $j \leq i_{n+1} < i_n$.

(3) *There is at least one $\bar{\alpha}_n$ which occurs as a component of γ in (i_{n+1}, b, γ), which is a j-subsection of α presuming that $\alpha_n \neq \alpha$.*

DEFINITION 26.39. Let $\bar{\eta}$ be a j-subsection of α. We say that $\bar{\eta}$ j-omits $\bar{\alpha}_n$ if $\bar{\eta}$ is not contained by any $\bar{\alpha}_n$, $\bar{\eta}$ does not contain any $\bar{\alpha}_n$, and $\bar{\eta}$ is not properly contained by any of $\bar{\alpha}_0, \bar{\alpha}_1, \ldots, \bar{\alpha}_{n-1}$.

PROPOSITION 26.40. I^n. *Let $\bar{\eta}$ be a j-subsection of α that contains $\bar{\alpha}_n$. Suppose for each occurrence of $\bar{\alpha}_n$ in $\bar{\eta}$ there is an element of I that has an occurrence in $\bar{\eta}$ that is connected to $\bar{\alpha}_n$ and is less than i_{n+1}. Let $\alpha_n^1, \ldots, \alpha_n^m$ be all the occurrences of $\bar{\alpha}_n$ in $\bar{\eta}$ and let q_k be the least element of I that has an occurrence connected to α_n^k. Let $q = q_n(\bar{\eta}) = \max(q_1, \ldots, q_m)$. Then $\eta <_p \alpha_n$ for every p such that $q < p \leq i_n$. (Note that $j \leq q < i_{n+1}$.)*

II^{n+1}. *If $\bar{\eta}$ j-omits $\bar{\alpha}_{n+1}$, then $\eta <_p \alpha_{n+1}$ for any p such that $j \leq p \leq i_{n+1}$.*

III^{n-1}. *Let α and β be connected ordinal diagrams and suppose that* $\operatorname{apr}(n, j, \alpha)$, $\operatorname{apr}(n, j, \beta)$ *and* $\operatorname{apr}(n+1, j, \alpha)$ *are defined. Suppose also that* $\alpha_n = \operatorname{apr}(n, j, \alpha) = \operatorname{apr}(n, j, \beta) = \beta_n$ *(hence* $\alpha_0 = \beta_0$, $\alpha_1 = \beta_1, \ldots, \alpha_{n-1} = \beta_{n-1}$*).*

(1) *If* $v_{n+1}(j, \beta) = i'_{n+1} < i_{n+1} = v_{n+1}(j, \alpha)$, *then* $\beta <_j \alpha$.

(2) *If*

$$v_{n+1}(j, \beta) = v_{n+1}(j, \alpha) = i_{n+1}$$

and

$$\beta_{n+1} = \operatorname{apr}(n+1, j, \beta) <_{i_{n+1}} \operatorname{apr}(n+1, j, \alpha) = \alpha_{n+1},$$

then $\beta <_j \alpha$.

PROOF. (By induction on n.) Note that I^0, II^0, II^1, III^0 and III^1 have been established (cf. Proposition 26.47, Lemma 26.45, Propositions 26.51, 26.43 and 26.46). First we will prove I^n by assuming III^r for all r such that $1 \leq r \leq n$. Then we will prove II^{n+1} from I^n and II^n. Finally we will prove III^{n+1} from II^n and II^{n+1}.

I^n. For any $p, j \leq p \leq i_n$ and any $r, 0 \leq r \leq n$, $v_r(p, \alpha_n) = v_r(j, \alpha_n) = v_r(j, \alpha)$; also $\operatorname{apr}(r, p, \alpha_n) = \operatorname{apr}(r, j, \alpha_n) = \operatorname{apr}(r, j, \alpha)$. Now let $q < p \leq i_n$. Then for some $r, 0 \leq r \leq n$, $\operatorname{apr}(0, p, \eta) = \operatorname{apr}(0, p, \alpha_n), \ldots,$ $\operatorname{apr}(r-1, p, \eta) = \operatorname{apr}(r-1, p, \alpha_n)$ and either $v_r(p, \eta) < v_r(p, \alpha_n)$ or $v_r(p, \eta) = v_r(p, \alpha_n) (= i_r)$ and $\operatorname{apr}(r, p, \eta) <_{i_r} \operatorname{apr}(r, p, \alpha_n) (= \alpha_r)$. So by III^r applied to p, $\eta <_p \alpha_n$ for every such p.

II^{n+1}. By induction on $l(\eta)$.

1) $\bar{\eta}$ is $\bar{\alpha}_n$ or $\bar{\eta}$ j-omits $\bar{\alpha}_n$. Then by II^n $\eta \leq_p \alpha_n <_p \alpha_{n+1}$ if $j \leq p \leq i_{n+1}$.

2) $\bar{\eta}$ contains $\bar{\alpha}_n$ properly and there is an occurrence of $\bar{\alpha}_n$ such that all the elements of I that have occurrences in $\bar{\eta}$ that are connected to it are $\geq i_{n+1}$. Then $\eta <_{i_{n+1}} \alpha_{n+1}$ by definition of α_{n+1}. Let $j \leq p < i_{n+1}$ and let $\bar{\delta}$ be a component of a p-section of $\bar{\eta}$. Then either $\bar{\delta}$ omits $\bar{\alpha}_n$ or $\bar{\delta}$ is a p-subsection of $\bar{\alpha}_n$. Therefore $\delta <_p \alpha_n <_p \alpha_{n+1}$. Then by induction on $\iota(p, \eta)$, $\eta <_p \alpha_{n+1}$ for all such p.

3) $\bar{\eta}$ properly contains $\bar{\alpha}_n$ and for every occurrence of $\bar{\alpha}_n$ in $\bar{\eta}$ there is an element of I which is less than i_{n+1} and has an occurrence in η that is connected to $\bar{\alpha}_n$. By I^n, $\eta <_p \alpha_n$ if $(q <) i_{n+1} \leq p \leq i_n$. In particular $\eta <_{i_{n+1}} \alpha_n <_{i_{n+1}} \alpha_{n+1}$. Let $j \leq p < i_{n+1}$ and let $\bar{\delta}$ be a component of a p-section of $\bar{\eta}$. Then, as in 2) incorporated with induction on $l(\eta)$, $\delta <_p \alpha_{n+1}$, which implies $\eta <_p \alpha_{n+1}$ for all such p.

III^{n+1}. (1) Let $\bar{\xi} = (\overline{i_{n+1}, b, \gamma})$ be any j-subsection of α whose outermost index is i_{n+1} and $\bar{\gamma}$ contains $\bar{\alpha}_n$ as a component. We shall show that, for every j-subsection of β, say $\bar{\eta}$, which either j-omits $\bar{\beta}_n$ or contains $\bar{\beta}_n$, $\eta <_p \xi$ if $j \leq p \leq i_{n+1}$. As a special case $\beta <_j \xi \leq_j \alpha$.

1) $\bar{\eta}$ is $\bar{\beta}_n$ or $\bar{\eta}$ omits $\bar{\beta}_n$. Then $\eta \leq_p \beta_n$ by II^n. So $\eta \leq_p \beta_n = \alpha_n <_p \xi$.

2) $\bar{\eta}$ properly contains $\bar{\beta}_n$. Recall that $\bar{\beta}_n$ occurs in $\bar{\eta}$ in the following context. There is a j-subsection of $\bar{\eta}$, say $(\overline{k, c, \rho})$, where $\bar{\beta}_n$ occurs in $\bar{\rho}$ as a component and $j \leq k \leq i'_{n+1} < i_{n+1}$.

We can show that

(*) There is a number r, $0 \leqslant r \leqslant n$, such that

$$\mathrm{apr}(r-1, i_{n+1}, \eta) = \alpha_{r-1} \,(= \mathrm{apr}(r-1, i_{n+1}, \xi))$$

and either

$$v_r(i_{n+1}, \eta) < i_r,$$

or

$$v_r(i_{n+1}, \eta) = i_r \quad \text{and} \quad \mathrm{apr}(r, i_{n+1}, \eta) <_{i_r} \alpha_r \,(= \mathrm{apr}(r, i_{n+1}, \xi)).$$

Applying (*) and III^r to η, ξ and i_{n+1} it follows that $\eta <_{i_{n+1}} \xi$. Let $j \leqslant p < i_{n+1}$ and let $\bar{\delta}$ be a component of a p-section of $\bar{\eta}$. If $\bar{\delta}$ satisfies the same condition as $\bar{\eta}$, then $\delta <_p \xi$ by the induction hypothesis. If $\bar{\delta}$ is a p-subsection of $\bar{\beta}_r$ for some r, $0 \leqslant r \leqslant n$, then $\delta <_p \beta_r \leqslant_p \beta_n = \alpha_n <_p \xi$. In any case, $\delta <_p \xi$, and by induction on $\iota(p, \eta)$, $\eta <_p \xi$.

The assertion (*) is a special case of the following.

(**) Let $\bar{\eta}$ be an arbitrary j-subsection of β that properly contains $\bar{\beta}_n$. Let m be any element of I such that $m > i'_{n+1}$ $(= v_{n+1}(j, \beta))$. Then there is a number r, $0 \leqslant r \leqslant n$, such that $\mathrm{apr}(r-1, m, \eta) = \beta_{r-1}$ $(= \alpha_{r-1})$ and either $v_r(m, \eta) < i_r = v_r(j, \beta)$ or $v_r(m, \eta) = i_r$ and $\mathrm{apr}(r, m, \eta) <_{i_r} \beta_r$ $(= \alpha_r)$. (Note that $m > i_{n+1}$ includes $m = i_{n+1}$.) When $r = 0$, $v_r(m, \eta) < (i, a) = v_r(m, \beta)$ or $v_r(m, \eta) = (i, a)$ and $\mathrm{apr}(0, m, \eta) <_i \beta_0$.

We prove (**) in the following way. Since $m \geqslant j$, $v_0(m, \eta) \leqslant v_0(j, \beta)$ $(= (i, a))$. If this is a strict inequality we are done. If equality holds, then consider $r = 1$. Continuing the same argument, suppose that we have reached $\mathrm{apr}(n-1, m, \eta) = \beta_{n-1}$ and $v_n(m, \eta) = i_n$. Then $\mathrm{apr}(n, m, \eta)$ must contain $\bar{\beta}_{n-1}$. This, $m > i'_{n+1}$, and the fact that $\bar{\eta}$ properly contains $\bar{\beta}_n$ imply that $\mathrm{apr}(n, m, \eta) \neq \beta_n$. So by definition of $\mathrm{apr}(n, j, \eta) (= \bar{\beta}_n)$, $\mathrm{apr}(n, m, \eta) <_{i_n} \beta_n$.

(2) We shall show that

(***) for any $\bar{\eta}$ a j-subsection of β which either omits $\bar{\beta}_{n+1}$, is $\bar{\beta}_{n+1}$ or properly contains $\bar{\beta}_{n+1}$, $\eta <_p \alpha_{n+1}$ for any p such that $j \leqslant p \leqslant i_{n+1}$.

As a special case of (***) we have $\beta <_j \alpha_{n+1} \leqslant_j \alpha$. The proof of (***) is by induction on $l(\eta)$.

1) $\bar{\eta}$ is $\bar{\beta}_{n+1}$ or $\bar{\eta}$ omits $\bar{\beta}_{n+1}$. By II^{n+1}, $\eta \leqslant_{i_{n+1}} \beta_{n+1} <_{i_{n+1}} \alpha_{n+1}$. Using an argument similar to one employed earlier we can prove by induction on $\iota(p, \eta)$ that $\eta <_p \alpha_{n+1}$ provided $j \leqslant p < i_{n+1}$.

2) $\bar{\eta}$ contains $\bar{\beta}_{n+1}$ properly and there is an occurrence of $\bar{\beta}_n$ in $\bar{\eta}$ such that every element of I that has an occurrence in $\bar{\eta}$ and is connected to it is $\geqslant_{i_{n+1}}$. Then, by definition of $\bar{\beta}_{n+1}$,

$$\eta <_{i_{n+1}} \beta_{n+1} <_{i_{n+1}} \alpha_{n+1}.$$

That $\eta <_p \alpha_{n+1}$ for $j \leqslant p < i_{n+1}$ can be shown as above.

3) $\bar{\eta}$ properly contains β_{n+1}, and, for every occurrence of $\bar{\beta}_n$ in $\bar{\eta}$, there is an element of I which has an occurrence connected to $\bar{\beta}_n$, and is $<_{i_n+1}$.

Then, by I^n, $\eta <_p \beta_n$ if $q < p \leq i_n$, where $j \leq q < i_{n+1}$. Therefore, letting $p = i_{n+1}$ we obtain

$$\eta <_{i_{n+1}} \beta_n <_{i_{n+1}} \beta_{n+1} <_{i_{n+1}} \alpha_{n+1}.$$

If $j \leq p < i_{n+1}$, then $\eta <_p \alpha_{n+1}$ is proved as before.

We next define refinements of approximations. We shall define $\bar{\alpha}_{(n,k)}$ by induction on k in such a way that $\bar{\alpha}_{(n,k)}$ is an i_{n+1}-subsection of $\bar{\alpha}_{n+1}$.

DEFINITION 26.41. $\bar{\alpha}_{(n,0)}$ is any occurrence of $\bar{\alpha}_n$ that is i_{n+1}-active in $\bar{\alpha}_{n+1}$.

Suppose $\bar{\alpha}_{(n,k)}$ has been defined so that $\bar{\alpha}_{(n,k)}$ is an i_{n+1}-subsection of $\bar{\alpha}_{n+1}$. Suppose $\bar{\alpha}_{(n,k)} \neq \alpha_{n+1}$. Let $\bar{\gamma}_1, \ldots, \bar{\gamma}_m$ be all the occurrences of i_{n+1}-subsections of $\bar{\alpha}_{n+1}$ that properly contain an occurrence of $\bar{\alpha}_{(n,k)}$. Let

$$\alpha_{(n,k+1)} = \max_{<_{i_{n+1}+1}} (\gamma_1, \ldots, \gamma_m).$$

Then $\bar{\alpha}_{(n,k+1)}$ denotes any such occurrence of $\alpha_{(n,k+1)}$ in $\bar{\alpha}_{n+1}$.

Note that although $\bar{\gamma}_1, \ldots, \bar{\gamma}_m$ above are determined relative to an occurrence of $\bar{\alpha}_{n+1}$, $\gamma_1, \ldots, \gamma_m$ (as ordinal diagrams) are determined uniquely from α_{n+1}. The same is true of $\alpha_{(n,k+1)}$.

$\bar{\alpha}_{(n,k)}$ is called the (n, k)th j-approximation of α and is denoted by $\text{apr}((n, k), j, \alpha)$.

PROPOSITION 26.42. *Either $\bar{\alpha}_{(n,k)}$ is $\bar{\alpha}_{n+1}$ or it occurs in $\overline{(i_{n+1}, c, \delta)}$ as a component of $\bar{\delta}$.*

PROOF. Suppose $\bar{\alpha}_{(n,k)}$ occurs in $\overline{(p, c, \delta)}$ as a component of $\bar{\delta}$. Then by the definition of $\bar{\alpha}_{n+1}$, $p \geq i_{n+1}$. If $p > i_{n+1}$, then $p \geq i_{n+1} + 1$ hence $\bar{\alpha}_{(n,k)}$ in $\bar{\delta}$ is an $i_{n+1}+1$-subsection of $\overline{(p, c, \delta)}$. So $\alpha_{(n,k)} <_{i_{n+1}+1} (p, c, \delta)$. Furthermore, $\overline{(p, c, \delta)}$ contains some occurrence of $\bar{\alpha}_{(n,k-1)}$ as an i_{n+1}-subsection, hence, by definition of $\bar{\alpha}_{(n,k)}$, $(p, c, \delta) <_{i_{n+1}+1} \alpha_{(n,k)}$, which is a contradiction. Therefore, $p = i_{n+1}$.

DEFINITION 26.43. An i_{n+1}-subsection of $\bar{\alpha}_{n+1}$, say $\bar{\eta}$, j-omits $\bar{\alpha}_{(n,k)}$, if it does not contain any occurrence of $\bar{\alpha}_{(n,k)}$, is not contained by $\alpha_{(n,k)}$, and is not contained by any occurrence of $\bar{\alpha}_{(m,p)}$, where $(m, p) < (n, k)$.

Note that if $k = 0$, it is possible that $\bar{\eta}$ j-omits $\bar{\alpha}_{(n,0)}$ but not $\bar{\alpha}_n$.

PROPOSITION 26.44. *Suppose that $\bar{\eta}$ is an i_{n+1}-subsection of $\bar{\alpha}_{n+1}$ that j-omits $\bar{\alpha}_{(n,k)}$. Then $\eta <_{i_{n+1}} \alpha_{(n,k)}$.*

PROOF. (By induction on k within which by induction on $l(\eta)$.)

$k = 0$: $\bar{\eta}$ omits $\bar{\alpha}_{(n,0)}$.

1) $\bar{\eta}$ omits $\bar{\alpha}_n$. Then $\eta <_{i_{n+1}} \alpha_n \; (= \alpha_{(n,0)})$ by II^n of Proposition 26.40.

2) $\bar{\eta}$ contains $\bar{\alpha}_n$. Then for each j-active $\bar{\alpha}_n$ in η there is an element of I that has an occurrence in $\bar{\eta}$ connected to $\bar{\alpha}_n$, and which is $< i_{n+1}$. So by I^n of Proposition 26.56, $\eta <_{i_{n+1}} \alpha_n = \alpha_{(n,0)}$.

$k > 0$: $\bar{\eta}$ omits $\bar{\alpha}_{(n,k)}$.

1) $\bar{\eta}$ omits $\bar{\alpha}_{(n,k-1)}$. Then $\eta <_{i_{n+1}} \alpha_{(n,k-1)}$ by the induction hypothesis. But $\bar{\alpha}_{(n,k-1)}$ is an i_{n+1}-subsection of $\bar{\alpha}_{(n,k)}$ (by definition). So $\bar{\eta} <_{i_{n+1}} \alpha_{(n,k)}$.

2) $\bar{\eta}$ is $\bar{\alpha}_{(n,k-1)}$. Then $\eta <_{i_{n+1}} \alpha_{(n,k)}$.

3) $\bar{\eta}$ properly contains $\bar{\alpha}_{(n,k-1)}$. Then $\eta <_{i_{n+1}+1} \alpha_{(n,k)}$ by the definition of $\bar{\alpha}_{(n,k)}$. Let $\bar{\delta}$ be a component of an i_{n+1}-section of $\bar{\eta}$. Since $\bar{\delta}$ omits $\bar{\alpha}_{(n,k)}$, $\delta <_{i_{n+1}} \alpha_{(n,k)}$ by the induction hypothesis.

PROPOSITION 26.61. *Suppose*

$$\bar{\alpha}_{n+1} = \overline{\mathrm{apr}(n+1, j, \alpha)}, \qquad \bar{\beta}_{n+1} = \overline{\mathrm{apr}(n+1, j, \beta)},$$

$$\bar{\alpha}_{(n,k)} = \overline{\mathrm{apr}((n,k), j, \alpha)}, \qquad \bar{\beta}_{(n,k)} = \overline{\mathrm{apr}((n,k), j, \beta)}$$

are defined for α and β. If

$$\beta_{(n,k-1)} = \alpha_{(n,k-1)},\; v_{n+1}(j,\beta) = v_{n+1}(j,\alpha) = i_{n+1} \textit{ and } \beta_{(n,k)} <_{i_{n+1}+1} \alpha_{(n,k)},$$

then $\beta_{n+1} <_{i_{n+1}} \alpha_{n+1}$, hence $\beta <_j \alpha$.

PROOF. We first claim that

(1) $\beta_{(n,k)} <_{i_{n+1}} \alpha_{(n,k)}$.

But (1) is a special case of the following.

(2) For any $\bar{\gamma}$ an i_{n+1}-subsection of $\beta_{(n,k)}$ which either contains $\bar{\beta}_{(n,k-1)}$ or j-omits $\bar{\beta}_{(n,k-1)}$, $\gamma <_{i_{n+1}} \alpha_{(n,k)}$.

We prove (2) by induction on $l(\gamma)$.

1) $\bar{\gamma}$ is $\bar{\beta}_{(n,k-1)}$. Then $\gamma = \alpha_{(n,k-1)} <_{i_{n+1}} \alpha_{(n,k)}$.

2) $\bar{\gamma}$ omits $\bar{\beta}_{(n,k-1)}$. Then by Proposition 26.44 $\gamma <_{i_{n+1}} \beta_{(n,k-1)}$ hence $\gamma <_{i_{n+1}} \beta_{(n,k-1)} = \alpha_{(n,k-1)} <_{i_{n+1}} \alpha_{(n,k)}$.

3) $\bar{\gamma}$ properly contains $\bar{\beta}_{(n,k-1)}$. Let γ be of the form (p, c, γ'), where $p \geqslant i_{n+1}$. Then $\gamma \leqslant_{i_{n+1}+1} \beta_{(n,k)}$ by definition of $\bar{\beta}_{(n,k)}$, and hence

$$\gamma \leqslant_{i_{n+1}+1} \beta_{(n,k)} <_{i_{n+1}+1} \alpha_{(n,k)}$$

by the hypothesis of the proposition. Let $\bar{\delta}$ be a component of an i_{n+1}-section of $\bar{\gamma}$. If $\bar{\delta}$ satisfies the same condition as $\bar{\gamma}$, then $\delta <_{i_{n+1}} \alpha_{(n,k)}$ by the induction hypothesis. The remaining possibility is that $\bar{\delta}$ is a $\bar{\beta}_{(m,p)}$, where $(m,p) < (n, k-1)$. Therefore $\delta =_{i_{n+1}} \beta_{(m,p)} = \alpha_{(m,p)} <_{i_{n+1}} \alpha_{(n,k)}$. From this it follows that $\gamma <_{i_{n+1}} \alpha_{(n,k)}$.

In order to finish the proof of the proposition we need only prove that

(3) for any $\bar{\eta}$ an i_{n+1}-subsection of $\bar{\beta}_{n+1}$ which either omits $\bar{\beta}_{(n,k)}$ or contains $\bar{\beta}_{(n,k)}$, $\eta <_{i_{n+1}} \alpha_{(n,k)}$.

This we prove by induction on $l(\eta)$.

1) $\bar{\eta}$ is $\bar{\beta}_{(n,k)}$. By (1), $\eta <_{i_{n+1}} \alpha_{(n,k)}$.

2) $\bar{\eta}$ omits $\bar{\beta}_{(n,k)}$. By Proposition 26.44, $\eta <_{i_{n+1}} \beta_{(n,k)} <_{i_{n+1}} \alpha_{(n,k)}$.

3) $\bar{\eta}$ properly contains $\bar{\beta}_{(n,k)}$. Let η be (p, c, η'), where $p \geqslant i_{n+1}$. By definition of $\bar{\beta}_{(n,k)}$, $\eta <_{i_{n+1}+1} \beta_{(n,k)}$ and hence

$$\eta <_{i_{n+1}+1} \beta_{(n,k)} <_{i_{n+1}+1} \alpha_{(n,k)}.$$

From this it follows in the same manner as 3) in the proof of (2) above that $\eta <_{i_{n+1}} \alpha_{(n,k)}$.

This completes the theory of approximations. As we have seen, this theory supplies a criterion for the evaluation of orderings between two connected ordinal diagrams.

§27. A consistency proof of second order arithmetic with the Π_1^1-comprehension axiom

In 1967, the author published consistency proofs of three systems of second order number theory by using ordinal diagrams. In this section and the next section we discuss consistency proofs of these systems.

The following lemma, concerning the system of ordinal diagrams $O(\omega+1, \omega^3)$ is essential for the consistency proof of this section.

LEMMA 27.1 (the Main Lemma). *Let p be a natural number and let γ and δ be ordinal diagrams for which there exist two finite sequences of ordinal diagrams $\gamma = \gamma_0, \ldots, \gamma_m$ and $\delta = \delta_0, \ldots, \delta_m$ which satisfy the following conditions* (1)–(4).

(1) *Each γ_i, $i < m$, is of the form $(k, 0, \gamma_{i+1})$ for some natural number $k \geqslant p$, or $(\omega, a+1, \gamma_{i+1} \# \eta)$.*

(2) *Each δ_i, $i < m$, is $(k, 0, \delta_{i+1})$ or $(\omega, a+1, \delta_{i+1} \# \eta)$ according as γ_i is $(k, 0, \gamma_{i+1})$ or $(\omega, a+1, \gamma_{i+1}, \# \eta)$.*

(3) *$\delta_m <_j \gamma_m$ for each j such that $p \leqslant j \leqslant \omega$.*

(4) *For each j such that $p \leqslant j < \omega$ and for each j-section $\bar{\alpha}$ of δ_m, there exists a j-section $\bar{\beta}$ of γ_m for which $\alpha \leqslant_j \beta$.*

Then $\delta <_j \gamma$ for each j such that $p \leqslant j \leqslant \omega$, and for each j with $p \leqslant j < \omega$ and each j-section $\bar{\alpha}$ of δ, there exists a j-section $\bar{\beta}$ of γ such that $\alpha \leqslant_j \beta$.

PROOF. (By double induction on m and $n = \iota(j, \gamma, \delta)$.)

1. If $m = 0$, then the result is obvious from (3) and (4).

2. Suppose $m > 0$, $\gamma = (k, 0, \gamma_1)$ and $\delta = (k, 0, \delta_1)$ where $k \geqslant p$.

2.1. Then $\delta_1 <_k \gamma_1$ by induction hypothesis so that $\delta <_\infty \gamma$.

2.2. If $k < j \leqslant \omega$, then $\delta <_j \gamma$ since there are no q-sections of γ or δ for $q \geqslant j$, and since by 2.1 $\delta <_\infty \gamma$.

2.3. If $j = k$, then 2.2 implies that $\delta <_j \gamma$ if $\delta_1 <_j \gamma$. Since $\delta_1 <_j \gamma_1$ by the induction hypothesis on m and since $\gamma_1 <_j \gamma$ because γ_1 is a j-section of γ it follows that $\delta_1 <_j \gamma$.

2.4. If $p \leqslant j < k$, then by the induction hypothesis $\delta <_{j_1} \gamma$, where $j_1 = j_0(j, \gamma, \delta)$. Therefore, $\delta <_j \gamma$ if, for each j-section $\bar{\alpha}$ of δ, $\alpha <_j \gamma$. Suppose $\bar{\alpha}$ is a j-section of δ. Then $\bar{\alpha}$ is a j-section of δ, so by the induction hypothesis on m there exists a j-section $\bar{\beta}$ of γ_1 such that $\alpha \leqslant_j \beta$. Furthermore, β is a j-section of γ as well. Therefore, $\alpha \leqslant_j \beta <_j \gamma$.

3. Suppose $m > 0$, $\gamma = (\omega, c+1, \gamma_1 \# \eta)$ and $\delta = (\omega, c+1, \delta_1 \# \eta)$.

3.1. Since, by the induction hypothesis on m, $\delta_1 <_\omega \gamma_1$ it follows that $\delta <_\infty \gamma$.

3.2. If $j = \omega$, then it is sufficient to show that $\delta_1 \# \eta <_\omega \gamma$, that is, $\delta_1 <_\omega \gamma$ and $\eta <_\omega \gamma$. Since $\delta_1 <_\omega \gamma_1$ and $\overline{\gamma_1 \# \eta}$ is an ω-section of γ, it follows that $\delta_1 <_\omega \gamma$ and $\eta <_\omega \gamma$.

3.3. If $p \leqslant j < \omega$, then by the induction hypothesis on n, $\delta <_{j_1} \gamma$, where $j_1 = j_0(j, \gamma, \delta)$. Hence $\delta <_j \gamma$ if for any j-section $\bar{\alpha}$ of δ, $\alpha <_j \gamma$. Let $\bar{\alpha}$ be a j-section of δ. Then $\bar{\alpha}$ is either a j-section of δ_1 or a j-section of η. If the former is the case, then by the induction hypothesis on m, there exists a j-section $\bar{\beta}$ of γ_1 such that $\alpha \leqslant_j \beta$. Moreover, $\bar{\beta}$ is a j-section of γ. Therefore, $\alpha <_j \gamma$. If $\bar{\alpha}$ is a j-section of η, then $\bar{\alpha}$ is a j-section of γ, and hence $\alpha <_j \gamma$.

We now present a second order logic system **ISL** called the isolated second order logic. In order to simplify the discussion we will use $\neg$, $\wedge$, and $\forall$ as primitive logical symbols. Other symbols will be used as abbreviations.

DEFINITION 27.2. (1) The language, formulas, abstracts, sequents and proofs of second order arithmetic are as in Definition 18.1.

(2) A semi-formula or a semi-abstract is respectively a formula-like, or an abstract-like expression, where a bound variable may occur free. The outermost logical symbol of a semi-formula or a semi-abstract is defined naturally.

(3) Let A be a semi-formula or a semi-abstract, let $\forall\phi B$ be a semi-formula in A and let $\#$ be the outermost $\forall$ in $\forall\phi B$, i.e., the $\forall$ which precedes ϕ. Let G be an arbitrary symbol in B. Then we say that $\#$ ties G and G is tied by $\#$ in A. If G is a $\forall$ on a second order variable in B and G ties ϕ in B, then we say that $\#$ affects G in A.

(4) Let $\#$ be a $\forall$ on a second order variable in A. We say that $\#$ is isolated in A if the following conditions are satisfied.

(4.1) No $\forall$ on a second order variable in A affects $\#$.

(4.2) $\#$ does not affect any $\forall$ on a second order variable.

(5) A semi-formula or a semi-abstract A is called isolated if every $\forall$ on a second order variable in A is isolated. Originally we used "semi-isolated" instead of "isolated".

The following result is easily shown.

PROPOSITION 27.3. *The class of isolated formulas (abstracts) is Π^1_1-in-the-wider-sense* (cf. (3) of Definition 18.1). *Therefore, if V is an isolated abstract and $F(\alpha)$ is isolated, then so is $F(V)$.*

From Proposition 27.3 we see that when we study isolated formulas, we are essentially dealing with Π^1_1-formulas.

DEFINITION 27.4. By the *isolated second order logic*, **ISL**, we mean a second order predicate calculus obtained from $\mathbf{G^1LC}$ in Definition 15.17 by restricting the second order $\forall$: left to only those

$$\frac{F(V), \Gamma \rightarrow \Delta}{\forall \varphi\, F(\varphi), \Gamma \rightarrow \Delta},$$

where either $\forall \varphi\, F(\varphi)$ or V is isolated.

We shall show the following

THEOREM 27.5 *The cut elimination theorem holds in* **ISL**.

The theory of relativization and this theorem immediately imply the following

COROLLARY. *Peano arithmetic together with the equality axiom $\forall \varphi\, \forall x\, \forall y\, (x = y \wedge \varphi[x] \supset \varphi[y])$ and mathematical induction*

$$\forall \varphi\, (\varphi[0] \wedge \forall x\, (\varphi[x] \supset \varphi[x+1]) \supset \forall x\, \varphi[x])$$

is consistent in **ISL**.

The second order number theory described in the Corollary was originally called **SINN**$'$. We now call it the isolated second order number theory **ISN**.

Theorem 27.5 will be proved using the system of ordinal diagrams $O(\omega+1, \omega^3)$. The proof will be presented stage by stage. We shall take over much of the terminology from first order arithmetic, for example, explicit and implicit bundles and formulas, end piece, boundary inferences, etc.

DEFINITION 27.6. Let A be a formula. We define the *γ-degree* of A, denoted $\gamma(A)$, as follows:

1) $\gamma(A) = 0$ if A is isolated.

In the following we assume that A is not isolated.

2) If A is of the form $\neg B$, then $\gamma(A) = \gamma(B) + 1$.
3) If A is of the form $B \wedge C$, then $\gamma(A) = \max(\gamma(B), \gamma(C)) + 1$.
4) If A is of the form $\forall x\, G(x)$, then $\gamma(A) = \gamma(G(a)) + 1$.

5) If A is of the form $\forall\phi\, F(\phi)$, then $\gamma(A) = \gamma(F(\alpha)) + 1$.
6) The γ-degree of an abstract $\{x_1, \ldots, x_n\}\, H(x_1, \ldots, x_n)$ is defined to be $\gamma(H(a_1, \ldots, a_n))$.

PROPOSITION 27.7. *If V is an isolated abstract, then $\gamma(F(V)) = \gamma(F(\alpha))$.*

PROOF. If $\gamma(F(\alpha)) = 0$, the proposition is evident (cf. Proposition 27.3). If $\gamma(F(\alpha)) \neq 0$, we shall prove the proposition by mathematical induction on the number of logical symbols in $F(\alpha)$. Since other cases are treated similarly we shall consider only the case where $F(\alpha)$ is of the form $\forall\phi\, G(\phi, \alpha)$. By the induction hypothesis, $\gamma(G(\beta, V)) = \gamma(G(\beta, \alpha))$. This implies that

$$\gamma(F(V)) = \gamma(G(\beta, V)) + 1 = \gamma(G(\beta, \alpha)) + 1 = \gamma(F(\alpha)).$$

PROPOSITION 27.8. *If V is isolated and $\gamma(F(V)) > 0$, then $\gamma(\forall\phi\, F(\phi)) = \gamma(F(V)) + 1$.*

PROOF. Let V be isolated and $\gamma(F(V)) > 0$. By Proposition 27.7, $\gamma(F(\alpha)) > 0$, that is, $F(\alpha)$ is not isolated. Hence $\forall\phi\, F(\phi)$ is not isolated either. Then $\gamma(\forall\phi\, F(\phi)) = \gamma(F(\alpha)) + 1 = \gamma(F(V)) + 1$.

DEFINITION 27.9. Let A be an occurrence of a formula in a proof P, in **ISL**. The *generalized grade* of A with respect to P, denoted by $g(A; P)$ or simply $g(A)$, is defined to be $\omega^2 \cdot \gamma(A) + \omega \cdot m_1 + m_0$, where m_1 is the number of second order free variables used as eigenvariables of second order $\forall$: right under the sequent containing A, and m_0 is the number of logical symbols in A. The number of occurrence, of logical symbols in A is called the *grade* of A denoted by $g_0(A)$. For a while we use only the generalized grade.

To prove Theorem 27.5, we shall modify the notion of proof in **ISL**, by introducing the following rule of substitution:

DEFINITION 27.10. Rule of substitution in **ISL**.

$$\frac{A_1, \ldots, A_n \to B_1, \ldots, B_m}{A_1\begin{pmatrix}\alpha\\ V\end{pmatrix}, \ldots, A_n\begin{pmatrix}\alpha\\ V\end{pmatrix} \to B_1\begin{pmatrix}\alpha\\ V\end{pmatrix}, \ldots, B_m\begin{pmatrix}\alpha\\ V\end{pmatrix}},$$

where α is a second order free variable and V is an arbitrary abstract with the same number of argument-places as α. Here α is called the eigenvariable of the substitution. The introduction of it helps us in the reduction of proofs in **ISL**.

DEFINITION 27.11. We say that an inference J, which is either a substitution or a second order $\forall$: right, *disturbs* a semi-formula A if the eigenvariable of J is tied by a second order $\forall$ in A.

DEFINITION 27.12. Let P be a proof in **ISL**. We call P a *proof with degree* if the following conditions are satisfied.

1) Every substitution is in the end-piece.

2) We can assign an ordinal number $\leqslant \omega$ to every semi-formula A or substitution J in P as follows. We denote this assigned number by $d(A; P)$ or $d(J; P)$, or, for short, $d(A)$ or $d(J)$ read "degree of A or J".

2.1) If A is explicit, then $d(A) = 0$. Suppose A is implicit.

2.2) If A is not isolated, then $d(A) = \omega$. Suppose A is isolated.

2.3) $d(A) = 0$ if A contains no logical symbol.

2.4) $d(A) = d(B) + 1$ if A is of the form $\neg B$.

2.5) $d(A) = \max(d(B), d(C)) + 1$ if A is of the form $B \wedge C$.

2.6) $d(A) = d(B(x)) + 1$ if A is of the form $\forall x\, B(x)$.

2.7) $d(A) = \max(d(F(\phi)), d(J_0)) + 1$ if A is of the form $\forall \phi\, F(\phi)$, where J_0 ranges over substitutions which disturb $\forall \phi\, F(\phi)$.

2.8) $d(B) < d(J)$ for every implicit formula B in the upper sequent of J.

2.9) $0 < d(J) < \omega$.

DEFINITION 27.13. Let P be a proof with degree and let S be a sequent in P. The *i-resolvent* of S is the upper sequent of the uppermost substitution under S whose degree is not greater than i, if such exists; otherwise, the i-resolvent of S is the end-sequent of P.

If a proof P has no substitution, then we can easily assign a degree to every semi-formula in P so that P becomes a proof with degree. Therefore it suffices to discuss only proofs with degree.

DEFINITION 27.14. Consider the system of ordinal diagrams $O(\omega + 1, \omega^3)$. We shall assign an ordinal diagram from $O(\omega + 1, \omega^3)$ to every sequent of a proof with degree, as follows:

1) The ordinal diagram of an initial sequent is 0.

2) If S_1 and S_2 are the upper sequent and the lower sequent, respectively, of a weak structural inference J, then the ordinal diagram of S_2 is equal to that of S_1.

3) If S_1 and S_2 are the upper sequent and the lower sequent respectively of $\neg$, $\bigwedge$: left, first order $\forall$, second order $\forall$: right or explicit second order $\forall$: left, then the ordinal diagram of S_2 is $(\omega, 0, \sigma)$, where σ is the ordinal diagram of S_1.

4) If S_1 and S_2 are the upper sequents and S is the lower sequent of an $\bigwedge$: right inference, then the ordinal diagram of S is $(\omega, 0, \sigma_1 \# \sigma_2)$, where σ_1 and σ_2 are the ordinal diagrams of S_1 and S_2, respectively.

5) If S_1 and S_2 are the upper sequent and the lower sequent respectively of an implicit, second order $\forall$: left of the form

$$\frac{F(V), \Gamma \rightarrow \Lambda}{\forall \phi\, F(\phi), \Gamma \rightarrow \Lambda},$$

then the ordinal diagram of S_2 is $(\omega, g(F(V))+2, \sigma)$, where σ is the ordinal diagram of S_1.

6) If S_1 and S_2 are the upper sequents and S is the lower sequent of a cut J, then the ordinal diagram of S is $(\omega, m+1, \sigma_1 \# \sigma_2)$, where m is the generalized grade of the cut formula and σ_1 and σ_2 are the ordinal diagrams of S_1 and S_2, respectively.

7) If S_1 and S_2 are the upper sequent and the lower sequent respectively of a substitution with degree i, then the ordinal diagram of S_2 is $(i, 0, \sigma)$, where σ is the ordinal diagram of S_1.

8) The ordinal diagram assigned to the end-sequent of a proof P with degree is called the ordinal diagram of P.

The ordinal diagram of a sequent S in P will be denoted by $O(S; P)$ or simply $O(S)$; the ordinal diagram of P will be denoted by $O(P)$.

DEFINITION 27.15. We shall define the notion of *reduction of proofs*.

1) Let $S_1, \ldots, S_m$ and S be sequents. S is *reducible* to $S_1, \ldots, S_m$ if S is provable without a cut presuming that $S_1, \ldots, S_m$ are provable without a cut.

2) Let $P_1, \ldots, P_m$ and P be proofs with degree. We say P is *reduced* to $P_1, \ldots, P_m$ if the following conditions are satisfied:

2.1) The ordinal diagram of each P_i is less than that of P (in the sense of $<_0$).

2.2) The end-sequent of P is reducible to the end-sequents of $P_1, \ldots, P_m$.

(1) Preparation for the reduction. Suppose that the sequent $\Gamma \rightarrow \Delta$ is provable in **ISL**. In the following we shall reduce a proof P of $\Gamma \rightarrow \Delta$ to another proof of $\Gamma \rightarrow \Delta$. Then by transfinite induction on $<_0$, we can prove that there exists a proof in **ISL** of $\Gamma \rightarrow \Delta$ of which the entire part is the end-piece, namely there exists a cut-free proof.

Without loss of generality we may assume that all free variables used as eigenvariables in a proof are distinct and are not contained in the sequents under the inference in which it is used as an eigenvariable.

Let P be a proof of $\Gamma \rightarrow \Delta$.

(2) We can eliminate all explicit logical inferences one by one from below. Do this in a similar way as in (4) below.

(3) We may assume that the end-piece of P contains no initial sequent. Let Q be a proof with degree whose end-sequent is not necessarily $\Gamma \rightarrow \Delta$ but which satisfies the same conditions as those required for P. We can define Q^*, obtained from Q by eliminating weakenings in the end-piece of Q, by induction on the number of inferences in the end-piece of Q as for **PA**. We deal with the following case only: If the last inference of Q is a substitution, say

$$\frac{\Gamma_1 \rightarrow \Delta_1}{\Gamma_1 \binom{\alpha}{V} \rightarrow \Delta_1 \binom{\alpha}{V}},$$

where Γ_1 and Δ_1 are $A_1, \ldots, A_m$ and $B_1, \ldots, B_n$, respectively, and $\Gamma_1\binom{\alpha}{V}$ and $\Delta_1\binom{\alpha}{V}$ denote $A_1\binom{\alpha}{V}, \ldots, A_m\binom{\alpha}{V}$ and $B_1\binom{\alpha}{V}, \ldots, B_n\binom{\alpha}{V}$, respectively, and the end-sequent of Q_0^*, where Q_0 is the proof of the upper sequent, is $\Gamma_1^* \rightarrow \Delta_1^*$, then Q^* is

$$\frac{\begin{array}{c} Q^* \\ \vdots \\ \Gamma_1^* \rightarrow \Delta_1^* \end{array}}{\Gamma_1^* \binom{\alpha}{V} \rightarrow \Delta_1^* \binom{\alpha}{V}}.$$

If the end-piece of P contains a weakening, we can reduce P to P^*, where every substitution in P^* has the same degree as the corresponding substitution in P.

(4) Suppose that the end-piece of P contains initial sequents. Suppose P is of the following form and $D \rightarrow D$ is one of the initial sequents in the end-piece of P:

$$\begin{array}{c} \dfrac{\begin{array}{c} \\ \vdots \\ \Gamma_1 \xrightarrow{\mu} \Delta_1, \tilde{D} \end{array} \qquad \begin{array}{c} D \rightarrow D \\ \vdots \\ \tilde{D}, \Pi \xrightarrow{\nu} \Lambda_1, \tilde{D}, \Lambda_2 \end{array}}{\Gamma_1, \Pi \rightarrow \Delta_1, \Lambda_1, \tilde{D}, \Lambda_2} \\ \vdots \\ \Gamma \rightarrow \Delta \end{array}$$

where two $\tilde{D}$'s in the right upper sequent of the cut denote the descendants of the D's occurring in the initial sequent which is explicitly written.

We shall consider a proof P' of the following form:

$$\begin{array}{c} \vdots \\ \Gamma_1 \xrightarrow{\mu} \Delta_1, \tilde{D} \\ \hline \text{some weakenings and exchanges} \\ \hline \Gamma_1, \Pi \rightarrow \Delta_1, \Lambda_1, \tilde{D}, \Lambda_2 \\ \vdots \\ \Gamma \rightarrow \Delta \end{array}$$

where every substitution in P' is assigned the same degree as the corresponding one in P.

$$O(\Gamma_1, \Pi \rightarrow \Delta_1, \Lambda_1, \tilde{D}, \Lambda_2; P) = (\omega, g(\tilde{D})+1, \mu \# \nu),$$

while

$$O(\Gamma_1, \Pi \rightarrow \Delta_1, \Lambda_1, \tilde{D}, \Lambda_2; P') = \mu <_j (\omega, g(\tilde{D})+1, \mu \# \nu)$$

for all $j \leq \omega$ and, if $j < \omega$, for each $\bar{\beta}$ a j-section of μ, $\bar{\beta}$ is also a j-section of $(\omega, g(D)+1, \mu \# \nu)$. Thus by Lemma 27.1, $O(P') <_0 O(P)$.

If P is of the form

$$\dfrac{\begin{array}{c} D \rightarrow D \\ \vdots \\ \Gamma_1, \tilde{D}, \Gamma_2 \rightarrow \Delta, \tilde{D} \end{array} \qquad \tilde{D}, \Pi \rightarrow \Lambda}{\Gamma_1, \tilde{D}, \Gamma_2, \Pi \rightarrow \Delta, \Lambda}$$

then the reduction is carried out similarly.

(5) In the following we shall assume that the end-piece of a proof with degree contains no logical inference, weakening or initial sequents. Moreover, we may assume that the proof is different from its end-piece, for if the entire proof is its end-piece, then there are no cuts in the proof.

Let P be a proof with degree. We repeat the definition of a suitable cut: A cut in the end-piece of a proof with degree, P, is called suitable if both of its cut formulas have ancestors which are principal formulas of boundary (hence logical) inferences. We can show, exactly the same way as for **PA**, that under those conditions there exists a suitable cut in the end-piece of P.

Now, let P be a proof with degree whose end-sequent is $\Gamma \rightarrow \Delta$ and let J be a suitable cut in P. To define the essential reduction, we must treat separately several cases according to the form of the outermost logical symbol of the cut formulas of J.

(6) We shall first treat the case where the outermost logical symbol of J is second order $\forall$. Let P be of the following form:

$$\begin{array}{lcc}
 & \vdots & \vdots \\
 & J_0\ \dfrac{\Gamma_1 \xrightarrow{\lambda} \Delta_1, F_1(\alpha)}{\Gamma_1 \xrightarrow{(\omega,0,\lambda)} \Delta_1, \forall\phi\, F_1(\phi)} & \dfrac{F_2(V), \Pi_1 \xrightarrow{\mu} \Lambda_1}{\forall\phi\, F_2(\phi), \Pi_1 \xrightarrow{(\omega,n+2,\mu)} \Lambda_1} \\
S_1 & & S_3 \\
 & \vdots & \vdots \\
S_2 & J\ \Gamma_2 \xrightarrow{\tau} \Delta_2, \forall\phi\, F(\phi) & S_4\ \forall\phi\, F(\phi), \Pi_2 \xrightarrow{\rho} \Lambda_2 \\
\hline
S_5 & \multicolumn{2}{c}{\Gamma_2, \Pi_2 \xrightarrow{(\omega,m+1,\tau\#\rho)} \Delta_2, \Lambda_2} \\
 & \multicolumn{2}{c}{\vdots} \\
S_6 & \multicolumn{2}{c}{\Gamma_3 \xrightarrow{\nu} \Delta_3} \\
 & \multicolumn{2}{c}{\vdots} \\
 & \multicolumn{2}{c}{\Gamma \xrightarrow{\sigma} \Delta}
\end{array}$$

where $m = g(\forall\phi\, F(\phi))$, $n = g(F_2(V))$, and S_6: $\Gamma_3 \to \Delta_3$ is the i-resolvent of S_5: $\Gamma_2, \Pi_2 \to \Delta_2, \Lambda_2$, i being $d(\forall\phi\, F_1(\phi))$. Here we should remark that the i-resolvent $\Gamma_3 \to \Delta_3$ will be used only for the case when $\forall\phi\, F(\phi)$ is isolated.

Case 1. $\forall\phi\, F(\phi)$ is isolated.

Let, in the above figure,

$$S_1:\quad \Gamma_1 \to \Delta_1, \forall\phi\, F_1(\phi), \qquad S_2:\quad \Gamma_2 \to \Delta_2, \forall\phi\, F(\phi),$$

$$S_3:\quad \forall\phi\, F_2(\phi), \Pi_1 \to \Lambda_1, \qquad S_4:\quad \forall\phi\, F(\phi), \Pi_2 \to \Lambda_2.$$

Here we should remark that $\forall\phi\, F_1(\phi)$ and $\forall\phi\, F_2(\phi)$ are $\forall\phi\, F(\phi)$ itself; that is, no substitution applies to those formulas, for if there were a substitution with degree k between S_1 and S_2 which applies to $\forall\phi\, F_1(\phi)$, then this substitution would disturb $\forall\phi\, F_1(\phi)$. But this implies that $k < i$, which contradicts 2.8) of Definition 27.12. Thus $\forall\phi\, F_1(\phi)$ is $\forall\phi\, F(\phi)$. By the same reasoning, $\forall\phi\, F_2(\phi)$ is $\forall\phi\, F(\phi)$. In the inference J_0, $d(F_1(\phi)) < i$ $(= d(\forall\phi\, F_1(\phi)))$. Let P' be the following:

$$
\begin{array}{ll}
S'_1\; \begin{array}{c} \vdots \\ \Gamma_1 \overset{\lambda'}{\to} \Delta_1, F_1(\alpha) \\ \hline\hline \Gamma_1 \to F_1(\alpha), \Delta_1, \forall\phi\, F_1(\phi) \end{array}
&
S_3\; \begin{array}{c} \vdots \\ F_2(V), \Pi_1 \overset{\mu}{\to} \Lambda_1 \\ \hline \forall\phi\, F_2(\phi), \Pi_1 \to \Lambda_1 \end{array}
\end{array}
$$

$$
\begin{array}{c}
\begin{array}{ll} \quad\vdots & \quad\vdots \\ S'_2\; \Gamma_2 \overset{\tau'}{\to} F(\alpha), \Delta_2, \forall\phi\, F(\phi) & S_4\; \forall\phi\, F(\phi), \Pi_2 \overset{\rho}{\to} \Lambda_2 \end{array} \\ \hline
S'_5\; \Gamma_2, \Pi_2 \overset{(\omega, m+1, \tau' \# \rho)}{\longrightarrow} F(\alpha), \Delta_2, \Lambda_2
\end{array}
$$

$$
\begin{array}{c}
\begin{array}{ll}
\begin{array}{c} \vdots \\ S'_6\;\; J_1\;\; \dfrac{\Gamma_3 \overset{\theta}{\to} \Delta_3, F(\alpha)}{S_7\;\; \Gamma_3 \overset{(i,0,\theta)}{\longrightarrow} \Delta_3, F(V)} \end{array}
&
\begin{array}{c} \vdots \\ S'_3\; F(V), \Pi_1 \overset{\mu}{\to} \Lambda_1 \end{array}
\end{array} \\ \hline
\Gamma_3, \Pi_1 \overset{(\omega, n+1, (i,0,\theta) \# \mu)}{\longrightarrow} \Delta_3, \Lambda_1 \\ \hline\hline
S_8\;\; \forall\phi\, F(\phi), \Pi_1, \Gamma_3 \to \Delta_3, \Lambda_1
\end{array}
$$

$$
\begin{array}{c}
\begin{array}{ll} \quad\vdots & \quad\vdots \\ S_1\;\; \Gamma_2 \overset{\tau}{\to} \Delta_2, \forall\phi\, F(\phi) & S_9\;\; \forall\phi\, F(\phi), \Pi_2, \Gamma_3 \overset{\rho'}{\to} \Delta_3, \Lambda_2 \end{array} \\ \hline
\Gamma_2, \Pi_2, \Gamma_3 \overset{(\omega, m+1, \tau \# \rho')}{\longrightarrow} \Delta_2, \Delta_3, \Lambda_2 \\ \hline\hline
S_{10}\;\; \Gamma_2, \Pi_2, \Gamma_3 \to \Delta_3, \Delta_2, \Lambda_2 \\
\vdots \\
\Gamma_3, \Gamma_3 \to \Delta_3, \Delta_3 \\ \hline\hline
S_{11}\;\; \Gamma_3 \overset{\nu'}{\to} \Delta_3 \\
\vdots \\
\Gamma \overset{\sigma'}{\to} \Delta
\end{array}
$$

where J_1 is a substitution whose eigenvariable is α and whose degree is defined to be i. Every substitution in this proof other than J_1 is assigned the same degree as the corresponding substitution in P. Here we should remark that, in the upper sequent of J_1, the descendant of $F_1(\alpha)$ in $\Gamma_1 \to \Delta_1, F_1(\alpha)$ is $F(\alpha)$. As was remarked, no substitution disturbs $F(\alpha)$ between $\Gamma_1 \to \Delta_1, F_1(\alpha)$ and $\Gamma_2, \Pi_2 \to \Delta_2, \Lambda_2$ in P. If there were such a substitution with degree k between $\Gamma_2, \Pi_2 \to \Delta_2, \Lambda_2$ and $\Gamma_3 \to \Delta_3$, it would disturb $\forall \phi\, F(\phi)$, i.e., $k < i$. But this contradicts the fact that $\Gamma_3 \to \Delta_3$ is the i-resolvent of $\Gamma_2, \Pi_2 \to \Delta_2, \Lambda_2$.

We shall show that P' is a proof with degree. For this it is sufficient to show that $d(F_1(\alpha); P') < i$. If there is an inference other than J_1 which disturbs $F_1(\alpha)$ (in P'), the corresponding substitution in P disturbs $F_1(\phi)$. J_1 does not disturb $F(\alpha)$ for otherwise the outermost $\forall$ of $\forall \phi\, F_1(\phi)$ affects another $\forall$ for a second order variable in $\forall \phi\, F_1(\phi)$. But this contradicts the fact that $\forall \phi\, F_1(\phi)$ is isolated. So $d(F_1(\alpha); P') = d(F_1(\alpha); P) < i$.

In order to prove $O(P') <_0 O(P)$, or $\sigma' <_0 \sigma$, we first prove $\nu' <_j \nu$, where $j \leq \omega$, and, for any j, $0 \leq j < i$, and a j-section of ν', say $\bar{\eta}$, there is a j-section of ν, say $\bar{\xi}$, such that $\eta \leq_j \xi$. This is shown below (cf. (6.5)).

(6.1) For any $j \leq \omega$, $\tau' <_j \tau$, where $\tau = O(S_2; P)$ and $\tau' = O(S_2; P')$. If $j < \omega$, and $\bar{\alpha}$ is any j-section of τ', then there exists a j-section of τ, say $\bar{\beta}$, such that $\alpha \leq_j \beta$.

PROOF. Since there is no substitution above S_1 we see by the Main Lemma, with $p = 0$, that it is sufficient to show that $\lambda' <_j (\omega, 0, \lambda)$ for all $j \leq \omega$.

1) $j = \omega$. Then $\lambda' \leq_\omega \lambda$, by the definition of P'. Obviously $\lambda <_\omega (\omega, 0, \lambda)$. Therefore $\lambda' <_\omega (\omega, 0, \lambda)$.

2) $j < \omega$. Since there is no j-section of λ', $\lambda' <_j (\omega, 0, \lambda)$ if $\lambda' <_\omega (\omega, 0, \lambda)$. But $\lambda' <_\omega (\omega, 0, \lambda)$ by 1).

(6.2) $\theta <_j \nu$ for each $j \leq \omega$. If $j < \omega$, then for each j-section $\bar{\alpha}$ of θ, there exists a j-section $\bar{\beta}$ of ν such that $\alpha \leq_j \beta$.

PROOF. By the Main Lemma, with $p = 0$, it is sufficient to prove

1) $(\omega, m+1, \tau' \# \rho) <_j (\omega, m+1, \tau \# \rho)$ for all $j \leq \omega$, and

2) for each $j < \omega$, and for each j-section $\bar{\alpha}$ of τ', there exists a j-section $\bar{\beta}$ of τ such that $\alpha \leq_j \beta$.

But 2) is part of (6.1). We therefore need only prove 1). This we will do by induction on the total number of indices greater than j (super indices of j) in $(\omega, m+1, \tau' \# \rho)$ and in $(\omega, m+1, \tau \# \rho)$.

i) Since $\tau' <_\omega \tau$ it follows from (6.1) that

$$(\omega, m+1, \tau' \# \rho) <_\infty (\omega, m+1, \tau \# \rho).$$

ii) If $j = \omega$, then $\tau' \# \rho <_\omega \tau \# \rho$ by (6.1). Since $\tau \# \rho <_\omega (\omega, m+1, \tau \# \rho)$, 1) follows from i).

iii) If $j < \omega$, then by the induction hypothesis and i),

$$(\omega, m+1, \tau' \# \rho) <_{j_1} (\omega, m+1, \tau \# \rho),$$

where $j_1 = j_0(j, (\omega, m+1, \tau' \# \rho), (\omega, m+1, \tau \# \rho))$. Let $\bar{\alpha}$ be a j-section of $(\omega, m+1, \tau' \# \rho)$. Then $\bar{\alpha}$ is also a j-section of $\tau' \# \rho$. If $\bar{\alpha}$ is a j-section of τ', then $\alpha <_j (\omega, m+1, \tau \# \rho)$ by (6.1). If $\bar{\alpha}$ is a j-section of ρ, then $\bar{\alpha}$ is a j-section of $(\omega, m+1, \tau \# \rho)$ and hence $\alpha <_j (\omega, m+1, \tau \# \rho)$. Thus

$$(\omega, m+1, \tau' \# \rho) <_j (\omega, m+1, \tau \# \rho).$$

(6.3) $O(S_8; P') <_j O(S_3; P)$ for $i < j \leq \omega$ or $j = \infty$.

PROOF. $O(S_3) = (\omega, n+2, \mu)$ and $O(S_8) = (\omega, n+1, (i, 0, \theta) \# \mu)$. The proof is by induction on $\iota(j, O(S_3), O(S_8))$.

1) Since $n+1 < n+2$, $O(S_8) <_\infty O(S_3)$.

2) If $j = \omega$, then since $\mu <_\omega O(S_3)$ and $O(S_8) <_\infty O(S_3)$, it is sufficient to prove that $(i, 0, \theta) <_\omega O(S_3)$. But this is clearly the case since $(i, 0, \theta)$ has no ω-section and $(i, 0, \theta) <_\infty O(S_3)$.

3) If $i < j < \omega$, then $O(S_8) <_j O(S_3)$ because neither $(i, 0, \theta)$ nor μ has a j-section and from 2) $O(S_8) <_\omega O(S_3)$.

(6.4) If $i < j \leq \omega$, then $\rho' <_j \rho$. If $i < j < \omega$ then for each j-section $\bar{\alpha}$ of ρ' there exists a j-section $\bar{\beta}$ of ρ such that $\alpha \leq_j \beta$.

PROOF. Let us regard $i+1$, $O(S_4) (= \rho)$, and $O(S_9) (= \rho')$ as p, γ, and δ, respectively, in the Main Lemma. Let $\gamma_0 (= O(S_4))$, $\gamma_1, \ldots, \gamma_m (= O(S_3))$ be the sequence of distinct ordinal diagrams of sequents from S_4 to S_3 in P and let $\delta_0 (= O(S_9))$, $\delta_1, \ldots, \delta_m (= O(S_8))$ be the sequence of distinct ordinal diagrams of sequents from S_9 to S_8 in P'. The proposition then follows from the Main Lemma and (6.3). Here we should recall that $O(S_8)$ has no j-section if $i < j < \omega$.

(6.5) $\nu' <_j \nu$ for $j \leq \omega$.

PROOF. We first show that $\nu' <_j \nu$ for any $j, i < j \leq \omega$. Let p be any number, $i < p \leq \omega$ and let $p \leq j \leq \omega$. Take $O(S_6) (= \nu)$, $O(S_4) (= \nu')$ and p as γ, δ, p respectively in the Main Lemma. Let $\gamma_0 (= O(S_6)), \ldots, \gamma_m (= O(S_5))$ be the sequence of distinct ordinal diagrams of sequents from S_6 to S_5 in P and let

$$\delta_0 (= O(S_{11})), \ldots, \delta_m (= O(S_{10}))$$

be the sequence of distinct ordinal diagrams of sequents from S_{11} to S_{10} in P'. We then only have to prove that the conditions of the Main Lemma are satisfied for $O(S_5)$ and $O(S_{10})$. This we prove by induction on $\iota(j, O(S_5), O(S_{10}))$, where $O(S_5) = (\omega, m+1, \tau \# \rho)$ and $O(S_{10}) = (\omega, m+1, \tau \# \rho')$.

1) From (6.4) $\rho' <_\omega \rho$. Therefore, $O(S_{10}) <_\infty O(S_5)$.

In 2)–3) we assume that $O(S_{10}) <_{j_1} O(S_5)$, where $j_1 = j_0(j, O(S_5), O(S_{10}))$.

2) If $j=\omega$, then $O(S_{10}) <_\omega O(S_5)$ provided $\tau \# \rho' <_\omega O(S_5)$ and $O(S_{10}) <_\infty O(S_5)$. But this follows from (6.4) and 1).

3) If $p \leq j < \omega$, then $O(S_{10}) <_j O(S_5)$ provided for each j-section $\bar{\alpha}$ of $O(S_{10})$, $\alpha <_j O(S_5)$. Let $\bar{\alpha}$ be a j-section of $O(S_{10})$, i.e., of $\tau \# \rho'$. If $\bar{\alpha}$ is a j-section of τ, then $\bar{\alpha}$ is a j-section of $O(S_5)$ as well. Therefore, $\alpha <_j O(S_5)$. If $\bar{\alpha}$ is a j-section of ρ', then $\alpha <_j O(S_5)$ by (6.4).

Having established $\nu' <_j \nu$, $i < j \leq \omega$, now consider an i-section of ν'. If it is not $\bar{\theta}$, then it is an i-section of ν. If it is $\bar{\theta}$, then $\theta <_i \nu$ has been established in (6.2). For $j < i$, let $\bar{\alpha}$ be a j-section of ν'. It can be easily shown that there is a j-section of ν whose o. d. is α. So $\alpha <_j \nu$. Thus $\nu' <_j \nu$ for any $j \leq i$. This completes the first objective, $\nu' <_j \nu$ for all $j \leq \omega$.

Next, recall that either $\Gamma_3 \rightarrow \Delta_3$ is the end sequent or $\Gamma_3 \rightarrow \Delta_3$ is the upper sequent of a substitution of degree $(= k_0) < i$. If the former is the case, then $\nu' <_0 \nu$ means $\sigma' <_0 \sigma$. Suppose the latter is the case. Then $\sigma' <_0 \sigma$ follows from (6.5) by virtue of the Main Lemma; notice that k_0 as above prevents $\bar{\theta}$ from being an i-section of an o. d. between ν' and σ'.

Case 2. $\forall \phi\, F(\phi)$ is not isolated.

Let P'' have the following form:

$$\vdots \qquad\qquad\qquad\qquad\qquad\qquad\qquad\qquad \vdots$$
$$\Gamma_1 \rightarrow \Delta_1, F_1(V) \qquad\qquad\qquad F_2(V), \Pi_1 \rightarrow \Lambda_1$$
$$\text{some exchanges and a weakening} \qquad \text{some exchanges and a weakening}$$
$$\Gamma_1 \rightarrow F_1(V), \Delta_1, \forall\phi\, F_1(\phi) \qquad\qquad \forall\phi\, F_2(\phi), \Pi_1, F_2(V) \rightarrow \Lambda_1$$
$$\vdots \qquad\qquad \vdots \qquad\qquad \vdots \qquad\qquad \vdots$$
$$\dfrac{\Gamma_2 \rightarrow F(V), \Delta_2, \forall\phi\, F(\phi) \quad \forall\phi\, F(\phi), \Pi_2 \rightarrow \Lambda_2}{\Gamma_2, \Pi_2 \rightarrow F(V), \Delta_2, \Lambda_2} \qquad \dfrac{\Gamma_2 \rightarrow \Delta_2, \forall\phi\, F(\phi) \quad \forall(\phi)\, F(\phi), \Pi_2, F(V) \rightarrow \Lambda_2}{\Gamma_2, \Pi_2, F(V) \rightarrow \Delta_2, \Lambda_2}$$
$$\text{some exchanges} \qquad\qquad \text{some exchanges}$$
$$\dfrac{\Gamma_2, \Pi_2 \rightarrow \Delta_2, \Lambda_2, F(V) \qquad F(V), \Gamma_2, \Pi_2 \rightarrow \Delta_2, \Lambda_2}{\Gamma_2, \Pi_2, \Gamma_2, \Pi_2 \rightarrow \Delta_2, \Lambda_2, \Delta_2, \Lambda_2}$$
$$\dfrac{\text{some exchanges and contractions}}{\Gamma_2, \Pi_2 \rightarrow \Delta_2, \Lambda_2}$$
$$\vdots$$
$$\Gamma \overset{\sigma''}{\rightarrow} \Delta$$

where every substitution is assigned the same degree as the corresponding substitution in P, and the proof of $\Gamma_1 \overset{\vdots}{\rightarrow} \Delta_1, F_1(V)$ is obtained from the proof of $\Gamma_1 \overset{\vdots}{\rightarrow} \Delta_1, F_1(\alpha)$ by substituting V for α everywhere. Since V is isolated, $\gamma(G(V)) = \gamma(G(\alpha))$ by Proposition 27.7. Hence the ordinal diagram of $\Gamma_1 \rightarrow \Delta_1, F_1(V)$ is not greater than λ in the sense of $<_j$ for every j.

P'' is clearly a proof with degree. Since $g(F(V)) < g(\forall\phi\, F(\phi))$, we can easily see that $\sigma'' <_0 \sigma$.

(7) Next we treat the case in which the outermost logical symbol of the

cut formula of J is $\wedge$. Let P be of the following form:

$$\begin{array}{c}
\dfrac{\overset{\vdots}{\Gamma_1 \to \Delta_1, A_1} \quad \overset{\vdots}{\Gamma_2 \to \Delta_2, B_1}}{\Gamma_1, \Gamma_2 \to \Delta_1, \Delta_2, A_1 \wedge B_1} \qquad \dfrac{\overset{\vdots}{A_2, \Pi_2 \to \Lambda_1}}{A_2 \wedge B_2, \Pi_1 \to \Lambda_1} \\
\vdots \\
\dfrac{\overset{\vdots}{\Gamma_3 \to \Delta_3, A \wedge B} \qquad \overset{\vdots}{A \wedge B, \Pi_2 \to \Lambda_2}}{\Gamma_3, \Pi_2 \to \Delta_3, \Lambda_2} \\
\vdots \\
\Gamma \to \Delta
\end{array}$$

We see that P can be reduced to a P' of the following form:

$$\begin{array}{c}
\dfrac{\overset{\vdots}{\Gamma_1 \to \Delta_1, A_1}}{\Gamma_1, \Gamma_2 \to A_1, \Delta_1, \Delta_2, A_1 \wedge B_1}\ \text{some exchanges and a weakening} \qquad \dfrac{\overset{\vdots}{A_2, \Pi_1 \to \Lambda_1}}{A_2 \wedge B_2, \Pi_1, A_2 \to \Lambda_1}\ \text{some exchanges and a weakening} \\
\vdots \\
\dfrac{\dfrac{\dfrac{\overset{\vdots}{\Gamma_3 \to A, \Delta_3, A \wedge B} \quad \overset{\vdots}{A \wedge B, \Pi_2 \to \Lambda_2}}{\Gamma_3, \Pi_2 \to A, \Delta_3, \Lambda_2}}{\Gamma_3, \Pi_2 \to \Delta_3, \Lambda_2, A}\ \text{some exchanges} \quad \dfrac{\dfrac{\overset{\vdots}{\Gamma_3 \to \Delta_3, A \wedge B} \quad \overset{\vdots}{A \wedge B, \Pi_2, A \to \Lambda_2}}{\Gamma_3, \Pi_2, A \to \Delta_3, \Lambda_2}}{A, \Gamma_3, \Pi_2 \to \Delta_3, \Lambda_2}\ \text{some exchanges}}{\dfrac{\Gamma_3, \Pi_2, \Gamma_3, \Pi_2 \to \Delta_3, \Lambda_2, \Delta_3, \Lambda_2}{\Gamma_3, \Pi_2 \to \Delta_3, \Lambda_2}\ \text{some exchanges and contractions}} \\
\vdots \\
\Gamma \to \Delta
\end{array}$$

Every substitution in P' is assigned the same degree as the corresponding substitution in P. Thus P' is a proof with degree whose ordinal diagram is less than that of P.

(8) The remaining cases, i.e., the case in which the outermost logical symbol of the cut-formula of J is $\neg$ and the case in which the outermost logical symbol of the cut-formula of J is $\forall$, for a first order variable, are treated in the same way as the above cases.

This completes the proof of Theorem 27.5.

Our consistency proof of **ISN** in the corollary of Theorem 27.5 is not satisfactory in the following sense. The optimal ordinal number for **ISN** is $O(\omega+1, 1)$ but not $O(\omega+1, \omega^3)$. As far as the foundational meaning of the consistency proof is concerned, $O(\omega+1, 1)$ and $O(\omega+1, \omega^3)$ do not make much difference since the complexity of the accessibility proof of $O(I, A)$ mainly depends on I but not on A. However as far as the ordinal of **ISN** is concerned, $O(\omega+1, 1)$ is the right one and $O(\omega+1, \omega^3)$ is greater than $O(\omega+1, 1)$. Recently T. Arai improved the author's method and proved the consistency of **ISN** by using $O(\omega+1, 1)$. Now we shall discuss Arai's new proof. First we reformulate **ISN** in a simpler form.

DEFINITION 27.16. By the isolated second order number theory, **ISN**, we mean from now on a system of second order arithmetic as in Definition 18.1, where the induction formulas are arbitrary, i.e., the system has full induction, and the second order $\forall$: left

$$\frac{F(V), \Gamma \rightarrow \Delta}{\forall \varphi F(\varphi), \Gamma \rightarrow \Delta}$$

are restricted to one of the following forms:

(1) the principal formula $\forall \varphi F(\varphi)$ is isolated, or

(2) the abstract V in $F(V)$ is a second order free variable.

It is easily seen that the old system **ISN** and the new system **ISN** are equivalent. We shall give a new proof of the following

THEOREM 27.17. **ISN** *is consistent.*

The proof will be again presented stage by stage. We shall take over much of the terminology from the proof of Theorem 27.5, for example, substitution, Lemma 27.1, and the definition of grade in Definition 27.9 etc.

DEFINITION 27.18. Let P be a proof in **ISN**. We call P a proof with degree if the following conditions are satisfied.

(1) Every substitution is in the end-piece and there is no ind under a substitution.

(2) We can assign an ordinal number $\leq \omega$ to every semi-formula A or substitution J in P as follows. We denote this assigned number by $d(A, P)$ or $d(J; P)$, or, for short, $d(A)$ or $d(J)$, and read "degree of A or J".

(2.1) If A is explicit, then $d(A) = 0$.

Suppose A is implicit.

(2.2) If A is not isolated, then $d(A) = \omega$.

Suppose A is isolated.

(2.3) $d(A) = 0$ if A contains no logical symbol.

(2.4) $d(\neg A) = d(A)$, $d(A_1 \wedge A_2) = \max(d(A_1), d(A_2))$, and $d(\forall x\, A(x)) = d(A(x))$.

(2.5) $d(\forall \varphi F(\varphi)) = \max(d(F(\varphi)) + 1, d(J))$, where J ranges over substitutions which disturb $\forall \varphi F(\varphi)$.

(2.6) $d(B) < d(J)$ for every implicit formula B in the upper sequent of J.

(2.7) $0 < d(J) < \omega$.

Notice that a proof with degree in **ISN** is quite different from a proof with degree in **ISL**.

If a proof P has no substitutions, then we can easily assign a degree to every semi-formula in P so that P becomes a proof with degree. Therefore it suffices to discuss only proofs with degree.

DEFINITION 27.19. Let P be a proof in **ISN** and S a sequent in P. The *height* of S in P, denoted by $h(S; P)$ or simply $h(S)$, is defined inductively from below to above as follows:

(i) $h(S) = 0$ if S is the end-sequent of P or S is the upper sequent of a substitution in P.

(ii) $h(S) = h(S')$ if S is an upper sequent of an inference except substitution, cut, ind, and second order $\forall$: left, and S' is the lower sequent of the inference.

(iii) $h(S) = \max(h(S'), g_0(D)+1)$, if S is an upper sequent of cut, ind, or second order $\forall$: left, and D is the cut formula, induction formula or auxiliary formula of the inference respectively, and S' is the lower sequent of the inference.

We are going to assign an ordinal diagram from $O(\omega+1, 1)$ to every sequent of a proof with degree. For simplicity, we write (i, μ) for a nonzero connected ordinal diagram $(i, 0, \mu)$. We also define the ordinal diagram $\omega(n, \mu)$ for every ordinal diagram μ from $O(\omega+1, 1)$ and natural number n as follows:

$$\omega(0, \mu) = \mu, \quad \omega(n+1, \mu) = (\omega, \omega(n, \mu)).$$

DEFINITION 27.20. We assign an ordinal diagram from $O(\omega+1, 1)$ to every sequent of a proof with degree, as follows:

(1) The ordinal diagram of an initial sequent is O.

(2) If S_1 and S_2 are the upper sequent and the lower sequent, respectively, of a weak structural inference, then the ordinal diagram of S_2 is equal to that of S_1.

(3) If S_1 and S_2 are the upper sequent and the lower sequent respectively of $\neg$, $\wedge$: left, first order $\forall$, second order $\forall$: right or explicit second order $\forall$: left, then the ordinal diagram of S_2 is (ω, α), where α is the ordinal diagram of S_1.

(4) If S_1 and S_2 are the upper sequents and S is the lower sequent of an $\wedge$: right inference, then the ordinal diagram of S is $(\omega, \sigma_1 \# \sigma_2)$, where σ_1 and σ_2 are the ordinal diagrams of S_1 and S_2, respectively.

(5) If S_1 and S_2 are the upper sequent and the lower sequent respectively of an implicit second order $\forall$: left or ind, then the ordinal diagram of S_2 is $\omega(h(S_2) - h(S_1) + 1, \sigma)$, where σ is the ordinal diagram of S_1.

(6) If S_1 and S_2 are the upper sequents and S is the lower sequent of a cut, then the ordinal diagram of S is $\omega(h(S_1) - h(S), \sigma_1 \# \sigma_2)$, where σ_1 and σ_2 are the ordinal diagrams of S_1 and S_2 respectively.

(7) If S_1 and S_2 are the upper sequent and the lower sequent respectively of a substitution with degree i, then the ordinal diagram of S_2 is (i, σ), where σ is the ordinal diagram of S_1.

(8) The ordinal diagram assigned to the end-sequent of a proof P with degree is called the ordinal diagram of P.

As before the ordinal diagram of a sequent S in P will be denoted by $O(S; P)$ or simply $O(S)$; the ordinal diagram of P will be denoted by $O(P)$.

DEFINITION 27.21. For each i such that $0 \leq i \leq \omega$, we define a relation $\ll_i$ between two ordinal diagrams from $O(\omega+1, 1)$ as follows:

(1) $\mu \ll_i \nu$ iff the following two conditions are satisfied.

(a) For each j with $i \leq j < \omega$ and for each j-section ρ of μ, there exists a j-section τ of ν for which $\rho \ll_j \tau$, and

(b) for each k with $i \leq k \leq \omega$, $\mu <_k \nu$.

(2) $\mu \leqq_i \nu$ iff $\nu \ll_i \nu$ or $\mu = \nu$.

As a special case of Lemma 27.1 we have the following

LEMMA 27.22. *(1) $\mu \ll_i \nu$ implies $\omega(n, \mu) \ll_i \omega(n, \nu)$ and $\omega(n, \mu \# \theta) \ll_i \omega(n, \nu \# \theta)$ for every natural number n and every ordinal diagram θ from $O(\omega+1, 1)$.*

(2) $\mu \leqq_i \nu$ iff $\nu \ll_i \nu$ or $\mu = \nu$.

(1) Preparation for the reduction. Suppose that the sequent $\rightarrow$ is provable in **ISN**. In the following we shall reduce a proof P of $\rightarrow$ to another proof of $\rightarrow$. Then by transfinite induction on $<_0$, we can prove that there exists a proof in **ISN** of $\rightarrow$ of which the entire part is the end-piece. Following the method of the consistency proof of **PA**, we can eliminate the cut inference from the proof of $\rightarrow$ so obtained. But this is impossible.

Without loss of generality we may assume that all free variables used as eigenvariables in a proof are distinct and are not contained in the sequents under the inference in which it is used as an eigenvariable.

Let P be a proof of $\rightarrow$.

1) We add the following rules of inference, called term-replacement.

$$\frac{\Gamma_1, F(s), \Gamma_2 \rightarrow \Delta}{\Gamma_1, F(t), \Gamma_2 \rightarrow \Delta}, \qquad \frac{\Gamma \rightarrow \Delta_1, F(s), \Delta_2}{\Gamma \rightarrow \Delta_1, F(t), \Delta_2},$$

where s and t are terms which do not contain any free variable and which express the same number. (These rules of inference are redundant in the original system.)

2) If S_1 and S_2 are the upper sequent and the lower sequent of an application of term-replacement, then the ordinal diagram of S_2 is equal to that of S_1.

3) We substitute 0 for every free variable of type 0 in P except if it is

used as an eigenvariable. In this alteration the proof remains correct and neither the end-sequent of P nor the ordinal diagram of P changes.

(2) Suppose that P contains an application of ind in its end-piece. Because of 3), immediately above, P contains no first order free variables in its end-piece other than those used as eigenvariables. Let J be a lowermost induction in the end-piece of P:

$$\begin{array}{c} Q(a) \\ \vdots \\ J\quad \dfrac{A(a), \Gamma \overset{\mu}{\rightarrow} \Delta, A(a')}{A(0), \Gamma \rightarrow \Delta, A(t)}\,, \\ \vdots \\ \rightarrow \end{array}$$

where t contains no free variables and $Q(a)$ is the proof of the upper sequent of J. We obtain a proof P' from P by replacing J by the following:

Case 1. $t=0$. Replace the part of P above $A(0),\ \Gamma \rightarrow \Delta,\ A(t)$ (inclusive) by

$$\begin{array}{c} A(0) \rightarrow A(0) \\ \hline \text{some weakenings and exchanges} \\ \hline A(0), \Gamma \rightarrow \Delta, A(0) \\ \hline A(0), \Gamma \rightarrow \Delta, A(t). \end{array}$$

Since the ordinal diagram of $A(0),\ \Gamma \rightarrow \Delta,\ A(t)$ is 0, it is obvious that $O(P') <_0 O(P)$.

Case 2. $t \neq 0$. Then $t=n$ for some numeral n. Consider the following proof P':

$$\begin{array}{c}
\dfrac{\begin{array}{c} Q(0) \\ \vdots \\ A(0), \Gamma \overset{\mu}{\rightarrow} \Delta, A(0') \end{array} \qquad \begin{array}{c} Q(0') \\ \vdots \\ A(0'), \Gamma \overset{\mu}{\rightarrow} \Delta, A(0'') \end{array}}{A(0), \Gamma, \Gamma \rightarrow \Delta, \Delta, A(0'')} \\
\hline \text{some exchanges and contractions} \\
\hline
\dfrac{A(0), \Gamma \rightarrow \Delta, A(0'') \qquad \begin{array}{c} Q(0'') \\ \vdots \\ A(0''), \Gamma \overset{\mu}{\rightarrow} \Delta, A(0''') \end{array}}{A(0), \Gamma, \Gamma \rightarrow \Delta, \Delta, A(0''')} \\
\hline \text{some exchanges and contractions} \\
\hline A(0), \Gamma \rightarrow A(0'''), \Delta \\
\dfrac{\begin{array}{c} \vdots \\ A(0), \Gamma \rightarrow \Delta, A(n) \end{array} \qquad \dfrac{A(n) \rightarrow A(n)}{A(n) \rightarrow A(t)}}{A(0), \Gamma \rightarrow \Delta, A(t)} \\
\vdots \\
\rightarrow
\end{array}$$

Every substitution in P' is assigned the same degree as the corresponding substitution in P. It is easily seen that P' is a proof with degree whose end-sequent is $\rightarrow$.

That $O(P') <_0 O(P)$ is shown as follows.

First it is easily seen that

$$\mu_0 = O(A(0), \Gamma \rightarrow \Delta, A(t); P) = \omega(h_1 - h + 1, \mu),$$

where h_1 and h are the heights of $A(a), \Gamma \rightarrow \Delta, A(a')$ and $A(0), \Gamma \rightarrow \Delta, A(t)$ respectively. Since all the heights of $A(0'^{\cdots'}), \Gamma \rightarrow \Delta, A(0'^{\cdots''})$ and $A(n) \rightarrow A(t)$ in P' are h_1 and the height of $A(0), \Gamma \rightarrow \Delta, A(t)$ in P' is h, it is shown that

$$\mu_1 = O(A(0), \Gamma \rightarrow \Delta, A(t); P') = \omega(h_1 - h, \mu \# \ldots \# \mu \# 0)$$

Thus $\mu_1 <_\omega \mu_0$. There are no substitutions above an ind, so this implies $\mu_1 <_j \mu_0$ for every j, where j is ∞ or $j \leq \omega$. Therefore by Lemma 27.21, $O(P') <_0 O(P)$.

(3) Because of the reduction in (2), we may now assume that there is no ind, hence no first order free variable, in the end-piece of P. Suppose that there occur axioms of the form $s = t, A(s) \rightarrow A(t)$ in the end-piece of P. Let $s = t, A(s) \rightarrow A(t)$ be one such. Then there are numerals m and n such that m and n are equal to s and t, respectively. Either $m = n \rightarrow$ or $\rightarrow m = n$ is a mathematical, initial sequent.

Case 1. If $m = n \rightarrow$ is an axiom, then replace that axiom by

$$\dfrac{\dfrac{\dfrac{m = n \rightarrow}{\text{weakenings and an exchange}}}{m = n, A(m) \rightarrow A(n)}}{\dfrac{\text{term replacements}}{s = t, A(s) \rightarrow A(t).}}$$

This does not change the ordinal diagram.

Case 2. If $m = n \rightarrow$ is not a mathematical, initial sequent, replace the initial sequent by:

$$\dfrac{\dfrac{\dfrac{A(m) \rightarrow A(n)}{\text{term replacements}}}{A(s) \rightarrow A(t)}}{s = t, A(s) \rightarrow A(t).}$$

(4) By virtue of (3), we may assume that there are no applications of ind and no equality axioms as initial sequents in the end-piece of P. Suppose that the end-piece of P contains logical, initial sequents. Suppose P is of the following form and $D \rightarrow D$ is one of the initial sequents in the end-piece of P:

$$\begin{array}{cc}
\qquad\qquad D \rightarrow D & \\
\vdots \qquad\qquad \vdots & \\
\dfrac{\Gamma \overset{\mu}{\rightarrow} \Delta, \tilde{D} \quad \tilde{D}, \Pi \overset{\nu}{\rightarrow} \Lambda_1, D', \Lambda_2,}{\Gamma, \Pi \rightarrow \Delta, \Lambda_1, \tilde{D}, \Lambda_2}, & \begin{array}{c} l \\ m \end{array} \\
\vdots & \\
\rightarrow &
\end{array}$$

where two $\tilde{D}$'s in the right upper sequent of the cut denote the descendants of the D's occurring in the initial sequent and l is $h(\Gamma \rightarrow \Delta, \tilde{D}; P)$.

We shall consider a proof P' of the following form

$$P_0 \left\{ \begin{array}{c} \vdots \\ \Gamma \overset{\mu}{\rightarrow} \Delta, \tilde{D} \end{array} \right. \qquad m$$
$$\overline{\overline{\Gamma, \Pi \rightarrow \Delta, \Lambda_1, \tilde{D}, \Lambda_2}}, \qquad m$$
$$\vdots$$
$$\rightarrow$$

where every substitution in P' is assigned the same degree as the corresponding one in P. We see easily that for every sequent S in P_0

$$O(S, P') \leq_0 \omega(h(S; P) - h(S; P'), O(S; P)).$$

In particular $\mu' \leq_0 \omega(l - m, \mu)$ and so

$$\mu' \leq_0 \omega(l - m, \mu \# \nu) = O(\Gamma, \Pi \rightarrow \Delta, \Lambda_1, \tilde{D}, \Lambda_2, P').$$

Thus by Lemma 27.22, we have $O(P', d') \leq_0 O(P, d)$.

(5) We may assume besides the condition in (3) that the end-piece of P contains no logical, initial sequent. Let Q be a proof with degree whose end-sequent is not necessarily $\rightarrow$ but which satisfies the same conditions as those required for P. We can define Q^*, obtained from Q by eliminating weakenings in the end-piece of Q, by induction on the number of inferences in the end-piece of Q as for **PA**. We deal with the following case only: If the last inference of Q is a substitution, say

$$\frac{\Gamma \rightarrow \Delta}{\Gamma\binom{\alpha}{V} \rightarrow \Delta\binom{\alpha}{V}},$$

where Γ and Δ are $A_1, \ldots, A_m$ and $B_1, \ldots, B_n$, respectively, and $\Gamma\binom{\alpha}{V}$ and $\Delta\binom{\alpha}{V}$ denote $A_1\binom{\alpha}{V}, \ldots, A_m\binom{\alpha}{V}$ and $B_1\binom{\alpha}{V}, \ldots, B_n\binom{\alpha}{V}$, respectively, and the end-sequent of Q_0^*, where Q_0 is the proof of the upper sequent, is $\Gamma^* \rightarrow \Delta^*$, then Q^* is

$$\frac{\begin{array}{c} Q^* \\ \vdots \\ \Gamma^* \rightarrow \Delta^* \end{array}}{\Gamma^*\binom{\alpha}{V} \rightarrow \Delta^*\binom{\alpha}{V}}.$$

If the end-piece of P contains a weakening, we can reduce P to P^*, where every substitution in P^* has the same degree as the corresponding substitution in P. Similarly in the case (4) we have $O(P^*, d^*) \leq_0 O(P, d)$.

(6) In the following we shall assume that the end-piece of a proof with

degree contains no logical inference, ind, weakening or axioms other than mathematical axioms. Moreover, we may assume that the proof is different from its end-piece, for if the entire proof is its end-piece, then we can eliminate cuts as mentioned at the beginning of (1).

Let P be a proof with degree. We repeat the definition of a suitable cut: A cut in the end-piece of a proof with degree, P, is called suitable if both of its cut formulas have ancestors which are principal formulas of boundary (hence logical) inferences. We can show, exactly the same way as for **PA**, that under those conditions there exists a suitable cut in the end-piece of P.

Now, let P be a proof with degree whose end-sequent is $\rightarrow$ and let J be a suitable cut in P. To define the essential reduction, we must treat separately several cases according to the form of the outermost logical symbol of the cut formulas of J.

(7) The case where the cut formula of J is of the form $\forall\phi\, F(\phi)$.

Case 1. $\forall\phi\, F(\phi)$ is isolated.

Let P be the following form:

$$
\begin{array}{ccr}
\vdots & \vdots & \\
\dfrac{\Gamma_1 \xrightarrow{\lambda} \Delta_1, F_1(\alpha)}{\Gamma_1 \xrightarrow{(\omega,\lambda)} \Delta_1, \forall\phi\, F_1(\phi)} & \dfrac{F_2(V), \Pi_1 \xrightarrow{\mu} \Lambda_1}{\forall\phi\, F_2(\phi), \Pi_1 \xrightarrow{\omega(l_4-l_3+1,\mu)} \Lambda_1} & \begin{array}{r} l_4 \\ l_3 \end{array} \\
\vdots & \vdots & \\
J\ \dfrac{\Gamma_2 \xrightarrow{\tau} \Delta_2, \forall\phi\, F(\phi) \qquad \forall\phi\, F(\phi), \Pi_2 \xrightarrow{\rho} \Lambda_2}{\Gamma_2, \Pi_2 \xrightarrow{\omega(l_2-l_1,\tau\#\rho)} \Delta_2, \Lambda_2} & & \begin{array}{r} l_2 \\ l_1 \end{array} \\
\vdots & & \\
\Gamma_3 \xrightarrow{\nu} \Delta_3 & & 0 \\
\vdots & & \\
\xrightarrow{\sigma} & & 0
\end{array}
$$

where both $\forall\phi\, F_1(\phi)$ and $\forall\phi\, F_2(\phi)$ are $\forall\phi\, F(\phi)$ up to term-replacement. Let i be $d(\forall\phi\, F_1(\phi))$. $\Gamma_3 \rightarrow \Delta_3$ is the i-resolvent of $\Gamma_2, \Pi_2 \rightarrow \Delta_2, \Lambda_2$. Let P' be the following:

$$
\begin{array}{ccr}
\vdots & \vdots & \\
\dfrac{\Gamma_1 \xrightarrow{\lambda} \Delta_1, F_1(\alpha)}{\Gamma_1 \rightarrow F_1(\alpha), \Delta_1, \forall\phi\, F_1(\phi)} & \dfrac{F_2(V), \Pi_1 \xrightarrow{\mu} \Lambda_1}{\forall\phi\, F_2(\phi), \Pi_1 \xrightarrow{\omega(l_4-l_3+1,\mu)} \Lambda_1} & \begin{array}{r} l_4 \\ l_3 \end{array} \\
\vdots & \vdots & \\
\dfrac{\Gamma_2 \xrightarrow{\tau'} F(\alpha), \Delta_2, \forall\phi\, F(\phi) \qquad \forall\phi\, F(\phi), \Pi_2 \xrightarrow{\rho} \Lambda_2}{\Gamma_2, \Pi_2 \xrightarrow{\omega(l_2-l_1,\tau'\#\rho)} F(\alpha), \Delta_2, \Lambda_2} & & \begin{array}{r} l_2 \\ l_1 \end{array}
\end{array}
$$

$$
\begin{array}{lll}
J_1 \quad \dfrac{\Gamma_3 \overset{\theta}{\rightarrow} \Delta_3, F(\alpha)}{\Gamma_3 \overset{(i,\theta)}{\rightarrow} \Delta_3, F(V)} \quad 0 & F(V), \Pi_1 \overset{\mu}{\rightarrow} \Lambda_1 & l_4 \\
\hline
\qquad \Gamma_3, \Pi_1 \overset{\omega(l_4 - l_3, (i,\theta) \# \mu)}{\rightarrow} \Delta_3, \Lambda_1 & & l_3 \\
\hline\hline
\qquad \forall\phi\, F_2(\phi), \Pi_1, \Gamma_3 \rightarrow \Delta_3, \Lambda_1 & &
\end{array}
$$

$$
\begin{array}{lll}
\Gamma_2 \overset{\tau}{\rightarrow} \Delta_2, \forall\phi\, F(\phi) & \forall\phi\, F(\phi), \Pi_2, \Gamma_3 \overset{\rho'}{\rightarrow} \Delta_3, \Lambda_2 & l_2 \\
\hline
\Gamma_2, \Pi_2, \Gamma_3 \overset{\omega(l_2 - l_1, \tau \# \rho')}{\rightarrow} \Delta_2, \Delta_3, \Lambda_2 & & l_1 \\
\hline\hline
\Gamma_2, \Pi_2, \Gamma_3 \rightarrow \Delta_3, \Delta_2, \Lambda_2 & & \\
\vdots & & \\
\Gamma_3, \Gamma_3 \overset{\nu'}{\rightarrow} \Delta_3, \Delta_3 & & \\
\hline\hline
\Gamma_3 \rightarrow \Delta_3 & & 0 \\
\vdots & & \\
\overset{\sigma'}{\rightarrow} & & 0
\end{array}
$$

where J_1 is a substitution with eigenvariable α. We define $d'(J')$ for a substitution J' except J_1 to be $d(J'')$, where J'' is the substitution corresponding to J' in P. We define $d'(J_1)$ to be i.

The following propositions (7.1)–(7.5) are easily verified by

LEMMA 27.21.

(7.1) $\tau' \preccurlyeq_0 \tau$,

(7.2) $\theta \preccurlyeq_0 \nu$,

(7.3) $\omega(l_4 - l_3, (i, \theta) \# \mu) \preccurlyeq_{i+1} \omega(l_4 - l_3 + 1, \mu)$,

(7.4) $\rho' \preccurlyeq_{i+1} \rho$,

(7.5) *$\nu' <_j \nu$ for all $j \leqslant \omega$, and for each k such that $k < i$ and for each k-section π' of ν', there exists a k-section π of ν for which $\pi' \preccurlyeq_k \pi$, and for each i-section η of ν', $\eta <_i \nu$. (Here note that ν and ν' are connected.)*

It follows from (7.5) that $\sigma' <_0 \sigma$.

Case 2. $\forall\phi\, F(\phi)$ is not isolated.

Let P be the following form

$$
\begin{array}{llll}
& \dfrac{\Gamma_1 \overset{\lambda}{\rightarrow} \Delta_1, F_1(\alpha)}{\Gamma_1 \overset{(\omega,\lambda)}{\rightarrow} \Delta_1, \forall\phi\, F_1(\phi)} & \dfrac{F_2(\beta), \Pi_1 \overset{\mu}{\rightarrow} \Lambda_1}{\forall\phi\, F_2(\phi), \Pi_1 \overset{(\omega,\mu)}{\rightarrow} \Lambda_1} & \begin{array}{l} m \\ m \end{array} \\
& \vdots & \vdots & \\
J & \Gamma_2 \rightarrow \Delta_2, \forall\phi\, F(\phi) & \forall\phi\, F(\phi), \Pi_2 \rightarrow \Lambda_2 & l \\
\hline
& \multicolumn{2}{c}{\Gamma_2, \Pi_2 \rightarrow \Delta_2, \Lambda_2} & \\
& \multicolumn{2}{c}{\vdots} & \\
& \multicolumn{2}{c}{\Phi \rightarrow \Psi} & n \\
& \multicolumn{2}{c}{\vdots} & \\
& \multicolumn{2}{c}{\rightarrow} &
\end{array}
$$

where $\Phi \to \Psi$ denotes the uppermost sequent below J whose height is less than l. Let P' be the following:

$$\dfrac{\vdots}{\dfrac{\Gamma_1 \xrightarrow{\lambda} \Delta_1, F_1(\beta)}{\Gamma_1 \to F_1(\beta), \Delta_1, \forall\phi\, F_1(\phi)}} \qquad\qquad \dfrac{\vdots}{\dfrac{F_2(\beta), \Pi_1 \xrightarrow{\mu} \Lambda_1}{\forall\phi\, F_2(\phi), \Pi_1, F_2(\beta) \to \Lambda_1}} \quad m$$

$$\dfrac{\Gamma_2 \to F(\beta), \Delta_2, \forall\phi\, F(\phi) \quad \forall\phi\, F(\phi), \Pi_2 \to \Lambda_2}{\Gamma_2, \Pi_2 \to F(\beta), \Delta_2, \Lambda_2} \quad \dfrac{\Gamma_2 \to \Delta_2, \forall\phi\, F(\phi) \quad \forall\phi\, F(\phi), \Pi_2, F(\beta) \to \Lambda_2}{\Gamma_2, \Pi_2, F(\beta) \to \Delta_2, \Lambda_2} \quad l$$

$$\dfrac{\dfrac{\dfrac{\vdots}{\Phi \to F(\beta), \Psi}}{\Phi \to \Psi, F(\beta)} \qquad \dfrac{\dfrac{\vdots}{\Phi, F(\beta) \to \Psi}}{F(\beta), \Phi \to \Psi}}{\dfrac{\Phi, \Phi \to \Psi, \Psi}{\dfrac{\Phi \to \Psi}{\dfrac{\vdots}{\to}}}} \quad \begin{matrix} l' \\ n \end{matrix}$$

And for every substitution J' in P', $d'(J')$ is defined to be $d(J'')$, where J'' is the substitution corresponding to J' in P. From $l' < l$ we see easily that $O(P', d') <_0 O(P, d)$.

(8) The cases where the cut formula of J is of the form $F_1 \wedge F_2$, $\neg F$ or $\forall x\, F(x)$ are treated in the same way as the Case 2 in (7).

This completes the proof of Theorem 27.17.

It is also possible to prove Theorem 27.5 by using $O(\omega+1, 1)$. We chose the proof of Theorem 27.5 because it is a kind of preparation for the work in §28.

Now we state another formulation of **ISN**.

DEFINITION 27.23. By *bar induction* denoted by **BI** we mean the following inference-scheme.

$$\frac{F(V), \Gamma \to \Delta}{\forall\varphi\, F(\varphi), \Gamma \to \Delta},$$

where $\forall\varphi\, F(\varphi)$ is a Π^1_1-formula.

Then the following obvious proposition provides us with another formulation of **ISN**.

PROPOSITION 27.24. **ISN** *is equivalent to the second order number theory with the full inductions, Π^1_1-comprehension axioms, and bar inductions.*

Now we shall consider a subsystem of **ISL** which is denoted by $\mathbf{ISL}^0$.

DEFINITION 27.25. $\mathbf{ISL}^0$ is obtained from $\mathbf{G^1LC}$ by restricting the second order $\forall$: left to only those

$$\frac{F(V), \Gamma \rightarrow \Delta}{\forall \varphi\, F(\varphi), \Gamma \rightarrow \Delta},$$

where $\forall \varphi\, F(\varphi)$ is isolated.

It is obvious that **ISL**^{0} is a subsystem of **ISL**. The author proved the cut elimination theorem for **ISL** by using $\forall n\,(O(n, \omega)$ is accessible). Then W. Pohlers proved the cut elimination theorem for a similar system by using $\theta_{\Omega_\omega O}$ which is equivalent to $O(\omega, 1)$. We remark that the order type of $O(\omega, 1)$ is the supremum of the order type of $O(n, \omega)$. Here we prove the theorem by using $\forall n\,(O(n, 1)$ is accessible) namely the accessibility of $O(\omega, 1)$.

THEOREM 27.26. *The cuts in the proof of* **ISL**^{0} *can be eliminated.*

This theorem is weaker than the usual cut elimination theorem in the sense that the cut-free proof obtained by this theorem is not guaranteed to be a proof in **ISL**^{0} (but of course in **G**^{1}**LC**). The proof is also due to T. Arai. Since the proof is a modification of a proof of Theorem 27.17, we shall only give an outline.

DEFINITION 27.27. Let P be a proof in **ISL**^{0}. We call P together with d a proof with degree in **ISL**^{0} if P and d satisfy the following conditions:

(1) Every substitution is in the end-piece.

(2) $d(A) = 0$ if A is explicit.

Let A be implicit.

(3) $d(A) = 0$ if A does not contain any logical symbol.

(4) $d(A) = d(B)$ if A is of the form $\neg B$.

(5) $d(A) = \max\{d(B), d(C)\}$, if A is of the form $B \wedge C$.

(6) $d(A) = d(G(x))$, if A is of the form $\forall x\, G(x)$.

(7) If A is of the form $\forall \varphi\, F(\varphi)$,

$$d(A) = \max\{d(F(\varphi)) + 1, d(J), d(A_0)\},$$

where J ranges over all substitutions which disturb $\forall \varphi\, F(\varphi)$, and A_0 ranges over all principal formulas of the implicit inference $\forall$ right on a second order variable which disturb $\forall \varphi\, F(\varphi)$.

(8) For every implicit formula B in the upper sequent of a substitution J,

$$d(B) < d(J).$$

DEFINITION 27.28. A proof P with degree in **ISL**^{0} is said to be of order n ($n > 0$) if $d(A) < n$ and $d(J) < n$ for every formula A and for every substitution J in P.

DEFINITION 27.29. Let P be a proof of order n. The height $h(S)$ of a sequent in P is defined as follows.

1. If S is the end sequent of P or the upper sequent of a substitution, then $h(S) = 0$.
2. Let I be an inference in P. If S and S' is the upper sequent and the lower sequent of I respectively, and if S is not a cut, an implicit $\forall$ left on a second order variable, or a substitution, then $h(S) = h(S')$.
3. Let I be a cut or an implicit, $\forall$ left on a second order in P and S and S' be the upper or the lower sequent of I respectively. Then $h(S) = \max\{h(S'), g_0(D)\}$, where D is a cut formula, if I is a cut, and D is an auxiliary formula if I is an implicit $\forall$ left on a second order variable. As before $g_0(D)$ is the number of occurrences of logical symbols in D.

DEFINITION 27.30. For a natural number k and $\mu \in O(n+1, 1)$, we define $n(k, \mu) \in O(n+1, 1)$ by the equations $n(0, \mu) = \mu$ and $n(k+1, \mu) = (n, n(k, \mu))$.

DEFINITION 27.31. Let P be a proof with degree of order n. To every sequent S and every inference in P we assign ordinal diagrams $O(S), O(J) \in O(n+1, 1)$ as follows.

1. $O(S) = O$ for an initial sequent S.
2. Let $[S' \quad (S'')]/SJ$ be an inference where $O(S')$ (and possible $O(S'')$) has (have) been defined. Then $O(J)$ and $O(S)$ are defined as follows.
   - 2.1. If J is a weakening, a contraction, or an exchange, then $O(J) = O(S')$.
   - 2.2. $O(J) = O(S') \# 0$ if J is a logical inference except $\wedge$ right and implicit $\forall$ left on a second order variable.
   - 2.3. $O(J) = O(S') \# O(S'')$ if J is $\wedge$ right or a cut.
   - 2.4. $O(J) = (n, 0) \# O(S')$ if J is an implicit $\forall$ left on a second order variable.
   - 2.5. $O(J) = (n, O(S'))$ if J is a substitution.
   - 2.6. $O(S) = n(h(S') - h(S), O(J))$ if J is not a substitution.
   - 2.7. $O(S) = (d(J), O(J))$ if J is a substitution. The ordinal diagram $O(P)$ of a proof P is defined by the equation $O(P) = (n, O(S))$, where S is the end-sequent of P.

With these definitions and the following remark, the proof of Theorem 27.26 goes in the same way as in the proof of Theorem 27.17.

DEFINITION 27.32. **ISN**^{0} is obtained from **ISN** by

(1) restricting mathematical induction only to those of the form

$$\frac{A(a), \Gamma \to \Delta, A(a')}{A(0), \Gamma \to \Delta, A(t)},$$

where $A(a)$ is isolated and

(2) replacing (1) and (2) in Definition 27.16 by the following: The abstract V in $F(V)$ is isolated (F is arbitrary). $\mathbf{ISN}^0$ is sometimes called Π^1_1-CA_0 since it is a second order arithmetic with Π^1_1-CA and a restricted mathematical induction.

The theory of relativization and Theorem 27.26 immediately imply the following

THEOREM 27.33. $\mathbf{ISN}^0$ *is consistent.*

H. Friedman proved a generalization of Kruskal's theorem in §12 and also proved that this generalized Kruskal theorem is not provable in $\mathbf{ISN}^0$. We shall discuss his result.

DEFINITION 27.34. (1) A finite tree T (in Definition 12.69) together with a function $l: T \to n$ is called a finite tree with levels less than n, where n is considered as a set $\{0, 1, 2, \ldots, n-1\}$. The set of all finite tree (T, l) with levels less than n is denoted by $\mathcal{T}_n$.

(2) Let (T_1, l_1) and (T_2, l_2) be in $\mathcal{T}_n$. A function $f: T_1 \to T_2$ is said to be an embedding if f is an embedding in the sense of finite tree in Definition 12.69 and satisfies the following conditions:

(a) for every $b \in T_1$, $l_1(b) = l_2(f(b))$ and

(b) if $a, b \in T_1$ and b is an immediate successor of a in T_1 and $c \in T_2$ satisfies $f(a) < c < f(b)$, then $l_2(c) \geq l_2(f(b))$.

The condition (b) is called the gap condition.

(3) Let (T_1, l_1) and (T_2, l_2) be in $\mathcal{T}_n$. $(T_1, l_1) \leq (T_2, l_2)$ iff there exists an embedding of (T_1, l_1) into (T_2, l_2).

We state Friedman's generalized Kruskal theorem without proof.

THEOREM 27.35. *Let n be a positive natural number and $\langle (T_i, l_i) \mid i < \omega \rangle$ be a sequence from $\mathcal{T}_n$. Then there exist $i < j$ such that $(T_i, l_i) \leq (T_j, l_j)$.*

We denote Theorem 27.34 by **GKT**. We shall prove the following

THEOREM 27.36 (H. Friedman). **GKT** *is not provable in* $\mathbf{ISN}^0$.

PROOF. It suffices to show that the accessibility of $O(n, 1)$ follows from **GKT**. Let α be a connected ordinal diagram in $O(n, 1)$. We assign to α a tree $T(\alpha)$ in $\mathcal{T}_n$. If α is 0, then $T(\alpha)$ is a single point with level 0. If α is $(k, \alpha_1 \# \ldots \# \alpha_i)$, then $T(\alpha)$ is the tree

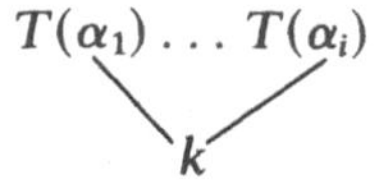

where k is the level of the point. It is easily seen that if $T(\alpha) \leq T(\beta)$, then $\alpha \leq_0 \beta$. Therefore **GKT** implies the accessibility of all connected ordinal diagrams in $O(n, 1)$. Hence **GKT** implies the accessibility of $O(n, 1)$.

Kruskal's theorem and Friedman's theorem can be considered as theorems on well-quasi-ordering. We shall develop the theory of ordinal diagrams as a theory on well-quasi-ordering.

M. Okada and the author developed the theory of quasi-ordinal diagrams as an extension of the theory of ordinal diagrams. We shall discuss this theory until the end of this chapter.

DEFINITION 27.37. (1) A structure $\langle D, \leq \rangle$ is said to be a quasi-ordered structure if it satisfies $\forall x \in D(x \leq x)$ and $\forall x, y, z \in D(x \leq y \wedge y \leq z \rightarrow x \leq z)$. In this case, $\leq$ is said to be a quasi-ordering on D.

(2) A quasi-ordering $\leq$ on D is said to be a well-quasi-ordering (abbreviated by wqo) if for every sequence $d_0, d_1, d_2, \ldots$, from D, there exist $i < j$ such that $d_i \leq d_j$.

(3) A quasi-ordering $\leq$ on D is said to be a weakly well-quasi-ordering (abbreviated by wwqo) if for every sequence $d_0 \geq d_1 \geq d_2 \geq \ldots$, from D, there exist $i < j$ such that $d_i \leq d_j$.

DEFINITION 27.38. Let S be quasi-ordered by $\leq$ and I be a well-ordered set whose ordering is also denoted by $\leq$. The system $Q(I, S)$ of quasi-ordinal diagrams (abbreviated by qod) is defined as follows.

(1) If $s \in S$, then s is a connected qod.

(2) If $i \in I$ and $\alpha \in Q(I, S)$, then (i, α) is a connected qod.

(3) If $\alpha_1, \ldots, \alpha_n$ $(n \geq 2)$ are connected qod, then $\alpha_1 \# \alpha_2 \# \ldots \# \alpha_n$ is a nonconnected qod.

A connected qod can be considered as a finite tree T together with level $l: T \rightarrow I \cup S$ such that l maps the endknots into S and the inner knots into I. More precisely if (T, l) is of the form

where i is the level of the root, and if $\alpha_1, \ldots, \alpha_n$ be qod corresponding to $T_1, \ldots, T_n$ respectively, then $(i, \alpha_1 \# \ldots \# \alpha_n)$ is the qod corresponding to (T, l).

We define the notions j-active, i-section, $j_0(j, \alpha, \ldots, \gamma)$, and $\tilde{I}$ in the same way as before.

DEFINITION 27.39. (I) Let α and β be qod's and $i \in \tilde{I}$. We define $\alpha \leq_i \beta$ as follows.

(1) If $s, t \in S$, then $s \leq_i t$ is equivalent to $s \leq t$.

(2) If $\alpha_1, \ldots, \alpha_n, \beta_1, \ldots, \beta_m$ be connected qod, then $\alpha_1 \# \ldots \# \alpha_n \leqslant_i \beta_1 \# \ldots \# \beta_m$ iff

(i) there exists $k \leqslant m$ such that $\alpha_1 \leqslant_i \beta_k$ and (if $n \geqslant 2$) $\alpha_2 \# \ldots \# \alpha_n \leqslant_i \beta_1 \# \ldots \# \beta_{k-1} \# \beta_{k+1} \# \ldots \# \beta_m$, or

(ii) there exists $k \leqslant m$ such that $\alpha_l \leqslant_i \beta_k$ and $\beta_k \not\leqslant \alpha_l$ for all $l \leqslant n$.

(3) If α and β are connected, if $i \neq \infty$ and if $j = j_0(i, \alpha, \beta)$ then $\alpha \leqslant_i \beta$ if one of the following holds.

(i) There exists an i-section σ of β such that $\alpha \leqslant_i \sigma$.

(ii) $\alpha \leqslant_j \beta$ and for every i-section σ of α, $\sigma \leqslant_i \beta$ and $\beta \not\leqslant_i \sigma$, where $j = j_0(i, \alpha, \beta)$.

(4) Let $\alpha = (j, \beta)$ and $s \in S$. Then $s \leqslant_\infty \alpha$ holds but $\alpha \leqslant_\infty s$ never holds.

(5) Let $\alpha = (j, \gamma)$ and $\beta = (k, \delta)$, then $\alpha \leqslant_\infty \beta$ if

(i) $j < k$ in $\tilde{I}$, or

(ii) $j = k$ and $\gamma \leqslant_j \delta$.

(II) Let α and β be qod's and $i \in \tilde{I}$. $\alpha \leqslant'_i \beta$ is the same as $\alpha \leqslant_i \beta$ except for the following changes:

In (2) we erase the condition (ii). In (3) the condition (ii) is replaced by (ii)′.

(ii)′ $\alpha \leqslant_j \beta$ and for every i-section σ of α, $\beta \not\leqslant_i \sigma$, where $j = j_0(i, \alpha, \beta)$.
We add (iii) in (3).

(iii) $\exists \gamma \alpha \leqslant_i \gamma \leqslant_i \beta$.

Recall that $\iota(i, \alpha, \beta, \ldots, \gamma)$ is the number of elements of I which are greater than i and occur α or β or ... or γ.

LEMMA 27.40. *For every $i \in \tilde{I}$, $\leqslant_i$ is a quasi-ordering on $Q(I, S)$.*

PROOF. Let $l(\alpha)$ be the total number of ( )'s and #'s in α. Then we prove the following proposition (∗) by induction on

$$\omega \cdot (l(\alpha) + l(\beta) + l(\gamma)) + \iota(i, \alpha, \beta, \gamma),$$

$$(*) \qquad \text{if } \alpha \leqslant_i \beta \text{ and } \beta \leqslant_i \gamma \text{ then } \alpha \leqslant_i \gamma.$$

Since other cases are trivial we only consider the case when α, β and γ are connected and $i \in I$.

Case (1). For some i-section σ of γ, $\beta \leqslant_i \sigma$. Then $\alpha \leqslant_i \beta \leqslant_i \sigma$. By induction hypothesis $\alpha \leqslant_i \sigma$. Therefore $\alpha \leqslant_i \gamma$.

Case (2). For all i-section δ of β, $\delta <_i \gamma$, i.e., $\delta \leqslant_i \gamma$ and $\gamma \not\leqslant_i \delta$, and $\beta \leqslant_{j'} \gamma$, where $j' = j_0(i, \beta, \gamma)$.

Subcase (2.1). For all i-section σ of α, $\sigma <_i \beta$, i.e., $\sigma \leqslant_i \beta$ and $\beta \not\leqslant_i \sigma$, and $\alpha \leqslant_{j''} \beta$, where $j'' = j_0(i, \alpha, \beta)$.

From $\sigma \leqslant_i \beta \leqslant_i \gamma$ and induction hypothesis $\sigma \leqslant_i \gamma$. On the other hand if $\sigma\, {}_i\!\geqslant \gamma$ then $\sigma\, {}_i\!\geqslant \gamma\, {}_i\!\geqslant \beta$, hence by induction hypothesis $\beta \leqslant_i \sigma$, which is a contradiction. Therefore $\sigma\, {}_i\!\not\geqslant \gamma$. On the other hand, from $\alpha \leqslant_{j''} \beta$ and $\beta \leqslant_{j'} \gamma$ we obtain $\alpha \leqslant_j \gamma$, where $j = j_0(i, \alpha, \beta, \gamma)$. Therefore $\alpha \leqslant_i \gamma$ holds.

Subcase (2.2). For some i-section δ of β, $\alpha \leqslant_i \delta$. Then $\alpha \leqslant_i \delta <_i \gamma$. Hence by induction hypothesis $\alpha \leqslant_i \gamma$.

DEFINITION 27.41. (1) If $\alpha = \beta_1 \# \dots \# \beta_n$ and $\beta_1, \dots, \beta_n$ are connected qod, then each β_i is called a component of α.

(2) Let $\{\alpha_n\}$ be a sequence of qod. Then a sequence $\{\beta_n\}$ of qod is said to be a component sequence of $\{\alpha_n\}$ if for each n, β_n is a component of α_n. A subsequence of a component sequence is called a component subsequence of $\{\alpha_n\}$.

(3) Let $\leqslant$ be a quasi-ordering on D. A sequence $d_0, d_1, d_2, \dots$, from D, is said to be bad with respect to $\leqslant$ if $d_0 \geqslant d_1 \geqslant d_2 \geqslant \dots$ and for every $i < j$, $d_i \nleqslant d_j$.

LEMMA 27.42. *Let $\{\alpha_n\}$ be a sequence of* qod. *If there are no bad component subsequence of $\{\alpha_n\}$ with respect to $\leqslant_i$, then $\{\alpha_n\}$ is not a bad sequence with respect to $\leqslant_i$.*

PROOF. In this proof, we call a bad sequence very bad if none of its component subsequences are bad. Suppose there exist a very bad sequence. Consider the first members of all very bad sequences. Let α_1 be a minimum among them in the sense that the number of component of α_1 is the smallest. Now consider the second member of all very bad sequences starting with α_1. Let α_2 be a minimum among them. Choose a very bad sequence $\alpha_1 \, {}_i\!\geqslant \alpha_2 \, {}_i\!\geqslant \alpha_3 \, {}_i\!\geqslant \dots$ in this way. We consider a suitable permutation of components of α_k such that $\alpha^1_k \, {}_i\!\geqslant \alpha^1_{k+1}$ for any $k \in \omega$, where $\alpha_k \equiv \alpha^1_k \# \alpha^2_k$ and α^1_k is a connected qod. By our hypothesis there are only finitely many empty α^2_k. Since any subsequence of $\{\alpha^1_k\}$ is not bad, by Ramsey's theorem there exists a subsequence $\{\alpha^1_{l_j}\}$ of $\{\alpha^1_n\}$ such that for every $j < k$, $\alpha^1_{l_j} \leqslant_i \alpha^1_{l_k}$. Now consider $\alpha^2_{l_1} \, {}_i\!\geqslant \alpha^2_{l_2} \, {}_i\!\geqslant \dots$. If this is a bad sequence, then $\alpha_1 \, {}_i\!\geqslant \alpha_2 \, {}_i\!\geqslant \dots {}_i\!\geqslant \alpha_{l_1 - 1} \, {}_i\!\geqslant \alpha^2_{l_1} \, {}_i\!\geqslant \alpha^2_{l_2} \, {}_i\!\geqslant \dots$ is a bad sequence. This contradicts to the choice of α_{l_1}. Therefore there exists $k_1 < k_2$ such that $\alpha^2_{l_{k_1}} \leqslant_i \alpha^2_{l_{k_2}}$. Then $\alpha_{l_{k_1}} \leqslant_i \alpha_{l_{k_2}}$ which is a contradiction.

DEFINITION 27.43. (1) If α and β are elements of S, then $\alpha = \beta$ iff $\alpha \leqslant \beta$ and $\beta \leqslant \alpha$ in S.

(2) If $\alpha \in S$ and $\beta \notin S$, then $\alpha = \beta$ does not hold.

(3) If $\alpha \equiv (i, \alpha_1)$ and $\beta \equiv (j, \beta_1)$ then $\alpha = \beta$ iff $i = j$ in I and $\alpha_1 = \beta_1$.

(4) If $\alpha \equiv \alpha_1 \# \dots \# \alpha_n$ and $\beta \equiv \beta_1 \# \dots \# \beta_m$, where $n + m > 2$ and each α_i or β_j is connected, then $\alpha = \beta$ iff $n = m$ and there is a permutation of $\{1, 2, \dots, m\}$, say $\{j_1, \dots, j_m\}$, such that

$$\alpha_1 = \beta_{j_1}, \alpha_2 = \beta_{j_2}, \dots, \alpha_m = \beta_{j_m}.$$

LEMMA 27.44. *For $\alpha, \beta \in Q(I; S)$, and for any $i \in \tilde{I}$, if $\alpha \leqslant_i \beta$ and $\beta \leqslant_i \alpha$, then $\alpha = \beta$.*

PROOF. By induction on $\omega \cdot l(\alpha, \beta) + \iota(i, \alpha, \beta)$.

THEOREM 27.45 (Okada–Takeuti). *For any well-ordered set I and for any weakly-well-quasi-ordered set S, $Q(I, S)$ is weakly well-quasi-ordered by $\leqslant_0$.*

PROOF. Let $\alpha_0, \alpha_1, \alpha_2, \ldots$ be a bad sequence with respect to $\leqslant_0$. Consider any component of 0-section α' of α_0. If there exists a bad sequence starting with α', say $\alpha'_0, \alpha'_1, \alpha'_2, \ldots$, then replace $\alpha_0, \alpha_1, \alpha_2, \ldots$ by $\alpha'_0, \alpha'_1, \alpha'_2, \ldots$. Repeating this method, we obtain a sequence $\alpha'_0, \alpha'_1, \ldots$ such that α'_0 is a subqod of α_0 and there are no bad sequences starting with any of 0-section of α'_0. Now consider any 0-section α'' of α'_1. If there exists a bad sequence starting with α'_0, α'', say $\alpha'_0, \alpha'', \alpha''_2, \ldots$, then replace $\alpha'_0, \alpha'_1, \alpha'_2, \ldots$ by $\alpha'_0, \alpha'', \alpha''_2, \ldots$. Repeating this method, we attain a sequence $\alpha^1_0, \alpha^1_1, \alpha^1_2, \ldots$ satisfying the following conditions:

(1) If $\alpha_0 \equiv \alpha^1_0, \ldots, \alpha_n \equiv \alpha^1_n$, and $\alpha_{n+1} \not\equiv \alpha^1_{n+1}$, then α^1_{n+1} is a subqod of α_{n+1}.

(2) $\alpha^1_0, \alpha^1_1, \alpha^1_2, \ldots$ is a connected bad sequence with respect to $\leqslant_0$.

(3) For every 0-section α''_n of α^1_n, there are no bad sequences starting with $\alpha^1_0, \ldots, \alpha^1_{n-1}, \alpha''_n$.

We claim that $\alpha^1_0, \alpha^1_1, \alpha^1_2, \ldots$ is also a bad sequence with respect to $\leqslant_1$. First we show that $\alpha^1_0 \,{}_1\!\geqslant \alpha^1_1 \,{}_1\!\geqslant \alpha^1_2 \,{}_1\!\geqslant \ldots$. If there exists a 1-section δ of α^1_i such that $\alpha^1_{i+1} \leqslant_0 \delta$, then $\alpha^1_0, \alpha^1_1, \ldots, \alpha^1_{i-1}, \delta, \alpha^1_{i+1}, \ldots$ is bad, or there exists j such that $\delta \leqslant_0 \alpha^1_{i+j}$. In the latter case we can see that $\delta = \alpha^1_{i+j}$ and that $\alpha^1_0, \alpha^1_1, \ldots, \alpha^1_{i-1}, \delta, \alpha^1_{i+j+1}, \ldots$ is bad. Either case is contradictory with the minimality of α^1_i. So there is no 1-section δ of α^1_i such that $\alpha^1_{i+1} \leqslant_0 \delta$. From $\alpha^1_i \,{}_0\!\geqslant \alpha^1_{i+1}$, we obtain $\alpha^1_i \,{}_1\!\geqslant \alpha^1_{i+1}$ by the definition of $\leqslant_0$. Next we show that there are no $i<j$ such that $\alpha^1_i \leqslant_1 \alpha^1_j$. Otherwise $\alpha^1_i = \alpha^1_j$, so $\alpha^1_i \leqslant_0 \alpha^1_j$. This is a contradiction to the badness of $\{\alpha^1_i\}$ with respect to $\leqslant_0$.

Now define $F_1 = \{\alpha \in Q(I, S) \mid$ every 0-section of α is a 0-section of some $\alpha^1_n\}$. Then $\alpha^1_0, \alpha^1_1, \alpha^1_2, \ldots$ is a bad sequence from F_1 with respect to $\leqslant_1$. Replacing $\alpha_0, \alpha_1, \alpha_2, \ldots$ and $\leqslant_0$ by $\alpha^1_0, \alpha^1_1, \alpha^1_2, \ldots$ and $\leqslant_1$, we obtain a sequence $\alpha^2_0, \alpha^2_1, \alpha^2_2, \ldots$ from F_1 satisfying the following conditions:

(1) If $\alpha^1_0 = \alpha^2_0, \ldots, \alpha^1_n = \alpha^2_n$, and $\alpha^1_{n+1} \neq \alpha^2_{n+1}$, then α^2_{n+1} is a subqod of α^1_{n+1}.

(2) $\alpha^2_0, \alpha^2_1, \alpha^2_2, \ldots$ is a connected bad sequence from F_1 with respect to $\leqslant_1$.

(3) For every 1-section β of α^2_n, there are no bad sequences starting with $\alpha^2_0, \ldots, \alpha^2_{n-1}, \beta$ from F_1 with respect to $\leqslant_1$.

In the same way as in the case of $\alpha^1_0, \alpha^1_1, \alpha^1_2, \ldots$ we can show that $\alpha^2_0, \alpha^2_1, \alpha^2_2, \ldots$ is a bad sequence with respect to $\leqslant_2$. We define $F_2 = \{\alpha \in F_1 \mid$ every 1-section of α is a 1-section of some $\alpha^2_i\}$. Repeating this method, we attain $\alpha^i_0, \alpha^i_1, \alpha^i_2, \ldots$ for every finite i satisfying the following properties:

(1) $\alpha^i_0, \alpha^i_1, \ldots$ is a bad sequence from F_i with respect to $\leqslant_i$.

(2) If $\alpha_0^i = \alpha_0^{i+1}, \ldots, \alpha_n^i = \alpha_n^{i+1}$, and $\alpha_{n+1}^i \neq \alpha_{n+1}^{i+1}$, then α_{n+1}^{i+1} is a subqod of α_{n+1}^i.

(3) $\alpha_0^{i+1}, \alpha_1^{i+1}, \alpha_2^{i+1}$ is a connected bad sequence with respect to $\leqslant_i$.

(4) For every i-section β of α_n^{i+1}, there are no bad sequences starting with $\alpha_0^{i+1}, \ldots, \alpha_{n-1}^{i+1}, \beta$ from F_i with respect to $\leqslant_i$.

(5) $F_{i+1} = \{\alpha \in F_i \mid$ every i-section of α is an i-section of some of $\alpha_n^{i+1}\}$.

Let ω be the first limit element of I. By the condition (2) above, there exists a sequence $\beta_0, \beta_1, \beta_2, \ldots$ from $\bigcap_{i<\omega} F_i$ such that

$$(*) \qquad \forall p\, \exists i\, \forall k\, (i \leqslant k < \omega \to \alpha_p^k \equiv \beta_p).$$

We claim that $\beta_0, \beta_1, \beta_2, \ldots$ is a bad sequence with respect to $\leqslant_\omega$. First we show that $\beta_0\, {}_\omega\!\geqslant \beta_1\, {}_\omega\!\geqslant \ldots$. By (*) for any p we can have $k < \omega$ such that $\beta_p \equiv \alpha_p^{k+1}\, {}_k\!\geqslant \alpha_{p+1}^{k+1} \equiv \beta_{p+1}$ and that if $k < h < \omega$ then h does not occur in β_p nor β_{p+1}. So $\beta_p \equiv \alpha_p^{k+1}\, {}_\omega\!\geqslant \alpha_{p+1}^{k+1} \equiv \beta_{p+1}$. Next we show that there are no $i < j$ such that $\beta_i \leqslant_\omega \beta_j$. Otherwise $\beta_i = \beta_j$, which implies $\alpha_i^{k+1} \leqslant_k \alpha_j^{k+1}$. This is a contradiction to the badness of $\{\alpha_i^{k+1}\}$ with respect to $\leqslant_k$.

Repeating this method, we can obtain $\alpha_0^i, \alpha_1^i, \alpha_2^i, \ldots$ for every $i \in \tilde{I}$ satisfying the following properties:

(1) $\alpha_0^i, \alpha_1^i, \alpha_2^i, \ldots$ is a bad sequence from F_i with respect to $\leqslant_i$.

(2) If $\alpha_0^i \equiv \alpha_0^{i+1}, \ldots, \alpha_n^i \equiv \alpha_n^{i+1}$, and $\alpha_{n+1}^i \not\equiv \alpha_{n+1}^{i+1}$, then α_{n+1}^{i+1} is a subqod of α_{n+1}^i.

(3) $\alpha_0^{i+1}, \alpha_1^{i+1}, \alpha_2^{i+1}, \ldots$ is a connected bad sequence with respect to $\leqslant_i$.

(4) For every i-section β of α_n^{i+1}, there are no bad sequences starting with $\alpha_0^{i+1}, \ldots, \alpha_{n-1}^{i+1}, \beta$ from F_i with respect to $\leqslant_i$.

(5) $F_{i+1} = \{\alpha \in F_i \mid$ every i-section of α is an i-section of some of $\alpha_n^{i+1}\}$. For a limit i in $\tilde{I}$, $F_i = \bigcap_{j<i} F_j$. Here ∞ is considered $i+1$, if i is the maximum element of I. Otherwise ∞ is considered the limit of I.

(6) If i is a limit element of $\tilde{I}$, then

$$\forall p\, \exists j < i\, \forall l\, (j \leqslant l \leqslant i \to \alpha_p^i = \alpha_p^l).$$

Now consider $\alpha_0^\infty, \alpha_1^\infty, \ldots$. Without loss of generality, we assume that all α_j^∞ are connected and $\alpha_j^\infty \equiv (i, \gamma_j)$ for the same i. Then we claim that $\gamma_0, \gamma_1, \ldots$ is a bad sequence with respect to $\leqslant_i$ and each γ_j is an i-section of some of α_n^{i+1}. Therefore each γ_j belongs to F_i. We may assume that γ_j is an i-section of $\alpha_{m_j}^{i+1}$ and $m_{j_1} < m_{j_2}$ for $j_1 < j_2$. Then $\alpha_0^{i+1}, \ldots, \alpha_{m_0-1}^{i+1}, \gamma_0, \gamma_1, \gamma_2, \ldots$ is a bad sequence from F_i with respect to $\leqslant_i$. We only have to show $\alpha_{m_0-1}^{i+1}\, {}_i\!\geqslant \gamma_0$ and $\alpha_l^{i+1} \not\leqslant_i \gamma_n$ for any $l < m_0$ and for any n. $\alpha_{m_0-1}^{i+1}\, {}_i\!\geqslant \gamma_0$ follows from the fact that γ_0 is an i-section of $\alpha_{m_0}^{i+1}$. Since γ_n is an i-section of some α_p^{i+1}, if $\alpha_l^{i+1} \leqslant_i \gamma_n$ then $\alpha_l^{i+1} \leqslant_i \alpha_p^{i+1}$ which is a contradiction. So $\alpha_0^{i+1}, \ldots, \alpha_{m_0-1}^{i+1}, \gamma_0, \gamma_1, \ldots$ is bad, which contradicts with the condition (4) above.

THEOREM 27.46 (Okada–Takeuti). *For every* $i \in \tilde{I}$, $Q(I, S)$ *is weakly well-quasi-ordered by* $\leqslant_i$.

PROOF. Let S_1 be the quasi-ordered system of $Q(I, S)$ together with $\leqslant_0$. Let $I^1 = I - \{0\}$. Then $Q(I, S)$ is isomorphic to $Q(I^1, S_1)$. Then the last theorem implies that $Q(I, S)$ is weakly well-quasi-ordered by $\leqslant_1$. Repeating this, the theorem is proved.

As a corollary, we have the usual well-ordering theorem for $Q(I, S)$.

COROLLARY 27.47. *Let I be an arbitrary well-ordered set and S an arbitrary well-ordered set. $Q(I, S)$ is well-ordered by $\leqslant_i$ for all $i \in \tilde{I}$.*

PROOF. If S is linear ordered, then $Q(I, S)$ is linear ordered. Here we should remark that in the case of linear ordering the notions of strong form and weak form of well-ordering are equivalent.

THEOREM 27.48 (Okada–Takeuti). *Let I be an arbitrary well-ordered set. For any well-quasi-ordered set S, $Q(I, S)$ is well-quasi-ordered by $\leqslant'_i$ for all $i \in \tilde{I}$.*

The proof is carried out in a similar way to the proof of Theorem 27.46. Here the notion of a bad sequence is changed as follows: Let $\leqslant$ be a quasi-ordering on D. A sequence $d_0, d_1, d_2, \ldots$, from D, is said to be bad with respect to $\leqslant$ if for every $i < j$, $d_i \not\leqslant d_j$.

§28. A consistency proof for a system with inductive definitions

In this section we will prove the consistency of a system obtained from **ISN** by adding inductive definitions with Π^1_1-clauses. This system we call the system of isolated inductive definitions, **IID**.

DEFINITION 28.1. **IID** is the system **ISN** with the following modifications.

1) **IID** contains a unary primitive recursive predicate I and a binary primitive recursive predicate $<^*$, where $<^*$ is a well-ordering of $\{a \mid I(a)\}$.

2) **IID** contains ternary predicate symbols $A_0, A_1, \ldots$ for which $A_n(s, t, V)$ is an atomic formula for s and t terms and V an abstract.

3) If $\forall \phi\, B$ is a semi-formula of **IID**, then the outermost quantifier $\forall$ affects A_n in B if there is a ϕ in an argument of A_n, i.e., if A_n occurs in B in the form $A_n(a, b, V)$ and ϕ occurs in V.

4) A semi-formula or abstract A of **IID** is isolated if no $\forall$ for a second order variable affects any other $\forall$ for a second order variable or $A_0, A_1, \ldots$, in A.

5) The initial sequents of **IID** are those in Definition 18.1, extended to include formulas with A_n's, and the sequents of the following forms:

$$I(s), A_n(s, t, V) \rightarrow G_n(s, t, V, \{x, y\}(A_n(x, y, V) \wedge x <^* s))$$

$$I(s), G_n(s, t, V, \{x, y\}(A_n(x, y, V) \wedge x <^* s)) \rightarrow A_n(s, t, V)$$

for $n = 0, 1, 2, \ldots$. Each $G_n(a, b, \alpha, \beta)$ is an arbitrary isolated formula containing none of $A_n, A_{n+1}, \ldots$, and no second order free variables except α and β, and V is an arbitrary abstract, which may contain $\forall$ for second order variables or $A_n, A_{n+1}, \ldots$.

(6) The rules of inference for **IID** are those of **ISL** together with full induction.

The purpose of this section is to prove the consistency of **IID**:

THEOREM 28.2. **IID** *is consistent.*

PROOF. This we will prove using the system of ordinal diagrams

$$O(\omega^{I_\infty}+1, \omega^{I_\infty} \cdot \omega \cdot \omega^{I_\infty}),$$

where $I_\infty = (2 \cdot |I| + 1) \cdot \omega$ and $|I|$ is the order-type of $<^*$. The proof is similar to the proof of Theorem 27.5, therefore we will present only new aspects of the proof.

PROPOSITION 28.3. *Let $F(\alpha)$ and V be an isolated formula and an isolated abstract, respectively. Then $F(V)$ is isolated.*

PROOF. (By induction on the number n of logical symbols contained in $F(\alpha)$.) If $n = 0$, the assertion is clear. Let $n > 0$. We shall treat several cases according to the outermost logical symbol of $F(\alpha)$. Since the other cases are easy, we shall consider the case where $F(\alpha)$ is of the form $\forall \phi\, G(\alpha, \phi)$. By the induction hypothesis $G(V, \beta)$ is isolated, where β is a free second order variable not contained in V. We have only to show that the outermost $\forall$ of $\forall \phi\, G(V, \phi)$ affects none of the $\forall$'s for second order variables or $A_0, A_1, \ldots$. But this is obvious since $\forall \phi\, G(\alpha, \phi)$ and $G(V, \beta)$ are isolated.

We next define several well-ordered systems.

DEFINITION 28.4. (1) Let $|I|$ be the ordinal of the well-ordering $<^*$. Let $\tilde{I}$ be $\{\tilde{\imath} \mid i \in I\}$ and let $I_* = I \cup \tilde{I}$. Then $<_*$ is the well-ordering of I_* defined as follows:

(1.1) If $i \in I$, then $i <_* \tilde{\imath}$.
(1.2) If $i <^* j$, then $i <_* j$.
(1.3) If $i <^* j$, then $\tilde{\imath} <_* \tilde{\jmath}$.
(1.4) If $i <^* j$, then $\tilde{\imath} <_* j$.
(1.5) If $i <^* j$, then $i <_* \tilde{\jmath}$.

The ordinal of $<_*$ is $2 \cdot |I|$.

(2) Let n be a natural number. Then $I_n = \{(i, n) \mid i \in I_*\} \cup \{\infty_n\}$ and $<_n$ is the well-ordering of I_n defined as follows:

(2.1) If $i <_* j$, then $(i, n) <_n (j, n)$.

(2.2) If $i \in I_*$, then $(i, n) <_n \infty_n$.

(3) $I_\infty = I_0 \cup I_1 \cup \ldots$ and $<_\infty$ is the well-ordering of I_∞ defined as follows:

(3.1) If $i \in I_n$, $j \in I_m$ and $n < m$, then $i <_\infty j$.

(3.2) If $i < j$ in I_n for some n, then $i <_\infty j$.

The order type of $<_\infty$ is $(2 \cdot |I| + 1) \cdot \omega$.

DEFINITION 28.5. Let A be a formula. The rank of A_n in A, denoted by $r(A_n: A)$, is an element of I_∞ defined as follows:

1) If $A_n(s, t, V) \wedge s <^* i$, occurs in A, where $I(i)$ is true and either s is a variable or s is a numeral for which one of $\neg I(s)$ or $i \leqslant^* s$ is true, then $r(A_n: A) = (i, n)$. Here i is constant.

2) If $A_n(j, t, V)$ occurs in A, where I(j) is true, and 1) does not hold, then $r(A_n: A) = (\bar{j}, n)$. Here j is a constant.

3) If A_n occurs in A and neither 1) nor 2) applies, then $r(A_n: A) = \infty_n$.

PROPOSITION 28.6. *Let B and C be two arbitrary formulas in which A_m and A_n occur, respectively. Then $r(A_m : B) <_\infty r(A_n : C)$ if $m < n$.*

DEFINITION 28.7. The γ-degree of a formula or an abstract, $\gamma(A)$, is a number less than ω^{I_∞}, defined in the following way. Here $<$ is the ordering of ω^{I_∞}.

1) If A is isolated, then $\gamma(A) = 0$.

In 2)–6), A is assumed not to be isolated.

2) If A is of the form $\neg B$, then $\gamma(A) = \gamma(B) + 1$.

3) If A is of the form $A_n(s, t, V) \wedge s <^* i$, then $\gamma(A) = \gamma(V) + \omega^{r(A_n : A)+1}$. If A is of the form $B \wedge C$ and not of the form just mentioned, then $\gamma(A) = \max(\gamma(B), \gamma(C)) + 1$.

4) If A is of the form $\forall x\, (G(x))$, then $\gamma(A) = \gamma(G(a)) + 1$.

5) If A is of the form $\forall \phi\, F(\phi)$, then $\gamma(A) = \gamma(F(\alpha)) + 1$.

6) If A is of the form $A_n(s, t, V)$, then $\gamma(A) = \gamma(V) + \omega^{r(A_n : A)}$.

7) If A is of the form $\{x_1, \ldots, x_n\} B(x_1, \ldots, x_n)$, then $\gamma(A) = \gamma(B(a_1, \ldots, a_n))$.

PROPOSITION 28.8. *Let $\{x_1, \ldots, x_n\} H(x_1, \ldots, x_n)$ be an abstract and let $s_1, \ldots, s_n$ be arbitrary terms. Then*

$$\gamma(H(s_1, \ldots, s_n)) \leqslant \gamma(\{x_1, \ldots, x_n\} H(x_1, \ldots, x_n)).$$

LEMMA 28.9. *If $G(\beta, \alpha)$ is an isolated quasi-formula (allowing other second order free variables as well) which contains none of $A_n, A_{n+1}, \ldots$, if s is a constant for which $I(s)$ is true, and if V is an arbitrary abstract which is not isolated, then*

$$\gamma(G(V, A_n^s(V))) \leqslant \gamma(V) + \sum_{l=1}^{k} \omega^{r(A_{j_l} : B_l)} + m$$

for some $j_1, \ldots, j_k \leq n$, *for some formulas* $B_1, \ldots, B_k$, *and for a number* m, *where* $A_n^s(V)$ *is an abbreviation for*

$$\{x, y\}(A_n(x, y, V) \wedge x <^* s), \quad \textit{and} \quad r(A_{j_l} : B_l) < r(A_n : A_n)$$

for $l \leq k$.

PROOF. By induction on the construction of G.

PROPOSITION 28.10. *If* s *is a constant for which* $I(s)$ *is true, if* V *is not isolated and if* $G_n(a, b, \alpha, \beta)$ *is as in Definition* 28.1, *then*

$$\gamma(G_n(s, t, V, A_n^s(V))) < \gamma(A_n(s, t, V)).$$

PROOF. As a special case of Lemma 28.9,

$$\gamma(G(V, A_n^s(V))) \leq \gamma(V) + \sum_{l=1}^{k} \omega^{r(A_{j_l} : B_l)} + m,$$

where $r(A_{j_l} : B_l) < r(A_n : A_n)$ and $m < \omega$. On the other hand,

$$\gamma(A_n(s, t, V)) = \gamma(V) + \omega^{\gamma(A_n : A_n)}.$$

The proposition then follows.

Next we add the rule of substitution to the system **IID** (cf. Definition 27.10).

DEFINITION 28.11. A substitution or a $\forall$: right for a second order variable, say J, is said to *disturb* a semi-formula A if the eigenvariable of J occurs in the scope of $\forall$ for a second order variable or in an argument of an A_n occurring in A.

We define a proof with degree to be a proof satisfying the following conditions.

1) Every substitution is in the end-piece, and no ind occurs under a substitution.

2) We can assign an element of $\omega^{I_\infty} + 1$ to every semi-formula or abstract A and every substitution J in the end-piece, which is called the degree of A or of J (written $d(A)$ or $d(J)$), respectively, so as to satisfy the following conditions:

2.1) If A is explicit, then $d(A) = 0$.

2.2) If A is implicit and not isolated, then $d(A) = \omega^{I_\infty}$.

2.3) Let A be implicit and isolated.

2.3.1) $d(A) = 0$ if A contains no logical symbol or $A_0, A_1, \ldots$.

2.3.2) $d(A) = d(B) + 1$ if A is of the form $\neg B$.

2.3.3) $d(A) = \max_J(d(V), d(J)) + \omega^{r(A_n : A)} + 1$, where J ranges over all the substitutions which disturb A, if A is of the form $A_n(s, t, V) \wedge s <^* i$.

$d(A) = \max(d(B), d(C)) + 1$, if A is of the form $B \wedge C$ and not of the form just mentioned.

2.3.4) $d(A) = d(B(x)) + 1$, if A is of the form $\forall x\, B(x)$.

2.3.5) $d(A) = \max_J(d(F(\phi)), d(J)) + 1$, where J ranges over all the substitutions which disturb $\forall \phi\, F(\phi)$, if A is of the form $\forall \phi\, F(\phi)$.

2.3.6) $d(A) = \max_J(d(V), d(J)) + \omega^{r(A_n : A_n)}$, where J ranges over all the substitutions which disturb A, if A is of the form $A_n(s, t, V)$.

3) $d(A) = d(B)$, if A is an abstract of the form $\{x_1, \ldots, x_n\} B$.

4) If J is a substitution in the end-piece, then $d(B) < d(J)$ for every formula B in the upper sequent of J.

(5) If J is a substitution, then $0 < d(J) < \omega^{I_\infty}$.

LEMMA 28.12. *Suppose $G(\beta, \alpha)$ is an isolated quasi-formula whose only second order free variables are β and α, and which contains none of $A_n, A_{n+1}, \ldots$. Assume also that i is a constant for which $I(i)$ is true. If V is isolated, then*

$$d(G(V, A_n^i(V))) \leq \max_J(d(V), d(J)) + \sum_{l=1}^{k} \omega^{r(A_{j_l} : B_l)} + m,$$

for some $j_1, \ldots, j_k \leq n$, some $B_1, \ldots, B_k$ and a number $m < \omega$, where $r(A_{j_l} : B_l) <_\infty r(A_n : A_n)$ and J ranges over all substitutions which influence V.

PROPOSITION 28.13. *Suppose $A_n(i, t, V)$ is isolated, i.e., V is isolated, and i is a constant for which $I(i)$ is true. If either*

$$I(i), A_n(i, t, V) \rightarrow G_n(i, t, V, A_n^i(V))$$

or

$$I(i), G_n(i, t, V, A_n^i(V)) \rightarrow A_n(i, t, V)$$

is an initial sequent in a proof with degree, in which $A_n(i, t, V)$ is implicit, then

$$d(G_n(i, t, V, A_n^i(V))) < d(A_n(i, t, V)).$$

PROOF. This is a special case of Lemma 28.12.

DEFINITION 28.14. Let A be a semi-formula or an abstract. We define the *norm* of A, $n(A)$, to be an element of ω^{I_∞} as follows:

1) If A contains no logical symbol or $A_0, A_1, A_2, \ldots$, then $n(A) = 0$.

2) If A is of the form $\neg B$, then $n(A) = n(B) + 1$.

3) If A is of the form $A_n(s, t, V) \wedge s <^* i$, then $n(A) = n(V) + \omega^{r(A_n : A)} + 1$. If A is of the form $B \wedge C$ and not of the above form, then

$$n(A) = \max(n(B), n(C)) + 1.$$

4) If A is of the form $\forall x\, B(x)$, then $n(A) = n(B(a)) + 1$.

5) If A is of the form $\forall \phi\, F(\phi)$, then $n(A) = n(F(\alpha)) + 1$.

6) If A is of the form $A_n(s, t, V)$, then $n(A) = n(V) + \omega^{r(A_n:A)}$.

7) If A is of the form $\{x_1, \dots, x_m\} H(x_1, \dots, x_m)$, then $n(A) = n(H(a_1, \dots, a_m))$.

LEMMA 28.15. *If $G(\beta, \alpha)$ contains none of $A_n, A_{n+1}, \dots$, if i is a constant for which $I(i)$ is true and if V is an arbitrary abstract, then*

$$n(G(V, A_n^i(V))) \leq n(V) + \sum_{l=1}^{k} \omega^{r(A_{j_l}:B_l)} + m,$$

where $j_l \leq n$, $r(A_{j_l} : B_l) < r(A_n : A_n)$ and $m < \omega$.

PROPOSITION 28.16. *If*

$$I(i),\, G_n(i, t, V, A_n^i(V)) \to A_n(i, t, V)$$

or

$$I(i),\, A_n(i, t, V) \to G_n(i, t, V, A_n^i(V))$$

is an initial sequent of our system, and i is a constant for which $I(i)$ is true, then

$$n(G_n(i, t, V, A_n^i(V))) < n(A_n(i, t, V)).$$

PROOF. A special case of Lemma 28.15.

DEFINITION 28.17. Let $N(I_\infty) = \omega^{I_\infty} \times \omega \times \omega^{I_\infty}$ and let $<$ be the lexicographical ordering of $N(I_\infty)$. The *grade* of a formula A, $g(A)$, is $\langle \gamma(A), a, n(A) \rangle$, where a is the number of eigenvariables in A for second order $\forall$: right under A, and $g(A)$ is an element of (I_∞).

PROPOSITION 28.18. *If*

$$I(i),\, A_n(i, t, V) \to G_n(i, t, V, A_n^i(V))$$

or

$$I(i),\, G_n(i, t, V, A_n^i(i, t, V) \to A_n(i, t, V)$$

is an initial sequent of a proof with degree, and i is a constant for which $I(i)$ is true, then

$$g(G_n(i, t, V, A_n^i(V))) < g(A_n(i, t, V)).$$

DEFINITION 28.19. We shall assign an element of $O(\omega^{I_\infty} + 1, \omega^{I_\infty} \times \omega \times \omega^{I_\infty})$ to every sequent of a proof P with degree as follows. We denote ω^{I_∞}, the maximum element of $\omega^{I_\infty} + 1$, by ξ.

1.1) The ordinal diagram of an initial sequent of the form

$$D \to D,\ s = t,\ A(s) \to A(t)$$

or a mathematical initial sequent is 0.

1.2) The ordinal diagram of an initial sequent of the form

$$I(i), A_n(i, t, V) \to G_n(i, t, V, \{x, y\}(A_n(x, y, V) \wedge x <^* i))$$

or

$$I(i), G_n(i, t, V, \{x, y\}(A_n(x, y, V) \wedge x <^* i)) \to A_n(i, t, V)$$

is $(\xi, \langle 0, 0, 0 \rangle, 0)$.

2) If S_1 and S_2 are the upper sequent and the lower sequent of a weak, structural inference, then the ordinal diagram of S_2 is equal to that of S_1.

3) If S_1 and S_2 are the upper sequent and the lower sequent of one of the inferences $\neg$, $\bigwedge$: left, $\forall$ for a first order variable, $\forall$: right for a second order variable and explicit $\forall$: left for a second order variable, then the ordinal diagram of S_2 is $(\xi, \langle 0, 0, 0 \rangle, \sigma)$, where σ is the ordinal diagram of S_1.

4) If S_1 and S_2 are the upper sequents and S is the lower sequent of $\bigwedge$: right, then the ordinal diagram of S is $(\xi, \langle 0, 0, 0 \rangle, \sigma_1 \# \sigma_2)$, where σ_1 and σ_2 are the ordinal diagrams of S_1 and S_2, respectively.

5) If S_1 and S_2 are the upper sequent and the lower sequent of an implicit $\forall$: left for a second order variable of the form

$$\frac{F(V), \Gamma \to \Delta}{\forall \phi\, F(\phi), \Gamma \to \Delta},$$

then the ordinal diagram of S_2 is $(\xi, \langle \mu, k, \nu \# 0 \# 0 \rangle, \sigma)$, where σ is the ordinal diagram of S_1 and $\langle \mu, k, \nu \rangle$ is $g(F(V))$.

6) If S_1 and S_2 are the upper sequents and S is the lower sequent of a cut, then the ordinal diagram of S is $(\xi, \langle \mu, k, \nu \# 0 \rangle, \sigma_1 \# \sigma_2)$, where $\langle \mu, k, \nu \rangle$ is the grade of the cut-formula and σ_1 and σ_2 are the ordinal diagrams of S_1 and S_2, respectively.

7) If S_1 and S_2 are the upper sequent and the lower sequent of a substitution J, then the ordinal diagram of S_2 is $(d(J), \langle 0, 0, 0 \rangle, \sigma)$, where σ is the ordinal diagram of S_1.

8) If S_1 and S_2 are the upper and the lower sequents of an ind, then the ordinal diagram of S_2 is $(\xi, \langle \mu, k, \nu \# 0 \# 0 \rangle, \sigma)$, where $\langle \mu, k, \nu \rangle$ is the grade of the induction formula and σ is the ordinal diagram of S_1.

9) The ordinal diagram of P is defined to be the ordinal diagram assigned to the end-sequent of P.

Suppose the sequent $\to$ is provable in this system. We shall reduce a proof P of $\to$ to another proof of $\to$. This reduction will be carried out in the same way as in §27. We can assume that the end-piece of P contains no first order free variable, ind, axiom of the form $m = n, A(m) \to A(n)$ or $D \to D$, or weakening and we assume that term-replacement has been introduced. Suppose that the end-piece of P contains an initial sequent of the form 5) of Definition 28.1, say

(*) $\quad I(i), A_n(i, t, V) \to G_n(i, t, V, \{x, y\}(A_n(x, y, V) \wedge x <^* i)),$

where we can assume without loss of generality that i and t are numerals. By our assumption either $I(i)\rightarrow$ or $\rightarrow I(i)$ is an initial sequent. We shall abbreviate $\{x, y\}(A_n(x, y, V) \wedge x <^* i)$ as $A_n^i(V)$.

Case 1. $I(i)\rightarrow$ is an initial sequent. Replace (*) by the following:

$$\cfrac{\cfrac{I(i)\rightarrow}{\text{weakenings and an exchange}}}{I(i), A_n(i, t, V)\rightarrow G_n(i, t, V, A_n^i(V)).}$$

The ordinal diagram of the proof is less than that of (*). Hence evidently I is reduced to the proof obtained by replacement.

Case 2. $\rightarrow I(i)$ is an initial sequent. Since every formula in P is implicity there exists a cut J where one of the cut-formulas is a descendant of $A_n(i, t, V)$ in (*). Let P be of the following form:

$$\begin{array}{ccc} A_n(i, t, V)\rightarrow A_n(i, t, V) & & I(i), A_n(i, t, V)\rightarrow G_n(i, t, V, A_n^i(V)) \\ \vdots & & \vdots \\ J \quad \Gamma \overset{\sigma_1}{\rightarrow} \Delta, A_n(i, t, V) & & A_n(i, t, V), \Pi \overset{\sigma_2}{\rightarrow} \Lambda \\ \hline & \Gamma, \Pi \overset{\sigma}{\rightarrow} \Delta, \Lambda & \\ & \vdots & \\ & \rightarrow & \end{array},$$

where $A_n(i, t, V)\rightarrow A_n(i, t, V)$ need not appear. Here we should note that no substitution applies to $A_n(i, t, V)$: in fact, if there were such a substitution J_0, it would disturb $A_n(i, t, V)$, i.e., $d(J_0) < d(A_n(i, t, v))$. But this contradicts 4) of Definition 28.11.

Consider the following proof P':

$$\begin{array}{ccc} \cfrac{I(i), A_n(i, t, V)\rightarrow G_n(i, t, V, A_n^i(V))}{A_n(i, t, V), I(i)\rightarrow G_n(i, t, V, A_n^i(V))} & & G_n(i, t, V, A_n^i(V))\rightarrow G_n(i, t, V, A_n^i(V)) \\ \vdots & & \vdots \\ \Gamma, I(i) \overset{\sigma_1'}{\rightarrow} \Lambda, G_n(i, t, V, A_n^i(V)) & & G_n(i, t, V, A_n^i(V)), \Pi \overset{\sigma_2'}{\rightarrow} \Lambda \\ \hline & \cfrac{\Gamma, I(i), \Pi\rightarrow \Delta, \Lambda}{\text{some exchanges}} & \\ \rightarrow I(i) & I(i), \Gamma, \Pi\rightarrow \Delta, \Lambda & \\ \hline \Gamma, \Pi \overset{\sigma'}{\rightarrow} \Delta, \Lambda & & \\ \vdots & & \\ \rightarrow & & \end{array}$$

Every substitution in P' has the same degree as the corresponding substitution in P. Then P' is a proof with degree by virtue of Proposition 28.10. Furthermore,

$$\sigma = (\xi, \langle \mu, j, \lambda \# 0\rangle, \sigma_1 \# \sigma_2)$$

and

$$\sigma' = (\xi, \langle 0, 0, 0 \# 0\rangle, \langle 0, 0, 0\rangle \# (\xi, \langle \nu, k, \delta \# 0\rangle, \sigma_1' \# \sigma_2')),$$

where $\langle \mu, j, \lambda\rangle = g(A_n(i, t, V))$, and $\langle \nu, k, \delta\rangle = g(G_n(i, t, V, A_n^i(V)))$. Proposition 28.18 implies that $\sigma' <_l \sigma (l \leq \xi)$, from which it follows that the ordinal diagram of P' is less than that of P. Thus P is reduced to P'. (For the computation of ordinal diagrams, one should refer to §27.)

Suppose that the end-piece of P does not contain a logical inference, ind, or initial sequents other than mathematical ones, or weakening. If P contains a logical symbol, we can find a suitable cut in P in the same way as for **PA** and define an essential reduction in the same way as in §27.

As an addendum to this section, as well as the previous section, we shall explain the general theory of γ-degree. We consider a second order language.

DEFINITION 28.20. A function γ from semi-formulas and abstracts to ordinals is called *monotone* if it satisfies the following conditions.

1) $\gamma(\neg A) \geq \gamma(A)$.
2) $\gamma(A \wedge B) \geq \max(\gamma(A), \gamma(B))$.
3) $\gamma(\forall x\, G(x)) \geq \gamma(G(x))$.
4) $\gamma(\{x_1, \ldots, x_n\} H(x_1, \ldots, x_n)) = \gamma(H(x_1, \ldots, x_n))$.
5) $\gamma(\forall \phi\, F(\phi)) \geq \gamma(F(\phi))$.
6) If A is an alphabetical variant of B, then $\gamma(A) = \gamma(B)$.
7) If $\gamma(V) = 0$ and $\gamma(\forall \phi\, F(\phi)) > 0$, then $\gamma(\forall \phi\, F(\phi)) > \gamma(F(V))$.

We say that A is *γ-simple* if $\gamma(A) = 0$.

A second order $\forall$: left, say

$$\frac{F(V), \Gamma \to \Delta}{\forall \phi\, F(\phi), \Gamma \to \Delta},$$

is called *γ-simple* if V is γ-simple; it is called *strictly γ-simple* if both V and $\forall \phi\, F(\phi)$ are γ-simple.

A proof P in $\mathbf{G^1LC}$ is called (strictly) γ-simple if every implicit, second order $\forall$: left in P is (strictly) γ-simple.

PROPOSITION 28.21. *Suppose γ is monotone and for every strictly γ-simple proof the cut-elimination theorem holds. Then the cut-elimination theorem holds for every γ-simple proof.*

PROOF. The grade of a formula in a proof, say A, is defined as $\omega^2 \cdot \gamma(A) + \omega \cdot m + l$, where m is the number of eigenvariables of the second order $\forall$: right introductions which occur under A, and l is the number of logical symbols in A. The grade of A will be denoted by $g(A)$. Let P be a γ-simple proof and let J be a cut in P. J is called "γ-simple" if the cut formula of J is γ-simple. The grade of J, $g(J)$, is defined to be the grade of the cut formula of J. The grade of P, $g(P)$, is defined to be $\sum_J \omega^{g(J)}$, where J ranges over all the cuts in P which are not γ-simple, and we assume that $\omega^{g(J)}$ in $\sum$ are arranged in the decreasing order.

If $g(P)=0$, then there is no implicit formula which is not γ-simple, in particular, the principal formula of every implicit $\forall$: left is γ-simple, which means that P is strictly γ-simple. Therefore, by the assumption of the proposition, the cut-elimination theorem holds for P. Suppose now that $g(P)>0$; hence there is a cut J in P which is not γ-simple and such that every cut above J is γ-simple. Since other cases are easily treated, we shall deal with the case where the cut formula is of the form $\forall\phi F(\phi)$:

$$J \quad \frac{\Gamma \to \Delta, \forall\phi F(\phi) \qquad \forall\phi F(\phi), \Pi \to \Lambda}{\Gamma, \Pi \to \Delta, \Lambda}.$$

Let P_0 be the proof ending with $\Gamma, \Pi \to \Delta, \Lambda$. Let A be the left cut formula of J and let B be the right cut formula of J. We may assume that the uppermost ancestor of $A(B)$ which is identical with $A(B)$ is the principal formula of a logical inference and $F(\alpha)$ is the auxiliary formula of such inference related to A. By replacing the ancestors of A which are identical with A by $F(\alpha)$, we obtain a proof P_1 ending with $\Gamma \to \Delta, F(\alpha)$.

Let $\Pi_1 \to \Lambda_1$ be an arbitrary sequent which occurs above the right upper sequent of J. We can construct a proof ending with a sequent of the form $\Pi_1^*, \Gamma \to \Delta, \Lambda_1$, where Π_1^* is obtained from Π_1 by eliminating all the ancestors of B which are identical with B. This can be done by induction on the number of inferences in the proof ending with $\Pi_1 \to \Lambda_1$. As an example, suppose $\Pi_1 \to \Lambda_1$ is the lower sequent of a cut:

$$\frac{\Pi_2 \to \Lambda_2, D \qquad D, \Pi_3 \to \Lambda_3}{\Pi_2, \Pi_3 \to \Lambda_2, \Lambda_3}$$

where Π_2, Π_3 is Π_1 and Λ_2, Λ_3 is Λ_1. Define the following:

$$\cfrac{\cfrac{\Pi_2^*, \Gamma \to \Delta, \Lambda_2, D \qquad D, \Pi_3^*, \Gamma \to \Delta, \Lambda_3}{\Pi_2^*, \Gamma, \Pi_3^*, \Gamma \to \Delta, \Lambda_2, \Delta, \Lambda_3}}{\Pi_2^*, \Pi_3^*, \Gamma \to \Delta, \Lambda_2, \Lambda_3.}$$

As another example, let $\Pi_1 \to \Lambda_1$ be the lower sequent of a second order $\forall$: left whose principal formula is an ancestor of B which is identical with B:

$$\frac{F(V), \Pi_2 \to \Lambda_1}{\forall\phi F(\phi), \Pi_2 \to \Lambda_1},$$

where $\forall\phi F(\phi), \Pi_2$ is Π_1. Consider the following:

$$\cfrac{\cfrac{\Gamma \to \Delta, F(V) \qquad F(V), \Pi_2^*, \Gamma \to \Delta, \Lambda_1}{\Gamma, \Pi_2^* \to \Delta, \Delta, \Lambda}}{\Pi_2^*, \Gamma \to \Delta, \Lambda,}$$

where $\Gamma \xrightarrow{\cdot\cdot} \Delta, F(V)$ is obtained from P_1 by substituting V for α everywhere.

By taking $\Pi_1 \rightarrow \Lambda_1$ to be $\forall\phi\, F(\phi), \Pi \rightarrow \Lambda$, we obtain $\Pi, \Gamma \rightarrow \Delta$, and hence $\Gamma, \Pi \rightarrow \Delta, \Lambda$. The grade of this proof, say Q, is less than $g(P_0)$, since $\gamma(F(V)) < \gamma(\forall\phi\, F(\phi))$ by assumption. Now replace P_0 by Q in P, obtaining a proof of the same end-sequent, but with a grade less than P. Then by the induction hypothesis the cuts can be eliminated.

DEFINITION 28.22. A set of semi-formulas and abstracts, say $\mathcal{F}$, is said to be closed if the following hold.

1) If A is atomic, then A belongs to $\mathcal{F}$.

2) If $\neg B$ belongs to $\mathcal{F}$, then B belongs to $\mathcal{F}$.

3) If $B \wedge C$ belongs to $\mathcal{F}$, then B and C belong to $\mathcal{F}$.

4) If $\forall x\, F(x)$ belongs to $\mathcal{F}$, then $F(s)$ belongs to $\mathcal{F}$ for every semi-term s.

5) If $\forall\phi\, F(\phi)$ belongs to $\mathcal{F}$, then $F(\alpha)$ belongs to $\mathcal{F}$ for every second order variable α.

6) If $\{x_1, \ldots, x_n\} H(x_1, \ldots, x_n)$ belongs to $\mathcal{F}$ then $H(a_1, \ldots, a_n)$ belongs to $\mathcal{F}$ for every $a_1, \ldots, a_n$; if $H(a_1, \ldots, a_n)$ belongs to $\mathcal{F}$ for some $a_1, \ldots, a_n$, then $\{x_1, \ldots, x_n\} H(x_1, \ldots, x_n)$ belongs to $\mathcal{F}$.

7) If B and C are alphabetical variants of one another, then B belongs to $\mathcal{F}$ if and only if C belongs to $\mathcal{F}$.

8) If $F(\alpha)$ and V belong to $\mathcal{F}$, then $F(V)$ belongs to $\mathcal{F}$.

We define a function γ relative to $\mathcal{F}$, which we call the γ determined by $\mathcal{F}$.

(1) $\gamma(A) = 0$ if A belongs to $\mathcal{F}$.

Assume A does not belong to $\mathcal{F}$.

(2) $\gamma(A) = \gamma(B) + 1$ if A is $\neg B$.

(3) $\gamma(A) = \max(\gamma(B), \gamma(C)) + 1$ if A is $B \wedge C$.

(4) $\gamma(A) = \gamma(F(x)) + 1$ if A is $\forall x\, F(x)$.

(5) $\gamma(A) = \gamma(F(\phi)) + 1$ if A is $\forall\phi\, F(\phi)$.

(6) $\gamma(\{x_1, \ldots, x_n\} H(x_1, \ldots, x_n)) = \gamma(H(x_1, \ldots, x_n))$.

In a manner similar to the proof of Proposition 27.7, we can easily prove the following.

PROPOSITION 28.23. *Suppose $\mathcal{F}$ is closed and γ is the function determined by $\mathcal{F}$. If V belongs to $\mathcal{F}$, then $\gamma(F(\alpha)) = \gamma(F(V))$.*

PROPOSITION 28.24. *Suppose $\mathcal{F}$ is closed and γ is the function determined by $\mathcal{F}$. Then γ is monotone.*

PROOF. Immediate from the definition of γ and Proposition 28.23.

CHAPTER 6

SOME APPLICATIONS OF CONSISTENCY PROOFS

In this chapter we discuss four applications of consistency proofs. Though we choose a specific system, our methods in the first three applications are general enough to apply for many other systems without any change.

§29. Provable well-orderings

We shall consider provable well-orderings of **ISN** and show that any provable well-ordering of **ISN** has order type less than that of the system of ordinal diagrams $O(\omega+1, 1)$, with respect to $<_0$. We will borrow much of the argument of §13. The results we will prove can be extended to **IID** with little modification.

DEFINITION 29.1. Let $<\cdot$ be a recursive linear ordering of the natural numbers which is actually a well-ordering. (Without loss of generality we may assume that $<\cdot$ is defined for all natural numbers and the least element with respect to $<\cdot$ is 0.) We use the same symbol $<\cdot$ to denote the formula in **ISN** which expresses the ordering $<\cdot$.

Let $\mathbf{TI}(<\cdot)$ be a formula expressing the principle of transfinite induction along $<\cdot$:

$$\forall \phi\, (\forall x\, (\forall y\, (y <\cdot\, x \supset \phi(y)) \supset \phi(x)) \supset \forall x\, \phi(x)).$$

If $\mathbf{TI}(<\cdot)$ is **ISN**-provable, then we say that $<\cdot$ is a provable well-ordering of **ISN**.

We assume that an arithmetization of the system of ordinal diagrams $O(\omega+1, 1)$ has been carried out. We use the same notation to denote both an object and its arithmetization.

THEOREM 29.2. *Let $<_0$ be the well-ordering of the system of ordinal diagrams $O(\omega+1, 1)$ with respect to* 0. (*Recall that the consistency of* **ISN** *was proved by using $<_0$.*) *If $<\cdot$ is a provable well-ordering of* **ISN**, *then there exists a recursive function from natural numbers to an initial segment of $<_0$ which is $<\cdot$-$<_0$ order-preserving. That is to say, there is a recursive function f such that $a <\cdot\, b$ if and only if $f(a) <_0 f(b)$ and there is an ordinal diagram μ in $O(\omega+1, 1)$ such that for every a, $f(a) <_0 \mu$.*

PROOF. We will follow the proof of Theorem 13.4; and shall only point out how to modify that proof so that the arguments fit **ISN**.

(1) **TJ**-proofs (for **ISN**) are defined as in 13.1); in particular the **TJ** initial sequents have the form

$$\forall x\,(x <\cdot\, t \supset \varepsilon(x)) \rightarrow \varepsilon(t),$$

and the end-sequents have the form

$$\rightarrow \varepsilon(m_1), \ldots, \varepsilon(m_n).$$

(2) $|m|_{<\cdot}$ and the end-number of a **TJ**-proof are defined as in 13.1).

(3) For 13.2), 13.5) simply read **ISN** in place of **PA**. We will, however, repeat the Fundamental Lemma:

LEMMA 29.3, the Fundamental Lemma (cf. Lemma 13.5). *The end-number of any* **TJ**-*proof is not greater than the order type of its ordinal diagram* (*with respect to* $<_0$).

(4) Ordinal diagrams are assigned to the sequents of the **TJ**-proofs as for **ISN**: The ordinal diagram of a **TJ**-initial sequent is

$$\omega(5, 0).$$

See 13.6).

(5) The proofs of 13.7) through 13.11) go through as before.

(6) In 13.12), the ordinal diagram of the proof presented there is

$$\omega(3, 0 \,\#\, (\omega, 0)),$$

regarding $A \supset B$ as an abbreviation for $\neg(A \wedge \neg B)$. This is less than the ordinal diagram of a **TJ**-initial sequent. By this, and obvious changes, P becomes a **TJ**-proof P' whose end-sequent is

$$\rightarrow \varepsilon(m), \varepsilon(m_1), \ldots, \varepsilon(m_n),$$

where $\rightarrow \varepsilon(m_1), \ldots, \varepsilon(m_n)$ is the end-sequent of P. The ordinal diagram of P' is less than that of P and the end number of P' is $|m|_{<\cdot}$.

(7) As in 13.13), we obtain the Gentzen-type theorem:

THEOREM 29.4 (cf. Theorem 13.6). *The order type of* $<\cdot$ *is less than the order type of* $O(\omega+1, 1)$ *with respect to* $<_0$.

(8) As in 13.14), we can define a proof P_k for every k, where the end-number of P_k is $|k|_{<\cdot}$. Then we define a function h as in 13.15), where $+$ is the ordinal sum.

(9) In order to claim that h is recursive, and that 13.16) holds, we need the following.

1) The ordinal sum, $+$, of ordinal diagrams is recursive.

2) If two ordinal diagrams, μ and ν, are connected (i.e., the last operations used to form μ and ν are not $\#$) and $\mu <_0 \nu$, then $\mu + \nu = \nu$.

(10) From (9), we conclude that h is recursive and 13.16) holds. This implies that h is order-preserving.

§30. The Π_1^1-comprehension axiom and the ω-rule

An analogue to Problem 13.9 can be proved for **ISN**, viz., the elimination of cuts in a system with the constructive ω-rule. We repeat some of the definitions which were given in Chapter 5.

DEFINITION 30.1. (1) We assume a standard Gödel numbering for axioms and for rules of finite inference. The ω-rule is expressed as follows:

$$\frac{\begin{matrix} P_0 & & P_n \\ \vdots & & \vdots \\ \Gamma \to \Delta, A(0) \ldots & & \ldots \Gamma \to \Delta, A(n) \ldots \end{matrix}}{\Gamma \to \Delta, \forall x\, A(x)}.$$

Here P_n is defined for every natural number n and is a proof of $\Gamma \to \Delta$, $A(n)$. To P_n assign a Gödel number of $\ulcorner P_n \urcorner$. If there exists a recursive function such that $f(n) = \ulcorner P_n \urcorner$ for every n, then the ω-rule is said to be constructive and $3 \cdot 5^e$ is assigned to the whole proof, where e is the Gödel number of f, i.e., $\{e\}(n) = \ulcorner P_n \urcorner$. Let **S** be any logical system. A proof, in the system obtained from **S** by adjoining the constructive ω-rule to it, is called an ω-proof in **S**.

(2) Let $S(a)$ and $a <\cdot\, b$ be primitive recursive predicates such that $<\cdot$ is a well-ordering of $\{a\colon S(a)\}$, whose first element is 0. A number-theoretic function Ψ is called $<\cdot$-recursive if it is defined by the following scheme which is a repetition of a previous definition.

(i) $$f(a) = a + 1.$$

(ii) $$f(a_1, \ldots, a_n) = 0.$$

(iii) $$f(a_1, \ldots, a_n) = a_i (1 \leq i \leq n).$$

(iv) $$f(a_1, \ldots, a_n) = g(\lambda_1(a_1, \ldots, a_n), \ldots, \lambda_m(a_1, \ldots, a_n)),$$

where g and λ_i $(1 \leq i \leq m)$ are $<\cdot$-recursive.

(v) $$f(0, a_2, \ldots, a_n) = g(a_2, \ldots, a_n)$$
$$f(a+1, a_2, \ldots, a_n) = \lambda(a, f(a, a_2, \ldots, a_n), a_2, \ldots, a_n),$$

where g and λ are $<\cdot$-recursive.

(vi) $$f(0, a_2, \ldots, a_n) = g(a_2, \ldots, a_n)$$
$$f(a+1, a_2, \ldots, a_n) = \lambda(a, f(\tau^*(a, a_2, \ldots, a_n), a_2, \ldots, a_n), a_2, \ldots, n),$$

where g, λ and τ are $<\cdot$-recursive and

$$\tau^*(a, a_2, \ldots, a_n) = \begin{cases} \tau(a, a_2, \ldots, a_n) & \text{if} \quad \tau(a, a_2, \ldots, a_n) <\cdot\, a+1, \\ 0 & \text{otherwise.} \end{cases}$$

We shall transform a proof in **ISN** whose end-sequent contains no first order free variables, into a proof of the same end-sequent in the system with the constructive ω-rule. In proving the consistency of **ISN** in §27 we defined reductions on a proof of $\rightarrow$. This notion, however, can easily be extended to any proof whose end-sequent has no first order free variables.

DEFINITION 30.2. (1) $O(\mathbf{ISN}) = O(\omega+1, 1)$ is the system of ordinal diagrams used to prove the consistency of **ISN** and $<$ is its well-ordering (namely $<_0$).

(2) For an ordinal diagram α, and natural number i, $\alpha^{(i)}$ is defined by $\alpha^{(0)} = \alpha$ and $\alpha^{(i+1)} = \alpha^{(i)} \# \alpha$.

(3) For an ordinal diagram μ and natural number m, $\langle \mu, m \rangle =_{\text{df}} (0, 0, \mu \# 0)^{(m)}$.

By an abuse of notation, we shall use the notations for proofs, ordinal diagrams and $<$ both for formal objects and their Gödel numbers.

To the sequents in a proof we make the same assignment of ordinal diagrams as in §27 and we define the ordinal diagram of a proof P to be $\langle \mu, m \rangle$, where μ is the ordinal diagram assigned to its end-sequent and m is the number of first order free variables in its end-piece (denoted $m(P)$). If the ordinal diagram of P is less than the ordinal diagram of Q, then we write $\ulcorner P \urcorner < \ulcorner Q \urcorner$, or simply $P < Q$.

REMARK. In the definition of a reduction, we may be asked to take, say, a lowermost inference satisfying a certain condition. Such an inference may not be uniquely determined; however, we may suppose that the inferences are Gödel-numbered, and then take an inference as required with smallest Gödel number.

THEOREM 30.3. *There exists a $<$-recursive function f such that, for every proof P in* **ISN** *whose end sequent contains no first order free variable, $f(\ulcorner P \urcorner)$ is the Gödel number of an ω-proof of the end-sequent of P which contains no cut and no application of mathematical induction or first order $\forall$: right.*

PROOF. Let P be a proof in **ISN** whose end-sequent contains no first order free variables. We define reductions $r(P)$ and $q(i, P)$ for each $i < \omega$ and a transformation $f(\ulcorner P \urcorner)$ by transfinite induction on the ordinal diagram of P.

1) The end-piece of P contains an application of induction or an explicit logical inference.

1.1) The end-piece of P contains a first order free variable which is not used as an eigenvariable. We define $r(P)$ to be the Gödel number of the proof obtained from P by substituting 0 for each of such first order free variables. Obviously, $r(P) < P$. We define $f(P)$ to be $f(r(P))$.

1.2) The end-piece of P does not contain a first order free variable which is not used as an eigenvariable. Let J be a lowermost induction or (explicit) logical inference. We consider several cases.

1.2.1) J is an induction. Let $r(P)$ be the proof obtained from P by applying to J the reduction in (2) of §27 and let $f(P)$ be $f(r(P))$. Then $r(P) < P$.

1.2.2) J is an explicit logical inference.

1.2.2.1) J is not a first order $\forall$: right. Since all the cases are treated similarly, we consider the case where J is $\bigwedge$: left. Let P be

$$\begin{array}{c} \vdots \\ \dfrac{A, \Gamma \to \Delta}{A \wedge B, \Gamma \to \Delta} \\ \vdots \\ \Gamma_0 \to \Delta_0. \end{array}$$

We define $r(P)$ to be the proof

$$\begin{array}{c} \vdots \\ A, \Gamma \to \Delta \\ \hline \text{some exchanges and weakening} \\ \hline A \wedge B, \Gamma, A \to \Delta \\ \vdots \\ \Gamma_0, A \to \Delta_0. \end{array}$$

Since $r(P) < P$, $f(r(P))$ has been defined by the induction hypothesis. We define $f(P)$ to be the following proof

$$\begin{array}{c} f(r(P)) \left\{ \begin{array}{c} \vdots \\ \Gamma_0, A \to \Delta_0 \end{array} \right. \\ \hline \text{some exchanges} \\ \hline A, \Gamma_0 \to \Delta_0 \\ \hline A \wedge B, \Gamma_0 \to \Delta_0 \\ \hline \text{some exchanges and a contraction} \\ \hline \Gamma_0 \to \Delta_0. \end{array}$$

We shall refer to this figure as $g(f(r(p)))$.

1.2.2.2) I is a first order $\forall$: right. Let P be the following form:

$$\begin{array}{c} \vdots \\ J\ \dfrac{\Gamma \to \Delta, A(a)}{\Gamma \to \Delta, \forall x\, A(x)} \\ \vdots \\ \Gamma_0 \to \Delta_0. \end{array}$$

For each i we consider the proof (referred to as $q(i, P)$):

$$\dfrac{\dfrac{\begin{array}{c}\vdots\\ \Gamma \rightarrow \Delta, A(i)\end{array}}{\text{some exchanges and a weakening}}}{\begin{array}{c}\Gamma \rightarrow A(i), \Delta, \forall x\, A(x)\\ \vdots\\ \Gamma_0 \rightarrow A(i), \Delta_0\end{array}}$$

where the proof of $\Gamma \rightarrow \Delta, A(i)$ is obtained from the proof of the upper sequent of J by substituting the numeral i for a. Obviously $q(i, P) < P$ for each numeral i. Thus $f(q(i, P))$ has been defined for each i. We define $f(P)$ to be the proof

$$\dfrac{\begin{array}{c} f(q(i,P))\left\{\begin{array}{c}\vdots\\ \Gamma_0 \rightarrow A(i), \Delta_0\end{array}\right. \\ \hline \text{some exchanges} \\ \hline \dots\ \ \Gamma_0 \rightarrow \Delta_0, A(i)\ \ \dots \text{for each } i \\ \hline \omega\text{-rule}\quad \Gamma_0 \rightarrow \Delta_0, \forall x\, A(x)\end{array}}{\dfrac{\text{some exchanges and a contraction}}{\Gamma_0 \rightarrow \Delta_0.}}$$

2) The endpiece of P contains no explicit logical inference or induction, but does contain an explicit, logical initial sequent. Then the end-sequent of P is obtained from it by some weakenings and exchanges. Let $f(P)$ be one such proof.

3) The end-piece of P contains no induction or logical inference or explicit, logical initial sequent. We define $r(P)$ to be the proof obtained from P by applying the reductions in (1) through (8) of §27, retaining explicit weakenings. Then $r(P) < P$. Since the end sequent is unchanged by the reductions, we define $f(P)$ to be $f(r(P))$.

We have identified many notions with their Gödel numbers, e.g., a proof P sometimes means its Gödel number. Thus we can consider the functions r, q, g, f to be number-theoretic functions. We can obviously take r, q and g to be primitive recursive. (Here we assume that **ISN** has only finitely many primitive recursive functions as function constants. If **ISN** has all primitive recursive functions as function constants, then r and q are ω^ω-recursive. Even so our proof goes through by embedding ω^ω in the ordering of $<$.) Let $P(a)$ be a primitive recursive predicate stating that a is a proof in **ISN** whose end sequent contains no first order free variables. Let P_0, P_1, P_2 and P_3 be defined by:

$P_0(m) \Leftrightarrow_{\text{df}} P(m)$ and one of the conditions 1.1), 1.2.1) or 2) applies.

$P_1(m) \Leftrightarrow_{\text{df}} P(m)$ and the end piece of m contains an explicit logical inference, other than first order $\forall$: right, to which the reduction applies.

$P_2(m) \Leftrightarrow_{\text{df}} P(m)$ and the reduction will apply to a first order $\forall$: right in the end piece of m.

$P_3(m) \Leftrightarrow_{df} \neg(P_0(m) \vee P_1(m) \vee P_2(m))$.

Obviously, P_0, P_1, P_2 and P_3 are primitive recursive and in the light of the consistency proof have the following properties:

$$\forall x\, \exists! i\, (i \leq 3 \text{ and } P_i(x));$$

$$P_0(m) \Rightarrow r(m) < m;$$

$$P_1(m) \Rightarrow r(m) < m;$$

$$P_2(m) \Rightarrow \forall n\, (q(n, m) < m).$$

With the help of recursion theory we shall show that f is recursive, in fact $<$-recursive. In fact,

$$f_0(e, m) \simeq \begin{cases} \{e\}(r(m)) & \text{if } P_0(m), \\ g(\{e\}(r(m))) & \text{if } P_1(m), \\ 3 \cdot 5^{S_1^2(c_0, e, m)} & \text{if } P_2(m), \\ m & \text{if } P_3(m), \end{cases}$$

where c_0 is the general recursive index $\Lambda n, e, m\{e\}(q(n, m))$ (i.e., an index for $\{e\}(q(n, m))$ as a function of n, e, m; see: Kleene, Introduction to Metamathematics (North-Holland, Amsterdam, 1967), p. 344. By the recursion theorem (op. cit., §66), there is a number c such that $f_0(c, m) \simeq \{c\}(m)$. Then define f by $f(m) \simeq \{c\}(m)$, i.e.,

$$f(m) \simeq \begin{cases} f(r(m)) & \text{if } P_0(m), \\ g(f(r(m))) & \text{if } P_1(m), \\ 3 \cdot 5^{S_1^2(c_0, e, m)} & \text{if } P_2(m), \\ m & \text{otherwise.} \end{cases}$$

Thus f is partial recursive. By transfinite induction on $<$ we can show that f is totally defined. It is also easy to see that f is $<$-recursive, that $f(P)$ has the same end-sequent as P, and that $f(P)$ has no cut or mathematical induction or first order free variable. The completes the proof.

DEFINITION 30.4. A number-theoretic function $f(a_1, \ldots, a_n)$ is called provably recursive in **ISN** if the following sequent is provable in **ISN**.

$$\rightarrow \forall x_1 \ldots \forall x_n\, \exists y\, T_n(e, x_1, \ldots, x_n, y),$$

where T_n expresses Kleene's primitive recursive predicate T_n (cf. §13; we can easily extend the definition in §13 to the case where there are more than one x) and e is a Gödel number of f.

As an application of our technique we can give an alternate proof of a theorem which was first proved by Kino. This is an analogue to Problem 13.8.

THEOREM 30.5. *Let ψ by a provably recursive function in* **ISN**. *Then we can find an ordinal diagram μ of* O(**ISN**) *such that ψ is $<^{\mu}$-recursive, where $<^{\mu}$ is $<$ restricted to arguments $<\mu$.*

PROOF. Without loss of generality we may assume that ψ is a function of one argument. Let e be a Gödel number of ψ such that the sequent $\rightarrow \forall x \, \exists y \, T_1(e, x, y)$ is provable in **ISN**. Let P be a proof of $\rightarrow \exists y \, T_1(e, a, y)$ whose ordinal diagram is μ. We define P_m to be the proof obtained from P by substituting the numeral m for a. The process of obtaining P_m from P is primitive recursive. To each P_m we apply the transformation f of the previous theorem. Then $f(P_m)$ is a proof without a cut. Since P does not contain any explicit $\forall$: right for a first order variable (which is the only inference which induces an application of the ω-rule in the transformation), it is easily proved by transfinite induction that $f(P_m)$ does not contain any application of the ω-rule. By checking the proof $f(P_m)$ we can find primitive recursively a numeral n satisfying $T_1(e, m, n)$. Since $n = \psi(m)$ and f is $<^{\mu}$-recursive by Theorem 30.3, we see that ψ is $<^{\mu}$-recursive.

In defense of the constructive infinite rule we submit the following argument. Many theorems in first order proof theory follow from the cut-elimination theorem. This is still true even for higher order proof theory in which the cut-elimination theorem is proved constructively. However, if one wishes to consider an extension of arithmetic, it is impossible to eliminate all cuts due to the fact that the formal proofs contain applications of mathematical induction. Schütte has introduced the ω-rule and eliminated all applications of the cut rule and ind in first order arithmetic. This is an excellent idea and can be considered an improved form of cut-elimination when ind is involved. However, since the main objective of our investigation is a finite proof, it is better if we can restrict the ω-rule so that the infinite proofs considered are possessed of some important properties of finite proofs. For this reason we consider the constructive ω-rule.

The adequacy of the constructive ω-rule has been proved by Shoenfield for first order arithmetic, and by Takahashi for second order arithmetic. Therefore, mathematically the constructive ω-rule is strong enough.

§31. Reflection principles

DEFINITION 31.1. (1) Let **P** be **PA** augmented with second order free variables which function as parameters.

(2) For the sake of technical convenience, we restrict the constants in **ISN** to the individual constants 0, 1; function constants $+$, $\cdot$; predicate

constants $=$, $<$; and we will use $\vee$, $\supset$ and $\exists$ as well as $\neg$, $\wedge$ and $\forall$ as logical symbols.

(3) A first order formula with second order parameters $\alpha_1, \ldots, \alpha_m$ is called rudimentary in $\alpha_1, \ldots, \alpha_m$ if every (first order) quantifier is bounded, that is, quantifiers occur in the form $\forall x\ (x < s \supset \ldots)$ or $\exists x\ (x < s \wedge \ldots)$ for a term s. These formulas will be denoted by $\forall x < s\,(\ldots)$ and $\exists x < s\,(\ldots)$, respectively.

We assume a standard Gödel numbering for expressions and notions concerning **ISN**. Because of (2) of Definition 31.1, we may assume that the mathematical initial sequents are those of Definition 9.3. The first purpose of this section is to prove the reflection principle in the following form.

THEOREM 31.2 (Takeuti and Yasugi). *Let $R(\alpha, a, b)$ be rudimentary in α and let* $\mathrm{Ind}_1(O(\mathbf{ISN}))$ *be the formula which expresses transfinite induction through $O(\mathbf{ISN})$ for the Σ_1^0-formulas (i.e., the formulas of the form $\exists x\, R(x, a)$, R recursive without second order parameters). Then*

$$\mathrm{Ind}_1(O(\mathbf{ISN})), \mathrm{Prov}(\ulcorner \forall x\, \exists y\, R(\alpha, x, y)\urcorner) \to \forall x\, \exists y\, R(\alpha, x, y)$$

is provable in **P**, *where $\ulcorner A \urcorner$ is the Gödel number of A and* $\mathrm{Prov}(\ulcorner A\urcorner)$ *means that "A is provable in* **ISN**".

In order to prove this we first observe the following.

PROPOSITION 31.3. *Let $R(a, \alpha)$ be rudimentary in α with one first order free variable a, and let $\exists x\, R(x, \alpha)$ be provable in* **ISN**. *Then there is a proof of $\exists x\, R(x, \alpha)$ in* **ISN** *containing no essential cut or induction. Moreover, this can be proved with the system of ordinal diagrams $O(\mathbf{ISN})$.*

The proposition could be stated for several parameters, $\alpha_1, \ldots, \alpha_m$, instead of just one α.

Let S be a sequent $A_1, \ldots, A_m \to B_1, \ldots, B_n$ of **ISN**. S is said to have the property (P) if the following conditions are satisfied:

p1. S contains no first order free variable.

p2. Every A_i, $1 \leq i \leq m$ is rudimentary in α.

p3. Every B_j, $1 \leq j \leq n$ is rudimentary in α or is of the form $\exists x\, R'(x, \alpha)$, where $R'(a, \alpha)$ is rudimentary in α.

We will prove the proposition in the following form:

PROPOSITION 31.4. *We can define a reduction, using $O(\mathbf{ISN})$, in such a way that if a sequent S has the property* (P) *and is provable in* **ISN**, *then its proof can be reduced to one with no essential cut or induction.*

Proposition 31.3 is only a special case of this proposition.

PROOF. The proof is for the most part the same as in §27. We introduce a new rule of inference bq, "bounded quantification":

$$\text{bq}\quad \frac{\Gamma \to \Delta, (0 < k \supset S(0)) \wedge \ldots \wedge (k-1 < k \supset S(k-1))}{\Gamma \to \Delta, \forall x\, (x < k \supset S(x))}.$$

where k is a numeral, $S(a)$ is rudimentary, and in which the formulas

$$(0 < k \supset S(0)) \wedge \ldots \wedge (k-1 < k \supset S(k-1)), \qquad \forall x\, (x < k \supset S(x))$$

are called, respectively, the auxiliary formula and the principal formula of the inference.

This rule is not regarded as one of the logical rules of inference but as a structural rule. (It is easily seen that the lower sequent of bq can be proved from its upper sequent without an essential cut or induction.) The ordinal diagram of the lower sequent of bq is defined to be the same as that of the upper sequent.

A proof is called a proof with degree if it contains applications of bq only in its end-piece as explicit inferences and is a proof with degree in the sense of §27.

We shall define the reduction of a proof P of a sequent satisfying (P). By a reduction-step we mean a process which decreases the ordinal diagram of the proof together with one or more preceding auxiliary processes which preserve the ordinal diagram of the proof. See the proof of Theorem 30.3.

Case 1. P contains an application of explicit logical inference or induction in its end-piece. We treat the cases according to the bottom most such inference.

Subcase1. Induction. As in §27.

Subcase 2. Explicit logical inference other than $\forall$: right for a first order variable. Since all the cases can be treated similarly, we give an example:

$$\frac{\Gamma \to \Delta, F(t)}{\Gamma \to \Delta, \exists x\, F(t)}$$
$$\vdots$$
$$\Gamma_0 \to \Delta_0.$$

We reduce this to P':

$$\vdots$$
$$\frac{\dfrac{\Gamma \to \Delta, F(t)}{\text{a weakening and some exchanges}}}{\Gamma \to F(t), \Delta, \exists x\, F(x)}$$
$$\vdots$$
$$\Gamma_0 \to F(t), \Delta_0.$$

The end-sequent of P' obviously satisfies (P) and P' has a smaller ordinal diagram than that of P, hence P' can be transformed to a proof without essential cuts and inductions. Then add some explicit inferences to obtain $\Gamma_0 \to \Delta_0$.

Subcase 3. Explicit $\forall$: right for a first order variable:

$$\dfrac{\begin{array}{c}\vdots\\ \Gamma\rightarrow\Delta,\ b<t\supset\tilde{R}(b,\alpha)\end{array}}{\begin{array}{c}\Gamma\rightarrow\Delta,\ \forall y\,(y<t\supset\tilde{R}(y,\alpha))\\ \vdots\\ \Gamma_0\rightarrow\Delta_1,\ \forall y\,(y<s\supset\tilde{R}'(y,\alpha)),\ \Delta_2,\end{array}}$$

where $\tilde{R}(b,\alpha)$ is rudimentary in α, t contains no variable, and s and $\tilde{R}'(y,\alpha)$ are obtained from t and $\tilde{R}(y,\alpha)$, respectively, by zero or more term-replacements. Let $t=n$ for a numeral n. If $n=0$, hence $s=0$, then, P is reduced to

$$\dfrac{c<s\rightarrow}{\Gamma_0\rightarrow\Delta_1,\ \forall y\,(y<s\supset\tilde{R}'(y,\alpha)),\ \Delta_2.}$$

If $n>0$, then for each $k<n$, let P_k be

$$\dfrac{\begin{array}{c}\vdots\\ \Gamma\rightarrow\Delta,\ k<n\supset\tilde{R}(k,\alpha)\end{array}}{\begin{array}{c}\Gamma\rightarrow k<n\supset\tilde{R}(k,\alpha),\ \Delta,\ \forall y\,(y<n\supset\tilde{R}(y,\alpha))\\ \vdots\\ \Gamma_0\rightarrow k<n\supset\tilde{R}'(k,\alpha),\ \Delta_1,\ \forall y\,(y<s\supset\tilde{R}'(y,\alpha)),\ \Delta_2.\end{array}}$$

where $\Gamma\rightarrow\Delta,\ k<n\supset\tilde{R}(k,\alpha)$ is the end-sequent of the proof obtained from that of $\Gamma\rightarrow\Delta,\ b<t\supset\tilde{R}(b,\alpha)$ by substituting k for b. Every substitution in P_k is assigned the same degree as the corresponding one in P. Then P is reduced to $P_0, P_1, \ldots, P_{n-1}$, for

$$\dfrac{P_0P_1\ldots P_{n-1}}{\dfrac{\Gamma_0\rightarrow\Delta',\ (0<n\supset\tilde{R}'(0,\alpha))\wedge(1<n\supset\tilde{R}'(1,\alpha))\wedge\ldots\wedge(n-1<n\supset\tilde{R}'(n-1,\alpha))}{\dfrac{\Gamma_0\rightarrow\Delta',\ \forall y\,(y<n\supset\tilde{R}'(y,\alpha))}{\Gamma_0\rightarrow\Delta_1,\ \forall y\,(y<s\supset\tilde{R}'(y,\alpha)),\ \Delta_2,}}\ \text{(bq)}}$$

where Δ' denotes $\Delta_1,\ \forall y\,(y<s\supset\tilde{R}'(y,\alpha)),\ \Delta_2$, is a proof of the end-sequent of P.

Case 2. P contains no explicit logical inference or induction but contains an axiom of the form $s=t,\ A(s)\rightarrow A(t)$ in its end-piece. Do the reduction as in §27.

Case 3. P contains no explicit logical inference or induction or axiom of the form $s=t,\ A(s)\rightarrow A(t)$, but contains either an explicit logical axiom or an implicit logical axiom of the form $D \to D$, e.g.,

$$\dfrac{\begin{matrix} \vdots \\ \Gamma \to \Delta, \tilde{D} \end{matrix} \qquad \begin{matrix} D \to D \\ \vdots \\ \tilde{D}, \Pi \to \Lambda_1, \tilde{\tilde{D}}, \Lambda_2 \end{matrix}}{\begin{matrix} \Gamma, \Pi \to \Delta, \Lambda_1, \tilde{\tilde{D}}, \Lambda_2 \\ \vdots \\ \Gamma_0 \to \Delta_0. \end{matrix}}$$

where $\tilde{D}$ and $\tilde{\tilde{D}}$ in the right upper sequent of the cut are the descendants of D's in the antecedent and succedent of $D \to D$, respectively. If the former, then the end-sequent of P is obtained from it by weakenings, exchanges and bq's. If the latter, and $\tilde{D}$ and $\tilde{\tilde{D}}$ are the same, up to term-replacement, we apply the corresponding reduction in §27. Otherwise, $\tilde{D}$ and $\tilde{\tilde{D}}$ are of the form

$$(s_0 < s \supset S(s_0)) \wedge \ldots \wedge (s_{n-1} < s \supset S(s_{n-1}))$$

and $\forall y\,(y < t \supset S'(y))$, respectively, where $s = n$ and $t = n$ for some numeral n, $s_i = i\ (i < n)$ and $S'(y)$ is either $S(y)$ itself or else obtained from it by term replacements. Then P is reduced to

$$\begin{array}{l} \quad \vdots \\ \quad \Gamma \to \Delta, \tilde{D} \\ \hline\hline \quad \Gamma \to \Delta, (0 < n \supset S'(0)) \wedge \ldots \wedge (n-1 < n \supset S'(n-1)) \\ \hline \text{bq} \quad \Gamma \to \Delta, \forall y\,(y < n \supset S'(y)) \\ \hline\hline \Gamma, \Pi \to \Delta, \Lambda_1, \tilde{\tilde{D}}, \Lambda_2 \\ \quad \vdots \\ \Gamma_0 \to \Delta_0 \end{array}$$

Case 4. Elimination of weakenings in the end-piece of P is defined as usual. If the last inference of a proof Q is a bq, say

$$Q \qquad Q_0 \left\{ \dfrac{\begin{matrix} \vdots \\ \Gamma \to \Delta, (0 < k \supset S(0)) \wedge \ \ldots \wedge (k-1 < k \supset S(k-1)) \end{matrix}}{\Gamma \to \Delta, \forall y < k\, S(y)} \right. ,$$

then the definition goes as follows.

If Q_0^* is $\begin{matrix} \vdots \\ \Gamma^* \to \Delta^* \end{matrix}$, then Q^* is Q_0^*. If Q_0^* is

$$\begin{matrix} \vdots \\ \Gamma^* \to \Delta^*, (0 < k \supset S(0)) \wedge \ldots \wedge (k-1 < k \supset S(k-1)), \end{matrix}$$

then Q^* is

$$\dfrac{Q_0^*}{\Gamma^* \to \Delta^*, \forall y\,(y < k\, S(y))} .$$

Case 5. In the following we assume that the end-piece does not contain any logical inference, induction, initial sequent other than mathematical initial sequents or weakening, while it may contain some applications of bq. We may also assume that the proof is different from its end-piece, for if the entire proof is the end-piece, then the end-sequent is provable from the mathematical initial sequents by bq, exchanges, contractions and non-essential cuts, and hence bq can be eliminated without use of an essential cut or induction. The existence of an essential cut and the essential reduction are carried out as usual, since applications of bq are all explicit.

This completes the proof of the proposition.

We consider an arithmetization of **ISN** in **P**. Let us introduce the following notational conventions:

Pf(⌜P⌝) for "P is a proof in **ISN**";

Prov(⌜P⌝, ⌜S⌝) for "P is a proof of a sequent S";

Prov(⌜S⌝) for "S is provable";

Prov(⌜A⌝) for "Prov(⌜$\rightarrow A$⌝)";

Pf*(⌜P⌝) for "P is a proof without an essential cut or induction";

Prov*(⌜P⌝, ⌜S⌝) for "P is a proof of S without an essential cut or induction";

Prov*(⌜S⌝) for "S is provable without an essential cut or induction";

Prov*(⌜A⌝) for "Prov*(⌜$\rightarrow A$⌝)".

It should be noted that under the assumption of this section **ISN** is axiomatizable, i.e., the set of the schemata for mathematical initial sequents is finite.

PROPOSITION 31.5. *Let $R(a, \alpha)$ be rudimentary in α. Then*

$$\mathrm{Ind}_1(O(\mathbf{ISN})), \mathrm{Prov}(\ulcorner \exists y\, R(y, \alpha)\urcorner) \rightarrow \mathrm{Prov}^*(\ulcorner \exists y\, R(y, \alpha)\urcorner)$$

is provable in **P**.

PROOF. This is proved by arithmetization of the proof of Proposition 31.3. We shall give only the outline of the proof that $\mathrm{Ind}_1(O(\mathbf{ISN}))$ is adequate.

First let us introduce some notational conventions. Assume that p denotes the Gödel number of a proof P in **ISN**. Then

ends(p) is the Gödel number of the end-sequent of P;

$Q(p)$ is true if and only if the end-sequent of P has the property (P);

$C(p)$ is true if and only if P is a proof which has no essential cut or induction;

$\tilde{o}(p)$ is defined by $\tilde{o}(p) = o(p) \# 0^{(p-1)}$, where $o(p)$ is the ordinal diagram of P and $0^{(p-1)}$ is as defined in Definition 30.2 (2). Note that $\tilde{o}(p)$ is an ordinal diagram of $O(\mathbf{ISN})$ and all these predicates and functions are primitive recursive.

Now from the proof of Proposition 31.4, we can define a primitive recursive function r as follows. Let p be the Gödel number of a proof P. If $C(p) \vee \neg Q(P)$, then define $r(p) = p$. If $\neg C(p) \wedge Q(p)$, define $r(p)$ to be the Gödel number of the resulting proof of the reduction of P. Then r is primitive recursive and satisfies the following.

1) $\tilde{o}(r(p)) < \tilde{o}(p)$ if $\neg C(p) \wedge Q(p)$.

2) $\tilde{o}(r(p)) = \tilde{o}(p)$ if $C(p)$.

Define $\tilde{r}(a, b)$ by

$$\tilde{r}(0, p) = p; \qquad \tilde{r}(n+1, p) = r(\tilde{r}(n, p)).$$

Then $\tilde{r}(a, b)$ is primitive recursive. Finally, define

$$p <\cdot\, q \Leftrightarrow_{\mathrm{df}} \tilde{o}(p) < \tilde{o}(q).$$

Then $<\cdot$ is a primitive recursive well-ordering of the natural numbers. Furthermore, the order type of $<\cdot$ is that of $O(\mathbf{ISN})$. So transfinite induction can be applied to the ordering $<\cdot$, with induction formula $Q(p) \supset \exists n\, C(r(n, p))$, or equivalently, $\exists n\, (Q(p) \supset C(r(n, p)))$, which is Σ^0_1.

DEFINITION 31.6. (1) A formula of **ISN** is said to have the property (Q) if it contains no second order quantifiers or first order free variables.

For every formula A having the property (Q) we define the subformulas of A as follows: A is a subformula of A; if $B \wedge C$ is a subformula of A then so are B and C. If $\neg C$ is a subformula of A then so is C; if $\forall x\, B(x)$ is a subformula of A, then so is $B(n)$ for every numeral n. Evidently, every subformula of A has the property (Q).

(2) We can give a truth-definition T_A for the subformulas of A and also for sequents consisting only of such subformulas. The truth definition is an arithmetical formula with second order parameters (i.e., free variables).

PROPOSITION 31.7. *Let A be a formula having the property* (Q). *Then the following are provable in* **P**.

(1) $\mathrm{T}_A(\ulcorner \neg B \urcorner) \leftrightarrow \neg \mathrm{T}_A(\ulcorner B \urcorner)$ *for every subformula B of A.*

(2) $\mathrm{T}_A(\ulcorner B \vee C \urcorner) \leftrightarrow \mathrm{T}_A(\ulcorner B \urcorner) \vee \mathrm{T}A(\ulcorner C \urcorner)$ *for every pair B and C subformulas of A.*

(3) $\mathrm{T}_A(\ulcorner \forall x_i\, B(x_i) \urcorner) \leftrightarrow \forall x\, \mathrm{T}_A(\ulcorner B(n(x)) \urcorner)$ *for every subformula $\forall x_i\, B(x)$ of A; here $n(a)$ denotes the ath numeral.*

(4) $\mathrm{T}_A(\ulcorner B(n(b_1), \ldots, n(b_k)) \urcorner) \leftrightarrow B(b_1, \ldots, b_k)$, *where $B(0, \ldots, 0)$ is an arbitrary subformula of A such that originally $B(y_1, \ldots, y_k)$ for some bound variables $y_1, \ldots, y_k$ occurred in A.*

(5) $P_A(a) \wedge \mathrm{Prov}^*(a) \rightarrow \mathrm{T}_A(a)$, *where $P_A(a)$ means "a is* (*the Gödel number of*) *a sequent consisting of subformulas of A".*

PROOF. (1) through (4) can be proved in the same manner as for the truth definition of **PA**.

(5) Assume $P_A(a)$ and $\mathrm{Prov}^*(a)$, and let P be a proof such that $\mathrm{Prov}^*(\ulcorner P \urcorner, a)$. We can show by induction on the number of inferences in

P, using (1)–(4), that

$$\mathrm{T}_A(\ulcorner \Gamma(n(c_1), \dots, n(c_k)) \to \Delta(n(c_1), \dots, n(c_k))\urcorner)$$

is provable in $\mathbf{P}$, where $n(c)$ denotes the cth numeral and $\Gamma \to \Delta$ is a sequent in $\mathbf{P}$.

PROOF OF THEOREM 31.2. Take $\forall x\, \exists y\, R(x, y, \alpha)$ as the A in Proposition 31.7 and let $T(a)$ denote $T_A(a)$. Then

(1) $\mathrm{Prov}(\ulcorner \forall x\, \exists y\, R(x, y, \alpha)\urcorner) \to \forall a\, \mathrm{Prov}(\ulcorner \exists y\, R(n(a), y, \alpha)\urcorner)$ is provable in $\mathbf{P}$.

By Proposition 31.5,

(2) $\mathrm{Ind}_1(O(\mathbf{ISN})), \mathrm{Prov}(\ulcorner \exists y\, R(n(a), y, \alpha)\urcorner) \to \mathrm{Prov}^*(\ulcorner \exists y\, R(n(a), y, \alpha)\urcorner)$ is provable in $\mathbf{P}$ for any free variable a.

By virtue of Proposition 31.7 the following are provable in $\mathbf{P}$:

(3) $\mathrm{Prov}^*(\ulcorner \exists y\, R(u(a), y, \alpha)\urcorner) \to \mathrm{T}(\ulcorner \exists y\, R(n(a), y, \alpha)\urcorner)$, since $P_A(\ulcorner \exists y\, R(n(a), y, \alpha)\urcorner)$ is provable in $\mathbf{P}$;

(4) $\forall a\, \mathrm{T}(\ulcorner \exists y\, R(n(a), y, \alpha)\urcorner) \to \mathrm{T}(\ulcorner \forall x\, \exists y\, R(x, y, \alpha)\urcorner)$;

(5) $\mathrm{T}(\ulcorner \forall x\, \exists y\, R(x, y, \alpha)\urcorner) \to \forall x\, \exists y\, R(x, y, \alpha)$.

The theorem follows from (1)–(5).

We can prove the uniform reflection principle by modifying the proof of Theorem 31.2.

THEOREM 31.8. *In* $\mathbf{P}$:

$$\mathrm{Ind}_1(O(\mathbf{ISN})) \to \forall m\, (\mathrm{Prov}(\ulcorner \forall x\, \exists y\, R(x, y, \alpha, n(m))\urcorner) \supset \forall x\, \exists y\, R(x, y, \alpha, m)).$$

PROOF. We first note that a modified version of Proposition 31.5:

$$\mathrm{Ind}_1(O(\mathbf{ISN})), \mathrm{Prov}(\ulcorner \exists x\, R'(x, \alpha, n(m)\urcorner) \to \mathrm{Prov}^*(\ulcorner \exists x\, R'(x, \alpha, n(m))\urcorner),$$

is provable, by replacing $\ulcorner \exists x\, R(x, \alpha)\urcorner$ by $\ulcorner \exists x\, R'(x, \alpha, n(m))\urcorner$ in Proposition 31.5. Taking $\forall z\, \forall x\, \exists y\, R'(x, y, \alpha, z)$ as A in Proposition 31.7, it follows that (1)–(5) in the proof of Theorem 31.2 are provable with $\ulcorner \forall x\, \exists y\, R'(x, y, \alpha, n(m))\urcorner$ instead of $\ulcorner \forall x\, \exists y\, R(x, y, \alpha)\urcorner$. With this observation, the theorem follows easily.

Now let $B(\alpha)$ be an arbitrary formula of $\mathbf{P}$ of the form

$$(*) \qquad \exists x_1\, \forall y_1 \dots \exists x_n\, \forall y_n\, B_0(\alpha, x_1, y_1, \dots, x_n, y_n),$$

where $B_0(\alpha, a_1, b_1, \dots, a_n, b_n)$ is a quantifier-free formula whose only free variables are $\alpha, a_1, b_1, \dots, a_n, b_n$.

The subformulas of $B(\alpha)$ are defined as in Definition 31.6.

PROPOSITION 31.9. *Given* $B(\alpha)$ *which satisfies* (*), *we can define the truth definition* $\mathrm{T}_{B(\alpha)}$ *for subformulas of* $B(\alpha)$ *in* $\mathbf{P}$ *with a* Σ^0_{2n}*-formula having the second order parameter* α. *It is obvious that* $\mathrm{T}_{B(\alpha)}$ *can be extended to sequents consisting of some subformulas of* $B(\alpha)$.

DEFINITION 31.10. Let $B(\alpha)$ be a formula satisfying (*). We define the condition $\mathscr{S}_{B(\alpha)}$ as follows; let $\ulcorner S \urcorner$ denote the Gödel number of the sequent S. We use quotes to mean that the quoted sentence is actually an arithmetized formula.

$\mathscr{S}_1(B(\alpha); \ulcorner S \urcorner)$: "Each formula of S is a subformula of $B(\alpha)$".

$\mathscr{S}_2(\ulcorner S \urcorner)$: "Each formula in the antecedent of S is quantifier-free".

$\mathscr{S}_3(B(\alpha); \ulcorner S \urcorner)$: $\neg \mathrm{T}_{B(\alpha)}(\ulcorner S \urcorner)$.

$\mathscr{S}_{B(\alpha)}(\ulcorner S \urcorner)$: $\mathscr{S}_1(B(\alpha); \ulcorner S \urcorner) \wedge \mathscr{S}_2(\ulcorner S \urcorner) \wedge \mathscr{S}_3(B(\alpha); \ulcorner S \urcorner)$.

From now throughout, $B(\alpha)$ shall be arbitrary but fixed so that it satisfies (*). For simplicity we shall abbreviate $\mathrm{T}_{B(\alpha)}$ and $\mathscr{S}_{B(\alpha)}$ as T and $\mathscr{S}$, respectively.

PROPOSITION 31.11.

$$\mathrm{Ind}_2(O(\mathbf{ISN})), \mathrm{Prov}(p, \ulcorner B(\alpha) \urcorner), \neg \mathrm{T}(\ulcorner B(\alpha) \urcorner) \\ \to \exists q \leqslant\cdot\, p(\mathrm{Pf}^*(q) \wedge \mathscr{S}(\mathrm{ends}(q)))$$

is **P**-*provable, where* $<\cdot$ *is the well-ordering of natural numbers defined in the proof of Proposition* 31.5 *and* $\mathrm{Ind}_2(O(\mathbf{ISN}))$ *is the schema which allows transfinite induction along the order* $<\cdot$ *applied to* Σ^0_{2n+1}*-formulas.*

The proposition is an immediate consequence of the following:

(**) $\mathrm{Ind}_2(O(\mathbf{ISN})), \mathscr{S}(\ulcorner S \urcorner), \mathrm{Prov}(p, \ulcorner S \urcorner) \to \exists q \leqslant\cdot\, p\, (\mathrm{Pf}^*(q) \wedge \mathscr{S}(\mathrm{ends}(q)))$ is **P**-provable.

Therefore, we shall prove (**). It is proved by applying $\mathrm{Ind}_2(O(\mathbf{ISN}))$ to the following formula:

$$\mathscr{S}(\mathrm{ends}(p)) \wedge \mathrm{Pf}(p) \supset \exists q \leqslant\cdot\, p\, (\mathrm{Pf}^*(q) \wedge \mathscr{S}(\mathrm{ends}(q))). \tag{1}$$

Since T and $\mathscr{S}$ are in Σ^0_{2n} and Π^0_{2n}, respectively, the induction formula is in Σ^0_{2n+1} with the parameter α.

It is obvious that in order to prove (1) it suffices to show

$$\mathscr{S}(\mathrm{ends}(p)) \wedge \mathrm{Pf}(p) \wedge \neg \mathrm{Pf}^*(p) \supset \exists q(\mathscr{S}(\mathrm{ends}(q)) \wedge \mathrm{Pf}^*(q) \wedge q <\cdot\, p), \tag{2}$$

for if $\mathrm{Pf}^*(p)$, then we may take p itself as q in (1).

Assume $\mathscr{S}(\mathrm{ends}(p)) \wedge \mathrm{Pf}(p) \wedge \neg \mathrm{Pf}^*(p)$ and find a q which satisfies (2). This is done in the same manner as in the consistency proofs of **ISN**, although, strictly speaking, the whole argument is developed in the arithmetized language.

Let P be the proof with Gödel number p.

1) Preparations for reduction as in §27 are applicable.

2) If there is an explicit logical inference or an induction in the end-piece of P, then the proof is carried out according to the bottom most such inference.

2.1) The last such inference is a first order $\exists$:right. Let P be of the form

$$\frac{\Gamma \to \Delta, \forall y_i \ldots \exists x_n \forall y_n B_0(\alpha, t_1, s_1, \ldots, t_i, y_i, \ldots, x_n, y_n)}{\Gamma \to \Delta, \exists x_i \forall y_i \ldots \exists x_n \forall y_n B_0(\alpha, t_1, s_1, \ldots, x_i, y_i, \ldots, x_n, y_n)}$$

$$\Pi \to \Lambda_1, \exists x_i \forall y_i \ldots \exists x_n \forall y_n B_0(\alpha, m_1, l_1, \ldots, x_i, y_i, \ldots, x_n, y_n), \Lambda_2.$$

Notice that t_i is a closed term which consists of 0, 1, + and $\cdot$. Therefore, t_i can be computed and is equal to a numeral m_i. P is reduced to the following.

$$\frac{\Gamma \to \Delta, \forall y_i \ldots \exists x_n \forall y_n B_0(\alpha, t_1, s_1, \ldots, t_i, y_i, \ldots, x_n, y_n)}{\begin{array}{r}\Gamma \to \forall y_i \ldots \exists x_n \forall y_n B_0(\alpha, t_1, s_1, \ldots, m_i, y_i, \ldots, x_n, y_n), \Delta, \\ \exists x_i \forall y_i \ldots \exists x_n, \forall y_n B_0(\alpha, t_1, s_1, \ldots, x_i, y_i, \ldots, x_n, y_n),\end{array}}$$

$$\begin{array}{r}\Pi \to \forall y_i \ldots \exists x_n \forall y_n B_0(\alpha, m_1, l_1, \ldots, m_i, y_i, \ldots, x_n, y_n), \Lambda_1, \\ \exists x_i \forall y_i \ldots \exists x_n \forall y_n B_0(\alpha, m_1, l_1, \ldots, x_i, y_i, \ldots, x_n, y_n), \Lambda_2\end{array}$$

$$\neg \mathrm{T}(\ulcorner \exists x_i \forall y_i \ldots \exists x_n \forall y_n B_0(\alpha, m_1, l_1, \ldots, x_i, y_i, \ldots, x_n, y_n)^\urcorner)$$

implies

$$\neg \mathrm{T}(\ulcorner \forall y_i \ldots \exists x_n \forall y_n B_0(\alpha, m_1, l_1, \ldots, m_i, y_i, \ldots, x_n, y_n)^\urcorner).$$

2.2) The last inference which satisfies the condition is a first order $\forall$: right. Let P be of the form

$$\frac{\Gamma \to \Delta, \exists x_{i+1} \ldots \exists x_n \forall y_n B_0(\alpha, t_1, s_1, \ldots, a, x_{i+1}, \ldots, x_n, y_n)}{\Gamma \to \Delta, \forall y_i \exists x_{i+1} \ldots \exists x_n \forall y_n B_0(\alpha, t_1, s_1, \ldots, y_i, x_{i+1}, \ldots, x_n, y_n)}$$

$$\Pi \to \Lambda_1, \forall y_i \exists x_{i+1} \ldots \exists x_n \forall y_n B_0(\alpha, m_1, l_1, \ldots, y_i, x_{i+1}, \ldots, x_n, y_n), \Lambda_2.$$

This is reduced to

(l_i)

$$\frac{\Gamma \to \Lambda, \exists x_{i+1} \ldots \exists x_n \forall y_n B_0(\alpha, t_1, s_1, \ldots, l_i, x_{i+1}, \ldots, x_n, y_n)}{\begin{array}{r}\Gamma \to \exists x_{i+1} \ldots \exists x_n \forall y_n B_0(\alpha, t_1, s_1, \ldots, l_i, x_{i+1}, \ldots, x_n, y_n), \Delta, \\ \forall y_i \exists x_{i+1} \ldots \exists x_n \forall y_n B_0(\alpha, t_1, s_1, \ldots, y_i, x_{i+1}, \ldots, x_n, y_n)\end{array}}$$

$$\begin{array}{r}\Pi \to \exists x_{i+1} \ldots \exists x_n \forall y_n B_0(\alpha, m_1, l_1, \ldots, l_i, x_{i+1}, \ldots, x_n, y_n), \Lambda_1, \\ \forall y_i \exists x_{i+1} \ldots \forall y_n B_0(\alpha, m_1, l_1, \ldots, y_i, x_{i+1}, \ldots, x_n, y_n), \Lambda_2,\end{array}$$

where (l_i) means the substitution of the numeral l_i for the free variable a in the proof and l_i is chosen so that

$$\mathrm{T}(\ulcorner \exists x_{i+1} \ldots \exists x_n \forall y_n B_0(\alpha, m_1, l_1, \ldots, l_i, x_{i+1}, \ldots, x_n, y_n)\urcorner)$$

holds, when

$$\neg T(\ulcorner \forall y_i \exists x_{i+1} \ldots \exists x_n \forall y_n B_0(\alpha, m_1, l_1, \ldots, y_i, x_{i+1}, \ldots, x_n, y_n)\urcorner)$$

is assumed.

2.3) All other cases of logical inferences are dealt with easily. By virtue of $\mathcal{S}_1(B(\alpha), \text{ends}(p))$ and $\mathcal{S}_2(\text{ends}(p))$, there is no first order $\forall$: left and no first order $\exists$: left.

2.4) The last inference which satisfies the condition is an ind. This case is proved as in §27.

3) Now we may assume that there is no explicit logical inference or induction in the end-piece of P. Hereafter we can follow exactly the consistency proof of §27. Thus we have proved (**).

PROPOSITION 31.12. $\mathrm{Ind}_2(O(\mathbf{ISN}))$, $\mathrm{Prov}(\ulcorner B(\alpha)\urcorner), \neg\mathrm{T}(\ulcorner B(\alpha)\urcorner) \rightarrow$ *is* **P**-*provable, where* $B(\alpha)$ *satisfies* (*).

PROOF. From the definition of T, $\mathrm{Pf}^*(q) \rightarrow \mathrm{T}(\text{ends}(q))$. But this contradicts $\mathcal{S}_3(\text{ends}(q), B(\alpha))$. Thus the proposition follows from Proposition 31.11.

Now we can present another form of the reflection principle for **ISN**.

THEOREM 31.13 (Takeuti and Yasugi).

$$\mathrm{Ind}_2(O(\mathbf{ISN})), \mathrm{Prov}(\ulcorner A(\alpha)\urcorner) \rightarrow A(\alpha)$$

is **P**-*provable for an arbitrary arithmetical sentence* $A(\alpha)$ *with a second order parameter* α, *where* $\mathrm{Ind}_2(O(\mathbf{ISN}))$ *applies to the formulas of* **P**, *that is, to the formulas arithmetical in some second order parameters.*

PROOF. It is well known that

(1) $$A(\alpha) \leftrightarrow B(\alpha)$$

is **P**-provable for some $B(\alpha)$ which satisfies (*).

(2) $$\mathrm{Ind}_2(O(\mathbf{ISN})), \mathrm{Prov}(\ulcorner B(\alpha)\urcorner) \rightarrow \mathrm{T}_{B(\alpha)}(\ulcorner B(\alpha)\urcorner)$$

and

(3) $$\mathrm{T}_{B(\alpha)}(\ulcorner B(\alpha)\urcorner) \rightarrow B(\alpha)$$

are **P**-provable from Propositions 31.12 and 31.7, respectively. It is also known that

(4) $$\mathrm{Prov}(\ulcorner A(\alpha)\urcorner) \leftrightarrow \mathrm{Prov}(\ulcorner B(\alpha)\urcorner)$$

is **P**-provable. Then (1)–(4) yield the theorem.

Here again we can prove the uniform reflection principle.

THEOREM 31.14.

$$\mathrm{Ind}_2(O(\mathbf{ISN})) \rightarrow \forall m\,(\mathrm{Prov}(\ulcorner A(\alpha, n(m))\urcorner) \supset A(\alpha, m)),$$

where $\mathrm{Ind}_2(O(\mathbf{ISN}))$ *applies to the formulas of* **P**.

PROOF. This is proved with modifications similar to those that have been carried out in the proof of Theorem 31.8: First apply (**) in the proof of Proposition 31.11 to $\ulcorner B(\alpha, n(m))\urcorner$ in the place of $\ulcorner B(\alpha)\urcorner$. Then take $\forall z\, B(\alpha, z)$ as $B(\alpha)$ and define the truth definition for $B(\alpha)$. The rest of the proof of Theorem 31.13 goes through after this alteration.

We now present another formulation of the reflection principle for the formulas $\forall \phi\, A(\phi)$, where $A(\alpha)$ is arithmetical in α. We shall state it in the form of the uniform reflection principle.

THEOREM 31.15. *Let* $A(\alpha, a)$ *be arithmetical in* α *and let* α *and* a *be the only free variables of* A. *Then*

(1) $$\mathrm{Ind}'(O(\mathbf{ISN})), \mathrm{Prov}(\ulcorner \forall \phi\, A(\phi, n(a))\urcorner) \rightarrow \forall \phi\, A(\phi, a)$$

is **ISN**-*provable, where* Ind′ *applies to* Σ_3^0-*formulas with a second order parameter.*

PROOF. First, with a slight extension of the language of **ISN** as specified in Definition 31.1, there exists a quantifier-free formula $R(\alpha, b, c, a)$ for which

(2) $$\forall \phi\, A(\phi, a) \leftrightarrow \forall \phi\, \exists x\, \forall y\, R(\phi, x, y, a)$$

is **ISN**-provable. Then (2) implies that

(3) $$\mathrm{Prov}(\ulcorner \forall \phi\, A(\phi, n(a))\urcorner) \leftrightarrow \mathrm{Prov}(\ulcorner \forall \phi\, \exists x\, \forall y\, R(\phi, x, y, n(a))\urcorner)$$

is **ISN**-provable. Finally, (2) and (3) guarantee that, in order to prove (1), we only have to prove

(4) $$\mathrm{Ind}'(O(\mathbf{ISN})), \mathrm{Prov}(\ulcorner \exists x\, \forall y\, R(\alpha, x, y, n(a))\urcorner) \rightarrow \exists x\, \forall y\, R(\alpha, x, y, a)$$

in **ISN**. But (4) follows from

(5)
$$\mathit{Ind}'(O(\mathbf{ISN})), \mathrm{Prov}(\ulcorner \exists x\, \forall y\, R(\alpha, x, y, n(a))\urcorner), \neg \mathrm{T}(\ulcorner \exists x\, \forall y\, R(\alpha, x, y, n(a))\urcorner) \rightarrow,$$

which is proved like Proposition 31.12.

Notice that T is the truth definition for $\exists x\, \forall y\, R(\alpha, x, y, a)$ so that we may assume it is a Σ_2^0-formula with the parameter α. But this implies that $\mathrm{Ind}'(O(\mathbf{ISN}))$ applies to Σ_3^0-formulas with the parameter α.

POSTSCRIPT

A very interesting recent development of proof theory is J. Y. Girard's Π^1_2-logic. He generalized the notion of proof by using categorical language. Therefore his notion proof is not purely syntactical as treated in this book. Nevertheless his work gives much information about the syntactical proofs. The reader is advised to read his book: Proof Theory and Logical Complexity, Bibliopolis, Napoli.

Now we give some references which are not mentioned in the course of the book.

CHAPTER 1.

Most material from §1 to §6 comes from a paper by

G. Gentzen: Untersuchungen über das logische Schliessen, Mathematische Zeitschrift 39 (1934) 176–210, 405–431.

In this paper Gentzen also introduced a system of the first order predicate calculus **NK** called natural deduction (**NJ** for the intuitionistic predicate calculus). The motivation of **NK** is to have a logical calculus close to actual reasoning. However this system is not convenient to formulate the cut-elimination theorem. So he invented **LK** and **LJ**.

D. Prawitz restored the theoretical value of natural deduction by proving a version of the cut-elimination theorem for **NJ**, called normalization theorem.

D. Prawitz: Natural Deduction, A Proof-Theoretic Study, Almquist and Wiksell, Stockholm (1965).

Natural deduction and normalization theorem are very successful also for higher order. For this, see

D. Prawitz: Ideas and Results in Proof Theory, Proc. of the 2nd Scandinavian Logic Symposium, Amsterdam (1971) 235–307.

A careful and detailed analysis of the correspondence between cut-elimination and normalization for the intuitionistic system including arithmetic was done by J. Zucker in his paper

J. Zucker: The Correspondence between Cut-elimination and Normalization, Annals of Math. Logic 7 (1974) 1–156.

The facts in Problem 3.12 are called Gentzen–Gödel double-negation interpretation. K. Gödel showed a similar interpretation of the classical number theory in the intuitionistic number theory in his paper

K. Gödel: Zur intuitionistische Arithmetik und Zahlen Theorie,

Ergebnisse eines mathematischen Kolloquium 4 (1933) 34–38.

Beth's definability theorem (Proposition 6.11) is proved by

E. W. Beth: On Padoa's Method in the Theory of Definition, Indag. Math. 15 (1953) 330–339.

Craig's interpolation theorem (Theorem 6.6) is proved by

W. Craig: Three users of the Herbrand-Gentzen Theorem in relating Model Theory and Proof Theory, J. Symbolic Logic 22 (1957) 269–285.

Craig also showed an elegant proof of Beth's definability theorem by using this interpolation theorem. Our proof of Craig's interpolation theorem is due to

S. Maehara: On the Interpolation Theorem of Craig (in Japanese), Sugaku 12 (1960–61) 235–237.

A semantical proof of Craig's interpolation theorem was given in

A. Robinson: A Result on Consistency and its Applications to the Theory of Definition, Indag. Math. 18 (1956) 47–58.

By using Maehara's method, K. Schütte extended the interpolation theorem for **LJ** in his paper:

K. Schütte: Der Interpolationssatz der intuitionistischen Pradikatenlogik, Math. Ann. 148 (1962) 192–200.

His result has further extended to the case with function symbols in **LJ** by

T. Nagashima: An Extension of the Craig–Schütte Interpolation Theorem, Ann. of Japan Assoc. Philos. Sci. 3 (1966) 12–18.

Harrop's theorem ((1) of Theorem 6.14) is proved by

R. Harrop; Concerning Formulas of the Types $A \rightarrow B \vee C$, $A \rightarrow (\exists x)B(x)$ in Intuitionistic Formal Systems, J. Symbolic Logic 25 (1960) 27–32.

A simpler proof of Harrop's theorem was given by

S. C. Kleene, Disjunction and Existence under Implication in Elementary Intuitionistic Formalisms, J. Symbolic Logic 27 (1962) 11–18.

The completeness theorem for the first order predicate calculus was proved by

K. Gödel, Dir Vollständigkeit der Axiome des logischen Funktionenkalkuls, Monatshefts für Mathematic und Physik 37 (1930) 349–360.

Our proof comes from

K. Schütte: Ein System des verknupfenden Schliessens, Archiv. Math. Logic Grundlagenf. 2 (1956) 55–67.

Similar methods were devised independently by

E. W. Beth: Semantical Entailment and Formal Derivability, Indag. Math. 19 (1956) 357–388

and by

J. Hintikka: Form and Content in Quantification Theory, Two Papers on Symbolic Logic, Acta Philosophica Fennica 8 (1955) 7–55.

Also see p. 50 in

A. Mostowski: Thirty Years of Foundational studies, Lectures on the development of mathematical logic and the study of the foundations of mathematics in 1930–1964, Acta Philosophia Fennica 17 (1965) 1–180.

Corollary 8.6 was proved independently by

A. Tarski: Contributions to the Theory of Models, I, II, Indag. Math. 16 (1954) 572–588

and by

J. Łos, The Extending of Models, I, Fund. Math. 42 (1955) 38–54.

Problem 8.4 and the proof of Corollary 8.6 were proved by

S. Feferman, Lectures on Proof Theory, Proceedings of the Summer School in Logic, Leeds, 1967, Lecture Notes in Math. 70 (1968) 1–107.

In this paper, his proof of Proposition 14.2 is also given. This can be also proved by using relativization. For this, see

N. Motohashi: Two Theorems on Mix-relativization, Proc. Japan, Acad. 49 (1973) 161–163.

Problem 8.7 is so-called eliminator of ε-symbols which was proved by

D. Hilbert and P. Bernays: Grundlagen der Mathematik, I, 1934 and II, 1939, Springer, Berlin.

Our proof is due to

S. Maehara: The Predicate Calculus with ε-symbol, J. Math. Soc. Japan 7 (1955) 323–344.

However Maehara's original proof is much better than our modification written in this book in the following sense. His proof can be easily modified to be valid for **LJ**. This is done in

G. Minc: Herbrand's Theorem, Mathematical Theory of Logical Deduction, Moskow (1967).

For the elimination of ε-symbol with equality axiom, see

S. Maehara: Equality Axiom on Hilbert's ε-symbol, J. Faculty of Science, Univ. Tokyo (1957) 419–435.

The following papers should be seen for ε-symbol in intuitionistic logic.

A. Dragalin: Intuitionistic Logic and Hilbert's ε-symbol, History and methodology of natural science, 16, Moscow State University (1974) 78–84.

G. Minc: Heytig Predicate Calculus with ε-symbol, Notes of scientific seminars of LOMI 40 (1974) 101–109.

G. Minc: Finite investigation of infinite derivation, Zapiski 49 (1975) 67–122.

D. Leivant: Existential Instantiation in a System of Natural Deduction for Intuitionistic Arithmetic, Stichting Mathematisch Centrum, Amsterdam (1973).

A generalized König's lemma in Proposition 8.11 comes from

G. Takeuti: A Generalization of König's Lemma, Proc. Japan Acad. 39 (1963) 331–332.

The proof of this book is due to J. Zucker.

Kripke structures for intuitionistic logic and his completeness theorem come from

S. Kripke: Semantical Analysis of Intuitionistic Logic I, Formal Systems and Recursion Functions, edited by J. N. Crosley and M. A. E. Dummett, North-Holland, Amsterdam (1965) 92–130.

Strong form of completeness theorem for intuitionistic logic using Kripke structure was proved by

R. H. Thomason: On the Strong Semantical Completeness of the Intuitionistic Predicate Calculus, J. Symbolic Logic 33 (1968) 1–7.

The proof in this book is due to the author. A formulation **LJ′** of the intuitionistic predicate calculus is due to

S. Maehara: Eine Darstellung der intuitionistic Logik und der Klassischen, Nagoya Math. J. 7 (1954) 45–64.

A good reference about complete Heyting algebras is

M. P. Fourman and D. S. Scott: Sheaves and Logic, Applications of Sheaves, Edited by Fourman, Mulvey and Scott, Springer Lecture Notes in Math. 753 (1979) 302–401.

The completeness theorem by using complete Heyting algebras was proved in

H. Rosiowa and R. Sikorski: The Mathematic of Metamathematics, Monografie Mathematyczne 41 (1963).

The completeness theorem for $\mathbf{LJ_I}$ (Theorem 8.35) was proved by

G. Takeuti and S. Titani: Intuitionistic Fuzzy Logic and Intuitionistic Fuzzy Set Theory, J. Symbolic Logic 49 (1984) 851–866.

CHAPTER 2.

Gödel's incompleteness theorems (Theorem 10.16 and Theorem 10.18) were proved by

K. Gödel: Über formal unentscheidbare Sätze der Principia Mathematica und verwandter Systeme, I, Monatshefte für Mathematik und Physik 38 (1931) 173–198.

Löb's theorem in Problem 10.21 was proved in

M. H. Löb: Solution of a Problem of Leon Henkin, J. Symbolic Logic 20 (1955) 155–118.

For some further results in this area, see

G. Kreisel and G. Takeuti: Formally Self-referential Propositions for Cut-free Classical Analysis and Related Systems, Dissertations Math. (118 (1974)) 1–55.

Rosser's improvement of Gödel's incompleteness theorem in Problem 10.22 was proved in

J. B. Rosser: Extensions of some Theorems of Gödel and Church, J. Symbolic Logic 1 (1936) 87–91.

For Hilbert's finite standpoint in §11, see

D. Hilbert and P. Bernays: Grundlagen der Mathematik I, Springer (1934).

I strongly put my belief in expressing finite standpoint in this section. See also

G. Takeuti: Consistency Proofs and Ordinals, Proof Theory Symposium, Kiel 1974, Springer (1975) 365–369.

For Gentzen's original accessibility proof of ordinal less than ε_0 see

G. Gentzen: Die Widerspruchsfreiheit der reinen Zahlentheorie, Math. Ann. 122 (1936) 493–565

and

G. Gentzen: Beweisbarkeit und Unbeweisbarkeit von Anfangsfällen der transfiniten Induktion in der reinen Zahlentheorie, Math. Ann. 119 (1943) 149–161.

The accessibility proof described in this section is based on my philosophical point of view.

For Gödel's consistency proof of Peano Arithmetic using higher type functionals see

K. Gödel: Über eine bisher noch nicht benützte Erweiterung des finiten Standpunktes, Dialectica 12 (1958) 280–287.

The consistency of Peano Arithmetic was first proved by G. Gentzen in his 1936 paper cited above. Our proof in §12 is based in Gentzen's paper

G. Gentzen: Neue Fassung des Widerspruchsfeiheitbeweis fur die reine Zahlentheorie, Forschungen Zur Logik und zur Grundlegung der exakten Wissenschaften, Neue Folge 4 (1938) 19–44.

Scarpellini's theorem in Problem 12.13 is in

B. Scarpellini: Some Applications of Gentzen's Second Consistency Proof, Math. Ann. 181 (1969) 325–344.

A better reference is the book

B. Scarpellini: Proof Theory and Intuitionistic Systems, Springer Lecture Notes in Math. 212 (1971).

The notion "ordinal recusive function" in Remark 12.15 and the ε_0-recursive functions (see Corollary 12.16) were found by G. Kreisel, in his paper

G. Kreisel: On the Interpretation of Non-finitist Proofs II, J. Symbolic Logic 17 (1952) 43–48.

It is done more elegantly in

G. Kreisel: Some Concepts Concerning Formal Systems of Number Theory, Math. Zeitschrift 57 (1952) 1–12.

The notion $\mathbf{PA}_k$ together with Corollary 12.16 and a characterization of the primitive recursive functions stated at the end of the Corollary are independently proved by G. Minc, and the author. See

G. Minc: Exact Estimates of Provability, Soviet Mathematics I.

The characterization of provably recursive function in **PA** by Hardy functions in Theorem 12.34 is proved by

S. S. Wainer: Ordinal Recursion, and a Refinement of the Extended Grzegorczyk Hierarchy, J. Symbolic Logic 37 (1972) 281–292.

A closely related result was obtained by

H. Schwichtenberg, Eine Klassification der ε_0-rekursiven Funktionen,

Zeitschrift für Math. Logik und Grundlagen der Math. 17 (1971) 61–74.
Goodstein's theorem in Theorem 12.36 is in
R. L. Goodstein: On the Restricted Ordinal Theorem, J. Symbolic Logic 9 (1944) 33–41.
Kirby-Paris Theorem on Goodstein's theorem in Theorem 12.36 is in
L. Kirby and L. Paris: Accessible Independence Results for Peano Arithmetic, Bull. London Math. Soc. 14 (1982) 285–293.
Our proof follows
E. A. Cichon: A short proof of two recently discovered independence results using Recursion Theoretic methods, Proc. Amer. Math. Soc. 87 (1983) 704–706.
Paris-Harrington's theorem in §12 is proved in
J. Paris and L. Harrington: A Mathematical Incompleteness in Peano Arithmetic, Handbook of Mathematical Logic, J. Barwise, ed., North-Holland, Amsterdam (1977) 1133–1142.
As for Ketonen–Solovay–Quinsey's improvement of Paris-Harrington's result in Theorem 12.68, see
Ketonen–Solovay's: Rapidly Growing Ramsey Functions, Ann. of Math. 113 (1981) 267–314.
The part from Definition 12.48 to Theorem 12.66 comes from Ketonen-Solovay's paper except that many theorems are replaced by weaker statements. Quinsey's improvement is in his thesis
J. E. Quinsey, Some Problems in Logic, Oxford (April 1980).
Paris–Harrington's work opened a very prosperous field in logic. However most of the works in the area are done model theoretically. Here we gave only the following reference which is done proof theoretically.
R. Kurata: Paris–Harrington Principles, Reflection Principles and Transfinite Induction up to ε_0, Ann. Pure and Applied Logic 31 (1986) 237–256.
Our discussion on a weaker version of Friedman's theorem on Kruskal's theorem in Theorem 12.73 and Friedman's theorem on generalized Kruskal theorem in Theorem 12.35 follows an excellent expository paper
S. G. Simpson: Unprovability of Certain Combinatorial Properties of Finite Tree, in: L. Harrington, M. Morley, A. Ščedrov and S. G. Simpson, eds., Harvey Friedman's Research in the Foundation of Mathematics, North-Holland, Amsterdam (1985).
The material of §13 comes from
G. Gentzen: Beweisbarkeit und Unbeweisbarkeit von Anfangsfällen der transfiniten Induktion in der Wahrheitsbegriff in der formalisierten Sprachen, Studia Philosophica 1 (1936) 261–405.
For a modern treatment see
C. Smorynski: The incompleteness theorem, Handbook of Math. Logic, J. Barwise ed., North-Holland, Amsterdam (1977) 822–865.
Proposition 14.2 is a theorem of A. Mostowski in his paper

A. Mostowski: On Models of Axiomatic Systems, Fund. Math. 3 (1952) 133–158.

Our proof is due to Feferman as we stated before. Exercise 14.3 is a theorem of R. Montague in his paper

R. Montague: Semantic Closure and Non-finite Axiomatizability I, Infinitistic Methods, Warsaw (1961) 45–69.

CHAPTER 3.

Most material in §15, §16, §17, and §20 comes from

G. Takeuti: On the Generalized Logic Calculus, Japan. J. Math. 23 (1953) 39–96.

In this paper the author proposed the cut elimination theorem of **GLC** as his fundamental conjecture. Problem 16.11 was independently proved by I. L. Novak and A. Mostowski. For Novak's proof see

I. L. Novak: Models of Consistent System, Fund. Math. 37 (1950) 87–110.

For Mostowski's proof see

J. B. Rosser and H. Wang: Non-standard Models for Formal Logic, J. Symbolic Logic 15 (1950) 113–129.

There is a proof theoretic proof by

J. R. Shoenfield: A Relative Consistency Proof, J. Symbolic Logic 19 (1954) 21–28.

Our proof comes from

G. Takeuti: A Metamathematical Theorem on Functions, J. Math. Soc. Japan 8 (1956) 65–78.

Problem 16.20 is a theorem of G. Kreisel in his abstract

G. Kreisel: The Status of the First ε-number in First Order Arithmetic, J. Symbolic Logic 25 (1960) 390.

Though the material in §18 and §19 comes from my lecture notes, 1964–65, University of Illinois, the contents belong to the common sense on truth definition.

As for the cut-elimination theorem for simple type theory in §21, W. W. Tait proved the cut-elimination theorem for send order in his paper

W. W. Tait: A Non-constructive Proof of Gentzen's Hauptsatz for Second Order Predicate Logic, Bull. Amer. Math. Soc. 72 (1966) 980–983.

The general case namely Theorem 21.15 was proved independently by

M. Takahashi: A Proof of Cut-elimination Theorem in Simple Type Theory, J. Math. Soc. Japan 19 (1967) 399–410, and

D. Prawitz: Hauptsatz for Higher Order Logic, J. Symbolic Logic 33 (1968) 452–457.

The cut-elimination theorem for simple type theory with extensionally (Theorem 21.3) was proved by M. Takahashi in his paper

M. Takahashi: A System of Simple Type Theory of Gentzen Style with

Inference on Extensionality and Cut-elimination in it, Comm. Math. Univ. Sancti Pauli 18 (1969) 129–147.

The works of J. Y. Girard, Martin–Löf, and Prawitz in Proceedings of the Second Scandinavian Logic Symposium, ed. J. E. Fenstad, North-Holland, 1971 mentioned in p. 175 are:

J. Y. Girard: Une Extension de l'Interprétation de Gödel à l'Analyse, et son Application à l'Elimination des Coupures dans l'Analyse et la Théorie de Types, 63–92.

P. Martin-Löf: Hauptsatz for the Intuitionistic Theory of Iterated Inductive Definitions, 179–216.

P. Martin-Löf: Hauptsatz for the Theory of Species, 217–234.

D. Prawitz: Ideas and Results in Proof Theory, 235–308.

There is an interesting simple proof of cut elimination theorem of $\mathbf{G^1LC}$ by P. Päppinghaus. See pp. 17–23 in §1, pp. 24–31 in §2 and pp. 39–42 in §4 of his paper

P. Päppinghaus: Completeness Properties of Classical Theorems of Finite Type and the Normal Form Theorem, Dissertations Math. 207 (1983) 1–66.

As related subject, also see

J. Y. Girard and P. Pappinghaus: A Result on Implications of Σ_1-sentences and its Application to Normal Form Theorems, J. Symbolic Logic 46 (1981) 634–642.

Using Päppinghaus' method, G. E. Minc proved cut-elimination for the second order logic with the axiom of choice in the form

$$\forall x\, \exists y\, A(x, y) \rightarrow \exists Y\, \forall x\, A(x, Yx).$$

With some new construction, he extended his result for the stronger form $\forall X\, \exists Y \rightarrow \exists Y\, \forall X$ of the axiom of choice formulated in the language with the (second-order) ε-symbol. (A private communication.)

CHAPTER 4.

The infinitary language in §22 was initiated by C. Karp in her monograph:

C. Karp: Language with Expressions of Infinite Length, Studies in Logic and the Foundation of Math. 32, North-Holland, Amsterdam (1964).

Most material in §22 comes from

S. Maehara and G. Takeuti: A Formal System of First-order Predicate Calculus with Infinitary Long Expressions, J. Math. Soc. Japan 13 (1961) 357–370.

Malitz's example in (3) of Example 22.2 is

J. Malitz: Problems in the Model Theory of Infinite Languages, Ph.D. Thesis, Berkeley (1966).

The theorem of Lopez–Escobar in Problem 22.21 is in

E. Lopez-Escobar: An Interpolation Theorem for Denumerably Long Sentences, F. Math. 57 (1965) 253–272.

Most material in §23 comes from

G. Takeuti: A Determinate Logic, Nagoya Math. J. 38 (1970) 113–138.

Infinite games were first studied by

D. Gale and F. Steward: Infinite Games with Perfect Information, Ann. of Math. Study 28 (1953) 245–266.

Axiom of Determinateness was proposed by

J. Mycielski: On the Axiom of Determinateness, Fund. Math. 53 (1964) 205–224.

The notion Henkin quantifier in §24 is due to

L. Henkin: Some Remarks on Infinitely Long Formulas, Infinitistic Methods, Warsaw, (1961) 167–183.

CHAPTER 5.

The theory of ordinal diagrams in §26 was first developed in the following papers.

G. Takeuti, Ordinal Diagram, J. Math. Soc. Japan 9 (1957) 386–394.

G. Takeuti: On the Formal Theory of Ordinal Diagrams, Ann. Japan Assoc. Philos. Sci. 3 (1958) 151–170.

G. Takeuti: Ordinal Diagrams II, J. Math. Soc. Japan 12 (1960) 385–391.

There is an interesting extension of ordinal diagrams by A. Kino in her paper

A. Kino: On Ordinal Diagrams, J. Math. Soc. Japan 13 (1961) 346–356.

The proof of the accessibility of ordinal diagrams in this book comes from

G. Takeuti: Proof Theory and Set Theory, Synthese 62 (1985) 255–263.

There is a new treatment of ordinal diagrams à la Girard by H. Jervell in his notes

H. Jervell: Ordinal Diagrams are Regular Bilators, Math. Reports, University of Tromsø 1983.

As for foundational discussion of the accessibility of ordinal diagrams, see

M. Yasugi: Groundness Property and Accessibility of Ordinal Diagrams, J. Math. Soc. Japan 37 (1985) 1–16.

and

M. Yasugi: Hyper-principle and the Functional Structure of Ordinal Diagrams: The opening part, Comm. Math. Univ. Sancti Pauli 34 (1985) 227–263, The concluding part, ibid., 35 (1986) 1–38.

Also see

T. Arai: An Accessibility Proof of Ordinal Diagrams in Intuitionistic Theories for Iterated Inductive Definitions, Tsukuba J. Math. 8 (1984).

In this paper he proved that the accessibility of each ordinal diagram of

$O(\xi+1, 1)$ is derivable in $\mathbf{ID}^i_\xi$ therefore $O(\omega+1, 1)$ is the ordinal of Π^1_1-CA + BI.

After ordinal diagrams appeared, several systems of proof-theoretic ordinals with similar strength have appeared. See Pohlers' notes at the end of Postscripts for the ordinal notations of Schütte's school and the relation between their ordinal notations and ordinal diagrams. H. Levitz first discussed such a relation in his paper

H. Levitz: On the Relationship between Takeuti's Ordinal Diagram $O(n)$ and Schütte's System of Ordinal Notation $\Sigma(n)$, Intuitionism and Proof Theory, Eds. Kino, Myhill and Vesley, North-Holland, Amsterdam (1970).

H. Pfeiffer discussed the relationship between Kino's system of ordinal diagrams and his system $W(X)$ in his paper

H. Pfeiffer: Vergleich zweier Bezeichnungssystem fur Ordinalzahlen, Arch. Math. Logik 15 (1972) 41–56.

Also see his monograph

H. Pfeiffer: Bezeichnungssysteme fur Ordinalzahlen, Communications of the Mathematical Institute Rijksuniversiteit Utrecht (1973).

Most material in §27 and §28 comes from

G. Takeuti: Consistency-proofs of Subsystems of Analysis, Ann. Math. 86 (1967) 299–348.

T. Arai's improvement of my results are mainly due to private communications. See also his paper

T. Arai: A Subsystem of Classical Analysis Proof to Takeuti's Reduction Method for Π^1_1-analysis, 9 (1985) 21–29.

Pohlers' result related to Theorem 27.25 is in

W. Pohlers: An Upper Bound for the Provability of Transfinite Induction in Systems with N-times Iterated Inductive Definitions, Proof-Theory Symposium Kiel 1974, Springer Lecture Notes in Math. 500 (1975).

For a further consistency-result related to §28, see

G. Takeuti and M. Yasugi: The Ordinals of the System of Second Order Arithmetic with Provable Δ^1_2-comprehension Axiom and with the Δ^1_2-comprehension Axiom respectively, Japan. J. Math. 41 (1963) 1–67.

Friedman's Theorem stated in Theorem 27.35 comes from Simpson's paper mentioned in the reference to Theorem 12.73.

The theory of quasi-ordinal diagram from Definition 23.37 to Theorem 17.42 is also discussed in

M. Okada and G. Takeuti, On the Theory of Quasi-ordinal Diagrams, to appear in Logic and Combinations, Amer. Math. Soc. (1986).

W. Buchholz has found a combinatorial statement which is independent from Π^1_1-CA + BI in his paper

W. Buchholz: An Independence Result for $(\Pi^1_1$-CA) + BI, in Ann. Pure and Applied Logic.

CHAPTER 6.

The result of §29 comes from

G. Takeuti: A Remark on Gentzen's Paper "Beweisbarkeit unt Unbeweisbarkeit von Anfangsfällen der transfiniten Induktion in der reinen Zahlentheorie", Proc. Japan Acad. 39 (1963) 263–269.

The result of §30 comes from

G. Takeuti: The Π_1^1-comprehension Schema and ω-rule, Springer Lecture Notes in Math. 70, (1968) 303–331.

For the result of Shoenfield and Takahashi discussed at the end of §30, see

J. Shoenfield: On a Restricted ω-rule, Bull. Acad. Polon. Sci. Ser. 7 (1959) 415–417,

and

M. Takahashi: A Theorem on the Second Order Arithmetic with ω-rule, J. Math. Soc. Japan 22 (1970) 15–24.

Theorem 30.5 is due to Kino in

A. Kino: On Provably Recursive Functions and Ordinal Recursive Functions, J. Math. Soc. Japan 20 (1968) 456–476.

The result of §31 comes from

G. Takeuti and M. Yasugi: Reflection Principles of Subsystems of Analysis, Contributions to Mathematical Logic, North-Holland, Amsterdam (1968) 255–273.

There are many aspects in proof theory which are not covered in this book. Though cut elimination is a central subject of this book, see the following paper for a different aspect of cut elimination theorem.

H. Schwichtenberg: Proof Theory: Some Applications of Cut-Elimination, Handbook of Mathematical Logic, J. Barwise, ed., North-Holland, Amsterdam, (1977).

An important subject which this book totally missed is functional interpretations. Functional interpretations were started by G. Kreisel's the no-counter-example-interpretation as is discussed in his article in the Appendix. Then Gödel's functional interpretation appeared as we have already discussed in Chapter 2. Gödel's work was extended to analysis by

C. Spector: Provably Recursive Functionals of Analysis, Proc. Symp. Pure Math., vol. 5, AMS, Providence (1962) 1–27.

The subject is closely related to intuitionism and Kleene's realizability. There are many nice works in this area by W. W. Tait, W. A. Howard, A. S. Troelstra, J. Diller, H. Luckhardt, and W. Friedrich.

Recently, many interesting works are being done on the fragment of Peano arithmetic relating to computer science. The following article is an excellent source for proof-theoretical study of the subject.

S. R. Buss: Bounded Arithmetic, Ph.D. dissertation, Princeton University (1985).

APPENDIX

PROOF THEORY: SOME PERSONAL RECOLLECTIONS

by

Georg Kreisel

§1. Hilbert's programme

Like many others (but particularly, Gödel [6] and Gentzen [4] (on p. 564) who expressed their reservations discretely) I was repelled by Hilbert's exaggerated claims for consistency as a sufficient condition for mathematical validity or some kind of existence. But unlike most others I was not only attracted by the logical wit of consistency proofs (which I learnt in 1942 from Hilbert-Bernays Vol. 2), but also by the so to speak philosophical question of making explicit the additional knowledge provided by those proofs (over and above consistency itself). My answers took 2 forms:

(i) particular applications to mathematical proofs, usually of Π^0_2-theorems, which mathematicians had wanted to unwind, but did not succeed (to their own satisfaction; cf. e.g. [28] on Littlewood's results about $\pi(x)$-$li(x)$ or [16] on Artin's results about sums of squares, etc.),

(ii) general formal criteria such as functional interpretations to replace the incomparable condition of consistency; 'incomparable' because the aim of functional interpretations is meaningful without restriction on metamathematical methods. In particular, as stressed already in my first paper [7], the addition of true Π^0_1-axioms does not affect the class of functions needed (key word: provably recursive functions), while for consistency *only* proofs of Π^0_1-statements are relevant.

The applications (i) used of course specific mathematical properties (just as applications of model theory do). But the early metamathematical results (ii) were formulated in terms of traditional logical categories such as type and complexity of (quantifier) prefixes. Incidentally, by the mid fifties I had come to assume—wrongly—that the potential for striking applications of (i) was low. In the late seventies I learnt of many areas (L-functions, Galois cohomology, ergodic theory, topological dynamics), where mathematicians wanted to unwind proofs of Π^0_2-theorems, but were not able to do so without logical guidance. At that time I also began to formulate general metamathematical results, for example, on Σ^0_2-theorems, by use of *additional* mathematical conditions, specifically the rate of growth of Herbrand's terms [33]; recently further exploited in [36].

Until the mid fifties I dismissed the general question of characterizing the notion of finitist proof (or of a variant better suited to my later experience) because it was irrelevant to the positive work I had been pursuing. Besides, for example, in Hilbert's *Über das Unendliche* not only Ackermann's function, but even functions of higher types defined by ordinal recursion (in the sketch concerning **CH**) were accepted as finitist without discussion; in other words, the idea of recursive function without restrictions on proofs justifying the defining equations seemed to be dominant.

§2. Informal notions of proof (in the logical tradition)

As it happened I was at the IAS in the mid fifties when Kleene's papers on hyperarithmetic hierarchies came out. Though the work did not contain the word 'predicative' its relevance was apparent (and stressed in my review [9]). At the same time Gödel brought up in conversation the one topic in his Princeton bicentennial lecture that had not been developed there, absolute provability. (Of course, the lecture had not been published, and he never mentioned it. He did mention that the letter '**L**' was intended to stand for 'lawlike', as an analysis of the totality of sets that can be defined by a law at all.) Being *sceptical* of absolute provability[1] as a rewarding subject of study, I turned to KINDS of provability familiar from the foundational literature; evidently, in the case of predicativity, if proofs were to be considered at all, Kleene's definability properties had to be sharpened. This led to what are now called 'autonomous progressions', more precisely, with suitable restriction on the number of iterations of the basic principle involved; in the case of predicativity it was numerical quantification (jump operator), in the case of finitist proofs it was enumeration (diagonalization). Obviously, a principal attraction of this kind of interpretation was that it also gave a meaning (interpretation) to the ordinal analyses which had come to dominate technical proof theory.

I described those analyses of informal notions of proof as a 'calculated risk' [13] because they left open to what extent the notions were adapted to the phenomena involved. But in the late fifties and early sixties I obtained some results which still seem to me to be of unqualified interest. They correct the false impression that (restrictions to) the kinds of proof mentioned would lead to *Insuperable Formal Complexity*; recall Hilbert's

[1] Finitists, predicativists etc. simply equate their favorite kind of provability with absolute provability, cf. [14]. Here 'absolute provability' refers of course to the outer limits of the mathematical imagination, not to some kind of contrast with—historical or other—relativity, say, w.r.t. reliability.

remark about keeping a boxer from using his fists. In the late forties I had verified this already in an exposition of most parts of standard function theory by Σ_1^0-induction. The trick used was to retain the familiar language of, say, impredicative analysis or abstract set theory, but restrict the axioms. Proofs in such systems code, in an obvious way, predicative proofs in—the language of predicative—analysis or set theory [18]. (A more systematic exposition of this idea, including the—obvious—interpretation of Gödel's work on **L** in similar terms, is in [19].)

As a result there was no obstacle to *internal* progress on traditional foundational classifications of proofs. (For internal limitations, that is, formal undecidability, recursion-theoretic or model-theoretic methods were used, for example, in the case of Cantor-Bendixson.) In fact a whole industry has developed along these lines, without, however, producing any evidence for the relevance of those classifications; cf. §7 for more on this.

§3. Intuitionistic logic

Until the mid fifties I found this subject distasteful because, as I said in [8], iterated implications made my head spin. They continue to do so, and the same is true of functions of all finite types. But Gödel's use of finite types, which he told me in 1955 and published in '58 (always applied only to arithmetic, but which I discussed at Cornell and Amsterdam in '57 primarily for analysis), impressed me because analysis has a functional interpretation by recursively continuous functions of finite types, but not of bounded type (cf. [15]); in particular, not the no-counterexample-interpretation.

N.B. This simple point still seems to me more memorable than the particular interpretation (by Spector) in terms of bar recursion, although the latter is mathematically incomparably more substantial.

Also I was impressed by the 'strength' of Heyting's first order arithmetic expressed by—what is now called—Markov's rule, **MR**, in [10], that is, the extension to Π_2^0 formulae of the old conservation results of Gödel and Gentzen for the negative fragment. The obvious extension of the latter to the impredicative theory of species was later noted in [21], and closure under **MR** was sketched in [23]. (Today I interpret this strength—tacitly, with respect to provability—as weakness for algorithmic purposes: it is as hard to unwind an intuitionistic proof of a Π_2^0-theorem, in order to extract a program, as a proof in the corresponding classical system; one appropriate 'correspondence' is provided by the negative translation, naturally, for those systems to which it applies.)

Another reason for my interest in intuitionistic logic was my success with completeness and incompleteness questions for propositional, resp. predicate logic. (N.B. The latter work was recursion-theoretic, not proof-theoretic.) The point of this work was that only very simple properties of the intended meaning were used, not some specific semantics. In this respect the work followed Gödel's early style for classical logic since he—unlike Tarski—did not bother to give a formal semantics. A difference: Gödel used predicates of natural numbers, and I had a parameter for choice sequences.

This was followed by the so-called elimination of choice sequences. As a corollary, the proof-theoretic strength of all current theories of choice sequences was seen to be $< \Pi^1_1$-CA or, more precisely, those theories were interpreted in the theory of (non-iterated) inductive definitions.

In the early sixties I was so preoccupied with intuitionistic logic that I decided to present Proof Theory in the Saaty article, entirely for intuitionistic systems; on the absurd ground that classical metatheorems followed easily from intuitionistic metatheorems, but generally not conversely. (This aberration was corrected in Survey of Proof Theory [21][2].) A more sensible theme in the Saaty volume was the following.

§4. Mainstream of logical foundations

By the early sixties I had come to recognize that the so-called rival foundational schemes were complementary. The main theme of the Saaty article was the internal coherence and fruitfulness of the different branches of foundations. (Model theory was left out because I had just completed the book with Krivine.)

N.B. I have retained the view of the complementary character of those schemes, but revised the characterization; cf. my article in Monist 67 (1984) [35] or the lecture at Stanford 4. VI. 84. However, the next step was a *failure*.

In 1968 I tried to write a text on Proof Theory in the general style of the book with Krivine (on Model Theory); circulated as my UCLA lectures: Elements of Proof Theory. The main topics were conservation results, equivalences between logical and mathematical principles (like reflection and induction), and between logical complexity and order type (of predicates to which, resp. on which induction was applied). The order types used were characterized up to definable isomorphism by appropriate functions on the orderings so that one could speak of 'natural' orderings. Also there were independence proofs, from finite subsystems of analysis, of such principles as ε_0 (or bar) induction applied to predicates of suitable complexity, that are still best obtained by proof theoretic methods [21] and useful (as in [1]); in contrast to more familiar independence results, for

---

[2] Naturally, [21] leaves out those metatheorems which do not have any counterpart for classical systems such as the hackneyed existence and disjunction theorems, not to speak of such imaginative theorems as those on transfinite induction by Friedman [2] and Friedman and Sčedrov [3] and [42].

example, of Σ^1_n-AC from Δ^1_n-CA ($n \neq 2$), which are valid also for ω-consequence ($n = 1$) or β-consequence ($n \geqslant 3$). The new material was published in the mid seventies under the title: Wie die Beweistheorie . . . [27].

The failure was this. Inspection of the numerous exercises showed that half of them did not properly belong to the main topics at all, which concern only PROVABILITY. In fact, more than half of the exercises concerned structural properties of PROOFS. So the lectures were never published.

§5. Operations on proofs

N.B. Of course, the whole of Proof Theory in the tradition of Gentzen manipulates proofs. But the results stated concern only provability. I now mean results which concern explicitly operations on proofs.

Towards the end of the 60's I learnt a bit about natural deduction from Prawitz, including so-called strong normalization which refers explicitly to operations on proofs, namely, the normal forms are independent of the order in which reductions are applied. For me this was of interest because of its structural character; not at all because of its use for so-called operational semantics, which excited Prawitz.

A first, albeit indirect, success of this new interest involved Girard who showed me a new functional interpretation of impredicative analysis by means of functions of a strange type. (The original form was contradictory.) I thought that his method also proved normalization—in particular, not only the existence of normal forms—for the theory of species. It thus proved Takeuti's conjecture reformulated—in fact, sharpened—in terms of operations on proofs. (Actually, Gentzen had stated that conjecture in a postcard to Bernays in 1934, with a clear distinction between a proof by brute force, that is, by model theory, and a proof of a more informative kind.) Nevertheless it was hard to find any uses at all for this solution of an old problem. One area of application was found, in connection with self-referential propositions, in collaboration with Takeuti [24].

I continued to search for structural properties of proofs till the mid seventies, partly in collaboration, partly in dissertation advice, and partly in the form of conjectures. For example, I published on relative consistency statements sharpened by bounding the relative lengths of hypothetical proofs of inconsistency [25], jointly with Mints and Simpson on continuous transformations of—not necessarily well-founded—proofs; Statman pursued the suggestion of studying the effect of explicit definitions (which is negligible measured in terms of length, but not—as he discovered—in terms of genus), and Parikh [37] solved a conjecture about $\forall x\, A$ if the lengths of the shortest proofs of $A[x/s^n 0]$ are bounded; at least for systems containing only finitely many schemata and only monadic function symbols[3].

The intention of all this work was to contribute to an understanding of proofs in the ordinary sense, including their intelligibility; in other words, their being easy to take in and remember; cf. [34]. But soon afterwards I came to the conclusion that logical proof theory had little to offer in this respect, and certainly less than, say, the axiomatic analysis of mathematics in the style of Bourbaki[4]: the choice between notions of the same logical complexity is more relevant (to the purpose mentioned above) than difference in logical complexity.

§6. Logical proof theory and computing

Preliminary distinction: automatic theorem proving (or proof checking) benefits more from axiomatic analysis as described in the last paragraph than from logical proof theory, while the latter can be used for automatic manipulation of proofs or programs, in particular, for program synthesis from (particular) proofs. The scheme was explained in lectures at Clermont-Ferrand (1975) [26] and Belgrade (1977) [29] with the title: From Foundations to Science (though 'Technology' is better than 'Science'). The dissertation by Goad (1980) [5] executed, and improved the scheme, the latter in the sense that the programs obtained (by eliminating redundancies) are more efficient than the programs implicit in the given proof. Formally speaking, normalization is replaced by so-called pruning; in contrast to normalization the end result depends on (a sensible choice of) the order in which reduction steps are applied.

Current interest: to modify the scheme for parallel computation by using—descriptions of—infinite proof trees, and applying continuous transformations. Since in such transformations only finite stumps of trees, albeit with many branches (not only binary trees), are used, a finite but large number of processors can work in parallel on the immediate

---

[3] He solved the problem only for the particular case of one monadic symbol (the successor in arithmetic). For finitely many, I appealed to Makanin's—complicated—decision method for word equations in semigroups, cf. [32]. Quite recently, a much simpler method was given by W. Farmer for the particular equations in the problem considered. In the meantime I had become interested, in collaboration with M. Baaz and P. Pudlák, in another aspect of [37] under the heading: short proofs of $A[x/s^n0]$ for large n can be generalized. (Not the numerical value of s^n0, but its large syntactic depth matters here; cf. [41, p. 146].) A little more explicitly, for systems related to those of [37], and for given c and A: there are bounds N and M s.t. if d proves $A[x/s^n0]$ in $\leq c$ steps and $n \geq N$, then there are $m \leq M$ and a derivation of similar logical form to d that proves $x \equiv s^m0 \pmod{s^n0} \rightarrow A$, for variable x. The old (global) result follows easily from the new formulation. Besides, the latter is related to *some* of the literature where proofs of general theorems are seen by inspection of computations; related to the results, not necessarily to the processes of abstraction involved.

[4] The 'manifesto' L'architecture des mathematiques is meant.

predecessors of any given node; c.f. [41, c(ii) on p. 142] concerning homogeneous proof trees.

§7. Philosophical progress: pursuing and examining logical idea(l)s

The story above began and ended with shifts away from traditional logical aims. Methods developed for the latter were turned into tools of science and technology. Experience has shown that, though wide, the use of these and other metamathematical tools is rarely central in any branch of science and technology. In this way the originally uninformed prejudice against logic has been properly adjusted.

There is a different area where mathematical logic generally, but especially Proof Theory seems to me to make truly central and highly effective contributions. This area, developed in our century as a new philosophical tradition, makes the examination, and not only the pursuit, of logical and other aims of academic philosophy its principal topic. In particular, it does not accept the hackneyed grail of refuting (dubious) sceptical doubts nor the dogma that the broad concerns behind so-called ontology and epistemology are better served by traditional dialectics than by the expansion of ordinary scientific knowledge.

The new tradition—incidentally, quite close to the popular meaning of 'philosophy'—is represented in writings of thoughtful scientists; but, with two provisoes, also in the professional philosophical literature. First, the latter is usually spoilt by exaggerations, for example, in Wittgenstein's "bewitchment by our language" or the "pernicious intrusion of logic in mathematics", where, in fact, the silent majority does not care at all. Secondly, the literature in the new tradition tends to be 'reinterpreted' by academics in their traditional terms; cf. [32] and [33, p. 87, footnote 3].

The contributions of mathematical logic to the new tradition may be compared to a familiar phenomenon: closely related technologies will create, identify and help to clean up pollution. The particular philosophical pollution associated with proof theory began with such pretentious—and therefore simple minded—claims as Frege's and Hilbert's that so-called complete formalization, especially with finitist metamathematics, is needed for reliability of proofs. Later there was Turing's suggestion, elaborated ad nauseam by epigones of Gentzen, that ordinals are the measure of depth for theorems, if not for all things. The flashy precision of such classifications pollutes the intellectual atmosphere by blinding the observer to their obvious inadequacy for understanding the (two) delicate aspects of mathematical reasoning just mentioned. As a corollary to the material in Section 2, the informal classifications of proofs considered there are similarly inadequate.

Of course, discredited (cl)aims need not be examined. They can simply be ignored as we ignore, for example, the principal dogmas of astrology or alchemy; cf. [40]. The two examples below, at opposite ends of the scale as it were, illustrate how logical pollution persists, and thus how the new philosophical tradition functions.

The first example is provided by the industry—incidentally, in both senses of the word—, alluded to already at the end of Section 2. Under the unwittingly appropriate trade name 'reverse mathematics' it proposes various formal equivalences in so-called basic systems as analyses of delicate mathematical phenomena; most brutally, at least on one occasion, as representing identity of ideas! True, the claim is suspect in view of the equivalence in ordinary formal arithmetic between $0=0$ and Mordell's conjecture. But the philosophical pollution becomes more blatant and thus more memorable when magnified by the paraphernalia of calibrations as precise and systematic as any astrological categories. More important still, the logic chopping provided by witty logical theorems locates the source: it lies in defects of the notions themselves, as opposed to lack of skill or of perseverance in their use.

The second example comes from personal experience, as is appropriate for the present personal record. It goes back to the equivalence in Section 1 between (i) formal derivability in S of Π_2^0-theorems from true Π_1^0-lemmas, and (ii) majorization of bounding functions by S-provably (total) recursive functions. In [8], 30 years ago, I stressed the direction (ii)$\rightarrow$(i), inferring formal underivability from 'mathematical' lower bounds. Later, I began [19] by emphasizing the problematic side in picking on any particular formal system, at least, without additional specific information. But I failed to apply to (i) the obvious corollary, namely the lack of significance of the kind of formal underivability involved. What makes this more than a merely personal blind spot is the popular interest in (ii)$\rightarrow$(i); cf. [38], 15 years after [19].

When writing the P.S. to [11] (for [11a], in 1978) I did stress the potentially greater market for (i)$\rightarrow$(ii), and told the story of—what I then felt to have been—a missed opportunity in pursuing some results on the infinite version $\mathbf{RT}_\infty$ of Ramsey's theorem[5]. But I failed to examine the problems of assessing the significance of (i)$\rightarrow$(ii) that are touched briefly in [27]. For one thing, there are proofs of (ii) without any detours via metamathematics; cf. [27] on Ketonen's original proof that was inspired by notions—not axioms of infinity!—in work on large cardinals, and the more standard combinatorial presentation in Ketonen-Solovay leading to

[5] At the time I remembered Ehrenfeucht's use in the early 70's of Kruskal's theorem for proving that Skolem's ordering-by-dominance of exponential terms was well founded. Since the ordinal is $\geqslant \varepsilon_0$ it was clear in the light of experience with $\mathbf{RT}_\infty$ that somewhere in the general area of Kruskal's theorem some fast growing bounding functions are liable to occur (but of course not, how fast they'd grow). As it happened, merely having realized the relevance of Ehrenfeucht's result came in useful a few years later when I could supply the reference effortlessly to Friedman and Leeb who needed it in their work. My reward is described below.

quantitative improvements. Another aspect noted in [27] is the trivial difference between the derivations of various finite versions of Ramsey's theorem from $\mathbf{RT}_\infty$.

The full implications of that aspect—for the philosophical tradition here considered—were driven home to me by the avalanche of later results on how sensitive the rate of growth of bounding functions is to tiny changes in proofs. The results are varied, occurring in some corners of ordinary and in some high spots of recreational mathematics. Without exaggeration: these results provide correct and memorable mathematical formulations of a familiar lesson of mathematical experience. Contrary to certain logical dogmas, realistically speaking, the algorithmic content of a proof is generally only a minor fraction of the information contained in the proof; and the bulk of the information is not proof theoretic at all in the sense that this word has acquired in logical research. Much more of that information is often made explicit by the delicate axiomatic analysis developed in mathematics during this century; cf. [35, p. 88, note 7].

Old results from around 1960, like Gödel's or Spector's, on the devastating algorithmic effects of mathematically trivial—and by Section 3, ideologically precious—extensions to higher types are seen in a new light. Obviously, they cannot be expected to be as entertaining as results in recreational mathematics. But they continue to serve as highly relevant warnings to computer scientists to beware of logical, that is, iterated types, and with luck as encouragement to look at less flashy, but more effective alternatives in the literature; cf. [41, 4b(ii), resp. A3].

References

[1] H. Friedman: Iterated inductive definitions and Σ^1_2-AC. in: J. Myhill, ed., Intuitionism and Proof Theory, North-Holland, Amsterdam (1970).

[2] H. Friedman: Programs and Results in Logic I, Notices A.M.S. 22 (1975) 476.

[3] H. Friedman and A. Sčedrov: Intuitionistically provable recursive well-orderings, Annals of Pure and Applied Logic 30 (1986) 165–171.

[4] G. Gentzen: Die Widerspruchsfreiheit der reinen Zahlentheorie, Math. Ann. 112 (1936) 493–565.

[5] C. A. Goad: Computational Uses of the Manipulation of Formal Proofs, Dissertation, Stanford University (1980).

[6] K. Gödel: Diskussion zur Grundlegung der Mathematik, Erkenntnis 2 (1931/32) 147–151.

[7] G. Kreisel: On the Interpretation of Non-finitist Proofs, I and II, J. Symbolic Logic, 16 (1951) 241–267; 17 (1952) 43–58.

[8] G. Kreisel: A Variant to Hilbert's Theory of the Foundations of Arithmetic, British Journal for the Philosophy of Science 4 (1953) 107–127.

[9] G. Kreisel: Review of S. C. Kleene, Hierarchies of Number-theoretic Predicates, Mathematical Review 17 (1956) 4.

[10] G. Kreisel: Mathematical Significance of Consistency Proofs, J. Symbolic Logic 23 (1958) 155–182.

[11] G. Kreisel: Elementary Completeness Properties of Intuitionistic Logic with a Note on Negations of Prenex Formulae, Ibid., 317–330.

[12] G. Kreisel: A Remark on Free Choice Sequences and the Topological Completeness Proofs, Ibid., 369–387.

[13] G. Kreisel: Hilbert's Programme, Dialectica 12 (1958) 346–372.

[13a] G. Kreisel: Revised Version of [13] with a Postscript, in: P. Benacerraf and H. Putnam, ed., Philosophy of Mathematics: Selected Readings, 2nd ed., Cambridge University Press, New York (1983) 207–238.

[14] G. Kreisel: Ordinal Logics and the Characterization of Informal Concepts of Proof, Address at the International Congress of Mathematics, Edinburgh (1958) 289–299.

[15] G. Kreisel: Interpretation of Classical Analysis by Means of Constructive Functionals of Finite Type, in: A. Heyting, ed. Constructivity in Mathematics, North-Holland, Amsterdam (1959) 101–128.

[16] G. Kreisel: Sums of Squares, in: Summary of talks presented at the Summer Institute for Symbolic Logic, 1957, Princeton, Institute for Defense Analysis (1960) 313–320.

[17] G. Kreisel: On Weak Completeness of Intuitionistic Predicate Logic, J. Symbolic Logic 27 (1962) 139–158.

[18] G. Kreisel: The Axiom of Choice and the Class of Hyper-arithmetic Functions, Dutch Academy A 65 (1962) 307–319.

[19] G. Kreisel: Mathematical Logic, in: T. L. Saaty, ed., Lectures on Modern Mathematics, Vol. 3, Wiley, New York (1965) 95–195.

[20] G. Kreisel: Lawless Sequences of Natural Number, Compositio 20 (1968) 222–248.

[21] G. Kreisel: A Survey of Proof Theory, J. Symbolic Logic 33 (1968) 321–388; cf. Zbl. 522 (1984) #03046.

[22] G. Kreisel: Formal Systems for some Branches of Intuitionistic Analysis, Annals of Mathematical Logic 1 (1970) 229–387, (with A. S. Troelstra).

[23] G. Kreisel: Church's Thesis: A Kind of Reducibility Axiom for Constructive Mathematics, in: J. Myhill, ed., Intuitionism and Proof Theory, North-Holland, Amsterdam (1970) 121–150; cf. Zbl. 199 (1971) 300.

[24] G. Kreisel: Formally Self-referential Propositions for Cut-free Classical Analysis and Related Systems, Dissertations Mathematicae 118 (1974) 1–50, with G. Takeuti, cf. Zbl. 336 (1977) #02027.

[25] G. Kreisel: The Use of Abstract Language in Elementary Metamathematics: Some Pedagogic Examples, in: R. Parikh, ed., Logic Colloquium, Symposium on Logic held at Boston, 1972–73, Springer Lecture Notes 453 (1975) 38–131, with E. G. Mints and S. G. Simpson.

[26] G. Kreisel: Some Uses of Proof Theory for Finding Computer Programs, in: M. Guillaume, ed., Logique. Colloques du CNRS, Editions du CNRS, Paris (1977) 123–134; cf. Zbl. 522 (1984) #03046.

[27] G. Kreisel: Wie die Beweistheorie zu ihren Ordinalzahlen kam und kommt, Jahresberichte D.M.V. 78 (1977) 177–223; cf. Zbl. 522 (1984) #03046.

[28] G. Kreisel: On the Kind of Data Needed for a Theory of Proofs, in: R. O. Gandy and J. M. Hyland, eds., Logic Colloquium 76, North-Holland, Amsterdam (1977) 111–128; cf. Zbl. 522 (1984) #03046.

[29] G. Kreisel: From Foundations to Science: Justifying and Unwinding Proofs, Recueil des travaux de l'Institut Mathematique, Belgrade, Nouvelle Serie 2 (1977) 63–72; cf. Zbl. 414 (1980) #03033.

[30] G. Kreisel: Neglected Possibilities of Processing Assertions and Proofs Mechanically: Choice of Problems and Data, in: P. Suppes, ed., University-level computer-assisted instruction at Stanford, 1968–1980, Stanford, CA, Stanford University, Institute for Mathematical Studies in the Social Sciences (1981) 131–147.

[31] G. Kreisel: Extraction of Bounds: Interpreting some Tricks of the Trade, Ibid., 149–165.

[32] G. Kreisel: Review of R. Statman, Zbl. 411 (1980) 26.

[33] G. Kreisel: Finiteness Theorems in Arithmetic: An Application of Herbrand's

Theorem for Σ_2-Formulas, in: Stern, ed., Proc. Herbrand Symposium (Marseilles 1981), North-Holland, Amsterdam (1982) 39–55; cf. Zbl., 499 (1983) 29#03045.

[34] G. Kreisel: Review of S. A. Kripke, Wittgenstein on Rules and Private Language, Canadian Philosophical Review 3 (1983) 287–289.

[35] G. Kreisel: Frege's Foundations and Intuitionistic Logic, Monist 67 (1984) 72–91.

[36] H. Luckhart: Auswertung von Σ_2-Herbrand Analysen: Polynomiale Anzahlschranken für den Satz von Roth (to appear).

[37] R. J. Parikh: Some results on the Length of Proofs, Transactions A.M.S. 177 (1973) 29–36.

[38] J. Spencer: Large Numbers, American Math. Monthly 90 (1983) 669–675; cf. Zbl. 539 (1985) #03042.

[39] R. Statman: Structural Complexity of Proofs, Dissertation, Stanford University (1974).

[40] G. Kreisel: Mathematical Logic: Tool and Object Lesson for Science, Synthese 62 (1985) 139–151.

[41] G. Kreisel: Proof theory and the Synthesis of Programs: Potential and Limitations, Springer Lecture Notes in Computer Science 203 (1985) 136–150.

[42] H. Friedman and A. Sčedrov: Arithmetic transfinite induction and recursive well orderings, Advances in Mathematics 56 (1985) 283–294.

CONTRIBUTIONS OF THE SCHÜTTE SCHOOL IN MUNICH TO PROOF THEORY

by

WOLFRAM POHLERS

§1. General aims

The main concern of the research of K. Schütte and his school was (and still is) the ordinal analysis of formal systems. The ordinal analysis of a formal system comprehends the computation of its proof theoretic ordinal. By the proof theoretic ordinal of a formal system one usually understands its least non provable ordinal. Strictly speaking, to deal with ordinals—represented by well orderings on the natural numbers—one has to use a second order language. But, since the well-foundedness of an order relation is expressible by a Π^1_1 formula, a first order system with free second order variables (or set parameters as we will call them) will suffice. So let **T** be a formal system, including Peano arithmetic, whose language contains set parameters. For an order relation $<$ we abbreviate the formula $\forall x\,(\forall y\,(y < x \supset y \in X) \supset x \in X) \supset \forall x\,(x \in \mathrm{Feld}(<) \supset x \in X)$ by $\mathbf{TI}(<, X)$. If $<$ is definable by a formula in the language of **T** also $\mathbf{TI}(<, X)$ is a formula in the language of **T** and the validity of $\mathbf{TI}(<, X)$ in the intended standard structure of **T** (where the set-parameter X is supposed to range over *all* sets of natural numbers) means that $<$ in fact is well-founded. Therefore we call an ordinal α a provable ordinal of **T** if there is an arithmetically definable well ordering $<$ of ordertype α such that **T** proves $\mathbf{TI}(<, X)$. Already Gentzen proved that ε_0 is the least ordinal which, in this sense, is not provable in formalized Peano arithmetic (**PA**). Let us assume that $<$ is a well-ordering of ordertype α. In general, what possibilities do we have to show the unprovability of $\mathbf{TI}(<, X)$ in **T**? Today, one first would try to find a model of **T** in which $\mathbf{TI}(<, X)$ fails. Since $<$ is a well-ordering in the 'real world', this necessarily has to be a nonstandard model. A more proof theoretic way is to give a consistency proof of **T** which uses besides an induction along $<$ only means which can be formalized in **T**. The unprovability of $\mathbf{TI}(<, X)$ then follows from Gödel's second theorem. In both cases, however, we do not yet have the unprovability of α since there might be another arithmetically definable well-ordering $<_0$ of ordertype α for which $\mathbf{TI}(<_0, X)$ holds. So one had to show that for any such well-ordering $<$ of ordertype α there is a nonstandard model for **T** such that $\mathbf{TI}(<, X)$ fails or that $\mathrm{Con}(\mathbf{T})$ can be proved by induction along any such well-ordering. In fact I don't know if such proofs exist.

The method used by Schütte and his school goes back to Gentzen's 1943 paper, although I'm going to describe it in more modern terms.

By Henkin [Henkin, 1954] and Orey [Orey, 1956] we know that ω-logic is complete for the structure $\mathcal{N}$ of natural numbers. Schütte himself gave a different proof of this fact in his book Beweistheorie [1960]. Schütte's proof, however, shows more. He proved the completeness of a cut-free system with ω-rule. We use this fact to introduce an ordinal-norm for valid arithmetical formulas with set parameters. In presence of the ω-rule a formula $\forall x\, A(x)$ may be inferred from "$A(n)$ for all $n \in \mathbb{N}$". So there is no need for eigenvariables and we may therefore dispense with free number variables at all. So we restrict our language to terms and formulas without occurrences of free number variables. We call an arithmetical formula $\mathfrak{A}[X_1, \ldots, X_n]$, whose set parameters are all in the list $X_1, \ldots, X_n$, valid in $\mathcal{N}$ (denoted by $\mathcal{N} \models \mathfrak{A}[X_1, \ldots, X_n]$) if $\mathcal{N} \models \mathfrak{A}[S_1, \ldots, S_n]$ holds for any choice of sets $S_1, \ldots, S_n$ of natural numbers. By $\mathbf{PA}_\infty^{\mathrm{CF}}$ we denote a cut-free system for Peano arithmetic with ω-rule[1]. By Schütte's completeness theorem we then have $\mathcal{N} \models \mathfrak{A}[X_1, \ldots, X_n] \Leftrightarrow \mathbf{PA}_\infty^{\mathrm{CF}} \vdash \mathfrak{A}[X_1, \ldots, X_n]$. Since the proof trees of $\mathbf{PA}_\infty^{\mathrm{CF}}$ are ω-branching trees they canonically represent countable (even recursive) ordinals.

By $\mathbf{PA}_\infty^{\mathrm{CF}} \vdash^\alpha F$ we denote, that there is a derivation of F in $\mathbf{PA}_\infty^{\mathrm{CF}}$ which represents the ordinal α. Using the completeness theorem we obtain the following norm

$$|F| := \min\{\alpha \colon \mathbf{PA}_\infty^{\mathrm{CF}} \vdash^\alpha F\}$$

for arithmetical formulas F which are valid in $\mathcal{N}$.

For a well-ordering $<$ on the natural numbers and z in the field of $<$ we denote by $|z|_<$ the ordertype of the $<$-predecessors of z. If β is less or equal to the ordertype of $<$ we denote by $<_\beta$ the segment of $<$ which has ordertype β. It's now an easy exercise to show the following lemma by induction on α.

LEMMA. *Let $\mathfrak{A}[X]$ be an arithmetical formula with only positive occurrences of X and $\beta = \sup\{|z_1|_<, \ldots, |z_n|_<\}$. If $\mathbf{PA}_\infty^{\mathrm{CF}} \vdash^\alpha \forall x\, (\forall y\, (y < x \supset y \in X) \supset x \in X), \underline{z}_1 \in X, \ldots, \underline{z}_n \in X \to \mathfrak{A}[X]$, then $\mathcal{N} \models \mathfrak{A}[<_{\beta+\omega^\alpha}]$.*

---

[1] The choice of $\mathbf{PA}_\infty^{\mathrm{CF}}$ is not essential. Usually one takes as axioms the logical tautologies $t \in X \to s \in X$ (t and s terms of same numerical value) together with the diagram of $\mathcal{N}$. The rules are the usual ones for introducing and eliminating the logical symbols.

As a specialization of this lemma we obtain the following theorem.

BOUNDEDNESS THEOREM. *Suppose that* $<$ *is a well-ordering and* $|\mathbf{TI}(<, X)| \leqslant \alpha$. *Then* $<$ *is of ordertype* $\leqslant \omega^{\alpha}$.

The boundedness theorem shows that an upper bound for the provable ordinals of a formal-system **T** may be obtained via the norm of its provable formulas. In fact, we want to redefine the proof theoretic ordinal of **T** via the norm. We put $|\mathbf{T}| := \sup\{|F|: F \text{ is arithmetic and } \mathbf{T} \vdash F\}$[2]. So the problem has reduced to determining the least upper bound for the norms of provable formulas of **T**. The natural way to obtain upper bounds for $|\mathbf{T}|$ is to transform a derivation in **T** into a derivation in $\mathbf{PA}_{\infty}^{\mathrm{CF}}$. This usually is done in two main steps. First we enlarge the system $\mathbf{PA}_{\infty}^{\mathrm{CF}}$ by additional rules and/or axioms such that we obtain a supersystem $\mathbf{T}_{\infty}$ of $\mathbf{PA}_{\infty}^{\mathrm{CF}}$ in which derivations of **T** can be easily embedded. Of course the embedding must give us upper bounds for the ordinals of the embedded derivations. In a second step we try to get rid again of all additional rules in the derivation of an arithmetical formula. This usually will alter the ordinal represented by the derivation tree. Therefore we have to find a reduction procedure for the derivation in $\mathbf{T}_{\infty}$ which allows us to keep control over the ordinal of the reduced derivation. The completely reduced derivation then is a derivation in $\mathbf{PA}_{\infty}^{\mathrm{CF}}$. Therefore its ordinal will give us an upper bound for the norm of the derived formula.

To show that an upper bound α is the least one, it suffices to show that for each $\beta < \alpha$ there is an arithmetical order relation $<$ of ordertype β such that **T** proves $\mathbf{TI}(<, X)$[3].

A good example for this method is given by **PA** itself. Here we just add the cut-rule to $\mathbf{PA}_{\infty}^{\mathrm{CF}}$ to obtain $\mathbf{PA}_{\infty}$. To translate the induction axiom (or the induction rule) of **PA** we really have to use derivations of transfinite length and we obtain an embedding theorem, saying that each provable formula of **PA** can be derived in $\mathbf{PA}_{\infty}$ by a derivation bounded by ω^2, say. (This depends a little upon the axiomatization of **PA**.) Then by the usual cut-elimination procedure we are able to transform it into a cut free derivation bounded by some $\alpha < \varepsilon_0$. So ε_0 is an upper bound for the norms of formulas provable in **PA**. Our task is threefold: First, to find the right system $\mathbf{T}_{\infty}$. Second, to develop a procedure which reduces $\mathbf{T}_{\infty}$-derivations of arithmetical formulas to $\mathbf{PA}_{\infty}^{\mathrm{CF}}$-derivations and finally, to show that the obtained bounds are precise, i.e., to give well-ordering proofs within **T**.

---

[2] Since all proof-theoretic ordinals of formal systems including **PA** are ε-numbers this definition is coherent to the previous one.

[3] Since **T** contains **PA**, α has to be an ε-number. Hence $\omega^{\beta} + 1 < \alpha$. For any arithmetic well-ordering $<$ of ordertype $\omega^{\beta} + 1$ we have $\beta < |\mathbf{TI}(<, X)|$ by the boundedness theorem. But **T** proves $\mathbf{TI}(<, X)$ for a suited well-ordering $<$.

In the next sections I will indicate how and for which system **T** this has been done by the Schütte school. But before I do so I want to say a few words about the foundational significance of our method.

Certainly the program, as we described it above, sounds to be far away from the original aims of proof theory coming out of Hilbert's program. We did hardly mention consistency proof and didn't speak about finitistic or constructive methods at all. But there is of course also a foundational feature in our program. The infinitary system $\mathbf{PA}_\infty^{\mathrm{CF}}$ trivially is consistent. Therefore an ordinal analysis of **T** will also give a consistency proof for **T**. The question is "how constructive is this proof?". First we should mention that we also can manage with a 'constructive' subsystem of $\mathbf{PA}_\infty^{\mathrm{CF}}$ in which all proof trees are given by primitive recursive trees. The proof predicate of this subsystem will be of complexity Π_1^0. Clearly one should try to keep the translation of **T** into $\mathbf{T}_\infty$ and the reduction procedure for $\mathbf{T}_\infty$ as constructive as possible. More precisely, that means that the interpretation and the reduction procedure should be formalizable in an intuitionistic system as small as possible. Of course we need sufficiently strong inductions in this system in order to show that the reduction procedure really terminates. It has turned out that a system for Heyting arithmetic without complete induction but augmented by transfinite induction up to all ordinals below $|\mathbf{T}|$ for restricted formulas suffices. Let us call this system $\mathbf{HA}_{<|\mathbf{T}|}$. So, if we succeed to keep the proof predicate of $\mathbf{T}_\infty$ formalizable in $\mathbf{HA}_{<|\mathbf{T}|}$, we will obtain the following result:

$$\mathbf{T} \vdash F \Rightarrow \mathbf{HA}_{<|\mathbf{T}|} \vdash \mathrm{Proof}_{\mathbf{PA}_\infty^{\mathrm{CF}}}(e, \ulcorner F \urcorner),$$

where the number e of the recursive proof tree in $\mathbf{PA}_\infty^{\mathrm{CF}}$ is recursively computable from the number of the proof tree in **T**. On the other hand we have the following reflection principle for formulas F which are stable with respect to the double negation translation:

$$\mathbf{HA}_{<|\mathbf{T}|} \vdash \mathrm{Proof}_{\mathbf{PA}_\infty^{\mathrm{CF}}}(e, \ulcorner F \urcorner) \rightarrow F.$$

Therefore every stable arithmetical formula which is provable in **T** also is provable in $\mathbf{HA}_{<|\mathbf{T}|}$ and vice versa, since $\mathbf{HA}_{<|\mathbf{T}|}$ is a subsystem of **T**. This shows that **T** is a conservative extension of $\mathbf{HA}_{<|\mathbf{T}|}$ with respect to stable arithmetical formulas (cf. [Pohlers, 1977]). In fact we can restrict the transfinite induction in $\mathbf{HA}_{<|\mathbf{T}|}$ to Π_1^0-formulas. So we obtain a proof theoretic reduction of **T** to a constructive system with transfinite induction over Π_1^0-formulas.

From a very strict constructive viewpoint our way of ordinal analysis might be regarded worse than a Gentzen style consistency proof dealing directly with the finite proof trees of **T**. Such a proof leads to a reduction of **T** to a constructive system with quantifier-free transfinite induction. We believe, however, that our method pays off in greater perspicuity. It allows us to look through proof theoretic results within our customary picture of the mathematical world. But there is another result which is immediately clear from a Gentzen style consistency proof for **T** using a transfinite induction along $<$. Since **T** has a primitive recursive proof-predicate we obtain by Gödel's second theorem that **T** does not prove PRWO($<$), i.e., the fact that there is no primitive recursive strictly $<$-descending sequence. Such results are of interest in connection with independence proofs for combinatorial theorems. We have to work a little harder to obtain the same result by ordinal analysis (but also gain greater perspicuity). This has to do with the fact that the norm definition for arithmetical formulas is a bit rough. It is easy to see that each valid arithmetical sentence (not containing set parameters) has a norm below ω independently from its provability in any formal system. So we can't state anything about sentences. But as recently shown by Buchholz [Buchholz 1984a], it is possible to refine the boundedness theorem in the following way.

Refined boundedness lemma. *If* $\mathbf{PA}_\infty^{\mathrm{CF}} \vdash^\alpha \forall x\, \exists y\, A(x, y)$ *for a* Π_2^0-*formula, then there is for each natural number* k *an* $n < H_{\omega^\alpha}(k)$ *such that* $\mathcal{N} \vDash A(k, n)$ [4].

Here H_α is the subrecursive hierarchy defined recursively by $H_0(n) = n$, $H_{\alpha+1}(n) = H_\alpha(n+1)$ and $H_\lambda(n) = H_{\lambda[n]}(n+1)$, where $\lambda[n]$ denotes the nth member of a (fixed) fundamental sequence for the limit ordinal λ.

The theorem then tells us that $\forall x\, \exists y\, (H_{|\mathbf{T}|}(x) = y)$ cannot be provable in **T** and so gives us a bound for the provably recursive functions of **T**. For convenience we define $(\alpha+1)[n] = \alpha$. Then if $\lambda[x][x+1] \ldots [x+y] = 0$ we obviously have $H_\lambda[x] = H_0[x+y]$. Clearly $|\mathbf{T}|[x], |\mathbf{T}|[x][x+1] \ldots$ is a strictly descending primitive recursive sequence. Now if we assume that **T** proves PRWO($|\mathbf{T}|$) we also obtain $\mathbf{T} \vdash \forall x\, \exists y\, (|\mathbf{T}|[x] \ldots [x+y] = 0)$, i.e., $\mathbf{T} \vdash \forall x\, \exists y\, (H_{|\mathbf{T}|}(x) = y)$, a contradiction to the refined boundedness theorem.

It is now tempting to define a norm for Π_2^0-formulas, and by it the proof theoretic ordinal for **T**, via the subrecursive hierarchy. Such a definition, however, is by far less canonical than the previous one, given for Π_1^1-formulas. There is too much arbitrariness in the definition of the subrecursive hierarchy. First, it is not compelling to put $H_{\alpha+1}(n) = H_\alpha(n+1)$. If we define $H_{\alpha+1}(n) = H_\alpha(n) + 1$ we obtain the so-called "slow growing hierarchy" which catches up with the previously defined "fast growing hierarchy" not before the Howard-Bachmann ordinal $\theta\varepsilon_{\Omega+1}0$ [Girard 1981]. On the other hand the definition of the subrecursive hierarchy is also very sensitive to the definition of fundamental sequences and we do not yet know how to characterize 'natural' fundamental

---

[4] The proof needs a more sophisticated ordinal assignment. So $\mathbf{PA}_\infty^{\mathrm{CF}} \vdash^\alpha F$ does not have exactly the same meaning as before. Nevertheless the new assignment generates the same ordinals as before.

sequences. Therefore it seems more reasonable to keep the old (and unique) definition for the proof theoretic ordinal.

§2. Predicative proof theory

The use of the ω-rule (then called "infinite induction") in proof theory has already been proposed by D. Hilbert [Hilbert 1931]. In his 1951 paper, K. Schütte introduces a system for pure number theory containing the ω-rule. (Later such systems are called "semiformal" by Schütte.) There he showed that this system allows cut-elimination and so re-obtained Gentzen's ordinal analysis for pure number theory in a much more perspicuous way. In his monography "Beweistheorie" he brought the technique of semiformal systems to a certain perfection. In more modern terms, the key point in the cut-elimination procedure for a semiformal system is the following

ELIMINATION LEMMA. *If* $\mathbf{T}_\infty \vdash^\alpha_\rho \Gamma \to \Delta, F$ *and* $\mathbf{T}_\infty \vdash^\beta_\rho \Gamma', F \to \Delta'$, *where F is a formula of complexity* ρ, *then* $\mathbf{T}_\infty \vdash^{\alpha\#\beta}_\rho \Gamma, \Gamma' \to \Delta, \Delta'$.

Here we express by $\mathbf{T}_\infty \vdash^\alpha_\rho \Delta \to \Gamma$ that there is a $\mathbf{T}_\infty$ proof-tree for the sequent $\Delta \to \Gamma$ whose ordinal is bounded by α and whose cut formulas are all of complexity strictly less than ρ. $\alpha \# \beta$ denotes the natural sum of the ordinals α and β.

The elimination lemma shows us how to avoid a cut of complexity ρ for the price of an increasing derivation length. Once the elimination lemma is established it is easy to obtain the following

ELIMINATION THEOREM. $\mathbf{T}_\infty \vdash^\alpha_{\beta+\omega^\rho} \Delta \to \Gamma$ *implies* $\mathbf{T}_\infty \vdash^{\varphi_\rho \alpha}_\beta \Delta \to \Gamma$[5], *by main induction on* ρ *and subsidiary induction on* α.

In the example of pure number theory we obtain that for each sentence F (possibly with set-parameters) provable in **PA** we have $\mathbf{PA}_\infty \vdash^{\omega^2}_n F$ for a finite ordinal n. By n-fold application of the elimination theorem we therefrom obtain $\mathbf{PA}_\infty \vdash^\alpha_0 F$ for some $\alpha < \varepsilon_0$. This shows that ε_0 is an upper bound for the norms of the provable formulas of **PA**. The same pattern should also apply to different examples of **T** producing different ordinals.

The presence of the ω-rule in $\mathbf{PA}_\infty$ renders a scheme or a rule for complete induction superfluous. Therefore the elimination lemma may be proved there essentially as in pure first order logic by a simple induction on

---

[5] φ_ρ denotes the following family of normal functions. φ_0 is the ordering function of the principal ordinals, i.e., $\varphi_0\alpha = \omega^\alpha$, $\varphi_{\alpha+1}$ enumerates the fixed-points of φ_α and for limit ordinals λ, φ_λ enumerates the intersection of the ranges of all φ_ξ for $\xi < \lambda$.

$\alpha \# \beta$. One should notice, however, that the proof depends heavily upon the fact that in $\mathbf{PA}_\infty$ all rules introducing logical connectives or quantifiers do have the subformula property, more precisely, the formula in which the symbol is introduced is interred from shorter ones. This allows us to reduce the complexity of the cut formula step by step. As we already know the situation becomes more involved when we turn to second order systems. There we need the rules

$$\forall^2 \text{ left}: \frac{F(\{x: A(x)\}), \Delta \to \Gamma}{\forall X F(X), \Delta \to \Gamma}$$

and

$$\exists^2 \text{ right}: \frac{\Delta \to \Gamma, F(\{x: A(x)\})}{\Delta \to \Gamma, \exists X F(X)},$$

which do not any longer have the subformula property.

This mirrors the fact that second order systems allow the treatment of impredicatively defined sets[6]. The usual way to avoid impredicative definitions is to define sets in stages. The sets in the basic stage $\mathscr{A}_0$ are the arithmetical definable subsets of $\mathbb{N}$. $\mathscr{A}_{\alpha+1}$ contains all subsets of $\mathbb{N}$ which are second order definable over $\mathscr{A}_\alpha$ (with parameters) and at limit stages we take the union of all previously defined stages. We so obtain a hierarchy $(\mathscr{A}_\alpha)_{\alpha \in O_n}$ which is known as ramified analytic hierarchy. The definition of this hierarchy is at least locally predicative, i.e., predicative in each step. Of course the iteration along arbitrary ordinals makes it impredicative again. There is even a countable ordinal β_0, such that the collection of all $\mathscr{A}_\alpha$ for $\alpha < \beta_0$ forms a model of second order arithmetic with full comprehension and axiom of choice.

To formalize the ramified analytic hierarchy we introduce ramified set variables X^α, $Y^\alpha, \ldots$. A formula F is of stage α if α is the largest superscript of a set variable occurring in F. A set $\{x: F(x)\}$ is of stage F if F is. The bound variables X^α, $Y^\alpha, \ldots$ are supposed to range over the sets of stages below α. So the real meaning of $\forall X^\alpha A(X^\alpha)$ is that of an infinite conjunction while $\exists X^\alpha A(X^\alpha)$ means an infinite disjunction. This proposes the following rules for the introduction of $\forall X^\alpha$:

$$\forall^\alpha_\infty\text{-left}: \frac{A(S), \Gamma \to \Delta}{\forall X^\alpha (X^\alpha), \Gamma \to \Delta}, \quad \text{where } \mathbf{S} \text{ is a set of stage } < \alpha$$

$$\forall^\alpha_\infty\text{-right}: \frac{\ldots \Gamma \to \Delta, A(S) \ldots}{\Gamma \to \Delta, \forall X^\alpha A(X^\alpha)} \quad \text{for all sets } \mathbf{S} \text{ of stage } < \alpha$$

and the dual ones for the introduction of $\exists X^\alpha$.

---

[6] We call a definition of a set **S** impredicative if it refers to an entirety to which **S** itself belongs.

So we obtain a new semiformal system **RA** with the new infinitary rules $\forall_\infty^\alpha$-right and $\exists_\infty^\alpha$-left. The interpretation of $\forall X^\alpha A(X^\alpha)$ and $\exists X^\alpha A(X^\alpha)$ as infinitely long formulas also suggests to measure their complexity by (possibly transfinite) ordinals measuring the length of the corresponding infinite formula. In this measurement the complexity of $A(S)$ will always be less than that of $\forall X^\alpha A(X^\alpha)$ or $\exists X^\alpha A(X^\alpha)$ whenever **S** is a set of stage $<\alpha$. But this means that in our new semiformal system also the rules $\exists_\infty^\alpha$-right and $\forall_\infty^\alpha$-left—and therefore all essential rules—have the subformula property and we may prove the elimination lemma for this system too. This enables us to compute an upper bound for the predicatively provable ordinals. Let us specify what we understand by a predicative ordinal. First we regard ω as a predicative ordinal. By $\mathbf{RA}_\alpha$ we denote the subsystem of **RA** in which all formulas and derivations are bounded by α. An ordinal β is then called predicative if there is an arithmetical well-ordering of ordertype β and a predicative ordinal $\alpha<\beta$ such that $\mathbf{RA}_\alpha \vdash \mathbf{TI}(<, X)$. From the elimination theorem and the boundedness theorem it is now clear that a bound for the predicative ordinals is given by the least ordinal γ such that $\alpha, \beta<\gamma$ also implies $\varphi_\alpha\beta<\gamma$. Today this ordinal is called Γ_0 following the suggestion of S. Feferman. It is easy to see that Γ_0 can be obtained as limit of the sequence $(\zeta_n)n<\omega$ where $\zeta_0=0$ and $\zeta_{n+1}=\varphi_{\zeta_n}0$. Since $\mathbf{RA}_{\zeta_n+\omega}$ proves $\mathbf{TI}(<, X)$ for an arithmetical well-ordering $\prec$ of ordertype ζ_{n+1} we see that Γ_0 is the precise bound for the predicative ordinals.

These results have been independently obtained by S. Feferman [Feferman 1964] and K. Schütte [Schütte 1964, 1965][7].

We do not want to go into a deeper discussion of predicativity. We just want to mention a few more facts which influenced the further research of the Schütte school. It is known from abstract recursion theory that the ramified hierarchy can be obtained by iterated jumps. It is possible to formalize the iteration of jumps in systems of iterated Π_1^0-comprehension. We denote the system of less than α times iterated Π_1^0-comprehension by $(\Pi_1^0\text{-CA})_{<\alpha}^-$. (The superscript $^-$ denotes that we do not allow set parameters in the comprehension formula.)

Due to the fact that one additional stage in the ramified analytic hierarchy is produced by ω-fold iterated jumps we may embed $(\Pi_1^0\text{-CA})_{<\omega\cdot\alpha}^-$ into $\mathbf{RA}_\alpha$. S. Feferman has shown that second order arithmetic with the Δ_1^1-comprehension rule, i.e.,

$$(\Delta_1^1\text{-CR}) \vdash \forall x\,(A(x) \leftrightarrow B(x)) \Rightarrow \vdash \exists X\, \forall x\,(x \in X \leftrightarrow A(x)),$$

where A is a Π_1^1- and B a Σ_1^1-formula, is proof theoretical equivalent to $(\Pi_1^0\text{-CA})_{<\omega^\omega}$. The above mentioned embedding then immediately gives $\varphi_\omega 0$ as an upper bound for the provable ordinals of $(\Delta_1^1\text{-CR})$. This bound can also be shown to be the precise one.

[7] Our exposition, however, is closer to the presentation given in [Tait 1968].

A similar technique gives us a bound for (Δ_1^1-CA), i.e., second order arithmetic with the comprehension scheme restricted to

$$\vdash \forall x\,(A(x) \leftrightarrow B(x)) \supset \exists X\, \forall x\,(x \in X \leftrightarrow A(x)),$$

where A and B are as before. This system is proof theoretical equivalent to (Σ_1^1-AC), i.e., second order arithmetic with the scheme

$$\forall x\, \exists X\, A(X, x) \supset \exists Z\, \forall x\, A(Z_x, x),$$

for Σ_1^1-formulas A. H. Friedman [Friedman, 1967] has proved that (Σ_1^1-AC) is equivalent to $(\Pi_1^0\text{-CA})_{<\varepsilon_0}$. The embedding then gives $\varphi_{\varepsilon_0}0$ as an upper bound, which again is the precise one. After knowing these results it is of course possible to embed (Δ_1^1-CR) or (Δ_1^1-CA) directly into $\mathbf{RA}_\alpha$. It is obvious how to do that for (Δ_1^1-CR) (cf. [Schütte, 1977]). For (Σ_1^1-AC) this had been done by W. Tait in [Tait, 1968]. A more recent version, using an intermediate semiformal system, can be found in [Jäger and Schütte, 1979].

§3. Beyond predicative proof theory

We have taken these examples to emphasize the fact that the formalized ramified analytical hierarchy provides a powerful and intelligible tool in the ordinal analysis of predicative subsystems of second order arithmetic. It has therefore been one of our most important aims to extend this or a similar method also to impredicative systems. Since S. Feferman and H. Friedman showed that various impredicative subsystems are reducible to systems for iterated inductive definitions, we first concentrated on the investigation of those systems.

A formal system for the theory of ν-fold iterated inductive definitions is obtained from a formal system for pure number theory by adding constants for fixed points of positive inductive definitions and their defining axioms. To get a system for ν-fold iteration we assume an arithmetical definable well-ordering of ordertype ν. We augment the language of pure number theory by a new constant P^A for each X-positive arithmetical formula $A(X, Y, x, y)$ only containing the indicated free variables. If we abbreviate the set $\{x\colon \langle x, z\rangle \in P^A\}$ by P_z^A and $\{x\colon \exists y\,(y < z \wedge \langle x, y\rangle \in P^A\}$ by $P_{<z}^A$, the defining axioms for the new constants are:

($\mathbf{ID}_\nu^1$) $$\forall z \in \text{field}(<)(\forall x\,(A(P_z^A, P_{<z}^A, x, z) \supset x \in P_z^A),$$

expressing that P_z^A is a fixed point of the monotone operator Γ_A given by

$$\Gamma_A(S) = \{x \in \mathbb{N}\colon A(S, P_{<z}^A, x, z)\},$$

and

($\mathbf{ID}_\nu^2$) $$\forall z \in \text{field}(<)(\forall x\,(A(F, P_{<z}^A, x, z) \supset F(x)) \supset \forall x\,(x \in P_z^A \rightarrow P(x)),$$

expressing that P_z^A is the least such fixed point.

For safety one often adds the scheme $\mathbf{TI}(<, F)$ for each formula F of the extended language.

By $\mathbf{ID}_{<\nu}$ we denote the union of the systems $\mathbf{ID}_\mu$ for $\mu < \nu$. There are manifold connections between systems of iterated inductive definitions and subsystems of second order arithmetic (often called 'analysis'). So Π^1_1-comprehension with the axiom (not the scheme) of complete induction, denoted by $(\Pi^1_1\text{-CA})_0$ is equivalent to $\mathbf{ID}_{<\omega}$; $(\Pi^1_1\text{-CA}) + (\text{BI})$, where (BI) denotes the scheme of classical bar induction,

$$\forall X\, \mathbf{TI}(<, X) \rightarrow \mathbf{TI}(<, F) \text{ for every formula } F'$$

is equivalent to $\mathbf{ID}_\omega$; $(\Delta^1_2\text{-CR})$ is equivalent to $\mathbf{ID}_{<\omega^\omega}$ and $(\Delta^1_2\text{-CA})$ is equivalent to $\mathbf{ID}_{<\varepsilon_0}$. A detailed discussion of these connections is presented in Chapter 1 of [Buchholz et al., 1981]. All this shows that an ordinal analysis of $\mathbf{ID}$, will also give an ordinal analysis for a lot of subsystems of second order arithmetic. The ordinal analysis for the simplest system $\mathbf{ID}_1$, the system for non-iterated inductive definitions, had been given about 1970 by W. Howard (from above) and H. Gerber (from below). Later it was re-obtained by several people using different methods. But there had been no obvious generalization to the iterated case.

The ordinal analysis of $\mathbf{ID}_1$, however, had not been the earliest treatment of an impredicative system. The first breakthrough in the treatment of impredicative systems was obtained by G. Takeuti who proved the consistency of $(\Pi^1_1\text{-CA}) + (\text{BI})$ in Gentzen style [Takeuti, 1967]. The first computations of the proof theoretic ordinals of the systems $\mathbf{ID}$ had later been obtained by a modification of Takeuti's method.

§3.1. Notations for impredicative ordinals

Takeuti in his consistency proof made use of his ordinal diagrams. The ordinal diagrams are given in a purely syntactical manner. So, in order to use Takeuti's proof in the light of our general aims, the ordinal diagrams had to be replaced by ordinal notations which could more easily be understood from a classical set-theoretic standpoint. In [Schütte, 1969] Schütte presented the notation systems $\Sigma(N)$ which still were purely syntactical but could more easily be interpreted in terms of normal functions. H. Levitz [Levitz, 1970] could prove that the systems $\Sigma(N)$ were equivalent to Takeuti's ordinal diagrams of finite order. Takeuti's consistency proof, however, needed ordinal diagrams of transfinite order. Therefore the systems $\Sigma(N)$ could not be sufficient for the ordinal analysis of $(\Pi^1_1\text{-CA}) + (\text{BI})$. Schütte himself noticed that the union of the systems $\Sigma(N)$ is not well-ordered but Pfeiffer [Pfeiffer, 1969] observed that it is possible to choose subsystems $\Sigma'(N)$ of $\Sigma(N)$ such that $\Sigma = \bigcup\{\Sigma'(N) : N < \omega\}$ is a well-ordering. Pohlers showed in [Pohlers, 1973] that the system Σ suffices for an ordinal analysis of $(\Pi^1_1\text{-CA})_0$.

For the treatment of stronger systems, however, there was still a demand for stronger canonical notation systems. Feferman, in an unpublished paper around 1970–71, proposed the use of quite simply definable ordinal functions, nowadays known as θ-functions, in notation systems. These functions had been further examined by Aczel, Weyhrauch and especially J. Bridge, who also obtained first results concerning the recursiveness of the ordinal segment accessible by the θ-functions.

In 1975 W. Buchholz succeeded in defining a recursive notation system based on the θ-functions. We define the functions θ_α and the α-Skolemhull $Cl(\alpha, \beta)$ of an ordinal β simultaneously by induction on α. $Cl(\alpha, \beta)$ is the least set of ordinals which contains 0, comprises β and is closed under addition and all functions θ_ξ for $\xi < \alpha$. An ordinal η is called α-inaccessible or α-critical if $\eta \notin Cl(\alpha, \eta)$, i.e., if η cannot be accessed from smaller ordinals by addition and/or applications of functions θ_ξ with $\xi < \alpha$. θ_α is the enumeration function of the class $Cr(\alpha)$ of α-critical ordinals.

It turns out that $Cl(\alpha, 0)$ contains an entire segment of the ordinals. Since every ordinal in $Cl(\alpha, 0)$ can be expressed by a term only containing the symbols 0, + and θ, this gives us a notation system for the segment. As far as we have defined it now, the largest segment will be in $Cl(\Omega_1, 0)$ where Ω_1 denotes the first uncountable ordinal. Its ordertype is Γ_0, the Schütte-Feferman ordinal and it is not too difficult to see that the functions θ_ξ coincide with the previously used functions φ_ξ on this segment. We may, however, enlarge the denoted segment by throwing uncountable cardinals into the sets $Cl(\alpha, \beta)$. We denote by $\Omega_1, \Omega_2, \ldots$ the sequence of uncountable cardinals. By closing $Cl(\alpha, \beta)$ under $\Omega_1, \ldots, \Omega_\nu$ we obtain a segment which is sufficient for the proof theoretic treatment of $\mathbf{ID}_\nu$. Of course things become the more complicated the more closure properties $Cl(\alpha, \beta)$ has. The details of the development have been worked out by W. Buchholz [Buchholz, 1974, 1975]. There he also proved the recursiveness of the obtained segment.

It turned out that it was easier not to deal with the functions θ_α itself, because of their fixed points, but to introduce fixed point free functions $\bar{\theta}_\alpha$. The function $\bar{\theta}_\alpha$ nearly is the enumeration function of the class $Cr(\alpha)\backslash Cr(\alpha+1)$ of at most α-critical ordinals. The function which enumerates $Cr(\alpha)\backslash Cr(\alpha+1)$ will no longer preserve number classes. To save this important property one has to shift the domain of the enumeration function such that argument and value belong to the same number class. The resulting functions are the $\bar{\theta}_\alpha$-functions. We obtain Buchholz's notation system $\bar{\theta}(\{g\})$, which is based on the $\bar{\theta}$-functions, by closing $Cl(\alpha, \beta)$ under a function g defined in the following way. g maps pairs of ordinals into the cardinals. Viewed as a function of its second argument, g_0 enumerates the uncountable cardinals, $g_{\alpha+1}$ enumerates the fixed points of g_α, while g_λ for limit λ enumerates $\bigcap\{rg(g_\xi)\colon \xi < \lambda\}$. $\bar{\theta}(\Omega)$ is the subsystem of $\theta(\{g\})$, where $Cl(\alpha, \beta)$ only has to be closed under the function $\lambda x \Omega_x$ which enumerates the uncountable cardinals and $\bar{\theta}(\tau)$ is the subsystem which is obtained by just throwing $\Omega_1, \ldots, \Omega_\tau$ into all sets

$Cl(\alpha, \beta)$. In a joint paper [Buchholz and Schütte, 1976] W. Buchholz and K. Schütte investigated the match-ups between the previously mentioned system Σ and $\bar{\theta}(\omega)$ and Buchholz [Buchholz, 1976] compared notation systems developed by Pfeiffer as well as Kino's modification of Takeuti's ordinal diagrams with subsystems of $\bar{\theta}(\{g\})$. $\bar{\theta}(\{g\})$ was the most far reaching notation system at that time but it turned out that it was still too weak for an ordinal analysis of $(\Delta^1_2\text{-CA}) + (\text{BI})$. It will later become plausible that any notation system which is sufficient for the ordinal analysis of such a strong system, somehow has to internalize the first (at least recursively) inaccessible ordinal. It was, however, easy to see how the system $\bar{\theta}(\{g\})$ could be extended. It is quite clear that the closure of the α-Skolemhulls $Cl(\alpha, \beta)$ under any reasonably effectively given function which maps ordinals into cardinals will not spoil the recursiveness of the notation system. So the problem was to find functions from the ordinals into the cardinals which on the one side are sufficiently effective and on the other side enumerate enough cardinals. The idea was to use the same process as we have used in the definition of the θ-function. For this purpose we defined a higher α-Skolemhull, say $Cl_1(\alpha, \beta)$ and equip it with additional closure properties which are strong enough to force the ordinals in $Cr_1(\alpha) = \{\eta\colon \eta \notin Cl_1(\alpha, \eta)\}$ to be cardinals. Then their enumeration functions $\theta_{1,\alpha}$ will map ordinals into cardinals. This is achieved by closing $Cl_1(\alpha, \beta)$ under number classes, i.e., whenever $Cl_1(\alpha, \beta)$ contains an ordinal η it already has to contain the entire number class of η. This ensures that only initial ordinals, i.e., cardinals, belong to $Cr_1(\alpha)$. Now we close the original α-Skolemhulls under θ_1 and proceed as before. Till now the functions θ_1 and g coincide. But again we may enlarge the segment in $Cl_1(\alpha, \beta)$ by throwing more ordinals into $Cl_1(\alpha, \beta)$. In the case of $Cl(\alpha, \beta)$ we obtained more by putting regular ordinals into $Cl(\alpha, \beta)$. The ordinals which play the role of regular ordinals relative to $Cl_1(\alpha, \beta)$ are the regular fixed points of the function Ω (which enumerates the uncountable cardinals), i.e., weakly inaccessible cardinals. For the treatment of $(\Delta^1_2\text{-CA}) + (\text{BI})$ it suffices just to put the first weakly inaccessible cardinal into $Cl_1(\alpha, \beta)$. Until today this is the strongest system which is proof theoretically meaningful. But there is of course no need to stop here. We may iterate the procedure, define $2-\alpha$-Skolemhulls $Cl_2(\alpha, \beta)$, close $Cl_1(\alpha, \beta)$ under the resulting θ_2-functions, etc. It is even possible to extend this iteration into the transfinite. The result is a 3-place θ-function and a really far-reaching notation system based on the corresponding 3-place $\bar{\theta}$-functions. The details of this approach are in [Pohlers, 1983].

More recently it has been observed by W. Buchholz that the approach to notation systems may be simplified. Instead of regarding the α-Skolemhull $Cl(\alpha, \beta)$ of an arbitrary ordinal β it should suffice just to look at the α-Skolemhulls of initial ordinals (where we take 0 as an initial ordinal). This leads to different sets $C_\kappa(\alpha)$ where κ is an initial ordinal and functions $\psi_\kappa(\alpha) := \min\{\xi\colon \xi \notin C_\kappa(\alpha)\}$. The sets $C_\kappa(\alpha)$ are supposed to comprise κ, to be closed under addition and to contain $\psi_\sigma(\xi)$ whenever $\sigma, \xi \in C_\kappa(\alpha)$ and $\xi < \alpha$. The resulting notation systems are in fact simpler than the $\bar{\theta}$-systems. An exposition of ψ-systems is given in [Buchholz, 1984b].

Jäger in [Jäger, 1984a] showed that it is possible to establish a recursive notation system, based on the ψ-functions, which incorporates a hierarchy of inaccessible cardinals. Its strength should be the same as that of the corresponding $\bar{\theta}$-system.

We conclude this section by remarking that we may replace the cardinals used in the definition of the notation systems by their recursive counterparts, i.e., by admissible ordinals and limits of admissibles. This becomes important for the interpretation of the higher number classes which arise in proof theoretic applications of the notation systems.

§3.2. Ordinal analysis of iterated inductive definitions by Takeuti's reduction procedure

In terms of the θ-functions the proof theoretic ordinal of $\mathbf{ID}_1$ is $\theta\varepsilon_{\Omega_1+1}0$. It has been conjectured by different people that the ordinal of $\mathbf{ID}_n$ for finite n is $\theta\varepsilon_{\Omega_n+1}0$. As an obvious generalization one should expect that the ordinal of $\mathbf{ID}_\nu$ is $\theta\varepsilon_{\Omega_\nu+1}0$. Fixed points of inductive definitions are expressible by Π^1_1-formulas and, vice versa, Π^1_1-relations may be expressed by (projections of) such fixed points. S. Feferman [Feferman, 1970] used this to show that formal systems for iterated inductive definitions are proof theoretical equivalent to formal systems for iterated Π^1_1-comprehension. Since there already existed a reduction procedure for (Π^1_1-CA) by Takeuti, it was an obvious idea to use this procedure in the computation of the proof theoretic ordinals of $\mathbf{ID}_\nu$. However, the procedure introduced in [Takeuti, 1967] was a bit too rough to deliver the precise ordinals. But it was possible to refine it. This refinement heavily rested on the fact that no set parameters are needed in the comprehension formula. It has been presented in [Pohlers, 1975], where he used it to prove the consistency of $\mathbf{ID}_n$ by a transfinite induction along a well-ordering of the conjectured ordertype $\theta\varepsilon_{\Omega_n+1}0$. In [Pohlers, 1978] this was extended to transfinitely iterated inductive definitions. The latter paper already contained the result: "if $\mathbf{ID}_\nu \vdash F$ for an arithmetical F, then there is an $\alpha < \theta\varepsilon_{\Omega_\nu+1}0$ such that $\mathbf{PA}^{\mathrm{CF}}_\infty \vdash^\alpha F$", which gives an ordinal analysis of $\mathbf{ID}_\nu$ in the spirit of Section 1.

On the other hand Buchholz and Pohlers showed in [Buchholz and Pohlers, 1978] that for each $\alpha < \theta\varepsilon_{\Omega_\nu+1}0$, there is a primitive recursive well-ordering $<$ of ordertype α such that $\mathbf{ID}^i_\nu$ (i.e., $\mathbf{ID}_\nu$ based on intuitionistic logic) proves $\mathbf{TI}(<, X)$. This also settled the open problem of the proof theoretic equivalence of $\mathbf{ID}_\nu$ and $\mathbf{ID}^i_\nu$ [8]. In the light of our

[8] The proof theoretic equivalence of $\mathbf{ID}_{<\nu}$ and $\mathbf{ID}^i_{<\nu}$ for limit ordinals ν had been independently proved by W. Sieg. cf. [Buchholz et al., 1981].

general aims, the result was satisfying. The proof, however, was not. This for several reasons. First it was not a generalization of the methods of predicative proof theory. The ω-rule in [Pohlers, 1978] only served to get rid of complete induction. The handling of comprehensions was the same as in finite derivations. What we really had been looking for was an infinitary rule which, on the one hand, does not conflict with the canonical cut elimination procedure and, on the other hand, makes visible why the ordinal analysis of $\mathbf{ID}_\nu$ needs higher number classes [9].

Second, the methods of the proof were not strong enough to solve the remaining questions. We could treat autonomously iterated inductive definitions. The first ordinal which is inaccessible by autonomously iterated inductive definitions is $\theta\Omega_{\Omega_1}0$. This ordinal already formally resembles the ordinal $\Gamma_0 = \theta\Omega 0$ which is inaccessible by autonomously iterated jumps. But—in contrast to predicative theories—the possibilities of iterating inductive definitions are not exhausted by autonomous iterations. In a formal theory for inductive definitions (or comparable theories) we may define the accessible part $<^*$ of any arithmetically definable order relation $<$. Iteration of inductive definitions along $<^*$ really leads to a new dimension of impredicativity (compared to iteration along a fixed previously given well-ordering). Feferman called the resulting formal theory $\mathbf{ID}_{<^*}$. Already the ordinal analysis of $\mathbf{ID}_{<^*}$ resisted the methods used in [Pohlers, 1978].

§3.3. Buchholz's $\Omega_{\nu+1}$-rules

The first who succeeded in finding such an infinitary rule was W. Buchholz. Since fixed points of inductive definitions and hyperjumps are essentially the same he tried to find an adequate infinitary rule by formalizing the hyperjump. Already W. Howard [Howard, 1972] used such a formalization to introduce an abstract ordinal notation system which comprises notations for the ordinal of $\mathbf{ID}_1$. Buchholz extracted the proof theoretic content of Howard's paper and formulated his $\Omega_{\nu+1}$-rules formalizing $\nu+1$-fold iterated hyperjumps. To characterize the idea we describe the Ω_1-rule. We sketch a semiformal system $\mathbf{ID}_1^*$ which is sufficient for the treatment of $\mathbf{ID}_1$. The language of $\mathbf{ID}_1^*$ is that of $\mathbf{ID}_1$ but without free individual variables. $\mathbf{ID}_1^*$ is based on a sequent calculus and we assume the usual logical rules including ω-rule. As additional rules we have the closure rule

$$(Cl) \qquad \Gamma \to \Delta, A(P_A, \underline{n}) \Rightarrow \Gamma \to \Delta, \underline{n} \in P_A$$

and the Ω_1-rule. The Ω_1-rule is supposed to formalize the hyperjump. A

[9] Since $\mathbf{ID}_\nu$ allows the definition of higher constructive number classes this was intuitively clear. But it became not visible in the proof.

representative for the hyperjump of the set ω of natural numbers is Kleene's $\mathcal{O}$, the set of codes for constructive ordinals. The essential step in the definition of $\mathcal{O}$ is the rule: "If $\forall x\,(x \in \omega \supset \{e\}(x) \in \mathcal{O})$ then $3 \cdot 5^e \in \mathcal{O}$". Here $3 \cdot 5^e$ is just a code for the function $\{e\}$. If we do not insist on $\mathcal{O}$ being a subset of ω, we may reformulate the hyperjump rule to: "If $\forall x\,(x \in \omega \supset f(x) \in \mathcal{O})$, then $\langle f(x) \colon x \in \omega \rangle \in \mathcal{O}$, where f is any recursive function."

We are trying to transfer this rule into a rule of $\mathbf{ID}_1^*$. First we notice that we have two classes of formulas in the language of $\mathbf{ID}_1^*$. One class contains the formulas with at most positive occurrences of the new symbol P_A—we call them formulas of level 0—the other formulas of level 1 which are supposed to have an actual negative occurrence of some P_A. Since $\underline{n} \in P_A$ can be described by an arithmetical formula with free set parameters the formulas of level 0 are in fact arithmetical. Only formulas of level 1 exceed the expressiveness of the arithmetical language. Sequents of level 0 and level 1 are defined analogously. The subsystem $\mathbf{IA}_\infty$ of $\mathbf{ID}_1^*$ contains only sequents of level 0. So $\mathbf{IA}_\infty$ is a slight generalization of $\mathbf{PA}_\infty$ and it likewise allows cut-elimination. We therefore start with the cut-free derivations in $\mathbf{IA}_\infty^{\mathrm{CF}}$ as basic set—corresponding to the set ω in the hyperjump rule. Let us denote derivations by lower case greek letters $\pi, \pi_1, \ldots$. By $(\pi \vdash \Gamma \to \Gamma')$ we express that $\Gamma \to \Gamma'$ is the end sequent of π. Now the Ω_1-rule says:

$$(\Omega_1)\begin{cases}\text{if } \forall \pi((\pi \vdash \Gamma \to \Gamma', \underline{n} \in P_A) \in \mathbf{IA}_\infty^{\mathrm{CF}} \supset (F(\pi) \vdash \Gamma, \Delta \to \Gamma', \Delta') \in \mathbf{ID}_1^*),\\ \text{then } (\langle F(\pi) \colon \pi \in \mathbf{IA}_\infty^{\mathrm{CF}} \rangle \vdash \underline{n} \in P_A, \Delta \to \Delta') \in \mathbf{ID}_1^*,\end{cases}$$

where F belongs to a collection of effectively given functions from derivations into derivations (which we will not define precisely).

This sounds confusing. But we have written it down in this way in order to emphasize the close connection of the Ω_1-rule to the hyperjump rule. The derivations in $\mathbf{ID}_1^*$ can be understood as the hyperjump of the derivations in $\mathbf{IA}_\infty$. To make the rule clearer we repeat it in words: If we want to derive $\Delta, \underline{n} \in P_A \to \Delta'$ we have to show that each $\mathbf{IA}_\infty^{\mathrm{CF}}$ derivation of $\Gamma \to \Gamma'$, $n \in P_A$ can be effectively transformed into a derivation of $\Gamma, \Delta \to \Gamma', \Delta'$. As an example we show how the axiom ($\mathbf{ID}_1^2$) can be derived in $\mathbf{ID}_1^*$. The axiom ($\mathbf{ID}_1^1$) is apparently an immediate consequence of the closure rule. We have to show that $\mathbf{ID}_1^*$ proves $\forall x\,(A(F, x) \supset F(x))$, $\underline{n} \in P_A \to F(\underline{n})$ for arbitrary n. By the Ω_1-rule it suffices to show that every $\mathbf{IA}_\infty^{\mathrm{CF}}$ derivation of $\to \underline{n} \in P_A$ can be transformed into a derivation of $\forall x\,(A(F, x) \supset F(x)) \to F(\underline{n})$. This is easily done by replacing $\underline{m} \in P_A$ by $F(\underline{m})$ all over the derivation and by adding $\forall x\,(A(F, x) \supset F(x))$ to the antecedents of all sequents of the derivation. The only rule which will be violated by this substitution is the closure rule. But in this case we obtain $\forall x\,(A(F, x) \supset F(x))$, $\Gamma \to \Gamma'$, $A(F, \underline{m})$, from which $\forall x\,(A(F, x) \supset F(x))$, $\Gamma \to \Gamma'$, $F(\underline{m})$ is easily derived.

This shows that $\mathbf{ID}_1$ can be embedded into $\mathbf{ID}_1^*$. We shall now indicate that the additional rules (Cl) and (Ω_1) do not disturb the canonical cut-elimination procedure. In a first step we eliminate all cut-formulas which are not of the shape $\underline{n} \in P_A$. We can do it in such a way that the remaining cuts have the form

$$\left.\begin{array}{c} \begin{array}{cc} \pi_0 \vdots & \langle F(\pi) : \pi \in \mathbf{IA}_\infty^{CF} \rangle \\ \Gamma \rightarrow \Gamma', \underline{n} \in P_A & \Delta, \underline{n} \in P_A \rightarrow \Delta' \end{array} \\ \hline \Gamma, \Delta \rightarrow \Gamma', \Delta' \end{array}\right\} \rho$$

If we presume that $\Gamma, \Delta \rightarrow \Gamma', \Delta'$ is a sequent of level 0 we may eliminate also these cuts by induction on the length of ρ. $\Gamma \rightarrow \Gamma', \underline{n} \in P_A$ again is of level 0. By the induction hypothesis we may therefore assume that π_0 already is cut-free. But then it is a derivation in $\mathbf{IA}_\infty^{CF}$ and we obtain a derivation $F(\pi_0)$ of $\Gamma, \Delta \rightarrow \Gamma', \Delta'$ by the Ω_1-rule. Since $F(\pi_0)$ is a strict subderivation of ρ we obtain a cut-free derivation of $\Gamma, \Delta \rightarrow \Gamma', \Delta'$ by the induction hypothesis.

This last step is often called 'collapsing' because it collapses an $\mathbf{ID}_1^*$-derivation, belonging to a higher constructive tree class, into an $\mathbf{IA}_\infty^{CF}$-derivation belonging to a lower tree class.

The ordinal of an $\mathbf{ID}_1^*$-derivation will in general be above ω_1^{CK}, the first nonrecursive ordinal, while the ordinals of $\mathbf{IA}_\infty^{CF}$-derivation will be below ω_1^{CK}. According to the concluding remark of Section 3.2 we may interpret Ω_1 by ω_1^{CK}. The first step—elimination of all cuts except those of the form $n \in \mathbf{P}_A$—will lead us to ε_{Ω_1+1} the first ε-number above Ω_1. The second step will collapse this ordinal to $\theta\varepsilon_{\Omega_1+1}0 < \Omega_1$. An $\mathbf{IA}_\infty^{CF}$ derivation of an arithmetical formula is obviously a $\mathbf{PA}_\infty^{CF}$ derivation. So we re-obtain $\theta\varepsilon_{\Omega_1+1}0$ as an upper bound for the provable ordinals of $\mathbf{ID}_1$.

The generalization to the iterated case is not too difficult. You use the $\Omega_{\nu+1}$-rule to define the $\mathbf{ID}_{\nu+1}^*$ derivations as the hyperjump of the $\mathbf{ID}_\nu^{*,CF}$ derivations. The details of this approach can be found in Chapters IV and V of [Buchholz et al., 1981]. Using his rule W. Buchholz could also give an ordinal analysis of $\mathbf{ID}_{<^*}$. The ordinal of $\mathbf{ID}_{<^*}$ is $\theta\varepsilon_{\Omega_{\Omega_1}+1}0$. So the relation of $\mathbf{ID}_{<^*}$ to autonomously iterated inductive definitions (with ordinal $\theta\Omega_{\Omega_1}0$) resembles that of $\mathbf{ID}_1$ to autonomously iterated jumps (with ordinal $\theta\Omega_1 0$). This result is presented in Buchholz's Habilitationsschrift [Buchholz, 1977]. Here he introduced the $\Omega_{\nu+1}$-rules for the first time.

Buchholz's rule not only allowed a more perspicuous ordinal analysis of iterated inductive definitions, it also can be applied directly to subsystems of second order arithmetic. The rule sounds more complicated there. Its formulation needs quite a lot of prerequisites. Therefore we are forced to renounce its presentation. The rule can be found in [Buchholz and

Schütte, 1980]. There they give new proofs for

$$|(\Delta^1_2\text{-CA})| \leq \theta\Omega_{\varepsilon_0}0 \ (\mathrm{D_2CA}),$$

$$|(\Pi^1_1\text{-CA}) + \mathrm{BI}| \leq \theta\varepsilon_{\Omega_\omega+1}0 \ (\mathrm{GP_1CA}),$$

$$|(\Pi^1_1\text{-CA})| \leq \theta(\Omega_\omega\varepsilon_0)0 \ (\mathrm{P_1CA})^{10}, \text{ and}$$

$$|(\Delta^1_2\text{-CR})| \leq \theta\Omega_{\omega^\omega}0,$$

not using inductive definitions.

The first and the last result are originally due to Takeuti and Yasugi, the second to Pohlers (who computed the ordinal of $\mathbf{ID}_\omega$ which, according to Feferman, coincides with that of $(\Pi^1_1\text{-CA})+\mathrm{BI}$). The third result has been proved for the first time by Buchholz in his Habilitationsschrift. A comprehensive presentation is given in [Buchholz et al., 1981]. It is an obvious question to ask if it is also possible to collapse derivations into finite one (i.e., with ordinals below ω). W. Buchholz showed in [Buchholz, 1984a] that this in fact can be done. There he proved that every deduction tree in $\mathbf{ID}^*_\omega$ of a Σ^0_1-formula can be collapsed into a finite one. As a consequence he obtained the refined boundedness lemma already mentioned in Section 10. He used this result to show that a combinatorial principle concerning labeled finite trees (a generalization of Kirby's and Paris' hydras) is independent of $(\Pi^1_1\text{-CA})+(\mathrm{BI})$.

§3.4. The method of local predicativity

Buchholz's rule was a great progress. The cut elimination procedure became perspicuous, the need of higher constructive number classes in the ordinal notations got plausible and it made the treatment of $\mathbf{ID}_{<*}$ (and even further iterations) feasible. Nevertheless we had not been completely satisfied by it. First the $\Omega_{\nu+1}$-rule itself is obtained by an inductive definition and therefore has no immediate formalization in $\mathbf{HA}_{<|\mathbf{ID}_{\nu+1}|}$. So the result $\mathbf{ID}_\nu \equiv \mathbf{HA}_{<|\mathbf{ID}_\nu|}$ (cf. Section 1) needs further considerations. On the other hand the resulting semiformal systems were no immediate generalizations of the system $\mathbf{RA}$ of predicative proof theory. There we had a completely uniform method of computing the proof theoretic ordinal of a formal system $\mathbf{T}$. We looked for an optional α such that $\mathbf{T}$ became embeddable into $\mathbf{RA}_\alpha$. The rest was an application of cut-elimination and boundedness. Why shouldn't it be possible to embed also impredicative formal systems by choosing α large enough? In other words we wanted to replace the hyperjump rule by adequately often iterated jump rules. We finally succeeded in doing this. Again the starting point were the formal theories for inductive definitions. The fixed point $\mathbf{I}_A$ of an X-positive inductive definition $A(X, x)$ is obtainable from below as the

---

[10] The notations in brackets are the notations of [Buchholz and Schütte, 1980].

union of its stages $\mathbf{I}_A^{\xi} = \{x: A(\mathbf{I}_A^{<\xi}, x)\}$ where $\mathbf{I}_A^{<\xi}$ stands for the union of all $\mathbf{I}_A^{\eta}$ for $\eta < \xi$. The hierarchy of stages collapses at ω_1^{CK} and we obtain $\mathbf{I}_A = \mathbf{I}_A^{<\omega_1^{\mathrm{CK}}}$. We see that this definition from below is at least locally predicative, i.e., each step is predicative, only iteration along ω_1^{CK} makes it impredicative. We had already interpreted the ramified quantifiers in **RA** as infinite disjunctions or conjunctions. It is easy to see how to formalize the stages of an inductive using infinite formulas. We use the ordinals of our notation system and define $\underline{n} \in \mathbf{I}_A^{<\xi}: \Leftrightarrow \bigvee \{\underline{n} \in \mathbf{I}_A^{\eta}: \eta < \xi\}$ and $\underline{n} \in \mathbf{I}_A^{\eta}: \Leftrightarrow A(\mathbf{I}_A^{<\eta}, \underline{n})$. It is clear how to modify the infinitary rules of **RA** to obtain a semiformal system $\mathbf{ID}_1^{\infty}$. But then it is also clear that this system will not go beyond $\theta\Omega_1 0$ $(=\Gamma_0)$. So it is still too weak for an embedding of $\mathbf{ID}_1$. What we have not yet formalized is the fact that the hierarchy of stages collapses at Ω_1 (which is supposed to play the role of ω_1^{CK}). So we add a new rule

$$(Cl_{\Omega_1}) \qquad \Gamma \to \Delta, \underline{n} \in \mathbf{I}_A^{\Omega_1} \Rightarrow \Gamma \to \Delta, \underline{n} \in \mathbf{I}_A^{<\Omega_1}.$$

This rule immediately proves the translation of $(\mathbf{ID}_1^1)$. $(\mathbf{ID}_1^2)$ is just a special case of transfinite induction up to Ω_1, which is canonically provable in $\mathbf{ID}_1^{\infty}$ (of course with a derivation tree of length $\geqslant \Omega_1$). It is exactly the rule Cl_{Ω_1} which makes the system $\mathbf{ID}_1^{\infty}$ impredicative. The impredicativity of this rule is again manifested by the fact that its premise is more complex than its conclusion. This usually is fatal for the cut elimination procedure. But here we are able to overcome the difficulties. The crucial case in the elimination procedure will be a cut of the following form:

$$\dfrac{\dfrac{\vdash^{\alpha_0}_{\Omega_1} \Gamma \to \Delta, \underline{n} \in \mathbf{I}_A^{\Omega_1}}{\vdash^{\alpha}_{\Omega_1} \Gamma \to \Delta, \underline{n} \in \mathbf{I}_A^{<\Omega_1}} \qquad \dfrac{\ldots \vdash^{\beta_\xi}_{\Omega_1} \underline{n} \in \mathbf{I}_A^{\xi}, \Gamma' \to \Delta' \ldots \quad \text{for all } \xi < \Omega_1}{\vdash^{\beta}_{\Omega_1} \underline{n} \in \mathbf{I}_A^{<\Omega_1}, \Gamma' \to \Delta'}}{\vdash^{\gamma}_{\Omega_1+1} \Gamma, \Gamma' \to \Delta, \Delta'}$$

Here the standard reduction step will not apply. We need further considerations to handle this case. In a first step we notice that the system $\mathbf{ID}_1$ has the

Boundedness property. If $\mathbf{I}_A^{<\eta}$ occurs positively in $F(\mathbf{I}_A^{<\eta})$ and $\vdash^{\alpha}_{\rho} \Delta \to \Gamma, F(\mathbf{I}_A^{<\eta})$ for some $\alpha < \Omega_1$, then we also have $\vdash^{\alpha}_{\rho} \Delta \to \Gamma, F(\mathbf{I}_A^{<\beta})$ for all β such that $\alpha \leqslant \beta \leqslant \Omega_1$.

For $n \in \mathbf{I}_A$ we have the norm $|n| = \min\{\xi: n \in \mathbf{I}_A^{\xi}\}$. As a consequence of the boundedness property we obtain that we need a derivation of length $\geqslant \alpha$ in order to prove $\underline{n} \in \mathbf{I}_A^{<\eta}$ for an n of norm α.

The boundedness property is easily proved by induction on α. The remarkable case in the proof is that of an inference (Cl_{Ω_1}). Here we have the premise

$$\vdash^{\alpha_0}_{\rho} \Delta \to \Gamma, \underline{n} \in \mathbf{I}_A^{\Omega_1} \text{ which yields } \vdash^{\alpha_0}_{\rho} \Delta \to \Gamma, \underline{n} \in \mathbf{I}_A^{\alpha_0}$$

by the induction hypothesis. For any $\beta \geqslant \alpha > \alpha_0$ we may now infer $\vdash_\rho^\alpha \Delta \to \Gamma,\ n \in \mathbf{I}_A^{<\beta}$ by an $\bigvee$-right inference.

We have emphasized this case because it shows that the impredicative (Cl_{Ω_1})-rule is eliminable in derivations of length $<\Omega_1$. This helps us to manage the crucial case. If we succeed to collapse α_0 to α_0' below Ω_1 we may use the boundedness property to conclude

$$\vdash_{\Omega_1}^{\alpha_0'} \Gamma \to \Delta,\ \underline{n} \in \mathbf{I}_A^{\alpha_0'}$$

and then single out the α_0'th premise of

$$\vdash_{\Omega_1}^{\beta} \underline{n} \in \mathbf{I}_A^{<\Omega_1},\ \Gamma' \to \Delta'.$$

Thus we obtain a cut

$$\frac{\vdash_{\Omega_1}^{\alpha_0'} \Gamma \to \Delta,\ \underline{n} \in \mathbf{I}_A^{\alpha_0'} \qquad \vdash_{\Omega_1}^{\beta\alpha_0'} \underline{m} \in \mathbf{I}_A^{\alpha_0'},\ \Gamma' \to \Delta'}{\vdash_{\Omega_1}^{\delta} \Gamma,\ \Gamma' \to \Delta,\ \Delta'}$$

of reduced cut rank. But we are not content with a terminating cut-elimination procedure, we also want to have an estimate for the length of the cut-free derivation. That means that we need δ as a function of γ. In general, however, γ will be above Ω_1 while δ might be below Ω_1. This requires a function which collapses ordinals below Ω_1.

On the basis of the $\bar{\theta}$-functions it is possible to define such a collapsing function $\mathcal{D}_0$, whose range is sufficiently large. $\mathcal{D}_0$ is defined in such a way that it becomes the identity of the ordinals below Ω_1. Once we have the collapsing function we define a relation $\alpha \ll \beta :\Leftrightarrow \alpha < \beta$ and $\mathcal{D}_0\alpha < \mathcal{D}_0\beta$. Then we label a derivation tree with ordinals which are increasing in the sense of $\ll$. Such an ordinal assignment still provides an upper bound for the canonical ordinal-length of a derivation and it has the advantage of remaining an increasing assignment also after an application of $\mathcal{D}_0$ [11]. Of course such an assignment is not globally possible. In the case of an inference

$$(\bigvee\text{-left})\quad \vdash_\rho^{\beta_\xi} \underline{n} \in \mathbf{I}_A^{\xi},\ \Delta \to \Gamma \text{ for all } \xi < \Omega_1 \Rightarrow \vdash_\rho^{\beta} \underline{n} \in \mathbf{I}_A^{<\Omega_1},\ \Delta' \to \Gamma'$$

or the dual ($\bigwedge$-right) inference we do not have Ω_1-many different $\beta_\xi \ll \beta$. Here practice taught us that it suffices to require $\beta_\xi < \beta$ and $\beta_\xi \ll \sigma$ for all σ such that $\xi \ll \sigma$ and $\beta \leqslant \sigma$.

It is now obvious that the ordinal assignment of a derivation without Ω_1-branching inferences is collapsible. A sequent of level 0 again is a sequent without negative occurrences of $\mathbf{I}_A$. So if we have

$$\vdash_{\Omega_1}^{\alpha} \Gamma \to \Delta$$

for a sequent $\Gamma \to \Delta$ of level 0 we obtain

$$\vdash_{\Omega_1}^{\mathcal{D}_0\alpha} \Gamma \to \Delta$$

[11] This technique has first been introduced in [Howard, 1972].

since the derivation does not contain Ω_1-branching rules. Now let us choose $\delta := \omega^\gamma$ and look at the crucial case in the light of the new ordinal assignment. Then we have $\alpha_0 \ll \alpha \ll \gamma$ and $\beta \ll \gamma$. If we assume that $\Gamma, \Gamma' \rightarrow \Delta, \Delta'$ is of level 0 we may apply collapsing to obtain

$$\vdash_{\Omega_1}^{\mathscr{D}_0\alpha_0} \Gamma \rightarrow \Delta, \underline{n} \in \mathbf{I}_A^{\Omega_1}.$$

So we may put $\alpha_0' = \mathscr{D}_0\alpha_0 < \Omega_1$ and reduce the cut in the proposed way. We obtain $\mathscr{D}_0\alpha_0 \ll \alpha \ll \delta$ and therefore also $\beta_{\mathscr{D}_0\alpha_0} \ll \delta$ which shows that the ordinal assignment of the reduced cut is also increasing in the sense of $\ll$.

Thus $\mathbf{ID}_1^\infty$ allows cut-elimination for derivations with endsequents of level 0. An arithmetical sequent certainly has level 0. A cut-free $\mathbf{ID}_1^\infty$-proof neither contains the (Cl_{Ω_1}) rule nor Ω_1-branching rules. So it is in fact a $\mathbf{PA}_\infty$-derivation and we obtain the desired ordinal analysis [Pohlers, 1981].

This method is generalizable to the iterated case [Pohlers, 1982a]. The generalization, however, is not completely straightforward. An additional problem arises in the case of a cut

$$\frac{\dfrac{\vdash_\rho^{\alpha_0} \Gamma \rightarrow \Delta, \underline{n} \in \mathbf{I}_A^{\xi_0}}{\vdash_\rho^{\alpha} \Gamma \rightarrow \Delta, \underline{n} \in \mathbf{I}_A^{<\eta}} \qquad \dfrac{\ldots \vdash_\rho^{\beta_\xi} \underline{n} \in \mathbf{I}_\rho^{\xi}, \Gamma' \rightarrow \Delta' \ldots \text{ for all } \xi < \eta}{\vdash_\rho^{\beta} \underline{n} \in \mathbf{I}_\varphi^{<\eta}, \Gamma' \rightarrow \Delta'}}{\vdash_{\rho+1}^{\gamma} \Gamma, \Gamma' \rightarrow \Delta, \Delta'},$$

where η is not an initial ordinal. The standard reduction step delivers the derivation

$$\frac{\vdash_\rho^{\alpha_0} \Gamma \rightarrow \Delta, \underline{n} \in \mathbf{I}_A^{\xi_0} \qquad \vdash_\rho^{\beta_{\xi_0}} \underline{n} \in \mathbf{I}_A^{\xi_0}, \Gamma' \rightarrow \Delta'}{\vdash_\rho^{\gamma} \Gamma, \Gamma' \rightarrow \Delta, \Delta'},$$

but here we have no possibility to infer $\beta_{\xi_0} \ll \gamma$.

In the non-iterated case this caused no problems since the relations $\ll$ and $<$ coincide on ordinals below Ω_1. To overcome this problem in the iterated case we have to put more information into the ordinal assignment. For an inference

$$(\bigvee\text{-right}) \qquad \vdash_\rho^{\alpha_0} \Gamma \rightarrow \Delta, \underline{n} \in \mathbf{I}_A^{\xi_0} \Rightarrow \vdash_\rho^{\alpha} \Gamma \rightarrow \Delta, n \in \mathbf{I}_A^{<\eta}$$

and its dual ($\bigwedge$-right) inference we not only require $\alpha_0 \ll \alpha$ but also $\xi_0 \ll \alpha$. Then we obtain $\xi_0 \ll \alpha \ll \gamma$ and from this $\beta_{\xi_0} \ll \gamma$ and the ordinal assignment of the reduced cut is correct.

This is the basic idea. To overcome the technical problems of the iterated case all definitions have to be formulated more complicated. A detailed description of the whole procedure is in Chapter VI of [Buchholz et al., 1981]. The method of local predicativity also permits the ordinal analysis of $\mathbf{ID}_{<\varepsilon^*}$ as shown in [Pohlers, 1982b].

Our original aim was the generalization of the semiformal system **RA** for the ramified analytical hierarchy to a system which also allows the embedding of impredicative formal systems. The above sketched systems $\mathbf{ID}_\nu^\infty$ are too much tailored to formal systems for inductive definitions as to serve as such a generalization. We need something like the closure rules in $\mathbf{ID}_\nu^\infty$ in order to strengthen the system **RA**. It turned out, however, that analogues of the closure rule are only artificially definable in **RA**. The situation changes when we regard an extension **RS** of **RA**.

RS is a system for ramified set theory which is supposed to formalize the constructible hierarchy over the natural numbers as urelements. From the set theoretical viewpoint the better behaviour of the system **RS** is not too astonishing since the ramified analytical hierarchy is just the trace of the constructible hierarchy on the powerset of the natural numbers. **RS** uses the language of set theory with constants for natural numbers and possibly also relations on natural numbers. The ramified quantifiers $\forall x^\alpha$ and $\exists x^\alpha$ are intended to range over elements of L_α.

Of course we will not be able to describe all stages of the constructible hierarchy completely in **RS**. In order to formulate **RS** as a semiformal system we have to take ordinals from a recursive notation system. Such a notation system cannot contain a segment up to ω_1^{CK}. So for all $\alpha \geqslant \Omega_1$ there will be gaps in the ordinal notations below α and the resulting $\mathrm{M}_\alpha := \{x^\alpha : \mathbf{RS} \vdash x^\alpha = x^\alpha\}$ will just be a subset of L_α. But we will be able to describe sufficiently many sets such that we are able to embed a lot of formal systems into **RS**. The power of **RS** stems from the following closure rules which formalize the initial ordinals $\Omega_{\nu+1}$:

$$(Cl_{\Omega_{\nu+1}}) \vdash \Gamma \to \Delta, \forall x^\alpha\, \exists y^{\Omega_{\nu+1}}\, A(x, y) \Rightarrow \vdash \Gamma \to \Delta, \exists z^{\Omega_{\nu+1}}\, \forall x^\alpha\, \exists y \in z A(x, y).$$

Here $A(x, y)$ must not contain quantifiers of stages $\geqslant \Omega_{\nu+1}$. We have seen before that the cut-elimination procedure by the method of local predicativity is based on the interplay of the boundedness and the collapsing property. We use the same ordinal assignment for **RS**-derivations as we used it for the systems $\mathbf{ID}_\nu^\infty$. This assures that **RS**-derivations again have the collapsing property. What about the boundedness property? In order to formulate the boundedness property we denote by $\Gamma^{[\alpha]} \to \Delta^{[\alpha]}$ a sequent which is obtained from $\Gamma \to \Delta$ by replacing all positive occurrences of a quantifier $\exists x^\gamma$ with $\alpha \leqslant \gamma$ by $\exists x^\beta$ for some β such that $\alpha \leqslant \beta \leqslant \gamma$.

BOUNDEDNESS PROPERTY. If $\mathbf{RS} \vdash_\rho^\alpha \Gamma \to \Delta$, then $\mathbf{RS} \vdash_\rho^\alpha \Gamma^{[\alpha]} \to \Delta^{[\alpha]}$.

The proof is by induction on α and the interesting case again is that of an inference $Cl_{\Omega_{\nu+1}}$. There we have the premise

$$\vdash_\rho^{\alpha_0} \Gamma \to \Delta_0, \forall x^\beta\, \exists y^{\Omega_{\nu+1}}\, A(x, y)$$

with $\alpha_0 < \alpha$ and obtain

$$\vdash_\rho^{\alpha_0} \Gamma^{[\alpha_0]} \to \Delta_0^{[\alpha_0]}, \forall x^\beta\, \exists y^{\alpha_0}\, A(x, y)$$

by the induction hypothesis. But this is equivalent to

$$\vdash_{\rho}^{\alpha_0} \Gamma^{[\alpha_0]} \rightarrow \Delta_0^{[\alpha_0]}, \forall x^{\beta}\, \exists y \in \mathrm{M}_{\alpha_0} A(x, y)$$

and, since $\mathrm{M}_{\alpha_0} \in \mathrm{M}_{\alpha}$, we obtain

$$\vdash_{\rho}^{\alpha} \Gamma^{[\alpha]} \rightarrow \Delta^{[\alpha]}, \exists z^{\alpha}\, \forall x^{\beta}\, \exists y \in z^{\alpha} A(x, y)$$

by an $\exists$-right-inference.

This shows that for $\alpha < \Omega_{\nu+1}$ the impredicative $Cl_{\Omega_{\nu+1}}$-rule is eliminable in a derivation of $\Gamma^{[\alpha]} \rightarrow \Delta^{[\alpha]}$. Formulas in the language of second order arithmetic are easily translated into the language of **RS**. The translation of a sequent which is provable in any predicative system will not contain quantifiers of stages $\geqslant \Omega_1$. An **RS** derivation of such a sequent therefore neither needs any closure rule nor a Ω_1- or stronger branching rule. This means, however, that such a derivation is essentially an **RA** derivation. In this sense the system **RS** is a straightforward generalization of the predicative semiformal system **RA**. The system **RS** provides a powerful and uniform tool for the ordinal analysis of various subsystems of second order arithmetic. To obtain an ordinal analysis for a formal system **T** one has to formalize the fact that **T** has a model of L_{α}. (Indeed one has to look at the least α such that every Π_2-formula stemming from a **T**-provable formula becomes true in L_{α}.) This formalization yields $\vdash_{\rho}^{\beta} \Gamma \rightarrow \Delta$ for every (translation of a) **T**-provable sequent and bounds for β and ρ depending on α. For arithmetical sequents we may eliminate the cuts and obtain a bound for the length of the resulting cut-free derivation. Since this is essentially a $\mathbf{PA}_{\infty}^{\mathrm{CF}}$-derivation this bound is also an upper bound for the proof theoretic ordinal of **T**.

What formal theories can be treated in this way? A closer look to the rule $Cl_{\Omega_{\nu+1}}$ shows that it is nothing but an axiomatization of the fact that $\Omega_{\nu+1}$ represents an admissible ordinal. Therefore we may treat all theories **T** for which the proof of $\mathrm{L}_{\alpha} \models \mathbf{T}$ depends on the admissibility of α or of certain ordinals below α. But the system **RS** also opens the possibility of treating subsystems of set-theory directly.

In recent years the system **KPu** of Kripke-Platek set theory with urelements had gained great importance in generalized recursion theory. In the late 70's G. Jäger started to investigate this system proof-theoretically. First he looked at a system **KPN**. This system has axioms which formalize that the intended set of urelements is the set of natural numbers. It allows complete induction for arbitrary formulas. The set theoretic axioms are those for pair, union, Δ_0-separation and the Δ_0-collection scheme. The foundation scheme, however, is restricted to Δ_0 formulas. In [Jäger, 1980] he proved that **KPN** is predicative. Its ordinal is $\theta\varepsilon_0 0$, the ordinal of (Δ_1^1-CA). If one drops the restriction of the foundation scheme one obtains an impredicative system **KPu**. In [Jäger, 1982a] it is shown that the ordinal of **KPu** is that of $\mathbf{ID}_1$. This proof already needed cut elimination for **RS** by the method of local predicativity. In [Jäger, 1982b] he introduces theories for iterated admissibles. Those theories are obtained by introducing a predicate constant Ad and its defining axioms such that $Ad(a)$ expresses that a is an admissible set. We obtain the theory **KPl** if we add to the defining axioms for Ad the axiom (Lim) $\forall x\, \exists y\, (Ad(y) \wedge x \in y)$ which expresses that the universe of a standard model of **KPl** is a limit of admissible sets. The ordinal of **KPl** is $\theta\varepsilon_{\Omega_\omega+1}0$, the ordinal of $(\Pi^1_1\text{-CA}) + (\text{BI})$. If we restrict the foundation scheme in **KPl** to sets (i.e., take the axiom of foundation) we obtain the theory **W-KPl** whose ordinal is $\theta(\Omega_\omega \cdot \varepsilon_0)0$, the ordinal of $(\Pi^1_1\text{-CA})$. If we restrict both the scheme of foundation and the scheme of complete induction we obtain the theory **KPl′**, with ordinal $\theta\Omega_\omega 0$. This is the ordinal of $(\Pi^1_1\text{-CA})_0$ with the axiom of complete induction. In this paper he also announces the ordinal for a theory **KPi**. In **KPi** we have axioms which require that the universe of its standard model is an admissible limit of admissible sets. It is quite obvious that the least model of this theory in the constructible hierarchy will be at $L_{\mathbf{I}_0}$ where $\mathbf{I}_0$ denotes the first recursively inaccessible ordinal. This explains why the ordinal analysis of **KPi** needs a notation system which internalizes the first recursively inaccessible ordinal. The ordinal analysis of **KPi** was obtained by joint work of Jäger and Pohlers [Jäger and Pohlers, 1982]. Its ordinal is the collapse of the first ε-number beyond $\mathbf{I}_0$. When we use the 3-place θ-functions mentioned in Section 3.1 it is $\theta_0\varepsilon_{\mathbf{I}_0+1}0$. In the original paper we used a different notation system in which this ordinal is expressed as $\theta_0(\theta_1\varepsilon_{\mathbf{I}_0}0)0$ [12]. Till now this is the strongest system which can be treated by the method of local predicativity. The ordinal analysis of **KPi** comprises an ordinal analysis of $(\Delta^1_2\text{-CA}) + \text{BI}$. Any formula in the language of second order arithmetic is easily translated into a formula of **KPi**. We just replace first order quantifiers $\forall x$, $\exists x$ by $\forall x \in N$, $\exists x \in N$ and second order quantifiers $\forall X$, $\exists X$ by $\forall x \supset N$, $\exists x \subset N$. Using the facts that **KPi** proves the axiom β and the Spector-Gandy-theorem one easily obtains an embedding of $(\Delta^1_2\text{-CA}) + (\text{BI})$. The proof that every segment of $\theta_0\varepsilon_{\mathbf{I}_0+1}0$ is well-ordered can be done in $(\Delta^1_2\text{-CA}) + (\text{BI})$ without complications. But one can do even better. As shown by Jäger in [Jäger, 1983b], this well-ordering proof can also be formalized in Feferman's theory $\mathbf{T}_0$ which is based on intuitionistic logic. Feferman himself had shown that $\mathbf{T}_0$ is embeddable into $(\Delta^1_2\text{-CA}) + (\text{BI})$. So we have $|\mathbf{T}_0| \leqslant |(\Delta^1_2\text{-CA}) + (\text{BI})| \leqslant |\mathbf{KPi}| \leqslant \theta_0\varepsilon_{\mathbf{I}_0}0 \leqslant |\mathbf{T}_0|$, which gives a positive answer to the problem whether $\mathbf{T}_0$ and $(\Delta^1_2\text{-CA}) + (\text{BI})$ are proof theoretically equivalent.

K. Schütte has developed a method to treat $(\Delta^1_2\text{-CA}) + (\text{BI})$ without the use of ramified set theory. In a forthcoming paper he will present a semiformal system which is mainly based on the modification of Buchholz's rule for subsystems of analysis. Only for the collapse below $\mathbf{I}_0$ he

---

[12] $\theta_1\varepsilon_{\mathbf{I}_0+1}$ collapses $\varepsilon_{\mathbf{I}_0+1}$ below $\mathbf{I}_0$ and $\theta_0(\theta_1\varepsilon_{\mathbf{I}_0+1}0)0$ is then the collapse below Ω_1.

needs a rule which stems from the method of local predicativity. An ordinal notation system, based on the ψ-functions, which is sufficiently strong for this analysis has already appeared in [Buchholz and Schütte, 1983].

In his Habilitationsschrift [Jäger, 1984b] G. Jäger gave a systematic treatment of subsystems of set theory. This work gives evidence that subsystems of set theory provide the most general and the most uniform framework for the proof theoretical analysis of various systems. One of the most striking results is that the theory $\mathbf{KPi}^0$, i.e., **KPi** without foundation and complete induction restricted to sets, has the proof theoretic ordinal Γ_0 and allows a very natural embedding of all known predicative theories.

I want to close this report by two apologies. First, it was my task to inform about the contributions of the Schütte school to proof theory. Therefore I only have cited those papers which influenced our work immediately. Of course this does not mean that we ignore the work of other people. Second, I am aware that the more technical parts of this report will be really understandable only to readers already somewhat familiar with our work. I tried to exhibit the basic ideas behind our work. The reader who really wants to understand the whole work in all its technicalities will have to study the original papers.

References

Abbreviations

JSL: The Journal of Symbolic Logic.
AMLG: Archiv für Mathematische Logik und Grundlagenforschung.
APAL: Annals of Pure and Applied Logic (Former: Annals of Mathematical Logic).
MA: Mathematische Annalen.
SdBA: Sitzungsberichte der Bayerischen Akademie der Wissenschaften, Mathematische-Naturwissenschaftliche Klasse.
IPT: Intuitionism and Proof Theory, Proceedings of the summer conference at Buffalo, N.Y., 1968, North-Holland, Amsterdam, (1970).
SLN: Springer Lecture Notes in Mathematics.

W. Buchholz
1974 : Rekursive Bezeichungssysteme für Ordinalzahlen auf der Grundlage der Feferman-Aczelschen Normalfunktionen, Dissertation München.
1975 : Normalfunktionen und konstruktive Systeme von Ordinalzahlen, SLN 500.
1976 : Über Teilsysteme von $\bar{\theta}\{g\}$, AMLG 18.
1977 : Eine Erweiterung der Schnitteliminationmethode, Habilitationsschrift München.
1984a): An independence result for $(\Pi^1_1\text{-CA})+(\text{BI})$, to appear in APAL.
1984b): A new system of proof theoretic ordinals, to appear in APAL.

W. Buchholz and K. Schütte
1976 : Die Beziehungen zwischen den Ordinalzahlsystemen Σ and $\bar{\theta}(\omega)$, AMLG 17.
1980 : Syntaktische Abgrenzungen von formalen Systemen der Π^1_1-Analysis und Δ^1_2-Analysis, SdBA.

1983 : Ein Ordinalzahlensystem für die beweistheoretische Abgrenzung der Π^1_2-Separation und Bar-Induktion, SdBA.

W. Buchholz and W. Pohlers
1978 : Provable well-orderings of formal theories for transfinitely iterated inductive definitions, JSL 43.

W. Buchholz, S. Feferman, W. Pohlers and W. Sieg
1981 : Iterated Inductive Definitions and Subsystems of Analysis: Recent Proof Theoretical Studies, SLN 897.

S. Feferman
1964 : Systems of predicative analysis, JSL 29.
1970 : Formal theories for transfinite iterations of generalized inductive definitions and some subsystems of analysis, IPT.

H. Friedman
1967 : Subsystems of set theory and analysis, Thesis, MIT.
1970 : Iterated inductive definitions and Σ^1_2-AC, IPT.

J. Y. Girard
1981 : Π^1_2-logic, APAL 21.

L. Henkin
1954 : A generalization of the concept of ω-consistency, JSL 19.

D. Hilbert
1931 : Die Grundlegung der elementaren Zahlentheorie, MA 104.

W. Howard
1972 : A system of abstract constructive ordinals, JSL 37.

G. Jäger
1980 : Beweistheorie von **KPN**, AMLG 20.
1982a): Zur Beweistheorie der Kripke-Platek-Mengenlehre über den natürlichen Zahlen, AMLG 22.
1982b): Iterating admissibility in proof theory, Proceedings of the Herbrand Logic Colloquium'81, North-Holland, Amsterdam.
1983a): The strength of admissibility without foundation, to appear in JSL, Preprint, 1983.
1983b): A well-ordering proof for Feferman's theory $\mathbf{T}_0$, AMLG 23.
1984a): ρ-inaccessible ordinals, collapsing functions and a recursive notation system, AMLG 24.
1984b): Theories for admissible sets: A unifying approach to proof theory, Habilitationsschrift München.

G. Jäger and K. Schütte
1979 : Eine syntaktische Abgrenzung der (Δ^1_1-CA) Analysis, SdBA.

H. Levitz
1970 : On the relationship between Takeuti's ordinal diagrams O(n) and Schütte's system of ordinal notations $\Sigma(n)$, IPT.

S. Orey
1956 : On ω-consistency and related properties, JSL 21.

H. Pfeiffer
1969 : Ein Bezeichnungssystem für Ordinalzahlen, AMLG 12.

W. Pohlers
1973 : Eine Grenze für die Herleitbarkeit der transfiniten Induktion in einem schwachen Π^1_1-Fragment der klassischen Analysis, Dissertation, München.

1975 : An upper bound for the provability of transfinite induction in systems with N-times iterated inductive definitions, SLN 500.

1978 : Ordinals connected with formal theories for transfinitely iterated inductive definitions, JSL 44.

1977 : Beweistheorie der iterierten induktiven Definitionen, Habilitationsschrift München.

1981 : Cut elimination for impredicative infinitary systems, Part I: Ordinal analysis of $\mathbf{ID}_1$, AMLG 21.

1982 : Cut elimination for impredicative infinitary systems, Part II: Ordinal analysis for iterated inductive definitions, AMLG 22.

1983 : Ordinal functions and notations based on a hierarchy of inaccessible cardinals, to appear in APAL.

K. Schütte

1951 : Beweistheoretische Erfassung der unendlichen Induktion in der Zahlentheorie, MA 122.

1960 : Beweistheorie, Springer.

1964 : Eine Grenze für die Beweisbarkeit der transfiniten Induktion in der verzweigten Typenlogik, AMLG 7.

1965 : Predicative Well-orderings, in Formal Systems and Recursive Functions, Proceedings of the 8th Logic Colloquium, Oxford 1963. North-Holland, Amsterdam.

1969 : Ein konstructives System von Ordinalzahlen, AMLG 11 and 12.

1977 : Proof Theory, Springer.

W. Tait

1968 : Normal derivability in classical logic, SLN 72.

G. Takeuti

1967 : Consistency proofs for subsystems of classical analysis, Annals of Mathematics 86.

SUBSYSTEMS OF $\mathbf{Z}_2$ AND REVERSE MATHEMATICS

by

STEPHEN G. SIMPSON *

§1. Introduction

By $\mathbf{Z}_2$ we mean the formal system of second order arithmetic. One can weaken the comprehension scheme of $\mathbf{Z}_2$ in various ways so as to define a great many different subsystems of $\mathbf{Z}_2$. Such subsystems are distinguished from each other by their stronger or weaker set existence axioms.

Gentzen-style proof theory is largely concerned with a structural analysis of proofs in certain subsystems of $\mathbf{Z}_2$. Such an analysis leads to a computation of the provable ordinals and provable recursive functions of the subsystems.

The purpose of this Appendix is to study subsystems of $\mathbf{Z}_2$ from a perspective which is somewhat different from that of Gentzen-style proof theory. Namely, our purpose here is to isolate and classify certain subsystems of $\mathbf{Z}_2$ according to the theorems of "ordinary mathematics" which can be proved within them.

The phrase "ordinary mathematics" is difficult to define precisely. For purposes of this Appendix, we define *ordinary mathematics* to consist of those parts of mathematics which do not essentially depend on the abstract theory of uncountable ordinal and cardinal numbers. Thus ordinary mathematics includes: geometry, number theory, calculus, differential equations, real and complex analysis, countable algebra, countable combinatorics, the topology of complete separable metric spaces, and the theory of separable Banach and Frechet spaces. On the other hand, ordinary mathematics does not include: uncountable algebra, uncountable combinatorics, general topology, or the theory of nonseparable topological vector spaces. A borderline area is descriptive set theory, i.e., the study of Borel and projective subsets of complete separable metric spaces.

With the above qualifications, we may say that most or all theorems of ordinary mathematics can be stated and proved within $\mathbf{Z}_2$. This basic

---

* Preparation of this paper was partially supported by NSF grant MCS-8317874.

discovery can be attributed to Hilbert and Bernays (see Supplement IV of [29]).

Somewhat more delicate is the question of which subsystems of $\mathbf{Z}_2$ suffice for the development of which parts of ordinary mathematics. We concentrate on the following

MAIN QUESTION. Given a theorem τ of ordinary mathematics, which set existence axioms (formulated in the context of subsystems of $\mathbf{Z}_2$) are needed in order to prove τ?

The present Appendix is essentially a report on some precise answers which have been obtained for certain special cases of the above question.

§2. The formal systems $\mathbf{Z}_2$ and $\mathbf{ACA}_0$

The language of $\mathbf{Z}_2$, denoted L_2, is a two-sorted first-order language with number variables $i, j, k, m, n, \ldots$ and set variables $X, Y, Z, \ldots$. The number variables are intended to range over the set of natural numbers $\omega = \mathbb{N} = \{0, 1, 2, \ldots\}$. The set variables are intended to range over subsets of ω. Numerical terms are built up as usual from number variables, constant symbols 0 and 1, and binary operations $+$ and $\cdot$. Atomic formulas are $t_1 = t_2$, $t_1 < t_2$, and $t_1 \in X$, where t_1 and t_2 are numerical terms. Formulas are built up from atomic formulas by means of propositional connectives, number quantifiers $\forall n$ and $\exists n$, and set quantifiers $\forall X$ and $\exists X$.

A formula is said to be *arithmetical* if it contains no set quantifiers. (Note however that an arithmetical formula may contain free set variables. Such free set variables are known as *set parameters*.)

All of the formal systems which we shall consider include the familiar ordered semiring axioms for $\mathbb{N}, +, \cdot, 0, 1, <$ as well as the induction axiom

$$(0 \in X \wedge \forall n\, (n \in X \rightarrow n+1 \in X)) \rightarrow \forall n\, (n \in X)\,.$$

At all times we assume the law of the excluded middle. The formal system $\mathbf{Z}_2$ consists of the above axioms plus the *comprehension scheme*

$$\exists X\, \forall n\, (n \in X \leftrightarrow \varphi(n)),$$

where $\varphi(n)$ is any L_2-formula in which X does not occur freely.

By a *subsystem of* $\mathbf{Z}_2$ we mean, of course, any formal system in the language L_2 whose theorems are included in those of $\mathbf{Z}_2$.

One especially important subsystem of $\mathbf{Z}_2$ is obtained by restricting the comprehension scheme to formulas $\varphi(n)$ which are arithmetical. This subsystem is known as $\mathbf{ACA}_0$. The letters **ACA** stand for Arithmetical Comprehension Axiom.

Clearly $\mathbf{ACA}_0$ has a minimum ω-model consisting of the arithmetical subsets of ω. It is also easy to see that $\mathbf{ACA}_0$ is a conservative extension of the formal system $\mathbf{Z}_1 = \mathbf{PA}$ = first-order Peano arithmetic. Thus, from the viewpoint of Gentzen-style proof theory, $\mathbf{ACA}_0$ is equivalent to $\mathbf{Z}_1$. In particular, the proof-theoretic ordinal of $\mathbf{ACA}_0$ is ε_0 (see Chapter 2).

§3. Ordinary mathematics within $\mathbf{ACA}_0$

We shall now sketch how some basic concepts of ordinary mathematics, including the number systems, continuous functions, countable algebraic structures, and complete separable metric spaces, can be developed within $\mathbf{ACA}_0$. Later, in §5, we shall point out that much of this development goes through in a much weaker system known as $\mathbf{RCA}_0$.

We begin with the numerical pairing function

$$(m, n) = (m+n)^2 + m.$$

If X and Y are subsets of $\mathbb{N} = \{n \colon n = n\}$, we define

$$X \times Y = \{(m, n) \colon m \in X \wedge n \in Y\}.$$

Thus $\mathbb{N} \times \mathbb{N} \subseteq \mathbb{N}$.

In order to define the set $\mathbb{Z}$ of *integers*, we first define an equivalence relation $=_{\mathbb{Z}}$ on $\mathbb{N} \times \mathbb{N}$ by putting $(m, n) =_{\mathbb{Z}} (p, q)$ if and only if $m + q = n + p$. We then put $\mathbb{Z} = \{(m, n) \colon (m, n)$ is minimal in its $=_{\mathbb{Z}}$-class$\}$. We define $+, -, \cdot, \ldots$ on $\mathbb{Z}$ so that $(m, n) + (p, q) =_{\mathbb{Z}} (m+p, n+q)$, $-(m, n) =_{\mathbb{Z}} (n, m)$, $(m, n) \cdot (p, q) =_{\mathbb{Z}} (mp + nq, mq + np)$, etc. One then proves in $\mathbf{ACA}_0$ (actually in $\mathbf{RCA}_0$) that the system $\mathbb{Z}, +, -, \cdot, 0, 1, <$ has the usual properties of an ordered integral domain, the Euclidean property, etc.

In order to define the set $\mathbb{Q}$ of *rational numbers*, we let $\mathbb{Z}^+ = \{a \in \mathbb{Z} \colon a > 0\}$ and define an equivalence relation $=_{\mathbb{Q}}$ on $\mathbb{Z} \times \mathbb{Z}^+$ by putting $(a, b) =_{\mathbb{Q}} (c, d)$ if and only if $ad = bc$. We then put $\mathbb{Q} = \{(a, b) \in \mathbb{Z} \times \mathbb{Z}^+ \colon (a, b)$ is minimal in its $=_{\mathbb{Q}}$-class$\}$. We define $+, -, \cdot, \ldots$ on $\mathbb{Q}$ so that $(a, b) + (c, d) =_{\mathbb{Q}} (ad + bc, bd), -(a, b) =_{\mathbb{Q}} (-a, b), (a, b) \cdot (c, d) =_{\mathbb{Q}} (ac, bd)$, etc. One then proves in $\mathbf{ACA}_0$ (actually in $\mathbf{RCA}_0$) that the system $\mathbb{Q}, +, -, \cdot, 0, 1, <$ has the usual properties of an ordered field, etc.

If X and Y are subsets of $\mathbb{N}$, a function $f \colon X \to Y$ is defined to be a set of pairs $f \subseteq X \times Y$ with the usual properties. A *sequence of rational numbers* is defined to be a function $f \colon \mathbb{N} \to \mathbb{Q}$. We denote such a sequence as $\langle q_n \colon n \in \mathbb{N} \rangle$ or simply $\langle q_n \rangle$, where $q_n = f(n)$.

A *real number* is defined to be a sequence of rational numbers $x = \langle q_n \rangle$ such that $\forall n\, \forall i\, (|q_n - q_{n+i}| \leq 2^{-n})$. We use $\mathbf{R}$ informally to denote the set of all real numbers. (Of course formally this set does not exist within $\mathbf{Z}_2$.) We define $\langle q_n \rangle = \langle q'_n \rangle$ if and only if $\forall n\, (|q_n - q'_n| \leq 2^{-n+1})$. We define

$\langle q_n \rangle < \langle q'_n \rangle$ if and only if $\exists n\,(q_n + 2^{-n+1} < q'_n)$. Also

$$\langle q_n \rangle + \langle q'_n \rangle = \langle q_{n+1} + q'_{n+1} \rangle,$$

$$-\langle q_n \rangle = \langle -q_n \rangle, \qquad 0 = \langle 0 \rangle, \qquad 1 = \langle 1 \rangle,$$

and

$$\langle q_n \rangle \cdot \langle q'_n \rangle = \langle q_{k+n} \cdot q'_{k+n} \rangle,$$

where $k =$ least j such that $2^{j-1} \geqslant \max(|q_0|, |q'_0|) + 1$. One then proves within $\mathbf{ACA_0}$ (actually $\mathbf{RCA_0}$) that the system $\mathbb{R}, +, -, \cdot, 0, 1, <, =$ has the usual properties of an Archimedean ordered field, etc.

A *sequence of real numbers* is defined to be a double sequence of rational numbers $\langle q_{mn}\colon m, n \in \mathbb{N} \rangle$ such that for each m, $\langle q_{mn}\colon n \in \mathbb{N} \rangle$ is a real number. Such a sequence of real numbers is denoted $\langle x_m\colon m \in \mathbb{N} \rangle$, where $x_m = \langle q_{mn}\colon n \in \mathbb{N} \rangle$. Within $\mathbf{ACA_0}$ (but not $\mathbf{RCA_0}$) one can prove that every bounded sequence of real numbers has a least upper bound.

A *continuous function code* is a function $\Phi : \mathbb{N} \to \mathbb{Q}^4$, i.e., a sequence of ordered quadruples of rational numbers, with certain properties to be described below. We write $\langle c, d, u, v \rangle \in \Phi$ to mean that $\exists n\,(\Phi(n) = \langle c, d, u, v \rangle)$. We require that

(i) $\langle c, d, u, v \rangle \in \Phi \to c < d, u \leqslant v$;
(ii) $\langle c, d, u, v \rangle, \langle c, d, u', v' \rangle \in \Phi \to \langle c, d, u', v \rangle \in \Phi$;
(iii) $\langle c, d, u, v \rangle \in \Phi,\ c \leqslant c' < d' \leqslant d \to \langle c', d', u, v \rangle \in \Phi$;
(iv) $\langle c, d, u, v \rangle \in \Phi,\ u' \leqslant u,\ v \leqslant v' \to \langle c, d, u', v' \rangle \in \Phi$.

Here $c, d, u, v, c', d', u', v'$ range over $\mathbb{Q}$.

The idea here is that Φ encodes a continuous partial function ϕ from $\mathbb{R}$ into $\mathbb{R}$. A quadruple $\langle c, d, u, v \rangle \in \Phi$ is a piece of information to the effect that, for all real numbers x in the interval $c < x < d$, one has $u \leqslant \phi(x) \leqslant v$ provided $\phi(x)$ is defined. Formally, we say that $\phi(x)$ *is defined* if for all $\varepsilon > 0$ there exists $\langle c, d, u, v \rangle \in \Phi$ such that $c < x < d$ and $|u - v| < \varepsilon$. In this case we put $\phi(x) = y$, where y is the unique real number such that $u \leqslant y \leqslant v$ for all $\langle c, d, u, v \rangle \in \Phi$ such that $c < x < d$. The existence of y is provable in $\mathbf{ACA_0}$ (actually $\mathbf{RCA_0}$).

With these definitions, all of the basic properties of continuous, real-valued functions of a real variable can be proved in $\mathbf{ACA_0}$. The definitions and results can be routinely extended to $\mathbb{R}^n$, $n \in \mathbb{N}$. In this case the appropriate code is a sequence of $(2n+2)$-tuples of rational numbers. On this basis, the theory of ordinary and partial differential equations, the calculus of variations, etc can be developed within $\mathbf{ACA_0}$. (Many of the results go through in the weaker system $\mathbf{RCA_0}$.)

We now turn to a discussion of countable algebra within $\mathbf{ACA_0}$. A *countable commutative ring* is defined to be a structure $R, +, -, \cdot, 0, 1$, where $R \subseteq \mathbb{N}$, $+ : R \times R \to R$, etc., and the usual commutative ring axioms, including $0 \neq 1$, are satisfied. An *ideal* in R is a set $I \subseteq R$ such that $0 \in I$, $1 \notin I$, $\forall a\, \forall b\,(a, b \in I \to a + b \in I)$ and $\forall a\, \forall r\,(a \in I, r \in R \to ar \in I)$.

We then put $R/I = \{r \in R: r$ is minimal in its $=_I$-class$\}$, where $R =_I s$ if and only if $r - s \in I$. With the appropriate operations R/I becomes a commutative ring. I is said to be *prime* if R/I is an integral domain, and *maximal* if R/I is a field. The basic results of ideal theory can be proved in $\mathbf{ACA}_0$. (Some of these results go through in $\mathbf{RCA}_0$.) Other countable algebraic structures, e.g. countable abelian groups, can be defined and discussed in a similar manner.

Next we indicate how some basic results of abstract analysis can be developed within $\mathbf{ACA}_0$. We define a *code for a complete separable metric space* to be a nonempty set $A \subseteq \mathbb{N}$ together with a pseudometric $d: A \times A \to \mathbb{R}$ satisfying $d(a, a) = 0$, $d(a, b) = d(b, a) \geq 0$, and $d(a, c) \leq d(a, b) + d(b, c)$. A *point of the complete separable metric space* $\hat{A}$ is defined to be a sequence $x = \langle a_n : n \in \mathbb{N}\rangle$, $a_n \in A$, satisfying $\forall n\, \forall i\, (d(a_n, a_{n+i}) \leq 2^{-n})$. For example $\mathbb{R} = \hat{\mathbb{Q}}$ under the appropriate metric. If $\hat{A}$ and $\hat{B}$ are two complete separable metric spaces, a continuous function $\phi: \hat{A} \to \hat{B}$ is encoded as an appropriate sequence of ordered quadruples $\langle a, u, b, v\rangle$, where $a \in A$, $b \in B$, $u, v \in \mathbb{Q}^+$. The idea is that $d(a, x) < u$ implies $d(b, \phi(x)) \leq v$. A *closed separable subspace* of $\hat{A}$ is defined to be a complete separable metric space $\hat{S}$ together with an isometry $i: S \to \hat{A}$. With these definitions, the usual proofs of basic topological results, e.g. the Tietze extension theorem for complete separable metric spaces, can be carried out within $\mathbf{ACA}_0$. (Some of the results, e.g. a version of the Baire category theorem, go through in $\mathbf{RCA}_0$.)

A *separable Banach space* is defined to be a complete separable metric space $\hat{A}$ such that A is a pseudonormed vector space over $\mathbb{Q}$. For example, let $A = \mathbb{Q}[x]$ = the ring of polynomials in one variable x over the rational field $\mathbb{Q}$. With the metric

$$d(f, g) = \left(\int_0^1 |f(x) - g(x)|^p\, dx\right)^{1/p},$$

$1 \leq p < \infty$, we have $\hat{A} = L_p[0, 1]$. Similarly $C[0, 1] = \hat{A}$, with

$$d(f, g) = \sup\{|f(x) - g(x)|: 0 \leq x \leq 1\}.$$

As suggested by these examples, the basic theory of separable Banach and Frechet spaces can be developed within $\mathbf{ACA}_0$. (Some of the results go through in $\mathbf{RCA}_0$.)

The observation that a great deal of ordinary mathematics can be developed formally within a system something like $\mathbf{ACA}_0$ is due to H. Weyl [41]. See also Takeuti [40] and Zahn [42].

§4. Arithmetical versus Π_1^1 comprehension

It is easy to see that $\mathbf{Z}_2$ has infinitely many subsystems which are stronger than $\mathbf{ACA}_0$. For each $k \in \omega$ we define the Π_k^1 formulas of L_2 to be those of the form $\forall Y_1\, \exists Y_2 \ldots Y_k\, \theta$, where θ is arithmetical. Let Π_k^1-CA_0

be the subsystem of $\mathbf{Z}_2$ obtained by restricting the comprehension scheme to formulas $\varphi(n)$ which are Π^1_k. It can be shown that for each $k \in \omega$, Π^1_{k+1}-CA$_0$ is properly stronger than Π^1_k-CA$_0$. Thus we have a hierarchy of subsystems of $\mathbf{Z}_2$. Note also that $\mathbf{ACA}_0 = \Pi^1_0$-CA$_0$ while

$$\mathbf{Z}_2 = \Pi^1_\infty\text{-CA}_0 = \bigcup_{k\in\omega} \Pi^1_k\text{-CA}_0.$$

We have remarked in §3 above that a large part of ordinary mathematics can be developed within $\mathbf{ACA}_0$. However, there are certain exceptional theorems of ordinary mathematics which cannot be proved in $\mathbf{ACA}_0$. These exceptional theorems come from several branches of ordinary mathematics including countable algebra, the topology of the real line, countable combinatorics and classical descriptive set theory.

What most of the exceptional theorems have in common is that they somehow involve countable ordinal numbers. Within $\mathbf{ACA}_0$ we define a *countable linear ordering* to be a structure A, $<_A$, where $A \subseteq \mathbb{N}$ and $<_A \subseteq A \times A$ is an irreflexive linear ordering of A. This is called a *well-ordering* if there is no infinite descending sequence $\langle a_n : n \in \mathbb{N}\rangle$, $a_{n+1} <_A a_n$. If A is a countable well ordering, we view A as encoding a countable ordinal number α. Two countable well orderings are said to encode the same countable ordinal number if they are isomorphic.

Two countable well orderings are said to be *comparable* if they are isomorphic or if one is isomorphic to a proper initial segment of the other. The fact that any two countable well orderings are comparable can be proved in Π^1_1-CA$_0$ but not in $\mathbf{ACA}_0$. Thus Π^1_1-CA$_0$ (but not $\mathbf{ACA}_0$) is strong enough to develop a good theory of countable ordinal numbers. Because of this, Π^1_1-CA$_0$ is strong enough to prove almost all theorems of ordinary mathematics which are provable in $\mathbf{Z}_2$.

As an example let us mention the structure theory for countable abelian groups (Kaplansky [32]). Let $G, +, -, 0$ be a countable abelian group. We say that G is *divisible* if for all $a \in G$ and $n > 0$, there exists $b \in G$ such that $nb = a$. We say that G is *reduced* if G has no nontrivial divisible subgroup. Within Π^1_1-CA$_0$ (but not $\mathbf{ACA}_0$), one can prove that every countable abelian group is the direct sum of a divisible group and a reduced group. Now let $G, +, -, 0$ be a countable abelian p-group, where p is a prime number. One defines a transfinite sequence of subgroups $G_0 = G$, $G_{\alpha+1} = pG_\alpha$, and at limit stages $G_\delta = \bigcap_{\alpha<\delta} G_\alpha$. Thus G is reduced if and only if $G_\infty = 0$. The *Ulm invariants* of G are the numbers $\dim(P_\alpha/P_{\alpha+1})$, where $P_\alpha = \{a \in G_\alpha : pa = 0\}$ and the dimension is taken over the integers modulo p. Each Ulm invariant is either a natural number or ∞. *Ulm's Theorem* states that two countable reduced abelian P-groups are isomorphic if and only if their Ulm invariants are the same. Using the theory of countable ordinal numbers which is available in Π^1_1-CA$_0$, one can carry out the construction of the Ulm invariants and the usual proof of Ulm's Theorem within Π^1_1-CA$_0$.

Turning to another branch of mathematics, we point out that Π^1_1-CA$_0$ is strong enough to prove all of the basic results of classical descriptive set theory (Mansfield–Weitkamp [34]). Let $\hat{A}$ be a complete separable metric space. A *code for a Borel subset of* $\hat{A}$ is defined to be a certain kind of tree $T \subseteq \mathbb{N}^{<\mathbb{N}}$ which is required to be *well founded*, i.e., have no infinite path. The end nodes of T are labeled with codes for basic open neighborhoods of $\hat{A}$. The interior nodes of T are labeled with symbols denoting the operations of complementation, countable union, and countable intersection. If x is a point of $\hat{A}$, the relation "$x \in$ (the Borel set coded by) T" is defined in the obvious way. *Analytic sets* may be defined as projections of Borel sets or by means of the Souslin operation applied to closed sets. Because Π^1_1-CA$_0$ includes a good theory of countable well orderings and countable well-founded trees, it is straightforward to carry out within Π^1_1-CA$_0$ the standard proofs of all of the classical separation and uniformization results for Borel, analytic, and coanalytic sets. For details see Simpson [20].

With the above examples, Π^1_1-CA$_0$ emerges as being of considerable interest with respect to the development of ordinary mathematics. We note in passing that Π^1_1-CA$_0$ has a minimum β-model consisting of all subsets of ω which are recursive in the sets $\mathcal{O}^{(n)}$, $n \in \omega$, where $\mathcal{O}^{(0)} =$ the empty set and $\mathcal{O}^{(n+1)} =$ hyperjump of $\mathcal{O}^{(n)}$. Also, the proof-theoretic ordinal of Π^1_1-CA$_0$ is given by Takeuti's ordinal diagrams of finite order ([39]; see also Chapter 5). This ordinal is the same as Schütte's $\theta\Omega_\omega 0$ [37] or Buchholz-Schütte's $\psi_0(\Omega_\omega)$ [25].

Other examples of ordinary mathematical theorems which are provable in Π^1_1-CA$_0$ are: the determinacy of open subsets of $\mathbb{N}^\mathbb{N}$; the Ramsey property for open subsets of $[\mathbb{N}]^\mathbb{N}$; and the Cantor–Bendixson Theorem (every closed set is the union of a countable set and a perfect set). These theorems, like Ulm's Theorem and the classical theorems on Borel and analytic sets, are exceptional in that they are not provable in $\mathbf{ACA}_0$.

It is perhaps worth mentioning that there are a small number of even more exceptional theorems. These are ordinary mathematical theorems which, for instance, are provable in Zermelo-Fraenkel set theory but not in $\mathbf{Z}_2$. As an example, consider the following corollary, due to Friedman [6], of a theorem of Martin [35]. Given a symmetric Borel set $E \subseteq I \times I$, $I =$ closed unit interval, there exists a Borel function $\phi : I \rightarrow I$ such that the graph of ϕ is either included in or disjoint from E. Friedman [6] has shown that this theorem is not provable in $\mathbf{Z}_2$ or even in simple type theory. This is related to Friedman's earlier result [5] that Borel determinacy is not provable in simple type theory. For further results of this kind, see [6], [7], [22].

§5. A weak base theory

We now introduce a certain subsystem of $\mathbf{Z}_2$ known as $\mathbf{RCA}_0$. $\mathbf{RCA}_0$ is proof-theoretically much weaker than $\mathbf{ACA}_0$ and yet strong enough to develop at least some portion of ordinary mathematics.

If t is a term of L_2, we abbreviate $\forall n\,(n<t\rightarrow\varphi)$ by $(\forall n<t)\varphi$ and $\exists n\,(n<t\wedge\varphi)$ by $(\exists n<t)\varphi$. The expressions $\forall n<t$ and $\exists n<t$ are known as *bounded number quantifiers*. An L_2-formula is said to be Σ^0_0 if it is arithmetical (i.e., contains no set quantifiers) and all of its number quantifiers are bounded. An L_2-formula is said to be Σ^0_1 (respectively Π^0_1) if it is of the form $\exists m\,\theta$ (respectively $\forall m\,\theta$), where θ is Σ^0_0. By Σ^0_1 *induction* we mean all axioms of the form

$$(\varphi(0)\wedge\forall n\,(\varphi(n)\rightarrow\varphi(n+1)))\rightarrow\forall n\,\varphi(n),$$

where φ is Σ^0_1. By Δ^0_1 *comprehension* we mean all axioms of the form

$$\forall n\,(\varphi(n)\leftrightarrow\psi(n))\rightarrow\exists X\,\forall n\,(n\in X\leftrightarrow\varphi(n)),$$

where φ is Σ^0_1, ψ is Π^0_1, and X does not occur in φ. The system $\mathbf{RCA}_0$ consists of the ordered semiring axioms for $\mathbb{N}, +, \cdot, 0, 1, <$ together with Δ^0_1 comprehension and Σ^0_1-induction. The letters **RCA** stand for Recursive Comprehension Axiom.

Clearly $\mathbf{RCA}_0$ is a subsystem of $\mathbf{ACA}_0$. It is not hard to prove that $\mathbf{RCA}_0$ is a conservative extension of Σ^0_1-**PA**. (Σ^0_1-**PA** is just first-order Peano arithmetic **PA** with the induction scheme restricted to Σ^0_1 formulas of the language L_1.) Also, it is known that $\mathbf{RCA}_0$ is a conservative extension of **PRA** (Primitive Recursive Arithmetic) with respect to Π^0_2 sentences. Thus the proof-theoretic ordinal of $\mathbf{RCA}_0$ is ω^ω. These results show that $\mathbf{RCA}_0$ is proof-theoretically much weaker than $\mathbf{ACA}_0$.

From the viewpoint of ordinary mathematics, the best way to view $\mathbf{RCA}_0$ is as a kind of formalized recursive mathematics. The minimum ω-model of $\mathbf{RCA}_0$ consists of the recursive subsets of ω. Within $\mathbf{RCA}_0$ one can prove that the class of all total functions on $\mathbb{N}$ is closed under composition, primitive recursion, and the μ-operator. The development of mathematics within $\mathbf{RCA}_0$ is to a large extent parallel to the positive results of what are known as recursive analysis (cf. Aberth [24]) and recursive algebra (cf. Fröhlich–Shepherdson [28] and Metakides–Nerode [36]). For instance, using the definition of continuous function code, given in §3 above, one can prove within $\mathbf{RCA}_0$ that all of the usual special functions such as polynomials, e^x, $\sin x$, etc. are continuously differentiable, and that all continuous functions have the intermediate value property. On the other hand, one cannot prove within $\mathbf{RCA}_0$ that every continuous function on the closed unit interval is uniformly continuous or attains a maximum value. This is because the usual recursive counterexamples show that these results are false in the minimum ω-model of $\mathbf{RCA}_0$ (consisting of the recursive subsets of ω). Similarly, in the realm of countable algebra, $\mathbf{RCA}_0$ is strong enough to prove that every countable field has an algebraic closure, but not strong enough to prove uniqueness of the algebraic closure.

Thus $\mathbf{RCA}_0$ is a subsystem of $\mathbf{Z}_2$ which corresponds roughly to the positive content of recursive analysis and recursive algebra. In terms of the Main Question, $\mathbf{RCA}_0$ serves as a *weak base theory*, i.e., a vantage point from which we can appreciate the non-recursive content of other parts of mathematics. The value of such a vantage point will be seen in the next section.

§6. Reverse mathematics

In §3 and §4 we have sketched an approximate answer to the Main Question. Namely, we have suggested that most theorems of ordinary mathematics can be proved in $\mathbf{ACA}_0$, and that of the exceptions, most can be proved in Π^1_1-CA_0.

In the rest of this Appendix, we shall consider the following sharpened form of the Main Question: *Given a theorem* τ *of ordinary mathematics, what is the weakest natural subsystem* $S(\tau)$ *of* $\mathbf{Z}_2$ *in which* τ *is provable*?

Surprisingly, it turns out that for many specific theorems τ this question has a precise and definitive answer. Furthermore, $S(\tau)$ often turns out to be one of five specific subsystems of $\mathbf{Z}_2$. For convenience we shall list these systems as S_1, S_2, S_3, S_4 and S_5 in order of increasing ability to accommodate ordinary mathematical practice. The odd-numbered systems S_1, S_3 and S_5 have already been introduced as $\mathbf{RCA}_0$, $\mathbf{ACA}_0$ and Π^1_1-CA_0 respectively. The even-numbered systems S_2 and S_4 are intermediate systems which will be described in §7 below.

Our method for establishing results of the form $S(\tau) = S_j$, $2 \leq j \leq 5$ is based on the following empirical phenomenon: "When the theorem is proved from the right axioms, the axioms can be proved from the theorem" (Friedman [3]). In other words, let τ be an ordinary mathematical theorem which is not provable in the weak base theory S_1. Then very often τ turns out to be equivalent to S_j for some $j = 2$, 3, 4 or 5. The equivalence is provable in S_i for some $i < j$, usually $i = 1$.

For example let $\tau = \mathrm{BW}$ = the Bolzano–Weierstrass Theorem: every bounded sequence of real numbers has a convergent subsequence. Then we have

THEOREM 1 (Friedman [4]). BW *is equivalent to* $\mathbf{ACA}_0$, *the equivalence being provable in* $\mathbf{RCA}_0$.

PROOF. In one direction, just observe that the usual proof of BW goes through in $\mathbf{ACA}_0$. For the converse, we work within $\mathbf{RCA}_0$ and assume BW. We are trying to prove arithmetical comprehension. Note first that by relativization, arithmetical comprehension is equivalent to Σ^0_1-comprehension. So let $\varphi(n)$ be a Σ^0_1-formula, say $\varphi(n) \equiv \exists m\, \theta(m, n)$, where θ is Σ^0_0. For each $k \in \mathbb{N}$ define

$$c_k = \Sigma\{2^{-n}: n < k \wedge (\exists m < k)\theta(m, n)\}.$$

Then $\langle c_k: k \in \mathbb{N}\rangle$ is a bounded increasing sequence of rational numbers. By

BW the limit $c = \lim_k c_k$ exists. Then we have

$$\forall n\,(\varphi(n) \leftrightarrow \forall k\,(|c - c_k| < 2^n \rightarrow (\exists m < k)\,\theta(m, n))).$$

Hence by Δ^0_1 comprehension we conclude $\exists X\,\forall n\,(n \in X \leftrightarrow \varphi(n))$. This proves arithmetical comprehension.

Note that the above proof involved the deduction of an axiom (arithmetical comprehension) from an ordinary mathematical theorem (BW). This is the opposite of the usual pattern of mathematical practice, in which theorems are deduced from axioms. The deduction of axioms from theorems is known as *Reverse Mathematics*. Thus Reverse Mathematics is the key to obtaining precise answers to our sharpened version of the Main Question.

We shall now list some examples of Reverse Mathematics with respect to $\mathbf{ACA}_0$.

THEOREM 2. *Within* $\mathbf{RCA}_0$ *one can prove that* $\mathbf{ACA}_0$ *is equivalent to each of the following ordinary mathematical theorems.*

1. *Every bounded* (*or bounded increasing*) *sequence of real numbers has a least upper bound* (Friedman [4]).
2. *The Bolzano–Weierstrass Theorem*: *every bounded sequence of real numbers* (*or of points in* $\mathbb{R}^n$) *has a convergent subsequence* (Friedman [4]).
3. *The Ascoli Lemma*: *every bounded equicontinuous sequence of real-valued functions on a bounded interval has a uniformly convergent subsequence* (Simpson [18]).
4. *Every countable commutative ring has a maximal ideal* (Friedman–Simpson–Smith [12]).
5. *Every countable abelian group has a unique divisible closure* (Friedman–Simpson–Smith [12]).
6. *Every countable vector space over* $\mathbb{Q}$ (*or over any countable field*) *has a basis* (Friedman–Simpson–Smith [12]).
7. *Every countable field* (*of characteristic* 0) *has a transcendence base* (Friedman–Simpson–Smith [12]).
8. *König's Lemma*: *Every infinite, finitely-branching subtree of* $\mathbb{N}^{<\mathbb{N}}$ *has an infinite path* (Friedman [4]).
9. *Ramsey's Theorem for colorings of* $[\mathbb{N}]^3$ (*or* $[\mathbb{N}]^4, [\mathbb{N}]^5, \ldots$) (Simpson [20]; cf. Jockusch [31]).

Next we shall list some examples of Reverse Mathematics with respect to Π^1_1-CA_0.

THEOREM 3. *Within* $\mathbf{RCA}_0$ *one can prove that* Π^1_1-CA_0 *is equivalent to each of the following ordinary mathematical theorems.*

1. *The Cantor–Bendixson Theorem: every closed subset of* $\mathbb{R}$ *(or of any complete separable metric space) is the union of a countable set and a perfect set* (Friedman [4]; cf. Kreisel [33]).
2. *Every countable abelian group is the direct sum of a divisible group and a reduced group* (Friedman–Simpson–Smith [12]; cf. Feferman [26]).
3. *Every difference of two open sets in* $\mathbb{N}^{\mathbb{N}}$ *is determined* (Steel [23]).
4. *Every* $F_\sigma \cap G_\delta$ *set in* $[\mathbb{N}]^{\mathbb{N}}$ *has the Ramsey property* (Simpson [20]; cf. Solovay (unpublished)).
5. *Silver's Theorem: for every coanalytic (or even* F_σ*) equivalence relation with uncountably many equivalence classes, there exists a nonempty perfect set of inequivalent elements* (Simpson [20]; cf. Harrington and Sami (unpublished)).

We would like to mention one more example of Reverse Mathematics with respect to Π^1_1-CA$_0$. By Kondo's Theorem we mean the theorem of classical descriptive set theory, according to which every coanalytic set in the plane can be uniformized by a single-valued coanalytic set.

THEOREM 4 (Simpson [20]). *Kondo's Theorem is equivalent to* Π^1_1-CA$_0$. *This equivalence is provable in the system* **ATR**$_0$ (to be defined in §7).

Unfortunately, the above equivalence is not provable in, for instance, **ACA**$_0$.

§7. Intermediate systems

In order to round out our discussion of Reverse Mathematics, we must now define and discuss the two intermediate systems S_2 and S_4 which were mentioned above.

We first discuss S_2. By *Σ^0_1-separation* we mean all axioms of the form

$$\sim \exists n\,(\varphi_0(n) \wedge \varphi_1(n)) \rightarrow \exists X\, \forall n\,(\varphi_0(n) \rightarrow n \in X . \wedge . \varphi_1(n) \rightarrow n \notin X),$$

where φ_0 and φ_1 are Σ^0_1 formulas of L_2 in which X does not occur. The system S_2 consists of the ordered semiring axioms for $\mathbb{N}, +, \cdot, 0, 1, <$ together with Σ^0_1-separation and Σ^0_1-induction. Clearly Σ^0_1-separation implies Δ^0_1-comprehension and is implied by Σ^0_1-comprehension. Thus S_2 is intermediate between **RCA**$_0$ ($= S_1$) and **ACA**$_0$ ($= S_3$).

It is known [18] that Σ^0_1 separation is equivalent over **RCA**$_0$ to a principle known as Weak König's Lemma: every infinite subtree of $2^{<\mathbb{N}}$ has an infinite path. For this reason S_2 is known in the literature as **WKL**$_0$. The ω-models of **WKL**$_0$ are known in the literature as Scott systems [38]. The recursive subsets of ω do not constitute an ω-model of **WKL**$_0$ but are the intersection of all such ω-models. In unpublished work, Harrington has shown that **WKL**$_0$ is a conservative extension of **RCA**$_0$ with respect to

Π^1_1-sentences. Hence the proof-theoretic ordinal of $\mathbf{WKL}_0$ is the same as that of $\mathbf{RCA}_0$, namely ω^ω.

We shall now list some examples of Reverse Mathematics with respect to $\mathbf{WKL}_0$.

THEOREM 5. *Within* $\mathbf{RCA}_0$ *one can prove that* $\mathbf{WKL}_0$ *is equivalent to each of the following ordinary mathematical theorems*:

1. *The Heine–Borel Theorem*: *every covering of the closed unit interval by a sequence of open intervals has a finite subcovering* (Friedman [4]).
2. *Every continuous function on the closed unit interval is bounded, or uniformly continuous, or Riemann integrable* (Simpson [20]).
3. *Every uniformly continuous real-valued function on the closed unit interval has, or attains, a supremum* (Simpson [20]).
4. *The local existence theorem for solutions of (finite systems of) ordinary differential equations* (Simpson [18]).
5. *Gödel's Completeness Theorem*: *every countable* (*or finite*) *consistent set of sentences in the predicate calculus has a model* (Friedman [4]).
6. *Every countable commutative ring has a prime ideal* (Friedman–Simpson–Smith [12]).
7. *Every countable field has a unique algebraic closure* (Friedman–Simpson–Smith [12]).
8. *Every countable formally real field is orderable, or has a* (*unique*) *real closure* (Friedman–Simpson–Smith [12]).
9. *The separable Hahn–Banach Theorem*: *if f is a bounded linear functional on a closed separable subspace of a separable Banach space, and if $\|f\| \leq r$, then f has an extension $\tilde{f}$ to the whole space such that* $\|\tilde{f}\| \leq r$ (Brown–Simpson [2]).

We now turn to S_4. By *Σ^1_1-separation* we mean all axioms of the form

$$\sim \exists n\,(\varphi_0(n) \wedge \varphi_1(n)) \rightarrow \exists X\,\forall n\,(\varphi_0(n) \rightarrow n \in X . \wedge . \varphi_1(n) \rightarrow n \notin X),$$

where φ_0 and φ_1 are Σ^1_1-formulas (i.e., of the form $\exists Y\,\theta$, where θ is arithmetical) in which X does not occur. The system S_4 consists of $\mathbf{ACA}_0$ plus Σ^1_1-separation. Clearly Σ^1_1-separation implies Δ^1_1 comprehension and is implied by Π^1_1-comprehension. Thus S_4 is intermediate between S_3 and S_5.

It is known [20] that Σ^1_1 separation is equivalent over $\mathbf{RCA}_0$ to Arithmetical Transfinite Recursion, i.e., the axiom which says that arithmetical comprehension can be iterated along any countable well ordering. For this reason the system S_4 is known in the literature as $\mathbf{ATR}_0$. It is known [17] that the hyperarithmetical subsets of ω do not form an ω-model of $\mathbf{ATR}_0$ but are the intersection of all ω-models or β-models of $\mathbf{ATR}_0$. The proof-theoretic ordinal of $\mathbf{ATR}_0$ is the same as that of Feferman's predicative analysis, viz. Γ_0 (see [27], [11], [30]). For a comparison of $\mathbf{ATR}_0$ versus predicative analysis with respect to the development of ordinary mathematics, see [19].

We shall now list some examples of Reverse Mathematics with respect to $\mathbf{ATR}_0$.

THEOREM 6. *Within* $\mathbf{RCA}_0$ *one can prove that* $\mathbf{ATR}_0$ *is equivalent to each of the following ordinary mathematical theorems*:

1. *Any two countable well orderings are comparable* (Friedman [4]).
2. *Ulm's Theorem*: *any two countable reduced abelian p-groups which have the same Ulm invariants are isomorphic* (Friedman–Simpson–Smith [12]).
3. *Every uncountable closed* (*or analytic*) *set has a nonempty perfect subset* (Friedman [4], Simpson [20].
4. *Lusin's Separation Theorem*: *any two disjoint analytic sets can be separated by a Borel set* (Simpson [20]).
5. *The domain of any single-valued Borel set in the plane is a Borel set* (Simpson [20]).
6. *Every open* (*or clopen*) *subset of* $\mathbb{N}^{\mathbb{N}}$ *is determined* (Steel [23], Simpson [20]).
7. *Every open* (*or clopen*) *subset of* $[\mathbb{N}]^{\mathbb{N}}$ *has the Ramsey property* (Simpson [14], [11]).

References

We first list the publications which touch on Reverse Mathematics (items [1]–[23]). This part of the bibliography is intended to be fairly complete. We then list the other works which have been cited above (items [24]–[42]).

For a detailed exposition of subsystems of $\mathbf{Z}_2$ and Reverse Mathematics, including proofs of most of the results which were merely stated above, see Simpson's forthcoming monograph [20]. More than half of this monograph has already been written.

Works on Reverse Mathematics

[1] D. K. Brown, Ph.D. Thesis, Pennsylvania State University, 1986, in preparation.

[2] D. K. Brown and S. G. Simpson, Which set existence theorems are needed to prove the Hahn–Banach theorem for separable Banach spaces?, Ann. Pure and Applied Logic, to appear.

[3] H. Friedman, Some systems of second order arithmetic and their use, in: Proceedings of the International Congress of Mathematicians, Vancouver 1974, Vol. 1, Canad. Math. Congress (1975) 235–242.

[4] H. Friedman, Systems of second order arithmetic with restricted induction I, II (abstracts), J. Symbolic Logic 41 (1976) 557–559.

[5] H. Friedman, Higher set theory and mathematical practice, Ann. Math. Logic 2 (1971) 326–357.

[6] H. Friedman, On the necessary use of abstract set theory, Advances in Math. 41 (1981) 209–280.
[7] H. Friedman, Unary Borel functions and second order arithmetic, Advances in Math. 50 (1983) 155–159.
[8] H. Friedman, The analysis of mathematical texts and their calibration in terms of intrinsic strength, I–IV, informally distributed reports, State University of New York at Buffalo (April–August 1975) 70 pages.
[9] H. Friedman, The logical strength of mathematical statements I, preliminary report (August 1976) 21 pages.
[10] H. Friedman, Provable equivalents of induction I, unpublished abstract, handwritten (July 1982) 6 pages.
[11] H. Friedman, K. McAloon, and S. G. Simpson, A finite combinatorial principle which is equivalent to the 1-consistency of predicative analysis, in: G. Metakides, ed., *Patras Logic Symposion*, North-Holland, Amsterdam (1982) 197–230.
[12] H. Friedman, S. G. Simpson, and R. L. Smith, Countable algebra and set existence axioms, Ann. Pure and Applied Logic 25 (1983) 141–181; addendum.
[12a] V. Harnik, Stability theory and set existence axioms, J. Symbolic Logic 50 (1985) 17–31.
[12b] V. Harnik, Set existence axioms for general (not necessarily countable) stability theory, preprint (1985) 21 pages.
[13] S. G. Simpson, Notes on subsystems of analysis, unpublished lecture notes, Berkeley (1973) 38 pages.
[14] S. G. Simpson, Sets which do not have subsets of every higher degree, J. Symbolic Logic 43 (1978) 135–138.
[15] S. G. Simpson, Reverse Mathematics, in: A. Nerode and R. Shore, eds., Proceedings of the Recursion Theory Summer School, Proc. Symp. Pure Math., Amer. Math. Soc., 42 (1985) 461–471.
[16] S. G. Simpson, Σ^1_1 and Π^1_1 transfinite induction, in: D. van Dalen, D. Lascar, and J. Smiley, eds., Logic Colloquium '80, North-Holland, Amsterdam (1982) 239–253.
[17] S. G. Simpson, Set theoretic aspects of $\mathbf{ATR}_0$, in: D. van Dalen, D. Lascar, and J. Smiley, eds., Logic Colloquium '80, North-Holland, Amsterdam (1982) 255–271.
[18] S. G. Simpson, Which set existence axioms are needed to prove the Cauchy/Peano theorem for ordinary differential equations?, J. Symbolic Logic, 49 (1984) 783–802.
[19] S. G. Simpson, Friedman's research on subsystems of second order arithmetic, in: L. Harrington, M. Morley, A. Ščedrov and S. Simpson, eds., Harvey Friedman's Research on the Foundations of Mathematics, North-Holland, Amsterdam (1985) 137–159.
[20] S. G. Simpson, Subsystems of Second Order Arithmetic, in preparation.
[21] S. G. Simpson and R. L. Smith, Factorization of polynomials and Σ^0_1-induction, preprint (1984) 29 pages.
[22] L. J. Stanley, Borel diagonalization and abstract set theory: recent results of Harvey Friedman, in: L. Harrington, M. Morley, A. Ščedrov and S. Simpson, eds., Harvey Friedman's Research on the Foundations of Mathematics, North-Holland, Amsterdam (1985) 11–86.
[23] J. R. Steel, Determinateness and subsystems of analysis, Ph.D. Thesis, Berkeley (1976) 107 pages.

Other cited works

[24] O. Aberth, Computable Analysis, McGraw-Hill, New York (1980) 187 pages.
[25] W. Buchholz and K. Schütte, Proof Theoretic Ordinals of Impredicative Subsystems of Analysis, in preparation.
[26] S. Feferman, Impredicativity of the existence of the largest divisible subgroup of an abelian *p*-group, in: Model Theory and Algebra, Springer Lecture Notes in Mathematics 498 (1975) 117–130.

[27] S. Feferman, Systems of predicative analysis, J. Symbolic Logic 29 (1964) 1–30; 33 (1968) 193–220.

[28] A. Fröhlich and J. C. Shepherdson, Effective procedures in field theory, Trans. Royal Soc. London 248 (1956) 407–432.

[29] D. Hilbert and P. Bernays, Grundlagen der Mathematik, Vols. I, II, Springer, Berlin (1934), (1939); 2nd edition (1968), (1970) 473 + 561 pages.

[30] G. Jäger, The strength of admissibility without foundation, J. Symbolic Logic 49 (1984) 867–879.

[31] C. G. Jockusch Jr., Ramsey's theorem and recursion theory, J. Symbolic Logic 37 (1972) 81–89.

[32] I. Kaplansky, Infinite Abelian Groups, University of Michigan Press, 1954; revised edition (1969) 95 pages.

[33] G. Kreisel, Analysis of the Cantor–Bendixson theorem by means of the analytic hierarchy, Bull. Acad. Polon. Scie. 7 (1959) 621–626.

[34] R. Mansfield and G. Weitkamp, *Recursive Aspects of Descriptive Set Theory*, Oxford Logic Guides, Oxford University Press (1985) 127 pages.

[35] D. A. Martin, Borel determinacy, Ann. Math. 102 (1975) 363–371.

[36] G. Metakides and A. Nerode, Effective content of field theory, Ann. Math. Logic 17 (1979) 289–320.

[37] K. Schütte, Proof Theory, Springer, Berlin (1977) 299 pages.

[38] D. Scott, Algebras of sets binumerable in complete extensions of arithmetic, in: Recursive Function Theory, Proc. Symp. Pure Math. 5, Amer. Math. Soc. (1962) 117–121.

[39] G. Takeuti, Ordinal diagrams, J. Math. Soc. Japan 9 (1957) 386–394; 12 (1960) 385–391.

[40] G. Takeuti, Two Applications of Logic to Mathematics, Iwanami Shoten and Princeton University Press (1978) 139 pages.

[41] H. Weyl, Das Kontinuum: Kritische Untersuchungen über die Grundlagen der Analysis, Veit, Leipzig, 1918, iv + 84 pages; reprinted in: H. Weyl, E. Landau, and B. Riemann, Das Kontinuum, und andere Monographien, Chelsea (1960), (1973).

[42] P. Zahn, Ein konstruktiver Weg zur Masstheorie und Funktionalanalysis, Wissenschaftliche Buchgesellschaft (1978) 350 pages.

PROOF THEORY: A PERSONAL REPORT

by

SOLOMON FEFERMAN*

It takes all kinds of in and outdoor schooling
To get adapted to my kind of fooling.

Robert Frost

Proof theory is for me a tool in the foundations of mathematics, one which I first had to apply in my work on predicativity and which I came thereafter to use more and more frequently in different aspects of my research. In this respect — as in my foundational activity generally — my approach tends to be pragmatic rather than programmatic, but it is still guided by an overall view of the main goals of foundations (concerning which I shall speak later). As to method, I am no purist and am perfectly happy to use model-theoretic or recursion-theoretic interpretations in place of proof-theoretic reductions, when these seem more warranted. No doubt this is due in part to the fact that my graduate studies in Berkeley concentrated on the set-theoretical side of logic: model theory, set theory and, to a lesser extent, recursion theory and algebra. In particular, I formed a special taste for model theory (which I still savor), but in the end moved away from it to do my thesis on the arithmetization of metamathematics. In any case, lack of purity will be evident in the following account; though I emphasize that part of my work in which proof theory plays the predominant role, I cannot separate it neatly from the use of other methods.

---

* I greatly appreciate Professor Takeuti's generous invitation to write something about my approach to proof theory for the second edition of his book. He left completely open the form, content and length of my piece, except that it should concentrate on my work and related work of students and colleagues; I have taken him at his word. Much of my research reported herein has been supported by grants over the years from the National Science Foundation.

Predicativity

The immediate background to my work on predicativity was provided by the paper F 1962[1] on transfinite *progressions* $\{\mathbf{T}_a\}$ of recursively presented theories, indexed by notations for recursive ordinals $a \in \mathcal{O}$. These are generated from initial $\mathbf{T}_0$ containing **PA** using various extension procedures $\mathbf{T} \mapsto \mathbf{T}'$ by reflection principles, in particular:

(R1) $\mathbf{T}' = \mathbf{T} + \text{all}\,[\text{Prov}_{\mathbf{T}}(\ulcorner\phi\urcorner) \rightarrow \phi]$ for closed ϕ, and

(R2) $\mathbf{T}' = \mathbf{T} + \text{all}[\forall x\, \text{Prov}_{\mathbf{T}}(\ulcorner\phi(\bar{x})\urcorner) \rightarrow \forall x\, \phi(x)]$ for all $\phi(x)$.

One takes $\mathbf{T}_{a\oplus 1} = \mathbf{T}'_a$ and $\mathbf{T}_a = \bigcup \mathbf{T}_b[b < a]$ for limit a. Following Turing 1939, completeness for a progression based on (R1) was obtained with respect to Π^0_1-statements, which I then showed to be best possible. The main result of F 1962 was completeness for a progression based on (R2) with respect to all Π^0_∞ (arithmetical) statements. However, due to non-uniqueness (in general, $|a| = |b|$ does not imply $\mathbf{T}_a \equiv \mathbf{T}_b$), this result had to be treated with care. In F 1962, paths P recursive in $\mathcal{O}$ were found along which $\bigcup \mathbf{T}_a[a \in P]$ is complete for Π^0_∞ statements. But it was shown in Feferman and Spector 1962 that $\bigcup \mathbf{T}_a[a \in P]$ is incomplete even for Π^0_1-statements when P is Π^1_1 through $\mathcal{O}$ (and such paths exist).

The crucial question then became, what are natural conditions to impose on the choice of ordinal notations used to index theories in a progression? The first suggestion in response came for progressions in a language containing a predicate $I(x)$ such that provability of $I(\bar{a})$ in a correct theory implies $a \in \mathcal{O}$. (This is automatic for 2nd-order theories, in which $I(x)$ can simply be taken as the Π^1_1 definition of $\mathcal{O}$.) Assuming $\mathbf{T}_0$ is correct, and that $\mathbf{T} \mapsto \mathbf{T}'$ preserves correctness, one takes the class $\text{Aut}\{\mathbf{T}_a\}$ of *autonomous* notations for the progression to be the smallest class A containing 0, closed under predecessors and successor, and such that $a \in A$ and $\mathbf{T}_a \vdash I(b)$ implies $b \in A$. This notion of autonomy (or 'boot-strap' condition) was first introduced in Kreisel 1958 and used in his proposals to characterize *finitism* and *predicativity* by suitable autonomous progressions of theories $\{\mathbf{F}_a\}$ and $\{\mathbf{R}_a\}$, respectively. The main result of Kreisel's paper was the determination of ε_0 as the l.u.b. of the $|a|$ for a autonomous with respect to the $\{\mathbf{F}_a\}$. The question was raised to determine the least non-autonomous ordinal for the second of these progressions $\{\mathbf{R}_a\}$, a form of transfinite ramified analysis in which $\mathbf{R}_a$ uses 2nd-order variables of ranks $b \preccurlyeq a$ only. This problem was solved independently by Schütte (1964, 1965) and myself (F 1964) and the answer was established to be Γ_0, the least fixed-point for the Veblen hierarchy φ_α of critical functions ($\varphi_0(\beta) = \omega^\beta$, φ_α enumerates $\{\xi\colon \varphi_\gamma(\xi) = \xi$, all $\gamma < \alpha\}$ for $\alpha \neq 0$, and $\Gamma_0 =$ least α with $\varphi_\alpha(0) = \alpha$).

[1] 'F' refers to papers under my name in the bibliography.

Here was my first use of proof theory, but one which essentially came ready-made from Schütte's book (1960), where the proof-theoretic ordinals of the $\mathbf{R}_a$ were determined in general as a function of $\alpha = |a|$. For an ε-number, one simply has $|\mathbf{R}_a| = \varphi_\alpha(0)$, so the least non-autonomous ordinal is at most the limit of the α_n, where $\alpha_0 = 0$, $\alpha_{n+1} = \varphi_{\alpha_n}(0)$; but $\lim_n \alpha_n = \Gamma_0$. The more difficult part (for both Schütte and myself) was to show this bound best possible (which we did by different methods).

Schütte's work on predicativity was just one application of his overall development of proof theory and he went on to other aspects of the subject, to which he has contributed so much. However, predicativity continued to occupy my attention off and on in the following years, because I regarded its study as having special philosophical significance. Speaking informally, the structure of natural numbers is assumed as given or understood, and one wishes to see which definitions of sets of natural numbers can be successively recognized to be reduced to that of the natural numbers. This leads one directly to consider the ramified hierarchy at both finite and transfinite levels α. In previous work by Spector and Wang, the question of which ordinals to use was restricted only by the condition of their definability (as well-orderings) at earlier levels. For that the limit was shown to be ω_1^{rec}, the least non-recursive ordinal. Using a result of Kleene, the sets definable in the ramified hierarchy up to ω_1^{rec} are the same as the hyperarithmetic or Δ_1^1 sets. But the concept of ordinal or well-ordering was there used in its impredicative sense, and no restriction was placed on how one came to recognize an expression to be a definition of a well-ordering. It is just this problem which was met by dealing with autonomous progressions of ramified theories, thereby imposing conditions on provability as well as definability.

That there is a fundamental difference between our understanding of the concept of natural numbers and our understanding of the set concept, even for sets of natural numbers, seems to me undeniable [2]. The study of predicativity, as what is implicit in accepting the structure of natural numbers, is thus of special foundational significance. This is *not* to say that only what is predicative is 'justified'. What we are dealing with here are questions of relative conceptual clarity and foundational status; of this, more below.

Given the intrinsic interest of predicativity, I next turned to the question of how much mathematics could be developed predicatively. Since ramified theories were unsuitable as a framework for the development of analysis [3], I began by looking at which formally unramified theories could be shown to be (proof-theoretically) equivalent to the autonomous pro-

---

[2] The idea that sets are prior to natural numbers and that the latter are to be 'defined' in terms of the former ridiculously turns this on its head.

[3] Bertrand Russell had introduced his infamous "axiom of reducibility" just to get around this problem, thus turning predicative theories into basically impredicative ones.

gression of ramified theories. This was done in two steps for the language of analysis in F 1964 (and later extended to the language of set theory, as will be explained below). Following another suggestion of Kreisel, the first step was to replace the $\{\mathbf{R}_a\}$ by a progression $\{\mathbf{H}_a\}$ based on the Hyperarithmetic Comprehension Rule (HCR): from $\forall x\,[P(x) \leftrightarrow Q(x)]$ infer $\exists X\,\forall x\,[x \in X \leftrightarrow P(x)]$, where $P(x)$ is any Π^1_1-formula and $Q(x)$ any Σ^1_1-formula. The motivation for this is the recognizable absoluteness (or invariance) of provably Δ^1_1-definitions, in the following sense. At each stage one has recognized certain closure conditions on the universe of sets, and the definitions $D(x)$ of sets one takes should be independent of what further closure conditions may be accepted. Thus if $\mathcal{U}$ represents a universe satisfying given closure conditions and is extended to $\mathcal{U}'$ (satisfying the same closure conditions), one wants $\forall x\,[D^{\mathcal{U}}(x) \leftrightarrow D^{\mathcal{U}'}(x)]$. This requirement is easily seen to hold for provably Δ^1_1-formulas, since Π^1_1-formulas $P(x)$ are provably persistent for extensions, $\forall x\,[P^{\mathcal{U}}(x) \rightarrow P^{\mathcal{U}'}(x)]$, and dually one has $\forall x\,[Q^{\mathcal{U}'}(x) \rightarrow Q^{\mathcal{U}}(x)]$ for Σ^1_1-formulas $Q(x)$. The problem of showing that all provably invariant definitions are provably Δ^1_1 was to lead me into another part of proof theory which will be taken up in the next section.

To return to the progression $\{\mathbf{H}_a\}$ (contained in ordinary 2nd-order analysis): $\mathbf{H}_0$, and each $\mathbf{H}_{a\oplus 1}$ is obtained by closing under the rule (HCR) and for limit a, $\mathbf{H}_a = \bigcup \mathbf{H}_b[b < a]$. It was shown in F 1964 that Γ_0 is also the l.u.b. of Aut$\{\mathbf{H}_a\}$, and $\bigcup \mathbf{H}_a[a < \Gamma_0]$ is proof-theoretically equivalent to $\bigcup \mathbf{R}_a[a < \Gamma_0]$. Locally (at each stage), this makes primary use of model-theoretic and recursion-theoretic arguments, partly related to Kleene's equivalence of the hyperarithmetic hierarchy and ramified hierarchy up to ω_1^{rec}. Each $\mathbf{H}_a$ is modelled in a certain $\mathbf{R}_{a'}$ and vice-versa.

The second step in comparing predicative systems with classical analysis was to replace the progression $\{\mathbf{H}_a\}$ by one simple 2nd order system, denoted **IR**. This is axiomatized by (HCR) together with the Bar-Rule (BR): from $I(\bar{a})$ infer $I(\bar{a}; \phi)$, the principle of transfinite induction on the $<$ relation restricted to the predecessors of a, for any property ϕ. (In place of notations a for ordinals satisfying $I(\bar{a})$ one can take provably Δ^1_1 definitions of well-orderings.) The main result concerning **IR** stated in F 1964 is that it proves the same theorems as $\bigcup \mathbf{H}_a[a < \Gamma_0]$.

Various technical questions that arose in the process of this work now led me off into some parts of general proof theory, as will be described next.

Infinitary proof theory and interpolation theorems

Some background, to begin with: in the hands of Schütte in the 1950's the proof theory of ramified systems was most simply treated by extensions of Gentzen's sequential calculus (or variants thereof) for the ordinary predicate calculus, to systems with infinitary rules of inference. Typically, a theory **T** with finitary language, axioms and rules of inference is embedded in some sort of infinitary sequential calculus $\mathbf{T}^+$, and ordinals are assigned as lengths to derivations in $\mathbf{T}^+$. Gentzen's cut elimination theorem is extended to $\mathbf{T}^+$ in an effective way and the l.u.b. $|\mathbf{T}^+_{\text{c.f.}}|$ of the ordinals of cut-free derivations in $\mathbf{T}^+$ turns out to be a bound for the proof-theoretic ordinal $|\mathbf{T}|$ of **T**, defined as the l.u.b. of the provably recursive well-orderings of **T**. As reference system for locating these ordinals, one starts with a sufficiently large 'natural' recursive well-ordering $\preccurlyeq$ whose initial segments of type α are denoted $\preccurlyeq_\alpha$. In a 2nd order language one has a Π^1_1-statement $I(\alpha) := \forall X I(\alpha; X)$ (or $I(\bar{a})$, where a is the notation in $\preccurlyeq$ for α), which expresses transfinite induction along $\preccurlyeq_\alpha$ with respect to every property $P(x)$: $x \in X$. Then $|\mathbf{T}| = \sup\{\alpha : \mathbf{T} \vdash I(\alpha)\}$; to deal with first-order languages as well one takes $|\mathbf{T}| = \sup\{\alpha : \mathbf{T} \vdash I(\alpha; P)\}$, where P is a 'free' predicate variable adjoined to **T**. Let $\alpha_0 = |\mathbf{T}^+_{\text{c.f.}}|$; the proof-theoretic work described above is broken up by showing

(i) $|\mathbf{T}| \leqslant \alpha_0$, and

(ii) for $\alpha < \alpha_0$ one has $\mathbf{T} \vdash I(\alpha; P)$.

In addition one is usually able to show that

(iii) $\mathbf{T} + I(\alpha_0; R) \vdash \mathrm{Con}_{\mathbf{T}}$ (the consistency of **T**), for a suitable recursive property R.

This is usually done by arithmetizing the proof-theory of $\mathbf{T}^+$ using recursively represented infinite derivations.

The simplest example of all this is provided by $\mathbf{T} = \mathbf{PA}$ (Peano Arithmetic), where the induction scheme is replaced by the ω-rule in $\mathbf{T}^+$:

$$\frac{\Gamma \to \Delta, A(\bar{n})\,(\text{each } n < \omega)}{\Gamma \to \Delta, \forall x\, A(x)}.$$

Each derivation d in **PA** is transformed into a derivation d^+ in $\mathbf{T}^+$ of finite cut-rank and with length $|d^+| < \omega \cdot 2$. Each lowering of cut-rank can be arranged with an (at most) $\exp_\omega$ increase in length. Hence for the system of these infinitary derivations, $|\mathbf{T}^+_{\text{c.f.}}| = \varepsilon_0$, and the above treatment recaptures Gentzen's consistency result for **PA** in a conceptually much simpler way. The next example is for ramified analysis of level 0, equivalently arithmetical analysis with the Π^0_∞-Comprehension Axiom

$$(\Pi^0_\infty\text{-CA}) \quad \exists X \forall x\, [x \in X \leftrightarrow A(x)],$$

for each arithmetical A. In place of this one takes in $\mathbf{T}^+$ the rule

$$\frac{\Gamma \to \Delta, \phi(\hat{x}\, A_n(x)) \quad (\text{each } n < \omega)}{\Gamma \to \Delta, \forall X\, \phi(X)},$$

where A_n is an enumeration of the arithmetical formulas. In this case it turns out that $|\mathbf{T}^+_{\text{c.f.}}| = \varepsilon_{\varepsilon_0}$. And so one can proceed through the various levels of ramified analysis, where every set variable is ranked: $X^\beta, Y^\beta, Z^\beta, \ldots$, (so for Π^0_∞-CA we use ranked variables $X^0, Y^0, Z^0, \ldots$).

In Schütte's treatment, the formulas of $\mathbf{T}^+$ were still taken as finite. But a further conceptional simplification takes place if one allows use of countably infinite conjunctions $\bigwedge$ and disjunctions $\bigvee$ in the language of $\mathbf{T}^+$. With each finitary formula ϕ is associated an infinitary formula ϕ^* by

$$(\forall x\,\phi(x))^* = \bigwedge_{n<\omega} \phi(\bar{n})^*, \qquad (\forall X^0\,\phi(X^0))^* = \bigwedge_{n<\omega} \phi(\hat{x}A_n(x))^*, \quad \text{etc.}$$

In place of a multiplicity of special rules for the different kinds of quantified variables, one simply takes the rule

$$\frac{\Gamma \to \Delta, \phi_n \quad (\text{each } n \in \omega)}{\Gamma \to \Delta, \bigwedge_{n<\omega} \phi_n}.$$

Everything is thus embedded in $\mathscr{L}_{\omega_1,\omega}$, the logic of countable conjunctions and disjunctions and ordinary quantification (actually only its propositional part need be used). This was the step taken by Tait in 1966 (see his 1968).

Another reason for moving to infinitary formulas has to do with some results that came out of proof theory but having nothing directly to do with ordinals and consistency proofs. One such is the interpolation theorem; it turned out that a form of this was needed to characterize the absolute properties in predicativity. But the interpolation theorem fails in ω-logic, at least as applied to ordinary finitary formulas; interpolants are found in $\mathscr{L}_{\omega_1,\omega}$.

In 1965, working on the problem of which 2nd order formulas are absolute (or invariant) for extensions of the universe of sets when the 1st order universe is regarded as fixed, I realized the need for (and obtained) a many-sorted interpolation theorem. This was then applied in a note with Kreisel (1966) in which the desired invariance and more general persistence results were stated for three languages: $\mathscr{L}_{\omega,\omega}$ (ordinary finitary language), $\mathscr{L}_{\omega_1,\omega}$ and an intermediate language $\mathscr{L}_{\mathrm{HYP}}$ of infinitary formulas with hyperarithmetic structure. However, the methods applied at this stage were still unsatisfactory, since a different argument for the main persistence result was required in each of the three languages.

The proper setting for a uniform treatment was soon provided by the work of Barwise in his 1967 Stanford doctoral dissertation on admissible languages $\mathscr{L}_A \subset \mathscr{L}_{\omega_1,\omega}$, i.e., languages whose formulas have set-theoretical structure represented in A, where A is an admissible set in the Kripke-Platek sense. The languages $\mathscr{L}_{\omega,\omega}$, $\mathscr{L}_{\mathrm{HYP}}$ and $\mathscr{L}_{\omega_1,\omega}$ are obtained by taking $A = L_\alpha$ for $\alpha = \omega$, ω_1^{rec} and ω_1, resp. (L_α = the sets constructible by stage α). Barwise established basic completeness and (in the countable case) compactness results for the $\mathscr{L}_A$, as well as a generalization of Craig's interpolation theorem (first extended to $\mathscr{L}_{\omega_1,\omega}$ by Lopez-Escobar); see Barwise 1969.

In my lectures on proof theory at the 1967 Leeds conference (F 1968) I took the framework of countably infinitary logic on admissible sets as one in which to formulate a general cut-elimination theorem; no essentially new arguments were needed for its proof [4]. Then following standard proofs of Craig's interpolation theorem using cut-free sequential systems, I obtained a new and strengthened interpolation theorem for many-sorted languages, in which conditions are also imposed on the sorts associated with essentially universal, resp. existential occurrences of bound variables. Finally this was used to give direct proofs characterizing the persistent and invariant formulas for various model-theoretic extension relations, uniformly for all the $\mathscr{L}_A$.

The point of all this in the study of predicativity was to show that if a formula ϕ in a 2nd order language is provably persistent, resp. invariant for extensions of the universe of sets (keeping the universe of individuals fixed) then ϕ is provably (essentially) Σ^1_1, resp. Δ^1_1. Analogous results for higher order languages also came from the general characterization theorem referred to above. For applications to set theory, a further extension of the many-sorted interpolation theorem was needed where restricted quantifiers could be given separate play; this was done subsequently in F 1968a. The main application there characterizes those formulas in the language of set theory provably persistent, resp. invariant for end extensions of the universe, namely as those which are provably (essentially) Σ_1, resp. Δ_1 (in the usual set-theoretical classifications modulo free use of restricted quantifiers).

Further applications of the many-sorted interpolation theorems were made in the area of model theory in F 1974. In fact, proof theory became completely dispensable here, since model-theoretic proofs had in the meantime been found by Stern for the same interpolation results [5]. In addition, the information on ordinal bounds provided by the cut-elimination theorem nowhere played a part. From this point of view, the role of proof theory in the above development simply appears as an historical accident. On the other hand, while I do not in general hold to proof theory as the preferred method, it still seems to me to be the best motivated approach in this case. The whole matter is hardly worth arguing about, but at least it should be pointed out that one standard model-theoretic method used to establish interpolation and persistence theorems for infinitary languages, that of *consistency properties*, is just a dual form of *derivability*

[4] While completeness of a cut-free system was also established by the work of Barwise, the cut-elimination theorem was needed to give this an effective form including ordinal bounds.

[5] This used the technique of model-theoretic forcing. Subsequently, different methods have been applied by others, notably that of consistency properties, concerning which see below.

properties for Gentzen (cut-free) systems, so these differences in method are only apparent; for a discussion see Nebres 1972 (where the latter are called validity properties)[6].

Predicativity (continued)

In my further publications on this subject, my aim was to break away from the rather artificial looking formulation in terms of ramified progressions of theories and move toward a more realistic logical model of predicativity as a body of mathematical thought growing through the process of active reflection on the choice of basic principles.

The first step in F 1967 concentrated on autonomy and invariant definitions, but still had vestiges of the impredicative notions of ordinal and well-ordering. Briefly the idea was to consider formally infinitary extensions ($\mathbf{A}^+$) of arithmetical analysis, allowing both infinitary derivations $\mathscr{D}$ and infinitary formulas $\mathscr{F}$. The infinitary system ($\mathbf{A}^+$) is obtained from the (infinitary) defining axioms Z for natural numbers by adding instances of the comprehension axiom:

$$(\mathrm{CA}_{\mathscr{A}}) \qquad \exists X\, \forall x\, [x \in X \leftrightarrow \mathscr{A}],$$

for each formula $\mathscr{A}$ of $\mathscr{L}_{\omega_1,\omega}$ without bound 2nd order variables ('X' not free in $\mathscr{A}$). Three successively stronger notions of autonomy of ordinals, formulas and derivations are then formulated (by a simultaneous inductive definition) within this setting. In the first, if the ordinal $|\mathscr{D}|$ of a derivation is autonomous then so also is $\mathscr{D}$ together with the end-formula of $\mathscr{D}$, and, further, if a relation $\preccurlyeq$ is provably invariant and the formula $I(\preccurlyeq)$ is autonomous, then so also is the ordinal $|\preccurlyeq|$. The main results for this first notion are that Γ_0 is the l.u.b. of the autonomous ordinals and that the set of autonomous theorems is closed under the hyperarithmetic comprehension rule.

The argument for closure under (HCR) in the preceding has proof-theoretical interest: Given a formula $\mathscr{P}$ in essentially Π_1^1 form (i.e., one built up from atomic formulas and their negations by infinite conjunctions, disjunctions and numerical quantification, but only universal 2nd order quantification) and a formula $\mathscr{Q}$, dually, in essentially Σ_1^1-form, if $(\mathscr{P} \leftrightarrow \mathscr{Q})$ has an autonomous derivation then there is a quantifier-free autonomous $\mathscr{A}$, with $(\mathscr{P} \leftrightarrow \mathscr{A})$ autonomously provable. To show this, I used an interpolation theorem for $\vdash(\mathscr{P} \rightarrow \mathscr{Q})$ in $\mathscr{L}_{\omega_1,\omega}$ which is due to Barwise in his 1967 dissertation, and which can be regarded as a kind of infinitary form of Herbrand's theorem for $(\neg\mathscr{P} \vee \mathscr{Q})$.

For the second notion of autonomy in F 1967 one also requires that the set $\mathrm{Tree}_{\mathscr{D}}$ of sequences of natural numbers which represents the tree of a

---

[6] Nebres' paper is based on part of his Ph.D. thesis.

derivation $\mathscr{D}$ is to be shown provably invariant before accepting $\mathscr{D}$ as autonomous. The main result above extends to this notion as well. Finally, for the third notion, it is also required that one has an autonomous derivation of the statement that $\text{Tree}_{\mathscr{D}}$ satisfies the defining conditions, including well-foundedness, for an $(\mathbf{A}^+)$ derivation, before accepting $\mathscr{D}$ itself as autonomous. The main result stated in F 1967 is that the set of $(\mathbf{A}^+)$-autonomous theorems in this third sense is a conservative extension of the theory (**IR**) of F 1964.

These results bolstered (in my view) the determination of Γ_0 as the least impredicative ordinal, and of **IR** as comprising the extent of predicative mathematics in 2nd order analysis. However, a further effort was needed to remove the prima-facie impredicative concepts of ordinal and well-ordering from the basic axiomatic formulations. This step was taken in F 1979. I shall not attempt to describe the system **P** of that paper here, but only indicate some of its main features. The language $\mathscr{L}_{\mathbf{P}}$ is 3d order, with variables for numbers, functions and predicates (or sets of natural numbers), and constants for functionals of type level 2. In $\mathscr{L}_{\mathbf{P}}$ only quantification over numerical variables is admitted. There is an auxiliary language $\exists/\mathbf{P}$ in which we may derive formulas of the form $\exists\beta\, A$ with 'β' a type 1 second order (function or predicate) variable and A in $\mathscr{L}_{\mathbf{P}}$. Associated with each formula $A(\underline{\alpha}; \beta)$ such that $(\exists/\mathbf{P}) \vdash \exists!\beta A(\underline{\alpha}; \beta)$ is a functional constant F_A with axiom $A(\underline{\alpha}; F_A(\underline{\alpha}))$ in **P**. The system $\exists/\mathbf{P}$ provides for explicit definition and unification: $\forall x\, \exists!\beta\, A(x, \underline{\alpha}; \beta) \to \exists\beta\, \forall x\, A(x, \underline{\alpha}; \beta(x))$, which is a weak form of choice (equality of type 1 objects being defined extensionally).

Finally, there is a rule $A(\alpha)/A(t)$, whenever t is of the same type as α, which allows one to derive a form of the bar-rule; $A(X)/A(\hat{x}B(x))$ for 'X' in right-of-ε position in A and A, B formulas of $\mathscr{L}_{\mathbf{P}}$. The logic can be intuitionistic or classical, yielding theories with the same proof-theoretical strength, for, the main result of this paper is that: $\mathbf{P}^{(\text{int})} \vdash I(\bar{a}; X)$ for each (standard) notation $|a| < \Gamma_0$, and $|\mathbf{P}^{(\text{class})}| \leq \Gamma_0$. The latter result is proved by showing

(i) **P** is interpretable in $(\Sigma^1_1\text{-AC}) + (\text{BR})$ (the bar rule), and

(ii) $|(\Sigma^1_1\text{-AC}) + (\text{BR})| \leq \Gamma_0$,

a fact established by methods first presented in F 1971, to be described below.

The third approach to predicativity extracts it from a much more general concept, that of the reflective closure of a schematic theory **S**. This work is still being prepared for publication, as F 198?, though it was realized over six years ago[7]. A schematic system **S** is given in a language

[7] The basic notions and results were first presented in a talk on Gödel's incompleteness theorems, as part of a symposium on the work of Kurt Gödel held at the meeting of the Association for Symbolic Logic in San Diego, March 1979. An elaboration of the ideas was the subject-matter of my retiring presidential address for the ASL on Dec. 30, 1983 in Boston, which is to appear as F 198?.

$\mathscr{L}_0 + \{X\}$ where X is a new unary predicate symbol. We can think of the axiom schemas induced by **S** in $\mathscr{L}_0$ as consisting of all substitution instances $A(\hat{x}B(x))$ for $A(X)$ in **S** and B in $\mathscr{L}_0$; the resulting system is denoted $\mathbf{S}_{\mathscr{L}_0}$. But more generally, for any $\mathscr{L} \supseteq \mathscr{L}_0 + \{X\}$ we can think of derivations from **S** in $\mathscr{L}$ as proceeding by the predicate substitution rule [8]: $A(X)/A(\hat{x}B(x))$ for A in $\mathscr{L}_0$, B in $\mathscr{L}$; this system is denoted $\mathbf{S}(X)_{\mathscr{L}}$. Familiar examples of schematic theories are Peano arithmetic, 2nd order arithmetic, Zermelo set theory, **ZF**, etc.

Now the reflection principles (R1) and (R2) discussed in §1 are just consequences in a given language of using a notion of truth outside that language. In a sense, underlying the iteration of reflection principles is the unrestricted iteration of the truth concept applied to itself. To set up a consistent theory of such, one uses either axioms for partial truth predicates T, F, à la Kripke and myself (see the respective papers in Martin 1984) or Aczel's notion of (definite) proposition D and truth T (Aczel 1980). Roughly speaking, the relationship between the two approaches, is given by $D(x) = [T(x) \vee F(x)]$ or $F(x) = D(x) \wedge \neg T(x)$. The axioms for partial truth are called the (T, F)-axioms. There are two senses of reflective closure for a schematic theory **S** in $\mathscr{L}_0 + \{X\}$, denoted $\mathbf{S}^{\supset}$ and $\mathbf{S}(X)^{\supset}$, resp.:

(i) take $\mathscr{L} = \mathscr{L}_0 + \{T, F\}$ and

$$\mathbf{S}^{\supset} := \mathbf{S}_{\mathscr{L}} + (T, F)\text{-axioms};$$

(ii) take $\mathscr{L} = \mathscr{L}_0 + \{T, F, X\}$ and

$$\mathbf{S}(X)^{\supset} := \mathbf{S}(X)_{\mathscr{L}} + (T, F)\text{-axioms in } \mathscr{L},$$

closing under the predicate substitution rule.

The main result of F 198?a is that

(i)* $|\mathbf{PA}^{\supset}| = \varphi_{\varepsilon_0}(0)$ and $\mathbf{PA}^{\supset} \equiv \bigcup_{a<\varepsilon_0} \mathbf{R}_a$, while

(ii)* $|\mathbf{PA}(X)^{\supset}| = \Gamma_0$ and $\mathbf{PA}(X)^{\supset} \equiv \bigcup_{a<\Gamma_0} \mathbf{R}_a \equiv \mathbf{IR}$.

The second of these results is proved once more by use of $(\Sigma^1_1\text{-AC}) +$ (BR), in which in this case, $\mathbf{PA}(X)^{\supset}$ is interpretable.

There are interesting open questions on the strength of the reflective closures of $\mathbf{S}^{\supset}$ and $\mathbf{S}(X)^{\supset}$ for other schematic theories **S**.

Proof-theoretic ordinals

Despite the fact that ordinals and systems of ordinal representation have played a central role in proof theory since Gentzen's consistency proof of arithmetic in 1936, they did not themselves begin to receive theoretical

[8] This is, again, a form of the bar rule.

attention until the late 1960's[9], and only the loosest characteristics were ordinarily mentioned concerning them. For example, in the Gentzen-style extension of Hilbert's program to treat a theory **T** embodying principles considered to be problematic (on philosophical grounds), the usual requirements are:

(i) to find an effectively decidable well-ordering (or well-founded relation) $\preccurlyeq$, for which

(ii) Con(**T**) is provable by induction $I(\preccurlyeq)$ on $\preccurlyeq$, using only elementary combinatorial (finitist) arguments in addition, and

(iii) $I(\preccurlyeq)$ itself is established by constructive methods[10].

For optimality, one wants also to show that

(iv) $\mathbf{T} \vdash I(\preccurlyeq_a)$ for each proper initial segment $\preccurlyeq_a$ of $\preccurlyeq$.

The main problem with this explanation is that the requirement (iii) is so vague. (One might also argue that the notion of finitist proof is vague, but in practice one can take Primitive Recursive Arithmetic (**PRA**) as the base, so that (ii) is replaced by: $\mathbf{PRA} + I(\preccurlyeq) \vdash \mathrm{Con}(\mathbf{T})$.) Now, if one drops (iii) in order to have a more precise notion (and modifies (ii) as just indicated), the conditions become too weak. For example, as pointed out in Kreisel 1968, pp. 333–334, one can cook up a primitive recursive ordering $\preccurlyeq$ of type ω^2 such that $I(\preccurlyeq)$ proves Con(**PA**) in an elementary way, and for each proper initial segment $\preccurlyeq_a$, $\mathbf{PA} \vdash (\preccurlyeq_a)$. But to recognize informally that $I(\preccurlyeq)$ holds one must invoke the consistency of **PA**, so there is no genuine reduction here.

Those wishing to detach themselves from doctrinal or philosophical questions while still pursuing Gentzen-style proof-theoretical work, simply talk about determining the ordinal $|\mathbf{T}|$, defined as the sup of the provably recursive well-orderings of **T**. The problem with this is that it leaves completely open what constitutes 'determining' an ordinal, i.e., in what terms or form an answer is to be given.

In addition to the foregoing requirements, it has long been recognized that systems of ordinal representation used in proof theory must also be *natural* in some way. Roughly speaking, this means that we are dealing with a system of terms generated from some symbols for certain specific ordinals by function symbols for certain ordinal functions, with the ordering relation $\preccurlyeq$ between terms induced by the ordering of the corresponding values[11]. The paradigm is Gentzen's use of Cantor's representation of ordinals $< \varepsilon_0$, where the terms are built from (the symbol for) 0 by the functions (symbols for) $+$ and $\omega^{(\cdot)}$. The requirement, that the

---

[9] In contrast, the general theory of recursive well-orderings and systems of ordinal notation, initiated by Church and Kleene in 1936, was developed very intensively beginning in the 1950's by Kleene, Spector and others, with work on recursion-theoretic hierarchies.

[10] See Takeuti 1975, p. 96, and Schütte 1977, pp. 2–3 for such expressions of this approach to the Hilbert-Gentzen program.

[11] Takeuti 1975 calls such systems *standard*; cf. op. cit. Ch. 2, §11.

system of ordinal representation used be natural, is nowhere explicitly necessary in meeting the conditions on $\preccurlyeq$ (i)–(iv) above for the Hilbert-Gentzen program, but in practice it plays a role in each of these. Again, there is no theoretical explanation as to how or why this is the case. Now one could hope to detach a theoretical study of ordinal systems used in proof theory from the vague requirement of constructivity, by supplying a precise explanation for the notion of natural system of ordinal representation. This has so far not been done, though there is a body of work which has been carried out that clearly bears on it; some of that is described below.

My own work on systems of ordinal representation as a theoretical subject began when I was writing up just what was needed for a system representing the ordinals up to Γ_0 in order to fill in the details of the various arguments sketched in F 1964. As would be expected, the resulting examination of this question in F 1968b carried me to systems for much larger ordinals; hence, predicativity was not an issue here. Indeed, the point of view taken at the outset of F 1968b was frankly set-theoretical: one assumes the class Ω of all countable ordinals and considers properties of systems $\vec{\varphi} = \langle \varphi_1, \ldots, \varphi_n \rangle$ of functions $\varphi_i : \Omega^{m_i} \to \Omega$ ($m_i \geqslant 0$). (Later on I would take the class ON of all ordinals for granted.) For any set $A \subseteq \Omega$, $Cl_{\vec{\varphi}}(A)$ is defined to be the closure of $A \cup \{0\}$ under the functions in $\vec{\varphi}$. Identifying each ordinal α with its set of predecessors, the $\vec{\varphi}$-closed ordinals (or $\vec{\varphi}$-inaccessibles) are just the fixed-points $\alpha = Cl_{\vec{\varphi}}(\alpha)$; then $(\vec{\varphi})'$ is defined to be the (normal) function which enumerates the class of $\vec{\varphi}\cdot$closed ordinals (denoted $\mathrm{In}(\vec{\varphi})$).

The most obvious requirement to place on $\vec{\varphi}$, when using it to represent ordinals up to a certain point, is that $Cl_{\vec{\varphi}}(0)$ be an ordinal; $\vec{\varphi}$ is called *complete* in this case. For example, the system $\vec{\varphi}$ consisting of the two functions $\varphi_1 = \lambda\alpha, \beta(\alpha+\beta)$ and $\varphi_2 = \lambda\alpha(\omega^\alpha)$ (or of the single function $\lambda\alpha, \beta(\omega^\alpha + \beta)$) is complete, with $Cl_{\vec{\varphi}}(0) = \varepsilon_0 = (\vec{\varphi})'(0)$; on the other hand the system consisting only of the φ_2 above is not complete, though $\langle\varphi_2\rangle'(0) = \varepsilon_0$, too. The idea of building ever larger systems by iterating the critical process goes back to Veblen 1908. It is not true in general that the property of being complete is inherited under this iteration, but a more general property is, namely that of *repleteness*, which I introduced in F 1968b: $\vec{\varphi}$ is called *replete* if for every ordinal α, $Cl_{\vec{\varphi}}(\alpha)$ is again an ordinal. The first main result of F 1968b is that if $\vec{\varphi}$ is replete then so also is $\vec{\varphi} * \langle\psi\rangle$ as well as $\vec{\varphi} * \langle\lambda\alpha, \beta(\psi_\alpha(\beta))\rangle$ for $\psi = (\vec{\varphi})'$; here ψ_α is the αth function in the Veblen hierarchy based on $\psi_0 = \psi$.

With each system $\vec{\varphi}$ is associated a system of representation using a symbol for 0 and function symbols f_i of m_i arguments for φ_i. Then each closed term t denotes an element $|t|_{\vec{\varphi}}$ of $Cl_{\vec{\varphi}}(0)$. The effectiveness requirement on systems of representation is here expressed by the condition: the relation $t_1 \leqslant_{\vec{\varphi}} t_2$, defined by $|t_1|_{\vec{\varphi}} \leqslant |t_2|_{\vec{\varphi}}$, is recursive. Again, this property is not necessarily inherited by $\vec{\varphi} * \langle(\vec{\varphi})'\rangle$. In F 1968b I defined a notion: $\vec{\varphi}$ is *effectively relatively categorical* (e.r.c.) to mean that there is a recursive relation $R(t_1, t_2, d)$ which for any open terms t_1, t_2 and assignment s to the free variables x_i of (t_1, t_2) into the class $\mathrm{In}(\vec{\varphi})$ of $\vec{\varphi}$-inaccessibles we have $|t_1|_{\vec{\varphi},s} \leqslant |t_2|_{\vec{\varphi},s}$ if and only if $R(t_1, t_2, \mathrm{Diag}(s))$. In other words, for such assignments s, the ordering relation is effectively reduced to the diagram of s, i.e., to $\{(i, j) | s(x_i) \leqslant s(x_j)\}$[12]. The second main result of F 1968b is that if $\vec{\varphi}$ is e.r.c. then so also is $\vec{\varphi} * \langle \psi \rangle$ and $\vec{\varphi} * \langle \lambda\alpha, \beta \cdot \psi_\alpha(\beta) \rangle$ for $\psi = (\vec{\varphi})'$. Application of the two main results of this paper to the initial system $\vec{\varphi} = \langle \lambda\alpha(1+\alpha) \rangle$, yields all the properties I needed for the system of representation $\langle \lambda\alpha, \beta(\varphi_\alpha(\beta)) \rangle$ of ordinals up to Γ_0. In summary, the paper F 1968b brought together, by the introduction of the two new notions above, several significant features of systems of natural representation: completeness, effectiveness, categoricity, and iteration of the critical process, in such a way as to insure preservation of the first three of these under that iteration.

If $\vec{\varphi}$ is relatively categorical (not necessarily effectively), then it induces a functor F on the category of sub-classes A of $In(\vec{\varphi})$ by $F(A) = (Cl_{\vec{\varphi}}(A), \leqslant, \vec{\varphi})$ (F 1968b, p. 202). The use of notions from category theory to treat certain aspects of ordinal functions met in proof theory had already been initiated by Aczel in his Oxford thesis work, briefly reported in Aczel 1967. In F 1972 I concentrated on the algebraic and logical aspects of the functors defined above. These functors preserve inclusions and direct limits (while Aczel's normal functors preserve initial segments and direct limits). In F 1972 I defined a more general notion of *κ-local functor* (κ a cardinal), for a rather general class of concrete categories; the ω-local functors are just those preserving $\subseteq$ and direct limits. The main result of F 1972 was that κ-local functors preserve the relations of elementary equivalence and elementary substructure in the language $\mathscr{L}_{\infty,\kappa}$. This showed in particular the strong logical consequences of dealing with relatively categorical systems $\vec{\varphi}$ of ordinal functions[13]. Further work on ordinal functors, due to Girard, will be discussed below.

After Γ_0, the next ordinal that stood out in proof-theoretical work was that found by Howard as a bound for $|\mathbf{ID}_1|$, where $\mathbf{ID}_1$ is the theory of one inductive definition, or more precisely of inductively defined subsets of ω given by arithmetical closure conditions (the set $\mathcal{O}$ of recursive ordinal notations being a typical example of such). The bound for $|\mathbf{ID}_1|$ was expressed in terms of a symbolism developed by Bachmann 1950 to go beyond the ordinals obtainable by Veblen's methods; Bachmann uses hierarchies of functions φ_α indexed by uncountable ordinals α in an initial

---

[12] The relationship of effective categoricity with existence of recursive isomorphism of suitable structures, studied earlier by Crossley and Parikh, is explained in F 1968b p. 204. Kreisel 1968 also emphasized this aspect of natural systems (cf. pp. 335–336).

[13] Using this, I recaptured, as corollaries, earlier results of Ehrenfeucht and Chang on elementary equivalence of certain ordinal systems with + and with +, ·.

segment of the ordinals up to Ω_2 (the least ordinal of cardinality $\aleph_2$). Howard's bound for $|\mathbf{ID}_1|$ expressed in this symbolism is $\varphi_{\epsilon_{\Omega+1}}(0)$, where the index $\alpha = \epsilon_{\Omega+1}$ is the first ε-number beyond Ω $(=\Omega_1)$. (Howard did not publish his proof until Howard 1972; that the bound is best possible was established in 1967 by Gerber.)

The essentially new feature of Bachmann's extension of the Veblen hierarchy was to first build a hierarchy of normal functions on Ω_2, then use these to define a system of representation up to certain portions of Ω_2 (such as $\varepsilon_{\Omega+1}$) in which he could associate particular fundamental sequences of type $\leqslant \Omega$ with limit ordinals. For a limit ordinal α of cofinality type $< \Omega$, with fundamental sequence $\langle \alpha_\xi \rangle_{\xi<\nu}$, $\nu < \Omega$, one then defines φ_α from the φ_{α_ξ} by the Veblen process of taking the enumeration of their common fixed points; but for α given as $\lim_{\xi<\Omega} \alpha_\xi$ one diagonalizes, taking $\varphi_\alpha(\xi) = \varphi_{\alpha_\xi}(0)$. Pfeiffer 1964 lifted this procedure to higher finite number classes by more and more complicated procedures; at each stage one steps down successively from portions of higher number classes Ω_{n+1} to Ω_n. Isles 1970 [14] was able to extend Bachmann's definitions to functions defined using portions of all accessible number classes (i.e., up to the first inaccessible cardinal).

The Bachmann–Pfeiffer–Isles definition procedures succeeded in providing recursive systems of representation for extraordinarily large ordinals, which turned out somewhat later to be needed in the proof-theory of finitely and transfinitely iterated inductive definitions. But to my mind, the lack of a clear conceptual basis for their approach and the complexity of details in its development was very unsatisfactory, and again called for a theoretical account. My first attempt at this was presented in F 1970 for the 1968 Buffalo IPT conference. Here I concentrated on repleteness as an essential characteristic. Using pairing, I reduced systems of functions to single functions $\varphi : \Omega \rightarrow \Omega$. Then one defines $Cr(\varphi)$ to be the pair $\langle \varphi, \varphi' \rangle$ (i.e., $\lambda\alpha\langle\varphi(\alpha), \varphi'(\alpha)\rangle$), where φ' enumerates $\mathrm{In}(\varphi)$ as before. Cr now appears as a functional of type 2 in the hierarchy of functionals of finite type over Ω, given by $\Omega^{(1)} = \Omega$, $\Omega^{(n+1)} = \{\varphi \mid \varphi : \Omega^{(n)} \rightarrow \Omega^{(n)}\}$. Then one defines $\mathrm{Rp}^{(n)} \subseteq \Omega^{(n)}$ as the *hereditarily replete functionals* of type n : $\mathrm{Rp}^{(1)}$ consists of the replete functions and $\mathrm{Rp}^{(n+1)} = \{\varphi \in \Omega^{(n+1)} \mid \varphi : \mathrm{Rp}^{(n)} \rightarrow \mathrm{Rp}^{(n)}\}$. In each type $n \geqslant 3$ there is a natural iteration functional $\mathrm{It}^{(n)} \in \Omega^{(n)}$, for which the first main result is that $\mathrm{It}^{(n)} \in \mathrm{Rp}^{(n)}$. Furthermore, $Cr \in \mathrm{Rp}^{(2)}$ and hence the application $\mathrm{It}^{(3)} Cr \in \mathrm{Rp}^{(2)}$, so that for any $\varphi \in \mathrm{Rp}^{(1)}$, also $\mathrm{It}^{(3)} Cr\varphi \in \mathrm{Rp}^{(1)}$ (application associated to the left); the first result of F 1968b is recaptured in this way. Using a natural covering relation $\trianglelefteq$, I conjectured that one could bound the functions in the Bachmann hierarchy starting with $\varphi_0 = \lambda\alpha(1+\alpha)$, by

[14] Presented at the important conference on Intuitionism and Proof Theory held in Buffalo in 1968, whose proceedings are in Myhill, Kino, Vesley 1970 (IPT below).

$\varphi_\Omega \trianglelefteq \mathrm{It}^{(3)} Cr\varphi_0$, $\varphi_{\Omega\Omega} \trianglelefteq \mathrm{It}^{(4)}\mathrm{It}^{(3)} Cr\varphi_0, \ldots$, and in general

$$\underbrace{\varphi_{\cdot_{\cdot_{\cdot_{\Omega}}}}\Omega}_{n} \trianglelefteq \mathrm{It}^{(n+2)} \ldots \mathrm{It}^{(3)} Cr\varphi_0.$$

Take $K_0 = \{\lambda\alpha(1+\alpha), Cr, \mathrm{It}^{(3)}, \ldots, \mathrm{It}^{(n+1)}, \ldots\}$; this is contained in the class PR_Ω of primitive recursive functionals of finite type over Ω, and Howard's work showed that the ordinals generated in PR_Ω satisfy $\mathrm{PR}_\Omega \cap \Omega \leq \varphi_{\varepsilon_{\Omega+1}}(0)$. Combining these led me finally to conjecture that if one closes K_0 under function application the resulting class of ordinals $Cl_{K_0}(0) \cap \Omega$ is exactly $\varphi_{\varepsilon_{\Omega+1}}(0)$. This conjecture was finally established by R. Weyrauch in his dissertation 1975 (completed in 1972).

One conceptual problem I saw with the Bachmann–Pfeiffer–Isles definition procedures was the essential use of the set-theoretical universe, in the sense that the ordinals $\Omega_1, \Omega_2, \ldots, \Omega_\nu, \ldots$ must be assumed regular in order to know that each Ω_ν employed is closed under the limits which arise in the definitions. Only when one is through and is able to establish recursion relations for the ordering between terms for countable ordinals does one see that Ω_1 can be 'collapsed' to ω_1^{rec}. In contrast, the finite type structure over Ω can be interpreted directly in recursion-theoretic terms, using (generalized) recursion theory on ω_1^{rec}, as the hereditarily effective operations over ω_1^{rec}. Furthermore, PR_Ω and thence K_0 may be established as hereditarily effective in this sense, thus leading one to see that $|Cl_{K_0} \cap \Omega| < \omega_1^{\mathrm{rec}}$ without any special calculation work (cf. F 1970, p. 295). In this way, one succeeded in trading a set-theoretic interpretation for a recursion-theoretic one. The latter being closer to constructive requirements, it thus seemed to me plausible to replace the use of finite and transfinite number classes to generate notations for recursive ordinals by the use of transfinite types over Ω, with Ω interpreted directly as ω_1^{rec}. This would suggest looking for a generalization of $\Omega^{(n)}$ to $\Omega^{(\alpha)}$ for α an ordinal, and of the $\mathrm{It}^{(n)}$ to $\mathrm{It}^{(\alpha)} \in \Omega^{(\alpha)}$. Such a step was taken by Aczel 1972, and indeed a match-up was made with certain higher reaches of the Bachmann hierarchy, but only to $\varphi_{F_{\Omega+1}(0)}(0)$, where $\langle F_\alpha \rangle_{\alpha<\Omega_2}$ is a Veblen-hierarchy on Ω_2 [15]. Aczel concluded that the use of transfinite types could not account the ordinals generated by Bachmann's approach let alone by the Pfeiffer–Isles extensions. However, in his 1972 paper Aczel only allowed access to $\Omega^{(\alpha)}$ and $\mathrm{It}^{(\alpha)}$ if α was already generated; it seems to me that one should be able to climb higher by the use, somehow, of 'impure' types so as to deal particularly with $\lambda\alpha\mathrm{It}^{(\alpha)}$ as an object. Whether this can be done in a reasonable way is an open matter; my own preliminary ideas in this direction did not arrive at any definite proposal.

---

[15] Aczel takes F_Ω for what Bachmann defined as $F_{\Omega+1}$; I follow the latter notation here. To take care of certain problems with the details of Aczel 1972, this was reworked in Höwel 1976 with the same end results.

Given this impasse, I thought still to make conceptual improvements in the Bachmann (et al.) approach by eliminating various complications in those definitions while accepting uncritically the use of the higher number classes Ω_α. What struck me was that a Veblen-style hierarchy was defined separately over portions of each number class, but that each followed the same pattern; it seemed reasonable to look instead for a definition which gives rise to a single sequence of functions $\theta_\xi : \mathrm{ON} \to \mathrm{ON}$ such that restricted to each Ω, $(\theta_\mu \restriction \Omega_\nu) : \Omega_\nu \to \Omega_\nu$ acts like the Veblen hierarchy on that number class. The problem with this, met at the outset, is that one can't take θ_Ω as enumerating the inaccessibles under the closure of $\langle\theta_\xi\rangle_{\xi<\Omega}$, since that takes one immediately out of Ω. But that problem arises only if one is allowed to use *each* θ_ξ for $\xi < \Omega$. If instead we identify $\langle\theta_\xi\rangle_{\xi<\Omega}$ with the single function $\lambda\xi < \Omega \cdot \lambda\eta\theta_\xi(\eta)$, then closure of a set A will only allow access to those θ_ξ for which ξ arises in the closure process itself; this is a kind of autonomous closure operation. Now if one wishes to 'name' larger cardinals, too, that can be accomplished by putting Ω_ξ in for each ξ generated in the closure process. In other words, one defines $Cl_{\langle\theta_\xi\rangle_{\xi<\alpha}}(\gamma)$ as the least class C containing each ordinal $< \gamma$, which contains with each $\xi \in C$ also Ω_ξ, and with each $\xi < \nu$ in C and each $\eta \in C$ also $\theta_\xi(\eta)$. Now if $\gamma < \Omega_{\nu+1}$, the cardinality of $Cl_{\langle\theta_\xi\rangle_{\xi<\alpha}}(\gamma)$ is $\leqslant \aleph_\nu$, no matter what α is and what Ω_ξ's arise in this closure. Hence the inaccessibles of $\langle\theta_\xi\rangle_{\xi<\alpha}$ under autonomous closure (i.e., the least ordinals not caught) are of type $\Omega_{\nu+1}$ in each class $\Omega_{\nu+1}$. Thus if one takes $\theta_\alpha : \mathrm{ON} \to \mathrm{ON}$ as enumerating $\mathrm{In}(\langle\theta_\xi\rangle_{\xi<\alpha})$ defined by the autonomous closure process, this maintains the requisite property that $(\theta_\alpha \restriction \Omega_\nu) : \Omega_\nu \to \Omega_\nu$ for each ν, no matter what α. In contrast to the definition of Bachmann-style hierarchies, this definition does not require simultaneous assignment of fundamental sequences to limit ordinals since the definition of θ_α is uniform for all α and is independent of the cofinality type of α.

I arrived at the preceding in 1970 and explained it to Aczel and Weyhrauch, among others, and advanced the general hypothesis that one could account for the ordinals generated by the Bachmann-style hierarchies just as well by this means. This was verified to begin with by Weyhrauch in his (1972, but 1975) thesis and independently and more extensively by Aczel in unpublished notes. Using Aczel's work as a basis, Bridge, in her 1972 Oxford thesis (with Gandy), pushed this much further, showing that all ordinals obtainable under Isles' definition on the accessible number classes could equally well be obtained using these θ_α's given suitable basic functions from ordinals to cardinals [16]. Bridge extended this even further, to accomplish a match-up with Isles 1971, by the assumption of certain Mahlo operations. Finally, she initiated work on calculations needed to show that the countable ordinals thus generated are

---

[16] See also Bridge 1975.

recursive. The latter was carried out systematically and in full by Buchholz 1975, further demonstrating the manageability of the new approach [17]. Since then, considerable extensions and variations of this idea have been carried out by Buchholz, Jäger and Pohlers (working both individually and in collaboration with each other); see in particular Buchholz 198?, Jäger 1984, and the appendix to this volume by Pohlers.

As mentioned earlier, the ordinal notations generated by 'long' hierarchies of ordinal functions using higher number classes have been employed in the determination of the ordinals $|\mathbf{ID}_\nu|$ for theories $\mathbf{ID}_\nu$ of iterated inductive definitions. Part of the story concerning these results will be traced below; the connection is mentioned here only to return to the question of whether the higher number classes have an essential role to play in this respect. Since the paradigm example of an $\mathbf{ID}_\nu$ is the theory of $\langle \mathcal{O}_x \rangle_{|x| \leq \nu}$ of 'constructive' higher number classes, which constitute recursive analogues of the classical $\langle \Omega_\xi \rangle_{\xi \leq \nu}$, the connection may be considered neither surprising nor troubling. But there is an essential difference: the Ω_α are linearly well-founded, and regularity also insures that one can choose the least ordinal not captured in a given closure process. But the $\mathcal{O}_x$'s are only partially ordered (and well-founded) and there is no obvious definition of hierarchies θ_a, like the θ_α, directly on the $\mathcal{O}_x$'s with the property that $\theta_a \restriction \mathcal{O}_x : \mathcal{O}_x \rightarrow \mathcal{O}_x$ for each x. If one had such, then in particular the property $\theta_a \restriction \mathcal{O}_1 : \mathcal{O}_1 \rightarrow \mathcal{O}_1$, where $\mathcal{O}_1 = \mathcal{O} =$ recursive ordinal notations, would ensure that all countable ordinals generated by Bachmann-style procedures are automatically recursive. In F 1978, p. 97, I suggested a framework in which this question might be answered, but no results have yet been achieved by means of it [18].

No discussion of systems of ordinal notations (for proof theory) would nowadays be complete without giving attention to the work of Girard and his school. However, a complete discussion of that would take me far beyond the bounds of the present personal report, and I can only try to indicate some connections. Beginning in 1975, Girard has explored a variety of concepts of a functorial character based on the category of ordinals (or ordinal-like objects). These efforts arrived at one very well-developed form in his theory of *dilators* (Girard 1981). Dilators are functors F from the category ON of ordinals to itself which preserve direct limits and (the categorical notion of) pull-backs. It is the latter preservation property which is an essentially new feature, not occurring in previous functoral approaches (of Aczel and myself, mentioned above).

---

[17] See also the exposition in Schütte 1977, Ch. IX, §§ 24–36.

[18] It has been suggested that, instead, one should be able to interpret the long hierarchies as operating directly on the (Kripke-Platek) admissible number classes τ_α, where $\tau_1 = \omega_1^{\text{rec}}$. However, no theory of such classes currently available allows one to 'name' higher admissibles in the definition of a function and have a given admissible such as τ_1 closed under it.

Girard 1981, pp. 76–79, shows how dilators are connected with systems of unique ordinal denotation, for each $x \in \mathrm{ON}$, of ordinals $< F(x)$; it is uniqueness which essentially insures the pull-back property. Unlike the functors which are associated with relatively categorical systems $\vec{\varphi}$ of functions from F 1968b described above, there is an essentially syntactic underlying component in the construction of dilators from systems of denotation, namely in the 'configurations' used for representations (Girard 1981, p. 77). A further analysis of the differences and relationships between these two approaches may be rewarding. Girard's approach has been applied in Girard and Vauzeilles 1984 to give an elegant functorial (re-)construction of the Veblen hierarchy, extracting it from a finite hierarchy by taking direct limits (since every ordering is a direct limit of finite orderings.) This work has been carried on in Girard and Vauzeilles 1984a to obtain a functorial account of the Bachmann hierarchy (up to something like the Howard ordinal). This second paper is technically dense and unrewarding from a conceptual standpoint; it does not explain what leads one to Bachmann-style constructions but rather builds in the ideas from the latter so as to fit the categorical framework.

For any two categories, there is a category of functors between them, and thence a subcategory of all functors preserving direct limits and pullbacks. The members of these functor categories built up successively from ON are called *ptykes* by Girard. These are analogous to the functionals of finite type over Ω [19].

Girard's approach has already had an impressive variety of applications in proof theory, infinitary logic, theory of sub-recursive hierarchies, generalized recursion theory and descriptive set theory. But so far, as concerns the role of ordinals in proof theory, it has not yet demonstrated that the questions which have concerned us above will be answered in a simple and conclusive way by its means. Perhaps one was grasping at straws and it was too much to hope that the subject (category theory) which gave a simple, convincing answer to the question — What is a natural map? — would do the same for the question — What is a natural system of ordinal representation? [20]

Theories of finite type; reduction of axioms of choice to (iterated) comprehension axioms

The material in this and the next two sections was developed since 1967–68 by pursuing what appeared to be separate lines, moving alter-

[19] Indeed, in his recent Habilitationsschrift, Päppinghaus 1985 has recast Howard's result $\mathrm{PR}(\Omega) \cap \Omega = \varphi_{\varepsilon_{\Omega+1}}(0)$ in a form concerning a subclass of ptykes corresponding to $\mathrm{PR}(\Omega)$. Päppinghaus also has generalizations and extensions of this result.

[20] So far as I know, Girard has nowhere suggested that this was his hope.

nately from one to another. But these paths were more closely related than I thought, and eventually converged in the period 1977–1981; Part of the story is recounted in my Preface to Buchholz, Feferman, Pohlers and Sieg 1981 (B, F, P, S, below), especially concerning the material in the next section on iterated induction definitions. Other parts, especially concerning the material in this section, can be extracted from my Chapters 1 and 2 with Sieg from B, F, P, S 1981, though not so easily. The following pulls this material together and also extends it by a report on the progress made since 1981.

The results here concern subsystems of analysis formulated both in 2nd order terms and using arbitrary finite types. The 2nd order principles receiving main attention are Comprehension Axioms for special classes $\mathcal{F}$ of formulas,

$$(\mathcal{F}\text{-CA}) \qquad \exists X\,\forall x\,[x \in X \leftrightarrow \phi(x)], \quad \text{for } \phi \text{ in } \mathcal{F},$$

and Axioms of Choice, for such classes,

$$(\mathcal{F}\text{-AC}) \qquad \forall x\,\exists f\,\phi(x\ f) \to \exists f\,\forall x\,\phi(x, (f)_x), \quad \text{for } \phi \text{ in } \mathcal{F},$$

where $(f)_x = \lambda y \cdot f(\langle x, y\rangle)$. Also suggested by hierarchy theory are the principles

$$(\Delta_{\mathcal{F}}\text{-CA})$$
$$\forall x\,[\phi(x) \leftrightarrow \psi(x)] \to \exists X\,\forall x\,[x \in X \leftrightarrow \phi(x)], \quad \text{for} \quad \phi, \sim\psi \text{ in } \mathcal{F}.$$

There is a corresponding rule ($\Delta_{\mathcal{F}}$-CR): from $\forall x\,[\phi(x) \leftrightarrow \psi(x)]$ infer $\exists X\,\forall x\,[x \in X \leftrightarrow \phi(x)]$. For $\mathcal{F} = \Pi^1_n$, one writes Δ^1_n for $\Delta_{\mathcal{F}}$. In finite type theory it is natural to consider

$$(\mathcal{F}\text{-AC}_{\sigma,\tau}) \qquad \forall x^\sigma\,\exists y^\tau\,\phi(x, y) \to \exists f^{(\sigma\to\tau)}\,\forall x^\sigma\,\phi(x, f(x))$$

for various $\mathcal{F}$, and the special case above is then labelled ($\mathcal{F}$-AC_{01}). Very often, a system is named by its main comprehension or choice principles. However, a distinction is made between theories **S** in which induction on ω is applied to arbitrary formulas and those, denoted $\mathbf{S}\restriction$, in which it is given by a single second order statement $\forall X \cdot I\,(<, X)$[21].

Kreisel 1962 showed that the systems (Σ^1_1-AC) and (Δ^1_1-CA) have the same least ω-model, namely the class HYP of all hyperarithmetic functions. In F 1964 I showed the system (Δ^1_1-CR) is predicative, being equivalent to $\bigcup \mathbf{R}_\alpha[\alpha < \omega^\omega]$, i.e., ramified analysis up to (but not including) level ω^ω. I conjectured there that the system Σ^1_1-AC is impredicative. This conjecture was shown to be wrong by H. Friedman in his MIT dissertation of 1967, where he showed $\Sigma^1_1\text{-AC} \equiv \Delta^1_1\text{-CA} \equiv \bigcup \mathbf{R}_\alpha[\alpha < \varepsilon_0]$. In place of ramified analysis $\mathbf{R}_\alpha$ up to level α, for $\alpha = \omega \cdot \alpha$, Friedman took the equivalent theory $(\Pi^0_1\text{-CA})_\alpha$, which expresses that ($\Pi^0_1$-CA) can be iterated (e.g. in the

---

21 '$\mathbf{S}_0$' is used instead of '**S**' by Friedman and others for theories with restricted induction; that notation is definitely *not* followed here.

form of the jump hierarchy) α times (along a natural well-ordering of type α). Then $(\Pi^0_1\text{-CA})_{<\alpha}$ denotes $\bigcup_{\beta<\alpha} (\Pi^0_1\text{-CA})_\beta$, and so Friedman's theorem is expressed as: $\Sigma^1_1\text{-AC} \equiv \Delta^1_1\text{-CA} \equiv (\Pi^0_1\text{-CA})_{<\varepsilon_0}$. (The proof-theoretic ordinal of this theory is $\varphi_{\varepsilon_0}(0)$.) Both his result and method of proof were surprising, the latter involving ingenious uses of non-standard models. My earlier result for $(\Delta^1_1\text{-CR})$ could be recast as: $(\Delta^1_1\text{-CR}) \equiv (\Pi^0_1\text{-CA})_{<\omega^\omega}$, obtained by a straightforward modelling argument. At the IPT meeting, Friedman 1970 generalized his theorem to show that $(\Sigma^1_{n+1}\text{-AC}) \equiv (\Delta^1_{n+1}\text{-CA}) \equiv (\Pi^1_n\text{-CA})_{<\varepsilon_0}$, and that $(\Sigma^1_{n+1}\text{-AC})$ is conservative over $(\Pi^1_n\text{-CA})_{<\varepsilon_0}$ for Π^1_k-statements where $k = \min(n+2, 4)$ and $\Pi^1_0 := \Pi^0_1$. Also at IPT in F 1970a, I gave a straightforward generalization of the theorem for Δ^1_1-CR to $(\Delta^1_2\text{-CR}) \equiv (\Pi^1_1\text{-CA})_{<\omega^\omega}$. The systems $(\Pi^1_1\text{-CA})_\alpha$ and $(\Pi^1_1\text{-CA})_{<\alpha}$ were of particular interest because of the proof-theoretic equivalences which I established there with the theories $\mathbf{ID}_\alpha$, $\mathbf{ID}_{<\alpha}$, resp., of iterated inductive definitions (see next section).

Since Friedman's results, but not his methods, were proof-theoretic, one was naturally stimulated to try to find alternative proof-theoretic methods to establish them. The first to do so — for$(\Sigma^1_1\text{-AC}) \equiv (\Pi^0_1\text{-CA})_{<\varepsilon_0}$—was Tait 1968, working with certain cut-free infinitary propositional systems. Then the proof-theoretical basis for an extension of this—to show $(\Sigma^1_2\text{-AC}) \equiv (\Pi^1_1\text{-CA})_{<\varepsilon_0}$—was laid in Tait 1970 (for IPT), where an interpretation was given of $(\Sigma^1_2\text{-AC})$ in $\mathbf{PL}_{<\varepsilon_0}$, a propositional calculus with infinitely long conjunctions and disjunctions ranging over abstract constructive number classes $\mathcal{O}_\alpha$ for $\alpha < \varepsilon_0$. However, it was not obvious how to obtain Friedman's results for $(\Sigma^1_{n+1}\text{-AC})$ for $n \geqslant 2$ by a further extension of these means.

The first step in that direction was taken in F 1971, by a different method, namely an adaptation of Gödel's functional ('Dialectica') interpretation of arithmetic in Gödel 1958, which gave an interpretation of Heyting arithmetic **HA** in a quantifier free theory **PR** of primitive recursive functionals of finite type over ω. Using Gödel's (1933) earlier ('negative' or 'double-negation') translation of **PA** in **HA**, one has that if $\mathbf{PA} \vdash \forall x\, \exists y\, R(x, y)$ with R primitive recursive then $\mathbf{HA} \vdash \forall x \neg\neg \exists y\, R(x, y)$ and hence **PR** proves the (Dialectica) interpretation of $\forall x \neg\neg \exists y\, R(x, y)$. Now a crucial point is that it happens Gödel's functional interpretation of $\forall x \neg\neg \exists y\, R(x, y)$ for R quantifier-free is the same as that of $\forall x\, \exists y\, R(x, y)$, namely $\exists f\, \forall x\, R(x, f(x))$. One concludes finally that there is a term t in **PR** such that $\mathbf{PR} \vdash R(x, t(x))$. Tait 1965 used a normalization procedure for infinitely long terms of finite type, analogous to cut-elimination for formulas of finite rank in $\mathcal{L}_{\omega_1,\omega}$, to reduce each term t to one in normal form of length $< \varepsilon_0$. Combining all these parts, Tait was able to recapture Kreisel's 1952 result on the provably recursive functions of **PA**, as just those which are ordinal recursive of order $< \varepsilon_0$.

A final observation needed for the following is that the negative translation and functional interpretation work just as well to interpret a finite type theory $\mathbf{PA}^\omega$ into $\mathbf{HA}^\omega$ and thence into **PR**, in such a way that $(\text{QF-AC}_{\sigma,\tau})$ (all types σ, τ) is verified in the process. For the interpretation of $\forall x^\sigma \neg\neg\exists y^\tau R(x, y)$ again turns out to be $\exists f^{(\sigma\to\tau)} \forall x^\sigma R(x, f(x))$ and this implies the interpretation of $\neg\neg\exists f^{(\sigma\to\tau)} \forall x^\sigma R(x, f(x))$. By formalizing the normalization of terms of length $< \varepsilon_0$ in **PA**, one can finally conclude that $\mathbf{PA}^\omega + (\text{QF-AC})^\omega$ (i.e., all $(\text{QF-AC}_{\sigma,\tau})$) is a conservative extension of **PA** for Π^0_2 sentences.

The main new realization in F 1971 was that all of the foregoing could be relativized to suitable type 2 functionals F, whether or not they are constructive. In particular, functionals which correspond to Skolem functions can be used to reduce certain formulas to quantifier-free form and thence various instances of AC in higher types can be derived from $(\text{QF-AC})^\omega$. Applying the negative translation and Dialectica interpretation lands one finally in a theory $\mathbf{PR}(F)$ of functionals of finite type primitive recursive in F. By normalization—which is uniform in F—the terms of type 1 can be described in terms of an F-jump hierarchy $\{H^F_\alpha\}_{\alpha<\varepsilon_0}$. Then one obtains conservation for formulas equivalent to $\forall\exists$(QF) form over any theory in which the normalization of terms and reduction to $H^F_\alpha (\alpha < \varepsilon_0)$ can be formalized.

The simplest example of this is provided by $F = \mu =$ the unbounded minimum operator. The associated axiom is:

$$(\mu) \qquad \forall f \forall x [fx = 0 \to f(\mu f) = 0].$$

The axiom implies that every Π^0_∞-(arithmetical) formula is equivalent to a QF-formula. Note in particular that Σ^1_1-AC, which is equivalent to Π^0_∞-AC, is a consequence of (QF-AC_{01}) in $\mathbf{PA}^\omega + (\mu)$. The first main result of F 1971 is that $\mathbf{PA}^\omega + (\mu) + (\text{QF-AC})^\omega$ is a conservative extension of $(\Pi^0_1\text{-CA})_{<\varepsilon_0}$, since we can formalize the μ-jump (= ordinary jump) hierarchy up to α ($\alpha < \varepsilon_0$) in the latter theory. Moreover, the conservation holds for Π^1_2-formulas, since these are equivalent to $\forall\exists$(QF) formulas under (μ).

Next in F 1971 I introduced a Skolem functional μ_1 for Σ_1-formulas which, over (μ), reduces them to QF-form; μ_1 picks the left-most descending branch in a non-well-founded tree, and is a form of the hyperjump operator. The second main result of F 1971 is that $\mathbf{PA}^\omega + (\mu) + (\mu_1) + (\text{QF-AC})^\omega$ is a conservative extension of $(\Pi^1_1\text{-CA})_{<\varepsilon_0}$; here one uses formalization of the hyperjump hierarchy up to α ($\alpha < \varepsilon_0$) in the latter theory. In this case conservation holds for Π^1_3 formulas, since these are equivalent to $\forall\exists$(QF) formulas under (μ) and (μ_1).

As corollaries of these results I obtained Friedman's conservation theorems for $(\Sigma^1_{n+1}\text{-AC})$ over $(\Pi^1_n\text{-CA})_{<\varepsilon_0}$ for $n = 0, 1$. It seemed likely then that the method of F 1971 could also be carried over to derive

Friedman's 1970 results for $n \geqslant 2$, but I did not pursue the matter at the time, having satisfied myself that one had arrived at a general procedural pattern. Incidentally, all of the foregoing works equally to give stronger conservation results announced by Friedman covering the axiom of dependent choices for various classes $\mathscr{F}$ of formulas,

($\mathscr{F}$-DC)

$$\forall f \exists g\, \phi(f, g) \rightarrow \forall h\, \exists f[(f)_0 = h \wedge \forall x\, \phi((f)_x, (f)_{x+1})], \quad \text{for } \phi \text{ in } \mathscr{F}.$$

For, (Σ^1_1-DC) is contained in $\mathbf{PA}^\omega + (\mu) + (\text{QF-AC})^\omega$ and (Σ^1_2-DC) is contained in $\mathbf{PA}^\omega + (\mu) + (\mu_1) + (\text{QF-AC})^\omega$. The derivations are straightforward using iteration of choice functionals G with $\forall f\, \phi(f, G(f))$, and the induction scheme on ω. Thus we have (Σ^1_{n+1}-DC) conservative over $(\Pi^1_n\text{-CA})_{<\varepsilon_0}$ for Π^1_{n+2}-statements, for $n = 0, 1$.

The arguments in F 1971 were only given in brief outline form; a fuller exposition was presented in the Handbook article F 1977. In the meantime, Friedman had announced some related conservation results for systems with restricted induction. My functional approach worked equally well for these, giving for example:

(i) $(\hat{\mathbf{PA}}^\omega\!\restriction) + (\mu) + (\text{QF-AC})$ is conservative over $(\Pi^0_1\text{-CA})\!\restriction$, and

(ii) $(\hat{\mathbf{PA}}^\omega\!\restriction) + (\mu) + (\mu_1) + (\text{QF-AC})$ is conservative over $(\Pi^1_1\text{-CA})\!\restriction$,

in both cases for $\forall\exists$(QF) sentences; basically, the ordinal ε_0 is replaced by ω, and one uses $(\Pi^i_1\text{-CA})_{<\omega}\!\restriction\, = (\Pi^i_1\text{-CA})\!\restriction$. Here $\hat{\mathbf{PA}}^\omega$ embodies Kleene's primitive recursive functionals $\hat{\text{PR}}$ instead of Gödel's PR; the difference is that in $\hat{\text{PR}}$ one can iterate only at type 0. In consequence of this and restriction of induction, the derivation described above of (Σ^1_1-DC) cannot be carried over to this restricted setting. As corollaries though, we obtain Friedman's results:

(i)′ $(\Sigma^1_1\text{-AC})\!\restriction$ is conservative over $(\Pi^0_1\text{-CA})\!\restriction$ for Π^1_2-sentences, and

(ii)′ $(\Sigma^1_2\text{-AC})\!\restriction$ is conservative over $(\Pi^0_2\text{-CA})\!\restriction$ for Π^1_3-sentences.

Furthermore, it is well known that $(\Pi^0_1\text{-CA})\!\restriction$ is conservative over arithmetic, so (i)′ also gives conservation of $(\Sigma^1_1\text{-AC})\!\restriction$ over **PA**.

A substantial methodological simplification (both conceptually and technically) in all this was initiated by Sieg (1981). He succeeded in replacing the passage through the negative translation, functional interpretation and normalization of terms by purely logical arguments in Gentzen-type systems. At the basis is a straightforward elimination of (QF-AC) in cut-free derivations $\mathscr{D}$ of sequents Γ, (QF-AC) $\vdash \Delta$, where Γ is purely universal and Δ purely existential. This is accomplished by inversion arguments which expand on Gentzen's derivation of Herbrand's theorem. In our joint work Feferman and Sieg 1981 (Ch. 2 of B, F, P, S), this was combined with the quantifier reduction method of Skolem functionals to recapture all of the conservation results mentioned above (op. cit., pp. 98–131); we also went on to show how to treat as well

(Σ^1_{n+1}-AC) for $n \geqslant 2$ by the same methods[22]. For values of $n \geqslant 2$, the required functionals are first defined under the assumption of the axiom of constructibility, and later that statement is eliminated by Gödel's inner model argument. Shoenfield's absoluteness theorem then guarantees conservation for Π^1_4-sentences, but not for Π^1_n with $n > 4$; this explains the general class of statements conserved as being Π^1_k for $k = \min(n+2, 4)$.

A further application of these methods was made in Feferman and Jäger 1983, where we dealt with elimination of (Σ^1_{n+1}-DC) in the presence of the Bar-Rule BR. The main result is that (Σ^1_{n+1}-DC) + (BR) is conservative over Aut(Π^1_n-CA), where the (Π^1_n-CA) is iterated autonomously. A corollary is that (Σ^1_1-DC) + (BR) is conservative over $(\Pi^0_1\text{-CA})_{<\Gamma_0}$ (or predicative analysis), a result first obtained in F 1979 by use of the functional methods described above. With the material of the next section, a second corollary is that (Σ^1_2-DC) + (BR) is conservative over Aut(Π^1_1-CA), which is equivalent to Aut(**ID**); the proof-theoretic ordinal of the latter theory has been determined by Buchholz and Pohlers 1978 to be $\theta_{\Omega_{\Omega_1}}(0)$.

In an opposite direction, proof-theoretically speaking, these methods have been employed in Sieg 1985 to provide a unified approach to subsystems (or fragments) of **PA** yielding a variety of results due to Minc, Parsons, Paris–Kirby, and Friedman as special cases. The use of Skolem functional operators has thus given to proof-theory, both Gödel-style and Gentzen-style, a much greater range of direct applicability. It would be of interest to see if either of these two styles may thus be brought to yield results not readily obtainable by the other. In any case, it is likely that further interesting applications of these methods are still to come.

Iterated inductive definitions

This section will begin with a summary of the relevant portions of my Preface to B, F, P, S 1981.

The study of theories for single and iterated inductive definitions (i.d.'s) with 1st order closure conditions was initiated by Kreisel (1963). He wanted to see whether Spector's bar-recursive functionals (used in Spector's 1961 consistency proof of analysis by an extension of Gödel's interpretation) could be modelled in such systems. Kreisel's conclusion was negative, these theories being much weaker than full analysis; in the end he conjectured that they would not go beyond (Σ^1_2-AC) in strength.

[22] The method of Sieg 1981 was for systems with restricted induction and works to give the results for (Σ^1_{n+1}-AC)$\restriction$; this had to be combined with infinitary proof theory to give the results for (Σ^1_{n+1}-AC).

The conjecture was not precise and turns out to be correct in one sense, but not another (see below).

$\mathbf{ID}_1$ is based on axioms of the form:

I. $A(P, x) \to P(x)$,

II. $\forall x\,[A(B, x) \to B(x)] \to \forall x\,[P(x) \to B(x)]$,

where A is arithmetical in P, but P is only to have positive occurrences in A, and in II, B is any formula in the language. The positivity condition ensures monotonicity: $A(P, x) \wedge P \subseteq P' \to A(P', x)$ (of which more below). A special kind of $\mathbf{ID}_1$ is for A of the form $A_0(x) \wedge \forall y\,[A_1(y, x) \to P(y)]$, whose least A-closed solution P is the accessible (or well-founded) part of the relation A_1, hereditarily in A_0. Examples of this are $\mathbf{ID}_1(\mathbf{W}_1)$ and $\mathbf{ID}_1(\mathcal{O}_1)$, where W_1 is the set of codes for recursive well-founded trees and $\mathcal{O}_1$ is the set of Church-Kleene recursive ordinal notations.

$\mathbf{ID}_\nu$ is defined in a way similar to $\mathbf{ID}_1$ by a sequence of formulas $\langle A_\alpha \rangle_{\alpha \leq \nu}$, where at each stage, A_α can make use unrestrictedly of $\langle P_\beta \rangle_{\beta < \alpha}$ (the P_β being introduced at earlier stages), but where P_α itself must have only positive occurrences in A_α; $\mathbf{ID}_{<\nu} = \bigcup \mathbf{ID}_\alpha$ $(\alpha < \nu)$. By iterating accessibility A_α's, we may speak of accessibility $(\mathbf{ID}_\nu)$'s and $(\mathbf{ID}_{<\nu})$'s. Iterating the generation of the W_α's, where at each stage one uses trees recursive in $\langle W_\beta \rangle_{\beta < \alpha}$, gives rise to the special accessibility theories denoted $\mathbf{ID}_\nu(W)$ and $\mathbf{ID}_{<\nu}(W)$. Similarly, iterating the build-up of the $\mathcal{O}_\alpha$'s, where at each stage one closes under recursive suprema over lower number classes $\mathcal{O}_\beta$ $(\beta < \alpha)$, gives accessibility theories denoted $\mathbf{ID}_\nu(\mathcal{O})$ and $\mathbf{ID}_{<\nu}(\mathcal{O})$, resp. Note that in this second case, recursiveness is *not* relativized to $\langle \mathcal{O}_\beta \rangle_{\beta < \alpha}$, but is taken in its absolute (ordinary) sense [23]. The basic logic of these theories may be classical or intuitionistic, with restriction to the latter indicated by a superscript 'i'.

The accessibility i.d.'s of sets P enjoy a privileged position in our informal conception, since we have a direct picture of how such P are generated 'from below'. But in general, justification of the $\mathbf{ID}$ axioms requires appeal to impredicative definition 'from above' (as the least P closed under A), or by appeal to a classical theory of ordinals. The theories $\mathbf{ID}_\nu(\mathcal{O})$ are given by accessibility i.d.'s, and are especially perspicuous, because each ordinal notation encapsulates the entire history (or derivation) of how it came to be placed in the generated set. Finally, the most constructive theories to deal with in this respect are the intuitionistic theories of accessibility i.d.'s, particularly the $\mathbf{ID}_\nu^{(i)}(\mathcal{O})$ and $\mathbf{ID}_{<\nu}^{(i)}(\mathcal{O})$.

Various work on the proof-theory of subsystems of analysis aimed explicitly or implicitly at the reduction of such systems to certain theories

---

[23] By work of Richter (1965) on 'constructive' notations for ordinals, one obtains notations for the same ordinals whether or not recursiveness is treated absolutely or relatively. However, the proof-theoretical equivalence of the corresponding theories does not fall out of Richter's work.

of iterated i.d.'s; to be as constructive as possible, one would want finally to arrive at intuitionistic theories of accessibility i.d.'s. A further problem would be to show this reduction best possible; for that, one measure would be the calculation of exact proof-theoretic ordinals.

The first such results were obtained for (Π^1_1-CA) by Takeuti 1967 and then for (Δ^1_2-CA) by Takeuti and Yasugi 1973. These papers used (partial) cut-elimination arguments, assigning ordinals in special systems of notation called *ordinal diagrams*, which are given by accessibility i.d.'s. Constructivity of the proofs insured reduction of the classical theories to intuitionistic theories of accessibility i.d.'s. However, no sharp ordinal bounds were obtained in natural systems of notation. Moreover, the ideas behind these consistency results were somewhat opaque, and there was call for a much more perspicuous treatment.

One link to such a treatment was provided by my IPT paper F 1970a. It was shown there by straightforward model-theoretic arguments that $(\Pi^1_1\text{-CA})_{<\nu} \equiv \mathbf{ID}_{<\nu} \equiv \mathbf{ID}_{<\nu}(W)$ for ν of the form ω^γ, γ a limit number [24]. Hence by Friedman's results concerning Σ^1_2-AC (described in the preceding section) one could immediately conclude that $(\Sigma^1_2\text{-AC}) \equiv \mathbf{ID}_{<\varepsilon_0}(W)$. It was hopeful at that point to look for direct constructive consistency proofs of theories of the form $\mathbf{ID}_\nu(W)$ and for calculation of their ordinals. I expected that this could be accomplished in part through a direct reduction to corresponding intuitionistic theories. Indeed, it was already known through work of Kreisel and Troelstra and by Howard (1972) that $\mathbf{ID}_1 \equiv \mathbf{ID}^i_1(\mathcal{O})$ and $|\mathbf{ID}^i_1(\mathcal{O})| = \varphi_{\varepsilon_{\Omega+1}}(0)$. Unfortunately, even in this case the arguments were not as straightforward as one would have liked, and it was by no means obvious how to extend them to the $\mathbf{ID}_\nu$ for $\nu > 1$. In fact, in his thesis (1971), Zucker found a definite obstacle to the expected reduction of $\mathbf{ID}_\nu$ to $\mathbf{ID}^{(i)}_\nu$. (The negative translation is of no help here, since it turns the i.d. of W_1 or $\mathcal{O}_1$ into a non-accessibility i.d.).

This obstacle was eventually overcome by Sieg in his 1977 thesis, at least for the reduction of $\mathbf{ID}_{<\nu}$ to $\mathbf{ID}^i_{<\nu}(\mathcal{O})$ in the case of limit ν, thus establishing a further link in the reduction ($\leqslant$) chain:

$$(*) \qquad (\Sigma^1_2\text{-AC}) \leqslant (\Pi^1_1\text{-CA})_{<\varepsilon_0} \leqslant \mathbf{ID}_{<\varepsilon_0}(W) \leqslant \mathbf{ID}^{(i)}_{<\varepsilon_0}(\mathcal{O}).$$

Sieg's method made use of the line of attack initiated in Tait 1970 (cf. Sieg 1981a).

In the meantime, Pohlers had succeeded in calculating bounds for the ordinals of the $\mathbf{ID}_\nu$'s as $\theta_{\varepsilon_{\Omega_\nu+1}}(0)$, first for finite ν in 1975 and then for general ν in 1977. These were shown best possible by Buchholz and

[24] For the same ordinals ν, these theories are also shown equivalent to $(\Pi^1_1\text{-CA})_{<\nu}$ + (BI), where BI is the principle of bar induction (i.e., applicability of transfinite induction to well-founded relations). Basic to my arguments in F 1970a is the equivalence of a weak form of (Π^1_1-CA) + (BI) with $\mathbf{ID}_1$ and of full (Π^1_1-CA) + (BI) with $\mathbf{ID}_\omega$. On the other hand, Zucker 1971 found a theory: Weak ($\mathbf{ID}_\omega$), which is equivalent to (Π^1_1-CA) without (BI).

Pohlers (1978). Initially, Pohlers adapted Takeuti's proof-theoretical arguments, but he was able later to replace these by the more perspicuous method he calls *local predicativity*. An alternative approach, generalizing Howard's method, was obtained by Buchholz using what he calls the $\Omega_{\nu+1}$-rules. All these approaches make use of prima-facie uncountably infinitary systems in one form or another. For a full exposition and detailed references see the chapters Sieg 1981a, Buchholz 1981, 1981a and Pohlers 1981 in B, F, P, S; for an extensive survey, see the Appendix by Pohlers in this volume.

A further link to the chain (*) above was established with the constructive theory $\mathbf{T}_0$ of functions and classes introduced in F 1975 (discussed also in the next section). I showed in F 1979 that $\mathbf{T}_0$ is interpretable in $(\Sigma^1_2\text{-AC}) + (\text{BI})$, and conjectured that these two theories are of the same strength. $\mathbf{T}_0$ contains a general axiom (IG) for inductive generation, whose principle of minimality is a form of BI. When minimality in IG is restricted to properties which define classes in $\mathbf{T}_0$, the resulting theory $\mathbf{T}_0(\text{IG}\restriction)$ is interpretable in $(\Sigma^1_2\text{-AC})$ (without BI). On the other hand, it is easily seen that $\mathbf{ID}^i_{<\varepsilon_0}(\mathcal{O})$ is interpretable in $\mathbf{T}_0(\text{IG}\restriction)$; hence one obtains from (*) the proof-theoretical equivalence

$$(**) \qquad (\Sigma^1_2\text{-AC}) \equiv \mathbf{T}_0(\text{IG}\restriction).$$

Later work by Jäger and Pohlers (1982) succeeded in establishing the conjectured equivalence

$$(***) \qquad (\Sigma^1_2\text{-AC}) + (\text{BI}) \equiv \mathbf{T}_0.$$

Moreover, their work gave an ordinal bound for these theories in a 'higher' system of θ functions, and Jäger 1983 showed that bound to be best possible. $(\Sigma^1_2\text{-AC}) + (\text{BI})$ (equivalently $(\Sigma^1_2\text{-DC}) + (\text{BI})$ is the strongest system to date for which we have reduction to a constructive theory, as well as calculation of its proof-theoretic ordinal in a natural notation system. The method of proof of Jäger and Pohlers does not explicitly involve theories of iterated i.d.'s but rather (the indirectly related) theories of iterated admissibles formulated in the language of set theory.

Technically speaking, the success of all this work on constructive reductions of strong impredicative classical theories via theories of iterated i.d.'s is far greater than it was reasonable to hope for at the time of the IPT meeting and that I expressed in F 1970a. It has now been broken up into more or less independent steps, each of which is separately understandable and manageable by fairly systematic means. My colleagues in the pursuit of this program who have done the main body of that technical work—Buchholz, Jäger, Pohlers and Sieg—have also responded in significant ways to my interminable nagging for conceptual simplification and elimination of ad hoc devices. But in this respect I think there is further work yet to be done [25].

I want to conclude this section with mention of two results related to theories of iterated i.d.'s, the first concerning a considerable weakening of the axioms and the second concerning a formal (though, as it turns out, not actual) strengthening.

By an iterated *fixed-point* theory $\widehat{\mathbf{ID}}_n$ is meant one with language $\mathscr{L}(P_1, \ldots, P_n)$ with basic axioms of the form

$$\forall x\,[A_k(P_k, x) \leftrightarrow P_k(x)], \quad (1 \leq k \leq n),$$

where P_k is positive in A_k and only $P_1, \ldots, P_k$ occur in A_k. Let $\theta(0|\alpha) = \alpha$, $\theta(n+1|\alpha) = \theta_{\theta_n(\alpha)}(0)$. The main result of F 1982 is that $|\widehat{\mathbf{ID}}_n| \leq \theta(n|\varepsilon_0)$. This was applied to settle Hancock's conjecture concerning the strength of Martin-Löf's theory $\mathbf{ML}_n$ with $n-1$ universes, by interpreting $\mathbf{ML}_n$ in a suitable $\widehat{\mathbf{ID}}_n$. Since previously Jervell had shown $\alpha_n \leq |\mathbf{ML}_n|$, it follows that $|\widehat{\mathbf{ID}}_n| = \alpha_n$ for $\widehat{\mathbf{ID}}_n$ based on suitable $A_1, \ldots, A_n$. These results also give the bounds $|\widehat{\mathbf{ID}}_{<\omega}| \leq \Gamma_0$ and $|\mathbf{ML}_{<\omega}| = \Gamma_0$. The method of proof of the main result is to interpret $\widehat{\mathbf{ID}}_{n+1}$ in (Σ^1_1-AC) over $\widehat{\mathbf{ID}}_n$ and successively eliminate (Σ^1_1-AC) by the methods of Feferman and Sieg 1981.

The paper F 1982a is concerned with iterated monotone i.d.'s, $\mathbf{ID}^{\mathrm{mon}}_n$ and $\mathbf{ID}^{\mathrm{mon}}_{<\omega}$. Here at each stage n we use axioms of the same form as for $\mathbf{ID}_n$, with $A_n(P_n, x)$ in $\mathscr{L}(P_1, \ldots, P_n)$ and with the positivity condition on P_n replaced by the requirement that $A_n(P, x)$ is to be provably monotone in P, i.e.., that $\mathbf{ID}_{n-1}$ proves $\forall x\,[A_n(P, x) \wedge P \subseteq P' \rightarrow A_n(P', x)]$. The main result of F 1982a is that each $\mathbf{ID}^{\mathrm{mon}}_n$ is a conservative extension of $\mathbf{ID}_n(W)$ and $\mathbf{ID}^{\mathrm{mon}}_{<\omega}$ is a conservative extension of $\mathbf{ID}_{<\omega}(W)$. The method of proof is by a straightforward modelling argument which is a variant of that given in F 1970a. The main result answers a question raised by Kreisel and is of interest because $\mathbf{ID}^{\mathrm{mon}}_n$ is at first justified only by definition from above. However since $\mathbf{ID}_n(W)$ is reducible to $\mathbf{ID}^{(i)}_n(\mathcal{O})$, this shows its justification can be reduced to that for constructive inductive definition from below.

A quite general formulation in a constructive setting of inductive definitions determined by monotonic operators can be given in the language of $\mathbf{T}_0$, as: every function f from classes to classes which satisfies $\forall X\,\forall Y[X \subseteq Y \rightarrow f(X) \subseteq f(Y)]$ has a least fixed point. At the end of F 1982a I raised the question of the strength of this principle when adjoined to $\mathbf{T}_0$ (or to one of its subtheories of interest). This question

[25] One should certainly look at alternative approaches, such as that of Girard 198?. As presently formulated, though, it is very difficult to establish the relationship of Girard's work with that described here.

has been investigated by my student S. Takahashi in his 1986 dissertation with partial but interesting results.

Theories for mathematical practice

Predicativity provided a clear case where a formal system **S** might be singled out for special attention as axiomatizing a foundationally significant part of mathematics, but itself be unsuitable for representing the practice implicitly resting on that foundation. To study the latter, one would want to find another system **S*** in which that part of practice could be naturally and directly represented and for which **S*** would be reducible to **S**, either by an interpretation or by proof-theoretic means. In the case of predicativity (in the form dealt with in F 1964) **S** is the system of ramified analysis up to Γ_0, $\mathbf{R}_{<\Gamma_0}$. Any form of ramified analysis, as Russell discovered to his great discomfort, is fairly hopeless for the formalization of even the most basic parts of classical analysis. One could make the move of Weyl 1918 to stick to the first system in the ramified sequence, namely arithmetical analysis (i.e., (Π^0_∞-CA)), in our present notation. Weyl showed that an impressive amount of 19th century analysis could be developed there without any appearance of ramification. The crucial observation is that every existence proof can be given by a construction which is explicitly defined (in terms of parameters) by purely arithmetical means.

It was natural to suppose (at the time of F 1964) that more analysis could be accounted for by stronger predicative means than provided by (Π^0_∞-CA). In particular, the aim would be to have a theory in which the positive theory of Lebesgue measure could be carried out[26]. Earlier work by Kreisel and Lorenzen showed this to have a predicative basis, at least in the informal conception. Thus I began in F 1964 to look for an unramified theory **S***, reducible to $\mathbf{R}_{<\Gamma_0}$, in which such more substantial parts of analysis could be carried out. As described in the opening section above, the first step was to get rid of ramification, replacing $\mathbf{R}_{<\Gamma_0}$ by the autonomous progression $\mathbf{H}_{<\Gamma_0}$, and then replacing the latter by the single theory **IR**, at each step maintaining proof-theoretic equivalence: $\mathbf{R}_{<\Gamma_0} \equiv \mathbf{H}_{<\Gamma_0} \equiv \mathbf{IR}$.

One still did not have with **IR** a sufficiently rich language to deal with actual practice. By its being 2nd order, talk about arbitrary real numbers could be carried on there, but not about arbitrary functions of real numbers, let alone more general function spaces used in modern analysis. To be sure, most applicable function spaces can be indexed by N^N, each function being prescribed by a countable amount of information. In this

[26] The adjective 'positive' here is used to indicate that portion of the theory which does not depend on the existence of non-measurable sets or functions.

way many special classes of function spaces (continuous, differentiable, measurable, etc.) can be dealt with in a 2nd order language. Still, a framework permitting the direct introduction of higher-order concepts and structures would be more natural and readier for the representation of practice. At first, I looked to set theory for a rich enough language in which to identify a predicative part suitable for a practical development. A system of predicative set theory **PS** was described in F 1966, with statement there of its basic properties. A full exposition for a closely related system $\mathbf{PS}_1$ was given in F 1974 (for the proceedings of the 1967 UCLA set theory conference). $\mathbf{PS}_1$ (or **PS**) is weaker than the theory **KP** of 'admissible set theory' due to Kripke and Platek, and is a conservative extension of **IR**. However, unlike ordinary set theory, $\mathbf{PS}_1$ lacks the ability to represent directly higher-order concepts, simply because it lacks the power-set (or function set) axiom. It *is* a suitable system in which to carry out a predicative development of non-analytic parts of mathematics, such as algebra. But for analysis, this move into set theory proved to be a dead-end.

The way through this impasse came with my work on systems of finite type theory in F 1971, described in the section before last. Once I realized that systems like $\mathbf{PA}^{\omega}+(\text{QF-AC})+(\mu)$ are predicatively reducible, it became obvious that here was a place where the higher-type concepts of analysis (functions, functionals, function spaces, etc.) could be directly represented, while at the same time be shown to rest on predicative foundations. Even more, by use of stronger closure conditions than (μ) one could provide a framework for the direct development of further portions of analysis geared to other more or less explicit—though impredicative—approaches, such as those of Borelian mathematics, projective descriptive set theory, etc. This direction of thought was spelled out in F 1977, where I outlined a program for the use of finite type theories for such purposes. Informally, one deals with universes of sets and functions satisfying various closure conditions (in the form of existence of specific functionals for selection and quantification) and shows what parts of mathematics can be carried out under progressively stronger conditions. On the formal side, one has a direct representation of the ⟨sets, functions, closure conditions⟩ framework in each case in a suitable theory $\mathbf{S}^*$ of finite type. Then one uses formal interpretations or proof-theoretic methods to reduce $\mathbf{S}^*$ to certain **S** whose foundational significance is more clear-cut; e.g. **S** might be (ramified) predicative, or a theory of (iterated) accessibility i.d.'s, etc. (In F 1977, the use of non-constructive functional interpretations coupled with the normalization of terms was stressed for such reductions, but this was later replaced by the methods of Feferman and Sieg 1981, as described in the section before last.)

In the period I was bringing F 1977 to completion, I became aware of work that Friedman was doing with systems $\mathbf{S}\upharpoonright$ in which induction on ω is restricted, i.e., the scheme of induction is replaced by the axiom: any set containing 0 and closed under successor contains all numbers [27]. In an abstract, Friedman announced results about the proof-theoretic strength of various restricted systems, including: $(\Sigma^1_1\text{-AC})\restriction \equiv (\Pi^0_1\text{-CA})\restriction \equiv \mathbf{PA}$ (a result independently found by Barwise and Schlipf), and $(\Sigma^1_2\text{-AC})\restriction \equiv (\Pi^1_1\text{-CA})\restriction$. As explained above, I was able to strengthen these by the methods of F 1977 to obtain $(\mathbf{PA}^\omega)\restriction + (\text{QF-AC}) + (\mu) \equiv \mathbf{PA}$ and $(\mathbf{PA}^\omega)\restriction + (\text{QF-AC}) + (\mu) + (\mu_1) \equiv (\Pi^1_1\text{-}CA)\restriction$; the first of these systems includes $(\Sigma^1_1\text{-AC})\restriction$ and the second $(\Sigma^1_2\text{-AC})\restriction$. (These results, and stronger ones for the systems $(\Sigma^1_{n+1}\text{-AC})$ were later recaptured equally well by the methods of Feferman and Sieg 1981 described above.)

The interest of such results for systems with restricted induction depends on the extent to which actual mathematics can be represented in them. What was surprising from Friedman's initiative was how little difference this restriction makes. Thus, for example, all the analysis that was checked to be carried out in $(\mathbf{PA}^\omega) + (\text{QF-AC}) + (\mu)$ (of strength $(\Pi^0_1\text{-CA})_{<\varepsilon_0}$) can already be carried out in the restricted form of that system (of strength **PA**). In fact, returning to Weyl 1918, one could just as well view his work as being carried on in $(\Pi^0_\infty\text{-CA})\restriction$. Further evidence for the mathematical strength of systems reducible to **PA** was offered by Takeuti 1978 (using a certain form of $(\Pi^0_\infty\text{-CA})\restriction$ extended to finite types) and Friedman 1980 (still using a 2nd order system). Indeed, I was able to verify that positive Lebesgue measure theory and substantial portions of functional analysis can be carried out in a theory conservative over **PA**. Thus, after all, the potential extra mathematical power offered by the full scope of predicativity is hardly used at all in those portions of mathematics which can be shown to rest on predicative foundations. I say "hardly", because one can still come up with isolated statements which do require more than **PA**. The simplest example of such is the infinite form of Ramsey's theorem, which is known to be provable in $(\Pi^0_\infty\text{-CA})$ but not in $(\Pi^0_\infty\text{-CA})\restriction$ (because it implies the Paris–Harrington statement, which is independent of **PA**). It would be of interest to try to see why predicativity in practice is so close to $(\Pi^0_\infty\text{-CA})$, and whether the greater theoretical power that it offers can actually be employed to advantage in practice.

The theories of finite type utilized in F 1977 were still not flexible enough to deal directly with all aspects of practice. In part this is due to the fact that the type structure is syntactically fixed. In order to deal with arbitrary topological or algebraic structures of various kinds whose domains are to be suitable types, or classes, one should have variables $X, Y, \ldots$ ranging over arbitrary types; these are to be closed under various type constructions such as $X \times Y$ (Cartesian product), Y^X (or, $X \to Y$, the class of all functions from X to Y), and $\{x \in X \mid \phi(x)\}$ (separation classes). In the years following F 1977, I worked up several

[27] For classical systems there are simple equivalents of this axiom formulated in terms of functions, but these are weaker in the constructive systems discussed below.

theories of variable types to meet this purpose. One such theory **VT** was presented in F 1985a, for the 5th Latin American Logic Symposium (Bogotá, 1981). **VT** is intermediate between ordinary typed and untyped theories. More recently, this was improved to a system **VFT** presented in my talk for the meeting of the Assoc. for Symbolic Logic held at Stanford in July, 1985; this most recent work is not yet prepared for publication.

In recent years there has been much work on the topic 'reverse mathematics' initiated by Friedman, which identifies the exact logical strength of various mathematical statements. For a review of this work, see Simpson 1984 and his Appendix to this volume. The results obtained are superior to mine with respect to measuring the logical strength of mathematical statements, since usually I only give upper bounds. However, Friedman's program is mainly carried out in 2nd order systems, and my theories of finite types, particular those of variable type **VT** and **VFT**, are much superior for the direct representation of mathematical practice. It would be surprising if the program of reverse mathematics only worked in a 2nd order setting.

One new system introduced by Friedman should be singled out for special attention; it is denoted by him as $\mathbf{ATR}_0$ and here as $\mathbf{ATR}\restriction$. Its main principle (**ATR**) is that of Arithmetic Transfinite Recursion, which states that given any well-ordering $<$ and arithmetical functional Φ, there is a hierarchy (starting with any X_0) on $<$ for which $X_a = \Phi(\langle X_b \rangle_{b<a})$ for each notation $a \neq 0$ in $<$. (**ATR**) thus takes the second rule of **IR** and makes it into an implication. Friedman showed that $|\mathbf{ATR}\restriction| = \Gamma_0$; his proof is found in Friedman, MacAloon and Simpson 1982, using non-standard model theoretic and recursion-theoretic methods. Jäger 1984a gives a direct proof of this result by embedding $\mathbf{ATR}\restriction$ in a theory $\mathbf{KPi}\restriction$ of sets which shares common axioms with **KP** but is in certain respects stronger and other respects weaker. Jäger then uses proof-theoretic methods to show $|\mathbf{KPi}\restriction| = \Gamma_0$. The result for $\mathbf{ATR}\restriction$ had previously resisted more standard proof-theoretic methods and had stood as a test-case for our subject. Jäger's success in applying such methods to set-theoretical systems is not an isolated one; in his Habilitationschrift 1984b, he uses admissible set theory to provide a unifying framework to obtain as corollaries many old and new results on 2nd order systems, of which the one on $\mathbf{ATR}\restriction$ is just one special case. My only reservation about this approach is that, just as with **PS** discussed earlier, $\mathbf{KPi}\restriction$ does not directly admit higher-type notions; still it must be granted that many interesting prima-facie impredicative mathematical facts can be established within it, and in fact already in $\mathbf{ATR}\restriction$. The predicativity-in-principle of $\mathbf{ATR}\restriction$ definitely depends on the restriction on induction, since $|\mathbf{ATR}| = \Gamma_{\varepsilon_0}$, as also shown by Jäger.

While working on systems for representing classical mathematical practice, I became interested (in the mid 1970's) in corresponding questions concerning Bishop-style constructive mathematics. In his 1967,

Bishop had succeeded in giving a constructive redevelopment of analysis in an informal way which could equally well be read in classical terms. In trying to understand from a logical point of view how this was managed, I was led to introduce a new system $\mathbf{T}_0$ in F 1975 with some novel conceptual features. I emphasized there the essentially differing roles of functions (informally, rule-governed operations) and classes (informally, classifications given by defining properties). However, functions can operate constructively on classes to give new classes. E.g. if f is any function defined on a class A such that for each $x \in A$, $f(x) = B_x$ is a class, then $\sum_{x\in A} B_x$ and $\prod_{x\in A} B_x$ are classes; these are given respectively by general join and product operations $j(A, f)$ and $p(A, f)$. Similarly, the accessible part of a class R (of pairs) hereditarily in A is the result of an operation $i(A, R)$. Explicit class constructions are also given for each suitable property $\phi(x, y_1, \ldots, y_n, A_1, \ldots, A_m)$ by an operation

$$c_\phi(y_1, \ldots, y_n, A_1, \ldots, A_m) = \{x \mid \phi(x, y_1, \ldots, y_n, A_1, \ldots, A_m\};$$

it then turns out that the product p is definable in terms of j and suitable c_ϕ. Operations themselves satisfy constructive closure conditions given by combinatory function axioms. The weakest subsystem of $\mathbf{T}_0$ considered is $(\mathbf{EM}_0)\restriction$, where $(\mathbf{EM}_0)$ provides for the function axioms and the class N of natural numbers. It is easily shown that $(\mathbf{EM}_0)\restriction$ with classical logic is conservative over **PA**; for the intuitionistic system, $(\mathbf{EM}_0)\restriction^{(i)}$ is conservative over **HA**, as shown by Beeson. This and other proof-theoretic results for subsystems of $\mathbf{T}_0$ are dealt with in F 1979, Feferman and Sieg 1981 and Beeson 1985. For example, $(\mathbf{EM}_0) + (\mathrm{J}) \equiv (\Sigma^1_1\text{-AC})$, where J is the join axiom and $(\mathbf{EM}_0) + (\mathrm{J}) + (\mathrm{IG}\restriction) \equiv (\Sigma^1_2\text{-AC})$, where IG is the inductive generation axiom and $(\mathrm{IG})\restriction$ is its restricted version. Finally, by the Jäger–Pohlers results mentioned in the previous section, $\mathbf{T}_0 = (\mathbf{EM}_0) + (\mathrm{J}) + (\mathrm{IG}) \equiv (\Sigma^1_2\text{-AC}) + (\mathrm{BI})$[28].

Again it is surprising how much mathematics can already be carried out in $(\mathbf{EM}_0)\restriction$; practically all of Bishop 1967 can be formalized directly there. Beyond this, the axioms for join and inductive generation provide for notions of transfinite type theory and constructive ordinal number theory, not used in ordinary constructive analysis.

In F 1978 and F 1979b I have shown that both constructive and non-constructive extensions of $\mathbf{T}_0$ are useful for encompassing various parts of mathematics not dealt with in $\mathbf{T}_0$ (even with classical logic); for example a system $\mathbf{T}_1^{(\Omega)}$ in the latter paper is used to generalize ordinary model theory and certain analogue theories on admissible sets.

Not to leave any stone unturned, in F 1982b I have presented two theories $\mathbf{FM}_0$ and **FM** for the direct formalization of metamathematics,

[28] Beeson 1985 gives an excellent presentation of work on the metatheory of various constructive theories in the 70's and compares different approaches due to Myhill, Friedman, Martin-Löf, Aczel and myself.

using finite inductive generation of classes as the basic principle; these theories are of strength **PRA** and **PA** respectively. Gödel's 2nd incompleteness theorem is generalized to extensions of $\mathbf{FM}_0$ there, while elementary model theory can be carried on in $\mathbf{FM}_1$. The idea of canonical presentations of axiomatic systems can also be explained in terms of the symbolism of $\mathbf{FM}_0$.

In my current work with **VFT** mentioned above I have found a common base from which to proceed in both classical and constructive directions. I intend to use it (or something close to it) as the point of departure for a long-promised book (already half-written in one form) on the metamathematics of formal theories related to mathematical practice.

The role of proof theory in foundational work

It is much harder to give good reasons for why one pursues a given line of work in logic (or for that matter, many other endeavors) than to do the work itself. For those in the swim of a subject, little justification is necessary; one result (idea, method, . . .) leads to another, and only the simplest of over-all motivations is necessary. The efficacy for proof theory of the Hilbert–Gentzen program of 'securing' mathematics by consistency proofs is undeniable. This has led, and continues to lead, to an enormous body of results and methods, especially concerning informative ordinal analysis of non-constructive theories and their reduction to constructive theories. But critical examination of this program leaves one feeling uncomfortable about just what is accomplished. In particular, is one's conviction about the consistency of **PA** or Σ^1_1-AC, or Π^1_1-CA, or Σ^1_2-AC increased by the proof-theoretical results for these systems? It would be stretching things to say that the consistency of such systems was an open question which was finally settled to one's satisfaction by the 'consistency proofs'. (Compare the questions: "Is (Σ^1_2-AC) consistent?" with "Is NF consistent?" [29].)

Hilbert's program is one of the large global foundational schemes, each of which has been riddled with criticism and is more suspect than the mathematics which it is supposed to assure. We have grown tired of them, but logicians have come up with no new schemes to take their place. A natural reaction to the malaise about the grand logical schemes was to look to mathematics itself for a cure (or philosophical pain-killer). For example, some mathematicians espouse category theory as a new, purely mathematical foundational scheme. But what has been done is to take relatively sophisticated structural notions (category, functor) defined

[29] See my comments along these lines in my reviews of Takeuti 1975 and Schütte 1977 in F 1977a and F 1979c, resp. Of course, such criticisms are by no means unique, but there is no agreement on what to offer as an alternative.

(informally) in terms of ordinary unstructured mathematical notions (class, function), and attempted to kick away the traces—'explaining' the latter in terms of the former. Whatever the organizational power of category theory in different parts of mathematics, this attempt at foundations is an exercise in self-deception.

In F 1984a and F 1985 I have argued against overreactions to the enterprise of logical foundations, by looking at it locally, in every-day terms, rather than globally. To quote from F 1984a (p. 148): "On my view, this is a direct continuation of work that mathematicians have carried on from the very beginning of our subject up to the present. The distinctive role of logic lies in its more conscious, systematic approach and its different ways of slicing up the subject." Further: "If one analyzes foundational activity (whether carried on by mathematicians as a matter of course or more consciously by logicians), one finds that it falls into one of five or six characteristic modes"—which I call foundational ways. The principal ones I identified are indicated by:

(i) conceptual clarification,

(ii) dealing with problematic concepts or principles by interpretations or models,

(iii) dealing with problematic concepts or principles by replacement or elimination,

(iv) dealing with problematic methods and results,

(v) organizational foundations and axiomatization, and

(vi) reflective expansion.

(The reader must look to F 1984a and F 1985 for detailed illustrations of each of these.)

The role of proof theory in foundational work lies, then, primarily as a tool in (iii). Roughly speaking, one eliminates problematic aspects of a system **S*** by reduction to more clearly understood **S**. The trouble with this aim is an apparent lack of objectivity—what is "problematic" varies from person to person (and from time to time). Obviously most mathematicians currently see nothing problematic about impredicative analysis or even set theory. What should be emphasized is that the problematic aspect lies not in a worry about possible inconsistency (though that is not to be neglected in some of the farther-out systems), but rather in not being able to explain clearly just what we are talking about. And here mathematicians' common experience is reasonably in agreement: we understand the concept of natural number much better than that of arbitrary set of natural numbers, or of arbitary function between natural numbers. We understand sets generated by specific constructions "from below" better than sets defined by intersections "from above", and so on. Our mathematical thought thus falls into bodies of mathematical understanding which are represented (no doubt only to a first approximation) by formal systems. Proof theory then provides a delicate language and tools to uncover both expected and unexpected relations between these bodies of thought. In the end, it endeavors to tell us what rests on what—if not directly then at least in principle. And *that* is where we ought now to rest.

References

Abbreviations

JSL: Journal of Symbolic Logic
LNM: Lecture Notes in Mathematics
B, F, P, S: Buchholz, Feferman, Pohlers and Sieg
IPT: Intuitionism and Proof Theory (Myhill, Kino, Vesley, 1970).

P. Aczel
1967 : Normal functors on linear orderings, JSL 32, 430.
1972 : Describing ordinals using functionals of transfinite type, JSL 37, 35–47.
1980 : Frege structures and the notions of proposition, truth and set, in: The Kleene Symposium, North-Holland, Amsterdam (1980) 31–59.

H. Bachmann
1950 : Die Normalfunktionen und das Problem des ausgezeichneten Folgen von Ordnungszahlen, Vierteljarschs. Nat. Ges. Zürich 95, 5–37.

J. Barwise
1969 : Infinitary logic and admissible sets, JSL 34, 226–252.
1977 : (ed.) Handbook of Mathematical Logic, North-Holland, Amsterdam (1978).

M. Beeson
1985 : Foundations of Constructive Mathematics. Metamathematical Studies, Springer, Berlin.

E. Bishop
1967 : Foundations of Constructive Analysis, McGraw-Hill, New York.

J. Bridge
1975 : A simplification of the Bachmann method for generating large countable ordinals, JSL 40, 171–185.

W. Buchholz
1975 : Normalfunctionen und konstruktive Systeme von Ordinalzahlen, LNM 500, 4–25.
1981 : The $\Omega_{\mu+1}$-rule, in: B, F, P, S (1981) 188–233.
1981a: Ordinal analysis of $\mathbf{ID}_\nu$, in B, F, P, S (1981) 234–260.
198? : A new system of proof-theoretic ordinals (to appear).

W. Buchholz, S. Feferman, W. Pohlers, W. Sieg
1981 : Iterated inductive definitions and subsystems of analysis: recent proof-theoretical studies, LNM 897. [B, F, P, S 1981].

W. Buchholz, W. Pohlers
1978 : Provable wellorderings of formal theories for transfinitely iterated inductive definitions, JSL 43, 118–125.

S. Feferman
1962 : Transfinite recursive progressions of axiomatic theories, JSL 27, 259–316.
1964 : Systems of predicative analysis, JSL 29, 1–30.
1966 : Predicative provability in set theory, Bull. A.M.S. 72, 486–489.

1967 : Autonomous transfinite progressions and the extent of predicative mathematics, in: J. J. Cohen, ed., Logic, Method., Phil. Sci. III, North-Holland, Amsterdam.
1968 : Lectures on proof theory, LNM 70, 1–107.
1968a: Persistent and invariant formulas for outer extensions, Compos. Math. 20, 29–52.
1968b: Systems of predicative analysis II. Representations of ordinals. JSL 33, 193–220.
1970 : Hereditarily replete functionals over the ordinals, in: IPT, 289–301.
1970a: Formal theories for transfinite iterations of generalized inductive definitions and some subsystems of analysis, in: IPT, 303–326.
1971 : Ordinals and functionals in proof theory, in: Proc. Int. Cong. Maths (Nice, 1970), 1, 229–233.
1972 : Infinitary properties, local functors and systems of ordinal functions, LNM 255, 63–97.
1974 : Applications of many-sorted interpolation theorems, in: Proc. Tarski Symposium (AMS Symp. Pure Math. XXV), 205–223.
1974a: Predicatively reducible systems of set theory, in: Axiomatic Set Theory, Part II (AMS Proc. Symp. Pure Math. XIII), 11–32.
1975 : A language and axioms for explicit mathematics, LNM 450, 87–139.
1977 : Theories of finite type related to mathematical practice, in: J. Barwise, ed., Handbook of Mathematical Logic, North-Holland, Amsterdam (1978) 913–971.
1977a: Review of Takeuti 1975, in Bull. A.M.S. 83, 351–361.
1978 : Recursion theory and set theory: a marriage of convenience, In: J.E. Fenstad et al., eds., Generalized Recursion Theory II, North-Holland, Amsterdam (1978) 55–98.
1979 : A more perspicuous formal system for predicativity, in: K. Lorenz, ed., Konstruktionen versus Positionen I, de Gruyter, Berlin 87–139.
1979a: Constructive theories of functions and classes, in: M. Boffa et al., eds., Logic Colloquium '78, North-Holland, Amsterdam (1979) 159–224.
1979b: Generalizing set-theoretical model theory and an analogue theory on admissible sets, in: J. Hintikka et al., eds., Essays on Mathematical and Philosophical Logic, Reidel, Dordrecht, 171–195.
1979c: Review of Schütte 1977, in Bull. A.M.S. (New Series) 1, 224–228.
1981 : Preface to B, F, P, S 1981.
1982 : Iterated inductive fixed-point theories, in: G. Metakides, ed., Patras Logic Symposion, 171–196.
1982a: Monotone inductive definitions, in: A.S. Troelstra et al., eds., The L.E.J. Brouwer Centenary Symposium, North-Holland, Amsterdam, 77–89.
1982b: Inductively presented systems and the formalization of metamathematics, in: D. Van Dalen et al., eds, Logic Colloquium '80, North-Holland, Amsterdam (1982) 95–128.
1984 : Toward useful type-free theories, JSL 49, 75–111 (also in Martin 1984, 237–287).
1984a: Foundational ways, in: Perspectives in Mathematics (Birkhäuser, Basel) 147–158.
1985 : Working foundations, Synthese 62, 229–254.
1985a: A theory of variable types, Revista Colombiana de Matemáticas 19, 95–105.
198? : Reflecting on incompleteness (retiring ASL Presidential Address, Dec. 1983, to appear).

S. Feferman and G. Jäger
1983 : Choice principles, the bar rule and autonomously iterated comprehension schemes in analysis, JSL 48, 63–70.

S. Feferman and G. Kreisel
1966 : Persistent and invariant formulas relative to theories of higher order, Bull. A.M.S. 72, 480–485.

S. Feferman and W. Sieg
1981 : Proof-theoretic equivalences between classical and constructive theories for analysis, in: B, F, P, S (1981) 78–142.

S. Feferman and C. Spector
1962 : Incompleteness along paths in progressions of theories, JSL 27, 383–390.

H. Friedman
1970 : Iterated inductive definitions and Σ^1_2-AC, in: IPT, 435–442.
1980 : A strong conservative extension of Peano arithmetic, in: J. Barwise et al., eds., The Kleene Symposium, 113–122.

H. Friedman, K. MacAloon and S. Simpson
1982 : A finite combinatorial principle which is equivalent to the l-consistency of predicative analysis, in: G. Metakides ed., Patras Logic Symposion, 197–230.

J.-Y. Girard
1981 : Π^1_2-logic, Part 1: Dilators, Ann. Math. Logic 21, 75–219.
198? : Proof-theoretic investigations of inductive definitions (Part 1), (E. Specker 60th birthday meeting).

J.-Y. Girard and J. Vauzeilles
1984 : Functors and ordinal notations, I: A functorial construction of the Veblen hierarchy, JSL 49, 713–729.
1984a: Functors and ordinal notations, II: A functorial construction of the Bachmann hierarchy, JSL 49, 1079–1114.

K. Gödel
1958 : Über eine bisher noch nicht benützte Erweiterung des finiten Standpunktes, Dialectica 12, 280–287.

W. A. Howard
1972 : A system of abstract constructive ordinals, JSL 37, 355–374.

K.-A. Höwel
1976 : Iteration von Funktionalen transfiniter Typen, Dissertation, Univ. Münster.

D. Isles
1970 : Regular ordinals and normal forms, in: IPT, 339–361.
1971 : Natural well-orderings, JSL 36, 288–300.

G. Jäger
1983 : A well-ordering proof for Feferman's theory $\mathbf{T}_0$, Archiv. f. Math. Logik u. Grundl. 23.
1984 : ρ-inaccessible ordinals, collapsing functions and a recursive notation system, Archiv. f. Math. Logik u. Grund. 24.
1984a: The strength of admissibility without foundation, JSL 49, 867–879.
1984b: Theories for admissible sets: a unifying approach to proof theory, Habilitationsschrift, Univ. München.

G. Jäger and W. Pohlers
1982 : Eine beweistheoretische Untersuchung von $(\Delta^1_2\text{-CA})+(\text{BI})$ und verwandter Systeme, Sitzungsber. Bayer. Akad. Wiss.

G. Kreisel
1958 : Ordinal logics and the characterization of informal concepts of proof, Proc. Int. Cong. Maths. (Edinburgh), 289–299.
1962 : The axiom of choice and the class of hyperarithmetic functions, Indag. Math. 24, 307–319.
1963 : Generalized inductive definitions, in: Stanford report on the foundations of analysis (mimeographed), Section III.
1968 : A survey of proof theory, JSL 33, 321–388.

R. L. Martin
1984 : (ed.) Recent Essays on Truth and the Liar Paradox, Oxford University Press, Oxford.

J. Myhill, A. Kino, R. E. Vesley
1970 : (eds.) Intuitionism and Proof Theory, North-Holland, Amsterdam.

B. F. Nebres
1972 : Infinitary formulas preserved under unions of models, JSL 37, 449–465.

P. Päppinghaus
1985 : Ptykes in Gödel's T und verallgemeinerte Rekursion über Mengen und Ordinalzahlen, Habilitationsschrift, Univ. Hannover.

H. Pfeiffer
1964 : Ausgezeichnete Folgen für gewisse Abschnitte der zweiten und weiteren Zahlklassen, Dissertation Tech. Hochschule Hannover.

W. Pohlers
1981 : Proof-theoretical analysis of $\mathbf{ID}_\nu$ by the method of local predicativity, in: B, F, P, S (1981) 261–357.

W. Richter
1965 : Extensions of the constructive ordinals, JSL 30, 193–211.

K. Schütte
1960 : Beweistheorie, Springer, Berlin.
1964 : Eine grenze für die Beweisbarkeit der transfiniten induktion in der verzweigten Typenlogik, Arch. f. Math. Logik u. Grundl. 7, 45–60.
1965 : Predicative well-orderings, in: J. Crossley et al., eds., Formal Systems and Recursive Functions, North-Holland, Amsterdam 280–303.
1977 : Proof Theory, Springer, Berlin.

W. Sieg
1981 : Conservation theorems for subsystems of analysis with restricted induction (abstract), JSL 46, 194–195.
1981a: Inductive definitions, constructive ordinals, and normal derivations, in: B, F, P, S (1981) 143–187.
1985 : Fragments of arithmetic, Ann. Pure and Applied Logic 28, 33–72.

S. G. Simpson
1984 : Reverse mathematics, in: Recursion Theory (Proc. Symposia Pure Maths., AMS 42) 461–472.

W. W. Tait
1965 : Infinitely long terms of transfinite type, in: J. Crossley et al., eds., Formal Systems and Recursive Functions, North-Holland, Amsterdam, 176–185.
1968 : Normal derivability in classical logic, in: LNM 72, 204–236.
1970 : Applications of the cut-elimination theorem to some subsystems of classical analysis, in: IPT, 475–488.

S. Takahashi
1986 : Monotone inductive definitions in $\mathbf{T}_0$, Dissertation, Stanford University.

G. Takeuti
1967 : Consistency proofs of some subsystems of analysis, Ann. Mathematics 86, 299–348.
1975 : Proof Theory, North-Holland, Amsterdam.
1978 : Two applications of logic to mathematics, Pubs. of the Math. Soc. of Japan (Iwanami Shoten and Princeton Univ. Press).

G. Takeuti and M. Yasugi

1973 : The ordinals of the systems of second order arithmetic with the provably Δ^1_2-comprehension axiom and with the Δ^1_2-comprehension axiom respectively, Japan. J. Math. 41, 1–67.

A. M. Turing

1939 : Systems of logic based on ordinals, Proc. London Math. Soc. Ser. 2, 45, 161–228.

O. Veblen

1908 : Continuous increasing functions of finite and transfinite ordinals, Trans. A.M.S. 9, 280–292.

R. Weyhrauch

1975 : Relations between some hierarchies of ordinal functions and functionals, Dissertation, Stanford University.

H. Weyl

1918 : Das Kontinuum, Veit, Leipzig.

J. I. Zucker

1971 : Proof-theoretic studies of systems of iterated inductive definitions and subsystems of analysis, Dissertation, Stanford University.

INDEX

A CATALOG OF SELECTED

DOVER BOOKS

IN SCIENCE AND MATHEMATICS

Mathematics–Probability and Statistics

BASIC PROBABILITY THEORY, Robert B. Ash. This text emphasizes the probabilistic way of thinking, rather than measure-theoretic concepts. Geared toward advanced undergraduates and graduate students, it features solutions to some of the problems. 1970 edition. 352pp. 5 3/8 x 8 1/2. 0-486-46628-0

PRINCIPLES OF STATISTICS, M. G. Bulmer. Concise description of classical statistics, from basic dice probabilities to modern regression analysis. Equal stress on theory and applications. Moderate difficulty; only basic calculus required. Includes problems with answers. 252pp. 5 5/8 x 8 1/4. 0-486-63760-3

OUTLINE OF BASIC STATISTICS: Dictionary and Formulas, John E. Freund and Frank J. Williams. Handy guide includes a 70-page outline of essential statistical formulas covering grouped and ungrouped data, finite populations, probability, and more, plus over 1,000 clear, concise definitions of statistical terms. 1966 edition. 208pp. 5 3/8 x 8 1/2. 0-486-47769-X

GOOD THINKING: The Foundations of Probability and Its Applications, Irving J. Good. This in-depth treatment of probability theory by a famous British statistician explores Keynesian principles and surveys such topics as Bayesian rationality, corroboration, hypothesis testing, and mathematical tools for induction and simplicity. 1983 edition. 352pp. 5 3/8 x 8 1/2. 0-486-47438-0

INTRODUCTION TO PROBABILITY THEORY WITH CONTEMPORARY APPLICATIONS, Lester L. Helms. Extensive discussions and clear examples, written in plain language, expose students to the rules and methods of probability. Exercises foster problem-solving skills, and all problems feature step-by-step solutions. 1997 edition. 368pp. 6 1/2 x 9 1/4. 0-486-47418-6

CHANCE, LUCK, AND STATISTICS, Horace C. Levinson. In simple, non-technical language, this volume explores the fundamentals governing chance and applies them to sports, government, and business. "Clear and lively ... remarkably accurate." – *Scientific Monthly*. 384pp. 5 3/8 x 8 1/2. 0-486-41997-5

FIFTY CHALLENGING PROBLEMS IN PROBABILITY WITH SOLUTIONS, Frederick Mosteller. Remarkable puzzlers, graded in difficulty, illustrate elementary and advanced aspects of probability. These problems were selected for originality, general interest, or because they demonstrate valuable techniques. Also includes detailed solutions. 88pp. 5 3/8 x 8 1/2. 0-486-65355-2

EXPERIMENTAL STATISTICS, Mary Gibbons Natrella. A handbook for those seeking engineering information and quantitative data for designing, developing, constructing, and testing equipment. Covers the planning of experiments, the analyzing of extreme-value data; and more. 1966 edition. Index. Includes 52 figures and 76 tables. 560pp. 8 3/8 x 11. 0-486-43937-2

STOCHASTIC MODELING: Analysis and Simulation, Barry L. Nelson. Coherent introduction to techniques also offers a guide to the mathematical, numerical, and simulation tools of systems analysis. Includes formulation of models, analysis, and interpretation of results. 1995 edition. 336pp. 6 1/8 x 9 1/4. 0-486-47770-3

INTRODUCTION TO BIOSTATISTICS: Second Edition, Robert R. Sokal and F. James Rohlf. Suitable for undergraduates with a minimal background in mathematics, this introduction ranges from descriptive statistics to fundamental distributions and the testing of hypotheses. Includes numerous worked-out problems and examples. 1987 edition. 384pp. 6 1/8 x 9 1/4. 0-486-46961-1

Browse over 9,000 books at www.doverpublications.com

Mathematics-Geometry and Topology

PROBLEMS AND SOLUTIONS IN EUCLIDEAN GEOMETRY, M. N. Aref and William Wernick. Based on classical principles, this book is intended for a second course in Euclidean geometry and can be used as a refresher. More than 200 problems include hints and solutions. 1968 edition. 272pp. 5 3/8 x 8 1/2. 0-486-47720-7

TOPOLOGY OF 3-MANIFOLDS AND RELATED TOPICS, Edited by M. K. Fort, Jr. With a New Introduction by Daniel Silver. Summaries and full reports from a 1961 conference discuss decompositions and subsets of 3-space; n-manifolds; knot theory; the Poincaré conjecture; and periodic maps and isotopies. Familiarity with algebraic topology required. 1962 edition. 272pp. 6 1/8 x 9 1/4. 0-486-47753-3

POINT SET TOPOLOGY, Steven A. Gaal. Suitable for a complete course in topology, this text also functions as a self-contained treatment for independent study. Additional enrichment materials make it equally valuable as a reference. 1964 edition. 336pp. 5 3/8 x 8 1/2. 0-486-47222-1

INVITATION TO GEOMETRY, Z. A. Melzak. Intended for students of many different backgrounds with only a modest knowledge of mathematics, this text features self-contained chapters that can be adapted to several types of geometry courses. 1983 edition. 240pp. 5 3/8 x 8 1/2. 0-486-46626-4

TOPOLOGY AND GEOMETRY FOR PHYSICISTS, Charles Nash and Siddhartha Sen. Written by physicists for physics students, this text assumes no detailed background in topology or geometry. Topics include differential forms, homotopy, homology, cohomology, fiber bundles, connection and covariant derivatives, and Morse theory. 1983 edition. 320pp. 5 3/8 x 8 1/2. 0-486-47852-1

BEYOND GEOMETRY: Classic Papers from Riemann to Einstein, Edited with an Introduction and Notes by Peter Pesic. This is the only English-language collection of these 8 accessible essays. They trace seminal ideas about the foundations of geometry that led to Einstein's general theory of relativity. 224pp. 6 1/8 x 9 1/4. 0-486-45350-2

GEOMETRY FROM EUCLID TO KNOTS, Saul Stahl. This text provides a historical perspective on plane geometry and covers non-neutral Euclidean geometry, circles and regular polygons, projective geometry, symmetries, inversions, informal topology, and more. Includes 1,000 practice problems. Solutions available. 2003 edition. 480pp. 6 1/8 x 9 1/4. 0-486-47459-3

TOPOLOGICAL VECTOR SPACES, DISTRIBUTIONS AND KERNELS, François Trèves. Extending beyond the boundaries of Hilbert and Banach space theory, this text focuses on key aspects of functional analysis, particularly in regard to solving partial differential equations. 1967 edition. 592pp. 5 3/8 x 8 1/2. 0-486-45352-9

INTRODUCTION TO PROJECTIVE GEOMETRY, C. R. Wylie, Jr. This introductory volume offers strong reinforcement for its teachings, with detailed examples and numerous theorems, proofs, and exercises, plus complete answers to all odd-numbered end-of-chapter problems. 1970 edition. 576pp. 6 1/8 x 9 1/4. 0-486-46895-X

FOUNDATIONS OF GEOMETRY, C. R. Wylie, Jr. Geared toward students preparing to teach high school mathematics, this text explores the principles of Euclidean and non-Euclidean geometry and covers both generalities and specifics of the axiomatic method. 1964 edition. 352pp. 6 x 9. 0-486-47214-0

Mathematics-History

THE WORKS OF ARCHIMEDES, Archimedes. Translated by Sir Thomas Heath. Complete works of ancient geometer feature such topics as the famous problems of the ratio of the areas of a cylinder and an inscribed sphere; the properties of conoids, spheroids, and spirals; more. 326pp. 5 3/8 x 8 1/2. 0-486-42084-1

THE HISTORICAL ROOTS OF ELEMENTARY MATHEMATICS, Lucas N. H. Bunt, Phillip S. Jones, and Jack D. Bedient. Exciting, hands-on approach to understanding fundamental underpinnings of modern arithmetic, algebra, geometry and number systems examines their origins in early Egyptian, Babylonian, and Greek sources. 336pp. 5 3/8 x 8 1/2. 0-486-25563-8

THE THIRTEEN BOOKS OF EUCLID'S ELEMENTS, Euclid. Contains complete English text of all 13 books of the Elements plus critical apparatus analyzing each definition, postulate, and proposition in great detail. Covers textual and linguistic matters; mathematical analyses of Euclid's ideas; classical, medieval, Renaissance and modern commentators; refutations, supports, extrapolations, reinterpretations and historical notes. 995 figures. Total of 1,425pp. All books 5 3/8 x 8 1/2.

Vol. I: 443pp. 0-486-60088-2
Vol. II: 464pp. 0-486-60089-0
Vol. III: 546pp. 0-486-60090-4

A HISTORY OF GREEK MATHEMATICS, Sir Thomas Heath. This authoritative two-volume set that covers the essentials of mathematics and features every landmark innovation and every important figure, including Euclid, Apollonius, and others. 5 3/8 x 8 1/2.

Vol. I: 461pp. 0-486-24073-8
Vol. II: 597pp. 0-486-24074-6

A MANUAL OF GREEK MATHEMATICS, Sir Thomas L. Heath. This concise but thorough history encompasses the enduring contributions of the ancient Greek mathematicians whose works form the basis of most modern mathematics. Discusses Pythagorean arithmetic, Plato, Euclid, more. 1931 edition. 576pp. 5 3/8 x 8 1/2. 0-486-43231-9

CHINESE MATHEMATICS IN THE THIRTEENTH CENTURY, Ulrich Libbrecht. An exploration of the 13th-century mathematician Ch'in, this fascinating book combines what is known of the mathematician's life with a history of his only extant work, the Shu-shu chiu-chang. 1973 edition. 592pp. 5 3/8 x 8 1/2. 0-486-44619-0

PHILOSOPHY OF MATHEMATICS AND DEDUCTIVE STRUCTURE IN EUCLID'S ELEMENTS, Ian Mueller. This text provides an understanding of the classical Greek conception of mathematics as expressed in Euclid's Elements. It focuses on philosophical, foundational, and logical questions and features helpful appendixes. 400pp. 6 1/2 x 9 1/4. 0-486-45300-6

BEYOND GEOMETRY: Classic Papers from Riemann to Einstein, Edited with an Introduction and Notes by Peter Pesic. This is the only English-language collection of these 8 accessible essays. They trace seminal ideas about the foundations of geometry that led to Einstein's general theory of relativity. 224pp. 6 1/8 x 9 1/4. 0-486-45350-2

HISTORY OF MATHEMATICS, David E. Smith. Two-volume history – from Egyptian papyri and medieval maps to modern graphs and diagrams. Non-technical chronological survey with thousands of biographical notes, critical evaluations, and contemporary opinions on over 1,100 mathematicians. 5 3/8 x 8 1/2.

Vol. I: 618pp. 0-486-20429-4
Vol. II: 736pp. 0-486-20430-8

Physics

Physics

INTRODUCTION TO MODERN OPTICS, Grant R. Fowles. A complete basic undergraduate course in modern optics for students in physics, technology, and engineering. The first half deals with classical physical optics; the second, quantum nature of light. Solutions. 336pp. 5 3/8 x 8 1/2. 0-486-65957-7

THE QUANTUM THEORY OF RADIATION: Third Edition, W. Heitler. The first comprehensive treatment of quantum physics in any language, this classic introduction to basic theory remains highly recommended and widely used, both as a text and as a reference. 1954 edition. 464pp. 5 3/8 x 8 1/2. 0-486-64558-4

QUANTUM FIELD THEORY, Claude Itzykson and Jean-Bernard Zuber. This comprehensive text begins with the standard quantization of electrodynamics and perturbative renormalization, advancing to functional methods, relativistic bound states, broken symmetries, nonabelian gauge fields, and asymptotic behavior. 1980 edition. 752pp. 6 1/2 x 9 1/4. 0-486-44568-2

FOUNDATIONS OF POTENTIAL THERY, Oliver D. Kellogg. Introduction to fundamentals of potential functions covers the force of gravity, fields of force, potentials, harmonic functions, electric images and Green's function, sequences of harmonic functions, fundamental existence theorems, and much more. 400pp. 5 3/8 x 8 1/2. 0-486-60144-7

FUNDAMENTALS OF MATHEMATICAL PHYSICS, Edgar A. Kraut. Indispensable for students of modern physics, this text provides the necessary background in mathematics to study the concepts of electromagnetic theory and quantum mechanics. 1967 edition. 480pp. 6 1/2 x 9 1/4. 0-486-45809-1

GEOMETRY AND LIGHT: The Science of Invisibility, Ulf Leonhardt and Thomas Philbin. Suitable for advanced undergraduate and graduate students of engineering, physics, and mathematics and scientific researchers of all types, this is the first authoritative text on invisibility and the science behind it. More than 100 full-color illustrations, plus exercises with solutions. 2010 edition. 288pp. 7 x 9 1/4. 0-486-47693-6

QUANTUM MECHANICS: New Approaches to Selected Topics, Harry J. Lipkin. Acclaimed as "excellent" (*Nature*) and "very original and refreshing" (*Physics Today*), these studies examine the Mössbauer effect, many-body quantum mechanics, scattering theory, Feynman diagrams, and relativistic quantum mechanics. 1973 edition. 480pp. 5 3/8 x 8 1/2. 0-486-45893-8

THEORY OF HEAT, James Clerk Maxwell. This classic sets forth the fundamentals of thermodynamics and kinetic theory simply enough to be understood by beginners, yet with enough subtlety to appeal to more advanced readers, too. 352pp. 5 3/8 x 8 1/2. 0-486-41735-2

QUANTUM MECHANICS, Albert Messiah. Subjects include formalism and its interpretation, analysis of simple systems, symmetries and invariance, methods of approximation, elements of relativistic quantum mechanics, much more. "Strongly recommended." – *American Journal of Physics.* 1152pp. 5 3/8 x 8 1/2. 0-486-40924-4

RELATIVISTIC QUANTUM FIELDS, Charles Nash. This graduate-level text contains techniques for performing calculations in quantum field theory. It focuses chiefly on the dimensional method and the renormalization group methods. Additional topics include functional integration and differentiation. 1978 edition. 240pp. 5 3/8 x 8 1/2. 0-486-47752-5

Physics

MATHEMATICAL TOOLS FOR PHYSICS, James Nearing. Encouraging students' development of intuition, this original work begins with a review of basic mathematics and advances to infinite series, complex algebra, differential equations, Fourier series, and more. 2010 edition. 496pp. 6 1/8 x 9 1/4. 0-486-48212-X

TREATISE ON THERMODYNAMICS, Max Planck. Great classic, still one of the best introductions to thermodynamics. Fundamentals, first and second principles of thermodynamics, applications to special states of equilibrium, more. Numerous worked examples. 1917 edition. 297pp. 5 3/8 x 8. 0-486-66371-X

AN INTRODUCTION TO RELATIVISTIC QUANTUM FIELD THEORY, Silvan S. Schweber. Complete, systematic, and self-contained, this text introduces modern quantum field theory. "Combines thorough knowledge with a high degree of didactic ability and a delightful style." – *Mathematical Reviews.* 1961 edition. 928pp. 5 3/8 x 8 1/2. 0-486-44228-4

THE ELECTROMAGNETIC FIELD, Albert Shadowitz. Comprehensive undergraduate text covers basics of electric and magnetic fields, building up to electromagnetic theory. Related topics include relativity theory. Over 900 problems, some with solutions. 1975 edition. 768pp. 5 5/8 x 8 1/4. 0-486-65660-8

THE PRINCIPLES OF STATISTICAL MECHANICS, Richard C. Tolman. Definitive treatise offers a concise exposition of classical statistical mechanics and a thorough elucidation of quantum statistical mechanics, plus applications of statistical mechanics to thermodynamic behavior. 1930 edition. 704pp. 5 5/8 x 8 1/4. 0-486-63896-0

INTRODUCTION TO THE PHYSICS OF FLUIDS AND SOLIDS, James S. Trefil. This interesting, informative survey by a well-known science author ranges from classical physics and geophysical topics, from the rings of Saturn and the rotation of the galaxy to underground nuclear tests. 1975 edition. 320pp. 5 3/8 x 8 1/2. 0-486-47437-2

STATISTICAL PHYSICS, Gregory H. Wannier. Classic text combines thermodynamics, statistical mechanics, and kinetic theory in one unified presentation. Topics include equilibrium statistics of special systems, kinetic theory, transport coefficients, and fluctuations. Problems with solutions. 1966 edition. 532pp. 5 3/8 x 8 1/2. 0-486-65401-X

SPACE, TIME, MATTER, Hermann Weyl. Excellent introduction probes deeply into Euclidean space, Riemann's space, Einstein's general relativity, gravitational waves and energy, and laws of conservation. "A classic of physics." – *British Journal for Philosophy and Science.* 330pp. 5 3/8 x 8 1/2. 0-486-60267-2

RANDOM VIBRATIONS: Theory and Practice, Paul H. Wirsching, Thomas L. Paez and Keith Ortiz. Comprehensive text and reference covers topics in probability, statistics, and random processes, plus methods for analyzing and controlling random vibrations. Suitable for graduate students and mechanical, structural, and aerospace engineers. 1995 edition. 464pp. 5 3/8 x 8 1/2. 0-486-45015-5

PHYSICS OF SHOCK WAVES AND HIGH-TEMPERATURE HYDRO DYNAMIC PHENOMENA, Ya B. Zel'dovich and Yu P. Raizer. Physical, chemical processes in gases at high temperatures are focus of outstanding text, which combines material from gas dynamics, shock-wave theory, thermodynamics and statistical physics, other fields. 284 illustrations. 1966–1967 edition. 944pp. 6 1/8 x 9 1/4. 0-486-42002-7

Astronomy

CHARIOTS FOR APOLLO: The NASA History of Manned Lunar Spacecraft to 1969, Courtney G. Brooks, James M. Grimwood, and Loyd S. Swenson, Jr. This illustrated history by a trio of experts is the definitive reference on the Apollo spacecraft and lunar modules. It traces the vehicles' design, development, and operation in space. More than 100 photographs and illustrations. 576pp. 6 3/4 x 9 1/4. 0-486-46756-2

EXPLORING THE MOON THROUGH BINOCULARS AND SMALL TELESCOPES, Ernest H. Cherrington, Jr. Informative, profusely illustrated guide to locating and identifying craters, rills, seas, mountains, other lunar features. Newly revised and updated with special section of new photos. Over 100 photos and diagrams. 240pp. 8 1/4 x 11. 0-486-24491-1

WHERE NO MAN HAS GONE BEFORE: A History of NASA's Apollo Lunar Expeditions, William David Compton. Introduction by Paul Dickson. This official NASA history traces behind-the-scenes conflicts and cooperation between scientists and engineers. The first half concerns preparations for the Moon landings, and the second half documents the flights that followed Apollo 11. 1989 edition. 432pp. 7 x 10. 0-486-47888-2

APOLLO EXPEDITIONS TO THE MOON: The NASA History, Edited by Edgar M. Cortright. Official NASA publication marks the 40th anniversary of the first lunar landing and features essays by project participants recalling engineering and administrative challenges. Accessible, jargon-free accounts, highlighted by numerous illustrations. 336pp. 8 3/8 x 10 7/8. 0-486-47175-6

ON MARS: Exploration of the Red Planet, 1958-1978--The NASA History, Edward Clinton Ezell and Linda Neuman Ezell. NASA's official history chronicles the start of our explorations of our planetary neighbor. It recounts cooperation among government, industry, and academia, and it features dozens of photos from Viking cameras. 560pp. 6 3/4 x 9 1/4. 0-486-46757-0

ARISTARCHUS OF SAMOS: The Ancient Copernicus, Sir Thomas Heath. Heath's history of astronomy ranges from Homer and Hesiod to Aristarchus and includes quotes from numerous thinkers, compilers, and scholasticists from Thales and Anaximander through Pythagoras, Plato, Aristotle, and Heraclides. 34 figures. 448pp. 5 3/8 x 8 1/2. 0-486-43886-4

AN INTRODUCTION TO CELESTIAL MECHANICS, Forest Ray Moulton. Classic text still unsurpassed in presentation of fundamental principles. Covers rectilinear motion, central forces, problems of two and three bodies, much more. Includes over 200 problems, some with answers. 437pp. 5 3/8 x 8 1/2. 0-486-64687-4

BEYOND THE ATMOSPHERE: Early Years of Space Science, Homer E. Newell. This exciting survey is the work of a top NASA administrator who chronicles technological advances, the relationship of space science to general science, and the space program's social, political, and economic contexts. 528pp. 6 3/4 x 9 1/4. 0-486-47464-X

STAR LORE: Myths, Legends, and Facts, William Tyler Olcott. Captivating retellings of the origins and histories of ancient star groups include Pegasus, Ursa Major, Pleiades, signs of the zodiac, and other constellations. "Classic." – *Sky & Telescope.* 58 illustrations. 544pp. 5 3/8 x 8 1/2. 0-486-43581-4

A COMPLETE MANUAL OF AMATEUR ASTRONOMY: Tools and Techniques for Astronomical Observations, P. Clay Sherrod with Thomas L. Koed. Concise, highly readable book discusses the selection, set-up, and maintenance of a telescope; amateur studies of the sun; lunar topography and occultations; and more. 124 figures. 26 halftones. 37 tables. 335pp. 6 1/2 x 9 1/4. 0-486-42820-6

Chemistry

MOLECULAR COLLISION THEORY, M. S. Child. This high-level monograph offers an analytical treatment of classical scattering by a central force, quantum scattering by a central force, elastic scattering phase shifts, and semi-classical elastic scattering. 1974 edition. 310pp. 5 3/8 x 8 1/2. 0-486-69437-2

HANDBOOK OF COMPUTATIONAL QUANTUM CHEMISTRY, David B. Cook. This comprehensive text provides upper-level undergraduates and graduate students with an accessible introduction to the implementation of quantum ideas in molecular modeling, exploring practical applications alongside theoretical explanations. 1998 edition. 832pp. 5 3/8 x 8 1/2. 0-486-44307-8

RADIOACTIVE SUBSTANCES, Marie Curie. The celebrated scientist's thesis, which directly preceded her 1903 Nobel Prize, discusses establishing atomic character of radioactivity; extraction from pitchblende of polonium and radium; isolation of pure radium chloride; more. 96pp. 5 3/8 x 8 1/2. 0-486-42550-9

CHEMICAL MAGIC, Leonard A. Ford. Classic guide provides intriguing entertainment while elucidating sound scientific principles, with more than 100 unusual stunts: cold fire, dust explosions, a nylon rope trick, a disappearing beaker, much more. 128pp. 5 3/8 x 8 1/2. 0-486-67628-5

ALCHEMY, E. J. Holmyard. Classic study by noted authority covers 2,000 years of alchemical history: religious, mystical overtones; apparatus; signs, symbols, and secret terms; advent of scientific method, much more. Illustrated. 320pp. 5 3/8 x 8 1/2. 0-486-26298-7

CHEMICAL KINETICS AND REACTION DYNAMICS, Paul L. Houston. This text teaches the principles underlying modern chemical kinetics in a clear, direct fashion, using several examples to enhance basic understanding. Solutions to selected problems. 2001 edition. 352pp. 8 3/8 x 11. 0-486-45334-0

PROBLEMS AND SOLUTIONS IN QUANTUM CHEMISTRY AND PHYSICS, Charles S. Johnson and Lee G. Pedersen. Unusually varied problems, with detailed solutions, cover of quantum mechanics, wave mechanics, angular momentum, molecular spectroscopy, scattering theory, more. 280 problems, plus 139 supplementary exercises. 430pp. 6 1/2 x 9 1/4. 0-486-65236-X

ELEMENTS OF CHEMISTRY, Antoine Lavoisier. Monumental classic by the founder of modern chemistry features first explicit statement of law of conservation of matter in chemical change, and more. Facsimile reprint of original (1790) Kerr translation. 539pp. 5 3/8 x 8 1/2. 0-486-64624-6

MAGNETISM AND TRANSITION METAL COMPLEXES, F. E. Mabbs and D. J. Machin. A detailed view of the calculation methods involved in the magnetic properties of transition metal complexes, this volume offers sufficient background for original work in the field. 1973 edition. 240pp. 5 3/8 x 8 1/2. 0-486-46284-6

GENERAL CHEMISTRY, Linus Pauling. Revised third edition of classic first-year text by Nobel laureate. Atomic and molecular structure, quantum mechanics, statistical mechanics, thermodynamics correlated with descriptive chemistry. Problems. 992pp. 5 3/8 x 8 1/2. 0-486-65622-5

ELECTROLYTE SOLUTIONS: Second Revised Edition, R. A. Robinson and R. H. Stokes. Classic text deals primarily with measurement, interpretation of conductance, chemical potential, and diffusion in electrolyte solutions. Detailed theoretical interpretations, plus extensive tables of thermodynamic and transport properties. 1970 edition. 590pp. 5 3/8 x 8 1/2. 0-486-42225-9

Engineering

FUNDAMENTALS OF ASTRODYNAMICS, Roger R. Bate, Donald D. Mueller, and Jerry E. White. Teaching text developed by U.S. Air Force Academy develops the basic two-body and n-body equations of motion; orbit determination; classical orbital elements, coordinate transformations; differential correction; more. 1971 edition. 455pp. 5 3/8 x 8 1/2. 0-486-60061-0

INTRODUCTION TO CONTINUUM MECHANICS FOR ENGINEERS: Revised Edition, Ray M. Bowen. This self-contained text introduces classical continuum models within a modern framework. Its numerous exercises illustrate the governing principles, linearizations, and other approximations that constitute classical continuum models. 2007 edition. 320pp. 6 1/8 x 9 1/4. 0-486-47460-7

ENGINEERING MECHANICS FOR STRUCTURES, Louis L. Bucciarelli. This text explores the mechanics of solids and statics as well as the strength of materials and elasticity theory. Its many design exercises encourage creative initiative and systems thinking. 2009 edition. 320pp. 6 1/8 x 9 1/4. 0-486-46855-0

FEEDBACK CONTROL THEORY, John C. Doyle, Bruce A. Francis and Allen R. Tannenbaum. This excellent introduction to feedback control system design offers a theoretical approach that captures the essential issues and can be applied to a wide range of practical problems. 1992 edition. 224pp. 6 1/2 x 9 1/4. 0-486-46933-6

THE FORCES OF MATTER, Michael Faraday. These lectures by a famous inventor offer an easy-to-understand introduction to the interactions of the universe's physical forces. Six essays explore gravitation, cohesion, chemical affinity, heat, magnetism, and electricity. 1993 edition. 96pp. 5 3/8 x 8 1/2. 0-486-47482-8

DYNAMICS, Lawrence E. Goodman and William H. Warner. Beginning engineering text introduces calculus of vectors, particle motion, dynamics of particle systems and plane rigid bodies, technical applications in plane motions, and more. Exercises and answers in every chapter. 619pp. 5 3/8 x 8 1/2. 0-486-42006-X

ADAPTIVE FILTERING PREDICTION AND CONTROL, Graham C. Goodwin and Kwai Sang Sin. This unified survey focuses on linear discrete-time systems and explores natural extensions to nonlinear systems. It emphasizes discrete-time systems, summarizing theoretical and practical aspects of a large class of adaptive algorithms. 1984 edition. 560pp. 6 1/2 x 9 1/4. 0-486-46932-8

INDUCTANCE CALCULATIONS, Frederick W. Grover. This authoritative reference enables the design of virtually every type of inductor. It features a single simple formula for each type of inductor, together with tables containing essential numerical factors. 1946 edition. 304pp. 5 3/8 x 8 1/2. 0-486-47440-2

THERMODYNAMICS: Foundations and Applications, Elias P. Gyftopoulos and Gian Paolo Beretta. Designed by two MIT professors, this authoritative text discusses basic concepts and applications in detail, emphasizing generality, definitions, and logical consistency. More than 300 solved problems cover realistic energy systems and processes. 800pp. 6 1/8 x 9 1/4. 0-486-43932-1

THE FINITE ELEMENT METHOD: Linear Static and Dynamic Finite Element Analysis, Thomas J. R. Hughes. Text for students without in-depth mathematical training, this text includes a comprehensive presentation and analysis of algorithms of time-dependent phenomena plus beam, plate, and shell theories. Solution guide available upon request. 672pp. 6 1/2 x 9 1/4. 0-486-41181-8